INSECT PHEROMONE RESEARCH

NEW DIRECTIONS

INSECT PHEROMONE RESEARCH

NEW DIRECTIONS

EDITED BY

RING T. CARDÉ

DEPARTMENT OF ENTOMOLOGY, UNIVERSITY OF CALIFORNIA
RIVERSIDE, CALIFORNIA

ALBERT K. MINKS

DLO RESEARCH INSTITUTE FOR PLANT PROTECTION,
WAGENINGEN, THE NETHERLANDS

CHAPMAN & HALL

ITP International Thomson Publishing

New York • Albany • Bonn • Boston • Cincinnati • Detroit • London • Madrid • Melbourne
Mexico City • Pacific Grove • Paris • San Francisco • Singapore • Tokyo • Toronto • Washington

Cover design: Trudi Gershenov
Cover photo: Charles Hedgcock (R.B.P.) Tucson, AZ

Printed in the United States of America

For more information, contact:

Chapman & Hall
115 Fifth Avenue
New York, NY 10003

Chapman & Hall
2-6 Boundary Row
London SE1 8HN
England

Thomas Nelson Australia
102 Dodds Street
South Melbourne, 3205
Victoria, Australia

Chapman & Hall GmbH
Postfach 100 263
D-69442 Weinheim
Germany

International Thomson Editores
Campos Eliseos 385, Piso 7
Col. Polanco
11560 Mexico D. F.
Mexico

International Thomson Publishing - Japan
Hirakawacho-cho Kyowa Building, 3F
1-2-1 Hirakawacho-cho
Chiyoda-ku, 102 Tokyo
Japan

International Thomson Publishing Asia
221 Henderson Road #05-10
Henderson Building
Singapore 0315

1 2 3 4 5 6 7 8 9 10 XXX 01 00 99 98 97

Library of Congress Cataloging-in-Publication Data

Insect pheromone research: new directions / edited by Ring T. Cardé, Albert K. Minks.
p. cm.
Includes bibliographical references and index.
ISBN 0-412-99611-1 (alk. paper)
1. pheromones. I.Cardé, Ring T. II.Minks, A. K.
QP572.P47P44 1996
595.7'0166--DC20 96-26762
CIP

British Library Cataloguing in Pubication Data available

To order this or any other Chapman & Hall book, please contact **International Thomson Publishing, 7625 Empire Drive, Florence, KY 41042.** Phone: (606) 525-6600 or 1-800-842-3636. Fax: (606) 525-7778. e-mail: order@chaphall.com.

For a complete listing of Chapman & Hall's titles, send your request to **Chapman & Hall, Dept. BC, 115 Fifth Avenue, New York, NY 10003.**

Edmund Arimantas Arbas (1950–1995)

With great sadness and sense of loss, we report the untimely death of our close colleague and dear friend, Dr. Edmund Arimantas Arbas, on 18 June 1995. Ed made important contributions to diverse fields of science, including the evolution of nervous systems and behavior, cellular physiology of neurons, pheromonal information processing in insects, and mechanisms underlying olfactory-guided locomotion.

Ed Arbas's life ended in a tragic accident at the seashore. He spent his last few days instructing his daughters Audra and Alia on the fine points of wind-surfing and boogie boarding, and revealing to them the mysteries of the strange and beautiful denizens of the tide pools. As difficult as it is for us to accept his death, it seems fitting that Ed left this world at the place he loved most, doing the things that he loved most. Ed was one of a kind—combining a brilliant analytical and synthetic mind, a vibrant and engaging personality, a dazzling quickness of wit, and a passionate love of science with a rare commitment to his beautiful and loving family. We will miss him always.

John G. Hildebrand and Mark A. Willis
Tucson, November 1995

Contents

III. Neuroethology of Pheromone-Mediated Responses

Organized by Thomas C. Baker

IV. Use of Pheromones in Direct Control

Organized by Albert K. Minks and Ring T. Cardé

Preface

This book contains the proceedings of the "First International Symposium on Insect Pheromones," which was held at Wageningen, The Netherlands, from March 6 to March 11, 1994. Eighty participants from 17 countries attended the symposium, which turned out to be a unique forum for the exchange of the latest worldwide findings on insect pheromones, an opportunity to discuss and debate unsettled issues, and a mechanism to define new directions in pheromone research and foster interdisciplinary collaborations. The meeting comprised five sessions representing the breadth of disciplinary interest in pheromones, a typical characteristic of this research area. In the sessions the following topics were presented: (1) control of pheromone production (organized by W. L. Roelofs), (2) sensory processing of pheromone signals (T. L. Payne), (3) neuroethology of pheromone-mediated responses (T. C. Baker), (4) use of pheromones in direct control (A. K. Minks and R. T. Cardé), and (5) evolution of pheromone communication (C. Löfstedt). All sessions started with a series of 30-minute lectures, after which ample time was reserved for discussion. In each session some participants were asked to serve as discussants and to initiate and stimulate discussion, and a rapporteur was recruited to make notes of these discussions and to summarize the general trends emerging from the session. The general programming of the symposium was in the hands of R. T. Cardé, A. K. Minks, and T. L. Payne.

This symposium was the outcome of intensive deliberations in the preceding years, initiated by the three persons mentioned just above. In the early 90s, researchers on insect pheromones realized that, although there were many national and international meetings at which this topic was considered, not a single specific meeting had taken place, where most workers in this field could convene. The only meetings in the pherome field that have perhaps come close to that ideal were the symposium "Mechanisms and Insect Olfaction," held in Oxford, United Kingdom, in 1984, the symposium "Chemical Ecology: Odour Communication in

Animals," held in Noordwijkerhout, The Netherlands, in 1978, and the conference "Chemical Control of Behavior: Theory and Application," held at Bellagio, Italy, in 1975. These meetings were relatively small in size and characterized by invited presentations with ample opportunity for informal discussions.

Out of these deliberations, gradually the plan came forward to organize a series of regular meetings on insect pheromones with a frequency of 3 or 4 years, taking the long-established insect–host plant relationships group as an example. It was considered as essential that topics represent the breadth of pheromone research, including biochemistry, integration of sensory signals, orientation of flying insects, and successes in communication disruption, so that different kinds of pheromone researchers could meet each other. Also, all insect taxa that have pheromone-mediated behavior should be presented, although care should be taken with including fully the vast area of social insects. Subsequent meetings could consider the same or new topics. Finally, it was considered as important to keep the size of the meetings limited to create more favorable conditions for open discussion.

Circulation of this plan by the initiators among colleagues in the pheromone area generated considerable enthusiasm, and therefore an organizing committee was established to provide suggestions on a productive format, topics, possible participants, and a site for the first meeting and also to supervise further meeting arrangements. Members of the committee are: H. Arn, T. C. Baker, J. H. Borden, R. T. Cardé (chairman), J. G. Hildebrand, C. Löfstedt, J. N. McNeil, A. K. Minks, H. Mustaparta, T. L. Payne, W. L. Roelofs, M. Tóth, and J. H. Tumlinson. From this group R. T. Cardé, A. K. Minks, and T. Payne were chosen to prepare the first symposium, which was realized at Wageningen in March 1994.

The Wageningen meeting became a formidable success and fully met the expectations of the organizers as well as those of the participants. It turned out to be an extraordinary opportunity to summarize the current advances in the insect pheromone field and to provide directions for further studies. The present book reflects the high standard of the meeting and presents a unique compilation of high-quality research data, state-of-the-art reviews, and ideas for future research.

We gratefully acknowledge the local organizers: A. K. Minks, J. J. A. van Loon, and J. H. Visser and staff members of the International Agricultural Centre (IAC) for all the work they have done. The IAC appeared to be an excellent venue for the meeting and provided a delightful catalyst for our deliberations. We are also much indebted to the sponsors of the meeting and are particularly grateful for the financial support which they provided to the participants from the eastern European countries. The sponsors are as follows: Royal Netherlands Academy of Sciences, Amsterdam; Directorate of Science and Technology of the Netherlands Ministry of Agriculture, Nature Management and Fisheries (LNV-DWT), the Hague; Agricultural Research Department (DLO-NL), Wageningen; Research Institute for Plant Protection (IPO-DLO), Wageningen; Wageningen Agricultural University; Syntech, Hilversum; International Agricultural Centre,

Wageningen; National Research Initiative, United States Department of Agriculture.

Ring T. Cardé
Albert K. Minks

A booklet of the discussions following each paper is available *gratis* from either editor.

Contributors

Miriam Altstein
Department of Entomology
Institute of Plant Protection ARO
The Volcani Center
Bet Dagan 50250, Israel

Larry Antilla
Arizona Cotton Research and Protection Council
2402 West Huntington Drive
Tempe, AZ 85282

Edmund A. Arbas (deceased)
Arizona Research Laboratories-Division of Neurobiology and Department of Physiology
University of Arizona
Tucson, AZ 85721

Heinrich Arn
Swiss Federal Research Station for Horticulture and Viticulture
CH-8820 Wädenswil, Switzerland

Thomas C. Baker
Department of Entomology
Iowa State University
Ames, IA 50011

Orna Ben-Aziz
Institute of Plant Protection ARO
The Volcani Center
Bet Dagan 50250, Israel

Gary J. Blomquist
Department of Biochemistry
Mail Stop 330
University of Nevada
Reno, NV 89557

John H. Borden
Centre for Pest Management
Department of Biological Sciences
Simon Frazer University
Burnaby, British Columbia V5A 1S6, Canada

Heinz Breer
University Stuttgart-Hohenheim
Institute of Zoophysiology
70593 Stuttgart, Germany

Roger K. Butlin
Department of Genetics
The University of Leeds
Leeds LS2 9JT, United Kingdom

Francisco Camps
Department of Biological Organic Chemistry CID-CSIC
Jordi Girona 18-26
08034-Barcelona, Spain

Ring T. Cardé
Department of Entomology
University of California
Riverside, CA 92521

Thomas A. Christensen
Arizona Research Laboratories
Division of Neurobiology and Center for Insect Science
University of Arizona
Tucson, AZ 85721

Michel Cusson
Natural Resources Canada
Canadian Forest Service—Quebec Region
P.O. Box 3800
1055 du P.E.P.S.
Sainte-Foy, Québec G1V 4C7, Canada

Johanne Delisle
Natural Resources Canada
Canadian Forest Service—Quebec Region
P.O. Box 3800
1055 du P.E.P.S.
Sainte-Foy, Québec G1V 4C7, Canada

Joseph C. Dickens
U.S. Department of Agriculture
Agricultural Research Service
Insect Chemical Ecology Laboratory
Building 007 Room 310 BARC-West
Beltsville, MD 20704

Gehua Du
Departments of Chemistry and Biochemistry & Cell Biology
University at Stony Brook
Stony Brook, NY 11794

John S. Dugdale
Manaaki Whenua-Landcare Research New Zealand Ltd.
Private Bag 92170
Auckland, New Zealand

Ezra Dunkelblum
Institute of Plant Protection ARO
The Volcani Center
Bet Dagan 50250, Israel

Osama El-Lissy
Arizona Cotton Research and Protection Council
2402 West Huntington Drive
Tempe, AZ 85282

Gemma Fabriàs
Department of Biological Organic Chemistry CID-CSIC
Jordi Girona 18-26
08034-Barcelona, Spain

Nianbai Fang
Insect Attractants, Behavior and Basic Biology Laboratory
U.S. Department of Agriculture
Agricultural Research Service
P.O. Box 14565
Gainesville, FL 32604

Andreas Färbert
Department of Physics
University of Kaiserslautern
D-67653 Kaiserslautern, Germany

Petra Färbert
Department of Animal Physiology
University of Kaiserslautern
D-67653 Kaiserslautern, Germany

Adrien Fónagy
Department of Zoology
Plant Protection Institute of the Hungarian Academy of Sciences
P.O. Box 102
H-1525 Budapest, Hungary

Stephen P. Foster
The Horticulture and Food Research Institute of New Zealand
Mt. Albert Research Centre
Private Bag 92 169
Auckland, New Zealand

Tal Gabay
Institute of Plant Protection ARO
The Volcani Center
Bet Dagan 50250, Israel

Yoav Gavit
Institute of Plant Protection ARO
The Volcani Center
Bet Dagan 50250, Israel

Carina Gileadi
Department of Stored Products ARO
The Volcani Center, P.O. Box 6
Bet Dagan 50250, Israel

Bill Hansson
Department of Ecology
University of Lund
Lund S-223-62, Sweden

Kenneth F. Haynes
Department of Entomology
University of Kentucky
Lexington, KY 40546

John G. Hildebrand
Division of Neurobiology
611 Gould-Simpson Building
University of Arizona
Tucson, AZ 85721

Russell A. Jurenka
Department of Entomology
New York State Agricultural Experiment Station
Geneva, NY 14456

Batya Kamensky
Department of Stored Products ARO
The Volcani Center
P.O. Box 6
Bet Dagan 50250, Israel

Ryohei Kanzaki
Institute of Biological Sciences
University of Tsukuba
Tsukuba Science City
Ibaraki 305, Japan

G. Karg
The Horticulture and Food Research Institute of New Zealand Ltd.
P.O. Box 51
Lincoln, New Zealand

Uwe T. Koch
Department of Animal Physiology
University of Kaiserslautern
D-67653 Kaiserslautern, Germany

Mikhail Kozlov
Laboratory of Ecological Zoology
Biological Faculty
University of Turku
Turku FIN-20500, Finland

Ernst Kramer
Max-Planck-Institut für Verhaltenphysiologie
D-82319 Seewiesen, Germany

Stuart B. Krasnoff
U.S. Department of Agriculture
Agricultural Research Service
Tower Road
Ithaca, NY 14853

Walter Soares Leal
Laboratory of Chemical Prospecting
National Institute of Sericulture and Entomological Science
1-2 Ohwashi
Tsukuba 305, Japan

Dangsheng Liang
Department of Entomology
Box 7631
North Carolina State University
Raleigh, NC 27695

Charles E. Linn, Jr.
Department of Entomology
New York State Agricultural Experiment Station
Geneva, NY 14456

Christer Löfstedt
Department of Ecology
University of Lund
Lund S-223-62, Sweden

Friedrich Louis
Federal Research Station for Agriculture, Viticulture and Horticulture (SLFA)
D-67435 Neustadt an der Weinstraße, Germany

Peter Ma
Department of Entomology
New York State Agricultural Experiment Station
Geneva, NY 14456

Agenor Mafra-Neto
Department of Entomology
University of California
Riverside, CA 92521

M. Pilar Marco
Department of Biological Organic Chemistry CID-CSIC
Jordi Girona 18-26
08034-Barcelona, Spain

Shogo Matsumoto
Institute of Physical and Chemical Research (RIKEN)
Hirosawa 2-1, Wako-shi
Saitama 351-01, Japan

Jeremy N. McNeil
Départment de biologie
Université Laval
Sainte-Foy, Québec G1K 7P4, Canada

Albert K. Minks
DLO Research Institute for Plant Protection
P.O. Box 9060
6700 GW Wageningen, The Netherlands

Takashi Mitsui
Insect Toxicology and Physiology Laboratory
Institute of Physical and Chemical Research (RIKEN)
Hirosawa 2-1, Wako-shi
Saitama 351-01, Japan

Susan J. Muggleston
The Horticulture and Food Research Institute of New Zealand
Mt. Albert Research Centre
Private Bag 92 169
Auckland, New Zealand

John Murlis
H.M. Inspectorate of Pollution
Department of the Environment
London SW1P 3PY, United Kingdom

Hanna Mustaparta
Department of Zoology
University of Trondheim
Trondheim, Norway

P. Larry Phelan
Department of Entomology
Ohio State University
Wooster, OH 44691

John A. Pickett
IACR-Rothamsted
Harpenden Hertfordshire AL5 2JQ, United Kingdom

Glen Prestwich
Department of Medicinal Chemistry
308 Skaggs Hall
The University of Utah
Salt Lake City, Utah 84112

Ada Rafaeli
Department of Stored Products ARO
The Volcani Center
P.O. Box 6
Bet Dagan 50250, Israel

Ashok K. Raina
Insect Neurobiology and Hormone Laboratory
U.S. Department of Agriculture
Agricultural Research Service
Beltsville, MD 20705

Sonny B. Ramaswamy
Department of Entomology and Plant Pathology
Mississippi State University
Mississippi State, MS 39762

Wendell L. Roelofs
Department of Entomology
New York State Agricultural Experiment Station
Geneva, NY 14456

C. J. Sanders
Canadian Forest Service
Saulte Ste. Marie, Ontario, Canada

Coby Schal
Department of Entomology
Box 7631
North Carolina State University
Raleigh, NC 27695

Victoria Soroker
Department of Zoology
Tel Aviv University
Tel Aviv 69978, Israel

Robert T. Staten
U.S. Department of Agriculture
Animal and Plant Health Inspection Service
4125 Broadway
Phoenix, AZ 85040

D. M. Suckling
The Horticulture and Food Research Institute of New Zealand Ltd.
P.O. Box 51
Lincoln, New Zealand

Akinori Suzuki
Department of Agricultural Chemistry
The University of Tokyo
Tokyo 113, Japan

Peter E. A. Teal
Insect Attractants, Behavior and Basic Biology Laboratory
U.S. Department of Agriculture
Agricultural Research Service
P.O. Box 14565
Gainesville, FL 32604

Adam J. Trickett
Department of Genetics
The University of Leeds
Leeds LS2 9JT, United Kingdom

John T. Trumble
Department of Entomology
University of California
Riverside, CA 92521

James H. Tumlinson
Insect Attractants, Behavior and Basic Biology Laboratory
U.S. Department of Agriculture
Agricultural Research Service
P.O. Box 14565
Gainesville, FL 32604

J.N.C. Van der Pers
VDP Laboratories
P.O. Box 1547
1200 BM Hilversum, The Netherlands

Neil J. Vickers
Department of Entomology
Iowa State University
Ames, IA 50011

Lester J. Wadhams
IACR-Rothamsted
Harpenden Hertfordshire AL5 2JQ, United Kingdom

Mark A. Willis
Arizona Research Laboratories-Division of Neurobiology
University of Arizona
Tucson, AZ 85721

Peter Witzgall
Department of Plant Protection Sciences
Swedish University of Agricultural Sciences
230 53 Alnarp, Sweden

Christine M. Woodcock
IACR-Rothamsted
Harpenden, Hertfordshire AL5 2JQ, United Kingdom

Tristram D. Wyatt
Department for Continuing Education and Department of Zoology
University of Oxford
1 Wellington Square
Oxford OX1 2JA, United Kingdom

Uzi Zisman
Department of Stored Products, ARO
The Volcani Center
P.O. Box 6 Bet Dagan 50250, Israel

INSECT PHEROMONE RESEARCH

NEW DIRECTIONS

PART I

Control of Pheromone Production

Organized by Wendell L. Roelofs

1

Neural and Endocrine Control of Pheromone Production and Release in Cockroaches

Coby Schal, Dangsheng Liang, and Gary J. Blomquist

1. Regulatory Mechanisms of Pheromone Production and Release

Integration of physiological and behavioral events is required to produce and emit sex pheromones in coordination with other reproductive events. Thus, insects usually produce and emit sex pheromones, and exhibit specific behaviors associated with sexual receptivity, only when they are reproductively competent. While the production of pheromones is regulated biochemically or developmentally, emission of pheromones, particularly of volatile attractants, can also be regulated behaviorally, involving specific central nervous (CNS)-generated motor patterns. Insects thus employ a variety of mechanisms in the production and emission of sex pheromones:

- Pheromone production in some insects depends upon availability of biosynthetic precursors from host plants. Some danaid butterflies and arctiid moths modify plant pyrrolizidine alkaloids into volatile derivatives that are used as male pheromones (e.g., Schneider et al. 1975). It has become accepted that bark beetles detoxify tree monoterpenes through allylic oxidation into corresponding alcohols, which are then used as pheromones (review: Vanderwel 1994). However, recent data have demonstrated that *Ips paraconfusus* and *I. pini* synthesize the monoterpene mercene and the hydroxylated pheromone products ipsenol and ipsdienol (*I. pini* synthesizes only ipsdienol) from labeled acetate injected into the insects (Seybold et al. unpublished results). This raises the question of the relative importance of the *de novo* pathway *versus* the use of plant derived precursors to form the hydroxylated pheromone components in this group of insects.
- Neural mechanisms are involved in regulating both production and emission of pheromone in the gypsy moth, *Lymantria dispar*; neural messages

from the anterior portion of the CNS descend to the terminal abdominal ganglion (TAG) through the ventral nerve cord (VNC) (Tang et al. 1987). In this, as well as in other insects, neural signals originating from the abdomen of mated females ascend the VNC and inactivate sex pheromone production and/or release (e.g., Foster and Roelofs 1994).

- The bulk of recent work on regulation of pheromone production in insects, as well as emphasis in the present volume, is on humoral pheromonotropins, particularly the pheromone biosynthesis activating neuropeptides (PBAN) and related myotropic peptides with homologous carboxyl-terminal amino acids (reviews: Raina 1993; Nagasawa et al. 1994). In some insects, PBAN may stimulate release of other factors, such as a bursa copulatrix factor in the redbanded leafroller moth, *Argyrotaenia velutinana*, which in turn stimulates pheromone production (Jurenka and Roelofs 1993).
- Work with the housefly, *Musca domestica*, has shown that ecdysteroids, produced by the vitellogenic ovary, stimulate synthesis of the sex pheromone *Z*-9-tricosene (review: Blomquist et al. 1993). On the other hand, evidence from the moth *Heliothis virescens* suggests the involvement of 20-hydroxyecdysone in suppression of pheromone production after mating (Ramaswamy and Cohen 1992).
- Juvenile hormone (JH), a product of the corpora allata (CA), was shown to induce pheromone production and release in several coleopterans, cockroaches, and some lepidopterans (reviews: Cusson et al. 1994a; Schal and Smith 1990; Vanderwel 1994). In some insects, however, indirect evidence suggests that JH may also exert pheromonostatic effects (Webster and Cardé 1984). Although evidence for a direct action of JH on the site of pheromone production is lacking in the majority of cases, recent evidence from the boll weevil suggests that JH III enhances the production of sex pheromone by isolated male fat body *in vitro* (Wiygul et al. 1990).
- Developmental regulation of pheromone-producing glands occurs in all insects in which a sex-specific gland appears in the adult. In some lepidopterans, functional competency of the gland may be regulated by changing ecdysteroid titers during organogenesis (e.g., Tang et al. 1991). In cockroaches, cyclic maturational changes in the gland coincide with the ovarian cycle (Liang and Schal 1993c).
- Multiple mechanisms probably occur in most insects. In cockroaches and lepidopterans that respond to JH, it is likely that peptides are released in response to elevated titers of JH. For example, while either JH or PBAN can stimulate sex pheromone production in allatectomized *Agrotis ipsilon* (black cutworm moth), JH alone fails to induce pheromone production in decapitated females; it is therefore suggested that JH induces

release of pheromonotropic factors from the CNS (Picombon et al. 1995). Also, neural inactivation of pheromone production in many lepidopterans conceivably operates through suppression of pheromonotropic peptide release.

2. Juvenile Hormone: A Master Regulator in Cockroach Reproductive Biology

Cockroaches are a diverse group. Some exhibit parthenogenesis, whereas most reproduce sexually with reproductive modes including oviparity, ovoviviparity, and viviparity (Roth 1970). They therefore offer excellent models for comparative studies of the mechanisms that regulate female reproductive physiology and behavior, including the synthesis and emission of pheromones. *Blattella germanica* (German cockroach) and *Supella longipalpa* (brown-banded cockroach) are both oviparous blattellids. Females of both species emerge as sexually unreceptive adults, undergo several days of sexual maturation (which can be extended indefinitely if inappropriate conditions prevail), become sexually receptive, recruit conspecific males with sex pheromones, mate, and oviposit their vitellogenin-laden basal oocytes into an ootheca. Females of the brown-banded cockroach deposit their oothecae within 36 hours of oviposition and immediately enter a new vitellogenic wave. In contrast, under identical rearing conditions in our laboratory, *B. germanica* females carry the ootheca externally for about 21 days until the young hatch. During this time, oocyte development is inhibited, as in ovoviviparous and viviparous cockroaches in which the ootheca is retracted for embryonic development within the mother. Thus, while *S. longipalpa* females exhibit relatively uninterrupted cycles of oocyte development, *B. germanica* females experience discrete patterns of oocyte development which are interrupted by long pregnancies. These complex adult reproductive life histories contrast with most moth species, which mate soon after emergence, do not feed, and have short adult lives.

In *Blattella* and *Supella*, as in all cockroaches studied to date, vitellogenesis and cyclic maturation of oocytes parallels and is dependent upon JH III synthesis by the CA (Feyereisen 1985; Tobe and Stay 1985; Scharrer 1987). Thus, in females of both species the size of basal oocytes serves as a reliable predictor of relative CA activity (Bellés et al. 1987; Gadot et al 1989b; Smith et al. 1989). Likewise, synthesis of oothecal proteins is JH-regulated, and in the absence of JH (e.g., surgical allatectomy) both oocyte maturation and colleterial (accessory) gland growth are arrested (Zalokar 1968; Burns et al. 1991). JH production increases as the oocytes grow after the imaginal molt; and it declines just before ovulation, as the oocytes become chorionated, and remains low during pregnancy, while oocyte development is arrested. The JH biosynthetic activity of the CA in adult female cockroaches is dependent upon and modulated by (a) intrinsic signals originating from the brain, ovary, mating, and nutrients and (b) extrinsic signals including temperature, pheromones, tactile cues, and social conditions, such as isolation and crowding, that act through sensory pathways (Engelmann 1970). The degree of dependence of allatal activity on any one of these inhibitory or stimulatory signals varies widely and appears to be species-specific. In *B. germanica*, activity of the CA is potentiated more by signals from grouping than by those from mating (Gadot et al. 1989a). Conversely, isolated and grouped adult *S. longipalpa* females exhibit similar patterns of oocyte development, and mating stimulates JH synthesis (Chon et al. 1990). In *Blattella*, diet quantity and quality also influence activity of the CA (Schal et al. 1993).

In all cockroaches studied to date, the CA have been shown to be restrained to varying degrees by neural signals from the brain (see Feyereisen 1985; Tobe and Stay 1985; Khan 1988). Recent evidence supports the idea that allatostatic peptides are in large measure responsible for this inhibition (see Woodhead et al. 1994 and references there). Specific relevant cues, such as those from mating in *Supella* or social interactions and food quality in *Blattella*, can, together or independently, lift the brain inhibition on the CA. Conversely, in gravid females, especially in ovoviviparous and viviparous females and in oviparous females with nonoverlapping growth of basal and penultimate oocytes, low JH titers are necessary to sustain pregnancies and to avoid premature expulsion of oothecae. This is accomplished through ascending neural signals from the uterus or vestibulum that suppress CA activity until "the right time." The brain integrates the multiplicity of stimulatory and inhibitory signals; and through a graded and gradual lifting of neural inhibition of the CA, it effects a cycle of JH biosynthesis. Therefore, transection of the nervous connections between the brain and the CA, which removes this brain inhibition and facilitates JH synthesis, has been used as an effective tool in studies of JH regulation of reproduction and, in turn, pheromone production.

Females of both cockroach species may mate once or multiple times, depending on the amount of sperm in their spermathecal reservoirs. Because adult cockroaches are long-lived and most of their gonadal maturation cycles do not require reinsemination of the female, it is expected that they would evolve mechanisms to effect pheromone release and associated behaviors only during specific gonotrophic stages. At other times, suppression of sexual receptivity would conserve energy and reduce exposure to predators and parasites. In this review we compare and contrast the hormonal and neural mechanisms that regulate sexual receptivity in *Blattella* and *Supella*. Since both share JH III as a common regulator of reproduction, we are particularly interested in examining the effects of modulators of CA activity upon synthesis and emission of sex pheromones. Our pivotal question is whether cockroaches that exhibit different patterns of CA activity (long *versus* short pregnancy) share common underlying regulatory mechanisms of pheromone production. Moreover, what mechanisms are involved in suppression of mate-recruitment during pregnancy? And lastly, we review the biosynthesis, regulation, and internal transport of a contact sex pheromone in the German cockroach.

Pheromones, Glands, and Calling Behavior in *Blattella* and *Supella*

In spite of the vast economic importance of pest cockroaches, to date, only two major attractant sex pheromones have been identified. In the American cockroach, *Periplaneta americana*, periplanone-B, a 10-membered germacranoid structure, was identified and confirmed by synthesis (Persoons et al. 1976; Still 1979). Recent work with related blattids has elucidated related compounds that serve in species-specific blends and thus act in species recognition and isolation.

In the scotophase, *S. longipalpa* females engage in calling behavior, characterized by elevated tegmina and wings, a recurved abdomen, and occasional expansion of the genital atrium (Hales and Breed 1983). Calling is expressed only by receptive females at least 5 days old, is under circadian control, and is associated with the release of a volatile sex pheromone that elicits anemotactic responses in males (Smith and Schal 1990a, 1991). Behavioral assays of males and electroantennographic (EAG) assays with hexane extracts of various female body parts identified the fourth and fifth tergites as the major sites of sex pheromone production (Schal et al. 1992). We extracted the pheromone from integumental secretory units, each consisting of a secretory cell that discharges into a cuticular pore via a long subcutaneous duct supported by a duct cell (Liang and Schal unpublished). These cuticular pores are distributed throughout all tergites, but their density is highest on the lateral margins of the fourth and fifth tergites (Schal et al. 1992). The morphological evidence, together with behavioral and electrophysiological data, strongly implicate these structures as the site of pheromone production and release while the female engages in calling behavior.

An arduous rearing, extraction (from the tergites of some 20,000 virgin females and from air blown over calling females), and purification effort recently yielded supellapyrone (Fig. 1.1A) as the *S. longipalpa* sex pheromone (Charlton et al. 1993b). A synthetic racemic blend of supellapyrone [5-(2,4-dimethylheptanyl)-3-methyl-2*H*-pyran-2-one] elicits strong behavioral (long-distance anemotaxis) and electrophysiological (EAG) responses in males, as does the natural pheromone. The stereochemistry of supellapyrone was shown to be (2′*R*,4′*R*) by dual electroantenographic and flame ionization detection coupled to chiral capillary gas chromatography (Leal et al. 1995). Elucidation of the chemical structure of this pheromone and availability of synthetic supellapyrone will now facilitate *in vivo* and *in vitro* studies of endocrine regulation of its biosynthesis.

Virgin adult *B. germanica* females also exhibit a characteristic calling behavior; but in this species, calling peaks just before the end of the scotophase and extends well into the photophase; the diel pattern of calling coincides with the periodicity of mating (Liang and Schal 1993b). During calling, the female emits a volatile

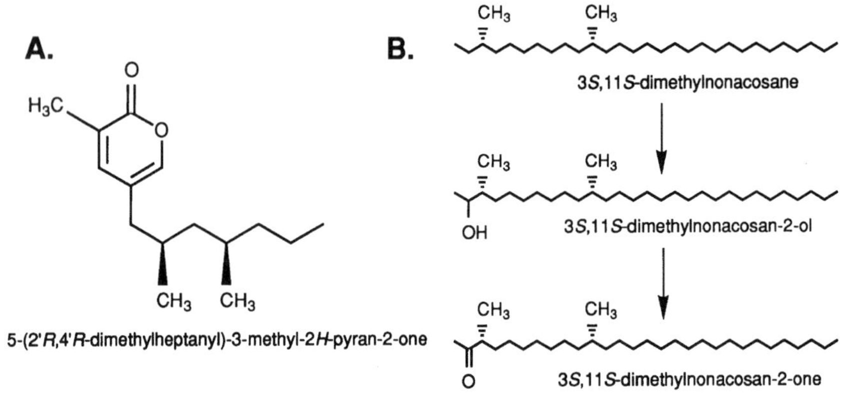

Figure 1.1. **(A)** Structure of supellapyrone, the sex pheromone of *Supella longipalpa*. **(B)** Biosynthesis of the contact sex pheromone of *Blattella germanica* showing conversion of the dimethylalkane to methyl ketone.

sex pheromone that serves to attract males from a distance. Olfactometer assays showed that the attractant is produced exclusively by adult females and that it elicits behavioral responses only in adult males. Behavioral and EAG assays localized the site of pheromone production in the tenth abdominal tergite, at the anterior of which an adult female-specific gland is found (Liang and Schal 1993a). This gland, too, is made of aggregates of integumental secretory cells that open via long ducts into orifices within deep depressions on the cuticular surface. The secretory cells of active glands (virgin day-6 females) are characterized by a large number of electron-lucid secretory vesicles, abundant rough and smooth endoplasmic reticulum (RER and SER), a large nucleus, and a long, convoluted end-apparatus which is surrounded by numerous long microvilli (Fig. 1.2) (Liang and Schal 1993c). The contents of the secretory vesicles are exocytosed into extracellular reservoirs at the base of microvilli and then transported to the cuticular surface through the long ducts. The chemical identity of the attractant sex pheromone of *B. germanica* remains unknown.

Figure 1.2. Schematic diagram showing maturation and developmental changes in the pheromone gland of *Blattella germanica* females in relation to the gonadal maturation cycle, juvenile hormone synthesis, and pheromone production. The immature secretory cell (SC) of a day-0 pheromone gland enlarges by approximately threefold by day 6, in relation to increasing JH production, basal oocyte size, and pheromone content. Active secretory cells contain numerous secretory vesicles (Sv), many of which discharge through large microvilli into an end-apparatus (EA). After oviposition, and throughout pregnancy, secretory cells regress, as do JH biosynthesis and pheromone content of the gland. Secretory cells enlarge again in the second gonotrophic cycle, coincident with an increase in JH

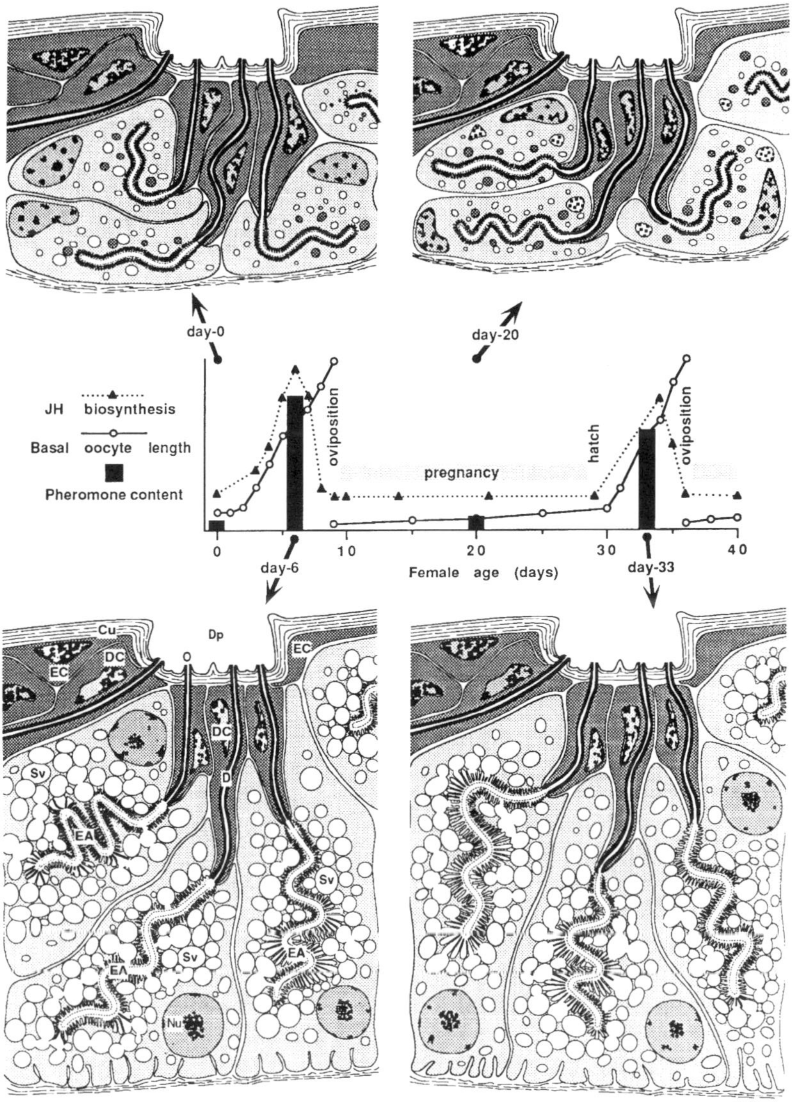

Figure 1.2. (Con't.) production, pheromone production, and oocyte growth. Rates of JH biosynthesis were determined with an *in vitro* radiochemical assay and were redrawn from Chiang et al. (1991b); data for basal oocyte length were redrawn from Chiang and Schal (1994). Pheromone content of the gland was determined by electroantennograms of gland extracts, as described in Liang and Schal (1993c). Schematics are based in part on descriptions in Liang and Schal (1993c). Schematics are based in part on descriptions in Liang and Schal (1993c). Cu, cuticle; D, duct; DC, duct cell; Dp cuticular depression; EC, epidermal cell; M, mitochondria; Nu nucleus; O, duct orifice.

4. Juvenile Hormone Induction of Sexual Receptivity and Pheromone Production and Emission in *Blattella* and *Supella*

JH involvement in the control of sexual behavior has been shown in several insects. Engelmann (1960) provided early evidence for the regulation of mating behavior by the endocrine system: Only 30% of 1-day-old *Leucophaea maderae* female cockroaches whose CA were removed mated during the next month. Subsequent work by Roth, Barth, and Bell (review: Barth and Lester 1973) showed that pheromone production in some cockroaches is under endocrine control. A major conclusion from several comparative studies of cockroaches (see reviews: Barth and Lester 1973; Truman and Riddiford 1974; Schal and Smith 1990) is that regulation of pheromone production, its release, and copulatory readiness by JH is species-specific. Thus, in *Byrsotria fumigata*, JH controls pheromone production but not mating readiness (Roth and Barth 1964); in some species (e.g., *Diploptera punctata*), newly eclosed females mate immediately, while still teneral, before the CA become active (Stay and Roth 1958; Stay and Tobe 1977), and yet in others (e.g., *B. germanica*) JH controls all three events (see below). Similarly, in a cricket, JH is responsible for the directed phonotactic responses of females to males, presumably by reducing the phonotactic threshold of an auditory neuron in the prothoracic ganglion (Stout et al. 1992; but see Loher et al. 1992). The regulation of copulatory readiness is much less clearly understood in this insect, since females with extripated CA mate when placed in contact with males (Koudele et al. 1987).

4.1. Sexual Receptivity

The close relationship between JH biosynthesis and the onset of pheromone production, release, and oocyte maturation in *Supella* and *Blattella* (Smith et al. 1989; Smith and Schal 1990a; Liang and Schal 1993b) suggests that the CA might be required for production of sexual signals and development or expression of female receptivity. Females emit these signals and mate only when JH titers reach a certain threshold level. Thus, teneral adult *B. germanica* females are sexually unreceptive; they routinely mount courting males, but by extending their legs, they move the abdomen out of the reach of males, or by mounting the male from the side, they avoid copulation. There is a clear relationship between copulatory readiness and size of the basal oocytes in virgin females with daily access to males: In a group of females of the same age, females that mate early have larger oocytes (Liang and Schal 1993b). Unobtrusive time-lapse video records showed that females mated on average 5.7 ± 0.13 (SEM) days after the imaginal molt (Schal and Chiang 1995), when their basal oocytes averaged 1.36 ± 0.03 mm, 54% of their maximal length at ovulation (see Fig. 1.2). Similarly, after virgin females abort their second infertile ootheca, they refuse to mate until their oocytes reach 1.28 ± 0.03 mm. This suggested a relationship between oocyte maturation, or factors controlling oocyte growth (i.e., JH), and sexual receptivity. We therefore exposed females to conditions that would uncouple the usually tightly linked factors of female age and physiological (endocrine) state. An association between sexual receptivity and JH, independent of age, would provide support for the hypothesis that JH affects sexual receptivity. Various experimental manipulations, including ablation of the CA, therapy with JH analogs, CA denervation, ovariectomy, and changing nutrient quality, coupled with time-lapse infrared video recording, support the hypothesis that JH controls readiness of the female German cockroach to copulate (Schal and Chiang 1995).

4.2. Onset of Pheromone Production and Calling

Because the chemical identity of pheromones of both species was not known, regulation of the onset of pheromone production was studied by employing behavioral and EAG assays of whole-body or gland extracts. Thus, the relative amount of pheromone was estimated, not *de novo* biosynthesis. In both *B. germanica* and *S. longipalpa*, JH is required for production of sex pheromones and the expression of calling. Virgin *Supella* females initiated pheromone production at 4 days old and were calling at 6.15 + 0.91 days (Smith and Schal 1990a), in relation to increasing JH biosynthesis by the CA *in vitro* (Smith et al. 1989). Likewise, in *Blattella*, onset of calling and pheromone production in relation to increasing rates of JH synthesis suggested involvement of the CA (Liang and Schal 1993c, 1994). In both species, denervation of the CA from the brain of teneral females through nervi corporis allati I (NCA-I) or nervi corporis cardiaci I and II (NCC-I and II) transection significantly accelerates the onset age of calling (Fig. 1.3) (Smith and Schal 1990a; Liang and Schal 1994). Since this operation accelerates the rate of JH synthesis by the CA (Gadot et al. 1989b; Schal et al. 1993), these results lend further support that JH is a regulator of calling in both species. Extirpation of the CA (allatectomy) of newly emerged adult females prevents pheromone production and calling in both species, and both events can be restored by reimplantation of active CA or by treatment with JH or JH analogs. Although growth of the basal vitellogenic oocytes is controlled by and highly correlated with JH biosynthetic rates, direct or even intermediary involvement of the ovaries in regulating calling and pheromone production was excluded by ovariectomies in both species (unpublished; Smith and Schal 1990a).

Thus, in both species, females synthesize pheromones and initiate calling when JH reaches a certain threshold titer. In the absence of JH, both events are not expressed. It remains to be determined whether JH exerts its actions directly on secretory cells of the pheromone gland, or whether it acts indirectly by inducing the synthesis and/or release of pheromonotropic neuropeptides. The latter model would be similar to that of several noctuid moths in which JH induces pheromone production presumably by promoting the release of PBAN (Cusson et al. 1994a; Picimbon et al. 1995). The primary site of JH action in this scheme is likely to be the brain; and, if so, JH receptors should be found in the CNS.

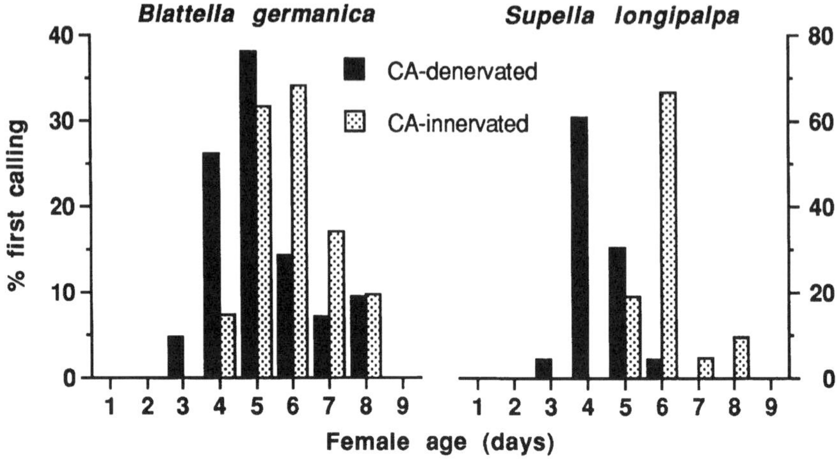

Figure 1.3. Effect of allatal nerve transection on calling behavior in *Blattella germanica* and *Supella longipalpa* females. Data were drawn from Liang and Schal (1994) and Smith and Schal (1990a), respectively.

5. Neural suppression of Pheromone Production and Calling in *Blattella* and *Supella*

Both pheromone production and calling are usually expressed only in receptive females seeking mates. In sexually immature adult females, absence of sexual signaling can be explained by low titers of JH since exogenous JH induces both calling and pheromone production in these females (see above). However, in older females, expression of both events (and sexual receptivity) is inhibited through specific signals to the CNS. Two important signals include cues associated with copulation and those from the ootheca. In spite of increasing rates of JH synthesis immediately after mating (Smith et al. 1989; Gadot et al. 1989a), the emission of sex pheromones is curtailed after copulation in both species. Termination of calling after copulation occurs through two successive events, with only minor differences between these two cockroach species. A first stage involves the physical insertion of a spermatophore into the bursa copulatrix, while a second stage involves presence of sperm in the spermathecae. In *Supella*, a brief and transient presence of a normal or spermless spermatophore (from a vasectomized male) in the female's genital atrium is sufficient to completely suppress calling in the first ovarian cycle; however, without migration of sperm into the spermathecae, females resume calling within several days, in relation to a new gonadal maturation cycle (Smith and Schal 1990b). While insertion of a spermatophore into the bursa copulatrix causes a temporary suppression of calling in *B. germanica*, sperm and/or associated secretions play a key role in the complete suppression of calling in mated females even in the first ovarian cycle (Liang and Schal 1994).

Mechanical pressure from the spermatophore appears to be the relevant signal in cockroaches, since implantation of artificial spermatophores also terminated calling (Smith and Schal 1990b). Also, chemical factors from the male appear to not be involved, since injection of sperm, spermatophores, spermatophore extracts, or hemolymph from mated females did not suppress calling in virgin females. Since females that mate with castrated males resume calling, sperm appear to be involved in the second stage of matedness. In both cockroach species, an intact ventral nerve cord is required for the inhibitory signals from mating to be effective suppressers of calling behavior (Smith and Schal 1990b; Liang and Schal 1994) as in *Nauphoeta cinerea* females, where an intact VNC is required to turn off sexual receptivity after mating (Roth 1962). Thus, it appears that nervous signals resulting from mechanical distention of sexual organs in the female ascend the VNC and inhibit calling.

Virgin females of both *Supella* and *Blattella* do not exhibit calling behavior while carrying infertile oothecae (Smith and Schal 1990a; Liang and Schal 1993b). This is also due to signals that ascend the VNC, since transection of the VNC in gravid virgin or mated females released the expression of calling behavior. However, the ventral nerve cord also transmits signals that suppress JH synthesis in gravid females (Roth and Stay 1962; Chiang et al. 1991a; Gadot et al. 1991). Therefore, the occurrence of calling behavior in gravid *B. germanica* after VNC transection could also be explained by activation of the CA and a rise in the titer of JH in the absence of VNC-transmitted inhibitory signals. To distinguish between direct suppression of calling by neural directives and indirect suppression through inhibiting a rise in JH, we treated gravid females with a JH analog and either removed the ootheca or transected the VNC. Intact virgin gravid females did not exhibit calling behavior even when treated with exogenous JH; however, such females initiated calling immediately after the ootheca was removed or the VNC cut (Liang and Schal 1994). Therefore, the ootheca plays a dual function in the control of calling behavior: It inhibits calling directly, as well as indirectly by suppressing JH synthesis. Both signals ascend the VNC, and CA inhibition is effected by the brain (Gadot et al. 1991).

6. Developmental Changes Control Pheromone Gland Function in *Blattella*

Mature endocrine or exocrine glands, wherein all the cellular synthetic machinery is present and functional, can be regulated through precise and rapidly reversible quantitative changes in rate-limiting enzymes along the biosynthetic pathway. PBAN appears to operate in this manner in many, primarily short-lived, lepidopterans. For example, the pheromone gland of *Helicoverpa armigera* and *H.*

zea, taken from virgin or mated females of various ages and during various photoperiods, were capable of responding to PBAN *in vitro* (Rafaeli 1994). However, probably more universal among insects is a slow, developmental regulation of gland function through changes in the cellular machinery available for hormone or pheromone synthesis. Such regulation can occur in three phases:

1. Organogenesis of a sex-specific gland in the adult must be developmentally regulated during the imaginal molt. In short-lived insects that emerge as reproductively competent teneral adults, it is expected that functional competence of the gland to produce pheromone might be regulated by the same factors that control its organogenesis (e.g., ecdysteroids—see Tang et al. 1991 for example).
2. In insects whose gland acquires functional competence during an imaginal maturational period, developmental regulation might be involved at the target tissue level and manifested as ultrastructural changes. Examples include many insects that mature sexually for several days after eclosion, as well as adult insects that enter diapause or migrate before the onset of reproduction. Competence to produce pheromone may be maintained for the life of the insect.
3. In insects in which reproduction is interrupted by long periods of sexual inactivity, developmental regulation at the target tissue level results in the cyclic acquisition and loss of competence through periodic maturation and atrophy of glandular machinery.

Nothing is known about organogenesis of the sex pheromone gland in *B. germanica*. However, based on studies of adult male-specific glands (e.g., Sreng and Quennedey 1976), we infer that each secretory unit in the female pheromone gland (glandular cell, duct cell, duct) is formed during the imaginal molt from four progenitor cells in response to low JH titer throughout the last stadium and a rise in ecdysteroids late in the instar.

The imaginal gland exhibits a clear pattern of developmental maturation in relation to sexual maturation of the female (Liang and Schal 1993c) and subsequent development cycles that correspond to reproductive cycles. In correspondence to a low titer of extractable pheromone, the secretory cells of newly formed glands in teneral females are small (about 5–10 μm in diameter), the nucleus is elongated and conspicuous, occupying about one-third of the cell volume, and only a few small secretory vesicles are evident (Fig. 1.2). The microvilli surrounding the end-apparatus are extremely short (10–20% of active) and few in number. As pheromone production increases with age in virgin females, so does the size of pheromone-secreting cells: Mature day-6 glands are characterized by large secretory cells (about threefold larger than on day 0) containing a large number of spherical or oval electron-lucid secretory vesicles, an end-apparatus lined with numerous long microvilli, and a large spherical nucleus (Liang and Schal 1993c).

Shortly after mating, while JH titer is high, the secretory cell begins to decrease in size, the secretory vesicles begin to lyse, and the microvilli shorten in relation to a rapid decline in pheromone content. In gravid virgin and mated females the pheromone gland regresses to the appearance of a day-0 gland and its pheromone activity is barely detectable (Liang and Schal unpublished). The glandular layer in gravid females appears even thinner than in day-0 females, the nuclei decrease in volume and become elongated and branched, and densely packed residual bodies, probably containing degenerated organelles, appear in the cytoplasm. After the infertile ootheca is aborted in virgin females, the secretory cells of the pheromone gland become hypertrophied again, as in day-6 virgins, in relation to a reinstatement of sexual receptivity, pheromone production, and calling. This suggests that the sex pheromone gland in *B. germanica* females may be developmentally regulated during the reproductive cycles of the adult female as well as during organogenesis in the preimaginal insect. Further research will be needed to verify this point and to delineate whether the pheromone gland responds to JH and ecdysteroids.

It is reasonable to presume that, in order to conserve cell energy, this type of regulatory mechanism should occur in insects requiring long-term arrestment of pheromone synthesis because of sexual inactivity (long gestation period, diapause, migration). Given that, developmental regulation of pheromone biosynthesis should play a less significant role in *S. longipalpa*, which exhibits more continuous JH cycles with very short periods of CA inactivity. However, our morphometric studies of the CA indicate that cyclic growth and atrophy of CA cells play a major role in the regulation of JH biosynthesis during reproductive cycles in both *Blattella* and *Supella* females (Chiang et al. 1991a; Chiang and Schal 1994). Thus, it is plausible that developmental regulation of pheromone biosynthesis is not restricted to species with long periods of gestation. In *Supella*, developmental changes in the pheromone gland might be involved in the long-term suppression of pheromone production after mating, rather than in tracking JH titers at each ovarian cycle as in *Blattella* (see Smith and Schal 1990a; unpublished).

7. Courtship-Inducing Contact Pheromone in *Blattella*: Synthesis and Release

To elucidate which step(s) of pheromone biosynthesis is (are) endocrine-regulated, it is necessary to understand the biosynthetic pathway. The courtship-inducing contact sex pheromone of *B. germanica* has been extensively studied in this regard. It consists of oxygenated derivatives of cuticular hydrocarbons (HCs) and includes 3*S*, 11*S*-dimethylnonacosan-2-one (C_{29} methyl ketone) as a major component, derivatives containing an alcohol or an aldehyde group at the 29 position, and the C_{27} methyl ketone homolog, 3,11-dimethylheptacosan-2-one (Nishida and Fukami 1983; Schal et al. 1990b). Since the cuticular HCs of *B.*

germanica include 3,7-, 3-9-, and 3,11-dimethylnonacosane as major components (Augustynowicz et al. 1987), and only the 3,11- branching patterns are seen in the pheromone components, we proposed that the 3,11- HC isomer might give rise in the adult female to the methyl ketone pheromone with the same methyl branching pattern (Jurenka et al. 1989). Radiotracer and carbon-13 nuclear magnetic resonance (NMR) studies of incorporation of valine, isoleucine, and methionine, as well as succinate, into methyl-branched HCs revealed that the branched-chain amino acids serve as precursors to the methylmalonyl-CoA, which serves as the methyl-branch donor (Chase et al. 1990). Together with carbon-13 NMR studies of propionate incorporation, these studies clearly show that the methyl branches are inserted early in chain elongation. Presence of methyl-branched fatty acids, including *n*-3,11-dimethyl fatty acids with 16–20 carbons, in the integument of female *B. germanica* and incorporation of [methyl-^{14}C]methylmalonyl-CoA into methyl-branched fatty acids by a microsomal fraction provided evidence for a microsomal fatty acid synthetase (FAS) (Juarez et al. 1992). The methyl-branched fatty acids formed by microsomal FAS are then presumably elongated and converted to methyl-branched hydrocarbons. Comparative kinetic studies on the microsomal and soluble FAS have been reviewed by Blomquist et al. (1993).

During the first two ovarian cycles, biosynthesis of methyl ketone pheromone *in vivo* (assayed with [1-^{14}C]propionate incorporation into methyl branches) and its accumulation on the epicuticle correspond to the pattern of JH biosynthesis by the CA *in vitro* (assayed with L-[methyl-^{3}H]methionine as a methyl donor) and oocyte maturation (Schal et al. 1990a, 1991, 1994). This suggested that the CA and JH were involved in regulating pheromone production. Removal of the CA reduced the amount of C_{29} methyl ketone on the cuticle, whereas the JHA hydroprene significantly accelerated both oocyte development and pheromone production (Schal et al. 1990a). Since JH analogs also restore pheromone production in decapitated and head-ligated females, factors from the corpora cardiaca (CC), such as PBAN or related factors, appear to not be involved.

While pheromone production is JH-mediated, the synthesis of the parent methyl-branched HC is dependent to a large extent on food availability, not JH (Schal et al. 1994). Normal females synthesize methyl-branched HC rapidly during their vitellogenic feeding stage and HC synthesis declines well before any appreciable decrease in JH biosynthesis, but concomitantly with reduced food intake. Conversely, allatectomized females eat less and synthesize less HC during the same time. While gravid females feed only sporadically and concurrently synthesize little HC, rates of HC synthesis in allatectomized females (without oothecae) remain relatively high. An important site for deposition of HC in normal females is the basal oocytes, because the amount of HC deposited into oocytes equals the amount on the epicuticle (see below). In the absence of ovarian uptake of HC in alltectomized females, the hemolymph titer of HC increases well beyond normal levels (Schal et al. 1994). As internal HCs increase, so do cuticular HCs and both internal (hemolymph) and cuticular methyl ketone pheromones. These patterns suggest that normally, stage-specific feeding regulates the amount of 3,11-dimethylnonacosane available for pheromone production. Thus, availability of *de novo* synthesized substrate is one mechanism by which *B. germanica* regulates contact pheromone production.

A second mechanism involves the stage-specific JH-mediated conversion of the HC to methyl ketone (Fig. 1.1B). Chase et al. (1992) concluded from radiolabeling studies that the pheromone arises from conversion of 3,11-dimethylnonacosane to 3,11-dimethylnonacosan-2-one via an alcohol intermediate, and they showed that JH analogs can stimulate the conversion of radiolabeled hydrocarbon to methyl ketone in females. Since [11,12-$^3H_2$]3,11-dimethylnonacosan-2-ol is readily metabolized to methyl ketone by both males and females, and only vitellogenic females efficiently hydroxylate the radiolabeled HC, it appears that formation of the 3,11-dimethyl alcohol occurs only in adult females and that this step might be under JH control (Chase et al. 1992).

To understand the mechanisms by which HCs and pheromones are compartmentalized among sites of synthesis, the hemolymph, the ovaries, and the cuticle, it is necessary to elucidate the sites of synthesis and dynamics of transport among target deposition sites. Both HC and contact sex pheromone in *B. germanica* females are synthesized primarily by abdominal integument, while incorporation of [^{14}C]propionate into HC and pheromone in other body parts (head, wings, legs, thorax, fat body, digestive tract, and vitellogenic ovaries) is negligible (Gu et al. 1995). Moreover, the sternites synthesize significantly more methyl-branched HC and methyl ketones than the tergites. These patterns strongly favor abdominal oenocytes or epidermal cells as the sites of synthesis of both HC and pheromone. However, occurrence of both lipids internally after extensive washing of the cuticle prompted us to examine the distribution of HC and pheromone in internal tissues. Most of the internal HCs and their methyl ketone derivatives can be recovered from ovaries and the integument. It appears that hemolymph carries and circulates HC and pheromone in the female.

In most insects the mechanisms of lipid transport involve lipophorin (reviews: Shapiro et al. 1988; Kanost et al. 1990; Van der Horst 1990; Law et al. 1992). Using *B. germanica* female hemolymph, we isolated and purified a high-density lipophorin (HDLp, 1.109 ± 0.002 g/ml) by KBr gradient ultracentrifugation and showed that it was the only hemolymph protein that carries HC and contact pheromone (Gu et al. 1995). Injected radiolabeled HC was recovered from HDLp, as was the radiolabeled pheromone to which it was converted. These results suggested the hypothesis that lipophorin loads newly synthesized HC and pheromone from the abdominal epidermis and transports them to various tissues, including the ovaries and epicuticle. A time course of incorporation of [^{14}C]propionate showed that newly synthesized methyl-branched HCs appear first in the epidermal fraction and hemolymph, and later on the epicuticle, supporting the Lp-mediated transport hypothesis. Experimental support for this idea was provided by severing the veins to the forewings, thus blocking hemolymph transport to the wings. The amount of newly synthesized HC on the wings was significantly lower than on intact forewings, and the amount of topically applied radiolabeled HC that was recovered in the hemolymph was greater than on the surface of the wings (Gu et al. 1995). These results, and the abundance of epicuticular HCs on nonsynthesizing body parts (e.g., wings, legs), clearly show that transport of HCs and their derivative pheromones is mediated by a HDLp, which shuttles newly synthesized lipids from the abdominal oenocytes to epicuticular and internal deposition sites, including the ovaries.

8. Conclusions

Often, the reproductive state of individual insects can be inferred from the sexual signals that they produce; they exhibit specific behaviors associated with sexual receptivity only when they are reproductively competent. Receptive virgin individuals produce songs, displays, pheromones, and other signals that attract conspecifics or elicit courtship in members of the opposite sex, while immature virgins and mated individuals do not solicit mates. In females with reproductive cycles, such as cockroaches, these behaviors usually coincide with ovarian maturation, suggesting that these signals may ensure courtship by males only at appropriate stages in the female's gonadal maturation cycle. We propose that cyclic acquisition and loss of sexual receptivity and emission of mate-recruiting signals might be related to the ecology of the insect. In some species, especially in solitary short-lived insects, the female may remain sexually receptive throughout the reproductive cycle, including periods during which she does not produce or emit sexual signals, because she is unlikely to contact and be courted during such periods. In contrast, coordination of copulatory readiness with mate-solicitation should be more important in gregarious long-lived species. Mate attraction by one receptive female may cause nonsoliciting females within the group to respond to male courtship if their mating receptivity is not physiologically restrained. In such species, especially those with repeating gonadotrophic cycles (e.g., cockroaches), it is likely (but not necessary) that hormones that regulate sexual physiology will also coordinate the expression of specific sexual behavioral responses. Thus, in insects in which JH regulates reproductive physiology, it would be efficient for JH also to coordinate the production of sex recruitment signals and sexual receptivity.

Juvenile hormone exerts major pleiotropic effects on insect development and reproduction. It acts as a repressor in larval growth and development, and it regulates some major gene products in adult females, including vitellogenin, a yolk protein synthesized by the fat body, and oothecin, an egg case protein synthesized in the accessory sex glands. This may be why a close relationship has evolved in some cockroaches between gonadal maturation and sexual behavior, as in many metazoans. Our results support the hypothesis that JH controls sexual receptivity and sexual signals emited by *Blattella* and *Supella* females, including sex pheromones. The activity of the CA in adult female cockroaches is dependent upon and modulated by both internal states (physiological, nutritional) and environmental stimuli (temperature, photoperiod, social interactions). Favorable stimuli result in a graded lifting of brain inhibition upon the CA, permitting the synthesis and release of JH. In addition to inducing protein synthesis in the female, in some species JH also stimulates the female cockroach to produce sexual signals, including both attractant and courtship-eliciting pheromones, and to become sexually receptive. Importantly, mated females also produce JH, but they neither emit sex pheromone nor regain sexual receptivity. It thus appears that while JH is required for the expression of copulatory readiness and pheromone production in *S. longipalpa* and *B. germanica*, signals associated with copulation (spermatophore, sperm, accessory secretions) can inhibit this behavioral state even when titers of JH are permissive for sexual receptivity.

Our results suggest that JH plays a "permissive" indirect role in the regulation of calling behavior. It is necessary for the expression of calling, but is not sufficient in the presence of ascending inhibitory signals from the terminal abdominal ganglion. These neural signals might inhibit the release of humoral signals from the brain, whose release is possible only when JH titers are high. Alternatively, neural signals that ascend the VNC might inhibit a JH-inducible motor program in the CNS. While signals associated with the ootheca are transient (up to 36 hours in *Supella*; 21 days in *Blattella*), inhibitory signals from mating are effective for the life of the female so long as enough viable sperm reside in the spermathecae. Clearly, the involvement of PBAN or PBAN-like peptides must be examined in production and emission of volatile pheromones in cockroaches. Conversely, JH appears to have direct effects on contact pheromone production since decapitated females treated with JH synthesize pheromone. The effect of JH needs to be confirmed *in vitro* using isolated epidermal tissues and oenocytes.

Developmental regulation of pheromone production and behavioral control of pheromone release are important mechanisms in *B. germanica*, and probably other insects that experience protracted periods of reproductive inactivity. Although we are aware of several reports on pheromone gland organogenesis during the imaginal molt, maturational phenomena, such as those we describe for the pheromone gland of *B. germanica* and for the CA of cockroaches, appear to be rare. This, we believe, is due to the scarcity of studies on long-lived insects with cyclic reproduction. It is likely that in such insects with long periods of reproductive inactivity (e.g., pregnancy, migration, diapause, nutrient deficiency), regression of pheromone glands will be found as a mechanism to conserve energy and ensure sexual inactivity. For example, our work with regulation of *B. germanica* CA reveals an early acquisition of competence to produce JH (CA maturation), followed by periodic volumetric regression of CA cells accompanied by loss of competence to biosynthesize JH, even in the presence of late precursors along the biosynthetic pathway (Gadot et al. 1989b; Chiang et al. 1991a; Chiang and Schal 1994). These volumetric changes in CA cells are synchronized and paced by various cues, including ovarian factors, resulting in a precise correlation between JH cycles and oocyte development. Thus, in *B. germanica*, rates of JH synthesis during ovarian cycles are mainly regulated through cyclic maturation and regression of cellular machinery in the CA, and these rates can be modified rapidly by allatostatic and allototropic modulation of rate-limiting enzymes.

To our knowledge, *B. germanica* is the only species in which lipophorin transports a sex pheromone. It remains to be seen whether Lp-mediated transport of sex pheromone is unique to this species or a general principle for long-chain cuticular lipids. Indirect evidence from other species suggests that an internal transport pathway, possibly involving Lp, might also occur in species that utilize medium- and short-chain HCs as sex pheromones (Blomquist et al. 1993; Charlton and Roelofs 1991; Schal et al. unpublished).

Acknowledgments

We thank our many colleagues and collaborators on these projects, and in particular R.A. Jurenka and W.L. Roelofs. W.L. Roelofs also offered critical comments on an earlier draft. Without G. Prestwich's and K. Touhara's (Stony Brook, N.Y.) generous gifts of radiolabeled 3,11-dimethylnonacosane and the alcohol and methyl ketone derivatives, much of the work on the synthesis and transport of the contact pheromone would not have been possible. Work in CS's laboratory was supported in part by the Blanton J. Whitmire Endowment and grants from the NSF (IBN-9407372) and USDA-NRI (9501922).

2

Control of Pheromone Production in Moths

Ashok K. Raina

1. Introduction

Most species of moths are nocturnal in behavior, and mating often requires the production and release of a sex pheromone by the females to attract conspecific males. Moths in general are short-lived; some species eclose with mature oocytes and may not even feed during their adult life. Such species mate, lay eggs, and die within a short period of time. Based on these facts, Barth (1965) hypothesized that the cyclic production of pheromone in moths may not require hormonal regulation. However, moths often exhibit a diel periodicity of pheromone production, suggesting the requirement of some regulatory mechanisms. In nocturnal species, peak pheromone production occurs during the scotophase; the resulting avoidance of mating in daylight may reduce the risk of predation. Additionally, moths such as the corn earworm, *Helicoverpa zea*, may also delay production of pheromone until a suitable host is located for oviposition (Raina 1988), further suggesting the need for regulation.

Neuroendocrine control of pheromone production and release was first proposed by Riddiford and Williams (1971) for two species of moths, *Hyalophora cecropia* and *Antheraea polyphemus*. They suggested that pheromone release in these species was induced by environmental cues integrated by the brain, which in turn stimulated release of a hormone from the intrinsic cells of the corpora cardiaca (CC). Based on results from experiments in which transfusion of hemolymph from calling into noncalling females of *A. polyphemus* stimulated calling in the latter, Riddiford (1974) speculated that a calling hormone released from the CC was possibly involved. When the experiments were subsequently repeated with the same species, it appeared the CC played no role in the induction of calling (Sasaki et al. 1983). However, injection of hemolymph or CC homogenates from calling females into isolated abdomens of *H. cecropia* did induce calling in 30–60% of cases. They further reported that the activity was not species-specific and suggested that a neurosecretory factor present in the CC and hemolymph may modulate the neural control of calling behavior.

These earlier studies laid the groundwork for the explosion of research in the 1980s on regulation of pheromone production and calling in moths. This chapter reviews some of the historical aspects of such studies on *H. zea* and other species and presents information on our current knowledge of regulation at the molecular level.

2. Control of Pheromone Production in the Corn Earworm

2.1. Isolation and Identification of PBAN

H. zea females exhibit a diel periodicity of pheromone production (Raina et al. 1986), and all mating related activities are confined to the scotophase. It was in this species that pheromone production was unequivocally shown to be under the control of a peptidic factor (Raina and Klun 1984). Subsequently, a 33-amino-acid residue peptide termed the *pheromone biosynthesis activating neuropeptide* (PBAN) was isolated from the brain–subesophageal ganglion (brain–SOG) complex of *H. zea* (Raina et al. 1989a). Kitamura et al. (1989) isolated and identified a similar pheromonotropic peptide from the heads of silk worm, *Bombyx mori*. Recently PBAN has been identified from the gypsy moth, *Lymantria dispar* (Masler et al. 1994). Structures of the three PBANs are shown in Fig. 2.1. The

5 10

Hez-PBAN **Leu**-Ser-Asp-**Asp-Met-Pro-Ala-Thr**-Pro-**Ala-Asp-**

Bom-PBAN **Leu**-Ser-Glu-**Asp-Met-Pro-Ala-Thr**-Pro-**Ala-Asp-**

Lyd-PBAN **Leu**-Ala-Asp-**Asp-Met-Pro-Ala-Thr**-Met-**Ala-Asp-**

15 20

Gln-Glu-Met-**Tyr**-Arg-Gln-Asp-**Pro-Glu**-Gln-Ile-

Gln-Glu-Met-**Tyr**-Gln-Pro-Asp-**Pro-Glu**-Glu-Met-

Gln-Glu-Val-**Tyr**-Arg-Pro-Glu-**Pro-Glu**-Gln-Ile-

25 30

Asp-**Ser-Arg**-Thr-Lys-**Tyr-Phe-Ser-Pro-Arg-Leu**-NH_2

Glu-**Ser-Arg**-Thr-Arg-**Tyr-Phe-Ser-Pro-Arg-Leu**-NH_2

Asp-**Ser-Arg**-Asn-Lys-**Tyr-Phe-Ser-Pro-Arg-Leu**-NH_2

Figure 2.1. Amino acid sequences of PBANs from *Helicoverpa zea* (Hez-PBAN), *Bombyx mori* (Bom-PBAN), and *Lymantria dispar* (Lyd-PBAN). Amino acids shown in bold letters represent sequence homology common to all three PBANs.

three PBANs exhibit ~80% sequence homology. Pheromonotropic activity has also been reported in brain–SOG from a number of other insect species (reviewed in Raina 1993).

2.2. *Structure and Activity of Hez-PBAN*

The C-terminal amide is essential for biological activity of PBAN; the free acid version of the peptide is almost 1000-fold less active. In an *in vivo* bioassay, 2–4 pmol of synthetic Hez-PBAN stimulated pheromone production to levels found in control females (Raina et al. 1989a). Oxidized PBAN (in which the two Met residues were converted to sulfoxides) was about fivefold more active than the unoxidized peptide (Raina et al. 1991). The C-terminal pentapeptide, common to all three identified PBANs and to several other myotropic peptides (Schoofs et al. 1993), is the minimum sequence with pheromonotropic activity (Raina and Kempe 1990). The C-terminal hexapeptide was slightly more active, and the substitution of L-Phe with D-Phe produced an analog with even higher activity. However, the activity was about 10-fold lower than the full-length PBAN. The conformation of this hexapeptide in a mixed solvent of water and DMSO revealed a β-turn at the Ser-Pro-Arg-Leu fragment, suggesting that this is the active conformation for the peptide (Wang et al. 1994). Recently, it was reported that modification of the N-terminal amino group of the hexapeptide by addition of acetyl, benzoyl, ethyl, benzyl, pyroglutamyl, or D-alanyl groups resulted in analogs that were 10–30 times more active than the unoxidized PBAN in *B. mori* pheromonotropic assay (Kuniyoshi et al. 1992a; Nagasawa et al. 1994). In a separate study, Kuniyoshi et al. (1992b) reported that Lem-PK, Lom-PK, Lom-MT-I, and Lom-MT-II (all of which share the C-terminal pentapeptide sequence FXPRL-NH_2 with PBAN, and where X is Thr, Val, or Ser) exhibited pheromonotropic activity in *B. mori* that was one order of magnitude more potent than that of unoxidized PBAN. Lom-MT-II was the most active, with activity similar to that of the oxidized PBAN. In both studies the greater potency of these peptides was attributed to their possible resistance to aminopeptidases. Recently, Lom-MT-II was tested for pheromonotropic activity in *H. zea* and found to be more active than Hez-PBAN at the 0.1 and 1 pmol doses (Fig. 2.2). Response at the higher doses was not significantly different from that of the 1 pmol dose. Based on structure–activity studies, Raina and Kempe (1992) had reported that an internal pentapeptide fragment of Hez-PBAN amidated at its C-terminus was very active at the low dose of 1 pmol but inactive at 100 and 1000 pmol doses. Since then we have repeated the synthesis and testing of this fragment. Despite our earlier finding, we are now unable to obtain biological activity with this analog.

2.3. *Immunocytochemistry and Release*

Immunohistochemical studies using polyclonal antibodies raised against Hez-PBAN revealed three clusters of PBAN immunoreactive cells along the ventral

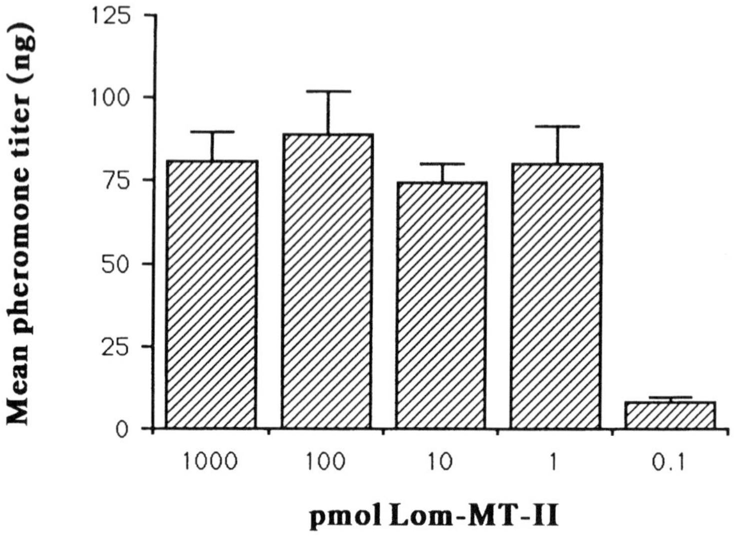

Figure 2.2. Pheromonotropic response of *Locusta migratoria* myotropin II (Lom-MT-II) tested in ligated *Helicoverpa zea* females. Values are means ± SEM, $N = 10$.

midline of SOG (Kingan et al. 1992). Cells in the most posterior cluster (labial) have axons projecting to the CC and into the esophageal nerve via the NCC-III. Immunoreactive axons from the other two clusters project into the maxillary nerve and then into the CC via the NCC V. In addition, axons from the maxillary cluster project into the ventral nerve cord (VNC); these axons arborize in all segmental ganglia. Based on the presence of pheromonotropic activity in the CC, a known neurohemal structure, Raina and Menn (1987) had proposed that PBAN is produced continuously in the SOG and released into the hemolymph from the CC. However, Teal et al. (1989a) proposed an alternate pathway for the release of PBAN resulting in pheromone production in *H. zea*. They suggested that pheromonotropic factors are transported via the VNC to the terminal abdominal ganglion (TAG) and that the nerves posterior to TAG had to be intact to stimulate pheromone production. This suggestion was based on the fact that pheromonotropic activity (Teal et al. 1989a) or PBAN immunoreactivity (Kingan et al. 1992) was not detected consistently in the hemolymph. Raina et al. (1991) have shown that ligation or cutting of the VNC or removal of TAG did not prevent pheromone production in intact *H. zea* females. Detection of PBAN in hemolymph will require more sensitive assays, because of the extremely minute quantities involved. Recently, we used HLPC-fractionated hemolymph collected from *H. zea* females during early scotophase and subjected the fractions to enzyme-linked immunosorbent assay (ELISA). Immunoreactivity was found in fractions that corresponded in retention time with PBAN (Kingan and Raina unpublished). While these results suggest that PBAN is released from the CC and the esophageal nerve into the hemolymph to activate pheromone production (Fig. 2.3A), direct release of PBAN into the hemolymph remains to be demonstrated along with experiments to correlate hemolymph levels of PBAN with pheromone accumulation.

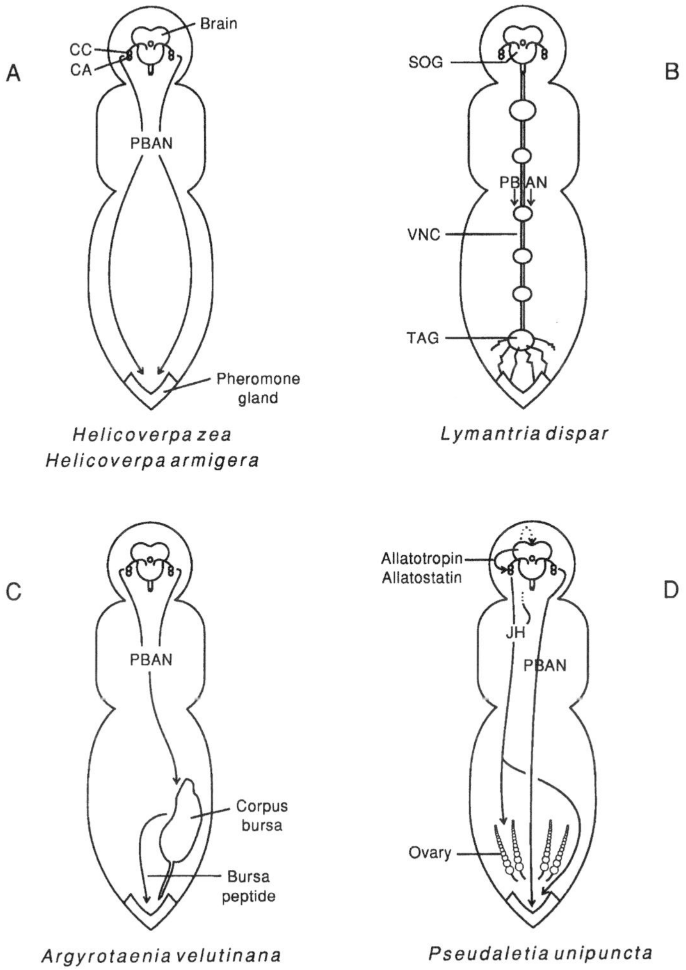

Figure 2.3. Various models showing probable modes of transport and sites of action of pheromonotropic factors in *Helicoverpa zea* and *Helicoverpa armigera* (**A**), *Lymantria dispar* (**B**), *Argyrotaenia velutinana* (**C**), and *Pseudaletia unipuncta* (**D**). Ca, corpora allata; CC, corpora cardiaca; JH, juvenile hormone; SOG, subesophageal ganglion; TAG, terminal abdominal ganglion; VNC, ventral nerve cord.

2.4. PBAN Gene and Expression in Baculovirus

The Hez-PBAN gene was isolated and partially characterized from a genomic library of *H. zea* (Davis et al. 1992b). Sequence analysis of the cloned gene showed that a 0.63-kb intron interrupts the PBAN gene in the codon encoding amino acid 14 (Met). An eight-amino-acid peptide, which shares the C-terminal pentapeptide sequence with that of PBAN, follows the PBAN gene. Upstream of the PBAN gene the codon GTG was suggested to code for Met (translation initiation). However, if GTG is instead designated to code for its usual amino acid Val, we find an 18-amino-acid peptide that has 83% sequence homology with a pheromonotropic peptide Pss-PT, isolated from the army worm, *Pseudaletia separata*, larvae (Matsumoto et al. 1992a). In *H. zea* we are designating this peptide as pheromonotropic-melanizing peptide (Hez-PMP). It is interesting to note that Hez-PMP has the same C-terminal pentapeptide sequence as Lem-PK and Lom-MT-II, both of which are highly pheromonotropic (Kuniyoshi et al. 1992b).

Sato et al. (1993) reported the characterization of a cDNA clone from the cDNA library derived from *B. mori* SOG. The mRNA was reported to encode an open reading frame consisting of 192 amino acid residues. Besides coding for Bom-PBAN, the gene codes for a diapause hormone (Bom-DH) and three other peptides (7, 17, and 8 amino acid residues, respectively). The 17-mer bears only about 33% homology to Pss-PT, most of it in the C-terminal region. The octapeptide has the same C-terminal pentapeptide sequence as PBAN. Recently, Ma et al. (1994) isolated and characterized the *H. zea* cDNA that codes for Hez-PBAN. In this case the open reading frame is reported to code for a 194-amino-acid precursor protein. Besides PBAN, the gene codes for four other peptides corresponding to Bom-DH, Hez-PMP, a heptapeptide, and an octapeptide. Both of these studies used Northern hybridization to show that the expression of this gene was confined to SOG only.

A synthetic gene of Hez-PBAN was recently cloned behind the polyhedrin promoter of AcMNPV and successfully expressed in cell culture but not in *H. zea* larvae or adults that were injected with the recombinant virus (Vakharia et al. 1995). The purpose of this study was to see if females of *H. zea* injected with the virus would produce sex pheromone during photophase. *H. zea* females were induced to produce pheromone during photophase by feeding starved females a sucrose solution containing PBAN or its penta- and hexapeptide analogs (Raina et al. 1994a). PBAN was also detected by ELISA in the hemolymph of fed females, indicating that some of the peptide survives intact the passage through the wall of the alimentary canal.

2.5. Pheromone Biosynthesis

In *Helicoverpa* and *Heliothis* species, the main component of female sex pheromones, (*Z*)-11-hexadecenal, is produced by oxidation of the corresponding alcohol

(Teal and Tumlinson 1986). In *H. zea*, using *in vivo* and *in vitro* gland cultures and radiolabeled precursors, Jurenka et al. (1991a) reported that presence of PBAN did not affect the amount of intermediates in the biosynthetic pathway, but greatly increased the amount of (*Z*)-11-hexadecenal. They concluded that PBAN controls pheromone biosynthesis in *H. zea* by regulating a step in or prior to fatty acid biosynthesis. *In vitro* studies conducted with pheromone glands from *H. zea* and *H. armigera* (Rafaeli et al. 1990a; Jurenka et al. 1991a) indicate that PBAN acts directly on the gland (Rafaeli 1994). An increase in intracellular cAMP in pheromone glands of *H. armigera*, incubated with PBAN, and a pheromonotropic response obtained with forskolin, 8-bromo-cAMP, and isobutyl-methylxanthine show that the response to PBAN may be mediated by cAMP (Rafaeli and Soroker 1989b; Rafaeli et al. 1990b). Further, extracellular Ca^{2+} was shown to be necessary for PBAN activity on isolated pheromone glands of *H. zea* (Jurenka et al. 1991a). In a recent study on *Helicoverpa* spp. it was reported that PBAN has a dual receptor mechanism, one activating adenylate cyclase and the other possibly activating phosphatidyl inositol hydrolysis, thereby activating intracellular Ca^{2+} release (Rafaeli 1994). Knowledge of the PBAN-receptor will help in further elucidating the steps in pheromone biosynthesis that are regulated by PBAN.

2.6. Mating-Induced Termination of Pheromone Production

Females of *H. zea* are known to mate several times during their life, but no more than once each night. Pheromone production is terminated following mating but is resumed during the subsequent night (Raina 1988). A factor present in the hemolymph of mated females was shown to terminate pheromone production in virgin females (Raina 1989). In *H. zea*, postmating pheromonostasis can be reversed by injecting PBAN in mated females, giving rise to the speculation that the mating factor must be disrupting the release or the action of PBAN (Raina 1989; Raina et al. 1994b). A peptide isolated from the male accessory glands was reported to be responsible for pheromonostasis but not termination of calling (Kingan et al. 1993). The pheromonostatic peptide (PSP) contains 57 amino acid residues with a single disulfide bridge (Kingan et al. 1994). Injection of as little as 0.3 male equivalents of PSP in decapitated *H. zea* females (that had previously been injected with PBAN) evoked a near-complete depletion of sex pheromone. In intact virgin females the activity of PSP was much less, indicating that continued release of PBAN may be responsible for the higher pheromone titer. Females whose VNC was cut in the abdominal region prior to mating did not show a decline in pheromone titer. Based on these observations, Kingan et al. (1994) suggested that the depletion of pheromone following mating may be caused by an ascending neural signal that stops PBAN-induced pheromone biosynthesis, whereas PSP may be responsible for clearing the pheromone present in the pheromone gland.

3. Control of Pheromone Production in the Gypsy Moth

Laboratory-reared females of the gypsy moth, *Lymantria dispar*, exhibit a daily rhythm of pheromone titer with peak production in the late afternoon (Charlton and Cardé 1982; Giebultowicz et al. 1992). However, females from a wild population held in outdoor cages under hot afternoon conditions (33°C) exhibited maximal pheromone titer in the early morning and almost negligible titer in the afternoon (Giebultowicz et al. 1992). During the isolation and identification of Hez-PBAN, it was observed that brain–SOGs of a number of species of moths exhibited pheromonotropic activity (Raina and Klun 1984). Neck ligation in females of all but one species, the cabbage looper, *Trichoplusia ni*, resulted in the termination of pheromone production within a few hours after ligation. Pheromone production in females of these species could be restored by injecting Hez-PBAN, except in the gypsy moth (Raina et al. 1989a). While ligation or cutting of the VNC in *H. zea* females had no adverse effect on pheromone production (Raina et al. 1991), such procedures in *L. dispar* resulted in termination of pheromone production (Thyagaraja and Raina 1994). Masler and Raina (1993) reported that a pheromonotropic peptide was present in the brain–SOG of *L. dispar*. The peptide was isolated and identified as a 33-amino-acid residue Lyd-PBAN (Masler et al. 1994). However, when this Lyd-PBAN was injected into ligated gypsy moth females, only a small increase in pheromone production was noticed, as had also been observed with Hez-PBAN. On the other hand, significant pheromone production was obtained if PBAN was administered to the cut end of VNC or injected directly into the TAG of decapitated females (Thyagaraja and Raina 1994). In addition, electrical stimulation of the TAG also resulted in significant pheromone production. In all these experiments, high pheromone titer was obtained only if the VNC was cut posterior to the thoracic ganglia, suggesting that the thoracic ganglia exert an inhibitory influence on pheromone accumulation. The above results suggest that in the gypsy moth, PBAN activates pheromone production via a neural route (Fig. 2.3B).

Postmating termination of pheromone production in the gypsy moth females is under a two-step neural control (Giebultowicz et al. 1991a). No hemolymph-borne factor was reported to be involved in the pheromonostatic response, but an intact neural connection of the spermatheca with the central nervous system appeared to be essential (Giebultowicz et al. 1991b).

4. Control of Pheromone Production in the Redbanded Leafroller

In the redbanded leafroller, *Argyrotaenia velutinana*, Jurenka et al. (1991c) reported that pheromone production was under the control of a neuropeptide produced in the brain. However, pheromone production was significantly higher in cultures of isolated abdomens treated with PBAN than in pheromone gland cultures. The higher pheromone titer in the abdominal preprations was attributed to the involvement of a second pheromonotropic factor originating in the corpus bursae. In a subsequent study, Fabriàs et al. (1992) reported that homogenates of the bursa elicited similar responses in both abdomen and isolated gland cultures, whereas PBAN elicited a very low response in gland cultures, in which the bursa was absent. Based on these results they suggested that PBAN is released into the hemolymph to stimulate the release of a bursa factor, which in turn stimulates pheromone biosynthesis in the pheromone gland (Fig. 2.3C). The factor, a peptide, has been partially characterized (W.L. Roelofs personal communication). In two related tortricid species, *Epiphyas postvittana* and *Planotortrix octo*, the female bursa copulatrix does not appear to contain a pheromonotropic factor (Foster and Roelofs 1994).

5. Control of Pheromone Production in the True Armyworm

Females of the true armyworm, *Pseudaletia unipuncta*, are relatively long-lived, and, depending on environmental factors, delay their sexual activity until after long-range migration to a suitable area (Cusson and McNeil 1989a). Pheromone release in this species was shown to be directly correlated with ovarian development. Further, allatectomized females failed to produce or release sex pheromone, and the effect could be reversed by treatment with JH. However, neck-ligated females could be stimulated to produce pheromone by injecting PBAN but not JH. Based on these results, Cusson and McNeil (1989a) and Cusson et al. (1994a) presented a model in which they proposed that JH plays an indirect role in stimulating pheromone biosynthesis by modulating either production or release of PBAN or increasing the responsiveness of the pheromone gland to PBAN (Fig. 2.3D). JH is also known to stimulate egg development in this and other lepidopterans, and reproductive maturity may affect pheromone production.

6. Control of Pheromone Production in the Cabbage Looper

Females of the cabbage looper, *Trichoplusia ni*, do not exhibit a diel periodicity of pheromone production; in addition, they do not show a marked decrease in pheromone titer following decapitation or mating (Tang et al. 1989). However, extracts of their brain–SOG were shown to contain high pheromonotropic activity when tested in *H. zea* females (Raina and Klun 1984). It was recently shown that the pheromone glands of *T. ni* females develop competency to produce pheromone prior to adult eclosion as a result of reduction in ecdysteroid titer (Tang et al. 1991). It is possible that pheromone biosynthesis is initiated at that time by a pheromonotropic factor, and the process does not need continuous activation in *T. ni* as in other species.

Acknowledgments

I thank Wendell Roelofs for organizing an excellent session on Control of Pheromone Production; he, together with Sonny Ramaswamy (whom I also thank), critically reviewed the manuscript. I also thank Timothy Kingan and Jadwiga Giebultowicz for many useful discussions during the course of writing this review.

3

Regulation of Pheromone Production in Lepidoptera: The Need for an Ecological Perspective

Jeremy N. McNeil, Johanne Delisle, and Michel Cusson

1. Introduction

Recently a graduate student in molecular biology, working on aspects of insect reproduction, was asked what he knew about the life history and ecology of the species he was studying. While the initial response "Oh, it lives in the rearing room down the hall" was accompanied by a telltale grin, it became clear after further discussion that the student's knowledge of his subject species did no extend far beyond the lab bench. The object of this story is not to slight molecular biologists—because behavioral ecologists and evolutionary biologists would not fare better if asked about the physiology or biochemistry of the species they work with each day—but rather to point out that we all tend to work within a somewhat narrow context.

In this chapter we wish to make a case for interdisciplinary research when addressing questions relating to both basic and applied aspects of semiochemicals. For example, a broader perspective could provide insight into our understanding of pheromone production, regardless of the level at which each scientist actually works. Even a rather superficial examination of several Lepidoptera currently or previously used to study the physiology of pheromone production shows not only a wide taxonomic diversity but also a marked variability with respect to a variety of life history traits. For example, in the three univoltine species listed in Table 3.1 the generally monandrous gypsy moth, *Lymantria dispar*, diapauses in the egg stage, while the polyandrous spruce budworm, *Choristoneura fumiferana*, and the presumed monandrous cecropia, *Hyalophora cecropia*, diapause in early larval and pupal stages, respectively. Furthermore, only spruce budworm adults are known to undertake long-distance migratory flight, generally by mated females in response to deteriorating stand quality (Greenbank et al. 1980). Similar differences in the life history parameters are seen with the multivoltine species

Table 3.1. Comparisons of certain life history traits of some lepidopteran species used in the study of physiological and biochemical aspects of pheromone biology

Family/species	Diapause[a]	Migratory flight	Voltinism	Frequency
Lymantriidae				
Lymantria dispar	Egg (O)	No	Univoltine	Monandrous
	Leonard (1968)	ODell and Mastro (1980)	Leonard (1968)	Doane (1968)
Noctuidae				
Heliothis zea	Pupal (F)	Yes	Multivoltine	Polyandrous
	Roach and Adkisson (1970)	Phillips (1979)	Hayes (1991)	Callahan (1958)
Pseudaletia unipuncta	Adult (F)	Yes	Multivoltine	Polyandrous
	McNeil (1987)	McNeil (1987)	Breeland (1958)	Callahan and Chapin (1960)
Saturnidae				
Hyalophora cecropia	Pupal (O)	No	Univoltine	Unknown[b]
	Williams (1946)	Williams (1946)	Willis et al. (1974)	
Sphingidae				
Manduca sexta	Pupal (F)	No	Multivoline	Unknown
	Rabb (1966)	Rabb et al. (1959)	Rabb (1966)	
Tortricidae				
Choristoneura	Larval (O)	Yes	Univoltine	Polyandrous
fumiferana	Jaynes and Speers (1949)	Greenbank et al. (1980)	Jaynes and Spears (1949)	Outram (1968)
Choristoneura	Larval (F)	No	Multivoltine	Polyandrous
rosaceana	Gangavalli and AliNiazee (1985)	Chapman et al. (1968)	Chapman et al. (1968)	Delisle (unpublished data)
Platynota stultana	No	No	Multivoltine	Monandrous
	Atkins et al. (1957)	Atkins et al. (1957)	Atkins et al. (1957)	Webster and Cardé (1984)

[a]Facultative (F) and obligatory (O) diapause.

[b]Probably monandrous, Riddiford (personal communication).

(Table 3.1). It is clear that each species has evolved to cope with very different suites of environmental conditions, and, when considered from an ecological perspective, it seems unlikely that the physiological processes involved in pheromone production will be identical. To elaborate on this point we will examine two phenomena associated with pheromone production: (i) the age at which virgin females first emit pheromone following emergence, a parameter that may permit the separation of migrant and resident species of Lepidoptera (McNeil 1986), and (ii) the degree of polyandry (multiple mating) in Lepidoptera.

2. The Age of Sexual Maturity

The conclusion of the initial research on the physiological regulation of pheromone production in Lepidoptera was that juvenile hormone (JH) did not play a significant role (Barth 1965; Hollander and Yin 1982; Itagaki and Conner 1986; Riddiford and Williams 1971; Sasaki and Riddiford 1984; Sasaki et al. 1983; Tang et al. 1987). However, these studies were carried out using nonmigrant species, such as the gypsy moth and the cecropia moth. In nonmigrant species the most effective reproductive strategy would be to mate as soon as possible following emergence, because any delay would only have potentially negative effects on overall lifetime reproductive success. In contrast, the rapid onset of reproduction could be detrimental for species that migrate in response to habitat deterioration, especially if migratory flight is generally initiated by sexually immature individuals. In resident species, such as the North American and Eurasian populations of the gypsy moth (Charlton and Cardé 1982), the oblique-banded leafroller moth, *Choristoneura rosaceana* (Delisle 1992), the bertha armyworm, *Mamestra configurata* (Howlader and Gerber 1986), and omnivorous leafroller moth, *Platynota stultana* (Webster 1987), calling behavior is usually observed within 24 hours following emergence, and its expression is not affected to any great extent by prevailing temperature conditions. A very different picture is observed in species such as the true armyworm, *Pseudaletia unipuncta* (Turgeon and McNeil 1982; Delisle and McNeil 1987a,b), the black cutworm, *Agrotis ipsilon* (Showers et al. 1989a,b) and the oriental armyworm, *Mythimna separata* (Han and Gatehouse 1991a), all believed to undertake seasonal migration in response to predictable habitat deterioration. In these species, even under conditions favorable for reproduction, the most precocious females do not express calling behavior for several days following emergence. Furthermore, when individuals are reared under low temperatures and short-day conditions, the delay in sexual maturation is considerably longer than one would expect as a direct physiological effect of the lower temperature on developmental rates. Work on the true armyworm has demonstrated that while PBAN is implicated in pheromone biosynthesis, JH is an essential component (Cusson and McNeil 1989a,b; Cusson et al. 1990, 1993, 1994a), as predicted in Barth's hypothesis relating to the role of JH in pheromone production (Barth 1965). Recent research on the black cutworm, another migrant noctuid, has demonstrated that both JH and PBAN are essential components involved in the process of pheromone synthesis (Gadenne 1993; Picimbon et al. 1995).

In species that migrate in response to unpredictable habitat deterioration, a somewhat different picture is seen with respect to the age at which females call for the first time. In the sunflower moth, *Homoeosoma electellum*, host plant pollen is an essential resource for neonate larvae (Rossiter et al. 1986) and contains an oviposition stimulant (Delisle et al. 1989). When pollen is present, calling, mating, and oviposition generally occur within 48 hours of emergence, a pattern very similar to that observed in nonmigratory species. However, in the absence of host plant pollen there is a significant delay in both ovarian development and the age at which females first express calling behavior (McNeil and Delisle 1989). A similar host plant effect has been reported for ermine moths, *Yponomeuta* spp. (Hendrikse and Vos-Bunnemeyer 1987), and, more recently, for *Helicoverpa zea* (Raina et al. 1992). This may be a common phenomenon. The black army cutworm, *Actebia fennica*, selectively oviposits at recently burned sites, and virgin females held in the presence of burned duff call at a significantly younger age than those held with unburned vegetation (Forcier and McNeil, in preparation). In all such cases, in the absence of suitable host plants, immature females could undertake extensive movement in search of acceptable oviposition sites.

It has been suggested that the delay in the onset of pheromone production and reproduction following emergence extends the time window available for migratory flight (Delisle and McNeil 1986, 1987a; Han and Gatehouse 1991b). However, one cannot rule out the possibility that, in species migrating in response to predictable habitat deterioration, this delay is associated with the acquisition of necessary resources and/or waiting for suitable meteorological conditions prior to the actual onset of migration (McNeil et al. 1995). Regardless of actual function, it seems evident, when seen from an ecological perspective, that the physiological mechanisms implicated in the timing of pheromone production in migrant species are unlikely to be identical to those observed in nonmigrant species.

It should not, therefore, be surprising that while the pheromone biosynthesis activating neuropeptide (PBAN) (Raina and Klun 1984), or PBAN-like substances, are involved in pheromone production of a number of moths (see references in Altstein et al. 1993), it does not play a role in pheromone synthesis of all species (Tang et al. 1991). Furthermore, even when pheromone biosynthesis is regulated by PBAN, interspecific differences have been reported with respect to the physiological mechanisms involved. PBAN is produced continuously in the subesophageal ganglion (SOG), but two possible release mechanisms have been proposed: into the hemolymph via the corpora cardiaca (CC) (Raina and Menn, 1987) or via the ventral nerve cord to the terminal abdominal ganglion (Teal et al. 1989). There has also been some debate concerning the actual target tissue of PBAN. Some studies have suggested that it stimulates the pheromone gland directly (e.g., Jurenka et al. 1991b; Rafaeli et al. 1990b), while others have reported that it acts on the terminal abdominal ganglion (TAG), which has nerve connections leading to the pheromone gland (Christensen et al. 1991b, 1994). Furthermore, in the redbanded leafroller it has been suggested that PBAN stimulates the corpus bursa, releasing a bursa factor responsible for stimulating the pheromone gland, although direct PBAN stimulation of the pheromone gland has not been ruled out (Fabriàs et al. 1992; Jurenka et al. 1991a; 1994a).

We suggest that these reported differences should not be seen as contradictions and that "someone has made a mistake." In fact, studies using RIA techniques to examine the distribution of PBAN in different tissues suggest that both humoral and neural pathways may be possible for the release of the neuropeptide (Rafaeli et al. 1991, 1993; Rafaeli, 1994). With additional information on an array of species, reared under different environmental conditions, one may see patterns associated with taxonomic relatedness and/or similar life history strategies. In addition, it is not inconceivable that the actual process by which pheromone production is governed may vary within the same species depending on the actual ecological context.

3. Mating Frequency

Species are generally classified as either monandrous (females mate once) or polyandrous (females mate with several males), although it is obvious that in monandrous species some individuals will mate more than once while single matings will occur in polyandrous species. Drummond (1984) proposed four possible benefits that lepidopterous females might obtain by mating more than once: (i) achieve an adequate sperm supply to fertilize her full egg complement, (ii) obtain resources other than sperm from males at the time of mating that increase her longevity and/or reproductive output, (iii) increase the genetic diversity of progeny, or (iv) minimize the time and energy lost while resisting persistent courting males. Additional benefits with respect to the quality of progeny may be accrued through competition of sperm from different males (Parker 1970) and by decreasing potential genetic defects that may result from prolonged storage of sperm (Halliday and Arnold 1987).

A higher-than-normal incidence of remating in Lepidoptera may result from male "harassment" under certain ecological conditions, such as strongly male-biased sex ratios (Pivnick and McNeil 1985). However, the probability that remating occurs to reduce the loss of time and energy rejecting persistent males is considerably lower in Lepidoptera than in other insects, especially in moths where males locate potential mates by responding to female sex pheromones that are only released by receptive individuals (Ridley 1990). Based on this, Ridley (1990) classified mating in the Lepidoptera as being under "female control."

Thus, the probability of remating will be dependent upon the processes that act to block pheromone production following mating. Corn earworm males transfer a homeostatic peptide, produced in the accessory glands of the reproductive system, to the female at the time of mating, and this passes via the hemolymph to inhibit the release of PBAN (Raina et al. 1994b and references therein). In the tobacco budmoth, *Heliothis virescens*, ecdysteroids from the testes and ejaculatory duct have been implicated in the loss of female receptivity (Ramaswamy et al. 1994 and references therein). In the gypsy moth there is a transient decline in pheromone production induced by mechanical signals at the time of mating, followed by a more permanent inhibition once viable sperm has entered the spermatheca. A neural message from the extended spermatheca passes to the terminal abdominal ganglion and terminates pheromone synthesis (Giebultowicz et al. 1991a). In the redbanded leafroller moth it is postulated that, following mating, a neural message passes via the ventral nerve cord to the brain and inhibits PBAN release (Jurenka et al. 1993), while other variants have been reported in two other tortricids, *Epiphyas postvittana* and *Planotortrix octo* (Foster and Roelofs 1994).

In many Lepidoptera, the first spermatophore produced by a male is significantly larger than subsequent ones (Pivnick and McNeil 1987; Svärd and Wicklund 1986; Royer and McNeil 1993b; Delisle and Bouchard, 1995). In species where stretch receptors in the bursa play a role, the bursa of a female that mates with a previously mated male will be less extended (Fig. 3.1A), due to the transfer of a much smaller spermatophore (Fig. 3.1C), so that the inhibition of pheromone production would be diminished. Also, the volume of sperm and/or accessory gland secretions transferred to the spermatheca may be insufficient to induce the appropriate neural impulses to result in permanent inhibition of pheromone production. However, it should be noted that considerable interspecific differences exist with respect to the effect of previous mating on ejaculate and spermatophore size. While the weight of the second spermatophore is about 40% of the first in the European skipper, *Thymelicus lineola* (Pivnick and McNeil 1987), the European corn borer, *Ostrinia nubilalis* (Royer and McNeil 1993b), and the oblique-banded leafroller (Delisle and Bouchard, 1996; Fig. 3.1C), in the true armyworm it is 70% (Marshall and McNeil 1989; Fig. 3.1D), resulting in a fully extended bursa (Fig. 3.1B). These differences in the relative change in spermatophore size may be reflected in the effect of previous male mating history on female reproductive success (Fig. 3.2). The reproductive output of once-mated females in the polyandrous true armyworm was not significantly affected, even when males had four previous matings (Fitzpatrick and McNeil 1989), whereas the reproductive output of once-mated females in the normally monandrous European corn borer declined with successive male matings (Royer and McNeil 1993b).

The size of the spermatophores produced may also be influenced by the type of host plant males ate during larval development. For example, the first spermatophore produced by oblique-banded leafroller males reared on maple is signifi-

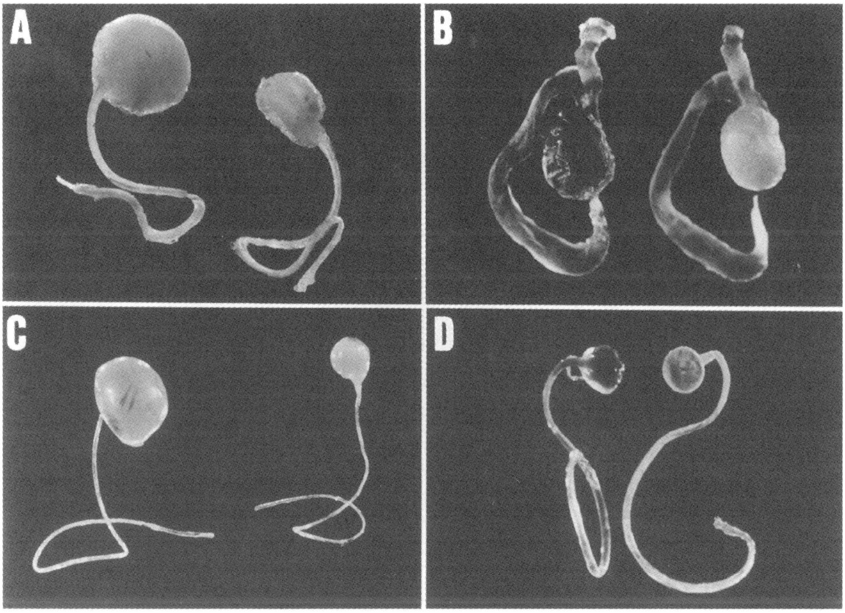

Figure 3.1. A comparison of bursa distention and the size of the spermatophore produced by the oblique-banded leafroller, *Choristoneura rosaceana* (**A** and **C**), and the true armyworm, *Pseudaletia unipuncta* (**B** and **D**), males during their first and second mating. The product of the first mating is on the left of each picture. Note that the photographs for the two species are not the same scale: Ejaculates from the armyworm are approximately four times that of the oblique-banded leafroller.

cantly larger (1.7 *versus* 1.2 mg wet weight) than those produced by males fed hazelnut (Delisle and Bouchard, 1995). Furthermore, while the second and third spermatophores were significantly smaller than the first produced by males reared on either larval diet, for any given subsequent mating (i.e., second or third), males fed maple produced a larger spermatophore than those fed hazelnut (Delisle and Bouchard, 1995).

A number of other factors must also be taken into account when examining the possible importance of spermatophore size or volume of material transferred at mating on neural messages affecting pheromone production. For example, the time elapsed from emergence to first mating in males (Delisle and Bouchard, 1995), or the time between successive matings (Svärd and Wiklund 1986, 1989; Royer and McNeil 1993b), is important, with spermatophore size increasing with time in both cases.

In the past there has been a tendency to equate spermatophore quality with size, especially when considering the importance of male investment on female reproductive output (Marshall and McNeil 1989, and references therein). In some

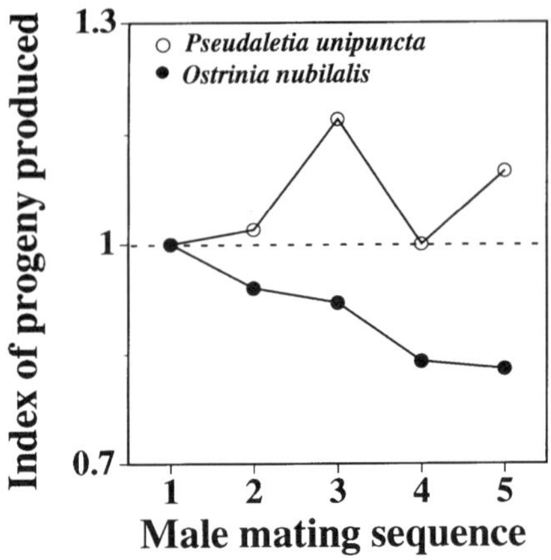

Figure 3.2. The effect of previous male mating history on the reproductive output of the polyandrous true armyworm, *Pseudaletia unipuncta*, and the monandrous European corn borer, *Ostrinia nubilalis*, females. The indices of reproduction, calculated by dividing the number of viable larvae produced at any given mating by the number produced from the male's first mating, was obtained for the armyworm and corn borer from data in Fitzpatrick and McNeil (1989) and Royer and McNeil (1993b), respectively.

cases, size and quality may be linked. As noted above, oblique-banded leafroller females, mated with males reared on maple, received a bigger spermatophore than those mated with males reared on hazelnut and had a significantly higher (480 *versus* 250 fertile eggs) reproductive success (Delisle and Bouchard, 1995). However, the dissociation of spermatophore size and quality is quite evident from the results of a preliminary experiment on the true armyworm. Females were mated on their second night of calling with virgin males and the fecundity and fertility, as well as the time spent calling, was determined for each individual for the next two nights. Females were considered as "well mated" if fertility was >50% and "poorly mated" if it was <50%. There were no differences in the duration of mating in the two categories (78.5 ± 13.8 *versus* 75.2 ± 8.1 min; $p = 0.83$, $F = 0.05$) and all females received similar-sized spermatophores. However, females considered as "poorly mated" laid few, mainly infertile eggs while "well mated" individuals laid many fertile eggs (Fig. 3.3). On both days, fecundity ($p < 0.01$, $F = 47.6$ and 11.9) and fertility ($p = 0.0001$, $F = 889.7$ and 1485.0) of "well mated" females were significantly higher than those of "poorly mated" individuals. Furthermore, following mating the females that produced few eggs called significantly more than those laying many eggs, even though both groups of mated females called significantly less than similar-aged virgins (Fig. 3.4; $p = 0.0001$, $F = 30.5$ and 30.1, for days 1 and 2, respectively).

These results support the hypothesis that, at least in some species, spermatophore size is not the dominant factor influencing postmating receptivity. The

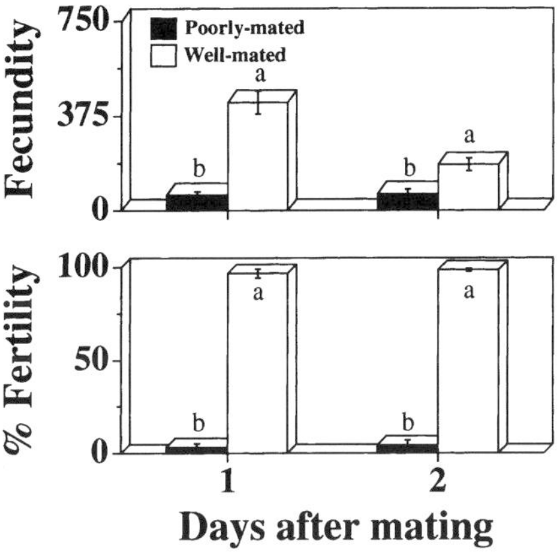

Figure 3.3. The fecundity and fertility of true armyworm, *Pseudaletia unipuncta*, females, considered as being "well" and "poorly" (>50 and <50% fertility, respectively) mated in the two days following mating. For any given day, values with different letters are significantly different ($p < 0.05$).

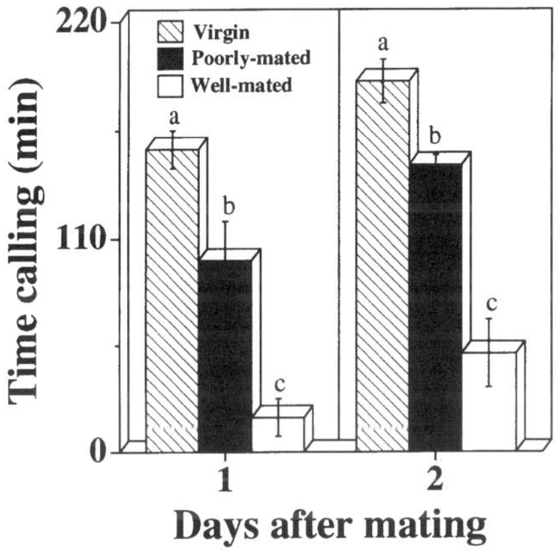

Figure 3.4. The time spent calling (minutes) by "well" and "poorly" (>50 and <50% fertility, respectively) mated true armyworm, *Pseudaletia unipuncta*, females during the two days following mating, as well as by virgin females of the same age. For any given day, values with different letters are significantly different ($p < 0.05$).

length of the refractory period may be related to the actual amount of sperm transferred, because many females that resume calling lay few, mainly infertile eggs, a situation that may be rectified with a second mating. Another cause may be the quantity of accessory compounds that are capable of inducing changes in female reproductive physiology, including pheromone production, following their transfer by the male. A third possibility relates to resource acquisition through which male-donated substances actually increase female reproductive success: Svärd and McNeil (1994) showed that a second mating by *P. unipuncta* females increased both the fecundity and longevity. Obviously, the three possible explanations are not mutually exclusive. Marshall and McNeil (1989) tested the hypothesis that spermatophore size does not necessarily reflect quantity of male donated resources and demonstrated that there was no relationship between the size of the first spermatophore produced by true armyworm males and either total lipid or hydrocarbon content. Furthermore, whereas total lipids declined with spermatophore size in successive matings, there were no significant changes in total hydrocarbons.

The idea that spermatophore size and quality are not related has been most strikingly demonstrated in a recent study on the effect of male larval nutrition in the spruce budworm (Delisle and Hardy, in press). Males reared on young balsam fir foliage, the insect's natural food source, produced spermatophores nearly 50% smaller than those reared on the standard McMorran artificial diet. However, <25% of females mated with males reared on foliage resumed calling after the first mating, significantly less than the >60% observed in females mated with males reared on artificial diet. Furthermore, those mated with foliage-fed males that resumed calling had both higher fecundity and fertility than those whose partners had been reared on artificial diet.

The relative importance of male-acquired nutrients will be affected by the female's own ability to acquire resources during both larval and adult development (Boggs 1990), and this will obviously change considerably depending on prevailing ecological conditions. For example, Boggs (1990) reported a significant, negative correlation between pollen feeding and the frequency of mating in *Heliconius cydno* females. The recent work on the spruce budworm (Delisle and Hardy, in press) and the oblique-banded leafroller (Delisle and Bouchard, 1995) provides clear evidence that male quality, and thus the probability of remating in females, will also be affected by resources acquired during larval development.

The exact manner in which nutrition influences aspects of male quality that affect different aspects of female reproduction remains to be clarified. For example, changes in both food quality and availability could explain why under outbreak conditions a high proportion of females in the normally monandrous gypsy moth (Doane 1968) remate (Richerson et al. 1976). One possible factor associated with male quality that merits consideration is JH, because males may transfer this hormone to females at the time of mating (Shirk et al. 1980). However, as noted above with respect to pheromone synthesis, there may be marked interspecific differences in the relative importance of JH in postmating pheromone production. Topical application of JH to virgin females reduced pheromone titers in the omnivorous leafroller (Webster and Cardé 1984) but did not affect this parameter in the corn earworm (Raina 1989) or the tobacco budworm (Ramaswamy and Cohen 1992). The extent, if any, to which males transfer JH to females at the time of mating (Shirk et al.1980) and the relationships between male-induced changes in JH titers (directly by transfer or indirectly through factors that modify female JH biosynthesis or JH esterase activity), vitellogenin production (Cusson et al. 1994b), and female receptivity following mating need to be examined in an array of species with very different life histories.

4. Concluding Remarks

Even though the use of pheromones in integrated pest management programs is often given as a justification for research dealing with pheromone physiology, with a few rare exceptions, these studies provide little or no information concerning the biology of the species under consideration. In fact the comparisons in Table 3.1 were limited, in large part, because we were unable to find information on the biological parameters selected. In the present chapter we propose that the approach taken by Barth (1965) be expanded to encompass a much broader ecological perspective than just the length of reproductive life. A clear understanding of the biochemical and physiological processes modulating pheromone production is required (Tumlinson 1988), and this information must be placed within behavioral, ecological, and evolutionary contexts (McNeil 1991, 1992) if we are to increase our basic understanding of the different aspects of pheromone biology and improve the use of pheromones in the control of insect pests.

Acknowledgments

Research discussed in this chapter has been supported by grants from NSERC, FCAR, and the Green Plan Programme. We also thank Pierre Therrien and Guy Plante for their assistance in preparing Fig. 3.1.

4

Interaction of PBAN with Biosynthetic Enzymes

Wendell L. Roelofs and Russell A. Jurenka

It has been well-documented that PBANs (pheromone biosynthesis activating neuropeptides) and PBAN-like compounds stimulate pheromone production in many moth species. The pheromone biosynthetic pathways have been elucidated for many of these species (Bjostad et al. 1987), and so it has been possible to investigate the pathway with respect to stimulation by PBAN. The biosynthetic pathways in most of the moth species in which PBAN is active involve the synthesis of saturated fatty acids via the fatty acid synthesis enzymes, and then a series of specific enzymes, including unique desaturases, limited chain-shortening enzymes, reductases, acetyltransferases, and oxidases. PBAN has been found to interact with different aspects of this pathway in different species, as discussed in this chapter, but examples discussed here all involve an effect on or before the first step of fatty acid synthesis.

1. Redbanded Leafroller (RBLR), *Argyrotaenia velutinana*

A study (Tang et al. 1989) involving labeled precursors was carried out with female RBLR moths to determine the interaction of its brain neuropeptide (PBAN homology) with the pheromone biosynthetic pathways. A bioassay was developed in which females were decapitated for 3 days to deplete their pheromone titer, and then labeled compounds were applied to the pheromone gland by topical application in dimethyl sulfoxide. The effect of a brain extract then was tested by injection into these same females. After a 4-hr incubation period the amount of label was determined in the main pheromone component (*Z*)-11-tetradecenyl acetate (Z11-14:OAc) produced in the sex pheromone gland at the end of the abdomen.

When [1-^{14}C]acetate was topically applied to the pheromone gland, significantly higher levels of labeled acetate were incorporated into the pheromone component and into all abundant biosynthetic intermediates when the brain hormone was injected into the female. This was true in scotophase as well as in photophase. If the brain hormone was not present in the assay, the amounts of labeled acetate incorporated into pheromone and intermediates were hardly above background. These results imply that the brain hormone acts by stimulating synthesis of the first long-chain acids (octadecanoic and hexadecanoic) in the sequence.

Additional evidence was provided by experiments with deuterium-labeled hexadecanoic acid (16:COOH). The results showed that decapitated females processed similar amounts of deuterium-labeled precursor into pheromone and intermediates, whether or not the brain hormone was present. Other studies were conducted with *in vitro* assays using pheromone glands, and similar results were found. No differences were found in the incorporation of labeled 16:COOH into pheromone with or without the presence of PBAN (Jurenka et al. 1991a). Therefore, enzyme abundance and activity states for the series of reactions that convert 16:Acyl into pheromone were unaffected by decapitation or the presence of brain extract. PBAN evidently did not influence these steps in the biosynthetic pathway.

Attention was then focused on the synthesis of fatty acids from acetate relative to fatty acid synthetase activity. An *in vitro* assay was developed for activity of this enzyme complex, and it revealed that there were no differences in quantity or quality of product formed in the presence or absence of the brain hormone. These data indicate, therefore, that PBAN in RBLR females stimulates either substrate supply for fatty acid synthesis or the first enzyme in the pathway, acetyl-CoA carboxylase.

2. Corn Earworm, *Helicoverpa zea*

A similar study (Jurenka et al. 1991b) was conducted with *H. zea* using both isolated abdomen and isolated gland incubations. The effect of PBAN on pheromone production was particularly dramatic with isolated gland incubations in which the difference in pheromone titer in the presence of PBAN was 90 ± 16 ng compared to 8 ± 1 ng in its absence. Likewise, the incorporation of [1-^{14}C]acetate into the main pheromone component, (*Z*)-11-hexadecenal (Z11-16:ALD), was 12,730 ± 662 cpm/gland in the presence of PBAN compared to 410 ± 36 cpm/gland in its absence. These experiments show that PBAN can interact directly with the gland and that it must activate either the availability of acetate or some enzyme in the biosynthetic pathway.

The biosynthetic pathway from saturated fatty acid to aldehyde was investigated with deuterium-labeled fatty acids. With isolated gland cultures it was found that the same amount of labeled fatty acid precursors was incorporated into pheromone with or without administration of PBAN. This indicated that PBAN is not affecting those enzymes involved in converting 16:COOH to Z11-16:ALD. In other words, the series of specialized enzymes—Δ11-desaturase, reductase, and oxidase—in the pathway are all active, regardless of the presence of PBAN.

The data on *H. zea* are similar to that obtained with RBLR and indicate a possible interaction with the first step in fatty acid biosynthesis involving acetyl-CoA carboxylase. A model incorporating these data and other data obtained on the possible mode of action of PBAN on the *H. zea* pheromone gland (Jurenka et al. 1991c; Jurenka and Roelofs 1993) is shown in Fig. 4.1. The accumulated data indicate that PBAN binds to a receptor on the pheromone gland cell membrane, the PBAN–receptor complex activates a calcium channel that increases intracellular calcium, which, in turn, activates adenylate cyclase to produce cAMP, and then cAMP stimulates other enzymes in the signal cascade that ultimately could stimulate pheromone production by phosphorylating or dephosphorylating acetyl-CoA carboxylase.

3. Cabbage Armyworm, *Mamestra brassicae*

A similar study was carried out with *M. brassicae* because it had been reported in the literature (Bestmann et al. 1989) that the key enzyme under control of PBAN in that species is Δ11-desaturase. The main pheromone component is (*Z*)-11-hexadecenyl acetate (Z11-16:OAc), which is biosynthesized via the fatty acid cycle and action of a series of specialized enzymes including Δ11-desaturase, reductase, and acetyltransferase. The purported involvement of PBAN with the

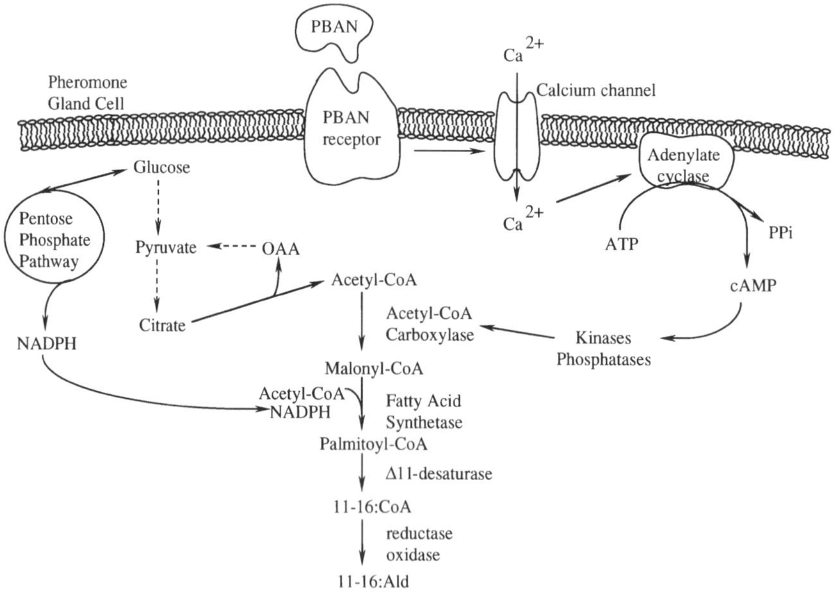

Figure 4.1. Proposed model for the stimulation of pheromone production by PBAN in *Helicoverpa zea*. PPi, pyrophosphate; OAA, oxalacetate.

Δ11-desaturase was unusual in view of the previous studies and so an investigation was undertaken with female *M. brassicae*.

Acetate labeled with both carbon-14 and carbon-13 was used to monitor fatty acid biosynthesis from the beginning of the pathway, and deuterium-labeled 16:COOH and Z11-16:COOH were used to monitor the enzyme activity after fatty acid biosynthesis (Jacquin et al. 1994). A large increase in the amount of incorporation of both the radio- and stable isotope-labeled acetate was obtained in the presence of PBAN. However, there was no significant differences in the amounts of label incorporation into pheromone with and without PBAN when the deuterium-labeled 16:COOH was used. Likewise, PBAN did not increase the amount of label incorporation into pheromone when deuterium-labeled Z11-16:COOH was used as the precursor. The results indicated that PBAN did not affect the portion of the pathway from 16:COOH to pheromone, but did affect the portion from acetate to fatty acids. This is in contrast to the previous report on this species, but is consistent with the findings on the above species that PBAN probably acts on the first step of fatty acid synthesis in these moths.

As discussed in other chapters, PBAN has homology with a family of myotropic peptides that have in common the last five C-terminal amino acids. These amino acids (Phe-X-Pro-Arg-Leu-amide, where X can be Ser, Thr, or Val) are required and sufficient for pheromonotropic activity, and so many insect peptides exhibit pheromonotropic activity in the bioassays. Although the different peptides fulfill various functional roles in the life of the insect, there may be similarities in the receptors and second messenger systems, as well as in the type of enzymes affected. In that case, one might expect that the enzymes affected by PBAN would be of the more general type, such as acetyl-CoA carboxylase and reductase (see other chapters), rather than a specialized one, such as Δ11-desaturase, which is found only in the moth pheromone gland.

5

PBAN Regulation of Sex Pheromone Biosynthesis in *Spodoptera littoralis*

M. Pilar Marco, Gemma Fabriàs, and Francisco Camps

The first experimental evidence that a neuroendocrine factor from the brain–subesophageal ganglion complex (Br–SOG) is involved in the control of sex pheromone production was provided by Raina and Klun (1984) using *Helicoverpa zea*. Subsequent research revealed that pheromonotropic activity was also present in the Br–SOG of other moths (Altstein et al. 1989; Ando et al. 1988; Bestmann et al. 1989; Cusson and McNeil 1989a; Martinez and Camps 1988; Rafaeli and Soroker 1989a; Tang et al. 1989). This factor, which was termed *pheromone biosynthesis activating neuropeptide* (PBAN), was later isolated and characterized from brains of both *H. zea* (Hez-PBAN) (Raina et al. 1989a) and *Bombyx mori* (Bom-PBAN I) (Kitamura et al. 1989) as a 33-amino-acid peptide with an amidated carboxyl terminal portion. An 80% primary sequence homology occurs between these two peptides. Subsequently, another pheromonotropic peptide (Bom-PBAN II) was isolated also from *B. mori* (Kitamura et al. 1990), which differs from Bom-PBAN I by an additional Arg-residue at the N-terminus. Recently, another pheromonotropic peptide with 18 amino acid residues has been found in *Pseudaletia separata* (Pss-PT) (Matsumoto et al. 1992a). Structure–activity studies revealed that the C-terminal portion of PBAN is crucial for eliciting biological activity, whereas the corresponding N-terminal portion is essential for achieving full potency (Kitamura et al. 1989; Raina and Kempe 1990).

Although PBAN appears to be involved in the endocrine control of sex pheromone production in most of the moths so far investigated, the conclusions from the studies addressing both the target site(s) and the enzyme(s) affected by PBAN are still controversial. According to some authors (Jurenka et al. 1991b; Raina et al. 1991; Soroker and Rafaeli 1989), PBAN is produced in the SOG and is subsequently transferred to the corpora cardiaca (CC) during scotophase, from which it is released into the hemolymph and acts directly on the pheromone gland. Alternatively, other researchers (Christensen et al. 1991b; Teal et al. 1989) have presented evidence suggesting that PBAN is transported through the ventral nerve cord (VNC) to the terminal abdominal ganglion (TAG), where it exerts its action. Neurons in this ganglion then would send a second signal, presumably octopamine (Christensen et al. 1991b), to the pheromone gland that could stimulate sex pheromone production. The fact that PBAN-like activity was present in the Br–SOG and CC of *H. zea* (Rafaeli et al. 1991; Raina et al. 1991; Raina and Klun 1984) and *Helicoverpa armigera* (Rafaeli et al. 1991) at selected times of the photoperiod, along with the finding that, in many species, PBAN activity can be achieved in isolated pheromone gland incubations (Arima et al. 1991; Fabriàs et al. 1994; Fonagy et al. 1992a; Jurenka et al. 1991b; Rafaeli et al. 1990a), supports the first hypothesis. Experimental evidence for the second hypothesis is found in the pheromonotropic action of octopamine (Christensen et al. 1991b, 1992), as well as the requirement that the TAG and its connectives are essential for the stimulatory response of the pheromone gland to PBAN in *H. zea* (Teal et al. 1989). Also in agreement with the neural approach is the pheromonotropic activity reported in thoracic ganglia and TAG at selected times of the photoperiod in *H. armigera* (Rafaeli et al. 1991) and in the VNC of *H. zea* (Teal et al. 1989) at the time of pheromone production. However, Raina et al. (1991) did not find any significant PBAN-like activity in thoracic ganglia, abdominal ganglia, and TAG of the same species.

Recently developed immunoassays have not helped to clarify these discrepancies. Thus, Rafaeli et al. (1991), using a radioimmunoassay, found PBAN-like immunoreactivity (PBAN-IR) in Br-SOG, CC, thoracic ganglia, and TAG of *H. armigera* at selected times of the photoperiod. However, no detectable levels of PBAN were observed in blood extracts during these sampling periods. An enzyme-linked immunosorbent assay (ELISA) has also been developed and used to quantify PBAN-IR in *Heliothis peltigera* (Gazit et al. 1992) and *H. zea* (Kingan et al. 1992). Although PBAN-IR was found in Br–SOG and CC, the amounts observed in the TAG and hemolymph were at or below the sensitivity level of the assay (Kingan et al. 1992).

As for the effect of PBAN on the different enzymes involved in sex pheromone biosynthesis, it has been reported that in *Argyrotaenia velutinana* (Tang et al. 1989), *H. zea* (Jurenka et al. 1991a), and *Mamestra brassicae* (Jacquin et al. 1994) this peptide acts on a step in or prior to fatty acid synthesis, However, in *Chrysodeixis chalcites* (Altstein et al. 1990) it has been proposed that PBAN affects a ΔT11 desaturation step, whereas in *Spodoptera littoralis* (Martinez et al. 1990) and *B. mori* (Arima et al. 1991) this neuropeptide appears to act on the reduction of fatty acyl moieties.

2. Physiological Mode of Action of PBAN in *Spodoptera littoralis*

2.1. Site of Action on PBAN

Previously, Martinez and Camps (1988) reported that *S. littoralis* virgin females produce their sex pheromone only during the dark period and that gland pheromone content peaks around 2 hours into the second and third scotophase. They also demonstrated that pheromone production could be stimulated during the photophase by injection of a Br–SOG extract. Similar results have been found recently in our laboratory by injection of synthetic Bom-PBAN into intact females in the photophase (Gosalbo et al. 1994). Likewise, we have recently observed that Bom-PBAN stimulates *S. littoralis* sex pheromone production in an isolated gland incubation system (Fabriàs et al. 1994). As shown in Fig. 5.1A, this effect is dose-dependent. The minimum dose tested that elicits a significant response is 0.01 pmol, whereas maximum stimulation is achieved with 0.5 pmol/gland. Similar responses are obtained with doses up to 10 pmol. Furthermore, pheromone glands incubated with PBAN (1 pmol/gland) showed a significant increase in pheromone titer after 45 minutes of incubation. Maximum stimulation is achieved after 3 hours, whereas incubation for longer periods does not result in increased pheromone titers up to 6 hours (Fig. 5.1B).

The fact that PBAN is active in the isolated pheromone gland preparation suggests that this gland might be a target organ for this peptide in *S. littoralis*. Similar results have also been reported in other moths (Arima et al. 1991; Fónagy et al. 1992a; Jurenka et al. 1991b; Rafaeli et al. 1990).

2.2. Determination of the Presence of PBAN in the Hemolymph

As mentioned in the introduction, some authors have speculated that PBAN stimulation of sex pheromone production occurs through the release of this peptide

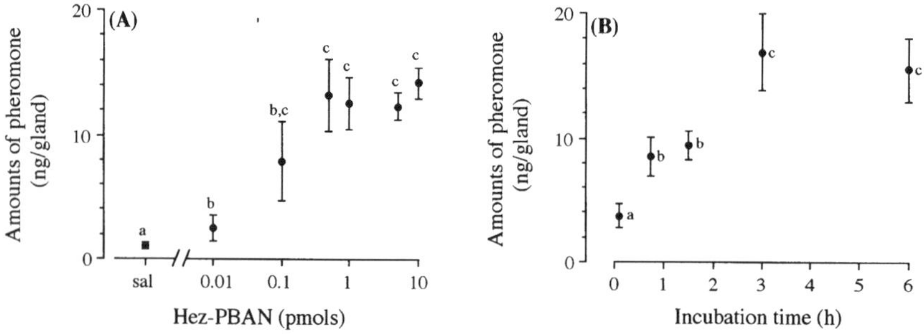

Figure 5.1. **(A)** Dose-dependence of PBAN induction of sex pheromone production in isolated gland incubations of *S. littoralis*. Glands were incubated for 3 hours in 10-μl drops of Grace's medium (sal) or medium containing PBAN. The doses of PBAN are given in the abscissa, and the amounts of (*Z,E*)-9,11-tetradecadienyl acetate are given in the ordinate. Data are mean ± SEM of 5–8 individual replicates. Different letters show statistical difference at $p < 0.05$ (unpaired, one-tail *t* test). **(B)** Pheromonotropic response to PBAN after different times of incubation. Isolated glands were incubated in 10-μl drops of Grace's medium containing 1 pmol of PBAN, and pheromone was extracted after a 0-, 1-, 1.5-, 3-, and 6-hr incubation. Each point represents mean ± SEM of nine individual replicates. Different letters indicate statistical difference at $p < 0.05$ (unpaired, one-tail *t* test). Extraction and quantification of pheromone was carried out as reported elsewhere (Fabriàs et al. 1994).

into the hemolymph. Although pheromonotropic activity has been detected in the blood of scotophase females (Jacquin et al. 1994; Raina et al. 1991), the existence of PBAN-IR could not be confirmed with the immunoassays currently available. This can be explained by considering that the times of sampling were not suitable or that the amounts of PBAN in the blood, which are sufficient to trigger the pheromonotropic response, are below the detection limit of the immunoassays reported. Consequently, both adequate sampling periods and high sensitivity of the analytical techniques employed might be critical. Additionally, no clear evidence exists that the pheromonotropic response is prompted by the complete 33-amino-acid PBAN, since it has been reported that smaller peptides with the C-terminal sequence of PBAN are also active (Fónagy et al. 1992b; Gazit et al. 1990; Raina and Kempe 1990).

In order to improve recognition of the C-terminal portion of PBAN, we prepared an immunogen protein conjugate by coupling Cys-Hez-PBAN to keyhole limpet hemocyanin (KLH) using *N*-succinimidyl 4-(*N*-maleidimidomethyl)cyclohexane-1-carboxylate (SMCC), an amino- and sulfhydryl-group-directed heterobifunctional reagent (Wong 1993), as cross-linker. Antisera obtained after the immunization protocol were used to develop a two-step competitive ELISA with a high degree of recognition for PBAN. The I_{50} value for Hez-PBAN was determined to be 50 pM (2.5 fmol/well), whereas the detection limit was around 1 fmol/well (Fig. 5.2). Cross-reactivity studies showed that this antiserum can also recognize smaller fragments of the carboxy-terminal portion of PBAN (see Table 5.1).

Hemolymph for ELISA determination was extracted with ethanol:water (1/1) centrifuged at 12,000 rpm for 5 minutes, and the supernatant was dried down in a Speed-Vac rotatory evaporator. This extraction procedure allows recoveries of

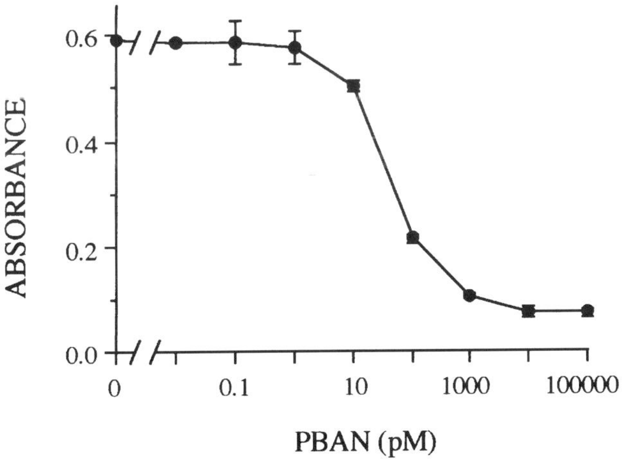

Figure 5.2. Standard curve for Hez-PBAN performed in PBST buffer. Analysis was performed by the two-step competitive ELISA using eight serial dilutions, in duplicates, of synthetic Hez-PBAN, ranging from 0.1 μM to 0.01 pM.

Table 5.1. Cross-reactivity of PBAN antiserum with selected peptides[a]

Peptide	Percent cross-reactivity
Hez-PBAN	100
Cys-Hez-PBAN	115
Cys-Hez-$PBAN_{(20-33)}$	148
Oxidized Hez-PBAN	64
Bom-PBAN	95
Leucopyrokinin	27
Adipokinetic hormone	nc[b]

[a]Cross-reactivity data were obtained by preparing standard curves in PBST (0.2 M phosphate buffer, 0.4% NaCl, 0.05% Tween 20, pH 7.5) with 10 different concentrations (10 μM to 0.01 pM) of selected peptides. Each I_{50} was determined in the competitive experiment and cross-reactivity values were calculated according to the equation: (I_{50} Hez-PBAN/I_{50} peptide) × 100.

[b]nc, no cross-reactivity under our experimental conditions.

PBAN around 80%. Furthermore, evaluation of parallelism and sensitivity of the standard curve for PBAN performed in the presence of hemolymph extracts diluted 1/10 showed no significant changes as compared to the curve in buffer. Therefore, the absence of a matrix effect makes this assay suitable to measure small amounts of PBAN-like material in the insect blood.

Amounts of PBAN-IR found in hemolymph of intact virgin females at different times of the scotophase and also during the last hour of the photophase are represented in Table 5.2. Concentrations of PBAN in samples taken 1 hour before

Table 5.2. PBAN-IR in hemolymph extracts of Spodoptera littoralis *virgin females at different times of the photoperiod*[a]

Hours into scotophase	PBAN-IR (pM)	n > DL[b]	n < DL[c]
−1	—	0	8
0	293 ± 49	3	9
1	411 ± 48	11	3
2	560 ± 63	13	0
3	540 ± 49	14	0
4	574 ± 53	12	0

[a]Analyses were performed by the two-step competitive ELISA. PBAN content was calculated from a standard curve of synthetic Hez-PBAN. A dilution of hemolymph extract of 1/10 was used, in duplicates. Data represent means ± SEM of those samples with concentrations of PBAN above the detection limit.

[b]Number of samples with levels of PBAN above the detection limit of the ELISA when measuring hemolymph (200 pM).

[c]Number of samples with levels of PBAN below the detection limit of the ELISA.

lights off were always below the detection limit, whereas 25% of samples taken just at the begining of the scotophase had already measurable amounts of PBAN. The concentration of PBAN-IR in 78% of samples obtained 1 hour into the scotophase was 411 pM, whereas amounts in all the samples taken 2, 3, and 4 hours into the dark period were around 550 pM. This pattern of PBAN distribution in the blood is similar to that of sex pheromone production (Martinez and Camps 1988). In contrast, the concentration of PBAN in samples of hemolymph taken 2 hours into the scotophase from females that had been decapitated 2 hours before darkness was always below the detection limit of the ELISA.

In summary, the findings that Bom-PBAN stimulates sex pheromone production in isolated glands (Fabriàs et al. 1994) and that PBAN-like IR is associated with the hemolymph of virgin females only during the period when pheromone is produced suggest that in *S. littoralis* PBAN follows a humoral route to its site of action, which is the pheromone gland itself.

3. Interaction of PBAN with the Biosynthetic Enzymes

In *S. littoralis*, the biosynthesis of the main pheromone component, (*Z*,*E*)-9,11-tetradecadienyl acetate (Z9,E11-14:Ac), involves the chain-shortening of palmitic acid followed by sequential E11 and Z9 desaturations and final reduction and acetylation (Martinez et al. 1990) (Fig. 5.3).

Previously, by injection of Br–SOG extracts into intact females, we hypothesized that reduction of acyl moieties might be activated by PBAN into the *S. littoralis* sex pheromone biosynthetic pathway (Martinez et al. 1990). We recently confirmed this hypothesis, using synthetic PBAN and the above-mentioned isolated gland incubation bioassay. Thus, under these conditions, pheromone glands incubated with [^{14}C]NaOAc (0.045 μCi/μl) in the presence of PBAN (2 pmol/gland) produced higher amounts of radiolabeled pheromone than did controls (Table 5.3) after 1 hour of incubation. Likewise, production of radioactive pheromone from either [1-^{14}C]myristic acid (0.045 μCi/μl) or [1-^{14}C]-(*Z*,*E*)-9,11-tetradecadienoic acid (0.032 μCi/μl) was enhanced in glands treated with PBAN (2 pmol/gland) as compared to controls (Table 5.3). On the other hand, no labeled (*E*)-11-tetradecenyl acetate (E11-14:Ac) or (Z9,E11-14:Ac) (Table 5.4) were detected in glands incubated with [13,13,14,14,14-$^2H_5$]-(*E*)-11-tetradecenoic acid

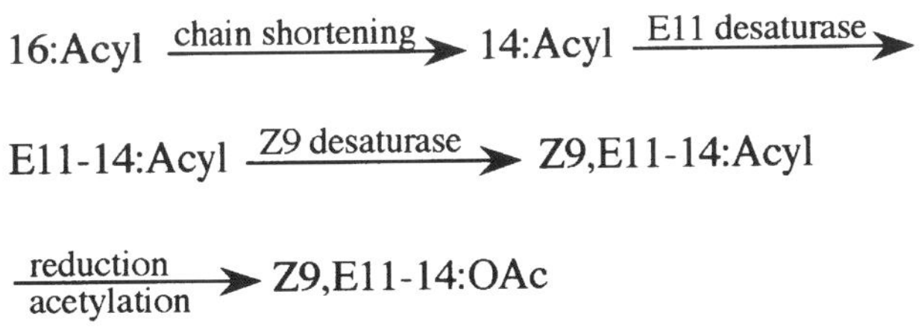

Figure 5.3. Biosynthetic pathway of the main component of *S. littoralis* sex pheromone.

Table 5.3. Effect of PBAN on incorporation of radiolabeled precursors into pheromone by the pheromone gland of Spodoptera littoralis in vitro

Precursor	Radioactive pheromone (dpm/gland)[a]	
	Saline	PBAN
[^{14}C]NaOAc	1376 ± 109 (10)	4647 ± 701 (10)
[1-^{14}C]Myristic acid	33 ± 4 (15)	426 ± 45 (15)
[1-^{14}C]-(*Z*,*E*)-9,11-Tetradecadienoic acid	490 ± 111 (8)	2342 ± 426 (8)

[a] Data are means ± SEM, and the number of individual replicates is given within parentheses. In all cases, differences between saline- and PBAN-treated glands are statistically significant at $p < 0.001$.

(1 mM) in the absence of PBAN, whereas both compounds were produced (Table 5.4) when PBAN (2 pmol/gland) was included in the medium. In these experiments, the amounts of labeled intermediates were also quantified. Whereas no differences in the amounts of methyl-[13,13,14,14,14-$^2H_5$]-(*E*)-11-tetradecenoate were observed between controls and treatments, amounts of labeled dienoate were lower in treated than in control preparations (Table 5.4). Thus, in all these experiments, production of labeled pheromone was enhanced in pheromone glands incubated in the presence of PBAN. In those experiments in which titers of natural pheromone were also measured, the amounts of pheromone components were 8- to 10-fold higher in PBAN treated glands than in controls. These results agree with our previous findings using Br–SOG extracts and whole females in that injection of the extract caused an increase in the production of both natural and labeled pheromone (Martinez et al. 1990). In contrast, the formation of d_5-labeled precursors occurred both in females that received the Br–SOG extract and in control insects. The amounts of each intermediate were not significantly different in both groups of animals, except for the labeled dienoate. Levels of

Table 5.4. Effect of PBAN on incorporation of d_5E11-14:acid into pheromone and its intermediates by the pheromone gland of Spodoptera littoralis in vitro

Treatment	Amounts of labeled compounds (ng/gland)[a]		
	E11-14:Me[b]	Z9,E11–14:Me[b]	Z9,E11–14:Ac[b]
Saline	4.0 ± 0.7 (10)	1.4 ± 0.2 (10)	n.d.
	[12 ± 2]	[34 ± 5]	[1.3 ± 0.1]
PBAN	4.0 ± 0.5 (9)	0.9 ± 0.1* (9)	0.9 ± 0.2 (7)
	[11 ± 2]	[34 ± 6]	[14 ± 1.5]

[a]Among all pheromone components, only Z9,E11-14:Ac was quantified. Data are means ± SEM, and the number of individual replicates is given within parentheses. Amounts of natural compounds are indicated within square brackets. Asterisk indicates statistical difference between saline- and PBAN-treated glands at $p < 0.05$. nd, not detected or peaks barely above background.

[b]E11–14:Me, methyl (*E*)-11-tetradecenoate; Z9,E11-14:Me, methyl (*Z*,*E*)-9,11-tetradecadienoate; Z9,E11-14:Ac, (*Z*,*E*)-9,11-tetradecadienyl acetate.

this precursor were lower in Br–SOG injected females than in controls, suggesting that this intermediate had been consumed to give rise to the corresponding acetate upon stimulation (Martinez et al. 1990). In agreement with these results, the experiments with [13,13,14,14,14-$^2H_5$]-(*E*)-11-tetradecenoate acid performed in the gland incubation bioassay revealed that the formation of labeled precursors occurred in isolated glands even in the absence of PBAN and also that the amounts of labeled dienoate were lower in PBAN-treated glands than in controls.

These studies indicate that PBAN acts on a step after the formation of the last intermediate. As indicated for other moth species (Bestmann et al. 1987b; Dunkelblum et al. 1989; Teal and Tumlinson 1987), the acetyl transferase is not under hormonal control in *S. littoralis*. Therefore, PBAN must activate a step prior to alcohol formation. All these results taken together indicate that, as we had previously suggested (Martinez et al. 1990), PBAN activation of *S. littoralis* sex pheromone biosynthetic pathway occurs at the reduction of acyl moieties. A similar conclusion was reached in a study conducted in the silkworm moth (Arima et al. 1991; Ozawa et al. 1993). Several authors have proposed that, since PBAN regulates the sex pheromone biosynthesis in different species with different pheromone blend composition, it should activate a step common to the different biosynthetic pathways. This step appears to be in (or prior to) fatty acid synthesis in some species but in (or prior to) the reduction of acyl precursors in others.

Acknowledgments

We thank Isabel Millán for rearing the insects used in our studies and thank CICYT for financial support (Grant AGF-92-0178). M.P.M. also thanks CSIC for a Contract as Research Associate in a Programme of the Spanish Ministry of Education and Science.

6

The Effect of PBAN on Conversion of Fatty Acyls to Pheromone Aldehydes in Female *Manduca sexta*

James H. Tumlinson, Nianbai Fang, and Peter E.A. Teal

The pheromone produced by females of *Manduca sexta* (L.) (Lepidoptera: Sphingidae) was identified as a blend of the 16-carbon aliphatic aldehydes hexadecanal, (Z)-9-hexadecenal, (Z)-11-hexadecenal, (*E*)-11-hexadecenal, (*E*,*Z*)-10,12-hexadecadienal, (*E*,*E*)-10,12-hexadecadienal, (*E*,*E*,*Z*)-10,12,14-hexadecatrienal, and (*E*,*E*,*E*)-10,12-14-hexadecatrienal (Starratt et al. 1979; Christensen et al. 1989a; Tumlinson et al. 1989). Subsequently, we showed that fatty acyl analogs of the pheromone aldehydes are present in the pheromone glands of females from about 3 or 4 days prior to adult eclosion until the adults are 8 days old or until death (Fang et al. 1992). By determining the fate of deuterium-labeled fatty acids applied to the gland, we also found that hexadecanoate is the precursor of the 16-carbon unsaturated pheromone aldehydes and fatty acids produced in the gland. Hexadecanoate is converted to (Z)-11-hexadecenoic acid and (*E*)-11-hexadecenoic acid by Δ11-desaturases, and (Z)-11-hexadecenoic acid is further desaturated and isomerized to the conjugated diene and triene 16-carbon fatty acids (Fang et al. 1995a). The deuterium labeling studies also showed quite clearly that the fatty acids in the gland were the precursors of the aldehydic pheromones. Normally, the aldehydes are produced and released only during the scotophase when the female is calling. However, injection of pheromone biosynthesis activating neuropeptide (PBAN) at any time of the photoperiod into the abdomens of females of any age resulted in production of the pheromone aldehydes after about an hour (Fang et al. 1992). Our labeling studies indicated, but did not prove, that the role of PBAN in regulating pheromone biosynthesis in this species was to activate the enzymes that convert the fatty acids stored in the gland into the analogous aldehydes.

In our previous studies with *M. sexta* we found that injection of synthetic Hez-PBAN did not affect the quantities or proportions of deuterated fatty acids in glands treated with deuterated hexadecanoic, (Z)-11-hexadecenoic, or (*E*)-11-hexadecenoic acids. Furthermore, during the photophase, endogenous and labeled fatty acids were converted to aldehydes only when PBAN was injected into the abdomens of females (Fang et al. 1992, 1995b). However, 24-hour incubations with labeled compounds were required to get sufficient incorporation of the label for analyses. Thus, we could not rule out the possibility that endogenous PBAN was affecting other steps in the biosynthetic pathway, particularly desaturation, in addition to conversion of fatty acids to aldehydes. It has been shown with other species that decapitation of females results in cessation of pheromone production, but that pheromone production can be induced by injection of PBAN as long as 3 days after decapitation (Tang et al. 1989). Therefore, we designed experiments with decapitated *M. sexta* females to determine which steps in the pheromone biosynthetic pathway in this insect are controlled by PBAN.

Manduca sexta females that were decapitated produced no pheromone during the scotophase following decapitation, indicating that they were free of PBAN. When deuterated hexadecanoic or (Z)-11-hexadecenoic acid was applied to the sex pheromone glands of decapitated or intact females of the same age, and allowed to incubate *in vivo* for 24 hours, deuterium labeled Δ11- and Δ10,12-unsaturated 16-carbon fatty acids were produced in both types of females. Injection of PBAN into intact or decapitated females 23 hours after application of labeled acids had no effect on the production of unsaturated labeled fatty acids. However, deuterium-labeled aldehydes were produced only in females that were injected with PBAN (Fang et al. 1995b). Although the results presented here do not eliminate the possibility that PBAN has a role in regulating the biosynthesis of hexadecanoate, they clearly indicate that, in this species, PBAN activates the process by which fatty acyl precursors in the pheromone gland are converted into the pheromonal aldehydes and has no effect on any of the desaturase enzymes in the pheromone biosynthetic pathway.

7

Structure–Function Relationship of PBAN/MRCH

Miriam Altstein, Orna Ben-Aziz, Tal Gabay, Yoav Gazit, and Ezra Dunkelblum

1. Introduction

Pheromone biosynthesis activating neuropeptides (PBANs) (Raina and Klun 1984) are C-terminally amidated neuropeptides that regulate sex pheromone biosynthesis in moths. To date, four different PBAN molecules were isolated and their primary structures have been determined. The first PBAN molecule (33 amino acids) was isolated from *Helicoverpa zea* (Hez-PBAN; Raina et al. 1989a). Two other molecules were isolated from *Bombyx mori*, Bom-PBAN-I (33 amino acids), and Bom-PBAN-II (34 amino acids) (Kitamura et al. 1989, 1990). A fourth molecule (33 amino acids) was recently isolated from *Lymantria dispar* (Lyd-PBAN, Masler et al. 1994). Hez-PBAN shares about 80% homology with Bom-PBAN and Lyd-PBAN. The C-terminal pentapeptide (Phe-Xxx-Pro-Arg-Leu-NH_2, Xxx = Ser) of all four PBAN molecules is identical and shares homology with the C-terminal pentapeptide of the pyrokinins, a family of myotropic peptides isolated from *Leucophea maderae* and *Locusta migratoria* (Xxx = Ser, Gly, Thr, Val; Schoofs et al. 1993). This C-terminal part is also homologous to the C-terminal region of two other neuropeptides: *Pseudaletia separata* pheromonotropin (Pss-PT, 18 amino acids) (Matsumoto et al. 1992a) and *B. mori* diapause hormone (Bom-DH) (Imai et al. 1991). The PBAN gene of *H. zea* (Davis et al. 1992b; Ma et al. 1994) and the c-DNA of *B. mori* PBAN (Kawano et al. 1992; Sato et al. 1993) have recently been cloned. In both moths the gene sequences contain, in addition to the PBAN sequence, peptides having the common C-terminal pentapeptide (Phe-Xxx-Pro-Arg-Leu-NH_2) (Sato et al. 1993; Ma et al. 1994).

The presence of PBAN has been demonstrated in a variety of moths as well as nonlepidopteran species, and its mode of action has been studied in many laboratories (for reviews, see Raina 1993 and Altstein et al. 1993). These studies have revealed that the neuropeptide is present in larvae and in adult male and female brain–subesophageal ganglia. The neuropeptide was also found in the corpora cardiaca–corpora allata complex of adult female moths as well as in the thoracic and abdominal ganglia. Natural and synthetic PBAN were found to activate pheromone biosynthesis in homologous and heterologous moth species during photophase and scotophase, and the activity of the neuropeptide was found to be mediated by the cyclic nucleotide adenosine 3′5′-monophosphate (cAMP) and dependent on the presence of Ca^{2+}. Some aspects, related to the potential target organ of PBAN, its route of transport, its mode of action on the pheromone biosynthetic pathway in the gland, and the involvement of additional peptide and nonpeptide hormonal factors in its activity, are controversial and are under investigation in various laboratories (Raina, 1993; Altstein et al., 1993).

Identification of the amino acid sequence of Hez-PBAN (Raina et al. 1989a) and Bom-PBAN (Kitamura et al. 1989, 1990) enabled structure–activity analysis using synthetic PBAN and fragments derived from its sequence. Studies performed in several laboratories using various species of Heliothinae and *B. mori* revealed that the C-terminal region is important for the onset of biological activity whereas the N-terminal is not (Kitamura et al. 1989; Gazit et al. 1990; Raina and Kempe 1990, 1992; Kuniyoshi et al. 1991, 1992a). Within this region, the C-terminal pentapeptide was found to be the minimal sequence required for any pheromonotropic activity. Structure–function studies were also performed using other insect neuropeptides containing the PBAN pentapeptide C-terminal sequence (the pyrokinin peptide family, Bom-DH, Pss-PT). All of the above peptides stimulated pheromone biosynthesis and confirmed the importance of the C-terminal region for the onset of the pheromonotropic activity (Fónagy et al. 1992d; Kuniyoshi et al. 1992b; Nachman et al. 1993b; Schoofs et al. 1993; Abernathy et al. 1995).

Although the C-terminal region was found to be essential, the activity of the C-terminal pentapeptide itself was much lower than that of the full-length PBAN (Raina and Kempe 1990, 1992; Kuniyoshi et al. 1991, 1992a). Substitution of L-amino acids with D-amino acids in the pentapeptide, along with blockage of its N-terminus, yielded high activities, which were in some cases higher than those of PBAN 1-33NH_2 (Kuniyoshi et al. 1992a; Raina and Kempe 1992). Furthermore, the N-terminus, which was not active by itself, was found to be essential for full potency. These results indicated that proteolytic enzymes might interfere with the bioassay and suggested that the presence of sequences within the PBAN molecule play a protective role against the enzymatic degradation *in vivo*. To test this hypothesis and to determine which sequences have a conformational role and which protect the peptide against proteolytic degradation, we performed a detailed structure function analysis using PBAN-derived peptides in the presence and absence of protease inhibitors.

Matsumoto et al. (1990) purified and revealed the primary structure of a neuropeptide from heads of adult *B. mori* which induces cuticular melanization. The peptide was termed *melanization and reddish coloration hormone* (Bom-

MRCH) and was found to be identical to Bom-PBAN-1. Both natural and synthetic Bom-MRCH were found to induce cuticular melanization in *Pseudaletia (= Leucania) separata* and *Spodoptera litura* larvae, as well as sex pheromone biosynthesis in adult *S. litura* females (Matsumoto et al. 1990), indicating that PBAN controls two different functions at different developmental stages. Currently, our understanding of the mode of action of PBAN with respect to the melanotropic activity is even more limited than our understanding of its pheromonotropic activity, and very little is known about the cellular and molecular basis of this process. In the present study, we report on the development of a quantitative analysis for the determination of the melanotropic activity of PBAN, and we discuss the use of this technique for studying the structure–function relationship between the peptide and the melanotropic activity.

2. Materials and Methods

C-terminal free and amidated fragments derived from the amino acid sequence of Hez PBAN (PBAN 9–33NH_2, PBAN 13–33NH_2, PBAN 17–33NH_2, PBAN 19–33NH_2, PBAN 28–33NH_2, PBAN 1–33COOH) were synthesized as described previously (Gazit et al. 1990). Assessment of the pheromonotropic activity of the full-length PBAN (PBAN 1–33NH_2) and the derived fragments was tested using the *Heliothis peltigera* bioassay as described by Gazit et al. (1992). Sex pheromone content was monitored with the major component, (*Z*)-11-hexadecenal (Z11-16:Ald) (Dunkelblum and Kehat 1989), 2 hours post-injection, by capillary gas chromatography as described previously (Gazit et al. 1992). Melanotropic activity was examined with fifth-instar *Spodoptera littoralis* larvae using the bioassay described by Matsumoto et al. (1981). The degree of cuticular melanization was determined by a quantitative computerized optical density scanning (CODS) analysis using an image analysis system (Biological Detection Systems, Pittsburgh, PA, USA) (Altstein et al. 1996).

3. Results and Discussion

3.1. Pheromonotropic Activity of PBAN: Structure–Function Analysis

Analysis of the activity of 1 pmol of the peptides (Table 7.1) 2 hours post-injection revealed that removal of the first eight N-terminal amino acids did not affect the pheromonotropic activity, and the amount of the major sex pheromone component that was formed in response to the injection of PBAN 9–33NH_2 was 96 ng/gland, compared with 108 ng obtained with PBAN 1–33NH_2. Further removal of four and eight amino acids from the N-terminus resulted in a loss of activity, and the amount of pheromone produced in response to the injection of

Table 7.1. Effect of synthetic Hez-PBAN and derived fragments on sex pheromone biosynthesis in Heliothis peltigera[a]

Injected peptide (1 pmol)	Z11–16:Ald content (ng/gland) ± SEM
PBAN 1–33NH_2	108 ± 10 (*n* = 44) a
PBAN 9–33NH_2	96 ± 19 (*n* = 14) ab
PBAN 13–33NH_2	4 ± 3 (*n* = 8) c
PBAN 17–33NH_2	33 ± 10 (*n* = 10) c
PBAN 19–33NH_2	74 ± 17 (*n* = 18) b
PBAN 26–33NH_2	6 ± 2 (*n* = 9) c
PBAN 28–33NH_2	5 ± 2 (*n* = 21) c
PBAN 1–33COOH	3 ± 2 (*n* = 9) c
Buffer	4 ± 1 (*n* = 54) c

[a]Peptide injection was performed using 3.5-day-old females in photophase. Numbers in parentheses indicate number of replicates. Statistical analysis was performed by ANOVA. Differences between means were tested for significance by the Newman–Keuls test at $P < 0.05$. Means with a common letter do not differ significantly.

PBAN 13–33NH_2 and PBAN 17–33NH_2 did not differ from that evoked by the injection of buffer (Table 7.1). Interestingly, injection of a shorter fragment lacking 18 amino acids from the N-terminus showed a marked activity (74 ng/gland), which did not differ significantly from that of PBAN 9–33NH_2 (Table 7.1). The C-terminally derived octa- and hexapeptides, PBAN 26–33NH_2 and PBAN 28–33NH_2, as well as the C-terminal free acid peptide, PBAN 1–33COOH, failed to stimulate pheromone biosynthesis, and the amounts of sex pheromone that were obtained in response to their injection were 6, 5, and 3 ng/gland, respectively (Table 7.1). These data indicated that the region between amino acids 9 and 13 plays an important role in maintaining the activity of the PBAN. The fact that the C-terminal free acid PBAN 1–33COOH was devoid of any activity, and the finding that the potency of the PBAN-derived fragments did not correlate directly with the length of the peptide (PBAN 19–33NH_2 was more potent than PBAN 13–33NH_2 or PBAN 17–33NH_2) raised the possibility that proteolytic activity affects the potency of PBAN and led to a further analysis of the activity in the presence and absence of protease inhibitors.

Examination of the biological activity of 1 pmol of the PBAN fragments PBAN 13–33NH_2, PBAN 17–33NH_2, PBAN 19–33NH_2, PBAN 28–33NH_2, and PBAN 1–33COOH in the presence of protease inhibitors, 2 hours post-injection, revealed that protease inhibitors do not affect the activity of these peptides, and the same stimulation of pheromone formation was observed with and without the inhibitors

(Table 7.2). Conversely, the same mixture of protease inhibitors had a significant effect on the activity of the full-length PBAN 1–33NH_2, elevating pheromone content from 108 to 193 ng/gland (Table 7.2).

The lack of activity of the PBAN-derived fragments in the presence of protease inhibitors may result from the activity of proteases that are not inhibited by the mixture used in this experiment. Alternatively, it may indicate that proteases do not affect the injected peptides. In the latter case, the results indicate that the C-terminal amide and/or the N-terminal sequence (between amino acids 9–13) play an important role in the bioactivity of the biopeptide.

3.2. Melanotropic Activity of PBAN: Quantitative Analysis

Quantitation of the melanotropic activity of synthetic Hez-PBAN involved the use of the CODS technique, which enabled us to determine the size and intensity of the pigmented and non-pigmented areas in a given cuticle. The melanization pattern of *S. littoralis* larvae consists of three pigmented stripes (two outer stripes and an inner stripe) which run along the larvae. The nonpigmented region occupies the area in between these stripes (Altstein et al. 1996). The effect of PBAN on cuticular melanization was examined by a comparison of the size of the melanized and nonmelanized areas in cuticles of untreated and treated (ligated and buffer- or PBAN-injected) *S. littoralis* larvae.

Using the melanotropic bioassay with the CODS technique, we found that the nonpigmented area in untreated sixth-instar larvae occupies 12% of the cuticle (Table 7.3). Ligation of fifth-instar larvae at head capsule slippage (which prevents access of released cerebral PBAN to epidermal target regions in the thorax and

Table 7.2. Effect of PBAN 1–33NH_2 and derived fragments on sex pheromone biosynthesis in the presence and absence of protease inhibitors[a]

Injected peptide (1 pmol)	Z11–16:Ald content (ng/gland) ± SEM	
	No inhibitors	With inhibitors
PBAN 1–33NH_2	108 ± 10 (n = 44)	193 ± 33* (n = 10)
PBAN 13–33NH_2	4 ± 3 (n = 8)	12 ± 4 (n = 8)
PBAN 17–33NH_2	33 ± 10 (n = 10)	51 ± 27 (n = 9)
PBAN 19–33NH_2	74 ± 17 (n = 18)	123 ± 29 (n = 10)
PBAN 28–33NH_2	5 ± 2 (n = 21)	0.3 ± 0.2 (n = 10)
PBAN 1–33COOH	4 ± 1 (n = 9)	3 ± 2 (n = 10)

[a]Injection of protease inhibitors (phenylmethylsulfonyl fluoride, N,α-p-tosyl-L-lysine chloromethyl ketone, aprotinin, and bacitracin, 0.5 μg each) was performed 30 minutes prior to the injection of the peptides. All other experimental details are as described in footnote *a* in Table 7.1. Numbers in parentheses indicate number of replicates. Statistical analysis was performed for each peptide separately, comparing activity in the presence and absence of inhibitors. The asterisk (*) indicates a significant difference in pheromone content at $p < 0.05$.

Table 7.3. Comparison of the extent of melanized and nonmelanized areas in cuticles of treated and untreated larvae of Spodoptera littoralis[a]

Treatment		Area (% of total)	
		Nonmelanized	Melanized
Untreated	(n = 26)	12 ± 1	88 ± 1 b
Ligated	(n = 48)	32 ± 1	68 ± 1 d
Ligated + buffer	(n = 61)	25 ± 2	75 ± 1 c
Ligated + PBAN	(n = 59)	7 ± 1	93 ± 1 a

[a]The size of the nonmelanized and the melanized area is expressed as a percentage of the total area of the cuticles ± SEM. PBAN 1–33NH_2 was injected at a dose of 10 pmol. Numbers in parentheses indicate number of replicates. Statistical analysis was performed by ANOVA. Differences between means were tested for significance by the Newman–Keuls test at $p < 0.05$. Means with a common letter do not differ significantly.

abdomen) resulted in a 2.6-fold increase in the nonpigmented area compared with untreated larvae. Injection of 0.1 M phosphate buffer (which is used as a vehicle for PBAN injection) resulted in a 2.1-fold increase in the nonpigmented area, and the injection of 10 pmol PBAN 1–33NH_2 causes a marked elevation in pigmentation, reducing significantly the size of the nonpigmented area to less than the level of the untreated insects (Table 7.3).

3.3. Melanotropic Activity of PBAN: Structure–Function Analysis

The development of a quantitative melanotropic bioassay enabled performance of a detailed structure–function analysis of synthetic Hez-PBAN and its derived fragments. The effect of all the tested peptides on cuticular melanization was examined by comparing the size of the melanized and nonmelanized areas in peptide- *versus* buffer-injected larvae. Removal of eight amino acids from the N-terminal region of PBAN 1–33NH_2 resulted in a decrease in the melanotropic activity, and the nonpigmented area increased from 7% to 12% (for PBAN 1–33NH_2 and PBAN 9–33NH_2, respectively, Table 7.4). Despite the decrease, the activity of PBAN 9–33NH_2 did not differ significantly from that of the full-length PBAN. These data indicated that the first eight amino acids do not play a role in the melanotropic activity. Removal of 12 amino acids from the N-terminus further decreased the melanotropic activity; the nonmelanized area increased to 25% of the total, similar to the size obtained in control larvae injected with buffer (Table 7.4). The drop in the activity obtained with this peptide was similar to that obtained in the pheromonotropic assay, and indicated the importance of the sequence between amino acids 9–13. Shorter peptides (PBAN 17–33NH_2 and PBAN 19–33NH_2) lacking 16 and 18 amino acids from their N-terminus were also inactive, increasing the size of the nonpigmented area to 18% and 20%, respectively (Table 7.4). The C-terminal derived peptides, PBAN 26–33 NH_2 and PBAN 28–33NH_2, exhibited a higher activity, similar to that of PBAN

Table 7.4. Effect of synthetic PBAN and derived fragments on the melanization pattern of Spodoptera littoralis *larvae cuticles*[a]

Injected peptide (10 pmol)	Nonpigmented area (%)
PBAN 1–33NH_2	7 ± 1 (n = 59)
PBAN 9–33NH_2	12 ± 6 (n = 5)
PBAN 13–33NH_2	25 ± 4* (n = 5)
PBAN 17–33NH_2	18 ± 5* (n = 5)
PBAN 19–33NH_2	20 ± 3* (n = 10)
PBAN 26–33NH_2	15 ± 5 (n = 4)
PBAN 28–33NH_2	14 ± 2 (n = 14)
PBAN 1–33COOH	21 ± 4* (n = 9)
Buffer	25 ± 1* (n = 61)

[a]Injection of peptides was performed into ligated fifth-instar larvae. The size of the nonmelanized area is expressed as a percentage of the total area of the cuticle. Numbers in parentheses indicate number of replicates. Statistical analysis was performed by ANOVA. The activity of each peptide was compared with that of PBAN 1–33 NH_2. Differences between means were tested for significance by the Newman–Keuls test. An asterisk (*) indicates significant differences at $p < 0.01$.

1–33NH_2 and PBAN 9–33NH_2 (Table 7.4). Removal of the C-terminal amide resulted in a complete loss of activity, suggesting that the C-terminal amide is essential for melanotropic activity.

The structure–function relationship studies of the pheromonotropic and melanotropic activities revealed similarities and dissimilarities. Neither activity depended on the presence of the first N-terminal 8 amino acids; and both were most dependent on the presence of the sequence between amino acids 9–13, as well as the presence of the C-terminal amide. The C-terminally derived fragments, PBAN 26–33NH_2 and PBAN 28–33NH_2, exhibited different activities in the two bioassays, being active in the melanotropic but not in the pheromonotropic assay. These data suggest that the C-terminal region plays a major role in the melanotropic activity and that it correlates well with the results of Matsumoto et al. (1992b), who demonstrated that the melanotropic activity resided in the C-terminus of another melanotropic neuropeptide, Pss-PT, and depends on the presence of the C-terminal amide. Two other peptides (Lom-MT-II and MT-III, Schoofs et al. 1993) belonging to the pyrokinin family and having the pentapeptide C-terminal sequence Phe-Xxx-Arg-Pro-Arg-Leu-NH_2 were also found to induce cuticular melanization, emphasizing the importance of this sequence in the melanotropic activity (Matsumoto et al. 1993).

4. Conclusion

The role of the C-terminal sequence in the pheromonotropic activity is not yet fully understood. The inability of the C-terminal short peptides to induce sex pheromone biosynthesis may result from either (a) the absence of sequences which are important conformationally for activation of the pheromonotropic receptor or (b) the absence of sequences which protect the peptide from proteolytic degradation. Our attempts to protect the short C-terminally derived fragments from proteolytic degradation, which may occur in the course of the pheromonotropic assay (Kuniyoshi et al. 1992a), were unsuccessful. This, however, does not exclude the possibility that the C-terminal sequences comprise the active site of the PBAN with respect to its pheromonotropic activity. The fact that the C-terminal amide was essential for activity, as well as the fact that the activity of the PBAN-derived peptides did not correspond directly with the length of the peptide, strengthened the notion that proteolysis may affect the ability of the shorter fragments to induce pheromone production. Additional experiments using different protease inhibitor mixtures, as well as more detailed time- and dose-dependent studies, have to be carried out in order to determine precisely the sequences which contain the pheromonotropically active site within the neuropeptide. The dependence of the melanotropic activity on the presence of the C-terminal amide, as well as the dependence on the presence of the sequences near the N-terminus of the molecule, suggests that proteases play a role in the melanotropic bioassay as well and that these sequences protect the neuropeptide from degradation. However, the conditions under which the melanotropic bioassay is conducted render the use of protease inhibitors ineffective; and alternative approaches, such as the use of proteolytically stable derivatives, have to be employed in order to determine the precise sequences which are conformationally necessary to activate the melanotropic receptor.

Acknowledgments

We thank Dr. Jacob Barg of the Cardiovascular and Hypertension Research Laboratory, Wolfson Medical Center, Tel Aviv University Medical School, Holon, Israel, for his assistance in the computerized analysis of cuticular melanization; Prof. Zvi Vogel of the Neurobiology Department, Weizmann Institute of Science, Rehovot, Israel, for critical evaluation of the results; and Mrs. Irit Schafler for her part in GC analysis. This research was supported by the Endowment Fund for Basic Research in Life Sciences, Charles H. Revson Foundation, administered by the Israel Academy of Sciences and Humanities.

8

Multifunctionality of PBAN-Related Neuropeptides: Melanotropic Activity of FXPRLamide Peptides

Shogo Matsumoto, Adrien Fónagy, Akinori Suzuki, and Takashi Mitsui

1. Introduction

Insect neurohormones have a significant role in regulating diverse physiological events according to environmental conditions or to specific developmental stages. Recent advances in insect neuropeptide chemistry have made it possible to compare sequence homology between various neurohormones/neuropeptides and elucidate the core sequence of each neuropeptide required for biological activity.

The pheromone biosynthesis activating neuropeptide (PBAN) in lepidopteran insects is an amidated 33 to 34-amino-acid peptide (Raina et al. 1989a; Kitamura et al. 1989, 1990) and has a specific C-terminal sequence, Phe-Ser-Pro-Arg-Leu-NH_2 (FSPRLamide), responsible for the pheromonotropic activity (Raina and Kempe 1990; Kuniyoshi et al. 1991). Interestingly, accumulated data have revealed that a group of neuropeptides having FXPRLamide at the C-terminus are widely distributed in insects (Fig. 8.1) and regulate diverse physiological events

Hez-PBAN:	L S D D M P A T P A D Q E M Y R Q D P E Q I D S R T K Y F S P R L amide
Bom-PBAN-I:	L S E D M P A T P A D Q E M Y Q P D P E E M E S R T R Y F S P R L amide
Bom-PBAN-II:	R L S E D M P A T P A D Q E M Y Q P D P E E M E S R T R Y F S P R L amide
Pss-PT:	K L S Y D D K V F E N V E F T P R L amide
Bom-DH:	T D M K D E S D R G A H S E R G A L C F G P R L amide
Lem-PK (LPK):	< Q T S F T P R L amide
Lom-MT-I:	G A V P A A Q F S P R L amide
Lom-MT-II:	E G D F T P R L amide
Lom-MT-III:	R Q Q P F T P R L amide
Lom-PK:	< Q D S G D G W P Q Q P F V P R L amide
Lom-PK-II:	< Q S V P T F T P R L amide
Lom-MT-IV:	R L H E N G M P F S P R L amide

Figure 8.1. Primary structures of insect neuropeptides having an FXPRLamide at the C-terminus (see Schoofs et al., 1993). < Q, pyroglutamyl; PBAN, pheromone biosynthesis activating neuropeptide; PT, pheromonotropin; DH, diapause hormone; PK, pyrokinin; MT, myotropin. Nomenclature of insect neuropeptides is according to Raina and Gade (1988).

such as larval pigmentation (Matsumoto et al. 1990), embryonic diapause (Imai et al. 1991), and muscle contraction (Holman et al. 1986) as well as sex pheromone biosynthesis. We here describe the functional diversity of PBAN-related neuropeptides, especially from the aspect of physiological roles in the larval pigmentation of some armyworm species.

2. Hormonal Control of Larval Pigmentation of Armyworms

A morphological color change is displayed by many insects according to environmental factors, such as background color, temperature, photoperiod and crowding, or specific developmental stages. Such color changes are caused by selective pigment breakdown or synthesis and are regulated by the endocrine system.

In the common armyworm, *Pseudaletia (= Leucania) separata*, the larvae are black when crowded and are yellow green, dull orange, or reddish brown when isolated (Iwao 1962). There is continuous variation in the intensity of black coloration depending on larval density. The black coloration in crowded conditions is due to deposition of indole melanin (Ikemoto 1971) and the morphological color change of *P. separata* is one of the characteristics of phase polymorphism of the armyworm (Iwao 1962). By contrast, the larvae of the cabbage armyworm, *Mamestra brassicae*, change color from green to brown with black markings after the last larval molt so that the morphological color change of *M. brassicae* is associated with a specific developmental stage (Hiruma et al. 1984). The principal brown pigment in the epidermis of this species is xanthomatin, and the black pigment in the cuticle is melanin.

Several workers have investigated the hormonal control of larval pigmentation of the armyworms by using microsurgical techniques, and it is believed that the cuticular melanization is regulated by a neurohormone mainly produced by the subesophageal ganglion (Ogura 1975a; Hiruma et al. 1984; Morita et al. 1988). In addition, this neurohormone also regulates epidermal reddish brown pigmentation in the presence of juvenile hormone in *M. brassicae* (Hiruma et al. 1984). Thus the hormone is named *melanization and reddish coloration hormone* (MRCH) (Ogura 1975a).

3. Distribution of Neurohormones from the Subesophageal Ganglion

In lepidopteran insects, the subesophageal ganglion (SOG) produces several distinct neurohormones in response to environmental conditions or to specific developmental stages. In such neurohormones, PBAN works at the adult stage and regulates sex pheromone production in the pheromone glands of several species of female moths (Raina and Klun 1984). MRCH works at the larval stage and regulates the morphological color change of several armyworm species (Ogura 1975a; Matsumoto et al. 1981). Another neurohormone produced by the

SOG, the diapause hormone (DH), works at the pupal stage and regulates embryonic diapause of the silkworm, *Bombyx mori* (Hasegawa 1951; Fukuda, 1951). Interestingly, a wide distribution of such neurohormone activities has been demonstrated in various insects which do not seem to require these neurohormones in their life cycles. For example, DH activity is found in the giant silkmoth, *Antheraea yamamai* (larval diapause) (Fukuda 1951), the common armyworm, *P. separata* (nondiapause) (Ogura and Saito 1973), and even in the cockroach, *Periplaneta americana* (nondiapause) (Takeda 1977). Likewise, MRCH activity is found in insects which do not show any morphological color change, such as *B. mori* (Ogura 1975b), and the cockroach, *P. americana* (N. Ogura, personal communication). Furthermore, PBAN activity is also seen in larvae or male moths (Raina et al. 1987; Ando et al. 1988), although PBAN regulates sex pheromone production in female moths. It has also been demonstrated that PBAN activity is in the cephalothoracic organs of the cockroach, *Blattella germanica* (Raina et al. 1987), and the African migratory locust, *Locusta migratoria* (Sreng et al. 1990). At this point, the question arises, What are the functions of such neurohormone activities in these insects and what is the structural relationship between these active substances?

4. *Bombyx mori* PBAN: Functional Dimorphism

In the course of chemical characterization of the silkworm neurohormones, we isolated a melanotropic neuropeptide from a head extract of *B. mori* by monitoring its MRCH activity with *P. separata* larvae (Matsumoto et al. 1986). The sequence analysis of the purified peptide revealed that this was a 33-amino-acid peptide, and, interestingly, this peptide consisted of exactly the same molecule as Bom-PBAN-I (Fig. 8.1) separately isolated from the same head extract (Kitamura et al. 1989; Matsumoto et al. 1990). This result suggests the multifunctionality of PBAN or PBAN-like neuropeptides.

Purified *B. mori* PBAN induced cuticular melanization in the armyworms *P. separata, Pseudaletia loreyi, Spodoptera litura*, and *M. brassicae* (Fig. 8.2) (Matsumoto et al. 1981), and also in the African armyworm, *Spodoptera exempta* (Yagi 1980). *B. mori* PBAN also induced epidermal reddish brown pigmentation in addition to cuticular melanization when it was injected into the yellow solitary larvae of *P. separata* (Matsumoto et al. 1988). In addition, when the abdomens of *M. brassicae* larvae were isolated at the time of head capsule slippage, subsequently treated with juvenile hormone, and then injected with *B. mori* PBAN, both cuticular melanization and epidermal brown pigmentation were induced (Hiruma et al. 1984). Since *B. mori* PBAN induced sex pheromone production in *S. litura* (Matsumoto et al. 1990) as well as in *B. mori* itself (Kitamura et al. 1989), these results indicate that PBAN actually has a dual action on larval pigmentation and sex pheromone biosynthesis (Matsumoto et al. 1990). At this

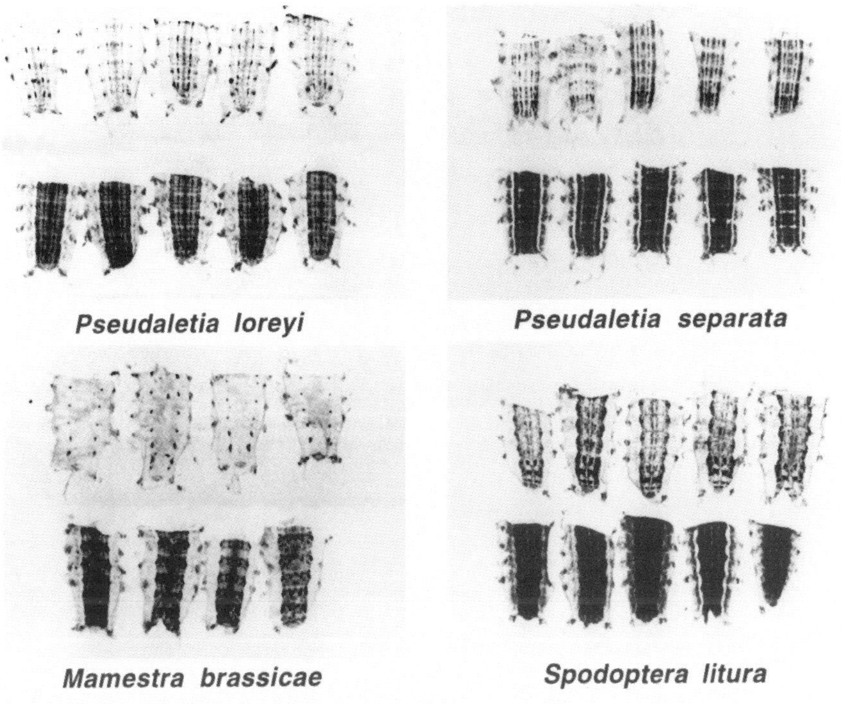

Figure 8.2. Induction of cuticular melanization in armyworm larvae by injection of *B. mori* PBAN. In each species, purified *B. mori* PBAN (*lower panel*) or distilled water (*upper panel*) was injected into the isolated abdomens of the penultimate instar larvae at the time of head capsule slippage (Matsumoto et al., 1981).

point, however, we still do not know if the same molecule as PBAN also regulates larval pigmentation in the armyworm, *P. separata,* itself.

Pseudaletia separata Pheromonotropin: Sequence Homology

To make sure if the larval pigmentation of armyworms is also regulated by the 33 to 34-amino-acid PBAN-like peptide, we attempted to purify the neuropeptide having melanotropic activity from a head extract of the fifth-instar *P. separata* larvae in the molting stage just before MRCH was released from the SOG. Purification was carried out starting with a head extract of 31,800 *P. separata* larvae (Fig. 8.3). In the third step, the fraction eluted from the activated charcoal column induced cuticular melanization in *P. separata* larvae as well as bombykol production in decapitated *B. mori* female moths, and the following purification proceeded by detecting the pheromonotropic activity on *B. mori.* The final step

<u>Purification of Pss-PT</u>

31,800 Heads of <u>Pseudaletia</u> larvae

1st	**Acetone powder**
2nd	**80% EtOH extract**
3rd	**Activated Charcoal column**
4th	**DEAE-Toyopearl column**
5th	**RP-HPLC on Senshu Pak VP-318**
6th	**RP-HPLC on Senshu Pak VP-318**
7th	**RP-HPLC on Senshu Pak VP-304**

Purified <u>Pseudaletia</u> pheromonotropin (Pss-PT)

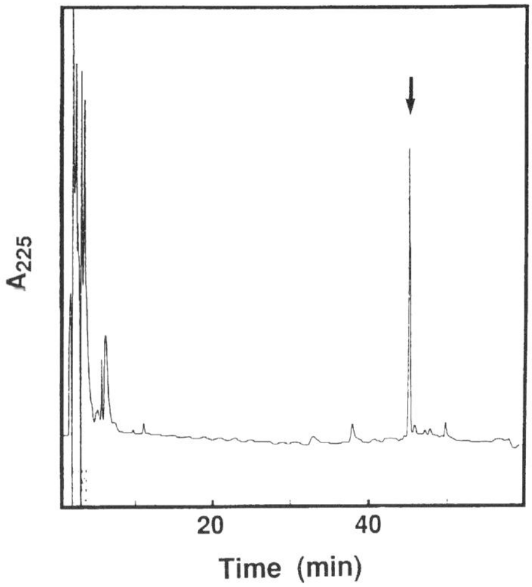

Figure 8.3. Purification procedure of *P. separata* pheromonotropin (Pss-PT) and the elution profile at the final step on RP-HPLC. A peak eluting at 45.0–45.7 minutes (*arrow*) showed pheromonotropic activity. [From Matsumoto et al. (1992a). Permission from Academic Press.]

of the purification on RP-HPLC is also shown in Fig. 8.3, and the purified peptide was named *P. separata* pheromonotropin (Pss-PT) (Matsumoto et al. 1992a).

Amino acid sequence analysis revealed that Pss-PT was an 18-amino-acid peptide which was obviously different from *B. mori* or *Helicoverpa zea* PBAN that consists of 33–34 amino acid residues (Fig. 8.1). The effect of synthetic Pss-PT on sex pheromone production and cuticular melanization was examined (Fig. 8.4). The C-terminal amide form, but not the free acid form, of Pss-PT showed both pheromonotropic and melanotropic activity (Matsumoto et al.

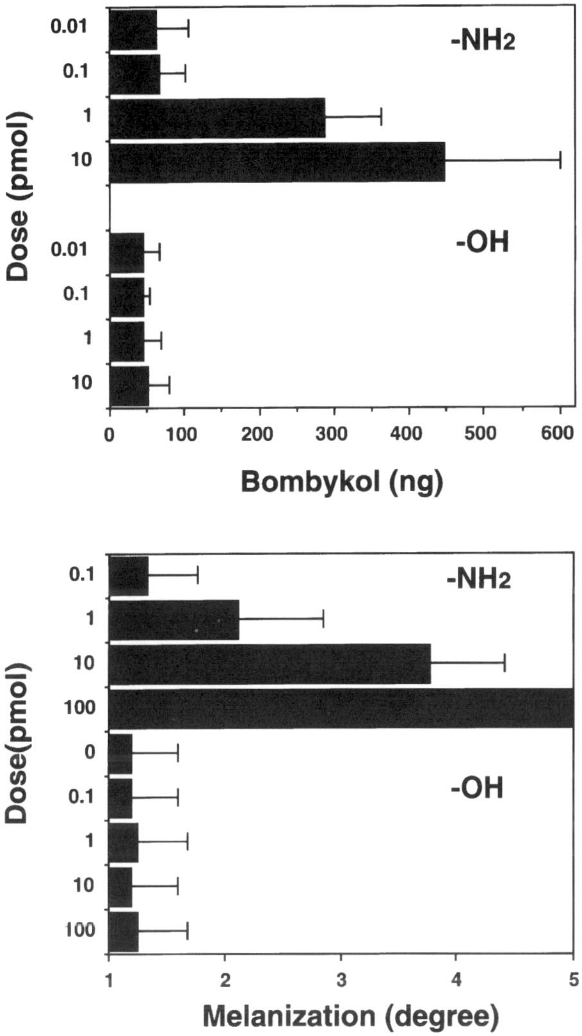

Figure 8.4. Pheromonotropic/melanotropic activity of synthetic Pss-PT. For the pheromonotropic assay, each peptide was injected into decapitated female moths of *B. mori* and the pheromone glands dissected 90 minutes after injection were extracted with *n*-hexane. The amount of the sex pheromone bombykol produced in each female was measured by HPLC (Matsumoto et al. 1992a). For the melanotropic assay, each peptide was injected into *P. separata* larvae at the time of head capsule slippage. The degree of melanization induced in the new cuticle was classified into five degrees; degree 5 corresponds with the most intense black coloration, and degree 1 corresponds with the least (Matsumoto et al., 1981). $-NH_2$, C-terminal amide; –OH, C-terminal free acid. [From Matsumoto et al. (1992a,b). Permission from Academic Press and Pergamon Press.]

1992b). As we mentioned before, Bom-PBAN-I also revealed a dual action on sex pheromone production in *B. mori* and cuticular melanization in *P. separata.* However, Bom-PBAN-I and Pss-PT have a homologous sequence only in the C-terminal portion, FXPRLamide, where X = S in Bom-PBAN-I and X = T in Pss-PT. This suggests that the active core for both pheromonotropic and melanotropic activity of these neuropeptides resides right in the common sequence of the FXPRLamide.

6. Core Sequence Required for Pheromonotropic/Melanotropic Activity

To confirm the core sequence required for pheromonotropic/melanotropic activity, we tried to synthesize fragment peptides of Pss-PT and to test their biological activity (Fig. 8.5). The N-terminal truncated peptides induced sex pheromone production in *B. mori* and also induced cuticular melanization in *P. separata.* In addition, the C-terminal pentapeptide, FTPRLamide, still retained both activities, but these activities disappeared in the C-terminal tetrapeptide TPRLamide. Since the C-terminal free acid form of Pss-PT was not positive in these assays (Fig. 8.4), these results indicate that the active core for both pheromonotropic and melanotropic activity of Pss-PT resides in the C-terminal pentapeptide FTPRLamide (Matsumoto et al. 1992a,b). It has also been demonstrated that the pheromonotropic activity of *H. zea* and *B. mori* PBAN resides in the C-terminal FSPRLamide (Raina and Kempe 1990; Kuniyoshi et al. 1991). In addition, since Pss-PT and Bom-PBAN-I reveal a similar specific activity for the induction of cuticular melanization or the sex pheromone production, the residue on the fourth position from the C-terminus seems tolerant for conservative amino acid substitution without substantial loss of biological activity. These results indicate that the core sequence of pheromonotropic/melanotropic activity is FXPRLamide. The remarkable pheromonotropic and melanotropic activity of insect myotropic neuropeptides having FXPRLamide (*vide infra*) also support the idea that the C-terminal FXPRLamide is the minimum sequence essential for both activities.

7. Functional Diversity of Pheromonotropic Neuropeptides

An 18-amino-acid pheromonotropic neuropeptide, Pss-PT, was isolated from *P. separata* larvae in the molting stage just before MRCH was released from the SOG. Since larval insects have little use for a sex-related pheromone-stimulating hormone, it seems that Pss-PT is functioning as MRCH and is involved in the regulation of larval pigmentation in the armyworm, rather than sex pheromone production which is likely to be regulated by a 33 to 34-amino-acid peptide, PBAN. Accordingly, we have proposed that Pss-PT should be named Pss-MRCH (Matsumoto et al. 1992b).

As we mentioned in the previous section, the C-terminal pentapeptide sequence, FTPRLamide, is the minimum sequence required for both pheromonotropic and

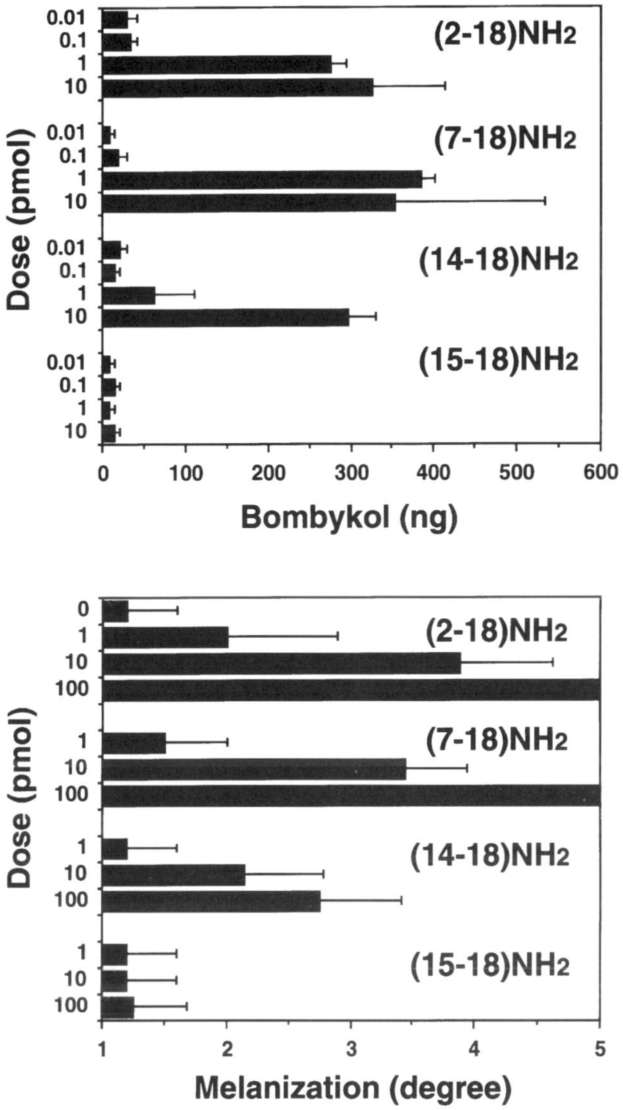

Figure 8.5. Pheromonotropic/melanotropic activity of Pss-PT fragment peptides. Peptides were injected as described in Fig. 4. [From Matsumoto et al. (1992a,b). Permission from Academic Press and Pergamon Press.]

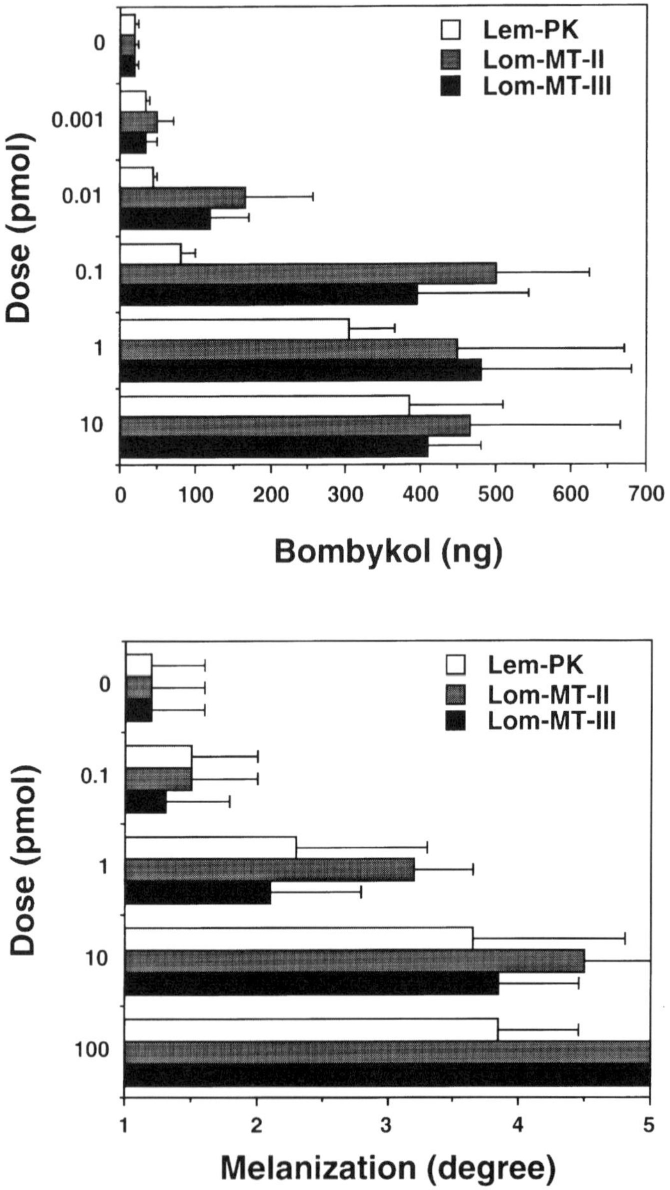

Figure 8.6. Pheromonotropic/melanotropic activity of insect myotropic neuropeptides having FXPRLamide at the C-terminus. Peptides were injected as described in Fig. 4. Nomenclature of insect neuropeptides is according to Raina and Gade (1988). [From Fónagy et al. (1992d) and Matsumoto et al. (1993). Permission from Pergamon Press and Pesticide Science Society of Japan.]

melanotropic activity of Pss-PT. It has been demonstrated that FTPRLamide is also the active core of a myotropic neuropeptide, leucopyrokinin (LPK), for its muscle contractile activity on the isolated cockroach hindgut (Nachman et al. 1986). In addition, it has been reported that a group of LPK-related myotropins are widely distributed in the cockroach and locust (see review by Schoofs et al. 1993). Since they all share the sequence of FXPRLamide (X = S, T, or V) at the C-terminus (Fig. 8.1), we tested the effect of such myotropins on both sex pheromone production in *B. mori* and cuticular melanization in *P. separata*. And, as seen in Fig. 8.6, the myotropins we tested here all revealed marked pheromonotropic and melanotropic activity (Fónagy et al. 1992; Matsumoto et al. 1993). It has also been demonstrated that PBAN has the muscle contractile activity on the *Leucophaea maderae* hindgut or *L. migratoria* oviduct (Fónagy et al. 1992d; Kuniyoshi et al. 1992b).

Among lepidopteran neurohormones produced by the SOG, *B. mori* diapause hormone also turned out to have the FXPRLamide sequence where X = Gly (Fig. 8.1) (Imai et al. 1991). Interestingly, Pss-PT also induced diapause egg production when it was injected into polyvoltine strain pupae of *B. mori* (Matsumoto et al. 1992b). Although the *B. mori* diapause hormone activity of Pss-PT was estimated at 1–5% of that of synthetic Bom-DH, it was also reported that a myotropic neuropeptide, Lom-PK, showed a comparable *B. mori* diapause hormone activity to synthetic Bom-DH (Nachman et al., 1993a). These findings also suggest the sequence significance of FXPRLamide even in the regulation of embryonic diapause of *B. mori*.

8. Conclusion

The wide distribution of hormonal activities seemed a tricky subject and provided us with several questions on the function and structural relationship of the neurohormones produced by the SOG in lepidopteran insects. However, advances in neuropeptide chemistry in this field are answering such questions and we now understand the following points: (1) A group of FXPRLamide peptides are widely distributed in insects and regulate diverse physiological events according to species or to specific developmental stages. (2) Pheromonotropic neuropeptides share the sequence of FXPRLamide, and this is a common core sequence for pheromonotropic, melanotropic, myotropic, and embryonic diapause-inducing activities. (3) The biological cross-activities of PBAN-related FXPRLamide peptides explain why PBAN, MRCH, and *B. mori* diapause hormone activities originating from the SOG are detected even in insects which do not seem to require such neurohormones in their life cycles.

Considering the significance as well as diverse functionality of the FXPRLamide sequence in insect physiology, it will be of interest to elucidate how these molecules have evolved and how they work in the regulation of diverse physiological events.

9

Physiological and Cellular Mode of Action of Pheromone Biosynthesis Activating Neuropeptide (PBAN) in the Control of Pheromonotropic Activity of Female Moths

Ada Rafaeli, Victoria Soroker, Batya Kamensky, Carina Gileadi, and Uzi Zisman

1. Introduction

With the identification, purification, and sequencing of a neuropeptide produced in the subesophageal ganglion (SOG) of *Helicoverpa zea*, termed Hez-PBAN (Raina et al. 1989a), progress in the field of pheromone neuroendocrinology in recent years has accelerated. PBAN was isolated from *Bombyx mori* adults (Kitamura et al. 1989) and showed an 80% homology to Hez-PBAN. PBAN-like activity was identified in a number of moth species as well as in several other insect species (Altstein et al. 1990; Bestmann et al. 1989; Cusson and McNeil 1989a; Foster et al. 1991a; Gazit et al. 1990; Martinez and Camps 1988; Ohguchi et al. 1985; Rafaeli and Soroker 1989a; Rafaeli and Klein 1994; Raina and Klun 1984; Tang et al. 1989). A peptide with pheromonotropic activity in *B. mori* female adults was subsequently isolated from larvae on *Pseudaletia separata* (Matsumoto et al. 1992a). Structure–function studies showed that the C-terminal pentapeptide sequence (Phe-X-Pro-Arg-Leu-NH_2, Xaa = Ser or Thr) is the minimum essential sequence for stimulation of pheromone production (Matsumoto et al. 1992a; Raina and Kempe 1990). A group of insect neuropeptides (pyrokinins and locustamyotropins) bear the same C-terminal sequence and have been demonstrated to have pheromonotropic activities (Fónagy et al. 1992a). All these studies were conducted using the original *in vivo* bioassay established for PBAN isolation (Raina and Klun 1984) whereby insects not producing pheromone (neck-ligated or decapitated females) were injected with the peptides. This bioassay has proved to be simple and efficient for isolation and identification. However, using this bioassay does not identify the target tissue and the mode of action at the receptor site. In addition, a weakness in such studies is that the injected PBAN may be degraded by the hemolymph.

Pheromone neuroendocrinology, although rapidly developing as a multidisciplinary science, is still a relatively new field and there are not enough studies on different species to permit the formulations of meaningful generalities. In some species, additional factors have been implicated as regulators of diel periodicity and pheromone production. In *Pseudaletia unipuncta* the corpora allata were shown to be essential for the release of PBAN (Cusson and McNeil 1989a). Therefore in this species an endocrine element is also implicated. In the redbanded leafroller, *Argyrotaenia velutinana*, (Jurenka et al. 1991a), PBAN induced pheromonotropic activity—but only in the presence of a bursal factor found in the corpus bursae, thereby implicating an additional, possibly hormonal, factor in the control of pheromone production in this species. In *Lymantria dispar*, for example, ligation between the head and thorax causes termination of pheromone production. However, injections of PBAN are only stimulatory when the ventral nerve cord is cut just before injection (Raina 1993). We observed a similar phenomenon in the pink bollworm moth, *Pectinophora gossypiella* (Rafaeli and Klein 1994) (see below). This indicates that, in these species, a possible neural element is involved in the control of pheromone production. The involvement of the ventral nerve cord (VNC) in the control of pheromone biosynthesis was implicated in *H. zea,* in which it was suggested that PBAN, produced in the SOG, was transported through the VNC to the terminal abdominal ganglion (TAG), thereby inducing a neural message to start pheromone production (Teal et al. 1989a). The pheromone gland was shown to be innervated, and electrical stimulation of the TAG caused a stimulation in pheromone biosynthesis (Christensen et al. 1991b). It was also suggested that the biogenic amine, octopamine, levels of which fluctuate diurnally (Christensen et al. 1992), can stimulate pheromone production when injected to intact photophase females (Christensen et al. 1991b). Levels of PBAN immunoreactivity as well as biological activity have been detected in TAG extracts from photophase females (Rafaeli et al. 1991, 1993). However, these were very low as compared to levels found in the corpora cardiaca. We also cannot rule out the possibility that immunoreactivity was due to cross-reactivity with peptides containing the C-terminal pentapeptide sequence. Contradictory reports showed that pheromonotropic activity was obtained in the absence of the TAG connectives in *H. zea, Heliothis virescens*, and several other moth species (Arima et al. 1991; Yi and Cao 1989; Rafaeli et al. 1990a,b; Raina et al. 1991; Soroker and Rafaeli 1989).

Through the development of a sensitive *in vitro* bioassay, we were able to present evidence, for the first time, that isolated pheromone glands of *Helicoverpa armigera* and *H. zea* are target tissues for PBAN (Soroker and Rafaeli, 1989; Rafaeli et al. 1990b). This bioassay showed that brain–subesophageal ganglia extracts, as well as synthetic Hez-PBAN, stimulate [^{14}C]acetate incorporation from the precursor sodium acetate into the *de novo* production of hexane extractable lipids. Subsequent analyses of the hexane extractable lipids, using three different separation techniques (TLC, HPLC, and radio-GC as well as GLC for unlabeled products), all showed a significant stimulation of a product coeluting with the main pheromone component, (Z)-11 hexadecenal. These observations were subsequently confirmed in other moth species (Arima et al. 1991; Jurenka et al. 1991b; Fónagy et al. 1992b). The response obtained was shown to be specific to the pheromone gland cultures independent of the presence of other abdominal tissues, including the TAG connectives (Rafaeli 1994). A significant stimulation in response to PBAN, not significantly different from the response obtained by intact females, was obtained in cultures *in vitro* (Rafaeli 1994). Pheromone gland cultures were responsive to PBAN irrespective of the physiological or chronological age and the photoperiod of the donor females (Rafaeli et al. 1993). Even mated females, which have ceased to produce pheromone *in vivo,* could be stimulated to produce normal pheromone levels *in vitro* on stimulation by PBAN (Rafaeli et al. 1993). Therefore we concluded from these studies that control of pheromone production most probably resides in the control of release and/or production of the neuropeptide and not in the competence of the tissue. The present investigation utilizes this bioassay in the study of the mode of action of PBAN at the physiological and cellular level.

2. Involvement of the Nervous System

2.1. Effect of K^+ Depolarization

It has been reported that the pheromone gland is innervated (Christensen et al. 1991b). These scientists argue that these terminals are the target tissue for PBAN and that pheromonotropic activity of the gland is attributed to the release of octopamine by these nerve terminals (Christensen et al. 1991b, 1992). To test this hypothesis we conducted a series of experiments in which pheromone gland cultures were subjected to high K^+ depolarization (100 mM) before treatment with PBAN. In theory, if these terminals contain the stimulatory compound, then depolarization should stimulate pheromonotropic activity, and, moreover, after depolarization the glands should not respond to PBAN. However, we found that K^+ depolarization did not stimulate pheromone production *in vitro* and that pheromone glands were equally stimulated by PBAN whether they were subjected to prior K^+ depolarization or not (Table 9.1).

Since octopamine has been implicated as the neurotransmitter involved in the transmission of the nervous signal for pheromone biosynthesis, it should be possible to stimulate pheromonotropic activity by local administration of octopamine or injection to photophase females. We therefore tested the effect of octopamine in cultures *in vitro* of *H. armigera.* No significant pheromonotropic effect was observed with octopamine (Fig. 9.1). In addition, the effect of injecting octopamine to photophase females was tested. These tests were undertaken on both *H. zea* and *H. virescens.* No increase of pheromone titers was observed (Fig. 9.2). Moreover, phentolamine, an α-adrenergic blocker, did not inhibit the

Table 9.1. Stimulatory effect of PBAN (5 pmol) after K^+ depolarization (100 mM)[a]

Treatment	*De novo* biosynthesis (cpm/gl) mean ± SEM (*n*)
Control	1411 ± 250.0 (27) a
K^+ alone (100 mM)	1045 ± 125.5 (28) a
PBAN (5 pmol)	15272 ± 2089.3 (33) b
PBAN after K^+ (30 min)	16373 ± 2760.0 (16) b

[a]Numbers followed by different letters are significantly different (Mann–Whitney, $p < 0.05$). Shorter depolarization times (10 min) showed similar results (data not shown).

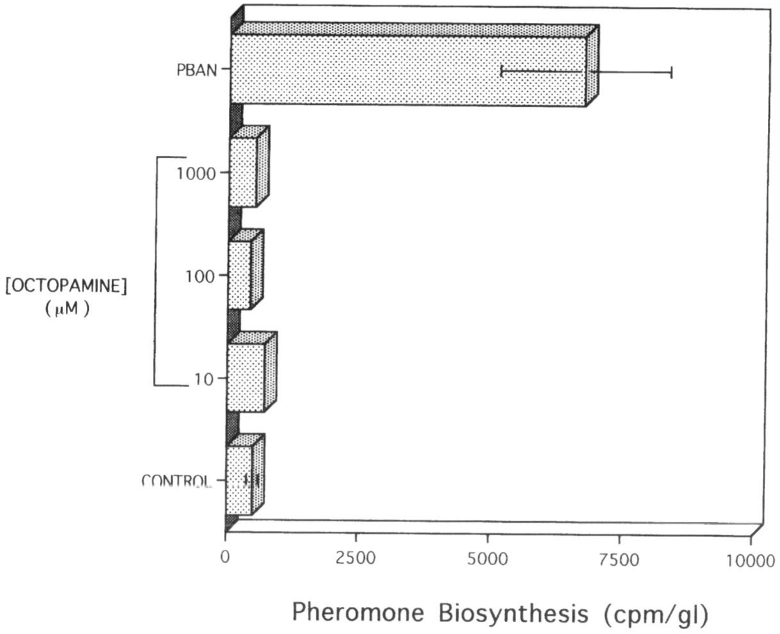

Figure. 9-1. The effect of various concentrations of octopamine on the pheromonotropic activity of pheromone gland cultures *in vitro* of *H. armigera.* Bars represent means ± SEM of at least six replicates for each treatment. A significant stimulation of pheromonotropic activity was only obtained in the presence of PBAN (5 pmol/gland in this experiment).

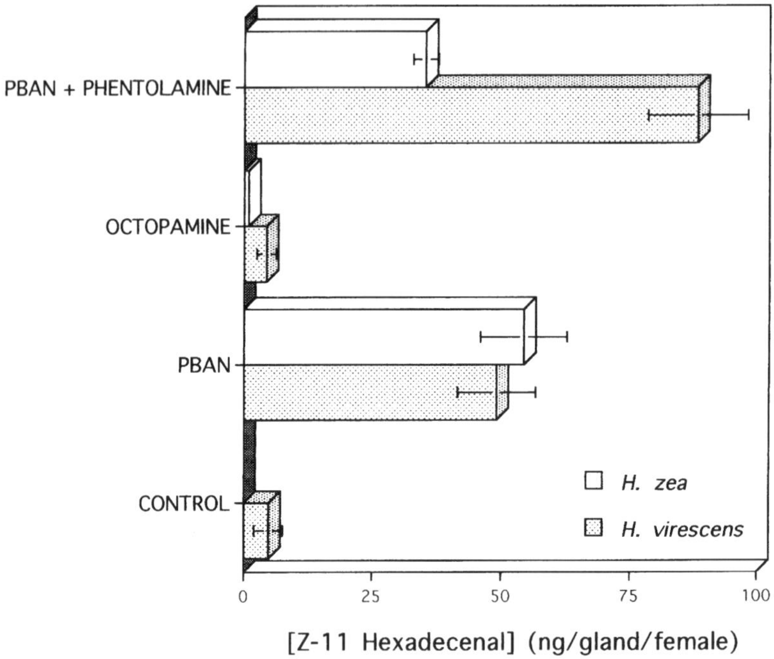

Figure 9.2. The effect of octopamine injections (10 μM) to photophase females on pheromonotropic activity *in vivo.* Bars represent means ± SEM of at least four replicates for each treatment. Levels of (*Z*)-11-hexadecenal were calculated using the internal standard method and GC analysis as reported previously (Rafaeli and Soroker, 1989a). PBAN was injected at 5 pmol/female; phentolamine was injected at a concentration of 1 mM.

normal response obtained with PBAN (5 pmol) and, in fact, enhanced stimulation in *H. virescens* (Fig. 9.2).

2.2. Effect of Ventral Nerve Cord Transection

Some evidence which does seem to point to an intervention by the nervous system was observed during a recent study on *P. gossypiella* (Rafaeli and Klein 1994). In this study we observed that females produce an average maximum of 22 ng pheromone/female (*Z,Z*-7,11 16:Ac + *Z,E,*-7,11 16:Ac) during the first hour of scotophase and that levels only drop to an average of 3.4 ng pheromone/female during the fourth hour of photophase, and thus the decline was very gradual. Decapitation for 24 hours did not replete pheromone glands from pheromone titers, and there was no significant response to PBAN in intact photophase females or those decapitated for 24 hours (Table 9.2). However, if they were decapitated for a longer period (72 hours), those surviving the ordeal had lowered their pheromone titers to a level which could subsequently be stimulated by PBAN

(Table 9.2). This stimulation, although significantly different from decapitated females, only reached 4 ng pheromone/female, a level much lower than that maximally obtained by scotophase females. A more significant level of stimulation was obtained when 72-hour decapitated females were challenged with PBAN immediately after the ventral nerve cord was severed (Table 9.2).

3. Transduction of the Pheromonotropic Response

Most neuropeptides act with surface receptors by binding and inducing a cascade of intracellular second messengers. We have shown that in *H. armigera* pheromone gland incubations the pheromonotropic action due to PBAN was associated with increases in intracellular cAMP levels (Rafaeli and Soroker 1989b). Pheromone gland stimulation could also be mimicked by elevation of intracellular cAMP levels as a result of treatment with pharmacological agents such as forskolin (adenylate cyclase activator), isobutyl methylxanthine (phosphodiesterase inhibitor), or 8-bromo-cAMP (cAMP analog) (Rafaeli and Soroker 1989b) (Fig. 9.3). Subsequently, Jurenka et al. (1991b) demonstrated the involvement of an additional second messenger, calcium. Recently, we implicated the involvement of another transducing pathway, phosphatidylinositol turnover (Rafaeli and Soroker 1994; Rafaeli 1994; Soroker and Rafaeli 1995).

3.1. Role of Cyclic-AMP, Calcium, and Protein Kinase C

PBAN induced significant levels of intracellular cAMP after 1 minute of stimulation, reaching saturation levels 10 minutes later (Soroker and Rafaeli 1995). Omission of calcium from the incubation medium completely abolished the PBAN-stimulated increase in intracellular cAMP levels (Soroker and Rafaeli 1995). Calcium dependence of pheromonotropic activity by PBAN (minimum concentrations of 0.5 pmol) was also observed. Calcium-dependence of phero-

Table 9.2. Pheromone titers of Pectinophora gossypiella *females as a result of decapitation, Hez-PBAN injections, and VNC severence*[a]

Treatment	Pheromone titers (ng *Z,Z*-7,11 16:Ac + *Z,E*-7,11 16:Ac/ female): Control	Hez-PBAN (5 pmol/female)
Photophase females	8.3 ± 0.85 (5)	6.9 ± 0.57 (5)
Decapitated 24 hr	5.2 ± 2.20 (5)	3.9 ± 0.81 (7)
Decapitated 72 hr	1.1 ± 0.63 (5)	4.1 ± 0.66 (5)[b]
Decapitated 72 hr and VNC severed	0 ± 0 (11)	8.7 ± 1.15 (8)[b]

[a]Results presented in a different form in Rafaeli and Klein (1994).

[b]Indicates a significant difference from the control value (*t* test, $p < 0.05$).

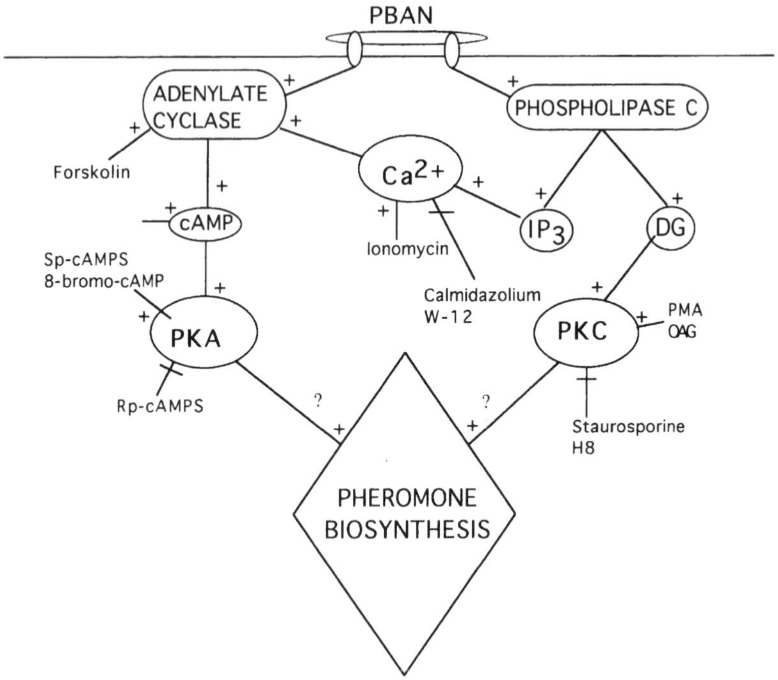

Figure 9.3. Schematic representation of the cellular events occurring at the pheromone gland cell membrane in response to the trigger by PBAN. For explanations refer to text. *Abbreviations:* 8-bromo-cAMP; 8-bromo-adenosine-3′:5′-cyclic monophosphate; cAMP; adenosine-3′:5′-cyclic monophosphate; H8; *N*-[2-(methylamino)ethyl]-5-isoquinoline sulfonamide; IBMX; isobutylmethylxanthine; OAG; 1,2-dioleolyl-*sn*-glycerol; PBAN; pheromone biosynthesis activating neuropeptide; PKA; protein kinase A; PKC; protein kinase C; PMA; phorbol-12-myristate 13-acetate; Sp-cAMPS; adenosine-3′,5′-cyclic monophosphathionate; W-12; *N*-(4-aminobutyl)-2-naphthalene sulfonamide.

monotropic activity has also been shown in *B. mori* (Fónagy et al. 1992c) and *H. zea* (Jurenka et al. 1991b). Using the analog Sp-cAMPS, calcium-independent pheromonotropic activity was obtained (Soroker and Rafaeli 1995). These experiments suggested that pheromone biosynthesis beyond the level of cAMP formation does not require extracellular calcium and that stimulation of protein kinase A leads to pheromonotropic activity (Fig. 9.3). On the other hand, the protein kinase A antagonist, Rp-cAMPS, did not inhibit the pheromonotropic activity obtained by PBAN (Soroker and Rafaeli 1995).

The pheromonotropic effects by PMA (protein kinase C activator) and by ionomycin (calcium ionophore) were also found to be dependent on the presence of calcium (Soroker and Rafaeli 1995). These results suggest that the activation pathway also operates via a calcium dependent protein kinase C (Fig. 9.3).

However, the protein kinase inhibitors, staurosporine and H8 (at concentrations of 10^{-5} M), failed to show any significant inhibition on the pheromonotropic effect by PBAN. Conversely, staurosporine, at a concentration of 10^{-5} M, stimulated pheromone biosynthesis. A similar phenomenon was reported in the corpora allata of the moth *Manduca sexta* (Reagan et al. 1992), where the rate of JH biosynthesis was increased by application of staurosporine.

The complex of calcium and calmodulin can alter the activity of enzymes such as adenylate cyclase and phosphodiesterase. If, in the pheromone gland, calcium calmodulin activates adenylate cyclase, inhibition of calmodulin would be expected to inhibit pheromone production. Intriguingly, in *H. armigera* (Rafaeli et al. 1993; Soroker and Rafaeli 1995) we observed that the calmodulin inhibitor, calmidazolium (10^{-3} M), did not interfere with PBAN stimulation. On the other hand, W-12 (10^{-3} M), another calmodulin blocker, inhibited the pheromonotropic response obtained with 20 pmol PBAN (Soroker and Rafaeli 1995). However, no effect on the pheromonotropic response obtained with 1 pmol PBAN was observed (Soroker and Rafaeli 1995).

3.2. "Cross-Talk" Between Calcium and Adenylate Cyclase

We found that ionomycin, in addition to its pheromonotropic effect, also caused significant elevation of intracellular cAMP levels. Moreover, the effects of ionomycin and PBAN on cAMP levels were additive (Soroker and Rafaeli 1995). Since intracellular calcium release, as a result of ionomycin, duplicates the pheromonotropic effect of PBAN on both pheromone production and cAMP synthesis, it appears that calcium activates adenylate cyclase (Fig. 9.3). This is further supported by the findings that forskolin stimulation of pheromone production is enhanced by calcium release (Fónagy et al. 1992c; Jurenka et al. 1991b).

Calcium release has been shown to be activated by inositol phosphate turnover. We have shown that the activation of pheromone production can operate via the stimulation of adenylate cyclase and/or the phosphatidylinositol pathway. Calcium is crucial for activation in both of these. The relationship between these three systems is schematized in Fig. 9.3. It could be hypothesized that PBAN receptor-binding stimulates phospholipase C, thereby activating calcium release which in turn activates both protein kinase C and adenylate cyclase with the subsequent activation of protein kinase A (Fig. 9.3). Another possibility is the activation of two separate receptors in a monodirectional manner (Nishizuka 1992) where adenylate cyclase and phospholipase C are activated independently after PBAN binds to the receptor site. Conclusive evidence awaits identification of the PBAN receptor(s). In addition, the nature of activation of the pheromone biosynthetic pathway by these protein kinases requires further elucidation.

4. Concluding Remarks

Two alternative transporting pathways for PBAN have been proposed, but neither have been unequivocally confirmed: PBAN may reach the pheromone gland cells via the hemolymph and act directly on them or via the ventral nerve cord where, by the secretion of a neurotransmitter such as octopamine, it can stimulate the cells. Evidence for both pathways has been proposed. It can be argued that these pathways depend on the species studied. However, in some cases the same species was studied at different laboratories. We should remember that reproductive behavior by female moths actually relies on two separate events: calling behavior accompanied by pheromone release and pheromone biosynthesis. These two elements of behavior may influence one another, but they are not necessarily interdependent on each other (Raina 1993). There is thus a need for further detailed physiological and biochemical studies of various moth species in order to separate the two elements of behavior and elucidate the exact role of PBAN in the control of these elements. It may well be that PBAN regulates the biosynthetic pathway whereas the timing of this event accompanied by calling is under the control of a neural message with octopamine acting as a modulator between the hormonal and neural control complex.

PBAN stimulates a complex multisignal, interrelated transduction cascade. The physiological significance of such a mechanism for PBAN activation may be in the precise regulation of the complex biochemical events involved in female pheromone production. The influence of PBAN on rate-limiting enzyme(s) in the biosynthetic pathway of pheromone production remains to be clarified. The availability of potential inhibitors as well as sensitive bioassays will greatly facilitate these explorations.

Acknowledgments

This research was supported by a Binational Agricultural Research Development Grant (BARD Research Project No. IS 1645-89) to A.R. and by a Foundation of Higher Council Planning and Grants Committee Fellowship to V.S. A.R. would like to thank the Insect Neurobiology and Hormone Laboratory (USDA, Beltsville, MD) for hosting her during her sabbatical when experiments were conducted on *H. zea* and *H. virescens,* and she would also like to thank Prof. A. Hefetz (Department of Zoology, Tel Aviv University) for hosting her during the study on *P. gossypiella.*

10

Mode of Action of Pheromone Biosynthesis Activating Neuropeptide in *Bombyx mori*

Adrien Fónagy, Shogo Matsumoto, and Takashi Mitsui

1. Introduction

The production of sex pheromone in a number of moth species is controlled by pheromone biosynthesis activating neuropeptide (PBAN), which is secreted by the subesophageal ganglion (Raina and Klun 1984; Ohguchi et al. 1985; Ando et al. 1988; Raina 1988; Rafaeli and Soroker 1989a). Recently, a growing number of investigations have focused on the identification of the target organ of PBAN for a clearer understanding of the mode of action of this hormone. In *Helicoverpa (Heliothis) armigera, Bombyx mori*, and *Spodoptera litura*, the target organ of PBAN has been demonstrated to be the pheromone gland itself (Rafaeli and Soroker 1989b; Soroker and Rafaeli 1989; Arima et al. 1991; Fónagy et al. 1992a). According to another model, in *H. zea, H. virescens,* and *H. subflexa, Helicoverpa-Heliothis* PBAN does not act on pheromone glands but on the terminal abdominal ganglion (TAG) and the abdominal nerve cord transports PBAN to the TAG, which sends a different message to the gland (Teal et al. 1989a). Moreover, direct, neural stimulation induced by PBAN is suggested via axonal branches innervating the pheromone glands in *H. zea* and *H. virescens* (Christensen et al. 1991b). In addition, reports say that the corpus bursa is the target organ of PBAN in *Argyrotaenia velutinana* and that it produces the "bursa factor" responsible for pheromone production (Jurenka et al. 1991a). These findings suggest diverse physiological mechanisms of PBAN action depending on the lepidopteran species.

To date, only a few studies have been conducted on the second messenger system involved in PBAN action. In *H. armigera,* it has been demonstrated that under *in vitro* conditions PBAN increased the intracellular level of cAMP in the pheromone gland, resulting in pheromone production (Rafaeli and Soroker 1989b; Rafaeli et al. 1990b). It has also been revealed for this species that the addition of 8-bromo-cAMP to the incubation media stimulated acetate incorporation into the pheromone (Rafaeli and Soroker 1989b). This effect was further increased by adding forskolin, a stimulant of adenylate cyclase (Seamon et al. 1981), and MIX, a potent phosphodiesterase inhibitor (Montague and Cook 1971), into the medium. It has been previously demonstrated that various enzyme activities in diverse systems depend on calcium ion and calmodulin concentrations (Cheung 1980) and that the synthesis or degradation of cyclic nucleotides in insect tissues also depends on calcium/calmodulin (Bodnaryk 1983; Smith and Combest 1985). Therefore, cyclic nucleotides could also be responsible for PBAN-induced pheromone synthesis, at rate-limiting step(s), in the pheromone gland of *B. mori*.

The main objective of our present study is to elucidate some further aspects in the second messenger system involved in the mode of action PBAN, in *B. mori*, by testing *in vivo* and *in vitro* effects of various cyclic nucleotides as well as some pharmacochemicals such as MIX and forskolin. Presuming the possible involvement of cAMP or other cyclic nucleotides, we investigated the *in vitro* effect of calcium ionophore, which is known to stimulate the penetration of divalent cations such as calcium ion (Pfeiffer et al. 1978), the influx of which might enhance the activation of adenylate cyclases. The effect of lanthanium ion, a calcium ion antagonist, which is known to displace specifically the divalent cations from the plasmamembrane (Weiss 1974), was also studied. Our investigations are based on the previous findings that the target organ of PBAN in *B. mori* is the pheromone gland (Arima et al. 1991; Fónagy et al. 1992a).

2. Materials and Methods

2.1. Insects

Bombyx mori (Kinshu & Showa strain from Katakura Co. Ltd) larvae were reared under a 16L:8D photoperiod at 25°C and fed on an artificial diet. The newly emerged females were decapitated within 3 hours and abdomens were kept for another 24 hours at room temperature.

2.2. Injection of Cyclic Nucleotides

In the *in vivo* cyclic nucleotide dose-dependence experiments, decapitated females of *B. mori* were injected with corresponding doses (50, 250, and 500 μg) of cAMP, cGMP, or dibutyryl cAMP (dbcAMP) (sodium salts; Sigma) dissolved in 10 μl of distilled water. The pheromone glands were excised 90–100 minutes after injection and extracted with 100 μl *n*-hexane. A 10-μl aliquot of the extracts was subjected to pheromone analysis. The amount of bombykol [(*E,Z*)-10,12-hexadecadien-1-ol] was measured by high-performance liquid chromatography (HPLC) using *n*-hexane:2-propanol (49:1) on a Senshu Pak NO_2 column at an absorbance of 230 nm.

2.3. Incubation of Pheromone Glands with Cyclic Nucleotides

Pheromone glands taken from decapitated females of *B. mori* were damped-dried and then cleaned of other tissue remnants, and the inside of the glands were carefully washed several times with the respective incubation medium using a microsyringe. The glands were then transferred as "hanging drops" into Eppendorf vials. The glands were incubated at room temperature in 150 μl Grace's medium containing various concentrations of cyclic nucleotides (1 μM–10 mM) for 90–100 minutes. After incubation, each gland was taken out from the medium and extracted with 100 μl *n*-hexane, and the amount of bombykol produced in the gland was measured as described above.

2.4. Incubation of Pheromone Glands with MIX or Forskolin in the Presence of Bom-PBAN-I

Stock solutions of both MIX (3-isobutyl-1-methylxanthine, Sigma) and forskolin (7-0-hemisuccinyl-7-deacetyl, Sigma) (100 mM) in DMSO were prepared and diluted in Grace's medium. Dose-dependence was studied applying different concentrations of MIX or forskolin, or both.

The stimulating effect of MIX and/or forskolin (for concentrations see Figs. 10-2 and 10-3) was studied in the presence of 0.5 ng or 1 ng Bombyx PBAN-I (Bom-PBAN-I)/150 μl Grace's medium. Bom-PBAN-I was synthesized by solid-phase method on an ABI model 430 peptide synthesizer using a *t*-butoxycarbonyl (Boc) protocol (Matsumoto et al. 1990). Crude Bom-PBAN-I was purified by reversed-phase HPLC on a Senshu Pak VP-318 column. The glands were prepared as described in the *in vitro* cyclic nucleotide assay and were incubated for 100–120 minutes. After incubation the glands were processed as described above.

2.5. Incubation of Pheromone Glands with Calcium Ionophore

A stock solution (10^{-1} M) of calcium ionophore A23187 (free acid) (Sigma) was prepared in DMSO and added to Grace's medium to obtain the desired concentrations. The *in vitro* dose-dependence studies were conducted at different ionophore concentrations (10–1000 μM in 150 μl Grace's medium; these concentrations are in a generally accepted working range). After 100–120 minutes of incubation the glands were processed as described before.

2.6. Incubation of Pheromone Glands with MIX or Forskolin in the Presence of Calcium Ionophore

Calcium ionophore was employed at different concentrations (333, 50, and 10 μM) in combination with MIX or forskolin (for exact arrangements see Figs. 10-5 and 10-6.). Following incubation they were processed as above.

2.7. *Incubation of Pheromone Glands with Lanthanium Ion in the Presence of Bom-PBAN-I*

The inhibitory effect of lanthanium ion ($LaCl_3$, Nacalai) was studied using a series of different doses of Bom-PBAN-I (2,5, and 10 ng/150 μl Grace's medium) along with a series of various concentrations of lanthanium ion (100 μM–10 mM). Glands were incubated and processed as previously described.

3. Results

3.1. *Effect of cAMP, cGMP, and dbcAMP* In Vivo *and* In Vitro

The injection of cAMP, cGMP, or dbcAMP into decapitated female *B. mori* induced bombykol production in a dose-dependent manner (Fig. 10.1). Among the tested cyclic nucleotides cAMP was found to be the most active one. Under *in vitro* conditions, however, when the pheromone glands were incubated with these cyclic nucleotides in a wide range of concentrations (1 μM–10 mM), they failed to induce measurable bombykol production (data are not given).

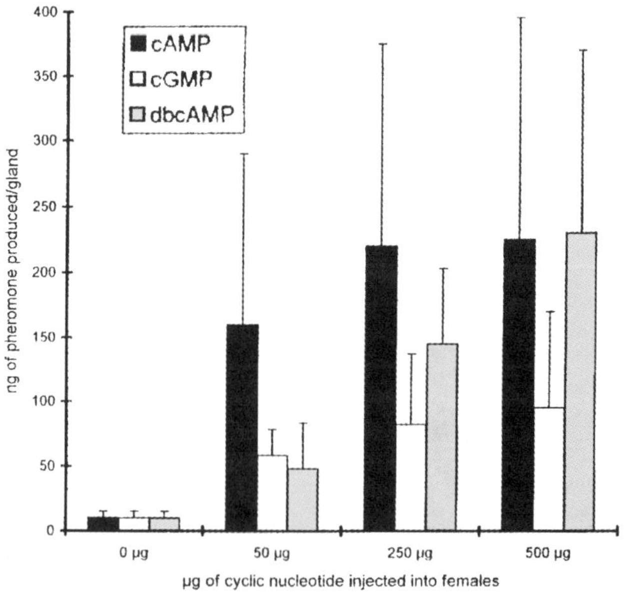

Figure 10.1. *In vivo* effects of cyclic nucleotides on pheromone production in *Bombyx mori*. Different doses of cyclic nucleotides were injected into females while controls received distilled water. The columns indicate a mean of minimum nine measurements of independent gland extracts. Thin bars represent standard deviations.

3.2. *The* In Vitro *effect of MIX or Forskolin in the Presence of Bom-PBAN-I*

As a preliminary experiment, pheromone glands were incubated with either MIX or forskolin (10 μM–1 mM), or both, but no pheromone production was induced [Figs. 10.2 and 10.3 (control and F control, respectively)].

The incubation of the pheromone glands with 0.5 ng Bom-PBAN-I/150 μl medium resulted in 102 ± 35 ng of bombykol production [using 10 ng Bom-PBAN-I, pheromone production shows maximum value of 430 ± 165 ng/gland as it was demonstrated before (Fónagy et al. 1992a)]. The effect of the hormone on bombykol production was enhanced by the addition of MIX to the incubation

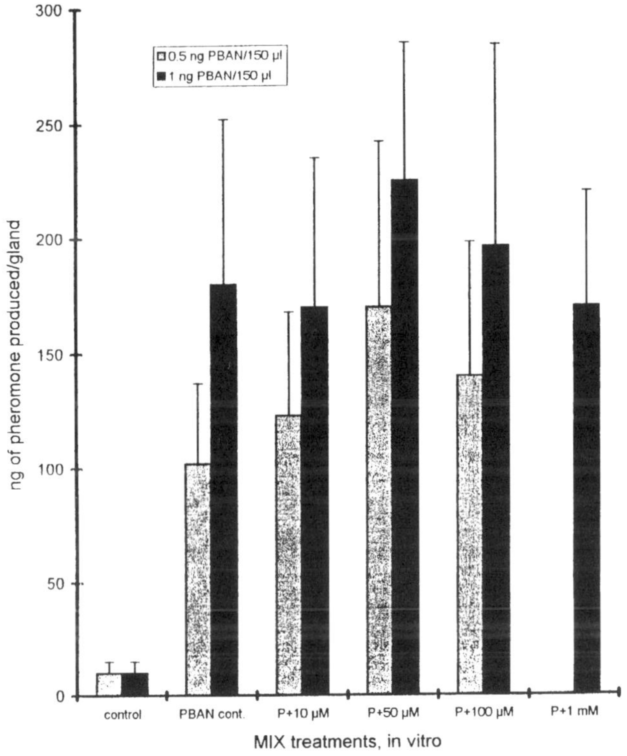

Figure 10.2. *In vitro* effect of MIX in the presence of 0.5 or 1 ng Bom-PBAN-1 in *Bombyx mori*. MIX, 3-isobutyl-1-methylxanthine; P, Bom-PBAN-I. Control: 100 μM MIX without P; PBAN cont.: 0.5 ng P/150 μl (gray column); 1 ng P/150 μl (black column); P + 10 μM, P + 50 μM, P + 100 μM, and P + 1 mM contain 0.5 or 1 ng PBAN + respective concentrations of MIX indicated. The columns indicate a mean of a minimum of 9 (but generally 15) independent gland extract measurements. Thin bars represent standard deviations. Result of P + 50 μM MIX differs significantly from that of PBAN control at $p < 0.05$ level with Student's t test.

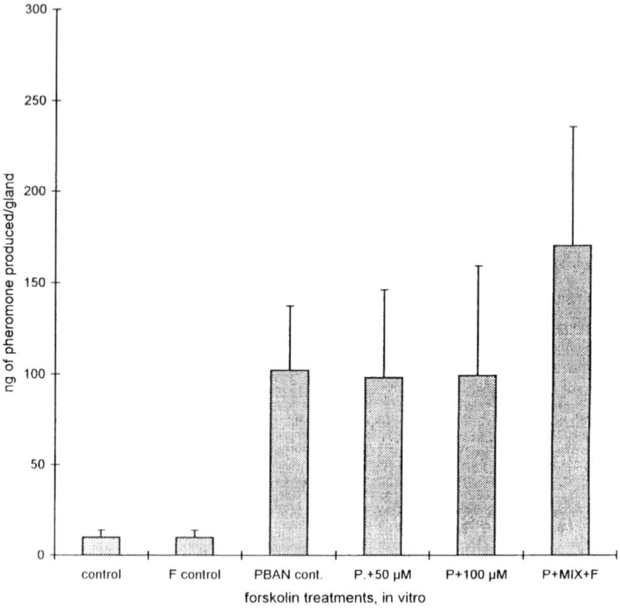

Figure 10.3. *In vitro* effect of forskolin in the presence of 0.5 ng Bom-PBAN-I in *Bombyx mori.* F, forskolin; for rest see control in legend to Fig. 10.2: glands in Grace's medium; F control: 100 μM forskolin without P; PBAN cont.: 0.5 ng P/150 μl; P + 50 μM and P + 100 μM contain 0.5 ng PBAN + respective concentrations of forskolin. P + MIX + F contains 0.5 ng PBAN + 50 μM of both MIX and F. Result of P + MIX + F differs significantly from that of PBAN control at $p < 0.05$ level in Student's t test. Otherwise refer to Fig. 10.2.

medium at concentrations between 10 and 100 μM. In this range, 50 μM MIX increased bombykol production by 60% (Fig. 10.2). The addition of MIX at a concentration of 50 μM to the incubation medium already containing 1 ng PBAN/150 μl increased pheromone production by about 30% (Fig. 10.2). In contrast, there was no enhancement of bombykol production when forskolin (50 or 100 μM) was added to the medium containing 0.5 ng PBAN/150 μl (Fig. 10.3). When both MIX and forskolin (50 μM) were added to the medium, the result was similar to the one when only MIX was employed (Fig. 10.2 and 10.3).

3.3. The In Vitro *Effect of Calcium Ionophore on Pheromone Production*

When glands of *B. mori* were incubated in a medium containing calcium ionophore at 10 μM concentration, pheromone production was remarkably induced, reaching the maximum at between 250 and 500 μM. Ionophore at higher concentrations decreased pheromone production slightly (Fig. 10.4).

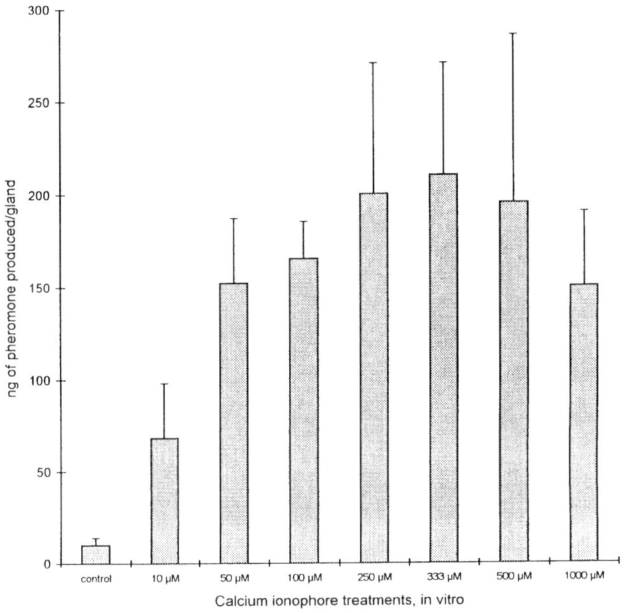

Figure 10.4. *In vitro* calcium ionophore dose-dependence in *Bombyx mori* pheromone glands. Glands were incubated in 150 μl of Grace's media containing respective concentrations of ionophore (or none in controls). Otherwise refer to Fig. 10.2.

3.4. *The* In Vitro *Effect of MIX or Forskolin in the Presence of Calcium Ionophore*

A twofold elevation was recorded when MIX (100 μM) was added to medium containing 50 μM calcium ionophore (Fig. 10.5). As shown above, calcium ionophore stimulated the pheromone gland to produce bombykol *in vitro,* and the effect was optimal at a concentration of 333 μM. Forskolin was added in different amounts to the above medium. When 100 μM concentration was applied, only a slight increase (20%) in bombykol production was achieved (Fig. 10.6). The stimulating effect, however, was much more significant when lower concentrations of calcium ionophore were employed. With 50 μM ionophore, an addition of 100 μM forskolin gave a twofold increase (Fig. 10.6). A medium which contained 100 μM forskolin and only 10 μM calcium ionophore could increase pheromone production more than threefold (Fig. 10.6).

3.5. *The* In Vitro *Effect of Lanthanum Ion in the Presence of Bom-PBAN-I*

The inhibitory action of lanthanum ion was studied using different doses of Bom-PBAN-I (Fónagy et al. 1992a) along with a series of various lanthanum ion

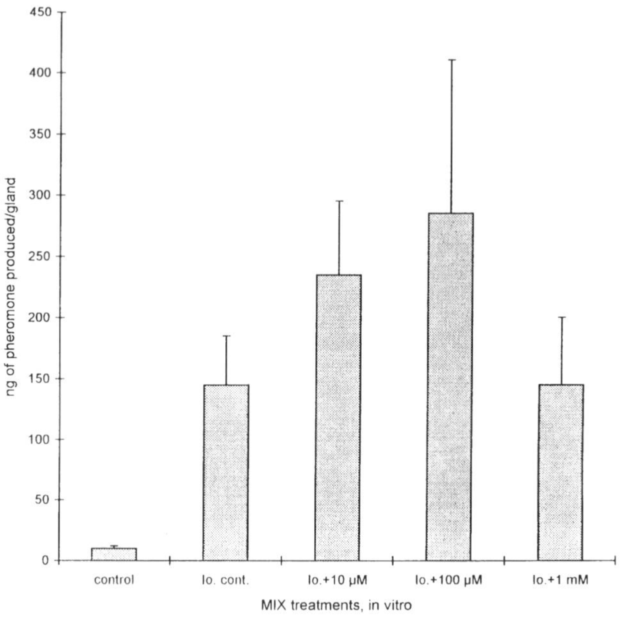

Figure 10.5. *In vitro* effect of MIX in the presence of 50 μM calcium ionophore on pheromone production, in *Bombyx mori.* Control: glands in Grace's medium; Io cont.: calcium ionophore control at 50 μM concentration; Io + 10 μM, Io + 100 μM, Io + 1 mM: 50 μM calcium ionophore + respective concentrations of MIX indicated. Result of Io 50 + 100 μM treatment differs significantly from that of control at $p < 0.01$ level in Student's t test. Otherwise refer to Fig. 10.2.

concentrations. Profound, total inhibitory effect was achieved with 1–10 mM lanthanum ion when a relatively low PBAN dose (2 ng/150 μl) was used, while 100 μM lanthanum ion resulted in 50% inhibition (Fig. 10.7). In the case of higher doses of Bom-PBAN-I (5 ng and 10 ng/150 μl) a concentration of 1 or 5 mM lanthanum ion, respectively, was necessary to reach 50% inhibition in pheromone production; 10 mM lanthanum ion induced about 75% inhibition with both doses (Fig. 10.7).

4. Discussion

As a part of extended studies related to the effect of Bom-PBAN-I in *B. mori,* a number of primarily *in vitro* studies were carried out in order to better understand the mode of action of the hormone in the target organ, the pheromone gland. Initially, the possible involvement of calcium ion and cyclic nucleotides (or both) in signal transduction process was studied.

Injection of different cyclic nucleotides into decapitated *B. mori* females re-

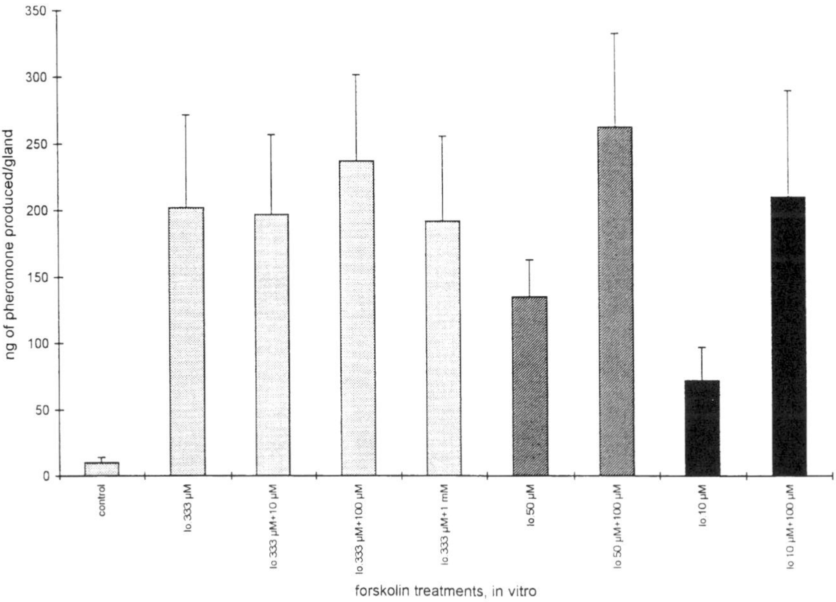

Figure 10.6. *In vitro* effect of forskolin in the presence of calcium ionophore in *Bombyx mori*. Control: glands in Grace's medium; Io 333 μM, Io 50 μM and Io 10 μM: calcium ionophore controls at 333, 50, or 10 μM concentration; Io 333 + 10 μM, Io 333 + 100 μM, Io 333 + 1 mM (stippled columns), Io 50 + 100 μM (hatched column) and Io 10 + 100 μM (black column) contain 50 or 10 μM calcium ionophore + respective concentrations of forskolin indicated. Results of Io 50 + 100 μM and Io 10 + 100 μM treatments significantly differ from those of controls at $P < 0.01$ level in Student's t test. Otherwise refer to Fig. 10.2.

vealed dose-dependent action in pheromone production. Cyclic AMP proved to be the strongest stimulant, and these observations already suggest that cAMP may play a role as a second messenger. The *in vivo* cyclic nucleotide tests, however, may serve only as guidelines, since direct *in vitro* action in *B. mori* could not be proved. The cAMP may act on the whole body indirectly and in a more complex way, or externally, as a hormone does, as was shown by Devroetes (1983). The negative results obtained with various nucleotides, MIX, and forskolin may be due to poor penetration into the target cells (Lehninger 1979), insufficient direct stimulation of *B. mori* pheromone glands under our experimental conditions, inadequate information from the measurement of pheromone production in which the chemicals are involved as second messenger (i.e., the stimulation of cAMP content may have been insufficient to initialize measurable bombykol production), or some other unknown factors. It must be noted that in *Manduca sexta,* the pupal prothoracic glands could not be stimulated by forskolin to produce ecdysone, although it has been proved that prothoracicotropic hormone (PTTH)

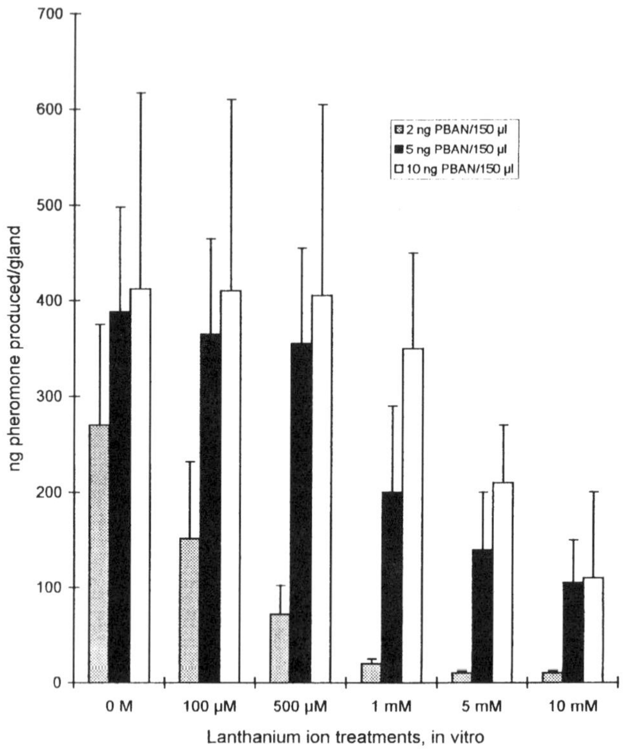

Figure 10.7. *In vitro* inhibitory action of lanthanum ion in the presence of Bom-PBAN-I on pheromone production in *Bombyx mori.* Various concentrations of Bom-PBAN-I were employed as indicated. On horizontal axis respective concentrations of lanthanum ion are given. The columns indicate a mean of minimum 13 independent gland extract measurements. Thin bars represent standard deviations.

induces cAMP formation and functions as second messenger for ecdysone production (Smith and Gilbert 1986). From these results it was concluded that the prothoracic glands apparently have a low level of forskolin-stimulatable adenylate cyclase activity and/or a high level of phosphodiesterase activity (resulting in the quick degradation of cAMP, if present). On the other hand, in *Locusta migratoria,* in which also cAMP was shown to be the second messenger of the antidiuretic hormone, MIX alone did not affect rectal cAMP levels and was ineffective on rectal water reabsorption, whereas forskolin was found to induce noticeable response (Herault and Proux 1987). Cyclic nucleotides also have been reported as the most likely second messengers of melanization and reddish coloration hormone (MRCH) (Matsumoto et al. 1979). It is worth noting that MRCH and PBAN-I of *B. mori* was found to be the same molecule possessing functional diversity depending on developmental stage (Matsumoto et al. 1990).

We may thus assume similar signal transduction mechanism for both hormones. The significant increase in cAMP levels after *H. armigera* pheromone glands were exposed to *Heliothis*-PBAN, *in vitro,* also favors this hypothesis (Rafaeli and Soroker 1989b; Rafaeli et al. 1990b).

Beside the possible involvement of cAMP in PBAN mode of action, the role of calcium ions was studied by using calcium ionophore and $LaCl_3$. The calcium ion influx might be relatively important in pheromone production induction since the ionophore mimicked the action of PBAN, and probably much more important than simply being necessary for PBAN binding to receptor. Furthermore, involvement of calcium ion may be direct, in hormone action, inducing respective enzyme(s) responsible for completion of pheromone biosynthesis. Based on several vertebrate and recent studies on insect hormonal mode of action (reviewed by Smith and Combest 1985) it could be also indirect as in the case of prothoracicotropic hormone (Gilbert et al. 1988). PBAN may act by stimulating calcium ion influx, with the cation in turn enhancing cAMP formation. The cAMP may activate sequential events responsible for possible enhancement of formation/activity of rate-limiting enzyme(s) involved in pheromone synthesis. Studies on isolated pheromone glands of *Helicoverpa (Heliothis) zea,* performed at the same time as ours but at a different laboratory, also proved that PBAN can act directly on the glands and that calcium ion and cAMP second messengers are involved in hormone signal transduction (Jurenka et al. 1991b).

Remarkable potentiations were observed, however, when MIX was added to medium containing Bom-PBAN-1 or when MIX or forskolin was applied in the presence of calcium ionophore. These results suggest that without initial stimulation or triggering of pheromone production, no enhancement could be achieved in *B. mori.* On the other hand, when low pheromone production is initiated (either by Bom-PBAN-I or by secondary means—that is, by calcium ionophore), elevated pheromone production can be measured in the presence of MIX or forskolin.

As far as the stimulation of glands are concerned, there may be differences between species as well. In *H. armigera,* both MIX and forskolin induced a significant increase in cAMP levels in glands and stimulated [^{14}C]acetate incorporation into the pheromone glands and the major pheromone component (Rafaeli and Soroker 1989b). When the glands of *S. litura* were incubated *in vitro* with MIX (1 mM), pheromone production increased twofold, while forskolin had no effect (Fónagy, unpublished observations). In the present experiments, it is unclear why forskolin did not exert a further effect when incubated with Bom-PBAN-I but affected pheromone production in a positive way when combined with calcium ionophore. A possible interpretation may be that a more precisely coordinated (hormone-second messenger-enzyme activation) system is switched on following PBAN action and that a less specific one (or a different one?) is switched on during only calcium ionophore incubation. The demonstrated stimulating effect of MIX in both cases, however, provides an overall indication for a possible

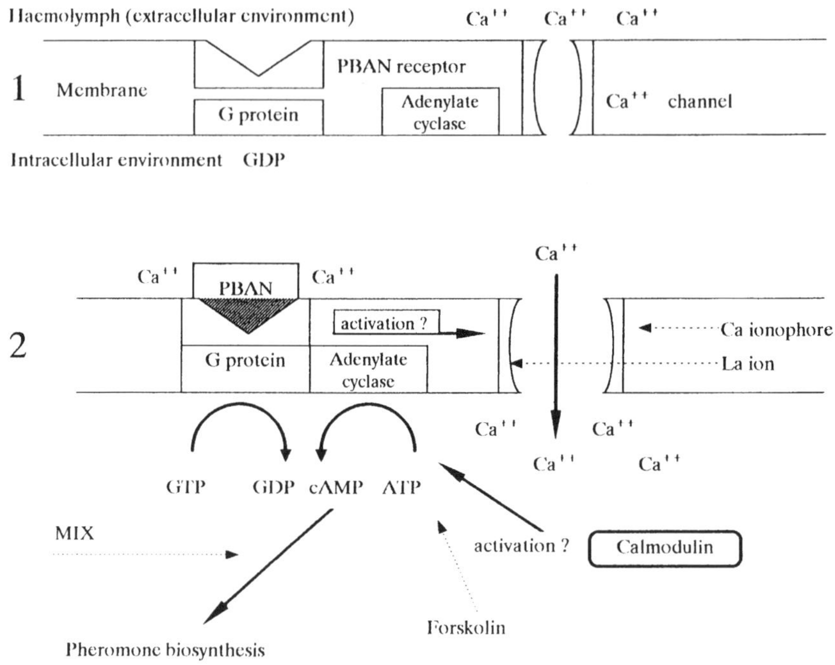

Figure 10.8. Suggested mode of action of PBAN hormone in pheromone biosynthesis in *Bombyx mori.*

cAMP involvement in the mode of action of PBAN. Actually, a similar enhanced effect was observed in pupal *M. sexta* prothoracic glands under *in vitro* conditions where PTTH (0.1 Unit) and MIX (50 μM) together induced cAMP accumulation, whereas separately these were relatively ineffective (Smith and Gilbert 1986).

Based on our findings, we believe that calcium ion has a decisive function instead of a simply permissive role. This is strengthened by the positive correlation found between the amount of Bom-PBAN-I and lanthanum ion employed. Results suggest a relation between PBAN receptor sites and calcium dependence of respective calcium channels due to the fact that they can be antagonized in a dose-dependent fashion. Similar dose-dependent antagonizing effect of lanthanum ion has been found in *M. sexta* prothoracic glands incubated with PTTH (Smith et al. 1985) or in *H. zea* pheromone glands when incubated with *Heliothis*-PBAN (Jurenka et al. 1991b).

Nevertheless, it may be concluded from the present study that different drugs, known to increase intracellular cAMP level, exerted positive/stimulating effect on *in vitro* pheromone production in the presence of PBAN or calcium ionophore. Further studies will be undertaken in order to more precisely understand and quantify the described potentiation effects. Direct measurement (on a time-course basis) of cAMP would be required in the same system, since the increased pheromone production observed may be regarded as indirect proof. Moreover, the measurement of cAMP may throw a light on the negative results obtained following only MIX and forskolin treatments.

Our experiments have provided some pieces of evidence regarding the sequence of events which might take place following PBAN action, like necessary influx of calcium ion, and suggested involvement of cAMP as second messenger. Our present hypothesis on events following various treatments is presented in Fig. 10.8. Further detailed studies are, however, required in this field, since the different physiological mechanisms involved in PBAN action in lepidopterans indicate diversity (Teal et al. 1989a; Christensen et al. 1991b; Jurenka et al. 1991a,b; Fónagy et al. 1992a) and the mediation of hormonal effect seems to be rather complex.

Acknowledgment

Dr. Adrien Fónagy gratefully acknowledges the receipt of an STA JISTEC research fellowship. This research was supported by STA (Japan) during 1990–1992.

11

Involvement of the Nervous System with PBAN

Wendell L. Roelofs, Russell A. Jurenka, and Peter Ma

It is generally accepted that PBAN (pheromone biosynthesis activating neuropeptide) and PBAN-like compounds possessing pheromonotropic activity are biosynthesized in the subesophageal ganglion (SEG). Immunocytochemical studies in several species have revealed discrete groups of PBAN-like immunoreactive cells in the SEG to support this conclusion (Davis et al. 1993; Kingan et al. 1992; Tips et al. 1993). In two lepidopterous species, *Helicoverpa zea* and *Bombyx mori*, a gene has been characterized from the SEG and found to encode PBAN and four other related peptides (Kawano et al. 1992; Ma et al. 1994). PBAN immunoreactivity also was found by these investigators in the corpora cardiaca (CC), as well as along the ventral nerve cord to the terminal abdominal ganglion. These data support a multiplicity of functions and means of transportation for these peptides, including PBAN. This chapter, however, is not intended as a review of all data that support either a mode of action for PBAN involving release from the CC into the hemolymph or involving transportation down the ventral nerve cord. The focus will be on research from our laboratory on several species and how some aspects of the mode of action of PBAN may or may not involve the nervous system in those species.

1. Redbanded Leafroller (RBLR), *Argyrotaenia velutinana*.

PBAN is involved in turning on pheromone production in RBLR females (Jurenka et al. 1991a; Tang et al. 1989) and appears to regulate the first step in fatty acid biosynthesis, acetyl-CoA carboxylase, or substrate supply for that step (see Roelofs and Jurenka this volume). However, it was found that in this species, another hormone released from the bursa copulatrix also is involved in stimulating pheromone production (Jurenka et al. 1991a; Fabriàs et al. 1992). The evidence for this was as follows: (a) PBAN did not stimulate sex pheromone production in isolated glands to the same extent as in isolated abdomens; (b) PBAN activity was abolished in isolated abdomens by removal of the bursa copulatrix; (c) pheromone titers also decreased in females that had been deprived of the bursa for 24 hours, although cutting the cervix bursae and leaving the bursae in the abdomen culture increased both titer and radiolabeled acetate incorporation into pheromone without the presence of PBAN; and (d) the bursa copulatrix of RBLR contains a substance with pheromonotropic activity. The data indicate that PBAN stimulates the release of the pheromonotropic bursa peptide (PBTP), and that PBTP then stimulates pheromone biosynthesis in the pheromone gland.

From the information above and data from other studies with RBLR (Jurenka et al. 1994a), we have developed a model for this species in which PBAN is released from the brain into the hemolymph and interacts with receptors on the bursa. The bursa, in turn, released PTBP into the hemolymph and interacts with the pheromone gland to stimulate pheromone production. The data supporting this mode of action in RBLR include the following: (a) The CC exhibit PBAN biological activity and immunoreactivity; (b) removal of the ventral nerve cord in isolated abdomens did not inhibit the action of PBAN; (c) severing the ventral nerve cord *in vivo* in normal females did not lower the pheromone titer; (d) PBAN-like activity was found in hemolymph collected from normal females but not from decapitated females; and (e) PBTP is active in isolated gland cultures and stimulation of pheromone production appears to be similar to the mode of action depicted in Roelofs and Jurenka (this volume) for *Helicoverpa zea* with a second messenger system involving extracellular calcium and cAMP.

Although in *H. zea* the primary route for PBAN appears to be by release from the CC, the ventral nerve cord does seem to play a role in inhibiting the release of PBAN in mated females (Jurenka et al. 1993). As is common in many moth species, the pheromone titer declines rapidly after mating. In RBLR it remains low for at least 4 days after mating. Studies to determine why the pheromone titer decreases after mating showed that both PBAN and PBTP are active when tested with mated females, and both brain and bursa extracts from mated females contained active material. Additionally, hemolymph from mated females exhibited no pheromonotropic activity, but hemolymph from virgin females did. The accumulated evidence suggested that all systems were functional but that PBAN was not being released in mated females. Studies in which the nerve cord was severed before mating showed that the pheromone titer did not decline in those females after mating. It is important to note, relative to discussions below, that a female with a severed nerve cord did produce pheromone and that a mated female produced pheromone similar to that of a virgin female. This supports a model in which a signal that inhibits the release of PBAN is sent via the ventral nerve cord to the brain (Fig. 11.1). This phenomenon was first described by Foster (1993) with the light brown apple moth, *Epiphyas postvittana*.

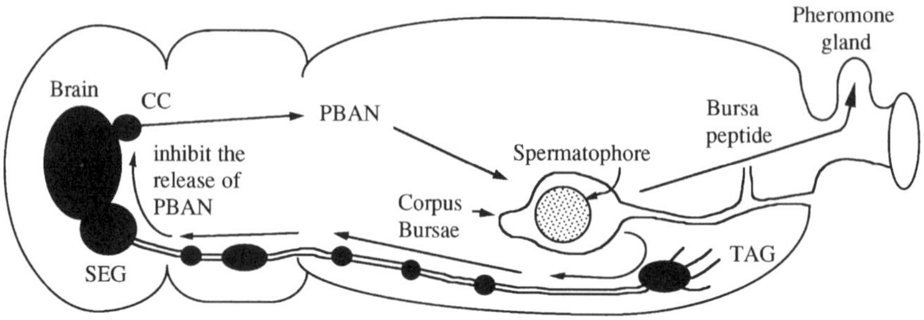

Figure 11.1. Proposed model for the control of pheromone biosynthesis in RBLR, *Argyrotaenia velutinana*. In virgin females, PBAN is released from the CC (corpora cardiaca) into the hemolymph, where it is transported to the corpus bursae, thereby stimulating the release of bursa peptide. The bursa peptide stimulates the pheromone gland to biosynthesize pheromone. In mated females, a signal is transmitted via the ventral nerve cord that inhibits the release of PBAN and the pheromone titer remains low. SEG, subesophageal ganglion; TAG, terminal abdominal ganglion; PBAN, pheromone biosynthesis activating neuropeptide.

2. Corn Earworm, *Helicoverpa zea*.

Much of the research on PBAN has been conducted on *H. zea* (see Raina this volume). Studies on the PBAN gene and immunoreactivity of various tissues have been cited above, and our model for its mode of action in this species has been presented in Roelofs and Jurenka (this volume). Since (a) PBAN activity is found in the male moths, as well as larvae, (b) it is expressed along with four other peptides, and (c) there is PBAN immunoreactivity in the CC and ventral nerve cord, it is assumed that there are multiple functions for these peptides and that they could be transported around the body by different methods.

Defining the interaction of PBAN with various nervous tissues is at best complicated. Therefore, experiments used to generate a knowledge base for the mode of action of these pheromonotropic peptides must be reproducible for that species. One debatable experiment that has been used to support a primary role for the ventral nerve cord involves the effect of severing the nerve cord on pheromone production. It was reported (Teal et al. 1989a) that severing the nerve cord in *H. zea* inhibits production of pheromone during the subsequent reproductive period. This experiment still is used in support of a conclusion that "some form of signal, be it neural or neuroendocrine, must be transmitted between the brain and the abdominal nerve cord for stimulation of pheromone production in this species" (Teal and Tumlinson 1991). The experiment has been repeated on this species by several investigators (Jurenka et al. 1991b; Raina et al. 1991), with the result that severing the nerve cord had no effect on the action of PBAN. In our hands, removal of or transection of the ventral nerve cord, either *in vivo* or in an incubated isolated abdomen, did not prevent the pheromonotropic action of PBAN. We found this to be true not only in *H. zea*, but also with RBLR (discussed above), with *Mamestra brassicae* (Jacquin et al. 1994), and with the European corn borer (Ma and Roelofs 1995a,b). Although these experiments indicate that the ventral nerve cord is not essential for the action of PBAN in females of these species, they cannot altogether rule out involvement of the nervous system. The important point here is that any data used in building the knowledge base should be accurate and reproducible.

In that respect, another debatable database concerns the possible involvement of octopamine as a second messenger in the stimulation of pheromone production. Again, before any meaningful hypotheses can be presented on the possible involvement of octopamine in pheromone production, the data on which these hypotheses are based must be convincing and reproducible. Although pheromonotropic activity for octopamine has been reported (Christensen et al. 1991b, 1992), attempts by other investigators to duplicate the exact experiments on the same species have not been successful. In our studies, which were conducted with a range of octopamine concentrations including those of the previous report and at the same time of day, we were unable to obtain any pheromonotropic activity with octopamine (Fig. 11.2). It is suggested that a reproducible assay be developed

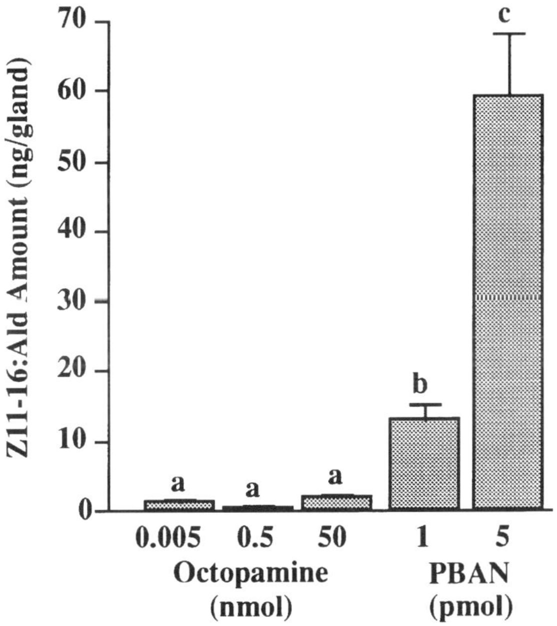

Figure 11.2. Octopamine or PBAN treatment of injected intact females of *Helicoverpa zea*. Bars represent the mean ± SEM; $n = 6$; different letters indicate statistical differences at $p < 0.05$.

that demonstrates pheromonotropic activity for octopamine before this second messenger is included in a model for PBAN mode of action.

3. European Corn Borer, *Ostrinia nubilalis*.

In similar studies with the European corn borer (Ma 1992; Ma and Roelofs 1995a,b), we found that PBAN-like factors are synthesized in three sets of neurosecretory cells in the SEG, and that release of these factors from the CC plays a more important role in pheromone production than the ventral nerve cord. This conclusion was derived from the following data: (a) Three cell clusters in the SEG of adult European corn borer females contain both PBAN-like biological activity and immunoreactivity; (b) cobalt backfilling from the CC filled these cell bodies; (c) most of the PBAN-like biological activity in the retrocerebral complex resides in the CC rather than CA (corpora allata), and PBAN-like immunoreactivity in bead varicosity is present in terminal arborizations only in the CC; (d) when the CC/CA complex was removed from the female moths and the pheromone titers determined 24 hours later, sham-operated females had the normal pheromone titer of 8 ng, whereas those with the CC/CA removed had little pheromone (<1 ng); and (e) females devoid of the ventral abdominal nerve cord were able to respond to exogenous PBAN.

Studies with the European corn borer also revealed another interesting point of difference from that reported with *H. zea*. With the latter species, it was reported (Christensen et al. 1991b) that the sex-pheromone gland is directly innervated by terminal nerves from the terminal abdominal ganglion. However, in the European corn borer (Ma and Roelofs 1995a), cobalt applied to cut ends of terminal nerves downstream of the terminal abdominal ganglion did not reveal any neuronal arborizations in the pheromone gland, but did reveal (a) afferent fibers from the ventral nervous system that innervate the median oviduct and (b) efferent fibers derived from sensory neurons of sensilla trichodea on the papillae anales. Also, in an ultrastructural study (Ma 1992) of the sex-pheromone gland, no neuronal elements were observed to be associated with the gland.

In summary, research on the complex functionality and mode of action for the pheromonotropic peptides produced in the SEG of male and female adult moths and larvae has only just begun. It now is clear that the peptides can be distributed from the SEG in various ways for various functions, and defining the biological role played by each in the different life stages and sexes will require a large effort involving creative bioassays.

12

Pheromonotropism in Insects: Molecules to Molecular Biology

Sonny B. Ramaswamy

1. Introduction

Since the identification of the first insect sex pheromone, bombykol, by Butenandt et al. (1959), sex pheromones and attractants of more than 1500 species of insects, predominantly lepidopterans, have been identified. The first two decades of research on sex pheromones were characterized by a spate of activity on (a) chemical identification, (b) behavior, and (c) development of methods to use pheromones, either directly or indirectly, in control of insects. Research was conducted also to determine the pathways involved in pheromone biosynthesis (e.g., Roelofs and Wolf 1988). Many of these studies had shown that sex pheromones were fundamental to the reproductive isolation of species (e.g., Roelofs and Cardé 1974).

Pheromone production, emission, and response to pheromones in insects were shown to be controlled by an endogenous circadian rhythm that is entrained by photoperiodic cues, which may be modified by other environmental factors such as temperature (Cardé and Webster 1980) and by the physiological status of the insects (Raina 1993; Ramaswamy et al. 1994; Schal et al. this volume; McNeil et al. this volume). It has been well established for many decades that the reproductive biology, in particular the pheromone biology, of a species is precisely timed and regulated (e.g., Thornhill and Alcock 1983). Such a phenomenon had suggested the requirement of neural as well as hormonal regulation (Cardé and Webster 1980). The first report providing evidence for the neurohormonal control of sexual communication and mating in insects was that of Barth (1965) in cockroaches. Barth hypothesized, however, that neurohormonal regulation of mating behavior was expected to occur only in long-lived, feeding adults and not in short-lived, nonfeeding species. Within a few years of this hypothesis, however, two nonfeeding adult moths were shown to require neurohormonal regulation of sex pheromone production and release (Riddiford and Williams 1971). The discovery of a cephalic neurohormone, the pheromone biosynthesis activating neuropeptide (PBAN) (Raina and Klun 1984) in relatively short-lived insects, served as an impetus for renewed efforts to explain regulation of pheromone production in insects. With the discovery of such hormones for the regulation of sex pheromone production has come the realization that these molecules can be used as tools to study the fundamental physiological and molecular bases of pheromone production.

2. Control of Pheromone Production

One can summarize that five general pathways appear to regulate sex pheromone biosynthesis and pheromone biology in insects: (1) species in which PBAN is involved, either directly or indirectly; (2) species in which juvenile hormone (JH) is involved solely or stimulates the release of PBAN; (3) species in which ecdysteroids are necessary for sex pheromone production; (4) neural mechanisms may regulate both production and emission of sex pheromone; or (5) the pheromone production machinery is activated due to developmental phenomena. In some species, there is evidence for multiple pathways.

Calling in moths, when pheromone is emitted, is controlled by the brain or other nerve centers (e.g., Itagaki and Conner 1987; Tang et al. 1987). However, in beetles and cockroaches, calling appears to require both neuronal and hormonal (JH) influences for calling (Schal et al. this volume). Pheromone biosynthesis in many species of moths requires the neurosecretion, PBAN, which is synthesized in the brain–subesophageal ganglion complex and stored in the corpora cardiaca (CC) (Raina and Klun 1984; Ohguchi et al. 1985; Rafaeli and Soroker 1989a; Tang et al. 1989; Jurenka et al. 1991a; Ramaswamy et al. 1994; Ma and Roelofs 1995a). Alternatively, in some insects such as *Argyrotaenia velutinana*, PBAN may stimulate factors such as the bursa factor, which in turn stimulates pheromone production (Fabriàs et al. 1992; Roelofs et al. this volume).

Some moth species either do not require PBAN (*Trichoplusia ni*) or require JH from the CA, which in turn may activate the release of PBAN for pheromone biosynthesis and also trigger calling behavior (*Pseudaletia unipuncta*) (Cusson and McNeil 1989a; Tang et al. 1989). Evidence with *T. ni*, which produces sex pheromone throughout the diel period and does not require PBAN, suggests that during the pharate adult stage, sex pheromone glands become competent to produce pheromone in the absence of 20-OH-ecdysone during a discrete and critical window of time; that is, competence is developed as the result of developmental/metamorphic events (Tang et al. 1991).

In contrast to the situation in lepidopterans, insects belonging to orders such as Dictyoptera and Coleoptera have been shown to require JH, and pheromone production in these species is closely allied with development of the reproductive system (Cardé and Webster 1980; Hughes and Renwick 1977; Barth and Lester 1973; Schal et al. this volume). Research on houseflies has shown that ecdysteroids of ovarial origin stimulate sex pheromone biosynthesis (e.g., Blomquist et al. 1993). Ecdysteroids have been shown to also suppress pheromone production in *Heliothis virescens*, mimicking the postcoital pheromonostasis observed in this species (Ramaswamy and Cohen 1992).

3. Pheromonotropic Molecules

The PBAN identified from *Helicoverpa zea* is a peptide (Hez-PBAN) with a molecular weight of 3900 and consisting of 33 amino acid residues (Raina et al. 1989a). Further, at a minimum, the C-terminal pentapeptide (-Phe-Ser-Pro-Arg-LeuNH_2) is sufficient to elicit pheromonotropic activity, although optimal activity is obtained with the entire molecule (Raina this volume). Two homologs of PBAN identified from *Bombyx mori* (Bom-PBAN I and Bom-PBAN II) have about 80% homology with Hez-PBAN (Kitamura et al. 1989, 1990), with sequences in the C-terminal pentapeptide similar to that of Hez-PBAN. Bom-PBAN II differs from Bom-PBAN I by the presence of an additional Arg at the N-terminus, possibly due to post-translational processing from precursor molecule. The PBAN from *Lymantria dispar* has been identified to be a 33-amino-acid peptide (Lyd-PBAN), sharing the same C-terminal pentapeptide sequence, but with ~75% sequence homology with the other PBANs (Masler et al. 1994; Raina this volume). Another shared characteristic of these molecules is that oxidation of the methionine residues affects pheromonotropic activity (Raina this volume).

To date, seven different homologs of JH have been identified from insects (Schooley et al. 1984; Yin 1994; Ho et al. 1995). The lepidopterans are unique among insects in that they are the only group in which six JH homologs have been found in males and females of the different species. Of these homologs, only JH I, II, and III have been found consistently in adult female lepidopterans, with different ones predominating in different species (Schooley et al. 1984). Which one of these compounds specifically is involved in regulation of sex pheromone production is unknown. It is possible that the different JHs have different roles in adult lepidopterans, in many of which JHs are known to regulate reproduction as well (Cusson and McNeil 1989a; Ramaswamy et al. 1990; Satyanarayana et al. 1992). In cockroaches and beetles, only JH III has been identified (Schooley et al. 1984). This homolog appears to regulate both reproductive and sex pheromone biologies in dictyopterans and coleopterans (Cardé and Webster 1980; Schal et al. this volume). Although several different ecdysteroids have been identified from insects (Hagedorn 1983), only 20-OH-ecdysone is known to regulate sex pheromone production in houseflies (Blomquist et al. 1993).

4. Modes of Action

In several species of moths, PBAN, synthesized in the brain–subesophageal ganglion complex (brain–SEG), is stored by the CC and released into the hemolymph, under the influence of circadian rhythms (Raina 1993; Raina this volume; Rafaeli et al. this volume; Marco et al. this volume; Ma and Roelofs 1995b). Its activity is mediated by Ca^{2+} via calmodulin and cAMP in the sex pheromone glands (Rafaeli and Soroker, 1989a,b; Rafaeli et al. 1990b, 1991; Rafaeli, 1994; Jurenka et al. 1991b; Ma and Roelofs 1995a; Rafaeli et al. this volume; Fónagy et al. this volume; Marco et al. this volume). Additionally, recent studies have suggested that phosphatidylinositol may regulate sex pheromone production in heliothine moths (Soroker and Rafaeli 1995; Rafaeli et al. this volume). The discovery of these alternative pathways for the mode of action of PBAN has suggested the possible occurrence of separate receptors for PBAN on the pheromone glands, one mediating the activity through Ca^{2+}/cAMP and the other mediating it through phosphatidylinositol; however, the physiological significance of this remains to be determined (Rafaeli et al. this volume).

Because injected PBAN stimulates pheromone production by female moths lacking an intact ventral nerve cord, it has been inferred that these species do not require the ventral nerve cord for mediation of PBAN activity (Rafaeli et al. 1990b; Jurenka et al. 1991a,b; Arima et al. 1991; Raina et al. 1991; Ramaswamy et al. 1995; Ma and Roelofs 1995a; Roelofs et al. this volume; Marco et al. this volume; Rafaeli et al. this volume). Additionally, octopamine was not involved as a mediator for the pheromonotropic activity of PBAN in *H. zea* (Jurenka et al. 1991b; Ramaswamy et al. 1995). These finds have been interpreted to suggest the sex pheromone gland to be the target organ for PBAN in some species, without the mediation of any other organs or neuromodulators (Rafaeli et al. 1990b, 1991, 1993; Jurenka et al. 1991a,b; Fabriàs et al. 1992; Rafaeli, 1994; Ramaswamy et al. 1995; Roelofs et al. this volume; Marco et al. this volume; Rafaeli et al. this volume).

In contrast, Teal et al. (1989a) and Christensen et al. (1991b, 1992) have reported that PBAN activity in *H. zea* and *Heliothis virescens* requires the presence of an intact ventral nerve cord, and that PBAN triggers the terminal abdominal ganglion to release octopamine via nerves innervating the sex pheromone gland; the octopamine in turn is suggested to stimulate pheromone production and or release. In ultrastructural analyses, Christensen et al (1991b) showed that sex pheromone glands in *H. zea* and *H. virescens* are innervated, and electrical stimulation of the ventral nerve cord by itself was sufficient to stimulate pheromone production. However, Roelofs et al. (this volume) discovered that sex pheromone glands of the European cornborer were not innervated. Another reason for interpreting the necessity of an intact nerve connection for pheromone production is that PBAN immunoreactivity is found, in addition to brain–SEG–CC complex, in other ventral ganglia and the ventral nerve cord, including the terminal abdominal ganglion (Kingan et al. 1992; Ma and Roelofs 1995b,c; Tumlinson this volume). However, many of these antibodies were developed to the C-terminal end of PBAN, and the immunohistochemical localization of PBAN may in fact be due to the cross-reactivity of the antibodies with other neuropeptides with sequence homologies, such as myotropins. Further, the discovery that the PBAN gene codes for five peptides, including PBAN (Ma et al. 1994, 1996), and that some of these peptides were localized in the ganglia suggests that transport of PBAN through nerves may be questionable.

Christensen (this volume) has suggested that octopamine and an intact nerve connection may be necessary not so much for pheromone biosynthesis but for pheromone release. This interpretation was based on their ability to elicit electroantennogram responses from male moths to female sex pheromone glands that were electrically stimulated as well as to glands that were incubated in saline under highly depolarizing conditions of high K^+ in the saline. In contrast, in other heliothine moths, Raina (this volume) and Rafaeli et al. (this volume) were unable to stimulate pheromone production or release by electrical stimulation or subjecting incubated glands to the highly depolarizing conditions of 100 mM KCl. These conflicting findings are intriguing and need to be resolved and suggest that there might be strain/species differences in the insects under study or that the conflicting observations may be due to methodological differences.

The lack of explicit evidence for the presence of PBAN in the hemolymph has been used as another reason for suggesting a neuronal route for transport of PBAN rather than a hemolymph route. Pheromonotropic activity was found in hemolymph from scotophase females injected into neck-ligated females (Raina and Klun 1984). Recently, Jacquin et al. (1994) demonstrated pheromonotropic activity and PBAN-like immunoreactivity (PBAN-ir) in the hemolymph of *Mamestra brassicae*. Using enzyme-linked immunosorbent assays (ELISAs) on hemolymph fractionated on a high-pressure liquid chromatograph, PBAN was detected in several heliothine species and *Spodoptera littoralis* (Ramaswamy et al. 1995; Raina this volume; Marco et al. this volume). Results of these studies suggested that hemolymph of females during scotophase has ~5–10 fmols of PBAN. Ramaswamy et al. (1995) have determined also that the pheromonotropic substance in the hemolymph of *H. zea* appears to be a molecule with a molecular weight that is less than the 3900 determined for PBAN (Raina et al. 1989a). These studies point to the possibility that pheromonotropic substances synthesized in the brain–SEG complex may likely be modified before release to the hemolymph.

Pheromonotropic activity has been found by bioassays and immunoassays in neuronal tissues as well as hemolymph. It is likely that depending on species, strains, and rearing and handling parameters, a combination of routes of transfer of pheromonotropic substances is occurring in moths. In insects such as cockroaches and *L. dispar*, both neuronal and hormonal routes appear to be important for sex pheromone production (Raina this volume; Schal et al. this volume). As a morphogenetic and gonadotropic hormone, JH activity is mediated by its interactions at the gene level (Jones 1995). JH activity in certain processes is also known to be mediated by protein kinase C-mediated phosphorylation (e.g., Sevala and Davey 1989). While the activity of JH as a pheromonotropic hormone is well established in several groups of insects (Schal et al. this volume), its mode of action in this physiological process is unknown.

5. Cross-Reactivity of PBAN with Other Neuropeptides

Owing to the C-terminal pentapeptide sequence of PBAN being similar to that of a family of peptides that include myotropins, pyrokinins, and melanotic hormones (Raina 1993), many investigators have shown the PBANs identified thus far exhibit also various levels of myotropic and melanotic activities. Alternatively, myotropins and melanotic hormones exhibit pheromonotropic activities (Altstein et al. 1994; Altstein et al. this volume; Matsumoto et al. 1992a; Matsumoto et al. this volume). These neuropeptides have been found in various parts of the central nervous system in larvae and adults. One may speculate from these recent studies on structure–activity relationships and homologies of PBANs and myotropins/pyrokinins that such peptides must have had a similar evolutionary origin. It is an intriguing possibility that PBANs may have originated as a myotropin for triggering calling behavior, which is dependent on neuronal control (Itagaki and Conner 1987; Tang et al. 1987), later taking on the role of regulating pheromone biosynthesis as well. Alternatively, this family of peptides may have similar molecular precursors. The recent discovery that the PBAN gene codes for five peptides, including PBAN, with functions for the other peptides yet to be determined (Sato et al. 1993; Ma et al. 1994, 1996; Roelofs et al. this volume), is exciting and should allow deciphering interactions between different peptides.

6. Pheromonotropic Substances and Pheromone Biosynthesis

Pheromone biosynthesis in the Lepidoptera that have been studied to date suggests that the process follows a similar scheme (e.g., Roelofs and Wolf 1988; Roelofs and Jurenka this volume; Marco et al. this volume; Tumlinson et al. this volume). One may summarize that fatty acyl CoA substrates are synthesized from acetyl CoA via malonyl CoA, catalyzed by acetyl CoA carboxylase and fatty acid synthase. The fatty acyl CoA is desaturated by specific desaturases, followed by the activity of microsomal chain-shortening enzymes and/or reductases, oxidases, acetylases yielding appropriate blends of alcohols, aldehydes, and acetates, respectively. In species that use monounsaturated compounds as their sex pheromone, PBAN appears to regulate pheromone biosynthesis at the initial two steps (acetyl CoA carboxylase and/or fatty acid synthase) (Roelofs and Jurenka this volume). In contrast, in species that use dienes or trienes in their pheromone blend, regulation by PBAN occurs at the level of the reduction enzymes (Marco et al. this volume; Tumlinson et al. this volume). Regulation of pheromone biosynthesis by PBAN appears to be mediated by cAMP activating protein kinases/phosphatases that activate or inactivate biosynthetic enzymes by phosphorylation or dephosphorylation (Rafaeli et al. this volume). While some strides have been made in determining the molecular role of PBAN in stimulating pheromone biosynthesis, very little is known of the molecular action of JH in pheromone biosynthesis in insects which utilize this hormone. For example, Chase et al. (1992) showed in the German cockroach that JH induces the hydroxylation of a parent alkane to an alcohol, which then is oxidized in a nonregulated step to a ketone contact sex pheromone.

7. Molecular Biology

Davis et al. (1992b) sequenced the PBAN gene partially in *H. zea* and suggested that it coded for three peptides. In *Bombyx mori*, Sato et al. (1993) determined that the PBAN gene codes for five peptides, including PBAN and diapause hormone. Similarly, in a complete sequencing of the PBAN gene in *H. zea*, Ma et al. (1994) and Roelofs et al. (this volume) describe that the gene codes for five peptides. As noted above, the physiological significance of some of these gene products is unknown. The identified peptide sequences have homologies of varying degrees to other neuropeptides such as myotropins, pyrokinins, melanization and reddish coloration hormone, and diapause hormone (Imai et al. 1991; Kuniyoshi et al. 1992b; Altstein et al. this volume; Matsumoto et al. this volume). The evolutionary significance of such relationships are unknown. However, with various DNA and immunological probes becoming available, along with studies on PBAN receptor antagonists and agonists (Roelofs et al. this volume; Raina this volume; Altstein et al. this volume), progress on explaining pheromonotropism at the molecular level should be forthcoming. Further, efforts to develop recombinant baculoviral DNA carrying the PBAN gene (Raina this volume), leading the way for future studies on potential modulation of pheromone production at population levels, are exciting possibilities.

8. Conclusions

Notwithstanding the controversy on the mode of action of PBAN, pheromonotropism appears to be common in many species of insects, both long-lived feeding and short-lived nonfeeding ones, counter to the hypothesis proposed by Barth (1965). Several pathways, including elements of developmental, neuronal, and/or hormonal influences, for regulation of pheromone production appear to be operational in different species of insects. The controversies seen thus far are probably the result of strain/species differences and/or methodological differences, requiring the need for more collaborative efforts by scientists in this area of endeavor. The neuronal versus hormonal routes of action for PBAN may not be mutually exclusive; they may actually be acting on the pheromone production machinery cooperatively in some species. The future in this field of endeavor will likely explain the molecular basis of the mode of action of such hormones, especially as it pertains to the finer intricacies in the intracellular activities in the pheromone producing cells. Additionally, explaining the molecular/physiological roles of the different peptides coded by the PBAN gene and the other hormones, including studies on receptor interactions, are exciting avenues of research. Several laboratories around the world are involved in research on determining the feasibility of using this knowledge of the physiological and molecular activities of pheromonotropic substances to control insects.

PART II

Sensory Processing of Pheromone Signals

Organized by Thomas L. Payne

13

Sensory Processing of Pheromone Signals

John G. Hildebrand

The detection and sensory processing of volatile pheromones in insects as diverse as moths, cockroaches, and honeybees are functions of the olfactory system. To understand how pheromones influence behavior, it is necessary to unravel mechanisms that operate at the molecular, cellular, and systems levels at various stages of information processing in the neural pathway—from olfactory receptor cells in the antenna and olfactory interneurons in the antennal lobe to higher-order interneurons in the olfactory foci of the protocerebrum and descending premotor neurons that convey olfactory influences to the thoracic flight motor. Research in this area of insect neurobiology seeks to reveal how olfactory receptor cells transduce qualitative, quantitative, and temporal information about the pheromonal signal into patterns of action potentials, how information about properties of the stimulus is encoded in patterns of activity in different types of interneurons in the central nervous system, how sensory information is integrated with inputs of other modalities, and how synaptically processed olfactory information ultimately influences activity in the neural circuits that control flight behavior.

The purpose of this overview is to introduce the olfactory system and the tasks it performs, tracing the processing of sex-pheromonal information in male moths from receptor cells in the antenna through the deutocerebrum and protocerebrum to descending premotor pathways. This introduction is meant to serve as an overture to the other chapters in this volume that deal with aspects of olfaction.

Research in my laboratory that forms the basis for this brief account has focused mainly on the giant sphinx moth *Manduca sexta*. Other reviews have dealt with aspects of that research in greater detail (Christensen and Hildebrand 1987a, 1990; Hildebrand 1985, 1995; Hildebrand and Christensen 1994; Hildebrand et al. 1992; Homberg et al. 1989).

A male moth's reception, analysis, and integration of the sex-pheromonal signal emitted by a receptive conspecific female constitutes an experimentally tractable "model" olfactory task carried out by a relatively simple olfactory system that bears striking similarities to its counterpart in other invertebrates and vertebrates. Likewise, the oriented flight of a male moth in response to the pheromonal message is a dramatic example of odor-regulated behavior. These sensory and behavioral models offer opportunities for insights into fundamental mechanisms of olfaction as well as sensory control of behavior.

For the sex-pheromonal communication system to be effective in enhancing reproductive success and isolation, the male moth must be able to discriminate the conspecific female's multicomponent pheromone blend from the blends released by females of other species. Key to that ability is the analytic power of the male's olfactory system. The capacity of the olfactory system to distinguish different odors depends on the response characteristics of olfactory receptor cells, which are exposed to the environment and responsible for initial recognition and transduction of odor components. Hence odor discrimination ultimately depends on the cascades of molecular and cellular events, leading from molecular recognition at the receptor site to the generation of action potentials in temporal (in each receptor-cell axon) and spatial (across the array of receptor-cell axons) patterns that represent features of the stimulus. Also essential for odor discrimination is the neural circuitry in the central nervous system through which afferent olfactory information is integrated, abstracted, and recognized. To begin to understand how chemical signals are analyzed by the olfactory system and ultimately affect behavior, it is necessary to consider the anatomical and functional organization of the olfactory pathway and the processing performed, and abstraction accomplished, by neural circuitry at each level in that pathway.

Specialized receptor cells in a male moth's antenna detect individual components of the sex pheromone and pass information about those substances on to the first-order processing center for pheromonal information in the brain—a specialized, male-specific glomerular structure or complex in the antennal lobe of the deutocerebrum. Within that structure (the macroglomerular complex), the axons of the pheromone-selective receptive cells terminate and male-specific deutocerebral interneurons have their arborizations. Some antennal-lobe interneurons are confined to the lobe (local interneurons), whereas others extend an axon into the protocerebrum and hence are called output or projection neurons. The projection neurons convey the synaptically shaped outputs of glomerular information processing to higher-order olfactory foci in the brain.

Antennal-lobe projection neurons that have arborizations confined to the macroglomerular complex and exhibit similar morphological characteristics give various classes of physiological responses to antennal stimulation with sex-pheromone components. These neurons encode information about stimulus quality, quantity, and temporal fluctuations. Different neurons encode different aspects of the stimulus. Some projection neurons respond similarly to several pheromone components and thus cannot discriminate one component from another. Other projection neurons are considered "pheromone specialists" because they can discriminate one pheromone component from another. Some of these specialists respond to only one pheromone component, and others give excitatory responses to one component and inhibitory responses to another component. These "mixed-response" projection neurons (also called "–/+/–" projection neurons) are especially effective at following behaviorally essential temporal discontinuities in the sex-pheromonal signal.

Thus, information about sex pheromone is relayed, by way of patterns of action potentials shaped by synaptic processing in the macroglomerular complex, via the axons of projection neurons to higher centers in the protocerebrum. Of the pheromone-responsive protocerebral neurons that have been studied to date, many have had arborizations in the lateral accessory lobes of the protocerebrum. Each lateral accessory lobe is linked, by neurons that have arborizations in it, to the ipsilateral superior protocerebrum as well as the lateral protocerebrum, where axons of antennal-lobe projection neurons terminate. The two lateral accessory lobes are also linked to each other by bilateral neurons with arborizations in each. Neuropil adjacent to the lateral accessory lobe contains branches of many neurons that descend in the ventral nerve cord. Local neurons link the lateral accessory lobe to this adjacent neuropil. Some descending neurons also have arborizations in the lateral accessory lobe. Thus, the lateral accessory lobe appears to lie in the pathway of olfactory information flow from the deutocerebrum through the lateral protocerebrum to descending neurons.

All protocerebral neurons investigated to date that responded to antennal stimulation with pheromone were excited by that input. Certain protocerebral neurons show "long-lasting excitation" that sometimes outlasts the olfactory stimulus by many seconds. In some other protocerebral neurons, pheromonal stimuli elicit brief excitations that recover to background firing rates within 1 second after stimulation. Long-lasting excitation is more frequently elicited by the sex-pheromone blend than by individual components and is more common in bilateral protocerebral neurons with arborizations in the lateral accessory lobes than in protocerebral local neurons innervating the lateral accessory lobes. Long-lasting excitation appears to be one important kind of physiological response that is transmitted to thoracic motor centers. How this persistent excitatory activity, evoked by the pheromone blend, might participate in the generation of the male moth's characteristic behavioral response is a subject for future investigation.

The long-term goal of this line of research is to understand the neurobiological mechanisms through which a specific olfactory signal, the female's sex-pheromone blend, is integrated with inputs of other sensory modalities and ultimately initiates and controls the characteristic responses of a conspecific male moth that lead to location of a mate.

Acknowledgments

This contribution is dedicated to the memory of my friend, colleague, and collaborator Ed Arbas, whose untimely death in June 1995 took from us one of the most gifted investigators and stimulating thinkers in our field.

I thank my many present and former co-workers and collaborators who contributed to the work reviewed in this chapter. Our research in this field is currently supported by NIH grants AI-23253 and NS-28495.

14

Molecular Mechanisms of Pheromone Reception in Insect Antennae

Heinz Breer

1. Introduction

In insects, olfaction plays a major role in the regulation of behavior; accordingly, most insects discover the source of airborne cues with great sensitivity and accuracy. This is particularly true for the detection of sex-specific pheromones by male moths; they display an incredible odor perception and discrimination when flying long distances to "calling" females through turbulent air carrying the pheromone molecules. The pheromone plume originating from a single female moth shows a high degree of internal discontinuity due to turbulence (Murlis and Jones 1981). During the searching flight a male encounters and loses numerous pheromone packets en route to the female, leading to switches in flight behavior from an oriented zigzag flight to lateral casting which may occur as rapid as 0.5 sec after pheromone loss (Kennedy et al. 1980; Kramer 1975). The task of monitoring the temporal, spatial, and intensity parameters of pheromone plume and conveying the messages of this ever-changing environment to the animal's brain is accomplished by the extremely sensitive olfactory receptor neurons located in antennal sensilla. The sensillum is a bristle-like structure forming a hollow cuticular hairshaft housing the dendritic processes of the sensory cells; the inner cavity is filled with a fluid, the sensillum lymph, produced by auxiliary cells, which protect the delicate dendrites. The cuticular wall of the sensillum is perforated by tiny pores that permit access of volatile odor molecules but are small enough to retard loss of fluid. A common olfactory sensillum of moths, the sensillum trichodeum, consists of three non-neuronal auxiliary cells and at least two bipolar neurons which are specifically tuned to distinct components of a pheromone blend (Schneider 1969, 1992).

The chemosensory neurons respond to specific chemical stimuli with a change in the membrane potential, a receptor potential, which in turn elicits a distinct pattern of action potentials, encoding the strength and duration of a pheromone stimulus. The afferent neuronal signals travel along the axons of the sensory neuron toward the olfactory lobe where the pheromonal information is processed in glomerular complexes. Refined neuroanatomical and electrophysiological techniques have led to a detailed knowledge about the structure, the development, and the physiological properties of pheromone-sensitive antennal receptor neurons as well as the processing of pheromonal information in the olfactory lobe and higher brain centers (Christensen and Hildebrand 1987a; Kaissling 1986a; Keil 1989). However, only recently have studies concerned many aspects of the primary process of pheromone perception, including recognition of odor molecules and mechanisms underlying the chemoelectrical transduction processes (Stengl et al. 1992); our knowledge in this field clearly lags behind the recent progress in understanding molecular mechanisms of olfactory signaling in vertebrates (Breer and Boekhoff 1992; Breer 1994a; Lancet and Ben-Arie 1993).

2. Perireceptor Events

After adsorption onto the cuticular surface of the antennae (Kanaujia and Kaissling 1985) the hydrophobic pheromones diffuse into the interior of the sensilla through microscopic pores in the cuticle of the hairshaft (Keil 1982). How the lipophilic molecules subsequently partition into the aqueous sensillum lymph and traverse the hydrophilic barrier before reaching the chemosensory membranes of the dendrites is unclear; this problem is analogous to the odorant's entry and passage through the mucus covering the vertebrate olfactory epithelium. The processes controlling the entry, exit, or residence time of odorant molecules in the receptor environment have been denoted as perireceptor events (Getchell et al. 1984). Biochemical analysis have shown that the sensillum lymph of pheromone-detecting sensilla of a male moth is not an ionic solution, but rather a proteinaceous medium containing soluble proteins that bind pheromones (Vogt and Riddiford 1981; Vogt 1987; Klein 1987) and enzymes degrading pheromones (Vogt et al. 1985; Kasang et al. 1988; Prestwich et al. 1989). These findings have led to the concept that the proteins in the sensillum lymph may control the access of pheromones to the receptive sites on the dendrites as well as the lifetime of pheromone molecules.

2.1. Pheromone-Binding Proteins

2.1.1. Identification of Pheromone-Binding Protein

The first sensillar protein identified was the pheromone-binding protein (PBP) of the large silkmoth *Antheraea polyphemus* (Vogt and Riddiford 1981; Klein 1987). This 16-kDa protein was found exclusively in the extracellular fluid of sensilla from male antennae. Meanwhile, PBPs have been identified in a number of additional species; they are expressed during adult development, just prior to adult emergence (Vogt et al. 1989). Recent immunological approaches provide evidence that the PBPs are synthesized by the auxiliary supporting cells of the pheromone-sensitive sensilla trichodea (Steinbrecht et al. 1992). The notion that PBPs may facilitate the translocation of pheromones from the pores toward receptive sites in the dendritic membranes is supported by recent studies, recording from individual sensory hairs that are perfused with isolated PBP (Van den Berg and Ziegelberger 1991). The greatly improved efficiency of pheromones in sensilla perfused with PBP compared to controls suggests that PBPs may indeed act as a shuttle for the hydrophobic pheromone molecules to traverse the fluid barrier, as evident for other small lipophilic ligands in the body fluid. A better understanding of the functional role of PBPs will be provided by a detailed knowledge of the structural features of these proteins.

2.1.2. Molecular Properties of PBP

Molecular cloning approaches have recently allowed the deciphering of the primary structure of PBPs from several insect species including *Manduca sexta* (Györgyi et al. 1988), *Antheraea polyphemus* (Raming et al. 1989), *Antheraea pernyi* (Raming et al. 1990a), *Heliothis virescens* (Krieger et al. 1993), and *Drosophila melanogaster* (Pikielny et al. 1994). A comparison of the sequence data indicates that the PBPs of various species are related to each other, although the moderate overall homology retains significant interspecies differences. The primary structures display the features of globular proteins comprising two major hydrophobic domains, which may form the binding pocket for the pheromones (Fig. 14.1). In a recent microsequencing analysis of a heterologously expressed and photoaffinity-labeled PBP, the covalent attachment site for pheromones has been identified in the major hydrophobic domain (Du et al. 1994). The fact that two genes coding for different PBPs, which differ mainly in the hydrophobic domains, are expressed simultaneously in *A. pernyi* (Krieger et al. 1991) suggests that a given PBP may be fine-tuned for recognizing distinct pheromones. Binding proteins may be tuned either to the different length of the carbon chain or to the functional group (e.g., aldehyde or ester) of the pheromone. The microdiversity of binding proteins support the concept that PBPs are not merely passive transporters for lipophilic ligands but rather selective for certain ligands; thus PBP may act as a selective signal filter binding and consequently translocating only distinct, physiologically relevant ligands. In fact, a defined PBP heterologously expressed in *Spodoptera frugiperda*-derived (Sf9) cell line and purified to homogeneity showed a specific binding only for distinct pheromones (Krieger et al. 1992).

Genomic DNA has been analyzed to elucidate the structural organization of the genes encoding PBPs; it was found that a gene spans about 3.5 kilobases comprising three exons and two intervening sequences (introns) that interrupt the protein-coding regions of the gene. Control elements for initiation of transcrip-

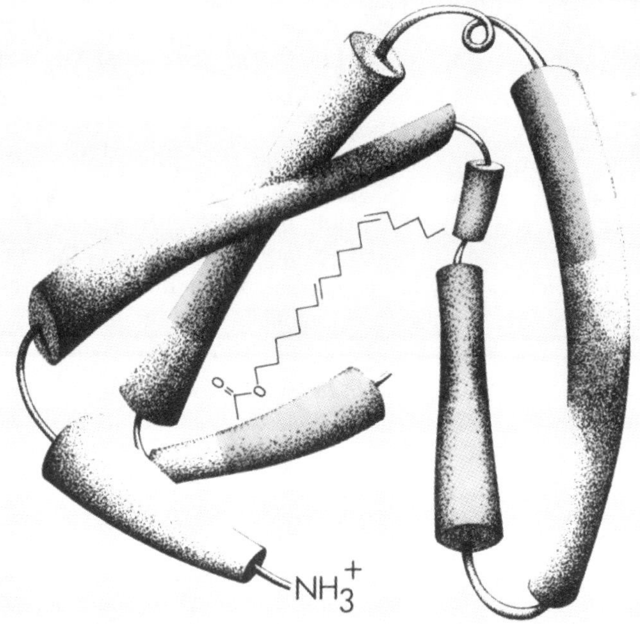

Figure 14.1. Hypothetical model of PBP based on structural properties and hydropathic features of the polypeptide chain resulting in a soluble globular protein with an interior hydrophobic binding pocket.

tion, such as "TATA" box and "CAAT" box, were identified in the 5′-flanking region, and sets of potential polyadenylation signals were located in the 3′-untranslated region. The genomic analysis revealed no evidence for alternative splicing sites which could give rise to different isoforms; instead, multiple genes appear to be the basis for binding protein heterogeneity (Krieger et al. 1991).

Besides the male-specific PBPs, a novel class of binding proteins has been identified in various moth species (Breer et al. 1990a; Vogt et al. 1991a; Krieger et al. 1993) which are expressed in male and female antennae and are supposed to bind general odorants (Vogt et al. 1991b). The primary structure of general odor-binding proteins (GOBPs) is highly conserved when compared among species. However, the overall homology between PBPs and GOBPs is rather low (–30%); nevertheless, the two classes of proteins share a number of sequence motifs and structural features. Besides having the same size, the polypeptide chains display very distinct hydrophilic and hydrophobic domains, thereby enabling the protein to form binding pockets for lipophilic ligands and to be soluble in the aqueous sensillum lymph. Furthermore, six cysteine residues are conserved and are located at the same position in all binding proteins from the same as well as from phylogenetically distant species (Fig. 14.2). The formation of disul-

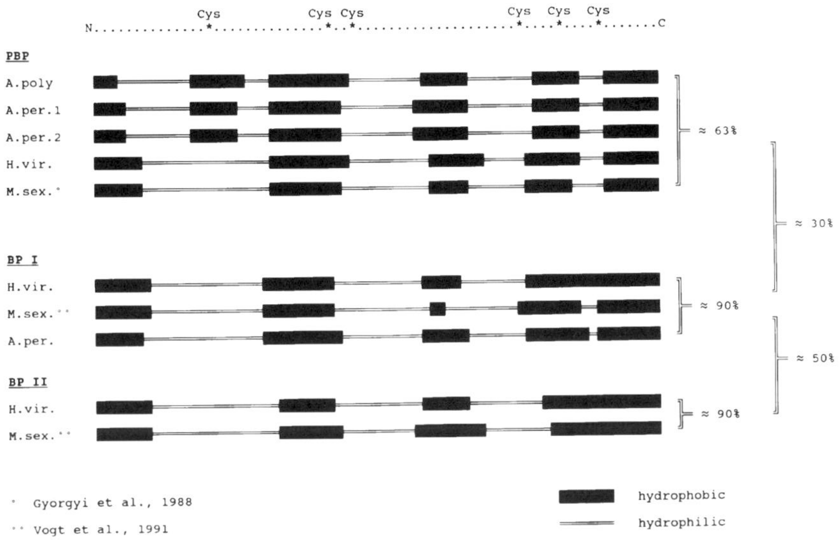

Figure 14.2. Sequence similarities and structural features of antennal binding proteins from various insect species. Groups of PBPs as well as general odorant bindings are compared. The average degree of structural homology is given as the percentage of sequence identity. The bars represent hydrophobic domains and the lines represent hydrophilic regions. Note the positions of the cysteines which are conserved in each of the proteins. A. poly, *Antheraea polyphemus*; A. per, *Antheraea pernyi*; H. vir, *Heliothis virescens*; M. sex, *Manduca sexta*. ° Györgyi et al. (1988); °° Vogt et al. (1991a,b).

fide bonds may play a crucial role in generating the tertiary structure of the binding proteins.

2.1.3. Evolution of Odorant-Binding Proteins

In spite of the low degree of sequence identity between the two classes of insect antennal binding proteins, none of the primary structures display any significant homology to the odorant-binding proteins (OBPs) identified in various vertebrate species (Pevsner and Snyder 1990). Since low-molecular-weight binding proteins for odorants have never been found in the olfactory system of aquatic animals, the invention of OBP appears to be an adaptation to terrestrial life, facilitating the probability for capturing volatile lipophilic signal molecules. The vertebrate OPBs are members of a large family of binding proteins for small lipophilic ligands (lipocalins) including retinol- and cholesterol-binding proteins. Interestingly, a carrier protein for lipophilic molecules (insectocyanin) from the hemolymph of *Manduca sexta* also belongs to this family (Riley et al. 1984). These observations favor the view that insects and vertebrates evolved binding proteins for odorants independently—that is, that OBPs in insects and vertebrates may represent an evolutionary convergence.

The phylogenetic relationship between PBPs and GOBPs in insects is currently unclear. However, based on the degree of sequence conservation it can be assumed that the multigene family encoding antennal binding proteins derived from a common ancestral gene. Selective pressure apparently forced PBPs to diverge, whereas GOBPs remain conserved during evolution.

2.1.4. Functional Roles of PBPs

Although a definitive proof for the function of PBPs in the sensillum lymph of insect antennae is still lacking, there is some evidence (Van den Berg and Ziegelberger 1991) supporting the original suggestion that PBPs may facilitate the transfer of hydrophobic pheromone molecules through the aqueous sensillar fluid surrounding the sensory dendrite (Vogt and Riddiford 1981; Vogt 1987). The translocation of pheromones through the sensillum lymph by binding proteins could be achieved either via facilitated diffusion as described for oxygen translocation in muscles by myoglobin (Gayeski et al. 1985) or via carrier bound delivery as evident for the transport of lipophilic molecules, such as fatty acids, cholesterol, or retinol, in the body fluid. In addition, it is conceivable that PBPs may play a dual role as binder of chemostimulants as well as initiators of signal transduction. Such a mechanism is realized in bacteria, like *Escherichia coli*, which contain a soluble maltose-binding protein (MBP) in the periplasmic space. Upon binding of maltose the conformation of MBP is changed so that the maltose–MBP complex can interact with a signal transducer protein in the plasma membrane, thus initiating the chemotactic response (Higgings et al. 1990). It is interesting to note that aphrodisin, a proteinaceous sex pheromone in hamster, appears to be a lipocalin-like protein loaded with a small hydrophobic ligand that elicits the transduction process (Singer and Macrides 1990). Furthermore, it is also possible that PBPs may act as scavengers by binding and thereby inactivating pheromone molecules after the receptor activation process (Kaissling 1986a). Despite the general agreement that PBPs play an important role in the perireceptor process in insect antennae, their exact function remains elusive (Fig. 14.3).

2.2. Inactivation of pheromones

Within a natural pheromone plume a flying moth encounters frequent and extensive changes in the concentration of the signal molecules; in fact repetitive pheromone pulses up to 3 per second are more effective stimuli for orientation behavior than a constant flow of odorants (Kaissling 1986a). It is therefore of particular importance that active pheromone molecules are eliminated very rapidly to maintain the sensitivity for incoming pheromones. Such a fast inactivation may be accomplished by binding to proteins or, more probably, via enzymatic

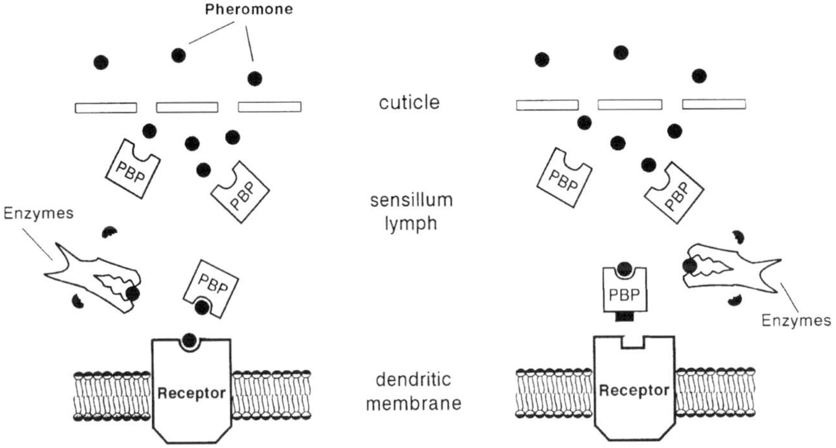

Figure 14.3. Perireceptor events in the sensillum lymph of a sensory hair controlling the entry and residence time of pheromone molecules in the vicinity of receptor sites. These auxiliary processes include the interaction of pheromones with specific soluble binding proteins which may act as a shuttle for the hydrophobic pheromones through the sensillum lymph and may participate in receptor activation. In addition, highly aggressive degrading enzymes inactivate the pheromone molecules, thus clearing the system between consecutive stimuli and protecting the sensory system from background pheromones.

modification of the pheromone molecules. In fact, two types of pheromone-degrading enzymes have been identified in the sensillum lymph of moth antennae: an esterase from *A. polyphemus* (Vogt and Riddiford 1981; Vogt et al. 1985; Klein 1987) and an aldehyde oxidase from *Manduca sexta* (Rybczynski et al. 1989). Based on the *in vitro* activity of the enzymes, an *in situ* half-life of pheromones between 0.6 and 15 msec has been predicted (Vogt et al. 1985; Vogt 1987). The data indicate that the enzymes can clear up to 10^6 pheromone molecules from the sensory hairs in less than 0.5 sec (Vogt 1987), the time required for a male to enter a new behavioral mode after on loss of pheromones (Kennedy et al. 1980). Thus, the kinetic parameter of the pheromone-degrading enzymes closely match the physiological and behavioral parameters observed after pheromone stimulation. Whether binding proteins also play a role in pheromone inactivation is currently a controversial issue, since degradation studies on whole antennae suggested half-life times of several minutes (Kasang et al. 1988). Such a slow degradation which averages over the entire antennae, rather than over the interior of the sensory hair, fails to match the physiological and behavioral time courses and suggests that a large proportion of the pheromone never enters the hair lumen where it should be rapidly inactivated, but may be lost elsewhere on or in the antennal cuticle (Vogt and Riddiford 1986). In fact, it has been suggested that only pheromone molecules striking at pore openings enter the sensillum rapidly enough to induce behavioral responses (Futrelle 1984). Thus, the rapid degradation of pheromones by inactivating enzymes in the sensillum lymph may not only ensure a short residence time for pheromones in the vicinity of the chemosensory membranes but may also contribute to prevent slowly entering background pheromones from being detected. These considerations suggest that the sensillar enzymes play a major role in pheromone clearance, which may be reminiscent of the functional role of biotransformation enzymes in the nasal epithelium of vertebrates (Lazard et al. 1991).

3. Signal Transduction in Chemosensory Cells

3.1. Electrophysiological Responses

It is the task of the sensory neurons in the antennal sensilla to recognize specific odorous compounds and to translate the information about strength, duration, and quality of a pheromone stimulus into the language of the nervous system, into a frequency of action potentials which are conveyed via the axonal process toward the deutocerebrum, the first relay station in the antennal afferent pathway. Inter- and intraspecific chemical stimuli that induce changes in the behavior and/ or physiology of individual insects are usually complex mixtures of several components; thus, each effective blend elicits cooperative activities of various sensory neurons. The peripheral olfactory coding for pheromones seems to be quite different from general odors. In most species, chemical information concerning sex pheromones is conveyed to the brain via separate nonoverlapping sensory axons; that is, the activity of distinct sensory cells encodes the quality of one particular pheromone. This coding is termed *labeled-lines pattern*, whereas general odors activate various types of less specialized cells, with the information encoded in an *across-fiber pattern* (Masson and Mustaparta 1990). Electrical responses from olfactory receptor cells in insects were first recorded in electroantennograms (EAGs), which are assumed to be the summated receptor potentials of the sensory cells (Schneider 1957). Quantitative evaluation of responses from single olfactory neurons are restricted to extracellular recordings, since the small sensory cells are tightly covered by auxiliary cells below a thick cuticle and thus are not readily accessible for intracellular or patch-clamp analysis *in situ*.

With electrodes put over the cut tip of a sensillum, slow potential changes with superimposed action potentials can be recorded after odor stimulation (Kaissling 1986); the electrical responses elicited by pheromone stimulation always occurred with a latency of at least 15 msec. The odor-induced potential changes detected in tip recordings may be considered as sensillar potential (Stengl et al. 1992) which possibly may include the electrical response of the different sensory cells, but also that of the non-neuronal cells. Because intracellular recordings from olfactory receptor cells have not been possible, it remains unclear to what extent the extracellularly measured summated potentials reflect pheromone-induced potential changes in a single olfactory neuron (receptor potential). While monitoring the frequency of nerve impulses, both excitatory and inhibitory responses correlating with the intensity of the stimulus were observed (Boeckh 1962; Boeckh et al. 1965).

3.2. Primary Reaction (Second Messenger Signaling)

In appropriate sensilla, the application of low pheromone doses usually elicit an electrical response in one of the specialized sensory neurons, indicating that the reactive cell is tuned to recognize the pheromone molecule. It is generally agreed that chemostimulants interact with highly specific receptor sites in the dendritic membrane of the olfactory neurons. Also the molecular entities responsible for pheromone recognition remain to be identified; circumstantial evidence suggests that the chemoelectrical transduction process is initiated by the interaction of pheromone molecules with specific receptor proteins, which are supposed to be members of the G-protein-coupled receptor superfamily; pheromone receptors may be related to the olfactory receptors recently identified in various vertebrate species, including rat (Buck and Axel 1991; Raming et al. 1993a), mouse (Ressler et al. 1993), and fish (Ngai et al. 1993).

The extreme sensitivity of pheromone responsive neurons and the latencies of pheromone-induced electrical reactions of these cells were initially considered as an indication for a possible involvement of intracellular second messenger cascades in the chemoelectrical transduction process. This would be in line with the primary processes of signal transduction in vertebrate olfactory neurons, where two second messenger systems, notably cyclic adenosine monophosphate (cAMP) and inositol trisphosphate (IP_3), are operating in two alternative pathways. In fact, early EAG recordings from antennae perfused with cyclic nucleotide analogs had suggested that elevated cAMP levels may mediate the generation of receptor potentials (Villet 1978). However, subsequent biochemical studies demonstrated that the specific activity of adenylate cyclase is rather low in insect antennae (Breer et al. 1990b) and that stimulation with various concentrations of pheromones did not affect the concentration of cAMP (Ziegelberger et al. 1990). In contrast, the activity of phospholipase C, the key enzyme of the phosphatidyl inositol pathway generating IP_3 and DAG (diacylglycerol), was found to be rather high in antennae of various insect species, and the enzyme activity was significantly stimulated by low doses of pheromone (Breer et al. 1990b; Boekhoff et al. 1990a).

Since appropriate changes in the concentration of second messengers, rather than changes in enzyme activity, are important for signaling pathways, the formation of IP_3 upon pheromone stimulation was analyzed, demonstrating that nanomolar concentrations of pheromone induced a more than fivefold increase of the IP_3 level. The generation of IP_3 induced by sex pheromones was found to be species- and tissue-specific and occurred only in male antennae (Boekhoff et al. 1990b, 1993).

The pheromone-induced stimulation of antennal phospholipase C and the formation of IP_3 appeared to be essentially dependent on the presence of guanosine triphosphate (GTP), suggesting that the activation is mediated by guanine nucleotide regulatory proteins (G-proteins). In fact, specific inhibitors of G-proteins, such as GDPβ-S or pertussis toxin, completely blocked the pheromone-induced response, indicating that the antennal phospholipase C is under the control of a pertussis-toxin-sensitive G-protein, like G_o- or G_i-proteins. Putative alpha-subunits of G_0-proteins have been identified in antennal preparations from various insect species by means of pertussis-toxin-catalyzed ADP-ribosylation and Western blot analysis (Breer et al. 1988; Boekhoff et al. 1990b). Antennal cDNA clones encoding an alpha-subunit of a G_o-protein have recently been cloned and sequenced (Raming et al. 1990b); however, it remains unclear if a specific G-protein isoform exists in pheromone-sensitive cells of insects as has been proposed for vertebrate olfactory neurons (Jones and Reed 1987). Nevertheless, the data indicate that the pheromone effect on phospholipase C is mediated by G-proteins and suggest that similar to the role of transducin in photoreceptors (Stryer 1986), the G-protein in the pheromone-receptive cells of insect antennae may play a pivotal role in signal amplification which may be the basis of the extreme sensitivity of pheromone detection in insects (Boeckh et al. 1965; Kaissling 1986a; Masson and Mustaparta 1990). Furthermore, the functional role of G-proteins in pheromone-induced second messenger signaling suggests that the G-protein may couple the phospholipase C to specific pheromone receptors; in lack of direct evidence, this notion may be considered as a further indication that the putative pheromone receptors may be members of the G-protein-coupled receptor superfamily (Dohlman et al. 1987).

The notion that a pheromone-induced increase of IP_3 concentration may be part of the chemoelectrical transduction cascade implies that the biochemical reaction precedes the onset of the electrical response which proceeds on a subsecond time scale. Employing a "stopped flow" methodology has allowed monitoring of pheromone-induced changes in second messenger levels on a physiologically relevant time scale—that is, in the subsecond time range (Fig. 14.4). These fast kinetic experiments demonstrated that upon application of picomolar doses of pheromone, the concentration of IP_3 rapidly rises and reaches a stimulus-dependent maximum after about 50 msec. Thereafter, the elevated IP_3 concentration declines to the basal level within a few hundred milliseconds (Breer et al. 1990c). Thus, the transient elevation in IP_3 correlates with the rapid rising phase of the sensillar potential. Such a rapid and transient molecular signal in response to a single, short pulse of a low pheromone dose seems ideally suited for mediating the chemoelectrical transduction in olfactory neurons that respond to repeated pheromone stimulation and are able to resolve several (3–10) stimuli per second (Rumbo and Kaissling 1989; Marion-Poll and Tobin 1992).

Using stronger pheromone stimuli increased the second messenger responses over several orders of magnitude depending on the concentration of pheromone.

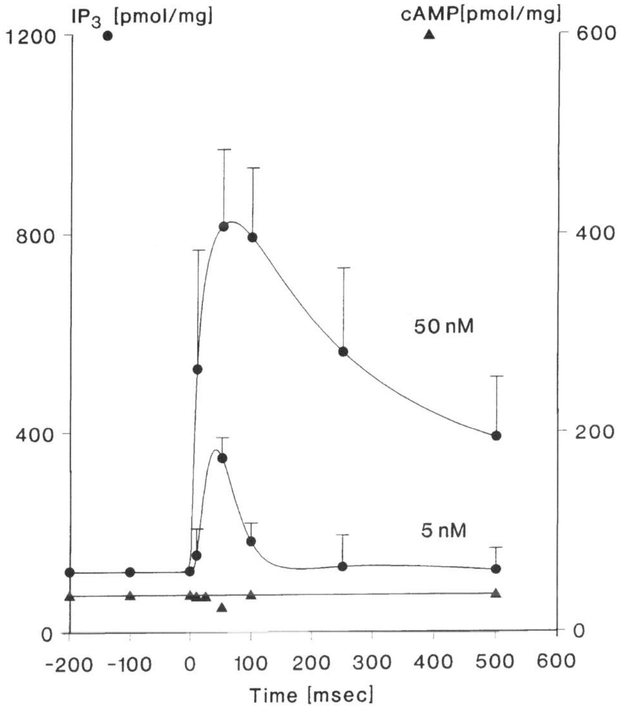

Figure 14.4. Rapid kinetics of second messenger responses elicited by sex pheromones in antennal preparation from *Heliothis virescens.* Note the fast rise in IP_3 concentration reaching a peak level after 50 msec, followed by a rapid decay of the signal; the concentration of cAMP was not changed. At higher pheromone doses the "off-set" kinetic is changed; resulting in a phasic–tonic response.

Furthermore, higher pheromone doses led to a different "off-set" kinetic: The decay resulted in an intermediate (tonic) IP_3 level between the basal and the maximal (phasic) levels (Fig. 14.4). The intermediate level slowly decayed. This discrepancy is probably due to the time required to inactivate high concentrations of pheromones (Breer et al. 1990c).

3.3. Signal Termination (Desensitization)

A remarkable feature of the pheromone-induced second messenger response in preparations from male antennae is the transient nature of this molecular signal, which is similar to odor-induced second messenger reactions in olfactory cilia preparations from vertebrates (Boekhoff and Breer 1992). Such "pulse"-like changes of second messenger levels to single, short odorant stimuli are supposed to be essential for sensory cells which can follow repetitive stimuli. This is particularly important for pheromone-detecting receptor cells which encounter the frequent concentration changes in a natural pheromone plume (Murlis and Jones 1981). Accordingly, pheromone receptor cells display a high temporal resolution of odor pulses; they are able to resolve repetitive odor pulses at frequencies of at least 3 Hz, and in some cases even up to 10 stimuli sec^{-1} are detected (Rumbo and Kaissling 1989; Marion-Poll and Tobin 1992). So besides a rapid clearance of background pheromones between stimuli (see above), it is essential that the intracellular reaction cascade initiated by pheromone–receptor interaction is rapidly turned off, resulting in a brief change of the IP_3 level and resetting the system for the next stimulus. Recent studies have demonstrated that in rat cilia the rapid termination is mediated by phosphorylation of activated receptor proteins in a negative feedback reaction (Boekhoff et al. 1992; Krieger et al. 1994). Similar experiments employing specific kinase inhibitors have shown that the turn-off reaction in antennal preparations is also mediated by protein kinases—specifically by protein kinase C, which is activated by diacylglycerol that is concomitantly generated with IP_3 (Schleicher et al. 1994). Accordingly, it was found that pheromone stimulation leads to a significant and specific phosphorylation of antennal proteins which is completely blocked by specific kinase C inhibitors, such as calphostin C. Analysis of the substrate proteins revealed that in antennal preparations from *Heliothis virescens*, stimulation with the main component of the pheromone blend for 1 sec induced a phosphorylation of a single polypeptide band (M_r 55 kDa) (Schleicher et al. 1994). By analogy to the vertebrate olfactory system, the phosphorylated antennal proteins may be considered as candidate receptors for pheromones (Fig. 14.5).

Whether some regulatory proteins, such as arrestin, play a role in governing the pheromone-induced second messenger signaling as found in other systems is currently unclear. However, two subtypes of arrestin-like proteins have recently been identified in insect antennae by molecular cloning (Raming et al. 1993b); their functional implication in pheromone signaling is now under investigation.

3.4. Chemoelectrical Transduction

The mechanisms how the "pulse" of IP_3 elicits the electrical response of the cell is not entirely clear; however, in recent electrophysiological studies on olfactory neurons from *Manduca sexta* (Stengl 1993, 1994) and *Locusta migratoria* (Wegener 1994), IP_3-induced currents have been recorded, which are apparently mediated by ion channels in the plasma membrane directly opened via IP_3 (Fig. 14.6). The view that the transduction of pheromone stimuli into an excitatory electrical response of sensory neurons is mediated via IP_3 which directly activates ion channels in the plasma membrane is reminiscent of the signal transduction mechanisms in the olfactory receptor neurons from lobster, another arthropod. Ache and co-workers have elegantly demonstrated that odor-induced stimulation of the IP_3 pathway in lobster olfactory cells leads to depolarization and excitation

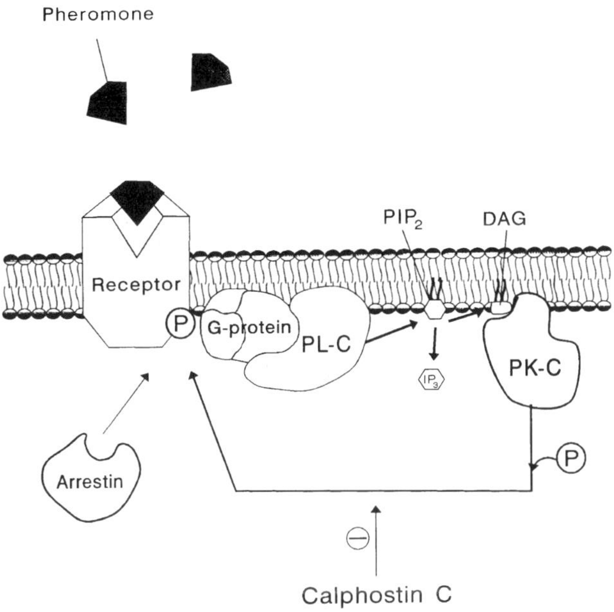

Figure 14.5. The transient nature of the pheromone-induced IP_3 signal is due to a negative feedback reaction catalyzed by protein kinase C (PK-C) which is activated via diacylglycerol (DAG) which is generated concomitantly with IP_3. Blockade of kinase C with specific inhibitors (Calphostin-C) abolishes the turn-off reaction. Pheromone-induced activation of kinase C leads to phosphorylation of specific proteins, possibly activated pheromone receptors, thus uncoupling the transduction cascade.

of the receptor neurons by activation of IP_3-gated ion channels (Fadool and Ache 1992), whereas stimulation of the cAMP pathway by odorants leads to hyperpolarization and inhibition of the cells by activation of cAMP gated K^+ channels (Michel et al. 1992). Interestingly, recent experiments have shown that odorants which elicit hyperpolarization and inhibition in insect receptor cells (Boeckh et al. 1965) induce the formation of cAMP in antennal preparations (C. Kroner personal communication), suggesting that dual transduction pathways similar to those in lobster olfactory cells may be operating in insect receptor neurons.

3.5. Functional Role of Cyclic GMP

The only rapid and transient molecular signal observed upon stimulation of male antennae from various species with the appropriate sex pheromone was the formation of IP_3. When antennal preparations were challenged with high doses of pheromone, after waning of the IP_3 signal, there was a delayed elevation of the cGMP concentration (Fig. 14.7), which was sustained over 10 sec (Boekhoff

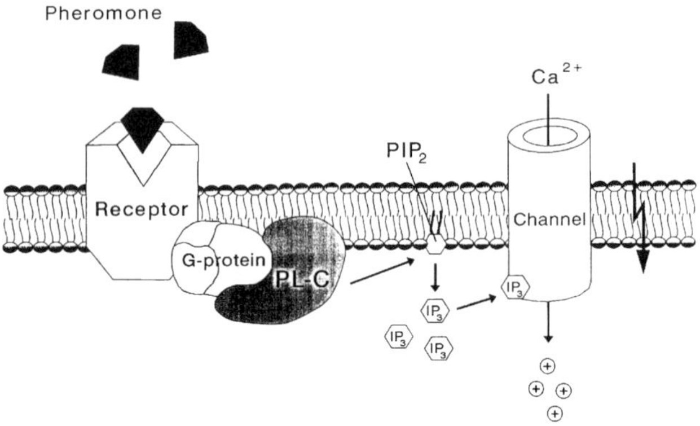

Figure 14.6. The chemoelectrical transduction process is initiated by interaction of pheromone molecules with specific receptor proteins in the dendritic membrane, leading to a G-protein-mediated activation of phospholipase C (Pl-C) and the generation of IP_3 pulses. Elevated IP_3 levels may activate IP_3-gated Ca^{2+} channels in the plasma membrane, thus converting the molecular signal into an electrical response.

et al. 1993). A long-lasting cGMP response has also been observed in experiments stimulating whole antennae (Ziegelberger et al. 1990). The necessity to apply high pheromone doses as well as the slow time course suggests that the cGMP response is not part of the primary transduction process, but rather may be part of the adaptation mechanism. This notion was supported in studies exploring the effect of exogenous cGMP on the responsiveness of antennal preparations (Fig. 14.8); it was found that elevated levels of cGMP significantly attenuated the pheromone-induced IP_3 response. Furthermore, rapid kinetic experiments revealed that elevated cGMP concentration particularly affected the phasic component of the IP_3 signal, whereas the tonic phase was only slightly reduced (Boekhoff et al. 1993). This differential effect is reminiscent of electrophysiological results demonstrating that the phasic response of moth antennal neurons is considerably attenuated during adaptation; pheromone-induced electrical responses of adapted receptor cells are less phasic and more tonic than unadapted responses (Kaissling 1987). The mechanisms underlying the generation of cGMP by strong pheromone stimuli are unclear; however, it is conceivable that cGMP formation may be a consecutive reaction to the IP_3 pulse of the primary reaction. Reaction cascades triggered by IP_3-mediated increase in intracellular Ca^{2+} concentration could be envisaged; elevated Ca^{2+} levels could either directly stimulate a calcium/calmodulin-dependent guanylate cyclase (Steinlen et al. 1990) or activate an NO-synthase leading to the generation of nitric oxide which in turn activates the cytoplasmic guanylate cyclase (Garthwaite 1991). The elevated cGMP levels may down-regulate the pheromone-activated phosphatidyl inositol signaling pathway via cGMP-dependent kinases, as recently found for human platelets (Guillon et al. 1992). In addition, cGMP may lead to closure of cation channels in receptor

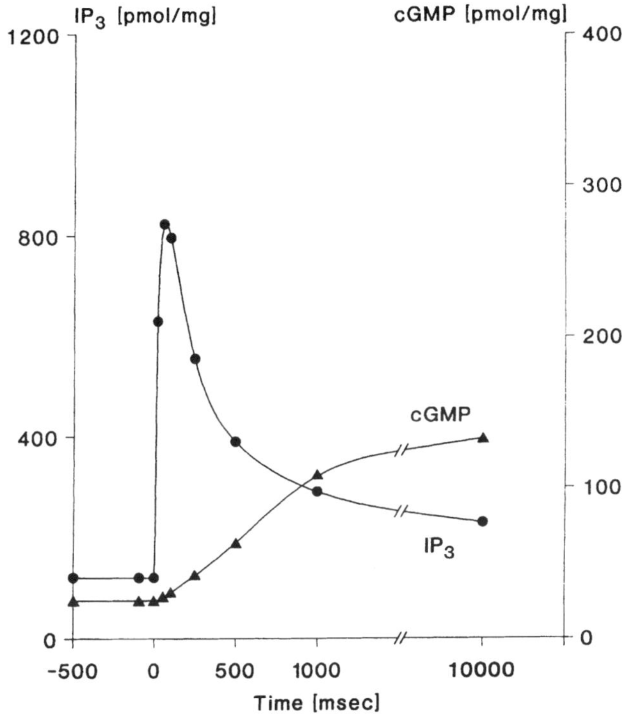

Figure 14.7. High doses of pheromones elicit a phasic–tonic IP_3 response followed by a delayed and sustained elevation of the cGMP concentration.

neurons (Zufall et al. 1991), thus reducing the electrical response. Although the molecular mechanisms are still elusive, experimental evidence suggests that the cGMP cascade may be involved in fine-tuning the sensitivity of pheromone receptor cells based on "cross-talks" between different signaling pathways.

4. Conclusion

The advent of modern biological techniques, such as patch-clamp and molecular cloning approaches, has opened new areas for research in insect olfaction. Studies exploring the molecular elements and processes in pheromone-sensitive sensilla of insects will eventually provide access to the molecular machinery underlying sensory transduction in sensory hairs and allow an important step toward an understanding of the principles of pheromone detection in insects to be deciphered. Comprehensive approaches by combination of aspects and methods in a multidisciplinary effort will contribute to turn concepts and theories about the molecular basis of biological events observed in physiological and behavioral experiments

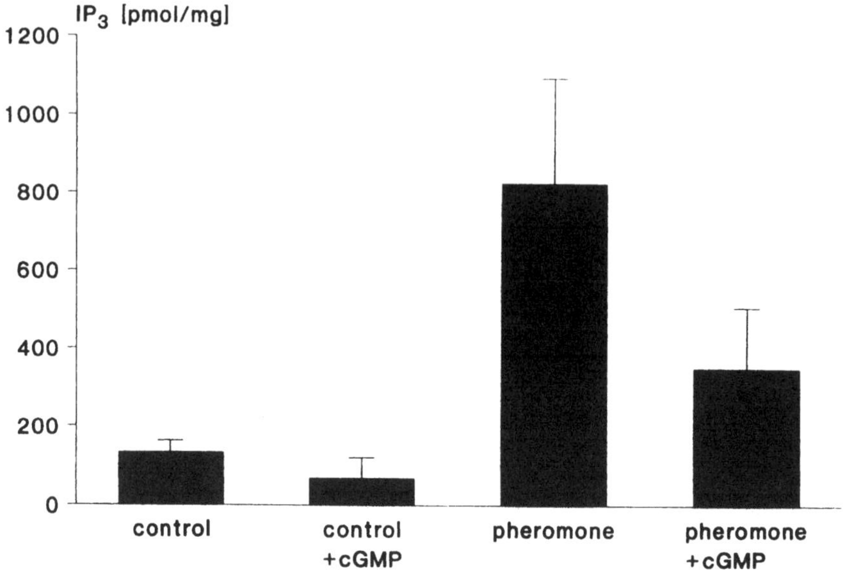

Figure 14.8. Increased concentrations of cGMP significantly attenuate the responsiveness of antennal preparations to pheromone stimulation, suggesting that cGMP pathways may be involved in adaption of olfactory receptor neurons.

of pheromone action into knowledge. A detailed exploration of the mechanisms underlying pheromone perception is also considered as an essential prerequisite before the unique opportunities of the sex pheromone system can be employed in more specific and safer insect management strategies.

Acknowledgment

The work of this laboratory was supported by the Deutsche Forschungsgemeinschaft and the Bayer AG, Leverkusen.

15

Pheromone-Binding Proteins, Pheromone Recognition, and Signal Transduction in Moth Olfaction

Glenn D. Prestwich and Gehua Du

1. Introduction to Pheromone-Binding Proteins

Odorant-binding proteins (OBPs) (Pelosi 1994) are small (ca. <20 kDa), soluble proteins that occur tissue-specifically at high concentrations in the nasal mucosa of vertebrates (Pevsner et al. 1988, 1990) and the antennal sensillum lymph of insects (Vogt 1987). Although the precise physiological role for these OBPs in sensory transduction is not completely understood, biochemical data strongly suggest roles in odorant transport, odorant degradation and clearance, and perhaps selective delivery to a membrane-associated receptor (Pelosi 1994; Pelosi and Maida 1990). Odorant-binding proteins are known in two insect orders, Lepidoptera and Diptera (McKenna et al. 1994; Pikielny et al. 1994), and have more distant homologs in the Coleoptera (Carlson 1993; McKenna et al. 1994).

In the Lepidoptera, the pheromone-binding proteins (PBPs) constitute one of three subfamilies (Krieger et al. 1993; Vogt et al. 1991a,b) of the insect OBPs, in which ligand specificity is postulated for the pheromone component(s) of the female-produced sex pheromone blend. The other two subfamilies are members of the so-called general OBP class, or GOBPs that occur primarily in female antennae (Breer et al. 1990b; Vogt et al. 1991a,b). PBPs are located in the sensillum lymph of the pheromone-responsive sensory hairs of the antennae of adult Lepidoptera, and they mediate the delivery of hydrophobic sex pheromones to specific receptor proteins located in the dendritic membrane (Prestwich 1993b). PBPs may also facilitate removal of the pheromone metabolites after signal transduction occurs (Vogt et al. 1985); indeed, apparently equal affinities have been observed for the pheromone acetate and its alcohol metabolite in photolabeling experiments with the *Antheraea polyphemus* PBP (Prestwich 1993a). To date, deduced protein sequences for five PBPs from four moth species (*A. polyphemus, A. pernyi, Heliothis virescens, and Manduca sexta*), and another five GOBPs (in two protein subfamilies) from three species have been described (Breer et al. 1990b; Györgi et al. 1988; Krieger et al. 1991, 1993; Raming et al. 1989, 1990a; Vogt et al. 1991a,b). In addition, cDNA sequences for two additional PBPs from *Lymantria dispar* and *Agrotis segetum* (S. LaForest, G. Du, and G. Prestwich, unpublished results) have been determined. PBPs and GOBPs share several highly conserved regions, including six conserved Cys residues (Fig. 15.1) that form intramolecular disulfide bridges (Prestwich 1993a).

In this chapter, we review the application of the techniques of molecular biology and protein biochemistry to the understanding of odorant discrimination and pheromone perception at the molecular level. First, we discuss the bacterial expression, refolding, and purification of the recombinant Apo-3 PBP (Prestwich 1993a), originally characterized from the antennae of adult male *A. polyphemus* (Vogt and Riddiford 1981). Second, we illustrate the use of photoaffinity labeling with two pheromone analogs to locate the pheromone-binding site in Apo-3 PBP (Du et al. 1994). Third, we present preliminary data for the development of a novel binding assay that delivers reliable dissociation constants for the interaction of hydrophobic ligands with the PBPs. Fourth, we elaborate the hypothesis that receptor activation may involve a PBP–pheromone complex. Finally, we discuss ongoing experiments to elucidate three-dimensional structures of the liganded and unliganded PBPs.

2. *Expression and Structural Studies of Recombinant Apo-3 PBP*

The first characterized PBP, Apo-3 PBP, was obtained from *A. polyphemus* male antennae (Klein 1987; Vogt 1987; Vogt and Riddiford 1981). The cDNA for this 142-residue protein was cloned and sequenced (Raming et al. 1989), and then the recombinant protein was expressed in a binding-active but potentially glycosylated form in baculovirus-infected insect cells (Krieger et al. 1992). The yield of protein from this methodology was insufficient to provide enough protein for biophysical characterization using nuclear magnetic resonance (NMR) spectroscopy and x-ray crystallography. Moreover, we anticipated that a high-resolution NMR method would require the expressed of uniformly ^{13}C, ^{15}N-labeled protein to allow full sequential assignments and establishment of sufficient geometrical constraints for a high-resolution three-dimensional structure. Thus, we employed expression-cassette polymerase chain reaction (ECPCR) (MacFerrin et al. 1993) to develop an *Escherichia coli*-based expression system (Prestwich 1993a). Thus, ECPCR was used to copy the open reading frame (ORF) of the Apo-3 cDNA (Raming et al. 1989) into dsDNA equipped with (5′ to 3′) an *Eco* RI site, a ribosome-binding-site, a start codon, the ORF, a stop codon, and a *Hind* III site. The restriction-digested, purified cassette was ligated into a complementary pBS-derived plasmid (pHN1+) equipped with a strong *tac* promoter upstream of an *Eco* RI site and a *rrn*BT_1T_2 transcription termination site downstream of the *Hind* III site. *E. coli* XA90 cells were transformed and

PBP:

SPEVMKNLCMNYGKAMDQ**C**KQ**E**LNLPDSVIADLYN**FW**KDD APR 2

SPEIIKNLSQNFCKAMDQ**C**KQ**E**LNIPDSVIADLYN**FW**KDD APR 1

SPEIMKNLSNNFGKAMDQ**C**KD**E**LSLPDSVVADLYN**FW**KDD APO 3

SPDVMKNLCLNFGKALDE**C**KA**E**MLNSDSIKDDFAN**FW**VEG MSEX

SKDVMHQMALKFGKPIKL**C**QQ**E**LGADDSVVKEFLD**FW**KDG LDIS 2

SQDVMKNLSMNFAKPLED**C**KK**E**MDLPDSVTTDFYN**FW**KEG HEL 1

GOBP I:

DVQVMKDVTLGFGQALEQ**C**RE**E**SQLTEEKMEEFFH**FW**RED GOBP-1

DVNVMKDVTLGFGQALDK**C**RE**E**SQLTEEKMEEFFH**FW**RDD HEL 11

GOBP II:

TAEVMSHVTAHFGKALEE**C**RE**E**SGLPVEVMDEFKH**FW**RED GOBP-2

TAEVMSHVTAHFGKALEE**C**RE**E**SGLSAEVLEEFQH**FW**RED HEL 10

TADIISHLCAHFGKALEE**C**RD**E**SGLSPEILNEFKH**FW**SEG APR 10

YVMTD**R**LA**GCA**IN**C**LSTKLDIVDPDGNL**HH**GNAKEFAMKH

YVMTD**R**LA**GCA**IN**C**MATKLDVVDPDGNL**HH**GNAKEFAMKH

YVMTD**R**LA**GCA**IN**C**LATKLDVVDPDGNL**HH**GNAKDFAMKH

YEVSN**R**DT**GCA**IL**C**LSKKLDMIDPDGKL**HH**GNAMEFAKKH

YVMKD**R**QT**GC**MLI**C**MAMKLELLDSAMEI**HH**GSTFAFAKAH

YEFTN**R**HT**GCA**SL**C**LSSKLELLDQEMKL**HH**GKAQEFAKKH

FKFEH**R**EL**GCA**LQ**C**MSRHFNLLTDSSRM**HH**ENTDKFIKSF

FKFEH**R**EL**GCA**IQ**C**MSRHFNLLTDSSRM**HH**DNTKKFIQSF

FEVVH**R**EL**GCA**II**C**MSNKFELLQDDTRI**HH**VKMHDYIKSF

FFVVH**R**EL**GCA**II**C**MSNKFSLLQDDSRM**HH**VNMHDYVKSF

FDVVH**R**EL**GCA**II**C**MSNKFSLLKDDTRI**HH**VNMHDYVKSF

Figure 15.1. Primary sequences of PBPs and GOBPs from five moths. **Key:** APR proteins (Breer et al. 1990b; Krieger et al. 1991; Raming et al. 1990a) from *A. pernyi*; APO-3 (Raming et al. 1989) from *A. polyphemus*; MSEX (Györgi et al. 1988) from *M. sexta*; HEL proteins from *H. virescens*; GOBPs (Vogt et al. 1991a,b) from *M. sexta*; LDIS from *L. dispar* (S. LaForest, R. G. Vogt, and G. D. Prestwich, unpublished results). Bold letters indicate amino acids conserved in all PBPs and GOBPs.

ampicillin-selected to give colonies containing the recombinant plasmid. Induction of growing cells with IPTG caused overproduction of a 14-kDa protein in both the soluble fraction and inclusion bodies.

The soluble recombinant PBP (rPBP) was readily purified by preparative isoelectric focusing (IEF) followed by gel filtration to give >95% homogeneous

```
GADDGMAHELVDIIHGCEKSSPPN**DDKCIKTMDIAMCFKK
GADASMAQQLVDIIHGCEKSAPPN**DDKCMKTIDVAMCFKK
GADETMAQQLVDIIHGCEKSAPPN**DDKCMKTIDVAMCFKK
GADEAMAKELVDIIHNCENSTPPN**DDACLKTLDIAKCFKK
GADEAMAQQIIDIVHGCTTTYPAAETNDPCQRAVNVAMCFKA
GADDAMAKQLVDMIHGCSQSTPDA*TDDPCMKALNVAKCFKA

PNGAVLSKTMVELIHNCELQHDAE**EDHCWRILRVAECFKI
PNGEVL-RQMVELIHSCEKQFDHE**EDHCWRISHLADCFKS

PNGQVLSEKMVQLIHNCEKQYDDI**ADDCDRVVKVAACFKK
PNGHVLSEKLVELIHNCEKKYDTM**TDDCDRVVKVAACFKV
PNGEVLSAKMVNLIHNCEKQYDDI**TDECDRVVKVAACFKV

EIHKLNWVPNMDLVVGEVLAEV
EIHKLNWVPDMDVVLGEVLAEV
EIHKLNWVPNMDLVIGEVLAEV
EIHKLNWAPNMDLVVGEVLAEV
DVHKLNWAPDVELLVADFLAESQ
KVHELNWAPSMELVVGEVLAEV

SCTKAGIAPSMEVMMAEFIMELKQ
SCVQRGIAPSMELMMTEFIMEAEAR

DAKKEGIAPEVAMIEAEIEKY
DAKAAGIAPEVTMIEAEMEKY
DAKKEGIAPEVAMIEAEIEKY
```

Figure 15.1. Continued

rPBP (Prestwich 1993a). This material was indistinguishable from the moth antenna-derived PBP in its gel mobility on native PAGE and SDS-PAGE and in its immunoreactivity with anti-PBP antisera. Moreover, rPBP was efficiently photoaffinity-labeled with [^{3}H] 6*E*,11*Z*-hexadecadienyl diazoacetate (6*E*,11*Z*-16:Dza), a photoactivatable analog of the naturally occurring acetate pheromone, and this labeling was displaceable by excess 6*E*,11*Z*-hexadecadienyl acetate (6*E*,11*Z*-16:Ac), the native pheromone component. However, only about 5–10 mg/liter of soluble recombinant protein was obtained, with the majority (over 100 mg/liter) of rPBP being located in an insoluble, denatured form in the inclusion bodies. Therefore, 0.2% Triton X-100-washed inclusion bodies were solubilized in 6 N guanidinium hydrochloride, reduced and unfolded by dilution into an equal volume of 10 mM DTT in 200 mM Tris-HCl at pH 8.0, and then refolded for 24 hours at room temperature following dilution into a cysteine–cystine mixture using air as the oxidant. After dialysis and concentration, the crude refolded protein was purified by IEF with pH 4–6 ampholytes using a 55-ml Rotofor® cell. Within 6 hours, the fractions in the pI 4.4–5.0 range (Apo-3 PBP has a calculated pI 4.7) were collected, analyzed by SDS-PAGE, dialyzed, and then purified by gel filtration. NMR, immunoreactivity, gel electrophoresis, and photolabeling all demonstrated that refolded rPBP behaved identically to the soluble rPBP. rPBP has proven to be incredibly robust, having survived unchanged after storage in aqueous solution for over 2 years at 8 °C, as well as after 48 hours in an NMR probe at temperatures up to 35 °C.

Photoaffinity labeling with [^{3}H]6*E*,11*Z*-16:Dza could be performed on rPBP in buffers between pH 6.0 and 8.5; outside this range, the tertiary structure was unstable and ligand binding was abrogated (Prestwich 1993a). Labeling was equally competable by 6*E*,11*Z*-16:Ac, 6*E*,11*Z*-16:OH, and 9Z-14:Ac, suggesting that the rPBP could not distinguish the major pheromone component from a nonpheromone component nor from a pheromone metabolite. A similar observation had been made earlier with rPBP from baculovirus-infected insect cells (Krieger et al. 1992). However, as we describe below, this apparent lack of selectivity is an artifact of the irreversibility (i.e., nonequilibrium nature) of the photoaffinity labeling experiment. Under equilibrium binding conditions developed recently, over 100-fold discrimination between pheromone and nonpheromone components can be readily quantified.

3. Photoaffinity Labeling and Active Site Mapping of Apo-3 PBP

Photoaffinity labeling allows identification of specific amino acids in the ligand binding site of a protein. The covalently modified amino acids are presumed to be proximal to the reactive moiety of the photoaffinity analog when it is properly situated in the binding site. Isolation and sequencing of smaller covalently modified peptide fragments of the PBP permitted the identification of the amino acids involved in the recognition of the acetate functional group for these two analogues.

The bacterially expressed rPBP of *A. polyphemus* was photoaffinity-labeled with [^{3}H]6*E*,11*Z*-hexadecadienyl diazoacetate, a photoactivatable analog of the naturally occurring acetate pheromone (Prestwich et al. 1984; Vogt et al. 1988). Radiolabeled peptides were separated from an endoproteinase Lys-C digestion by high-performance liquid chromatography (HPLC) and characterized by Edman degradation (Du et al. 1994). The label was exclusively found in the Asp^{39}-Lys^{58} fragment. Cleavage of this peptide (DDYVMTDRLAGCAINCLATK) with Arg-C gave a single radiolabeled peptide (DDYVMTDR), which was predicted to be α-helical; the adjoining LAGCAINCLATK fragment, which was highly conserved in PBP sequences, was predicted to contain a hydrophobic β-strand and has been proposed to be important in recognition of the pheromone chain. Edman degradation confirmed the location of the covalently attached ligand at Thr^{44} of the smaller hydrophilic peptide.

Another component of the pheromone blends of both *Antheraea* species, 4*E*,9*Z*-tetradecadienyl acetate (Kaissling 1986), and its photoaffinity analog, $[^3H]$4*E*,9*Z*-tetradecadienyl diazoacetate, were then synthesized (Du et al. 1994). Digestion of PBP photoaffinity labeled by $[^3H]$4*E*,9*Z*-14:Dza with Lys-C and separation by HPLC showed three peaks of radioactivity. Edman degradation of each fragment was performed. The major radioactive Lys-C fragment was assigned to Asp^{21}-Lys^{38}, the second peak corresponded to the same Asp^{39}-Lys^{58} fragment labeled by the 16-carbon homolog, and the last eluting peak was established as the C-terminal peptide. For the major peak, the radioactivity of each sequence cycle was measured, and the maximum labeling was detected in the twelfth cycle, corresponding to the residue Asp^{32}. The 14-carbon odorant molecule thus had two acceptable binding positions within the recognition site, while only a single binding position was available to the 16-carbon pheromone (Du et al. 1994). Moreover, it appeared that the C-terminal residues also participated in the formation of the PBP–pheromone-binding pocket.

The proposed binding of the two pheromone components in the pheromone binding site of Apo-3 PBP is diagrammed in Fig. 15.2. The ester carbonyl of

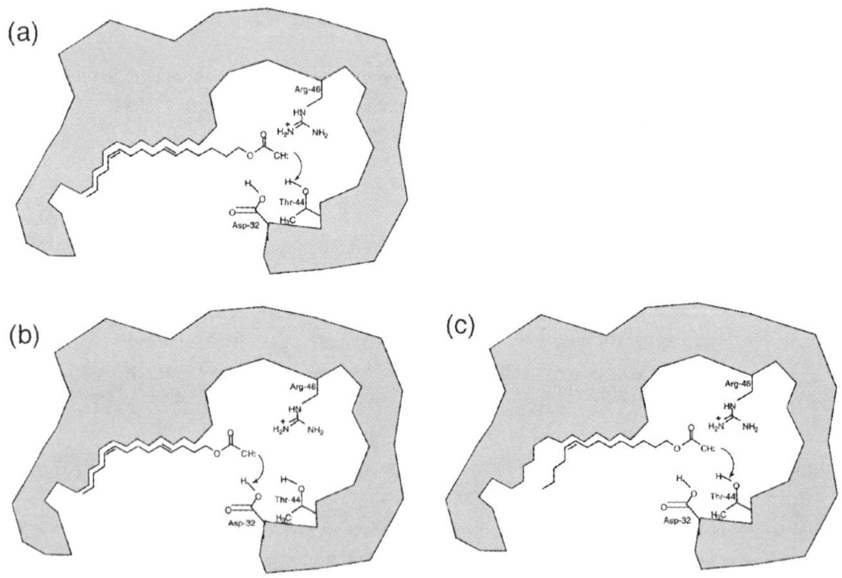

Figure 15.2. Binding of 6*E*,11*Z*-16:Ac and 4*E*,9*Z*-14:Ac to the Apo-3 PBP binding site. **A:** Binding interactions of 6*E*,11*Z*-16:Ac with PBP mediated by Arg^{44} electrostatic interaction with acetate carbonyl and by hydrophobic interaction with the kinked hexadecadienyl chain. **B** and **C:** Binding interactions of 4*E*,9*Z*-14:Ac with PBP. In **B**, the recognition of the *E,Z*-tetradecadienyl chain dominates the positioning of the ligand in the binding site. In **C**, ligand position is determined by the acetate carbonyl–Arg^{44} interaction.

acetate functionality of the pheromone could form an electrostatic interaction with the protonated guanidinium group of Arg^{46}, thereby anchoring the hydrophilic head of the pheromone in the hydrophilic helical region of the PBP. The hydrophobic hexadecadienyl chain would then associate with the hydrophobic β-sheet in the binding site. Photolytic production of the reactive carbene occurred several angstroms away from the Arg^{46}–carbonyl interaction; thus, the 16-carbon ligand [^{3}H]6*E*,11*Z*-16:Dza generated a carbene in the binding site that covalently modified Thr^{44} either by direct insertion into the O–H bond of Thr^{44} or by nucleophilic trapping of a rearranged ketene by this Thr^{44} side-chain hydroxyl.

In contrast, the 14-carbon ligand ([^{3}H]4*E*,9*Z*-14:Dza) would be approximately 2.6 Å shorter in its extended conformation, providing this ligand with two competing binding interactions that could determine its position in the binding site. In one state, the Arg^{46}–ester carbonyl interaction, which would not optimally satisfy the recognition of the hydrocarbon chain, would lead to modification of Thr^{44}. In the other state, hydrophobic recognition of the *E,Z*-diene and of the terminal butyl group would locate the reactive carbene near a different amino acid, Asp^{32}. Apparently, the 3:1 ratio of modification at Asp^{32} to Thr^{44} indicates an approximately 0.7-kcal/mol preference for hydrophobic backbone interactions. This region appears to be the major hydrophobic domain in all insect antennal binding proteins and appears to comprise a helix–sheet–helix motif. In Figure 15.1, the **RXXGCAXXC** motif is conserved in the primary sequences of the PBPs and GOBPs. The hydrophobic domains (amino acid 39–58) are more highly conserved in the GOBPs than in PBPs. This is consistent with the notion that this region of PBPs may associate with species-specific sex pheromones, while the GOBPs recognize one or more plant volatiles of common importance among moth species.

4. A New Binding Assay for PBPs and Hydrophobic Ligands

Early attempts to measure the binding affinity for a pheromone in aqueous solution were frustrated by the low solubility of long-chain acetates in water (estimated with [^{3}H]Z11–16:Ac to be <10 nM) and the tendency of the hydrophobic ligand to adsorb to the walls of the assay vessel (Prestwich 1987, 1991). The addition of bovine serum albumin or ovalbumin did not solve this problem (Prestwich 1987), nor did the use of polyethylene glycol coating or silanization procedures. One preliminary report (De Kramer and Hemberger 1987) used native Apo-3 PBP to solubilize [^{3}H]6*E*,11*Z*-16:Ac into an aqueous buffer from its adhesion to the glass vessel. "Affinities" and selectivities were determined with such an assay, and the K_D for 6*E*,11*Z*-16:Ac was estimated at 10^{-8}, but these values were not readily reproducible.

During experiments with J. Krieger and H. Breer at Stuttgart, one of us (G.D.P.) discovered that by coating a plastic microfuge tube with 1-dodecanol, the hydrophobicity of the plastic was reversed (Fig. 15.3). Thus, a solution of

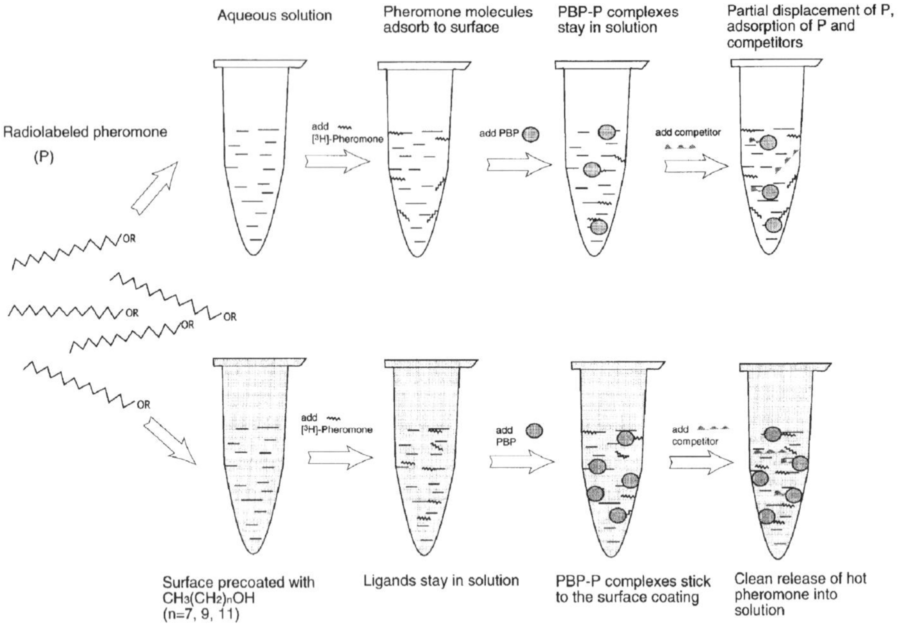

Figure 15.3. Development of novel binding assay for PBPs. **Top panel:** Hydrophobic pheromone molecules adsorb the vessel walls, leading to artifactual release of labeled pheromone into solution as competitors are added. **Bottom panel:** Precoating of the tubes with an *n*-alkanol allows development of a "reversed-phase" assay.

~50 nM [^{3}H]6*E*,11*Z*-16:Ac in aqueous buffer showed that virtually all (>95%) of the pheromone was still dissolved. We rationalized this as follows: The alkyl chain adhered to the vessel walls, as did the pheromones, while the hydroxy terminus was displayed into the aqueous phase. Basically, this takes the *liability* of nonspecific binding to the surface and converts it into an *asset*. Interestingly, when rPBP was added, the tritiated pheromone was removed from the solution, presumably because the PBP–pheromone complex stuck to the amphiphilic surface coating. These initial observations provided an excellent starting point for the development of a binding assay to determine binding affinities of pheromones and analogs to PBPs. The assay has now been developed and exploited to provide quantitative binding affinities for Apo-3 rPBP (Du and Prestwich 1995). The salient points of the assay will be described herein, and an indication of the type and quality of the data available from this assay will be presented.

Plastic microfuge tubes were coated (12 hr) with a saturated solution of an *n*-alkanol (n = 8, 10, 12), prepared by adding a 1% ethanolic solution of the alkanol to water. 1-Dodecanol gave the most reproducible data. The time course and saturability of binding were investigated using Apo-3 rPBP (25 nM) and [^{3}H]6*E*,11*Z*-16:Ac (Prestwich 1987; Prestwich et al. 1986) in 20 mM Tris-HCl, pH 6.8. Labeled pheromone was employed in the 5–200 nM range. Nonspecific binding was determined by addition of 300-fold excess of unlabeled 6*E*,11*Z*-16:Ac.

Binding to rPBP was linear with [PBP] between 5 and 100 μg/ml. Although specific binding was the same at 1 hr at 4°C and 25°C, the former gave more reproducible data. Half-maximal binding was observed in 15 min, with a plateau (saturation) at 45 min. Subsequent assays were incubated for 1 hr at 4°C. Scatchard analysis of the binding of [^{3}H]6*E*,11*Z*-16:Ac to rPBP showed a single saturable binding interaction with $K_D = 0.6$ μM. Displacement of [^{3}H]6*E*,11*Z*-16:Ac binding by other pheromone components (6*E*,11*Z*-16:Ald and 4*E*,9*Z*-14:Ac), a non-pheromone component (16:Ac), and pheromone catabolites (6*E*,11*Z*-16:OH and 4*E*,9*Z*-14:OH) showed a clear trend in relative binding affinity: 6*E*,11*Z*-16:Ac > 4*E*,9*Z*-14:Ac > 6*E*,11*Z*-16:Ald ≈ 16:Ac > 6*E*,11*Z*-16:OH > 4*E*,9*Z*-14:OH. These data clearly demonstrate a >1000-fold range of binding affinities among these very similar structures and unambiguously demonstrate the specificity of the PBP–pheromone interaction. Experiments are in progress to produce specific Apo-3 mutants with predicted alterations in binding affinity and to express additional *Antheraea* proteins with different ligand selectivities.

5. A PBP–Pheromone-Receptor Complex in Receptor Activation?

OBPs in general (Pelosi 1994; Pelosi and Maida 1990), and PBPs specifically, may play active roles in the initial biochemical recognition step leading to the perireceptor events of odorant perception (Prestwich 1993b; Stengl et al. 1992). Support for this notion comes from the known microheterogeneity of DNA encoding such proteins in moths (Krieger et al. 1991), the association of specific OBP subtypes with distinct classes of olfactory receptor neurons (Vogt et al. 1991a), the segregation of OBP types in *Drosophila* olfactory hairs (Pikielny et al. 1994), and multiple forms of vertebrate OBPs in the vertebrate nasal mucosa (Pelosi 1994). Ultrastructural (Steinbrecht 1992; Steinbrecht et al. 1992), developmental (Vogt et al. 1989), and electrophysiological studies (Van den Berg and Ziegelberger 1991) have established the cellular locations (Maida et al. 1993) of PBPs and have confirmed their importance in mediating pheromone detection (Kaissling et al. 1991). However, the involvement of PBPs in odorant discrimination and in odorant recognition at the putative dendritic membrane receptor proteins is as yet unknown.

Understanding the molecular recognition processes in PBP–pheromone interactions is necessary to unravel the sequential events leading to the activation of the odorant-receptor-coupled, G-protein-mediated signal transduction cascade in olfaction (Breer 1993; Breer et al. 1990a; Lancet et al. 1993; Prestwich 1993b; Raming et al. 1993a; Reed 1992; Ronnett and Snyder 1992). In contrast to the OBPs of vertebrates (Pelosi, 1994; Pevsner et al. 1990) that recognize a large number of odorants of diverse structure, the insect PBPs and GOBPs possess odorant specificities for a limited set of chemical structures and terminal functionalities. The GOBPs (Breer et al. 1990b; Vogt et al. 1991a,b), which are expressed in both male and female antennae, are over 95% conserved in the Lepidoptera. Male-specific antennal PBPs have highly conserved overall protein folds with 30–60% sequence identities, but retain significant interspecific differences. Importantly, the insect OBPs are members of a family of predominantly helical proteins completely unrelated to the eight-stranded beta-barrel structure of the vertebrate OBPs (Pelosi and Maida 1990; Pevsner et al. 1988). We believe that this macromolecular diversity corresponds to the small molecule diversity in the structures of lepidopteran pheromones.

Photoaffinity labeling studies with *A. segetum* and a variety of photoaffinity analogs, including [^{3}H]Z5-10:Dza and [^{3}H]Z7-12:Dza, and a host of displacing ligands have provided direct evidence that several PBPs with different chain-length specificities are present in insect antennae (G.D. Prestwich and C. Löfstedt, unpublished results). The binding data determined above are also strongly supportive of the notion that PBPs are endowed with specificity for recognition of the cognate pheromone. An additional data point (which is as yet unavailable) would be the segregation of PBPs for a given pheromone component to the sensillum lymph of a sensory hair responding to that molecule.

On the basis of the existing data in insects and in vertebrates (Pelosi 1994; Pelosi and Maida 1990), by analogy to the bacterial chemotaxis literature (Stewart and Dahlquist 1987), and borrowing from the mode of ligand display by immunophilins (Schreiber 1991), we propose that the PBPs and GOBPs serve three important roles.

- First, as high concentration components of the sensory hair, they are the proteins of first contact and transport for *all* hydrophobic molecules entering the sensory lymph via the pores and pore tubules. The ratio of protein to ligand is sufficiently high (Vogt 1987) to guarantee that even low-affinity (millimolar) ligands are completely bound. This prevents lipophilic molecules from adhering to hydrophobic surfaces and from having unimpeded access to lipid bilayers. Importantly, PBPs also deliver pheromones and nonpheromonal molecules to sensillar enzymes for inactivation (Prestwich et al. 1986, 1989; Vogt et al. 1985).
- Second, PBPs retain affinity for pheromone metabolites, and are likely participants in removal of these metabolites from the cellular lumen. Indeed, the rapid and continuous turnover of PBP in male *L. dispar* antennae (Vogt et al. 1989) supports such an active role.
- Third, we propose that the PBPs are active participants in the presentation of the pheromone ligand to the G-protein-coupled receptor located in the membrane of the pheromone-specific dendrite. In this model (Fig.

15.4), PBPs not only "solubilize" specific pheromones, but they present a combined pheromone–PBP surface to the receptor protein. Electrostatic and hydrophobic interactions from *both* the bound ligand and ligated protein are necessary and sufficient for receptor activation. Neither the PBP nor the pheromone-binding protein alone is sufficient to cause receptor activation. This is a testable hypothesis, requiring patch-clamp methodologies *in vitro* or specific PBP-knockout mutations *in vivo*. It also predicts a novel approach for the isolation of putative receptor proteins. It is worth noting that in both *A. segetum* (C. Löfstedt and G. D. Prestwich, unpublished results) and *A. polyphemus* (Vogt et al. 1988), putative dendritic membrane receptors were specifically labeled in the presence of PBPs.

6. Future Studies

The future of molecular studies of olfaction in insects will incorporate modern techniques of structural biology and protein biotechnology. For example, we have initiated studies of the circular dichroism (CD) of Apo-3 rPBP (Fig. 15.5). In contrast to the beta-barrel form of the vertebrate OBPs (Pelosi 1994), the PBPs are predominantly helical proteins. Deconvolution of the CD allows prediction of 49% α-helix and 8% β-sheet. Chou-Fasman and Garnier algorithms predict

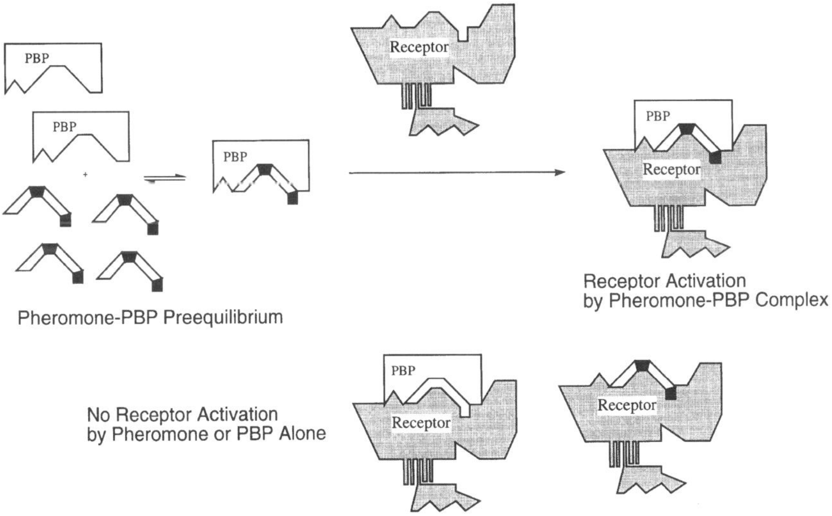

Figure 15.4. Schematic representation of the role of PBP in presentation of pheromone to a G-protein-coupled transmembrane receptor. The diagram emphasizes the notion that the receptor recognizes a composite surface formed by the specific pheromone–PBP complex.

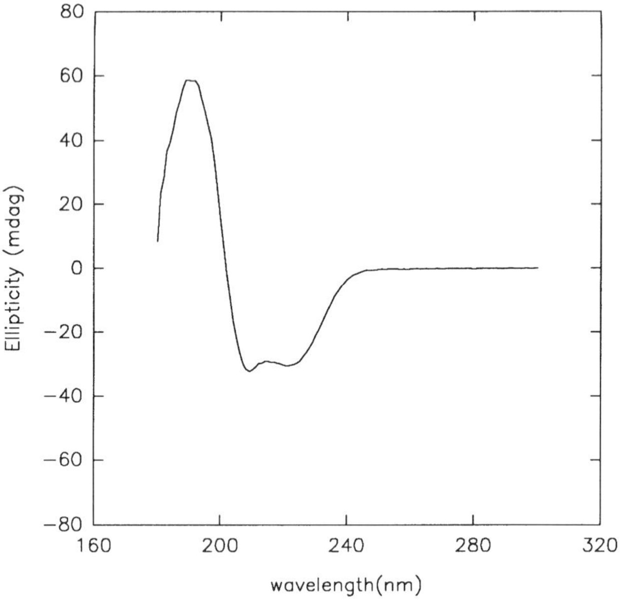

Figure 15.5. CD spectrum of Apo-3 rPBP. A 17 mM solution of rPBP in 50 mM phosphate, pH 6.5, was measured from 180 to 300 nm on a Jasco Model 720 spectrometer using a 1.0-mm cell. Estimates of secondary structure were made using the SSE-338 program with curve fitting of the 190–240 region.

47–48% helix content. Examination of PBP–pheromone complexes should be informative.

A high-resolution NMR structure is accessible for a soluble protein of 142 amino acid residues. Indeed, 1-D and 2-D spectra of Apo-3 rPBP at pH 6.0 show that sufficient dispersion and line sharpness would allow measurement of sufficient geometrical constraints to obtain structural data (Prestwich 1993a). Overexpression of Apo-3 rPBP has been accomplished from minimal medium with $^{15}NH_4Cl$ as the sole nitrogen source to give uniformly labeled ^{15}N-rPBP (Prestwich 1993a). Refolded, purified ^{15}N-PBP showed essentially all of the expected nitrogens in a single quantum coherence experiment (S. Mohanty and G. D. Prestwich, unpublished results).

Acknowledgments

We thank the NIH (NS29632 to G.D.P.) and the Herman Frasch Foundation for financial support of this research. A NATO research travel award to G.D.P. and

Professor H. Breer (Hohenheim-Stuttgart) made possible the collaboration leading to the development of the PBP expression and binding assays. We thank H. Breer and J. Krieger in particular for cDNAs, gemütlichkeit, and stimulating discussions. Dr. K.A. Lerro (Columbia University, Professor K. Nakanishi's laboratories) assisted in obtaining the CD spectrum.

16

Olfactory Coding Mechanisms for Pheromone and Interspecific Signal Information in Related Moth Species

Hanna Mustaparta

1. Pheromone Blends of Heliothine Moths

The subfamily Heliothinae consists of numerous moth species of which a number are serious pests in major agricultural staples, such as cotton and corn. The sex pheromones produced in the female's abdominal gland have been identified in several of these species, by the use of chemical analyses and behavioral studies. These signals consist of multicomponent mixtures, of predominantly unsaturated straight-chain aldehydes (C_{14}–C_{16}), but also of some acetates and alcohols (Fig. 16.1) (cf. Arn et al. 1992). Interestingly, two compounds (essential or principal pheromone components) in all the investigated species seem to be necessary and sufficient for eliciting the male attraction to the female or the full sequence of male sexual behavior, whereas the others (minor) components seem to play a role in close-range orientation. All, but one, heliothine species so far investigated produce (*Z*)-11-hexadecenal (Z11-16:Ald) (A) as the major component (Fig. 16.2). Species-specificity is realized by the addition of a second component differing in sympatric species. For example, in the American species *Heliothis virescens,* the principal pheromone components are Z11-16:Ald (A) and (*Z*)-9-tetradecenal (Z9-14:Ald) (B) produced in the ratio 16:1 and in the sympatric species *Helicoverpa zea,* Z11-16:Ald (A) and (*Z*)-9-hexadecenal (Z9-16:Ald) (C) in the ratio 16:0.1 (Roelofs et al. 1974; Tumlinson et al. 1975; Klun et al. 1980a,b; Pope et al. 1984; Teal et al. 1986). A third species, *Heliothis armigera,* distributed in other continents (Southern Europe, Asia, Africa, and Australia) utilizes the same principal components, Z11-16: Ald(A) and Z9-16:Ald (C) as *H. zea* (but in ratios between 10:1 and 100:1), supporting the hypothesis of a phylogenetic close relationship between the two species (Kehat et al. 1980; Kehat and Dunkelblum 1990). *Helicoverpa peltigera,* sympatric to *H. armigera,* has a similar pheromone blend of Z11-16:Ald (A) and Z9-14:Ald (B), as *H. virescens* (Dunkel-

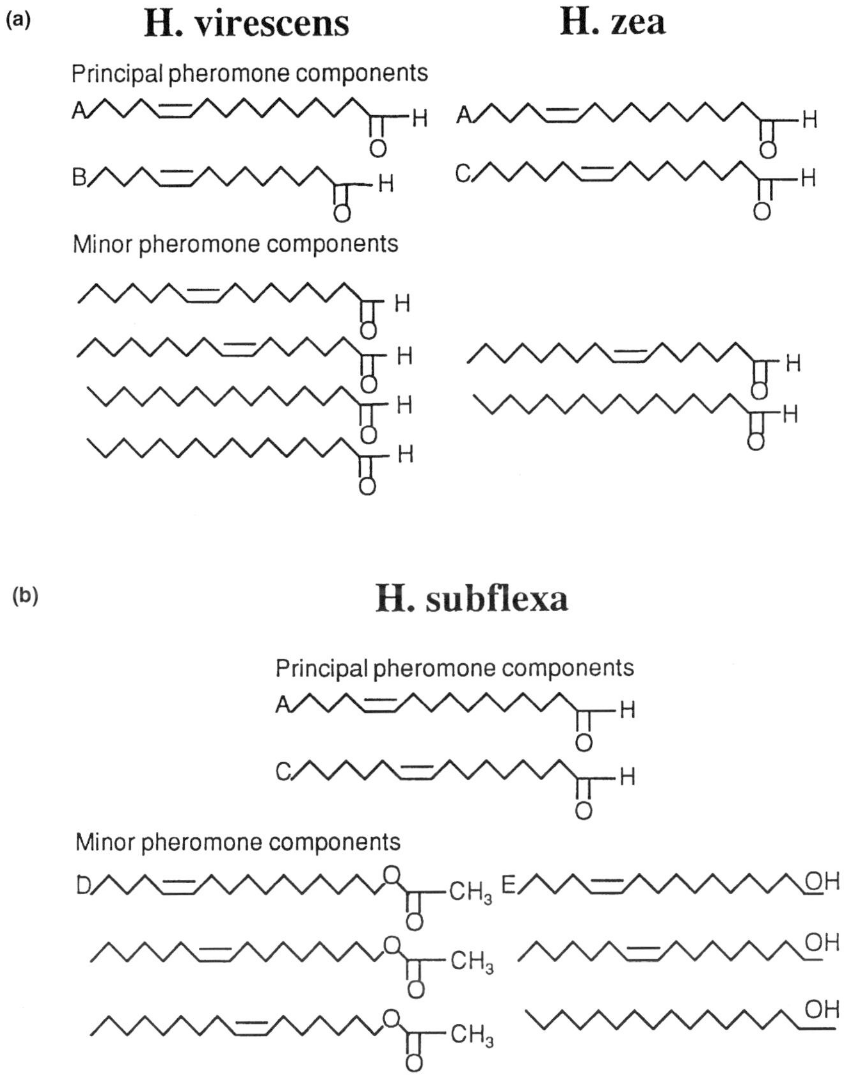

Figure 16.1. *(a)* Pheromones identified in *Heliothis virescens* and *Helicoverpa zea*. Two components A and B, the principal pheromone components, are necessary and sufficient for eliciting the full sequence of male sexual behavior, whereas the minor components seem to have subtle effects on the close-range orientation. *(b)* Pheromone components identified in *Heliothis subflexa*. The two principal pheromone components are A and C. The acetates, of which component D is present in largest amount, increase the captures of males when added to the principal pheromone components.

Heliothine species	Principal pheromone components	Interruption caused by
H. virescens	**A** + B	D, E
H. zea	**A** + C	B, E
H. subflexa	**A** + C	E
H. phloxiphaga	**A** + E	
H. peltigera	**A** + B	
H. armigera	**A** + C	B, E
H. punctiger	**A** + D	B
H. maritima	**A** + C	
H. assulta	**C** + A	B, E

Figure 16.2. All investigated species of heliothine moths, except for *Helicoverpa assulta,* uses the same compound A as the major pheromone component. The species-specificity is provided by the addition of a second component which varies in sympatric species, and of which some act interspecifically to inhibit attraction of sympatric species.

blum and Kehat 1989). In contrast to these and other heliothine species, producing A as major component, the oriental budworm moth *Helicoverpa assulta* present in Eurasia, Africa, and Australia utilizes C as major component and A as second pheromone component in a ratio 20:1 in the Korean population (Fig. 16.2) (Cork et al. 1991). The minor components in the various heliothine pheromone blends vary among species in number and analog structure (Fig. 16.1). In *H. virescens* and *H. zea* they seem to have only a subtle effect on close-range orientation when added to the principal components (Vetter and Baker 1983,1984), whereas in other species (e.g., *Heliothis subflexa*) the addition of some of them [the acetates of which (*Z*)-11-hexadecenylacetate (*D*) is present in the largest quantity] increases significantly trap captures of males (Teal et al. 1981a).

Another interesting aspect is the interspecific effect of pheromones, where one component of a certain species inhibits the attraction of a sympatric species (called *interspecific interruption*), which has previously been demonstrated among bark beetles as well as many lepidopterans (Roelofs and Comeau 1971; Birch et al. 1980; for review see Masson and Mustaparta 1990). Within the heliothine moths, Z9-14:A1 (B) interrupts the attraction of *H. zea* males (Shaver et al. 1982), which seems to ensure the isolation between this species and *H. virescens*

(Haile et al. 1973; Roach 1975; Carpenter et al. 1984; Lopez and Witz 1988). Likewise, the same compound interrupts the attraction of *H. armigera,* maintaining the isolation between this species and *H. peltigera* which utilizes B as the second principal pheromone component, as was described above (Kehat and Dunkelblum 1990). Compound B also seems to reduce pheromone attraction of males in the Australian species *Heliothis punctiger* (Rothschild et al. 1982). Other compounds known to interrupt pheromone attraction in several heliothine species are (Z)-11-hexadecenol (Z11-16:OH) (E) (*H. armigera, H. subflexa, H. zea,* and *H. assulta*) and (Z)-9-hexadecenol (Z9-16:OH) (F) (*H. assulta*), both reducing catches of males in pheromone traps or inhibiting upwind flight in wind tunnel experiments (Kehat et al. 1980; Teal et al. 1981a, 1984; Cork et al. 1991).

We have studied the olfactory coding mechanisms for pheromone and interspecific signal information in five species of heliothine moths, *H. virescens, H. zea, H. assulta, H. peltigera,* and *H. armigera.* The following three questions have been central in these studies: (1) How is pheromone information encoded by receptor neurons and central neurons? (2) What is the underlying mechanism for interspecific interruption? (3) How general are the coding mechanisms in related species? The third question also includes the ultimate goal to understand where in the olfactory system changes have occurred through evolution, resulting in different behavioral responses to the same compounds.

2. Receptor Neuron Specificities

2.1. Receptor Neuron Types and Labeled Lines

The coding of olfactory information starts peripherally with the specificities of the receptor neurons, defined as the neuron's dose–response relationship for the compounds it respond to (Kaissling 1971). The odor molecules apparently reach the dendritic surface of the receptor neuron by adsorption on the cuticle surface of the olfactory sensilla, diffusion through wall pores, and transport by odor-binding proteins (pheromone-binding proteins PBP in pheromone sensilla and general odor-binding proteins GOBP in plant odor sensilla) in the receptor lymph (Steinbrecht 1992). The PBPs which exhibit some specificities may act as a filter by transporting only certain molecules toward the membrane proteins of the dendritic surface (Prestwich 1995; Prestwich and Du this volume). It is obviously the putative membrane receptor proteins which determine the specificities of the receptor neurons. By interaction with the odor molecule or the complex of odor molecule and the binding protein, the intracellular transduction reaction is elicited, leading to opening of cation channels underlying the receptor potential and nerve impulses (cf. Stengl et al. 1992). Our knowledge about specificities of olfactory receptor neurons is derived from studies of structure–activity relationships where activity is measured as nerve impulses from an intact receptor neuron. We have extracellularly recorded nerve impulses from receptor neurons of heliothine moths, with tungsten microelectrodes penetrating the cuticle at the base of sensillum hairs. Recordings were made from the male-specific *sensilla trichodea* type 1 in all five species of heliothine moths and from *s. trichodea* type 2 in *H. virescens* and *H. zea.* Initially we screened the neurons for sensitivity to all relevant pheromone components of heliothine moths as well as some chemical analogs. The receptor neurons then were tested from low to high concentrations with the stimulatory compounds, resulting in dose–response relationships.

The receptor neurons so far found are activated by pheromones and other insect-produced compounds. No inhibitory effects have been detected. At this point, it should be kept in mind that inhibition might be difficult to detect because of the neurons' low spontaneous activity. However, some inhibitory effects of plant odors on pheromone receptor neurons have been recorded (Schneider et al. 1964; cf. Masson and Mustaparta 1990). Like in other insect species studied, we have found in the heliothine moths that the receptor neurons receiving information about pheromones and interspecific signals are tuned to one compound; that is, each neuron responds with high sensitivity to this key component with an increased firing rate as concentration is increased (Fig. 16.3). (Almaas and Mustaparta 1990, 1991; Almaas et al. 1991; Berg et al. 1995). However, when tested at high concentrations, these neurons also may respond to other pheromone components, often with a 2–3 log unit decrease in sensitivity. By adding these less effective compounds to the key compound, we have found that the receptor neurons respond equally to the mixture and the key compound alone, suggesting that no interaction of compounds takes place at the receptor level (Almaas and Mustaparta 1990; Berg and Mustaparta 1995). This implies that when males fly in a pheromone plume, each receptor neuron responds to its key compound as if it were the only compound present. At high concentrations any receptor protein may become activated by whatever structurally similar compound, a feature which applies to the various pheromone components. The results suggest that specificity of a given receptor neuron has evolved to detect one key component and "disregard" other related compounds; that is, each type of receptor neuron mediates information to the brain about one component, defined as a labeled-line mechanism. It implies that the message to the brain about pheromones and interspecific signals is mediated as the relative activities of the different receptor neuron types.

In spite of the relatively large number of pheromone components identified in heliothine moths, we so far found only four types of receptor neurons (Fig. 16.3). The two major groups of neurons tuned to the same two components, A and B, were found in all species, regardless of whether compound B is utilized as a pheromone component or not. When not produced by conspecific females, it appears to function as an interspecific signal. This suggests that in evolutionary respect the two compounds, A and B, have an early origin in the chemical communication of the heliothines.

In *H. virescens* we identified three types of receptor neurons (Almaas and Mustaparta 1990, 1991; Berg et al. 1995) (Fig. 16.3). In addition to the two types

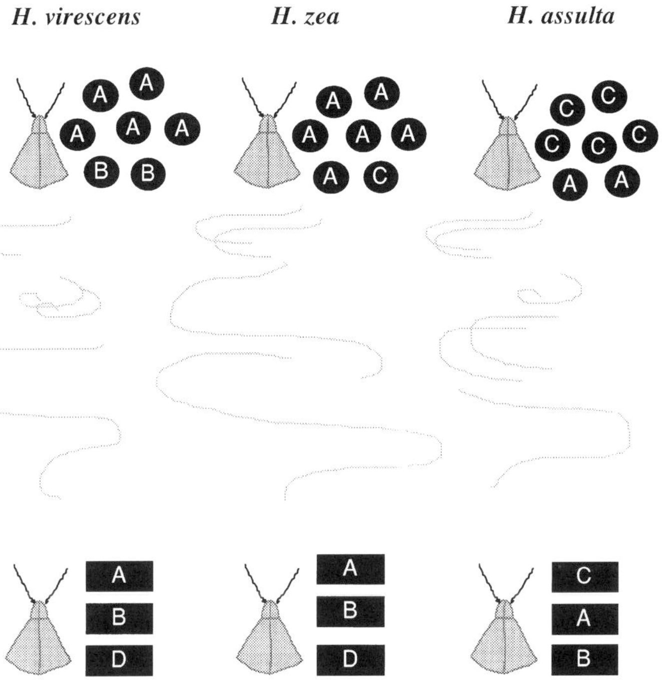

Figure 16.3. Principal pheromone components produced by females and receptor neuron types identified in males of *Heliothis virescens, Helicoverpa zea,* and *Helicoverpa assulta.* The A and B type of neurons are present in all three species, whereas the C-type neurons are found only in *H. assulta,* where C is the major pheromone component. Receptor neurons tuned to D are found in *H. virescens* and *H. zea,* where D may be an interspecific signal, demonstrating that the two sympatric species possess the same types of receptor neurons.

tuned to the principal pheromone components, A and B, the third type responded best to the acetate analog Z11-16:Ac (D). Thus, the information from these three compounds is conveyed to the brain via labeled lines. The message about the species-specific pheromone blend is mediated as the ratio of activities in the two receptor neuron types tuned to A and B, respectively. The receptor neurons tuned to the acetate D (Berg et al. 1995) were previously reported to respond best to the alcohol analog Z11-16:OH (E), but with a relative weak response, suggesting that another compound rather might be the key compound for these neurons (Almaas and Mustaparta 1991). In the latter study the acetates were not tested, and the recent study revealed that the alcohol E was the second best compound on the acetate neurons. These findings led to the assumption that the acetate neurons mediate interruption, either interspecifically (produced by *H. subflexa* females) or intraspecifically (produced by males of *H. virescens*) (Teal et al. 1981a; Teal and Tumlinson 1989). Support for the assumption is the recent observation that the acetates reduced upwind flight by *H. virescens* males when added to the pheromone blend (Vickers and Baker 1995). In *H. peltigera,* only the two types of receptor neurons tuned to A and B have been identified, suggesting also in this species a labeled-line mechanism for reception of the two principal pheromone components.

2.2. Dual Function of One Receptor Neuron Type

Since labeled lines for encoding pheromone information by receptor neurons seem to be the principal mechanism in most species investigated, it was surprising to find that the system in *H. zea* was different. In this species, no receptor neurons could be identified which were tuned to the second principal pheromone component C. One interpretation for this might be that the neurons are present, but in such a small number that they have not yet been found electrophysiologically. This seems possible in some species (Priesner 1979b). However, we propose another mechanism in the heliothine moth *H. zea,* that the information from C is mediated via the receptor neurons tuned to the interspecific signal B, since these neurons respond second best to compound C (Almaas et al. 1991; Christensen et al. 1991a; Vickers et al. 1991). It means that when the male is flying in a conspecific pheromone plume, the neuron type tuned to A will be strongly activated, whereas the neuron type tuned to the interspecific signal B will be only weakly activated by the second best compound C. B of course is not present in the conspecific plume. The activity ratio in the two neuron types (one strongly and one weakly activated) signify the pheromone information. However, when the males contact the plume of the sympatric species *H. virescens,* the receptor neurons tuned to B will be strongly activated, causing interruption of attraction. This implies that B and C might substitute for each other by decreasing and increasing the amount, respectively, in the mixture with the major component A. This was actually found in wind-tunnel studies, and it supports the hypothesis of a dual function of the B receptor neurons in *H. zea* (Vickers et al. 1991). As in *H. zea,* only two types of receptor neurons were found in *H. armigera,* tuned to A and B, respectively, the latter mediating interspecific interruption. No receptor neurons were identified which were activated by the second principal pheromone component C. Also in this species the neurons tuned to the interspecific signal B responded second best to compound C, suggesting that here also the B-neurons may have the dual function of mediating attraction and interruption of attraction. However, further behavioral tests are needed to verify this suggestion in *H. armigera.*

2.3. The Significance of the Major Pheromone Component and Interspecific Signals

Since no receptor neurons tuned to compound C were found in *H. zea* or in other heliothine species, it was interesting to study the pheromone reception in *H. assulta,* where C is the major component and A the second principal component.

The results showed that receptor neurons tuned to compound C are indeed present in *H. assulta* (Berg and Mustaparta 1995), indicating the great significance of the major components in pheromone communication. Here, the two largest groups of neurons were tuned to compound C and compound B, respectively. Actually, these two types of neurons were always recorded simultaneously from the same sensillum, with the large spikes originating from the neuron tuned to the major component C and the smaller spikes from the B neurons. A smaller group of neurons were tuned to A. The question as to the function of compound B was then raised. It was assumed that this compound acts as an interspecific signal, interrupting pheromone attraction in *H. assulta* as in *H. zea* and *H. armigera* males. This hypothesis was tested in a hot pepper field and in a nethouse in Korea. When compound B was added to the pheromone blend, it reduced trap catches of *H. assulta* males (Boo et al. 1995). Above a certain amount of B, males were no longer caught, supporting the hypothesis that B is an interspecific signal. Thus, in *H. assulta* the pheromone and interspecific signal information is mediated to the brain via a labeled-line system. We have here demonstrated that the three heliothine species, *H. assulta, H. zea,* and *H. armigera,* using compound B as an interspecific signal, all possess specific receptor neurons tuned to this compound. It further implies that interspecific interruption in these species is mediated via activation of the specific B neurons. This corresponds to results obtained in bark beetles and other insect species (Mustaparta et al. 1977; Priesner 1979a,b; for review see Masson and Mustaparta 1990).

2.4. Minor Components

Receptor neurons tuned to the minor components could not be found in any of the five species investigated. Further tests have been conducted with mixtures, excluding the possibility that minor components could interact with the major components on the receptor neurons (Almaas and Mustaparta 1990; Berg and Mustaparta 1995). One might wonder whether receptor neurons tuned to minor components are present in other sensilla (e.g., in *s. basiconicae*). It may be possible that the minor components influence the receptor neurons responding to plant odors and may thus play a less important role when plant odors are present. Another possibility is that these neurons have disappeared in cultured insects corresponding to changes detected in pheromone production (Raina et al. 1989b). This is in contrast to the fact that the influence of minor components on the behavior has been frequently demonstrated in insects from laboratory cultures.

3. Functional Similarities of Receptor Neurons

3.1. The A Type of Receptor Neurons

After having established that all heliothine species investigated possess the two groups of neurons either tuned to compound A or to B, our next question was whether these neurons are functionally similar—in other words, whether they exhibit a similar ranking of dose–response relationships for A and B, respectively, and their chemical analogs. The receptor neurons tuned to compound A, responded almost equally to the nonbiological formate (*Z*)-9-tetradecenylformate (Z9-14:FO), having an oxygen instead of a carbon atom in the C_2 position (Fig. 16.4)

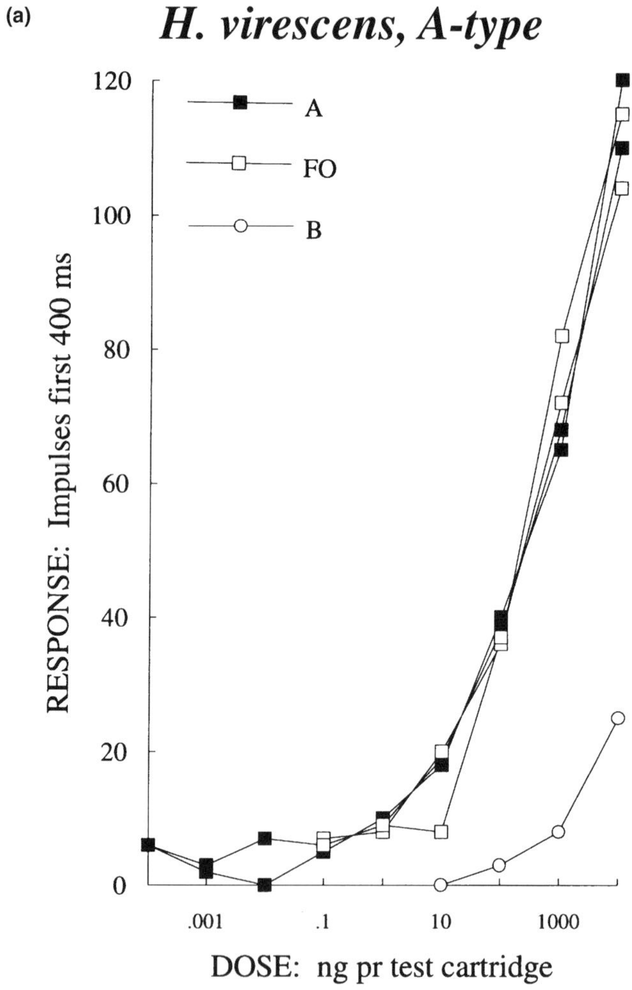

Figure 16.4. Dose–response curves of two receptor neurons tuned to compound A (*Z*)-11-hexadecenal in (**a**) *Heliothis virescens* and (**b**) *H. assulta.* The formate analog, (*Z*)-9-tetradecenylformate, has about the same effect as A on these neurons at the same quantity tested, whereas compound B has the second best effect. The other components have only minimal effect on the A neurons. [After Berg and Mustaparta (1995) Berg et al. (1995).]

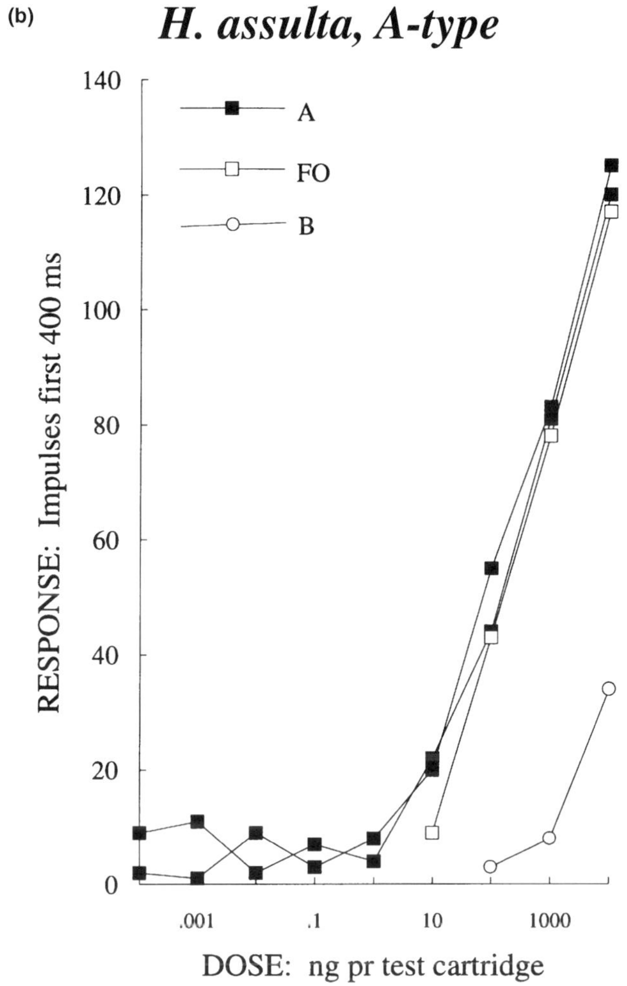

Figure 16.4. Continued

(Berg et al. 1995). The second best response was to compound B and had to be applied at 100–1000 times higher concentrations before it elicited the same strong response as did compound A. The other analogs had a minimal effect on these neurons. This ranking of dose–response relationships was characteristic for the A type of receptor neurons in all species investigated, suggesting that the membrane receptors are functionally similar—that is, that the receptor proteins have been functionally preserved through evolution (Almaas and Mustaparta 1990, 1991; Almaas et al. 1991; Berg et al. 1995; Berg and Mustaparta 1995).

Variations within this type of receptor neuron, which are similar in all species, were ascribed to different sensitivities. The most sensitive neurons responded according to the description above, whereas the less sensitive neurons responded only to A and the formate analog, simply because of the dose range. The sensitivities could roughly be correlated to the position on the antennae and the length of the sensillum hair. The sensitivity decreased from the lateral to the medial side and from the proximal to the distal annuli (Almaas and Mustaparta 1991). The receptor neurons located in the type 2 sensilla with short hairs had a marked lower sensitivity when compared to those of *s. trichodea* type 1 with long hairs.

3.2. The B Type of Receptor Neurons

The receptor neurons tuned to compound B showed similar dose–response relationships in *H. virescens* and *H. zea.* The second best pheromone component was C (as mentioned above), whereas the nonpheromonal compound (*Z*)-7-dodecenylformate (Z7-12:Fo) had a much lower effect on these neurons than did the corresponding formate on the A neurons (Fig. 16.5a) (Berg et al. 1995). The dose–response relationships of the B neurons in *H. assulta* differ from those of *H. virescens* and *H. zea* (Fig. 16.5b) (Berg and Mustaparta 1995). Here, compound C had no effect at any concentration, whereas the formate and the alcohol analogs had a relatively high effect. In contrast to the similarities of the B neurons in *H. virescens* and *H. zea,* suggesting a functional conservation of the receptor proteins in these two species, are the differences from those of *H. assulta,* suggesting the evolution of functionally different membrane receptors in the latter species. The ratio of pheromone compounds produced in conspecific females may explain why different specificities of the B neurons are evolved. When *H. zea* males are flying along a pheromone plume, the A neurons are strongly activated while the B neurons are weakly activated by compound C (Fig. 16.3). However, when *H. assulta* males are flying in its pheromone plume, the C-type receptor neurons and the A-type neurons are both strongly activated by their key compounds. If the B neurons would have become activated by compound C, being present here in a 200 times larger dosage than in *H. zea,* they might have become too strongly activated, causing interruption to its own pheromone plume. The specificities of the C-type neurons, present exclusively in *H. assulta,* exhibit similar dose–response relationships within the species, responding second best to compound B. When mixed with C, compound B did not influence the responses to compound C, suggesting that B has no biological function on these neurons. It means that nature is not more precise than necessary in producing functionally specific receptor proteins for a particular compound.

3.3. The D Type of Receptor Neurons

The acetate neurons identified in *H. virescens* showed similar dose–response relationships, responding second best to the nonbiological formate analog Z7-

(a)

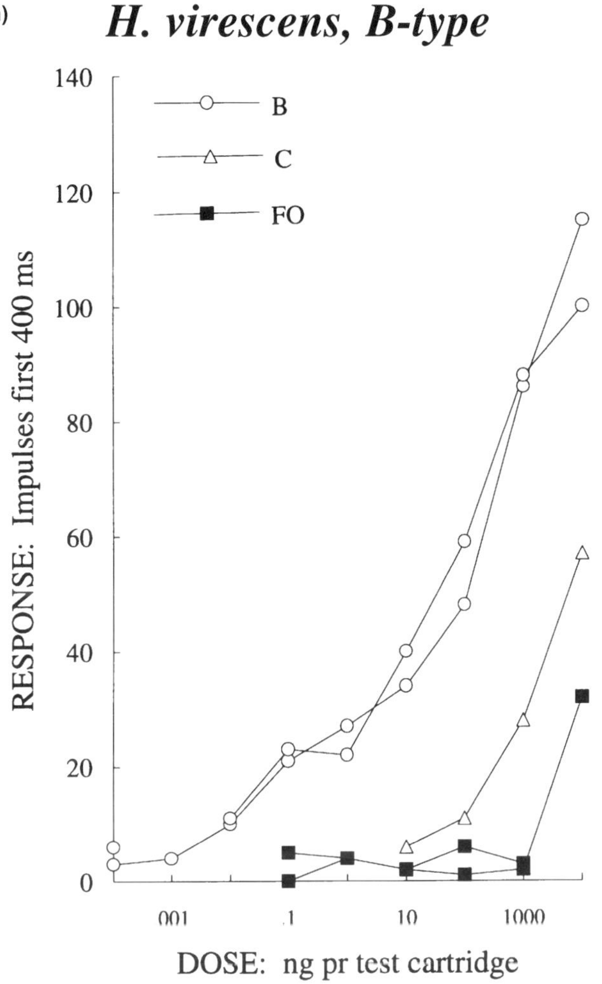

Figure 16.5. **(a)** Dose–response curves of a receptor neurons tuned to compound B in *Heliothis virescens*. The second best compound is C with a curve shifted about 2–3 log units to the right of that for compound B. The formate analog, (*Z*)-7-dodecenylformate, has much less effect on these neurons than does the corresponding formate analog on the A neurons.

12:FO and to the heliothine produced alcohol E. It is interesting that the receptor neurons tuned to interspecific signals, mediating interruption, seem to respond relatively strongly to two such behaviorally modifying compounds, as acetate D and alcohol E in *H. virescens* and as compound B and F in *H. assulta*. Furthermore, the response to the formate analog (Z7-12:FO) by the acetate neurons together

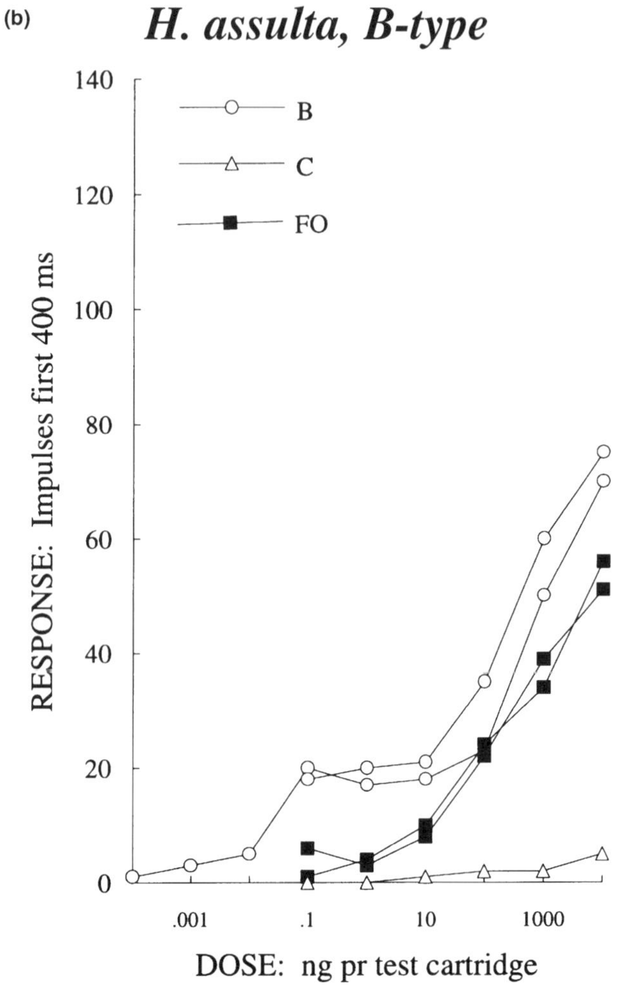

Figure 16.5. Continued **(b)** Dose–response curves of a receptor neuron tuned to compound B in *Helicoverpa assulta.* These neurons did not respond to compound C at any concentration. A higher effect of the formate analog and the alcohol E was also characteristic of the B neurons of *H. assulta,* as compared to the B-type neurons of the other species. [After Berg and Mustaparta (1995) and Berg et al. (1995).]

with the low response by the B neurons to the same formate can explain why this formate analog cannot substitute for compound B in *H. virescens,* whereas the corresponding formate Z9-14:FO can substitute for compound A (Berg et al. 1995).

The reviewed studies of the receptor neurons suggest that the labeled-line system seems to be general for encoding pheromone and interspecific signal information in heliothine moths as well as in other insect species (Kaissling 1979; Priesner 1977, 1979a,b; Mustaparta et al. 1979, 1980; for review see Mustaparta 1989). However, we have found an exception in *H. zea:* Neurons tuned to the interspecific signal B have a dual function, mediating pheromone information in one plume from an *H. zea* female and evidently interspecific interruption in a plume from *H. virescens.* Furthermore, the functional part of the membrane receptors seems in general to retain common features. However, we have also found an exception: Receptors evolved to detect the quality of an interspecific signal may also have become modified by the quantity of the major pheromone component that is produced.

4. Comparison of Input and Output of the Antennal Lobe

The information mediated to the brain by the four groups of receptor neurons represents the input to the antennal lobe (AL) about pheromones and interspecific signals. This information in insects is transmitted to the AL neurons in one particular glomerular structure, the macroglomerular complex (MGC), located close to the entrance of the antennal nerve (Hildebrand this volume; Christensen this volume). The MGC consists of three and four subcompartments in *H. zea* and *H. virescens,* respectively (Christensen et al. 1991a, 1995; Hansson et al. 1995; Hansson this volume). Two morphologically different types of AL neurons have been identified in heliothines (Christensen et al. 1991a, 1995) as well as in other moth species (Matsumoto and Hildebrand 1981; Christensen and Hildebrand 1987b). One type consists of local, anaxonal neurons, arborizing in almost all glomeruli. The other type, the projection neurons, exhibits dendritic arborizations in one glomerulus (those responding to pheromones) and has an axon projecting into the protocerebrum (Hildebrand this volume). Here, the axons terminate in two areas, the mushroom bodies and the lateral protocerebrum. The information from the receptor neurons is transmitted to the projection neurons via inhibitory input from the local interneurons, but may also be mediated directly (Christensen et al. 1993). The projection neurons usually respond by excitation to antennal stimulation with single pulses of pheromones. We have made intracellular recordings from the AL neurons during antennal stimulation with the various pheromone compounds (principal and minor components). "Lucifer Yellow" was used for staining of the neurons for subsequent reconstruction of their arborization and projection pattern. In this way, the structure and function of projection neurons have been studied so far in two sympatric species, *H. zea* and *H. virescens.*

4.1 *Heliothis virescens*

In *H. virescens,* the neurons could be classified in four categories (Christensen et al. 1995). One major group of neurons responded primarily to antennal stimulation with compound A and showed a wide range of response strengths to stimulation with the same dosages. The less sensitive neurons responded selectively to A, whereas the most sensitive neurons also showed weaker responses to B. Since these reactions reflect the responses of the receptor neurons tuned to compound A, it appears likely that they receive input exclusively from the receptor neurons tuned to compound A, where the receptor neurons of high sensitivity may converge on the most sensitive AL projection neurons. Another group of neurons responded to antennal stimulation with compound A as well as B, of which some neurons showed equal responses to these two principal pheromone components and obviously received input from both types of receptor neurons. Furthermore, they exhibited varying responses to stimulation with the mixture of A and B and the pheromone blend. Some neurons showed stronger and others weaker responses to these mixtures than to the single components. The neurons within this category also varied in their relative inputs from A and B, which often made distinguishing them from the previous neuronal type difficult. However, here the important point is that some neurons received input only from the major component A and others from both principal components. No AL-projection neurons in *H. virescens* received input exclusively or predominantly from compound B. The information from this component seems to be completely integrated with the information from the major component A. A synergistic effect of information from A and B was found for one neuron in *H. virescens,* which thus seemed to be a blend specialist detecting information about the conspecific pheromone blend. Of further interest is the finding that some *H. virescens* neurons responded selectively to the acetate, which interrupts pheromone attraction. This suggests that the information from the interspecific signal is conveyed in separate pathways throughout the antennal lobe.

Staining of the neurons responding selectively to stimulation with compound A in *H. virescens* revealed that these neurons arborized in the largest compartment of the MGC located closest to the antennal nerve entrance in the AL (Fig. 16.6). Unfortunately, we were not able to achieve staining for the few AL neurons responding selectively to the acetate. However, recent experiments with stain diffusion through the receptor neuron axons in *H. virescens* have revealed a functional organization in the MGC in this species (Hansson et al. 1995). Whereas the receptor neurons tuned to each principal pheromone components A or B, projected into the largest compartment of the MGC (corresponding to the arborizations of the AL neurons), the receptor neurons tuned to the acetate projected into a separate, more dorsomedial compartment (Fig. 16.6). Together these results suggest that in *H. virescens* the information from the pheromones and the interspecific signal is processed in separate compartments of the MGC.

4.2 Helicoverpa zea

In *H. zea,* with A as the major pheromone component and B as the interspecific signal, we identified two major classes of projection neurons (Christensen et al.

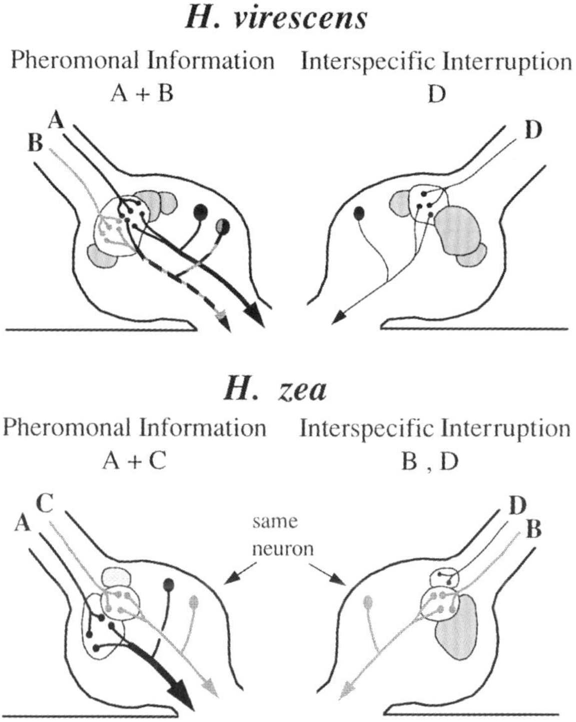

Figure 16.6. Scheme of the input and output information of the antennal lobes of *Helicoverpa virescens* and *Heliothis zea.* The same types of receptor neurons, tuned to the compounds A, B, and D, mediate the input information to the brain of both species. This information is differently processed in the antennal lobes of the two species. The major difference is the processing of information about compound B which mediates different messages, pheromone versus interspecific signal. Whereas the information about pheromones and interspecific signals seems to a large extent to be kept in separate pathways through the antennal lobe in both species (A and B versus D in *H. virescens* and A versus B and D in *H. zea*), the second pheromone component B seems to be completely integrated with the information from the major component in *H. virescens.* The system is more complicated in *H. zea,* where the second principal pheromone component and the interspecific signal is mediated via the same neurons.

1991a). One group of neurons was primarily activated by antennal stimulation with compound A, whereas the other group was activated primarily by compound B (Fig. 16.6). Within the former group, the neurons showed a wide range of sensitivities, with the less sensitive neurons responding selectively to compound A and the most sensitive neurons also weaker to compound B. When stimulated with a mixture of A and B or with the whole blend, the responses could not be distinguished from those elicited by A alone. These neurons corresponded to the A category of projection neurons in *H. virescens,* which seem to receive input exclusively from the receptor neurons tuned to A. It implies that the information from the major pheromone component is mediated via specific pathways to the protocerebrum. Neurons of the other major group was not encountered in *H. virescens*. These neurons in *H. zea* were activated by antennal stimulation with the interspecific signal B and did not respond to the major component A. However, when stimulated with high concentrations of compound C, these neurons became activated. Thus, the reactions of the B neurons in the AL of *H. zea* reflect the responses of the receptor neurons tuned to compound B, suggesting that the information from the interspecific signal B is mediated via separate pathways throughout the AL to protocerebrum (Fig. 16.6). It further indicates that the information from the interspecific signal is not integrated with the information from the major pheromone component in the antennal lobe (with exception of the synergist neurons, see below). No AL neurons were found in *H. zea* that were activated by the major pheromone component A and inhibited by the interspecific signal B. Instead, three AL neurons were encountered that responded synergistically to the two compounds when mixed in equal amounts of 10 ng. Since compound A and B are present in the pheromone blend of the sympatric species, these neurons may be used to signify the presence of the sympatric female, contributing to the interruption of attraction.

Staining revealed that the neurons had different patterns of arborizations in the MGC. The neurons specifically responding to compound A arborized in the whole MGC (i.e., in all three compartments), whereas the neurons responding to compound B arborized only in one dorsofrontal compartment of the MGC (Fig. 16.6). Recent results from stain diffusion of receptor neurons in *H. zea* have shown that the A-type receptor neurons terminated in the large ventral MGC compartment, whereas each of the receptor neurons tuned to the interspecific signals B and D projected to one of the two dorsal compartments, of which the dorsofrontal received the terminals of the B neurons (Almaas et al. 1995). Thus, also in *H. zea* the pheromone and interspecific signal information is to a large extent processed in different compartments of the MGC and kept in separate pathways throughout the AL. However, in this species the system is more complicated, where the information about the second principal pheromone component C and the interspecific signal B is mediated along the same receptor neurons and AL-projection neurons. This means that the message about conspecific versus sympatric pheromones is mediated to the protocerebrum as the ratio of activities in the two pathways of the A- and the B-projection neurons, where a high ratio between the A and B activity signifies a conspecific blend and a lower ratio (toward 1:1) indicates a blend of sympatric species.

5. Sensitivities and Dynamic Range

In general, the AL neurons show stronger responses to a given concentration than the receptor neurons, due to covergence (Boeckh and Boeckh 1979; Almaas and Mustaparta 1990, 1991; Almaas et al. 1991; Christensen et al. 1991a, 1995). However, both receptor neurons and AL neurons have individually different sensitivities varying by a factor of at least 1000, suggesting that the very sensitive AL neurons receive input from the most sensitive receptor neurons, whereas the less sensitive receptor neurons give input to the AL neurons of low sensitivity. We have proposed that the neuron's different sensitivities may be important for detecting changes of concentrations and increasing the dynamic range of the pheromone plume (Almaas and Mustaparta 1991). When concentration increases, the less sensitive receptor neurons are successively activated. This spatial recruitment of activity to the brain may be a mechanism by which males can detect the concentrations and estimate the distance from the odour source. This would be a more efficient mechanism than temporal summation of increased firing rate of each neuron, which would be problematic at low concentration levels in nature. It is difficult to estimate at which dosages the neurons are exposed in the natural pheromone plume. However, a strong response of a very sensitive AL neuron is elicited at concentrations below the rising phase of the receptor neuron's dose–response curves. It suggests that recruitments of receptor neurons at such concentration levels would be more efficient than temporal summations. It is, for instance, possible that the A neurons with their wide sensitivity range, rather than neurons tuned to minor components (and for which we have not identified any receptor neurons), may contribute to the close-range orientation.

Another aspect is the fact that the AL neurons of highest sensitivities receive input from only the A-type receptor neurons. This suggests that the pathways mediating information about the major component A is activated at lowest concentrations. This appears as contradictory to the results showing that the blend has the lowest behavioral threshold (Linn and Roelofs 1983). However, it might be that in nature the major component A elicits arousal or an across-wind searching behavior which is difficult to observe in wind-tunnel experiments. In this area we need more detailed studies on orientation mechanisms in order to understand why the major pheromone component is produced in a high quantity and why the males in addition possess a much larger number of receptor neurons tuned to the major component than to the second principal pheromone component (cf. Cardé and Charlton 1984).

6. Conclusions

1. Two major groups of receptor neurons, tuned to compounds A and B, respectively, are present in all heliothine moth species so far investigated, suggesting that the two compounds are evolutionary old in sex attraction of this noctuid group. The presence of major groups of receptor neurons tuned to compound A in all species and to C exclusively in *H. assulta* indicates the biological significance of the major pheromone components of these species. The major groups of receptor neurons

tuned to B in *H. zea, H. armigera,* and *H. assulta* and to D. in *H. virescens* and *H. zea* indicates the importance of interspecific signals for the reproductive isolation between sympatric species.

2. The receptor neurons in heliothine moths encoding information from pheromones and interspecific signals are, generally, tuned to one compound, mediating the information via a pure labeled line. However, one type of neuron with a dual function, mediating pheromone attraction when weakly activated and interspecific interruption when strongly activated, is an exception to this system.
3. As judged from the specificities of the receptor neurons present in the various heliothine species, all A-type and most B-type receptor neurons are functionally similar, suggesting that the functional part of the membrane receptor proteins are, in general, conserved through evolution. However, an exception from this is found in *H. assulta,* in which the B receptor is different by not responding to compound C, and may be an adaptation to the quantity produced of the major component.
4. Whereas the input to the AL is mediated by functionally similar receptor neurons tuned to A, B, and D, in the two species *H. virescens* and *H. zea* the processing of this information in their antennal lobes is different, leading to a different output. The information about conspecific pheromones and interspecific signals is to a large extent kept in separate pathways throughout the antennal lobes of both species, with exception of the information about B and C mediated by the same neurons in *H. zea.* The major differences between the two species is found for the processing of the input from the B receptor neurons. In *H. virescens,* the input from B serving as the second principal pheromone component is completely integrated with the information from the major component. In *H. zea,* the input from B serving as interspecific signals is processed in a separate compartment and is to a large extent kept separated from the input of the major pheromone component. The interspecific signals do not seem to cause inhibition of pheromone responses in the antennal lobe. The neuronal mechanisms for interspecific interruption must, therefore, take place at higher orders of neurons in protocerebrum. Together these results suggest that evolutionary changes in olfactory coding mechanisms take place in the central nervous processing of pheromone and interspecific signal information rather than peripherally by changes of receptor specificities.

Acknowledgments

The studies reported were supported by grants from the Norwegian Research Council (NFR/NAFV, project no. 452.89/006, 452.91/001452, 93/014) and

NATO Research Grants for International Collaborations (project no 880149). Dr. Tor J. Almaas cand.scient. Bente G. Berg and cand.sci Kari Andrésen, Trondheim, and Drs. Thomas A. Christensen and John G. Hildebrand, Arizona, are gratefully acknowledged for their collaborations on heliothine moths. Dr. James Tumlinson, Florida, is acknowledged for supply of purified pheromone compounds; and Dr. Max E. Angst, Basel, Dr. K. Boo, Seoul, and Dr. Ezra Dunkelblum, Bet-Dagan, are acknowledged for providing insect specimens. Bente Berg and Jon S. Kvaale are acknowledged for technical help with the figures.

17

Antennal Lobe Projection Patterns of Pheromone-Specific Olfactory Receptor Neurons in Moths

Bill S. Hansson

1. Introduction

The high sensitivity and specificity of pheromone-detecting olfactory receptor neurons, present on the antennae of the male moth, have attracted much interest as a model of chemoreception. In numerous publications, beginning with the pioneering studies of Professor Dietrich Schneider (e.g., Schneider 1962), male antennal sensory neurons have been shown to be able to detect minute amounts of the female produced sex pheromone. Also the morphology of the male olfactory sensilla has been scrutinized, showing how they are organized to allow passage of odor molecules into their internal dendritic processes (Steinbrecht 1987 and references therein).

Similarly, the primary olfactory centers of the moth brain, the antennal lobes, have been investigated. Extracellular studies by Boeckh and Boeckh (1979) and intracellular studies by Matsumoto and Hildebrand (1981) have shed light on the function and morphology of the antennal lobe. The sphinx moth, *Manduca sexta,* has become the model for these investigations. It has been shown that in the antennal lobe, an even higher sensitivity is encountered than in the antennal receptor neurons (Boeckh and Boeckh 1979; Hansson et al. 1991; Christensen and Hildebrand 1987b). Morphological investigations have demonstrated the presence of different types of antennal lobe interneurons (Homberg et al. 1988).

The peripheral olfactory organs and the primary olfactory centers of the brain are thus well studied in the male moth. This is contrary to the link between these two sites—that is, the axon and the axonal branches of the receptor neurons (Fig. 17.1). Mass filling of antennal neurons by bathing the cut antennal nerve in cobalt chloride or some other marker has been used to establish the morphology of different structures of the antennal lobe (Koontz and Schneider 1987); but the structure of individual receptor neurons, as well as correlations between physio-

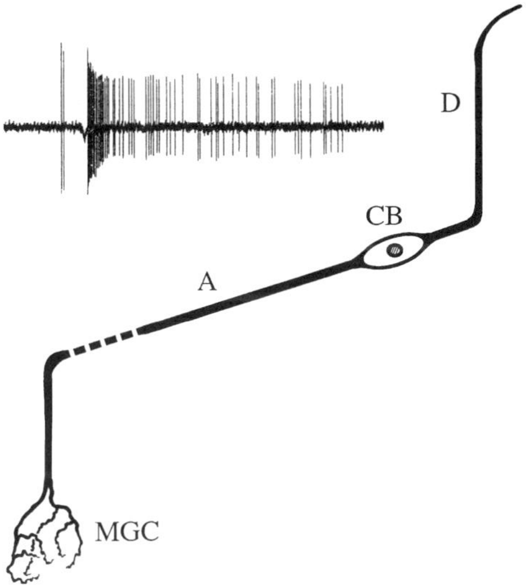

Figure 17.1. Schematic drawing of a pheromone-detecting receptor neuron in a moth. The pheromone molecules are detected by receptor sites on the outer dendritic segment (D). The signal passes the cell body (CB) and is transferred via the axon (A) to the axonal branches in the macroglomerular complex (MGC). The physiological response of such a neuron to the key stimulus is also shown.

logical and morphological characteristics at this level has not yet been described. This is mainly due to the lack of methods to conduct such investigations. It has been possible to observe a small selection of neurons in golgi preparations and in fills of restricted areas of the antennal surface, but the interest has been limited since no direct coupling between function and morphology could be established.

Recently, the first function–morphology correlations could be established through mass filling (Koontz and Schneider 1987), in conjunction with the staining of antennal lobe interneurons (Camazine and Hildebrand 1979). The axons of pheromone-specific receptor neurons projected to a large neuropil that is situated at the entrance of the antennal nerve into the antennal lobe of male moths. This neuropil, the male-specific macroglomerular complex (MGC), had already been described in 1924 by Bretschneider but was now "rediscovered." It was also established that non-pheromone-detecting receptor neurons sent their axons to the bulk of 40–70 sexually isomorphic, spheroidal neuropil glomeruli occupying the main part of the antennal lobe.

The antennal lobe consists of a large number of glomeruli. It has been the "holy grail" of many investigators to establish the function of these glomeruli. Why are they there? Is each glomerulus a relay station for information obtained from a specific type of receptor neuron? Or is the glomerular array a topographical map of the antenna? These questions could not be answered using the mass fill techniques. Other techniques, such as deoxyglucose staining (Teicher et al. 1980; Rodriguez 1988) and optical recordings using voltage-sensitive dyes (Cinelli and Kauer 1992; Lieke 1993), provided some evidence of functional properties of the glomeruli in other types of organisms, but in moths these methods never provided unambiguous information.

In *Drosophila,* Shanbhag and Singh (1992) showed an effect of cellular activation on the filling propensity of taste receptor neurons. By lacing a horseradish peroxidase solution with different solutions of taste stimulants, different staining patterns in the target areas of the axonal branches were characterized. By applying a similar technique to moth pheromone receptor neurons, it has been possible to stain single, physiologically identified neurons from the antenna extending into the antennal lobe. In this technique, a recording electrode filled with cobalt-lysine was slipped over the cut surface of a single olfactory sensillum, which subsequently was stimulated with the key stimulus for one of the receptor neurons present in the sensillum (Hansson et al. 1992).

In this chapter I will describe the general morphology of the antennal lobe, and particularly of the macroglomerular complex. I will also describe the method used to stain single, functionally defined receptor neurons and discuss the evolutionary significance of the results. Because the method was developed only 2 years ago, many of the studies referred to are still underway.

2. Methods

For study of the branching area and arborization pattern of antennal receptor neurons in the antennal lobe, it is important to establish the antennal lobe morphology (Fig. 17.2). This is done by impregnation of sections of the antennal lobe according to traditional staining methods. Methods often used are Bodian protargol staining, ethyl gallate staining, or Rowell's staining. Golgi preparations and mass fills of antennal neurons are other methods to gain insight into the gross antennal lobe morphology. For an overview of various neuroanatomical techniques, see Strausfeld and Miller (1980).

Once the antennal lobe morphology has been established, filling of single neurons can be initiated. I will describe here the method for moths of noctuid size or smaller. The method has also been used successfully on large moths and on locusts.

The moth is immobilized in a piece of plastic tubing or in a disposable pipette tip, depending on its size. The restraining device is modified to allow passage of the whole moth from one end, but only of the head through the other. An Ag/AgCl wire is inserted into the abdomen, to serve as a ground wire, and a moist ball of tissue paper is pressed in behind the animal to keep it and the ground wire in place. The head is then fixed in position with dental wax, which also covers the head capsule and fixes the antennae in the desired position. The whole

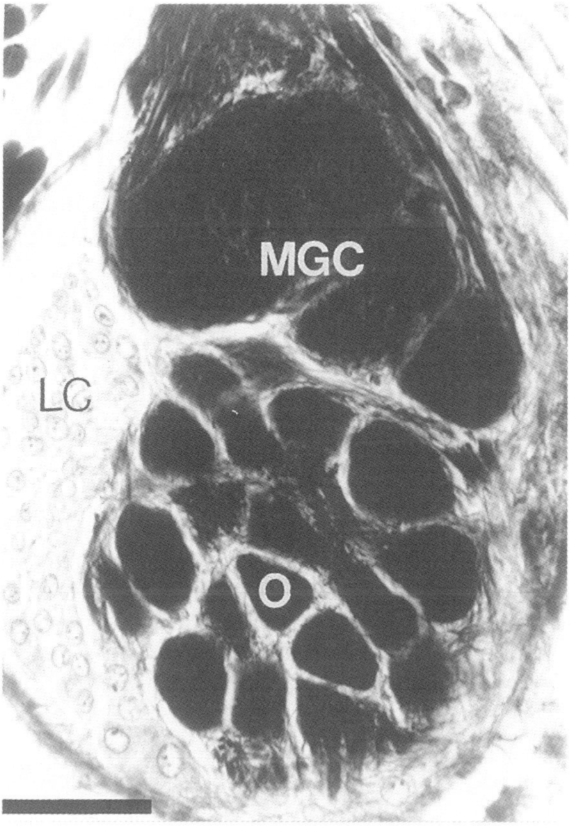

Figure 17.2. Frontal section through the antennal lobe of a male turnip moth, *Agrotis segetum.* Both ordinary glomeruli (O) and the prominent macroglomerular complex (MGC) can be observed. Note the glomeruli of the MGC. To the left, the lateral cell body cluster (LC) can be seen. Scale bar = 50 μm.

preparation is then transferred to a grounded micromanipulator, which is put in contact with the Ag/AgCl wire from the abdomen.

Now the recording and filling electrode can be prepared. A glass capillary electrode with an opening of approximately 4 μm is pulled on an electrode puller. The electrode is then filled with a cobalt lysine solution. This solution is made by adding 2.38 g of $CoCl_2$ and 5 g of l-lysine to 20 ml of distilled water. The solution must then be stirred for 24 hours, and finally set to pH 7.2–7.4 by addition of HCl. The produced solution has a deep purple color and is not optimal for staining. When the solution is left at room temperature, the complex between the metal ion and the amino acid is probably intensified, and the solution changes color to almost black. The optimal time to use the solution was found to be between 3 and 6 weeks after mixing. After 6 weeks a black precipitation can be observed in the bottle. When the electrode has been filled, it is connected, via a second Ag/AgCl wire, to a high-impedance AC/DC amplifier. The electrode is situated in a second micromanipulator.

To establish contact with the neurons in a sensillum, the tip of the sensillum is cut by means of microscopic glass knives; one of the knives is horizontal and placed beside the recording electrode, and the other one is vertical and beneath the antenna (Kaissling 1974; Van der Pers and Den Otter 1978). The electrode with the staining electrolyte is then put into contact with the cut surface, and with the receptor neurons in the sensillum. If the electrode is left in this position for a period up to several days, the neurons will absorb the marker and would most probably be stained. To achieve fast and selective staining of one of the neurons in the sensillum, the key stimulus for this neuron is pulsed into a charcoal-filtered and moistened air stream continuously flowing over the antenna at a speed of ~ 0.5 $msec^{-1}$. The stimulus should be applied at an amount giving a maximal response without adapting the cell. A suitable frequency for stimulation is 0.25–0.5 Hz. As cobalt is a toxic substance for the neuron, the firing of action potentials often ceases after 5–10 minutes. However, the receptor potential fluctuations do continue, and this is sufficient for staining. After 1 hour, the preparation is removed from the setup and transferred to a petri dish containing a moist tissue paper. The petri dish with the insect is then left at 8°C for 48 hours for postfilling.

After postfilling, the head capsule is opened carefully in front and back and is bathed in 1 ml of saline, 150 mM NaCl, 3 mM KCl, 2 mM $CaCl_2$, 9 mM TES (*N*-tris-{hydroxymethyl}-methyl-2-aminoethane-sulfonic acid), and 25 mM sucrose (pH 6.9), to which three drops of concentrated ammonium sulfide (38%) is added. The head is left in this solution for 10 minutes and is then rinsed in four changes of pure saline, each lasting 5 minutes. After the sulfide treatment, the head is put into an alcohol–formaldehyde–acetic acid (8:1:1) fixative for 24 hours. Subsequently, the brain is dissected out and is subjected to silver intensification (Timm 1958).

Finally, the preparation is dehydrated and cleared in methyl salicylate, so that stained neurons can be observed. If a neuron has been stained, the preparation is embedded in epoxy resin and sectioned for serial reconstruction.

3. The Antennal Lobe

The main part of the moth deutocerebrum is occupied by the antennal lobes. These are paired, conspicuous structures situated at the frontal–dorsal part of the brain. The antennal nerve, containing all the axons of the sensory neurons, reaches the lobe from the antenna. At the entrance into the lobe, the antennal nerve splits into two parts, one part bypassing the lobe laterally and one part entering the lobe and dividing into its constituent neural elements. The bypassing tract contains axons from mechano- and taste-sensitive neurons on the antenna, whereas all olfactory axons terminate within the antennal lobe.

The moth antennal lobe contains around 50 spheric ordinary glomeruli (Rospars 1988; Rospars and Hildebrand 1992) (Fig. 17.2). The glomeruli are glia-invested (Tolbert and Hildebrand 1981; Tolbert and Oland 1990) and consist of the axonal branches of receptor neurons and the branches of different types of antennal lobe interneurons (Koontz and Schneider 1987; Sun et al. 1993). Beside the glomeruli, the antennal lobe also houses the cell bodies of most of the interneurons arborizing in the lobe. The cell bodies are found in a lateral and a medial cell group. Some cell bodies are also found in a small anterior group (Homberg et al. 1988).

In moths utilizing long-distance sex pheromones, there is a strong sexual dimorphism in the antennal lobes (Bretschneider 1924). Just at the entrance of the antennal nerve into the antennal lobe, a greatly enlarged complex of glomeruli can be observed in the male moth, whereas no such structure is seen in the female. This male-specific MGC is the target for pheromone-detecting sensory neurons present on the male antenna (Fig. 17.2). The MGC is a true complex, because it consists of at least two and often more glomeruli.

The structure of the MGC has been described in several moth species, and a large interspecific variation in the morphology is characteristic, whereas the within species variation is minimal. The best investigated species is the sphinx moth, *M. sexta.* In males of this species, a large MGC glomerulus with a very inhomogeneous structure, resembling a cumulus cloud, is situated just at the entrance of the antennal nerve. It has therefore been named "the cumulus." Beneath the cumulus, a second MGC glomerulus is present. This forms a ring, like a donut or a toroid, on which the cumulus rests. These MGC glomeruli have been described in several different investigations (N.J. Strausfeld personal communication; Hansson et al. 1991) (Fig. 17.4b). The large, inhomogeneous glomerulus at the entrance of the antennal nerve has been found in all species investigated. To establish a consistent nomenclature, this glomerulus structure is therefore a good reference point. I suggest to name it "*a*", and that all satellite glomeruli around it are named alphabetically in an order starting with the glomerulus encountered first in a circular movement from the "*a*" glomerulus toward the midline of the brain, when the antennal lobe is viewed frontally (Fig. 17.3). This will give a comparable nomenclature in different species, whereas it does not exclude the use of more appealing names like cumulus or toroid at the same time. In my following discussion of MGC morphologies I will use this nomenclature and accordingly name the *M. sexta* cumulus *a* and the toroid *b.*

In a study of neuronal projections into the antennal lobe of three different moth species, *Bombyx mori, Lymantria dispar,* and *Antheraea polyphemus,* Koontz and Schneider (1987) reported on their MGC structure. The male *B. mori* displayed four different MGC glomeruli with the *a* glomerulus situated at the entrance of the antennal nerve as described above. The *b* and *c* glomeruli constituted ring-like structures at a similar location as the *b* glomerulus in *M. sexta.* Below the

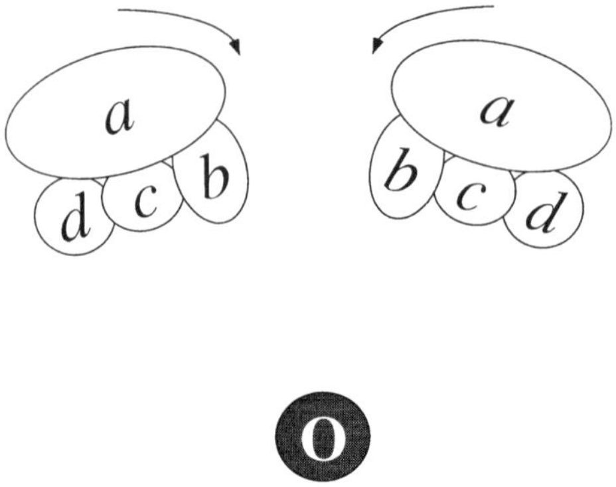

Figure 17.3. Nomenclature suggested for the macroglomerular complex glomeruli. The large glomerulus at the antennal nerve entrance is called *a,* and the satellite glomeruli are named alphabetically in a circular order, starting from the *a* glomerulus toward the midline of the brain in a frontal plane of section. O is the esophageal canal.

two ring structures, the *d* compartment was observed (Fig. 17.4a). In *L. dispar,* three MGC glomeruli were found, again with a typical *a* glomerulus and two smaller glomeruli situated below. In the *A. polyphemus* MGC, these authors state that three or possibly four glomeruli were observed. In a recent investigation of *A. polyphemus,* Williams and Hansson (in preparation) showed that the *a* glomerulus of this species is extremely complex and displays a wealth of invaginations. Below the *a* glomerulus, two other glomeruli are situated: the *b* glomerulus with a relatively homogeneous structure, and the *c* glomerulus with a very complicated and broken up morphology (Fig. 17.4c).

Among the noctuids, the MGC structure has now been established in several species. All of these have an *a* glomerulus at the nerve entrance. In the turnip moth, *Agrotis segetum,* three satellite glomeruli are surrounding the *a* glomerulus. A small *b* glomerulus is situated quite far down on the medial side of the antennal lobe, whereas the *c* and *d* glomeruli are somewhat larger and are situated medioventrally of the *a* glomerulus (Fig. 17.4d). To avoid confusion, it should be noted that Hansson et al. (1992) reverse the designations for the *a* and *b* glomeruli. In *Trichoplusia ni,* the most complex MGC, consisting of seven glomeruli, was found by Todd et al. (1992). In this case, the ever-present *a* glomerulus was surrounded by satellites on all sides except at the very entrance of the antennal nerve. The *b* glomerulus was found just beside the *a* part, on the mediodorsal side, and the *f* glomerulus was observed on the lateroventral side. Medioventrally of each of these three glomeruli, three additional glomeruli (*c, d,* and *e*) were present (Fig. 17.4e). In a third noctuid, *Spodoptera littoralis,* the MGC was found to be composed of three parts: one large *a* glomerulus and two satellites embracing it medially (*b*) and lateroventrally (*c*) (Ochieng et al. 1995)

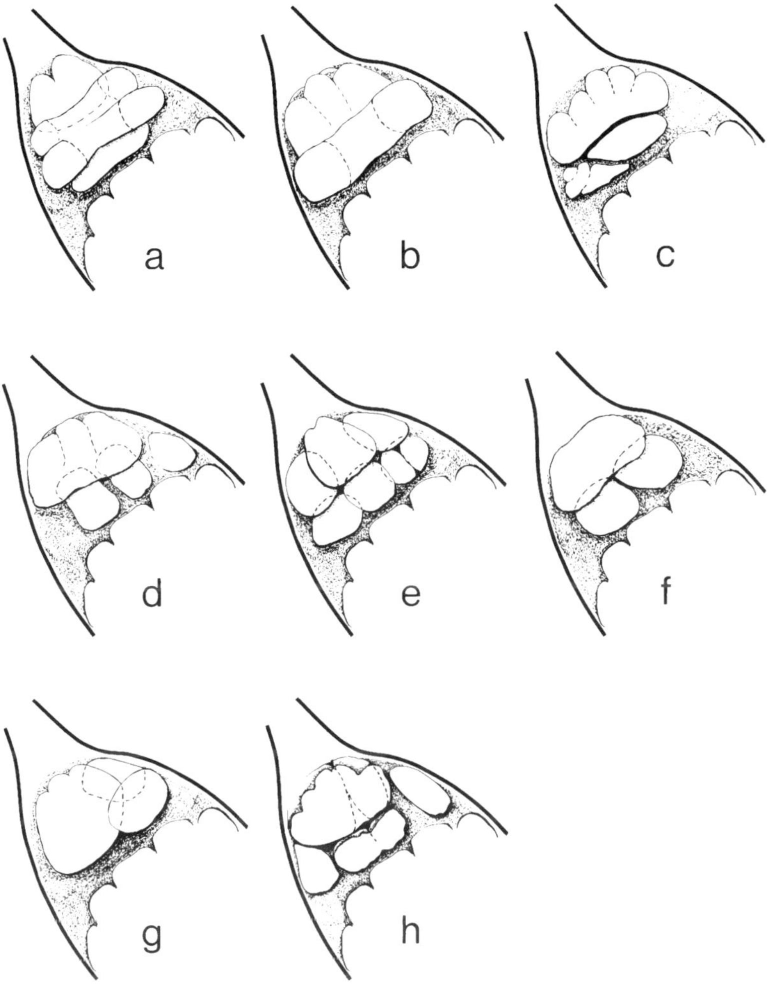

Figure 17.4. Graphic representation of the different macroglomerular complexes described. (**a**) *Bombyx mori,* (**b**) *Manduca sexta,* (**c**) *Antheraea polyphemus,* (**d**) *Agrotis segetum,* (**e**) *Trichoplusia ni,* (**f**) *Spodoptera littoralis,* (**g**) *Helicoverpa zea,* (**h**) *Heliothis virescens.*

(Fig. 17.4f). Among the heliothine noctuids, two species have been investigated: *Heliothis virescens* and *Helicoverpa zea.* In *H. zea,* the MGC consisted of three glomeruli: a large, ventrally situated *a* glomerulus with a mushroom-like appearance, along with two smaller, dorsally situated satellite glomeruli (Christensen et al. 1991) (Fig. 17.4g). In *H. virescens,* a MGC morphology very similar to that found in *A. segetum* was found. A large and inhomogeneous *a* glomerulus is surrounded at its medial, ventral, and lateral aspects by three smaller glomeruli. Of these, glomerulus *c* displayed invaginations (Hansson et al. 1995) (Fig. 17.4h).

All these examples of MGC morphology show that they are all variations on a common theme: one large central glomerulus surrounded by a number of smaller satellites. These satellites are in noctuids often spheric or ellipsoid, while in other families ring-formed structures have been observed.

4. Antennal Lobe Projections of Olfactory Receptor Neurons

Antennal olfactory receptor neurons have been stained in both male and female moths, even though the emphasis has clearly been put on male, pheromone-specific receptor neurons (Fig. 17.5). However, some general results, valid for both sexes, have emerged: (1) In both sexes it has been demonstrated that non-pheromone-specific neurons restrict their axonal branches to non-MGC, ordinary glomeruli. (2) Arborizations from the axon of a single receptor neuron are always restricted to a single glomerulus. (3) Pheromone-detecting receptor neurons project to a position just at the entrance of the antennal nerve into the antennal lobe.

Below, I will give an overview of receptor neuron projections in the moth species so far investigated. From here on, data will mainly stem from investigations of male moths—except in the case of *S. littoralis*, where both sexes are described.

4.1 Species Where Receptor Neurons Tuned to Different Pheromone Components Project to Separate MGC Compartments.

4.1.1. Agrotis segetum

The turnip moth, *A. segetum,* utilizes three major components in its sex pheromone: (Z)-5-decenyl acetate (Z5-10:OAc), Z7-12:0Ac, and Z9-14:OAc (Löfstedt et al. 1982). The male possesses receptor neurons tuned to these compounds on its antenna. Sensillum type I contains one neuron tuned to Z5-10:OAc and one neuron tuned to Z5-10:OH, a behavioral antagonist. Sensillum type II contains one neuron specifically responding to Z7-12:OAc and one neuron of unknown specificity. Sensillum type III contains one neuron tuned to Z9-14:OAc and one of unknown specificity. Sensillar types I, II, and III are present on the antenna at a 66:33:<<1 ratio, respectively (Löfstedt et al. 1982; Hansson et al. 1992).

Representatives of the different sensillum types were selected for cobalt staining. As mentioned before, the *A. segetum* MGC consists of four glomeruli. From sensillum type I, the Z5-10:OAc-specific neuron sends its axon exclusively to glomerulus *b*. Again the designations of *a* and *b* are reversed in the Hansson et al. (1992) article. The second neuron in sensillum type I, specific to Z5-10:OH, sends its axon to glomerulus *a*. The Z7-12:OAc-specific neuron in sensillum type

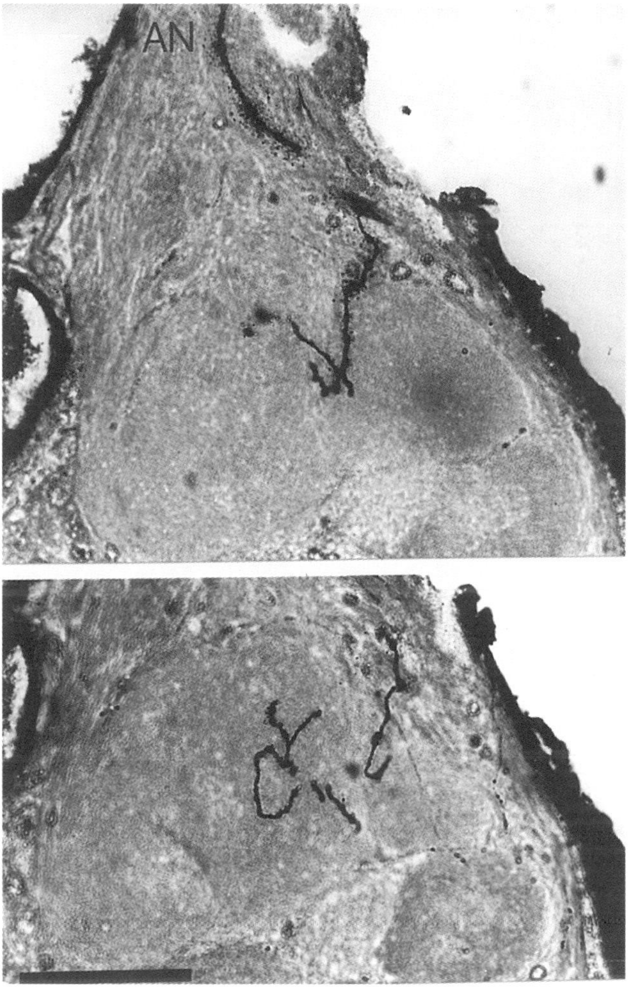

Figure 17.5. Example of a stained, pheromone-specific receptor neuron in the macroglomerular complex of a male moth. The neuron enters the antennal lobe through the antennal nerve (AN) and branches within the *a* glomerulus of the MGC. Scale bar = 50 μm.

II projects to glomerulus *c,* whereas the second cell in this sensillum projects to glomerulus *a.* The second neuron, of unknown specificity, was observed only in occasional double stainings. In 3000 contacts with different sensilla, only a single sensillum of type III was encountered. The filling attempt performed on this sensillum yielded a double staining, showing one neuron projecting to MGC glomerulus *d* and one projecting to glomerulus *a* (Hansson et al. 1992) (Fig. 17.6).

The selectiveness of the method could be tested in fills of sensillum type I,

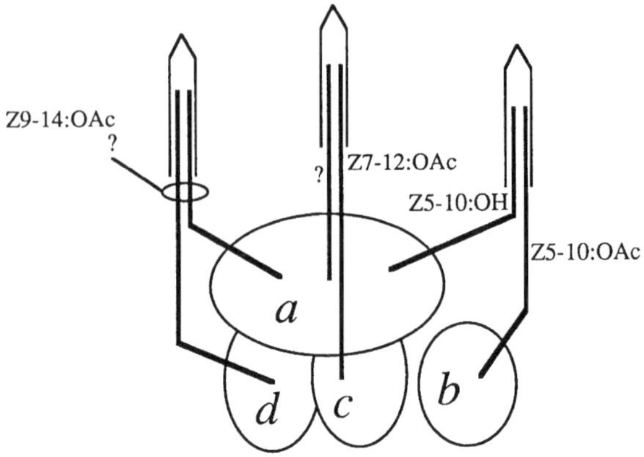

Figure 17.6. Schematic representation of the MGC branching patterns of the different types of pheromone-specific receptor neurons present on the male *Agrotis segetum* antenna. Three different sensillum types were found. A question mark indicates that the specificity of the receptor neuron was not known.

because both neurons are of known specificity. When the neuron is stimulated with Z5-10:OAc, five out of six single filled neurons project to glomerulus *b,* while eight out of nine single fills project to glomerulus *a* when the sensillum is challenged with Z5-10:OH. When tested statistically the selectivity is highly significant. The overall effect of stimulation on the probability of filling was tested in a separate experiment, where no nonstimulated neurons filled after being in contact with the cobalt solution for 1 hour.

The receptor neuron stainings in male *A. segetum* thus showed a separation of the input channels conveying information about two major pheromone components. The neuron detecting the behavioral antagonist projected to the largest MGC glomerulus, which also received input from two types of neurons of unknown specificity.

4.1.2. Spodoptera littoralis

The male cotton leafworm, *S. littoralis,* is attracted to a mixture of Z9,E11-14:OAc and Z9,E12-14:OAc in a 100:1 proportion (Campion et al. 1980; Kehat et al. 1976). These compounds are also produced in the female gland together with a host of other pheromone-like compounds. On the male antenna, two different types of sensilla are present. Long sensilla, present laterally on the antennal segments, contain two receptor neurons. One is tuned to the Z9,E12-14:OAc and one is tuned to a behavioral antagonist, Z9-14:OH. Shorter sensilla, situated more toward the midline of the antenna, contain one receptor neuron responding to Z9,E11-14:OAc (Ljungberg et al., 1993).

The receptor neurons in these different types of sensilla were subjected to cobalt lysine staining, to show the projection pattern of neurons tuned to the two pheromone components. The MGC contains three glomeruli. Neurons tuned to the major pheromone component send their axons exclusively to glomerulus *a*, whereas neurons tuned to the minor component projects only into glomerulus *c*. In all attempts to stain specifically by stimulation with Z9-14:OH, double stainings of two neurons, projecting to glomeruli *b* and *c*, respectively, were obtained. No clear reason for this phenomenon is at hand, but as the second neuron in this sensillum type had earlier unambiguously been shown to terminate in glomerulus *c*, the Z9-14:OH detecting neuron can only project to glomerulus *b* (Ochieng et al. 1995) (Fig. 17.7a).

The female cotton leafworm moth possesses a few long sensilla on each antennal segment. These sensilla contain two receptor neurons, of which one is tuned to the major component of the sex pheromone produced in her own pheromone gland. These neurons are just as sensitive as the same type of neurons found in the male (Ljungberg et al. 1993). The female has thus a very sensitive olfactory apparatus for autodetection of sex pheromone. After female, pheromone-detecting receptor neurons were stained, it became clear that they project into a glomerulus situated at a similar location as the male MGC. This glomerulus had,

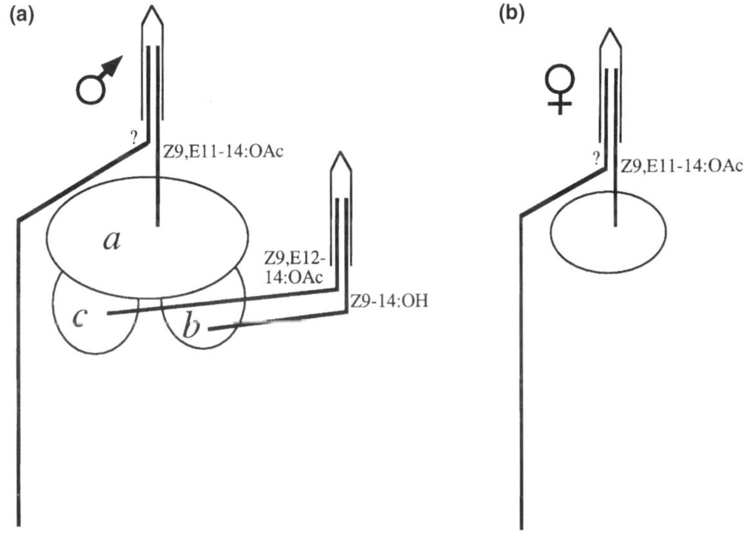

Figure 17.7. Schematic representation of the MGC branching patterns of the different types of pheromone-specific receptor neurons present on (**a**) the male and (**b**) the female *Spodoptera littoralis* antenna. Two different, pheromone-specific sensillum types were found in the male, and one in the female. In the sensillum containing a Z9,E11-14:OAc-detecting neuron, a second receptor neuron of unknown specificity was present both in the male and in the female. This neuron projected to a glomerulus situated deep and latero-ventrally in the lobe.

however, the appearance of an ordinary glomerulus—that is, a glomerulus not receiving input from pheromone-detecting receptor neurons. The second neuron in the same sensillum is of unknown specificity and could, in double stainings, be observed to project deep down on the lateral side of the antennal lobe (Ochieng et al. 1995) (Fig. 17.7b).

4.1.3. Manduca sexta

The sphinx moth, *M. sexta,* utilizes two major components in its sex pheromone: E10,Z12-16:Ald (bombykal) and E10, E12, Z14-16:Ald(EEZ) (Tumlinson et al. 1989). These two components are detected by two neurons in the same sensillum (Kaissling et al. 1989b). No selective-staining experiments have been performed on this moth, but because all sensilla contain two neurons responding to the major pheromone components, passive staining experiments have yielded valuable information. After staining of two neurons in such a sensillum, one neuron was seen to project to the *a* (cumulus) glomerulus and one to the *b* (toroid) glomerulus (T. A. Christensen personal communication) (Fig. 17.18). These results strongly suggest a separation of the neural elements carrying information of two different pheromone components. Recordings from antennal lobe projection interneurons corroborate the pattern found among the receptor neuron axonal arborizations, because projection interneurons specifically responding to bombykal restricted their dendritic arbors to the *b* glomerulus, whereas interneurons responding specifically to the second pheromone component sent dendritic arbours into the *a* glomerulus (Hansson et al. 1991).

4.2. Species Where Receptor Neurons Tuned to Different Pheromone Components Project to the Same MGC Compartment

4.2.1 Heliothis virescens

The male *H. virescens* orients to the female by a two-component sex pheromone consisting of Z11-16:Ald and Z9-14:Ald in a 16:1 ratio (Roelofs et al. 1974;

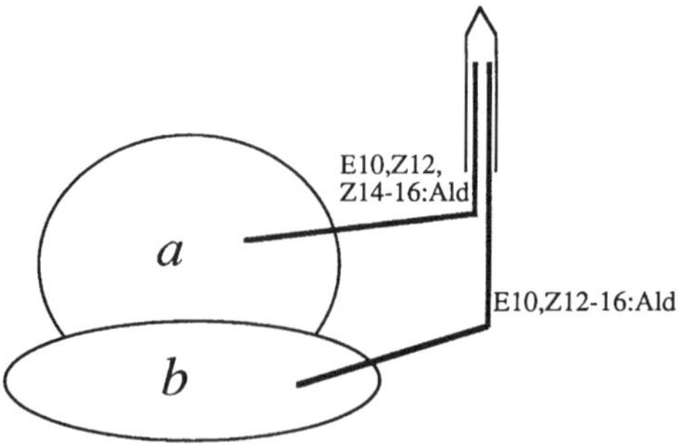

Figure 17.8. Schematic representation of the MGC branching patterns of the two different types of pheromone-specific receptor neurons present on the male *Manduca sexta* antenna as deduced from massfills. The physiological identity of the neurons could be reversed.

Tumlinson et al. 1975). Receptor neurons tuned to these compounds are present in different physiological sensillum types on the male antenna. In each of these two types, a silent receptor neuron is also present. Additionally, a third sensillum type contains a receptor neuron tuned specifically to a potential behavioral antagonist, Z11-16:OAc. Also in this sensillum a second neuron of unknown significance is found (Almaas and Mustaparta 1990; 1991; Berg et al. 1995).

The male-specific MGC in *H. virescens* consists of four glomeruli. Specific staining of the receptor neuron tuned to the major pheromone component showed that this neuron projects into the large *a* glomerulus. The silent neuron in the same sensillum terminates in the medial *b* glomerulus. Neurons tuned to Z9-14:Ald terminate in the large *a* glomerulus, as do the second neuron in the same sensillum as the Z9-14:Ald neuron. No spatial separation into different MGC glomeruli was thus observed between the receptor neurons detecting the different pheromone components. However, the neurons detecting the behavioral antagonist terminate in a separate compartment, *c*. On a single occasion, another type of sensillum containing a Z9-14:Ald-detecting cell was observed. In this sensillum, three types of receptor neurons were stained, projecting to glomerulus *a*, *b*, and *d*, respectively (Hansson et al. 1995) (Fig. 17.9).

4.2.2. *Antheraea polyphemus*

The most complicated projection pattern of pheromone-specific receptor neurons so far observed can be found in *A. polyphemus*. The female emits a two-component sex pheromone consisting of E6,Z11-16-16:OAc (Ac 1) and the corresponding aldehyde (Ald) (Kochansky et al., 1975a). Sensilla present on the

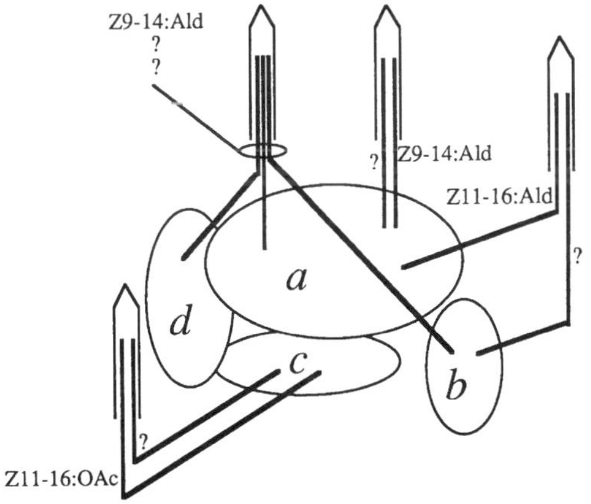

Figure 17.9. Schematic representation of the MGC branching patterns of the different types of pheromone-specific receptor neurons present on the male *Heliothis virescens* antenna. Four different sensillum types were found.

male antenna comprise receptor neurons tuned to these two and to a third substance, E4,Z9-14:OAc (Ac 2) (Meng et al., 1989), which is a pheromone component of the closely related species *Antheraea pernyi* (Bestmann et al., 1987a). The neurons are present in all possible combinations in the sensilla, but the Ac 2 detecting neurons are mostly found in a shorter sensillum type.

The key to establish the projection pattern of the different receptor neuron types in *A. polyphemus* turned out to be their arborization pattern in the MGC. The Ac 1 detecting neurons displayed a bifurcate pattern, whereas the neurons tuned to the Ald branched in a chandelier-like manner. These differential filling patterns were highly significant when tested statistically. The few Ac 2 neurons that were stained had a more stellate configuration of their axonal branches.

With these characteristics as a reference, it was possible to determine the targets of the different receptor neuron types within the MGC. The *A. polyphemus* MGC consists of three glomeruli. Arborizations of the Ac 1 type neurons were found in the *a* and in the *b* glomerulus. The Ald-type neurons were branching in the same two areas. Ac 2 neurons were found to send their axons to the *b* and to the *c* glomerulus. Glomerulus *a* did thus receive input from the two neuron types detecting the conspecific pheromone components, glomerulus *c* was targeted only by neurons detecting the interspecific signal, and glomerulus *b* contained arborizations from all three types of neurons (Williams and Hansson in preparation) (Fig. 17.10).

5. Discussion

5.1. Evolutionary Aspects of the Glomeruli, in Particular of the Macroglomerular Complex

If details of the MGC projections are studied, the investigated species fall into different groups depending on where receptor neurons tuned to different phero-

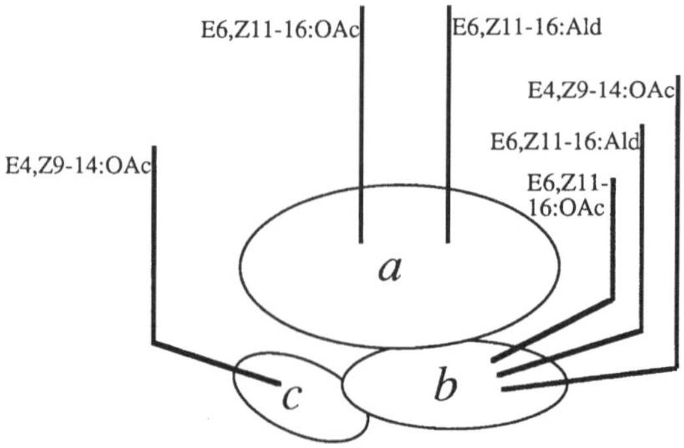

Figure 17.10. Schematic representation of the MGC branching patterns of the different types of pheromone-specific receptor neurons present on the male *Antheraea polyphemus* antenna. All combinations of neuron types within sensilla were observed.

mone components and behavioral antagonists project. As was mentioned earlier, the *a* part of the MGC is a conserved feature, which is present in all species so far studied. In the turnip moth, the receptor neurons detecting the behavioral antagonist were seen to project to the *a* compartment, whereas the major pheromone-component-specific neurons projected to satellite compartments. The *a* compartment also received input from two neuron types of unknown significance. In *M. sexta,* receptor neurons specifically responding to the component emitted from the female in the highest amount did not target the *a* compartment, whereas neurons tuned to the second pheromone component did. In the cotton leafworm, on the contrary, the neurons detecting the major pheromone component projected into the *a* compartment, whereas neurons detecting the minor component and a behavioral antagonist had axonal arborizations in the satellite compartments.

Even more differing MGC organizations are found in species where different physiological types of pheromone receptor neurons project axons to the same MGC compartment. This was the case in *H. virescens,* where both types of neuron projected to the *a* compartment, and in *A. polyphemus,* where the same pattern was observed. In male *A. polyphemus,* all types of pheromone detecting neurons also shared a common projection area in one of the satellite compartments. In both species, the interspecific signal was received in a separate MGC compartment. So, a common feature of all species studied is that receptor neurons detecting compounds acting as repellents always target a specific MGC glomerulus, which does not receive input from pheromone-detecting neurons.

MGC glomeruli receiving input from different receptor neuron types, particularly the *a* compartment, have always been found to be morphologically heterogeneous, often full of invaginations and folds, whereas the satellite glomeruli often display a more homogeneous structure. The heterogeneous structure is most probably a sign of a further compartmentilization within the single MGC glomerulus to accommodate different kinds of input. This raises the question whether there initially was only one MGC glomerulus, looking like an ordinary glomerulus, much like what has been documented in the female *S. littoralis.* This single MGC glomerulus might then have started to bud off additional compartments, giving rise to the structure we observe today. The heterogeneous structure of the *a* compartment could thus be a sign of a further subcompartmentalization, making the MGC morphologically more and more complex (Fig. 17.11a).

Conversely, the evolutionary development could also have gone in the opposite direction, where (a) a complex of individual glomeruli has formed what we now call the MGC and (b) the structure of the *a* compartment would thus be a remnant of this stage. In other words, the MGC is fusing over time (Fig. 17.11b). However, considering the structure in the female *S. littoralis,* where few pheromone detecting receptor neurons of only one type project into a single glomerulus of an "ordinary" size implies that this is probably also the structure that was present in the male before the sophisticated pheromone communication system and the elaborate structure of the male pheromone detecting apparatus had evolved. As

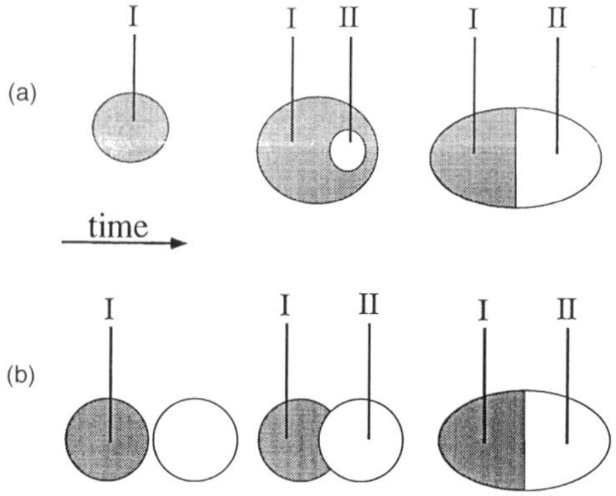

Figure 17.11. Two possible ways for the *a* glomerulus of the macroglomerular complex to reach its present, complex morphology over evolutionary time. (**a**) Through divergence from one original glomerulus of "ordinary" size, which expands when input neurons get more numerous, and which starts to diversify when neurons specific for additional pheromone components or behavioral antagonists occur. (**b**) Through fusion, where several ordinary glomeruli, receiving input from different receptor neuron types, fuse to form the MGC glomerulus.

more and more complex pheromone blends have developed, the male olfactory system—and thereby the MGC—has become more elaborate (Fig. 17.11a).

In a forum dealing with pheromone communication, the question "What relevance do the here studied features have to the efficient reproduction of these moth species?" must be asked. Why does the moth pheromone detecting system look as it does? The signals triggered by the presence of pheromone molecules in the ambient air are produced by the olfactory receptor neurons present on the antenna. These signals must be transferred to the brain. The lines performing this task are the axons, emanating from the cell bodies of the receptor neurons and traveling the length of the antennal nerve into the antennal lobe. The signal has then reached the central nervous system. In the antennal lobe, the signals are integrated and transferred to higher-order neurons. If the antennal lobe was a mere relay station, where the signal was transferred from receptor neuron onto neurons leading to higher brain levels, neurons with exactly the same specificity as the receptor neurons impinging on them, a glomerular organization would work very well. Receptor neurons of the same specificity could project to one glomerulus, where the second-level projection neurons could pick up the signal and lead it on, a pure labeled-line system. The real world is, of course, not that simple. In the antennal lobe, across-fiber patterning occurs parallel to labeled lines. Investigations of antennal lobe interneurons have shown that most receptor neurons contact wide-field, local interneurons (Christensen et al. 1993). The local interneurons then form synapses with each other and with the projection neurons.

In this way, inputs from many different types of receptor neurons can interact and mold the activity patterns of the projection neurons.

Despite the receptor neuron–local interneuron link, the axonal branches of receptor neurons and the dendritic branches of projection neurons, displaying the same physiological specificity, meet in the same MGC compartment in the sphinx moth, *M. sexta* (Hansson et al. 1991). In other species, however, no such pattern is observed. In the turnip moth, the inputs into the MGC glomeruli do not correspond at all with the specificity of the projection neurons entering the same compartments (Hansson et al. 1994).

Are the glomeruli remnants from ancient times when the antennal lobe was just a relay station, and were the glomeruli then formed as a *rendez-vous* site for receptor neurons and projection neurons with similar specificities? The stage was then set for a diversification of the lobe, where glomerular identities were broken up by local interneurons that could transfer information within the lobe. Projection neurons were no longer tied to the identity provided to the glomerulus by the receptor neurons aborizing within it. Again, the above outlined scenario is just one of several possible. However, it is now feasible to pose evolutionary questions regarding the structures and functions we observe in the moth olfactory system.

5.2. Studies of Receptor Neuron Projections

The staining of single, functionally defined olfactory receptor neurons in moths has proved to be very rewarding in all the case studies mentioned above. It has been possible to establish the projection patterns of the receptor neurons in several species. These patterns, however, are not as easily interpreted as the first study of the male turnip moth might have suggested. In this species, a clear separation between the channels carrying information about different pheromone components was observed, implying a clear functional significance to the MGC compartments. In *M. sexta* and *S. littoralis,* a similar organization of MGC input lines could be observed.

In *M. sexta,* olfactory neurons in the labial palp pit were stained with cobalt. In other moth species, these receptor neurons have been shown to detect minute changes in CO_2 concentration (Stange 1992). All the neurons in this accessory olfactory organ projected to one glomerulus, situated lateroventrally in the antennal lobe. This can be interpreted as another example where a certain type of receptor neuron targets a specific glomerulus. However, these neurons differ from the antennal neurons by having bilateral projection patterns (Kent et al. 1986).

In other organisms, the problem of receptor neuron projections and glomerular function has been studied using different techniques. In the fruit fly, *Drosophila melanogaster,* different patterns of activation among the antennal lobe glomeruli could be observed by utilizing deoxyglucose staining (Rodriguez 1988). Filling of antennal olfactory neurons by cobalt has also been performed in *Drosophila.* Filling of single, identified neurons was not possible, but specific types of sensilla, occupying distinct areas of the antennal surface, could be put in contact with the cobalt solution. In these staining attempts, different filling patterns in the antennal lobe could be observed when sensilla of different types were stained (Stocker et al. 1983).

In the cockroach *Periplaneta americana,* sensilla at different locations on the antenna were stained with DiI. The projections showed a topographical projection pattern into the MGC, a phenomenon never observed among moths (Hösl 1990).

In the rat olfactory bulb, a glomerular complex similar to the MGC in moths is present. This complex receives information about odors involved in nipple search by the young rat (Teicher et al. 1980). Here the staining of olfactory receptor neurons has been used in a reversed manner, compared to the studies in moths. A marker was injected into the area of the antennal lobe to be studied. This marker was allowed to travel up the length of the axons of olfactory receptor neurons, and in this way the location of receptor neurons projecting to the area in the olfactory bulb could be established (Jastreboff et al. 1984; Astic et al. 1987; Pedersen et al. 1986). Receptor neurons within the olfactory epithelium were stained in specific patterns depending on where the marker had been injected.

In the lobster, staining of individual, olfactory receptor neurons has revealed their projection patterns, but the specificity of the stained neurons was not established. Receptor neurons in the lobster were observed to send branches into several antennal lobe glomeruli (Schmidt and Ache 1992), a feature never seen in moths. A similar pattern of multiglomerular targeting by single receptor neurons has recently been observed also in the desert locust, *Schistocerca gregaria* (B. S. Hansson personal communication).

6. Concluding Remarks

Receptor neurons specifically tuned to compounds involved in attraction of males to conspecific females, or to compounds competing in the chemical communication channel and acting as behavioral antagonists, restrict their axonal arborizations to the MGC. Several patterns of receptor neuron projections have been observed in different species. In some species, a specific type of pheromone detecting receptor neuron targets one MGC glomerulus, which does not receive input from any other receptor neuron types. In other species, one MGC glomerulus can contain axonal branches from several different types of pheromone specific receptor neurons. MGC glomeruli invaded by more than one receptor neuron type display a more heterogeneous structure than glomeruli targeted only by one type. It seems that innervation by different types of receptor neurons triggers a differentiation of the glomerulus innervated, which, seen in proper evolutionary perspective, might lead to formation of new MGC glomeruli.

To obtain more evidence supporting the evolutionary hypotheses, carefully selected species should be made targets of future investigations. Primitive Lepidoptera could be such a group. How does the MGC look, and how do the antennal receptor neurons project among, for example, the eriocraniids? Another interesting group is formed by the caddis flies, in which a pheromone communication system has been shown to be present (Löfstedt et al., 1994b). First, does this primitive sister group of the Lepidoptera have an MGC or a corresponding structure? Second, how do the projection patterns of pheromone-detecting receptor neurons on the trichopteran antenna compare to those found in the Lepidoptera? Studies like these will provide vital data on the evolution of the antennal lobe and of the MGC.

The study of projection patterns of olfactory receptor neurons should also be diversified to include non-pheromone-specific neurons. Host odor-detecting neurons in moths are highly specialized in detecting specific chemicals emitted from the host plant (Todd and Baker 1993; Anderson et al. 1995). Is it possible to find correlations between receptor neuron function and antennal lobe projection patterns also among plant odor-specific neurons? Such investigations will require a much more elaborate mapping of the antennal lobe than has been needed in the MGC investigations, but techniques to perform such mapping are now available (Rospars 1988; Rospars and Hildebrand 1992). Finally, efforts should be made to identify the mechanisms of fast and selective staining of physiologically activated neurons.

Acknowledgment

I wish to express my thanks to all the colleagues involved in the different projects referred to in this chapter: Drs. T.J. Almaas, P. Anderson, S. Anton, T.C Baker, E. Hallberg, C. Löfstedt, J.L. Todd, and L. Williams and Mr. H. Ljungberg and Mr. S.A. Ochieng. I would also like to express my gratitude to colleagues at the Arizona Research Laboratories, Division of Neurobiology, University of Arizona, and especially to Drs. T.A. Christensen and N.J. Strausfeld, for very valuable discussions during the development of the staining method. Ms. L. Hansén, Mr. E. Jirle, and Ms. R. Wallén provided expert technical assistance during the different projects, and Dr. S. Anton and Mr. R. Tramontano commented on early versions of the manuscript.

18

Anatomical and Physiological Diversity in the Central Processing of Sex-Pheromonal Information in Different Moth Species

Thomas A. Christensen

1. Introduction

In all animals with a well-developed olfactory system, the first stages of information processing in the brain occur within an array of specialized units of synaptic neuropil—the olfactory glomeruli. This basic plan of organization is highly conserved across widely divergent species, which suggests that glomeruli play a fundamental role in the neural processing of olfactory information in the brain (Ache 1991; Hildebrand 1995). In insects, the glomeruli are situated within the paired antennal lobes (ALs), the first-order olfactory centers in the deutocerebrum (for reviews see Homberg et al. 1989; Masson and Mustaparta 1990; Boeckh and Tolbert 1993; Stocker 1994).

Sex pheromones have become important "tools" in the exploration of olfactory information processing and discrimination in the insect brain. The chemical identity and behavioral relevance of these essential signals have been verified in many species, and synthesized pheromones can be controlled in a systematic fashion and used in a number of different experimental situations. Moreover, like most environmental odors, a sex pheromone is typically not a single odorant, but a blend of several chemicals in a particular ratio (see Christensen et al. 1989b and references therein). Pheromones are therefore particularly well-suited for studies of how the brain discriminates individual odorants and integrates information about odor blends. Work from this laboratory has focused on characterizing neural circuits in the brain of the male moth that process pheromonal information from females (Fig. 18.1). Over the past few years, some basic principles of neural organization have emerged from this and other work that are helping us understand how information about environmental odors like sex pheromones are processed in olfactory glomeruli. The aim of this chapter is twofold: (1) to summarize the neural mechanisms operating at the first level of sensory processing in the AL

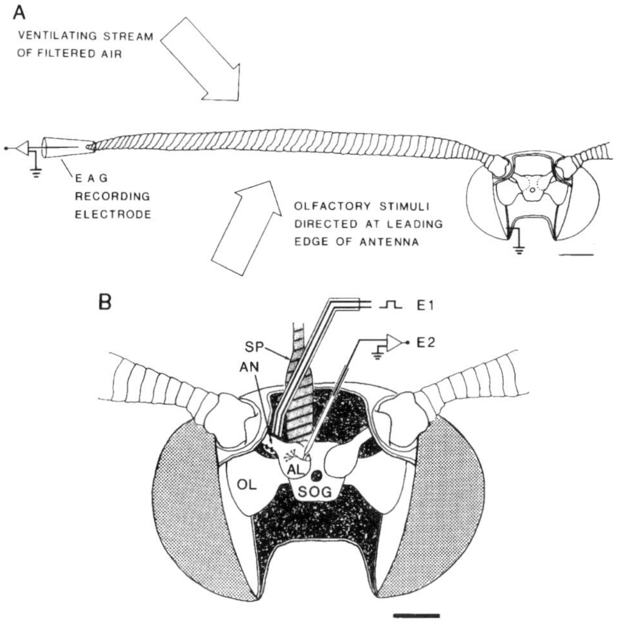

Figure 18.1. (**A**) Diagram illustrating the experimental setup for monitoring the electrophysiological activity of olfactory neurons in the moth brain. The electroantennogram (EAG) is recorded simultaneously to monitor responses of antennal sensory neurons to mechanosensory and olfactory stimulation. Scale bar = 2 mm. (**B**) Detail of the dissected head showing the positions of the antennal lobes (ALs), antennal nerves (ANs), optic lobes (OLs), and suboesophageal ganglion (SOG) in the brain. Antennal sensory-afferent axons are stimulated by presenting odors to the antenna, or electrically via silver-wire hook electrodes (EI) on the AN. Physiological responses of AL interneurons are recorded with intracellular microelectrodes (E2). SP = support platform; scale marker = 1 mm. [After Christensen and Hildebrand (1987a) with permission from Springer-Verlag.]

and (2) to develop a framework for understanding how these circuits ultimately contribute to the regulation of mate-seeking behaviors (see other chapters).

2. Organizing Principles of Pheromone-Information Processing

2.1. Glomeruli are Heterogeneous Structures

A very basic but important principle to emerge from the study of pheromone-information-processing centers in insects is that not all glomeruli in the AL are alike. Several lines of anatomical and physiological evidence have shown, for example, that only a subset of glomeruli in the ALs of certain male insects are responsible for processing information about female sex pheromones (Christensen and Hildebrand 1987b; Koontz and Schneider 1987; Rospars 1988; Boeckh et al. 1990; Hansson et al. 1991, 1992). These glomeruli are anatomically and physiologically distinct from those that process information about host odors (Boeckh et al. 1990; Anton and Hansson 1994, 1995; Christensen et al. 1991a, 1995). The macroglomerulus of cockroaches and the macroglomerular complex (MGC) of moths are examples of readily identifiable glomerular structures situated near the entrance of the antennal nerve as it enters the AL (Boeckh and Tolbert 1993; Fig. 18.2). These structures are also sexually dimorphic; in many species examined to date, the MGC is restricted to the male AL, but this is not necessarily a universal phenomenon (Anton and Hansson 1994).

Another basic principle to emerge from studies of pheromone information processing circuits is that the morphology of the MGC varies greatly from species to species, and that the glomeruli in each case are specialized for analyzing the particular constituents that comprise the pheromone blend of that species (Fig. 18.2). The MGCs of the silk moth *Bombyx mori* and the sphinx moth *Manduca sexta,* for example, are similar in that they both consist of a large central glomerulus and one or more ring-shaped glomeruli (Koontz and Schneider 1987; Hansson et al. 1991). In other species, like the corn earworm moth *Helicoverpa zea* and the tobacco budworm moth *Heliothis virescens,* the MGC appears as a simple cluster of three or four enlarged glomeruli (Christensen et al. 1991a, 1995; Mustaparta this volume; Hansson this volume). These distinct morphological differences between MGCs of different species might also reflect functional

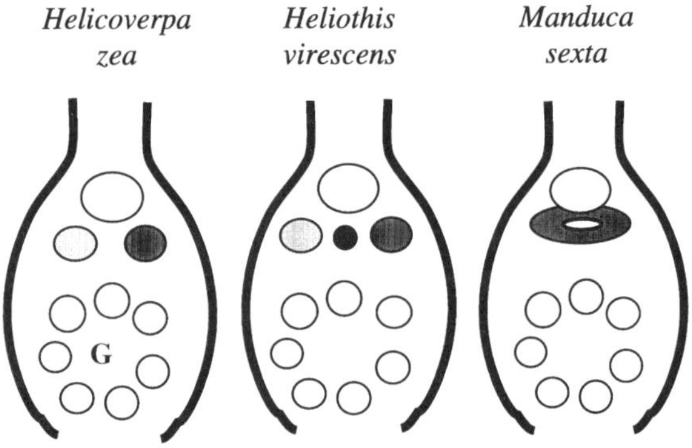

Figure 18.2. Schematic diagrams illustrating the anatomical diversity in the male antennal lobes of three lepidopteran species. In each drawing, the major subdivisions of the macroglomerular complex (MGC) are shown (shaded structures). Glomeruli that process nonpheromonal information (G) are anatomically and physiologically distinct from the MGC and appear as open circles. The MGCs of the two heliothine species (left and center) consist of an organized array of three or four glomeruli (spatial arrangements only approximate), whereas in the MGC of the sphingid moth *M. sexta,* one major subdivision (called the toroid) is very clearly a ring-shaped structure (Hansson et al. 1991). All antennal lobes are drawn to the same size for comparison.

differences in the organization of the MGC circuits and the mechanisms underlying pheromone information processing. One of our goals is to learn more about the functional roles of neural circuits in different MGCs and to relate these functions to the particular mate-seeking behaviors exhibited by different species.

2.2. Different Species Use Different Behavioral Strategies to Find a Mate

Two mechanisms (which are not mutually exclusive) have been proposed to help explain how a species-specific blend of pheromones might influence the sequence of behaviors that eventually leads a male moth to a "calling" female (Linn et al. 1986a,b; Teal et al. 1986). The "component" hypothesis assigns to the most abundant component of the blend the role in triggering long-distance attraction, whereas the "blend" hypothesis states that the complete blend is necessary for all phases of the behavior, at any distance from the source. At first glance, these may seem to be conflicting hypotheses, but if one scans the literature on pheromone-mediated behaviors in Lepidoptera, it is apparent that different species employ different strategies to locate their mates. To put it another way, they have different behavioral requirements for a pheromone blend. In some species, such as the oriental fruit moth *Grapholita molesta* and the redbanded leafroller moth *Argyrotaenia velutinana,* only a very small percentage of males will take flight in response to the major pheromone component presented alone at low dosages (Linn et al. 1986a,b). In other species, such as the cabbage looper *Trichoplusia ni* (Linn et al. 1986b; Mayer and McLaughlin 1992), the cotton bollworm *Helicoverpa armigera* (Kehat and Dunkelblum 1990), *H. zea,* and *Heliothis virescens* (Vetter and Baker 1984; Vickers et al. 1991), many more males will take flight, and some will even contact the source and attempt copulation if exposed only to the major component. In the corn earworm and tobacco budworm moths, these behaviors can be elicited by a low-dosage source with a stimulus release rate of less than 1 $ng\text{-}min^{-1}$ (Vetter and Baker 1984; Vickers et al. 1991). It should be pointed out that the natural pheromone blend evokes maximal levels of behavior in these species, but a minute amount of the major component is nevertheless sufficient to evoke flight and casting behaviors in a highly significant percentage (~ 50%) of males (Vickers et al. 1991). Clues to understanding these behavioral differences in sensitivity to the major pheromone component could lie within the neural circuits in the MGC that perform the first stages of processing in the insect brain. We have chosen to look at the level of the output pathways from the MGC because a major function of these particular neurons is to provide a route for essential pheromonal information from the MGC to higher-order centers in the protocerebrum (Homberg et al. 1989).

2.3. Output from the MGC Is Through Heterogeneous Pathways

Electrophysiological and anatomical studies have revealed a striking feature which is central to the organization of the MGC and how it processes pheromonal information in several species (summarized in Fig. 18.3). These studies show that the population of projection (output) neurons that first processes pheromonal information and then transmits this information to higher centers (see Kanzaki this volume) is not homogeneous, but anatomically and physiologically diverse (Homberg et al. 1989; Christensen et al. 1989b, 1991a, 1995; Christensen and Hildebrand 1987a; Hansson et al., 1991; Anton and Hansson 1994, 1995). This diversity furthermore reflects the chemical composition of the species-specific pheromone blend, which in each of the species we have studied consists of two principal components that are essential for optimal behavior. In both the sphinx moth *M. sexta* and in the corn earworm moth *H. zea,* information about the chemical components of the pheromone blend is transmitted through at least four functionally distinct output pathways from the MGC (Fig. 18.3). In summary, one pathway responds best to antennal stimulation with the major (or most abundant) component of the species-specific blend (for simplicity, component "A"). These output neurons are referred to as the "A" neurons in Fig. 18.3. Another set of neurons (the "B" neurons) responds best to the second essential component of the blend (but these cells can serve other functions as well (Christensen et al. 1991a; Christensen and Hildebrand 1994). The other two output channels appear to have more complex properties, and synaptic integration helps to shape their firing patterns. In one group of cells, responses to either component are qualitatively and quantitatively similar. These cells cannot, therefore, distinguish

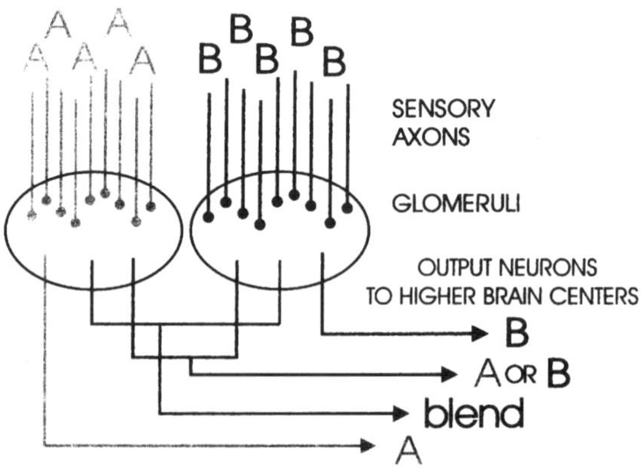

Figure 18.3. Summary diagram illustrating some of the key anatomical and physiological characteristics of the macroglomerular complex (MGC) observed in the male antennal lobes of several species. In this simplified diagram, the pheromone blend consists of two principal components, A and B. The first important feature is that the glomerular subdivisions of the MGC are functionally specified: Axons from receptor neurons that are selectively responsive to component A of the pheromone blend converge on one MGC glomerulus, and those specified for component B project to a different glomerulus. Secondly, output from the MGC is not homogeneous: At least four functionally different pathways transmit pheromonal information to higher centers in the brain (see text).

one component from the other and are thus referred to as the "A or B" neurons in Fig. 18.3. Another population of cells responds in a unique fashion to the blend of the two components (Christensen and Hildebrand 1987b, 1988; Christensen et al. 1989b, 1991a, 1995). These "blend" neurons appear to be among the most physiologically complex cell types in the MGC. Their responses may reflect the integration of both excitatory and inhibitory synaptic input from other interneurons in the MGC (as in *M. sexta:* Christensen et al. 1993), or they may result from central mechanisms that produce a synergistic response to the mixture (observed thus far in several noctuid species: Christensen et al. 1989b, 1991a, 1995; Anton and Hansson 1994, 1995).

In our detailed exploration of these four information-processing pathways in the ALs of male *M. sexta,* we found that the majority of the MGC-output neurons thus far encountered are "blend" neurons (Fig. 18.4) This result correlates well with the behavioral observations that *M. sexta* males, like some of the other species discussed earlier, will seldom take flight unless presented with the species-specific blend (Tumlinson et al. 1989). While fewer output neurons have thus

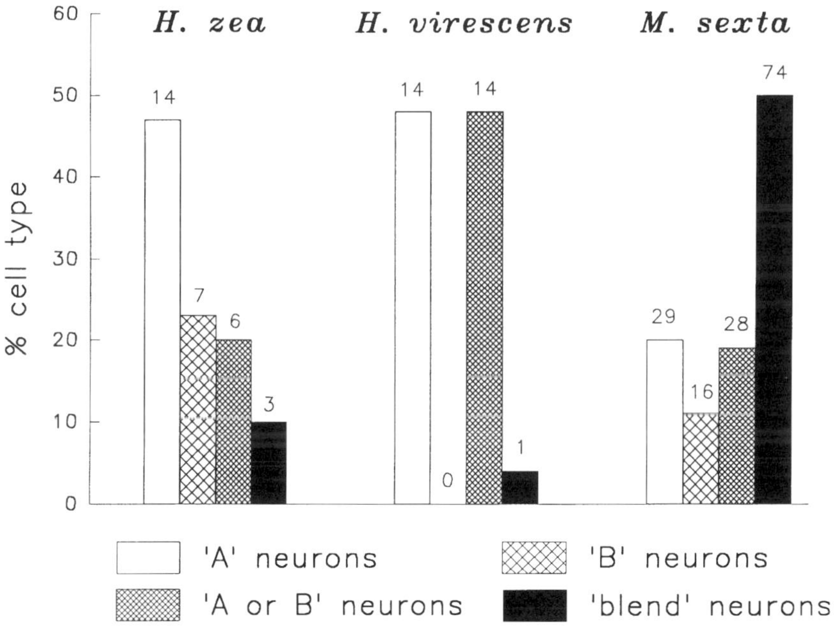

Figure 18.4. Frequency distributions of different types of output neurons thus far encountered in the MGCs of three lepidopteran species (numbers subject to change as more neurons are characterized). Note that most of the output neurons characterized in *M. sexta* are "blend" neurons, whereas most of the cells in the two heliothine species are either "A" neurons or the more generalist "A or B" neurons. The number of each cell type is shown at the peak of each bar.

far been characterized in *H. zea* and *H. virescens,* a different pattern appears to be emerging in the MGCs of these two noctuid species. Most of the neurons thus far encountered are not "blend" neurons, but neurons that show a strong excitatory response to the major component "A" (Fig. 18.4). Specifically, these include two neuronal populations: the "A" neurons and the "A or B" neurons mentioned previously. Both types of neurons respond to component "A" and to the species-specific blend in a qualitatively and quantitatively similar fashion. Thus an individual neuron cannot convey specific information about a blend. Similar findings were described more than 15 years ago in the saturniid moths *Antheraea pernyi* and *A. polyphemus* by Boeckh and Boeckh (1979). AL neurons in *A. pernyi* were most sensitive to the major aldehyde released by conspecific females, whereas cells in *A. polyphemus* were selectively stimulated by the major acetate of that species.

2.4. MGC-Output Pathways Do Not Have the Same Response Threshold

Another important physiological parameter that must be taken into account in behavioral assays, and that is now being examined very carefully in central processing circuits, is threshold sensitivity. Again, based upon the behavioral observations that response thresholds are often lowest for blends, we expected to find an enhanced sensitivity in the MGC neurons that respond in a unique fashion to the blend. To our surprise, we found that the most sensitive neurons in the two heliothine species we've studied thus far are not the "blend" neurons, but neurons that respond best to component "A." We now routinely find "A" neurons in the MGCs of heliothine males that are strongly depolarized and fire a burst of action potentials in response to brief pulses of only 0.01 ng of this single component of the pheromone blend. In the small sample of "blend" neurons encountered thus far, threshold levels are, on average, 2–3 orders of magnitude higher (Christensen et al. 1989b, 1991a, 1995). Based upon these initial observations, we believe that the "A" neurons in these species may participate in a neural pathway that initiates the observed behavioral "arousal" evoked by the sole presence of the major component of the blend. As mentioned previously, component-specific MGC-output neurons have also been studied in *M. sexta* (Fig. 18.5; Christensen and Hildebrand 1987b; Christensen et al. 1989b; Hansson et al. 1991), and analogous cells have been found in the cockroach *Periplaneta americana* (Boeckh et al. 1990; Boeckh and Tolbert 1993). There is also evidence that component-specific pathways exist in higher levels in the central nervous system. Some neurons that descend from the brain through the ventral nerve cord in several moths and in other invertebrate species exhibit component-specific olfactory responses (Olberg 1983a; Olberg and Willis 1990; Derby et al. 1984; Kanzaki et al. 1991b). All of this evidence suggests that information about single components is preserved to some degree even at the level of higher-order pathways in the central nervous system. With the development of new techniques that can

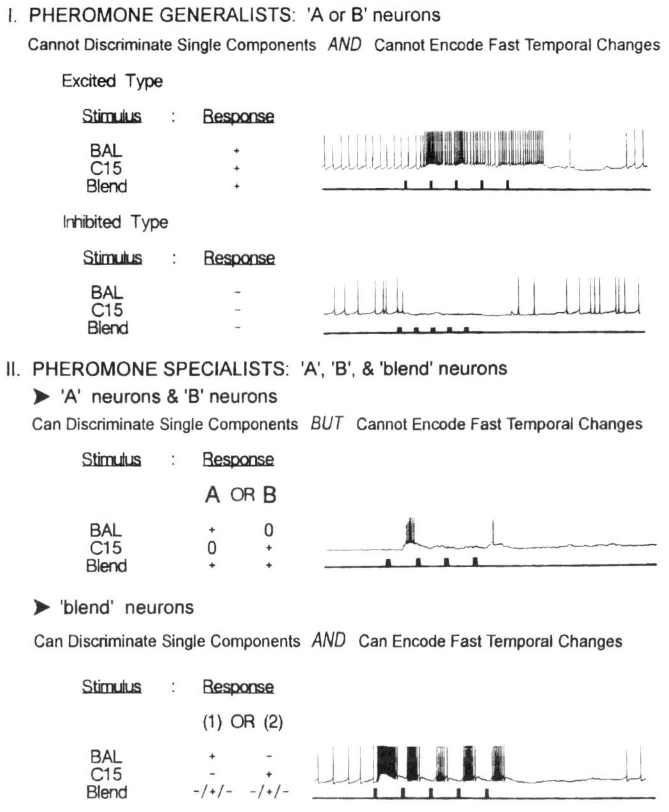

Figure 18.5. Summary of the electrophysiological response types recorded from various MGC-projection neurons in male *M. sexta.* Responses are listed as excitatory (+), inhibitory (–) or "mixed" (–/+/– = a combination of + and – synaptic input). The stimulus in each record on the right (lower trace) was one-half female-equivalent of *M. sexta* pheromone gland extract. The pheromone generalist ("A or B") neurons respond similarly to the complete natural blend and to the individual components. We use mainly two synthetic compounds to characterize these responses: bombykal (BAL), the principal component of the *M. sexta* pheromone blend, and (*E,Z*)-11,13-pentadecadienal (C15), a potent and stable mimic of a second blend component. Pheromone generalists thus cannot discriminate one component from the other. The pheromone specialists, on the other hand, discriminate the components in one of two ways: Some projection neurons receive input from only one class of antennal receptor neuron (either "A" or "B"-selective) and therefore cannot respond to the other component; the second type responds differentially to A and B and responds in a unique fashion to the blend of A + B. Some neurons are excited by BAL and inhibited by C15 (response type 1), whereas in others the polarity of responses is reversed (response type 2). Of all the types of MGC projection neurons characterized thus far, only the "blend" neurons can accurately encode the most rapid changes in the temporal pattern of the pheromonal stimulus. [After Christensen et al. (1989), with permission from Oxford University Press.]

monitor the electrophysiological activity of individual muscles in free flight (see Willis and Arbas this volume), we now have the means of assessing the threshold of activation for single motor units that are involved in flight (or even pre-flight) behaviors. It will be interesting to know for different species if any of these motor units can be activated by single components or if they are dependent upon antennal stimulation with the species-specific blend. Such an analysis is a necessary correlate to studies of behavioral thresholds to blends, because behavioral observations alone cannot reveal the neuromuscular events underlying flight or pre-flight movements. These electrophysiological recordings can now provide a high-resolution measure of neuromuscular activity evoked by experimentally controlled pheromonal and nonpheromonal stimuli.

The precise functional roles of AL neurons in controlling or modulating flight behaviors are presently unknown. Our initial results in two noctuid (heliothine) moths, however, point to a possible role for the abundant and sensitive "A" neurons in mediating the initial behavioral arousal observed in response to a calling female. At some distance far downwind of the female, a male will encounter pheromone concentrations that are just at the level of detection (Murlis et al. 1989; Murlis this volume). At this concentration, the most sensitive neurons in the MGC, the "A" neurons, are below the activation threshold for all components but the major component of the blend. Thus, at the edge of the "active space" for these heliothine males, activity evoked in the "A" neurons cannot signal any information about the pheromone blend. Because these neurons can respond to other components at other concentrations, however, it is certainly possible that blend information could be encoded *across* the complete ensemble of these sensitive output neurons at higher concentrations, thus creating a specific pattern of activity in the protocerebrum that signals the presence of the blend (Christensen et al. 1995). These output neurons could therefore serve more than one physiological function depending upon the ambient pheromone concentration. In neural systems that consist of relatively few neurons (like these AL output pathways; Homberg et al. 1989), such mechanisms would add considerable diversity to the system and would broaden the information-processing capabilities of these circuits.

3. Summary

We now see emerging a general pattern of organization in pheromone-processing circuits that appears to be conserved from species to species. If we examine these circuits more closely in several species, however, we find differences in the numbers and sensitivities of the different neuronal types, and these characteristics may help explain species-typical differences in behavior as well. Increasing evidence from single-cell analysis in the ALs of several moth species shows that the neural circuits in the MGC can accommodate some aspects of both the "blend" and "component" hypotheses of behavioral activation (see summary of *M. sexta* data in Fig. 18.5). Many of the output neurons from the MGC encode information about individual components of the pheromone blend, whereas the complete blend is encoded in other outputs. Likewise, some MGC outputs encode the natural temporal fluctuations that occur in a pheromone plume, whereas others cannot. These different but parallel processing pathways lend a great deal of flexibility to the analysis of the olfactory environment at early stages of processing and ensure that the fundamental features of the pheromone blend will be preserved for later stages of processing. In some respects, the output neurons in the MGCs of these moths are like the parallel processing pathways in early stages of the visual system, each carrying a basic feature of the stimulus to higher centers in the brain. Some output neurons carry information about a single component of a sex-pheromone blend, just as some retinal ganglion cells respond optimally to a single wavelength of light. Other output neurons from the MGC serve to enhance the signal contrast that is characteristic of an intermittent pheromone plume. Their responses are "sharpened" by the presence of a blend of pheromone components, and in *M. sexta* it is these neurons that can effectively track the spatiotemporal pattern of the sexual signal released by a calling female (Fig. 8.5; Christensen and Hildebrand 1988; Murlis this volume). It is evident, therefore, that one function of this heterogeneous population of output pathways from the MGC is to extract and analyze the different features of the stimulus that are important to the insect's perception of a pheromone signal under varying environmental conditions. Further behavioral and neurobiological studies are needed to determine if the general principles of organization described herein are applicable to other species as well.

Acknowledgments

The author wishes to thank Drs. T.C. Baker, B.S. Hansson, H. Mustaparta, and especially J.G. Hildebrand for more than 5 years of enjoyable and fruitful collaborations. Some of the work described in this chapter was supported by USDA Grant #92-37302-7634 to T.A.C. and T.C. Baker.

19

Neuroendocrine Factors in the Photoperiodic Control of Male Moth Responsiveness to Sex Pheromone

Charles E. Linn, Jr.

1. Introduction

Successful mate location, recognition, and syngamy in many insects is dependent on a chemical communication system. Males typically respond from long distance to a female released signal, and over the past 20 years great strides have been made in our understanding of the signal itself as well as the mechanisms involved in male orientation to the olfactory stimulus (Baker 1989c; Linn and Roelofs 1989, 1994; Mafra-Neto and Cardé 1994; Vickers and Baker 1994b; other chapters this volume). Efficient mate location also requires (a) temporal coordination of behaviors exhibited by sender and receiver and (b) the occurrence of these behaviors at a time most favorable with respect to habitat conditions. This coordination involves an endogenous clock that is entrained by environmental cues, principally photoperiod and temperature. In most moth species these coordinated activities take place at night, or late in the photophase, and a number of studies have demonstrated the importance of light intensity, photoperiod, and temperature cues in regulating the timing of female release of pheromone (Delisle and McNeil 1986; Dreisig 1986), and in fewer cases male responsiveness (Cardé and Webster 1981; Dreisig 1986).

The rhythmic expression of a behavioral event implies that there are mechanisms to modulate neuronal/physiological processes releasing the behavior. In the hawkmoth, *Manduca sexta,* a great deal is now known about the pathways involved in transmission of pheromone information from the antennal sensory units to descending pathways from the brain to motor centers in the thoracic ganglia (Christensen and Hildebrand 1987b; Arbas et al. 1990; Kanzaki et al. 1991a,b). This pathway is presented in Fig. 19.1. Presumably these pathways can be modulated, via inputs from visual pathways that interact with the clock, in ways to alter the threshold parameters affecting the probability of a male

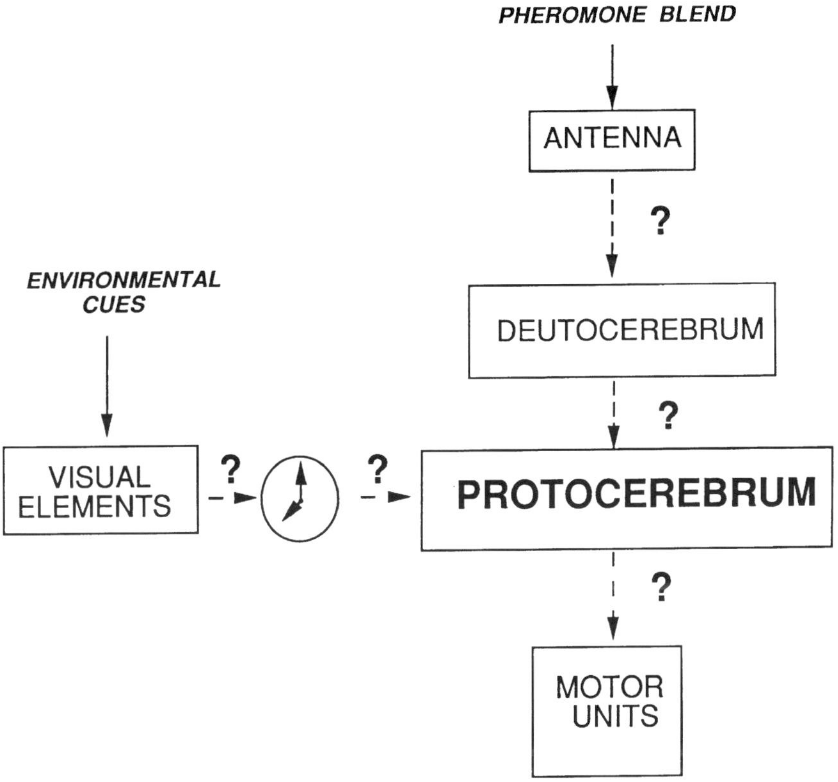

Figure 19.1. Flow diagram of essential elements involved in transduction of chemical communication and environmental signals involved in photoperiodic control of male responsiveness. Visual elements can include the compound eyes, ocelli, or photosensitive areas of the brain itself. The pathway to the right of the clock represents a simplified flow of information in the central nervous system for processing the pheromone signal. Inputs from the clock to the protocerebrum are emphasized here because it is in this area that multimodal units, which access the mushroom bodies, are found and are critical in the processing of pheromone blend information. Dashed arrows indicate neural pathways involved in transmitting information, and the question marks indicate potential sites where modulatory activity can occur.

exhibiting a response to the sex pheromone. One group of compounds that can function in a modulatory way on various levels of neural processing are the biogenic amines (Evans 1985; Brown and Nestler 1985; Orchard et al. 1993). Of particular interest in the present context are effects on central pathways involved in processing sensory stimuli, and those that are influenced by an endogenous circadian oscillator. Examples include proboscis extension reflexes in adult blowflies (Long and Murdock 1983), aggressive behavior in ants (David and Verron 1982), sensitization of medial giant interneurons to sound in crickets (O'Shea and Murphy 1978), learning in *Aplysia* (Kandel et al. 1987), short-term memory and sensitivity to olfactory stimuli in the honey bee (Mercer and Menzel 1982; see also Menzel 1983), phase-shifts of the locomotor rhythm in *Leucophaea,* (Page 1987), and regulation of retinal sensitivity in the eyes of *Limulus* (Barlow et al. 1977; Barlow 1979; Kass and Barlow 1984; Kass et al. 1988), *Aplysia* (Corrent et al. 1978; Eskin et al. 1982) *Hermissenda* (Crow and Bridge 1985), and *Gryllus bimaculatus* (Tomioka et al. 1993).

A number of additional reports have addressed the question of whether biogenic amine levels display diel rhythms, and whether the changes are correlated to overt rhythmic behaviors. These include moth flight activity (Hinks 1967), cricket song (Muszynska-Pytel and Cymborowski 1978a,b), locomotor activity in *Drosophila* (Fowler et al. 1972), an arachnid (Fowler and Goodnight 1966), a cockroach (Pandey and Habibulla 1980, 1982; Davenport and Evans 1984a,b; Owen et al. 1987), the locust (Davenport and Evans 1984a,b), and crickets (Cymborowski 1970a,b; Woodring et al. 1988). In a number of these examples it was stated that the diel changes were circadian in nature, but only in the case of Woodring et al. (1988) was the pattern observed in continuous dark conditions.

As part of a research program to understand the response specificity of male moths to multicomponent pheromone blends, we have been investigating the potential role of biogenic amines in the photoperiodic control of male responsiveness. Here I will present an overview of studies we have conducted with octopamine (OA) and serotonin (5-hydroxytryptamine, 5HT).

2. Effects of the Amines on Moth Behavior

We used the cabbage looper, *Trichoplusia ni,* for our studies because this noctuid moth displays a clear pattern of pheromone response and random locomotor activity only during the dark (Shorey and Gaston 1965; Linn and Roelofs 1992). We also have considerable information on the six-component pheromone produced by this species, as well as on a very sensitive flight-tunnel assay for male flight response to the pheromone (Bjostad et al. 1984, Linn et al. 1984). In a 16:8 L:D photoperiod, response levels typically rise from <5% at the start of scotophase to maximal levels (>85%) during the mid- to latter part of the dark. A similar pattern of random locomotor activity (here referring to non-pheromone-stimulated activity) occurs over this period, although maximum levels rarely exceed 60%. In our initial experiments OA, 5HT, and dopamine (DOP) were injected into the dorsal region of the head, close to the brain, with the injections occurring just prior to lights-off. The random locomotor activity and response to pheromone were then recorded over the course of the 8-hour dark period (Linn and Roelofs 1986b). Both behaviors were monitored because it has been hypothesized that photoperiod regulation of locomotor activity might be the underlying factor that results in the release of other behaviors that occur during the dark, such as feeding, egg laying, long-distance dispersal, and mating (see Brady 1974).

With a dosage of 10 μg OA/μl saline, males exhibited heightened sensitivity to pheromone, evidenced by a shift in the dose–response curve (as much as four- to fivefold; Linn and Roelofs 1986b, 1987). A 10 μg 5HT/μl saline dosage resulted in (a) increasing duration of bouts of random locomotor activity and (b) a greater probability of males responding to a normal dose of pheromone throughout the dark period. Injection of DOP did not affect either behavior.

In a comparative study similar effects of OA and 5HT were found in the gypsy moth, *Lymantria dispar,* (Linn et al. 1992). This moth was selected because it displays a prominent pattern of locomotor activity and response to pheromone during the photophase. Interestingly, we found no effect of the amines when males were treated during the early photophase hours, but rather the effects occurred when males were injected just prior to lights-off. Surprisingly, we also found that male *L. dispar* display a nocturnal pattern of locomotor activity and response to pheromone that were very different from those observed in photophase, as was the magnitude of the effects of OA and 5HT on male behavior (see Linn et al. 1992).

3. Photoperiod Cues and Amine Action

The overt effects of OA on pheromone response thresholds was of considerable interest to us, but perhaps the most striking aspect of this study was the observation that the effect lasted for the duration of the dark period. Additional studies on the effect of OA in *T. ni* demonstrated a link between the action of the amines and timing of the injections relative to the onset of the dark period (Linn and Roelofs 1992). First, maximal effect of OA was observed when injection occurred within 1 hour of the anticipated lights-off, with injection at progressively earlier or later times resulting in less of an effect on behavior. Second, shifting the time of lights-off earlier or later, with or without injection of the amines, resulted in loss of response to pheromone and diminished effect of the injected material on male response thresholds. Third, maximal response to pheromone in a population of moths, as well as maximal effect of OA on male sensitivity to pheromone, was dependent on a low light intensity (0.3 lux) characteristic of the dark period. Fourth, it was shown that a 1-hour period of dark at the anticipated time of lights-off was sufficient to alter male sensitivity and response several hours later, in lighted conditions under which males would not normally respond.

These investigations suggested that the action of OA during the scotophase is initiated by physiological changes that occur during the transition from light to dark (Linn and Roelofs 1992). We hypothesized that the time of this event in any given photoperiod constitutes a "critical period" during which target tissues/

pathways are most sensitive to modulatory action. The timing of this "critical period" was presumed to be regulated by an endogenous oscillator.

4. Response of Male *T. ni* In Different Photoperiods

Our next experiments were designed to test the hypothesis concerning a critical period, and they moved along two lines of investigation. One, to be discussed in the next section, dealt with analysis of tissue levels of the amines over the photoperiod. The second concerned a reanalysis of photoperiod responses of male *T. ni* to their pheromone (Shorey and Gaston 1965; Leppla et al. 1979). In our previous studies with *T. ni* a 16:8 L:D photoperiod was used, with maximal response to pheromone in the flight tunnel occurring during the fifth and sixth hours of dark. This relatively narrow period of peak response suggested the possibility of a gated window of maximal sensitivity and responsiveness, an idea that was also prompted by results of published studies, especially with respect to female release of pheromone (see McNeil 1991). Our photoperiod experiments were designed to address the question of whether there is a clock-regulated "window" of peak response to pheromone occurring during the dark period that has some characteristic phase relationship to the lights-off signal.

The pheromone-mediated upwind flight response of male *T. ni* (2–3 days old) was measured over a series of photoperiods: 6:18, 10:14, 12:12, 14:10, 16:8, and 18:6 Light:Dark, at a constant 25°C. For these tests we selected a source dose of pheromone (2 μg on a rubber septum) that would elicit 40–50% source contacts from a population of males tested during the mid-part of an 8-hour dark period (Linn and Roelofs 1992). We reasoned that if peak response occurs during a narrow period of maximal sensitivity to pheromone, this lower dose would allow a clear determination of that peak. However, the results did not support the proposed hypothesis; instead they showed that over the entire range of photoperiods tested, male *T. ni* displayed a broad period of responsiveness to pheromone, rising from very low levels just after lights-off to a plateau during the second to third hour of the dark, then decreasing in the last 1–2 hours of dark (Fig. 19.2; Linn et al. 1995). While somewhat surprising, in terms of our expectations based on previous observations of male response in a 16:8 L:D photoperiod, these results, in fact, supported earlier studies with *T. ni* by Shorey and Gaston (1965) and by Leppla et al. (1979). In both studies, nocturnal pheromone response in males did not display a clear peak during a narrow period of the dark, but instead displayed a broad pattern of response throughout the dark. The results also were similar to observations made by Haynes and Birch (1984), with the artichoke plume moth. In both 16:8 and 12:12 L:D photoperiods, males exhibited broad periods of response, essentially covering the dark period.

The photoperiod data were important for two reasons. First, they removed the problem of explaining the phase delay between injection of the amines and the

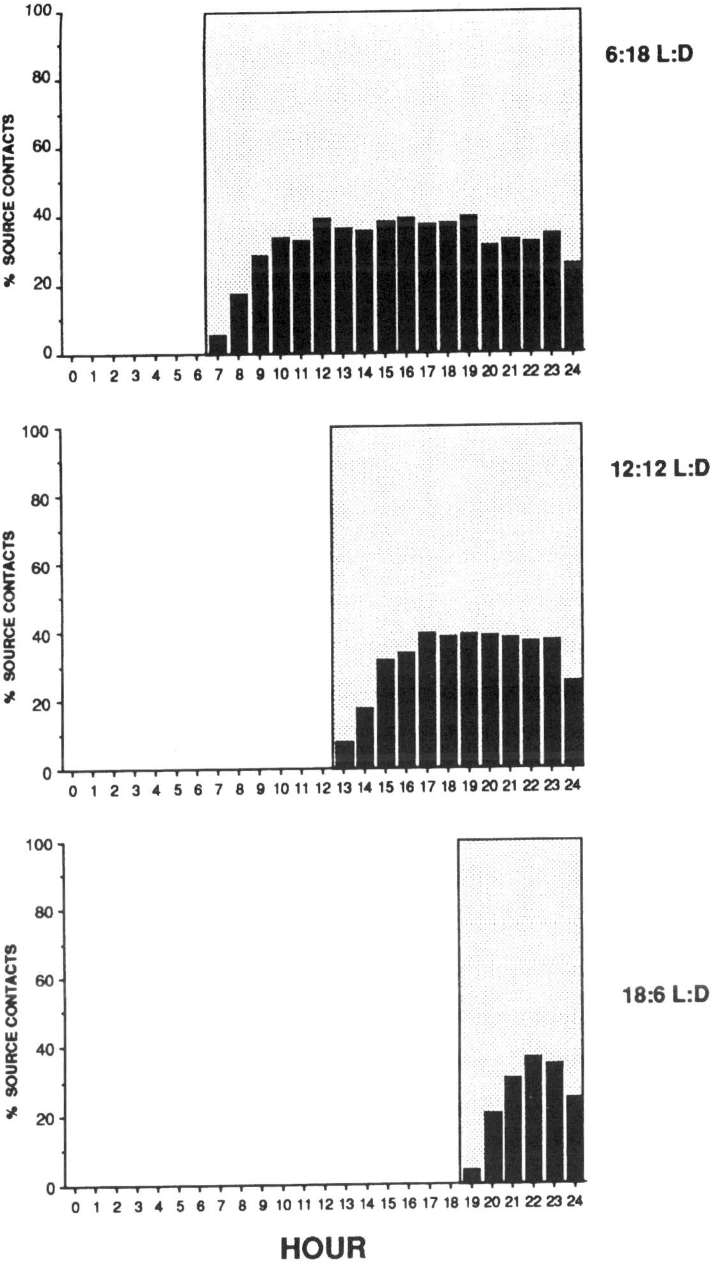

Figure 19.2. Percentage source contacts of male *T. ni* in a sustained-flight tunnel during their third scotophase in three different photoperiods: 6:18, 12:12, or 18:6 Light:Dark. Each hour represents 50 individuals.

time of maximal effectiveness. We realized that what had been interpreted as a "peak" of responsiveness in earlier studies was really a truncated plateau of response due to the tests being conducted in a short 8-hour dark period (Linn and Roelofs 1992). Second, they supported the hypothesis that male sensitivity is linked to events at the time of lights-off, because in all cases the plateau of response was reached within 2–3 hours after lights-off. If one considers that the early part of the dark is associated with a gradual decline in light intensity (or twilight period), rather than the abrupt change in our laboratory situation, the gradual rise in male activity and responsiveness can then be associated with a latency that is related to physiological processes associated with dark adaptation occurring during this period. Dreisig (1980,1986) showed that under natural twilight conditions, increased locomotor activity commences when the light intensity reaches a critical threshold. Initiation of locomotor activity by any individual is dependent on a species' characteristic threshold, the individual's threshold, and the rate of change in light intensity. He further emphasized the coupling of these activities to a circadian oscillator, stressing that in most nocturnally active moth species the transition from light to dark serves as the primary Zeitgeber phase setting the timing of the rhythms expression.

Concerning the role of an endogenous clock, previous work (Shorey and Gaston 1965) with *T. ni* suggested that the diel rhythm of responsiveness was circadian in nature, evidenced by the fact that at least one cycle persisted in continuous dark conditions, a minimum criterion for such a conclusion (see Brady 1974). We also addressed this problem in an expanded way, using our flight tunnel assay top test males over several days and in different photoperiods. Males were placed in continuous dark conditions after their first two days of adult life after being entrained from the pupal stage in 6:18, 12:12, or 18:6 Light:Dark regimes. Over the following 3 days they were tested to pheromone beginning with the last 2 hours of expected photophase through the expected dark period. The results showed that in continuous dark a period of responsiveness occurred during the anticipated dark and that the duration of the response period was well correlated with the previously experienced dark in one of the three photoperiods. The results also showed, however, that by the second and third 24-hour period in continuous dark the response "rhythm" was significantly damped. Response patterns of control males indicate that it is unlikely that the damped response in constant dark conditions was due to age-related effects. Rather, the results suggest that both the lights-off and lights-on signal have an impact on expression of the rhythm (Saunders 1982). The damping observed here has also been reported in several other species, including female release of pheromone in *Holomelina immaculata* (Cardé and Roelofs 1973), *T. ni* (Sower et al. 1970), and *Grapholita molesta* (Baker and Cardé 1979b). Damping was not observed, however, for male response to pheromone in constant light with *G. molesta,* nor was it observed for female release of pheromone in *Anagasta Kühniella* (Traynier 1970), *Holomel-*

ina lamae (Schal and Cardé 1986), *Cydia pomonella* (Castrovillo and Cardé 1979), or *Dioryctria abietella* (Fatzinger 1973).

As noted at the beginning of this chapter, it is generally held that circadian control of mating activities provides a mechanism for ensuring efficient mate location at a favorable time with respect to habitat conditions. With respect to *T. ni,* Sower et al. (1970) showed that the timing of pheromone release by female *T. ni* displays a clear phase relation to the lights on signal, a somewhat unexpected result because it is generally considered that circadian activities of nocturnally active organisms are more sensitive to the lights-off cue (Brady 1974; see Sanders and Lucuik 1992; and Dreissig 1986). Photoperiod studies with males have not demonstrated such a relationship, and our experiments indicate a strong relationship between male response and the lights-off cue, at least with respect to male sensitivity to the pheromone (see Linn et al. 1995). Studies with other moths indicate a variety of patterns, and thus it is not clear that either of these cues necessarily functions as an explicit time-keeper (Delisle and McNeil, 1986). Rather, the entrainment of a particular rhythm in most cases is a function of the photoperiod itself, with both cues influencing the phase of the rhythm. In the case of *T. ni* we would propose that the broad period of responsiveness to pheromone exhibited by males enhances the chances for mating because it anticipates peak female release of pheromone and yet also includes all of the time that females would be active. Future studies will hopefully address detailed studies for male response in a variety of species, as has been done with females, to determine whether the pattern observed for *T. ni* is a general one.

5. Photoperiod Changes in Amine Levels in Nervous Tissues

Another strategy in our analysis of the actions of biogenic amines on behavior has been to measure levels of amines and their metabolites in specific nervous tissues and hemolymph over the photoperiod. The objective here was to demonstrate the presence of the amines in relevant neural tissues and to determine whether changes in amine levels, implying utilization of the compounds, was correlated with specific behavioral states exhibited by males. From the extensive literature in this area (see Downer et al. 1985; Downer and Martin 1987) we developed a protocol utilizing high-performance liquid chromatography with electrochemical detection (HPLC-ECD) and have been able to obtain measurements from tissues in individual moths (Linn et al. 1994a–c).

5.1. Chromatography

The apparatus and testing protocols are described in detail in Linn et al. (1994a). Tissues (corpora cardiaca/corpora allata; optic lobes; the thoracic ganglia; and brain) from individual 2-to 3-day-old adult males were dissected at specific times in the photoperiod in ice cold saline, disrupted by sonication in ice cold 0.05 N perchloric acid, and centrifuged at 12,000*g* at +4.0°C. Fifteen microliters of the supernatant was injected onto an HPLC column [a Vydac C-18 HS-54 reversed-phase HPLC column (25-cm × 4.6-mm I.D., 3-μm particles, Vydac, Inc., Hesperia, CA, USA) protected by a Supelguard LC-18-DB guard column (Supelco, PA, USA)] for analysis. Hemolymph was collected by removing one of the antennae, squeezing the head gently, and collecting 0.5–1.5 μl into ice cold 0.05 N perchloric acid. The mobile phase (after Downer and Martin 1987) contained 70 mM monobasic sodium phosphate, 0.5 μM EDTA, 0.1 mM 1-octanesulfonate (sodium salt), 15% methanol, and 5% acetonitrile, pH 4.0. The HPLC pump, autosampler, guard cell, analytical cell, and dual channel coulometric detector were from ESA Inc., Chelmsford, MA, USA.

5.2. *Photoperiod Changes in Amine Levels*

Our initial survey of amine levels as a function of photoperiod showed several interesting points (Linn et al. 1994a). First, peak amine levels occurred during the photophase and were relatively unchanged over this period. Second, with the lights-off signal, levels of OA, 5HT, and DOP decreased, with the decreases different for each of the amines and within each tissue as well as among individuals (Fig. 19.3). Of particular interest were the changes in OA that occurred in the brain and thoracic ganglia, two areas where potential modulatory actions on pheromone response pathways could occur. Third, the *N*-acetylated metabolite of each amine was measured, but only in the dark period.

We interpreted the observed decreases in amine levels during the dark to suggest that the three amines were being metabolized within each tissue during the dark period. This followed from the appearance of the metabolites of the amines in relevant tissues at corresponding times to the decreases. Both *N*-acetyl-OA and *N*-acetyl-5HT were quantified only during the dark period, corresponding to the time when levels of OA and 5HT were decreasing.

5.3. *Amine Levels and Behavior*

The differences in changes in amine levels among individuals was examined further in tests where the random locomotor activity and response of males to pheromone was correlated with changes in levels of amines during the dark (Linn et al. 1994b). Random activity was recorded for each male at 30-minute intervals over the first half of the dark period, and then using our sustained-flight tunnel assay each male was tested to a low dose of pheromone (0.1 μg), one that elicits 10–15% completed flights to the source in a population of males. Males were then selected for dissection if they completed the response, or if they did not react at all. Following tissue analysis for amine levels, correlations were made with the amount of random activity exhibited during the dark period.

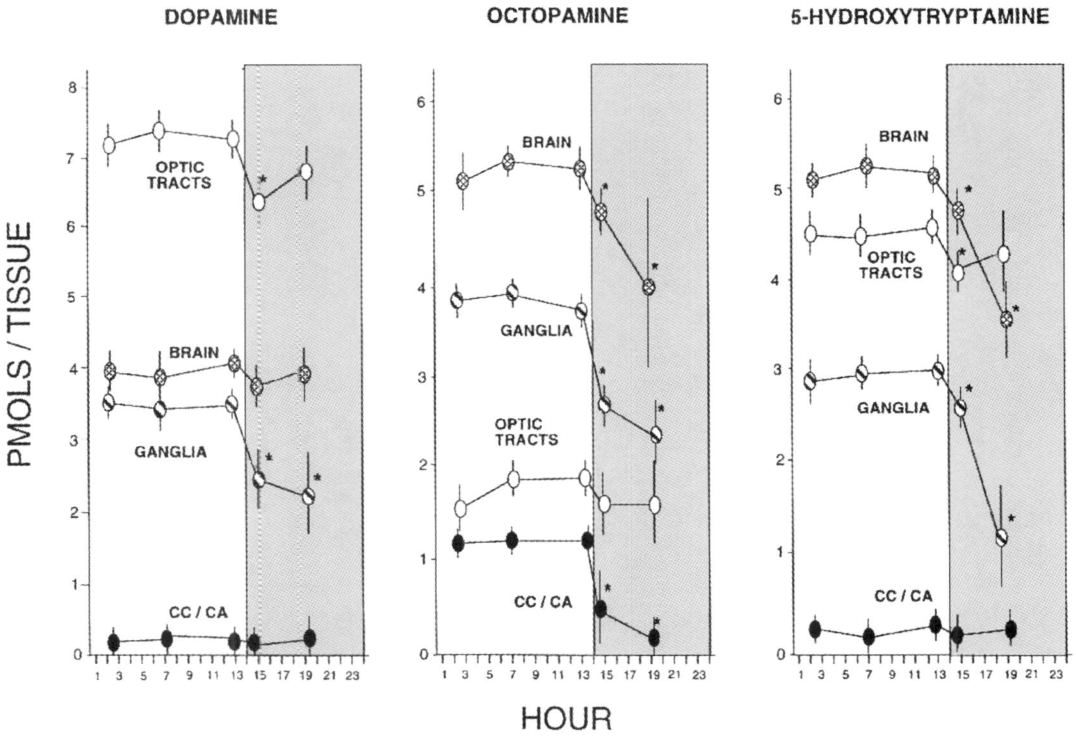

Figure 19.3. Quantitative measures (as pmol/tissue) of DOP, OA, and 5HT at five different times in the photoperiod from the corpora cardiaca/corpora allata, thoracic ganglia, optic tracts, and brains of individual *T. ni.* Values are means ± SD. Those values marked with an asterisk indicate a significant difference from the mid-photophase levels at hour 7, according to ANOVA with Dunnett's procedure for comparing means with a control group (see Linn et al. 1994a). Shaded box in each graph designates the scotophase. *N* = 43,42,45,43, and 41 for each time interval.

The following points were demonstrated. First, there was a significant correlation between decreased levels of OA in the brain and the probability of a male completing the upwind flight sequence to the low dose of pheromone (Fig. 19.4). Second, there was no direct correlation between decreased levels of 5HT in the brain or thoracic ganglia and sensitivity to pheromone or the level of random activity. Third, there was no correlation between the level of random activity and the probability of a response to pheromone. Fourth, there was a strong correlation between decreased levels of OA and of DOP in the thoracic ganglia of individuals, and there was also a strong correlation between decreased levels of these amines and the level of random activity exhibited (Fig. 19.4). Fifth, within individuals there was no correlation in the decrease in brain OA levels and those observed in the thoracic ganglia.

These results supported the proposed endogenous modulatory effect of OA on pheromone-response pathways, and they also showed the important point that performance of individuals in a biologically relevant and sensitive assay can be correlated with changes in the activity of neuroactive compounds. It should be noted, however, that a change in amine levels is not a necessary condition for demonstrating a neurotransmitter or modulatory function. Other studies (see Linn et al. 1994a) have shown increases during the dark, no changes, or decreases as seen in *T. ni*. Active uptake mechanisms for biogenic amines have been demonstrated, and in most cases recycling and synthesis of the compounds would maintain normal levels (Harris-Warrick et al. 1992). In our study, males were selected that exhibited the upwind flight sequence to a very low dose of pheromone, and it was in these males that the most significant changes were seen in brain levels of OA. However, it should be remembered that >90% of the males from the population tested would have flown upwind to a slightly higher dose. Thus we were testing the limits of the system by assaying those males who had the lowest response thresholds for pheromone.

6. Octopamine, General Arousal, and the Expression of a Rhythmic Behavior

Our studies discussed above on the effects of OA indicate that there is a relationship between the expression of a rhythmic photoperiod-regulated behavior, namely pheromone responsiveness, and the activity of this biogenic amine in the nervous system of male *T. ni*. The combined results support our original hypothesis that the time of the lights-off signal is a "critical period" during which physiological processes are initiated that result in threshold changes in the probability of the overt expression of this, and other, behaviors. Information from photoperiod studies and tissue analysis of amine levels indicate that during the early dark, dynamic changes are occurring in the CNS, changes that are reflected in response patterns throughout the remaining scotophase.

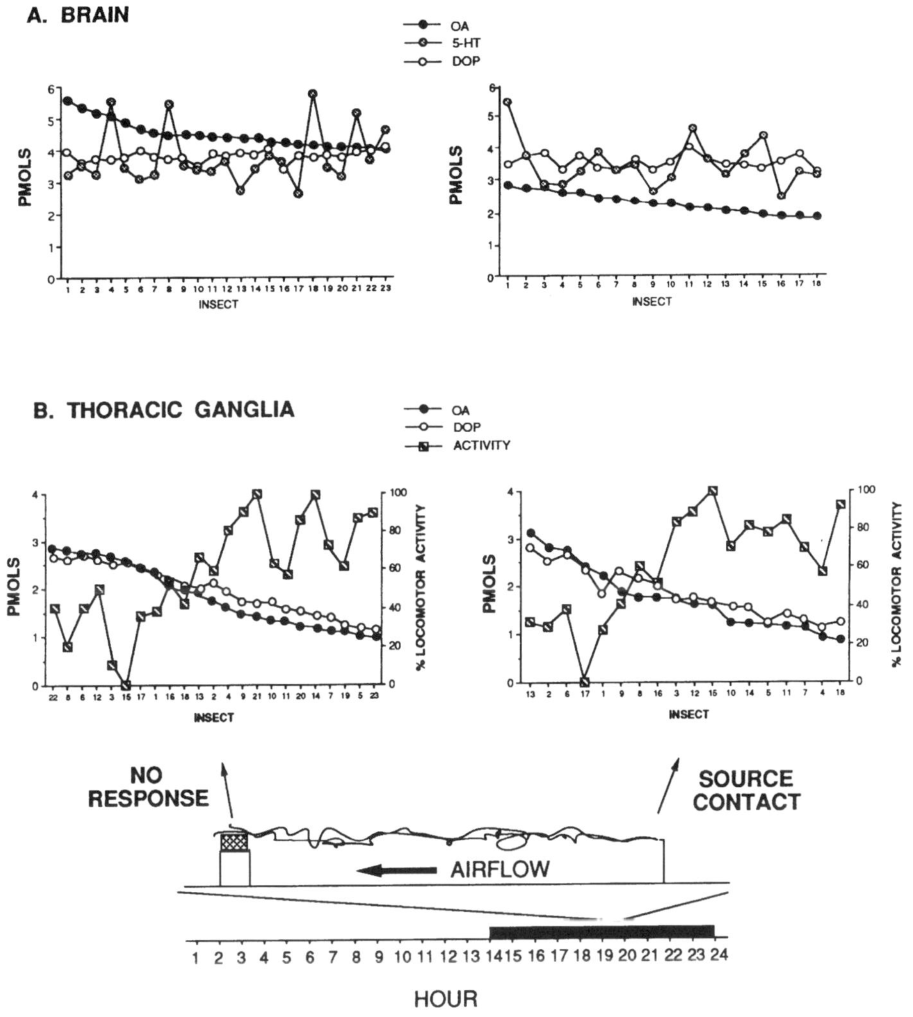

Figure 19.4. Distributions of brain (**A**) and thoracic ganglia (**B**) levels of OA, DOP, and 5HT in individuals exhibiting two behaviors in the flight tunnel—no response and source contact—to a 0.1-μg pheromone source. Total percentage locomotor activity for each male is also shown in **B.** Insects in each behavioral category are arranged on the ordinate on the basis of decreasing level of OA. Hour on the bottom of figure refers to the time in the photoperiod, with the black bar designating the scotophase period.

The results also support the idea that a clock mechanism is involved, in that the probability of overt behavior occurring, as well as the effectiveness of the amines, was dependent on the onset of dark occurring at the time anticipated from the previous day's photoperiod. Additionally, the duration of response in continuous dark reflected the length of dark in the photoperiod to which males had been entrained. However, the significant dampening of the "rhythm" in constant dark indicates that whereas an endogenous oscillator may be involved, there is a requirement for exposure to the respective photoperiod cues as well.

One of the conclusions from the many studies on the modulatory actions of amines on central pathways affecting perception of sensory stimuli (see Linn et al. 1994b) is that the effects involve a form of sensitization, producing a central excitatory state, thereby enhancing the effectiveness of sensory inputs and increasing responsiveness to stimuli. Orchard et al. (1993), in an important review of OA activity in insects, characterized the role of this and other amines as part of a "general arousal mechanism that transforms the insect from a quiescent into a dynamic state." Octopamine, for example, has been implicated in the "fight or flight" responses of several insects and has a demonstrated role in the mobilization of energy stores during the early stages of flight. The actions we have found for OA and 5HT also fit with this theme, in that the action of the amines is correlated with a number of changes that occur at the transition from light to dark, the transition from a behaviorally quiescent to an active state. During this period, local and neurohemal release of the amines can affect a variety of physiological systems, including energy metabolism, feeding, host location, and mating.

One question that remains, with respect to OA, is Where in the neural pathways that process pheromone information is this amine acting? Whereas a great deal of information is now available on the neural pathways involved in processing of pheromonal information (Christensen and Hildebrand 1987b; Homberg et al. 1988, 1989; Homberg and Hildebrand 1989a,b; Kanzaki et al. 1991a,b; Arbas et al. 1990), we believe that interneuronal units in the lateral accessory lobes of the brain are particularly interesting because they are located in regions where integration of other sensory information takes place, including the optic lobes and mushroom bodies (see Menzel 1983; Mercer and Menzel 1982; Kanzaki et al. 1991a,b; Homberg et al. 1988, 1989). Referring to Fig. 19.1, this argument places the modulatory action of OA in the brain and downstream from the endogenous clock. It is hoped that future immunocytochemical (Eckert et al. 1992) and neurophysiological studies will hopefully address these questions.

7. Melatonin in the Nervous System of Male *T. ni*

Our HPLC-ECD analysis of 5HT and its role in locomotor activity showed that there was no clear correlation between 5HT levels in the CNS and male activity, and the results in general with 5HT and locomotor activity are less clear than those with OA and pheromone responsiveness. Three points can be made. First, it could be due to the manner in which the behavior was measured, in that males were scored for locomotor activity at half-hour intervals, then exposed to pheromone in the flight tunnel, and then selected for dissection on the basis of their response to pheromone. Second, it may be that the effect of injected 5HT on overt behavior could be related to stimulation of peripheral pathways involved in motor function. Third, CNS levels of 5HT may be replenished in such a way that the observed level for any individual is not an accurate measure of the activity of the amine.

Our results with 5HT also did not indicate a specific action on pheromone responsiveness; however, it is known that in *M. sexta* there is a large 5HT cell associated with the protocerebrum that projects to the antennal lobe (Mercer et al. 1993). When 5HT is applied to this preparation it leads to spike broadening and enhanced excitability of antennal lobe interneurons. Thus 5HT also may have an important modulatory action on pheromone processing.

However, because of the action of this amine in other circadian and photoperiodic systems, we investigated the potential for the hormone melatonin (MEL; *N*-acetyl-5-methoxytryptamine) to be synthesized by male *T. ni.* In vertebrates, especially mammals, MEL has a well-established role in influencing both circadian and seasonal activities (Yu and Reiter 1993). Melatonin is synthesized from 5HT via *N*-acetyl-5HT and is produced predominantly in the pineal gland, with peak expression in all cases examined occurring during the dark. The photodependent synthesis and metabolism of MEL provides information on the photoperiodic cycle to the neuroendocrine system, allowing for coordination of reproductive physiology with an appropriate seasonal environment. The nature of the MEL message is explained by three different models (Reiter 1987). According to the "duration" hypothesis, the extent of the MEL peak during the dark period informs the organism about day length (e.g., in sheep, Djungarian hamsters, and the tammar wallaby). Alternatively, the "coincidence" model states that there is an important relationship between the timing of the MEL peak and sensitivity of target organs that determines the physiological response to photoperiod (as in the white-footed mice). The third model proposes that the "amplitude" of the MEL signal is a function of the photoperiod and provides the information concerning the length of the dark period.

Despite the extensive research on MEL, and its photoperiodic influence on behavior and reproduction in vertebrates, there has until recently been very little done with invertebrates. However, MEL is synthesized in the brains of a number of invertebrates (Vivien-Roels and Pevet 1993). Given our success with HPLC-ECD of biogenic amines, combined with evidence for MEL production and activity in other vertebrates, we developed an HPLC procedure for quantification of MEL, as well as 5HT and *N*-ac-5HT, the latter compounds involved in the synthesis of MEL (Linn et al. 1994d). Initially, males were held in a 14:10 L:D photoperiod and selected tissues were dissected at hourly intervals over the dark

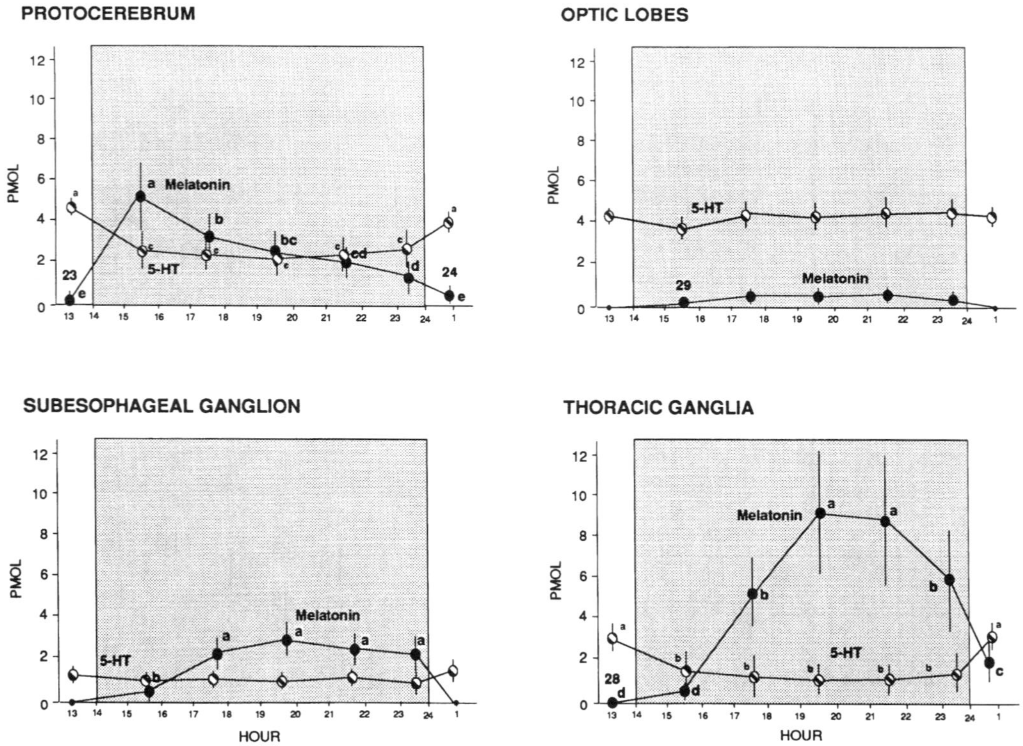

Figure 19.5. Levels of 5HT and MEL (pmol) in male *T. ni* over the dark period in four tissues of the central nervous system. Each point is the mean ± SD. Forty individuals were dissected at each time, and for each point $N = 40$ unless indicated by a number above the data point. The smaller dots for some photophase values of MEL refer to undetectable levels in all individuals. In each panel for each compound mean values with different letters are significantly different ($p < 0.05$; see Linn et al. 1995).

period. The following points were demonstrated. First, MEL was measured in the brain, thoracic ganglia, and hemolymph, but with very different profiles over the dark (Fig. 19.5). Second, maximal levels in the brain were observed within 1.5 hours after lights-off, whereas maximal levels in the thoracic ganglia and hemolymph occurred later in the dark period. Third, the level of MEL in the thoracic ganglia was higher than that observed in the brain. Fourth, high levels of MEL in the brain were not associated with the optic lobes. Fifth, brain levels of MEL produced throughout the dark period were significantly greater than the amount of decrease in 5HT, suggesting that active synthesis of the latter compound was occurring. Sixth, cycling of MEL was observed in continuous dark but not in continuous light. Seventh, MEL levels in brain and thoracic ganglia were quantified in males held at 14:10 and 6:18 L:D. In each case, maximal levels of MEL in the brain were observed within 2 hours after lights-off, with levels remaining high throughout the dark.

These results confirmed that large quantities of MEL are produced in the nervous system of male *T. ni,* and that this is a function of the dark period. Results also indicate that production of MEL accurately tracks the dark period of a given LD cycle, suggesting that the molecule could provide a time-keeping mechanism, similar to its demonstrated role in vertebrates (Reiter 1987).

Acknowledgments

I wish to thank Marlene Campbell, Kathy Poole, Marion O'Connor, Henry Facundo, Wen-Q, Wu, and Wendell Roelofs for contributions to this research. Various aspects of the project were funded by NSF and USDA competitive grants programs.

20

Neurobiology of Pheromonal Signal Processing in Insects

Joseph C. Dickens

1. Introduction

The basic mechanisms by which an insect detects an odorous signal has intrigued investigators since Fabre (1879) reported the attraction of male moths to females. Modern progress in the elucidation of these mechanisms has closely followed advances in electrophysiological, morphological, biochemical, and molecular biological techniques. At the *peripheral level,* development of electrophysiological techniques allowed for recording of odor-stimulated responses from whole antennae (Schneider and Hecker 1956; Schneider 1957) and from single sensilla either by using sharpened tungsten electrodes (Boeckh 1962) or by clipping tips of individual sensilla and recording electrical responses with electrolyte-filled glass capillaries (Kaissling 1974). Concurrent morphological studies allowed investigators to correlate whole antennal odor responses with the presence or absence of sensillar types and also to correlate the number of action potential types recorded with the number of neurons within a given sensillum (Boeckh et al. 1965). In male moths, trichoid sensilla were shown to have individual receptor neurons specialized for detection of a female-produced sex attractant pheromone component (Boeckh et al. 1965) or interspecific inhibitor (Grant et al. 1988). The ability of pheromone receptor neurons to respond to pulsed stimuli correlated with the ability of the insect to detect and respond to the temporal and spatial nature of filamentous odor plumes (Kaissling 1986b; Rumbo and Kaissling 1989; Borroni and O'Connell 1992) as revealed from the study of charged ions in nature (Murlis 1986; Murlis et al. 1992).

Investigations of the *central nervous system* have sought to record electrical responses to pheromonal stimuli from identified neurons. These results have progressed with development of improved methods of filling neurons with dyes in the central nervous system and combining dye staining with previously devised techniques for clipping and recording from antennal sensilla (Hansson et al. 1992).

The questions asked in studies of both peripheral receptor neurons and interneurons within areas of the central nervous system are similar. *How does the nervous system detect important chemicals signals (e.g., individual components and pheromonal blends) and abstract from them information on their quality, quantity, and spatiotemporal character sufficient for the insect to orient to them?*

The purpose of this symposium was to determine where we have made progress in our studies since the last pheromone symposium (Payne et al. 1986) and to suggest where we can progress in future studies.

2. Peripheral Events

At the time of the last symposium in 1984 (Payne et al. 1986), biochemical studies had revealed two proteins, a pheromone-binding protein (PBP) of molecular weight 15,000 and a sensillar esterase of molecular weight 75,000 in abundance in the receptor lymph of the pheromone-receptive trichoid sensilla of the silk moth, *Antheraea polyphemus* (Vogt and Riddiford 1981). A model was proposed in which pheromone was transported across the receptor lymph by PBP for interaction with the membrane receptor, following which it was deactivated by the sensillar esterase (Vogt and Riddiford 1986). This was in conflict with established theories which considered diffusion through pore tubules as the mechanism which transported odor molecules directly to receptor sites on the olfactory dendrites (Steinbrecht and Müller 1971; Kaissling 1974).

Currently, odorant-binding proteins (OBPs) are considered as the conveyance for pheromone molecules to membrane receptors. Further studies have revealed PBPs in several species of moths, and molecular techniques have provided either partial amino terminal sequences or in some instances complete sequences for them (Vogt et al. 1991a,b; Breer this volume; Prestwich and Du this volume). Knowledge of PBP sequences revealed homologies among those found in moths, as well as similarities to antennal proteins in *Drosophila* (Pikielny et al. 1994; McKenna et al. 1994). Due to the lack of homology with vertebrate OBPs, OBPs in insects were thought to be independently derived (Breer this volume). The argument was put forth that sequences for OBPs in insects were conserved in a multigene family. However, an antennal-specific protein in a hemimetabolous insect, the tarnished plant bug, *Lygus lineolaris* (Heteroptera: Miridae), of similar distribution and molecular weight to PBPs of moths has a novel amino-terminal sequence (Dickens and Callahan 1996; Dickens et al. 1995). This may indicate that OBPs in insects may be independently derived or widely divergent. More recent immunocytochemical studies show that antennal-specific protein of *L. lineolaris* associates with olfactory sensilla (Dickens and Vogt unpublished). Immunocytochemistry has localized PBP to pheromone-sensitive sensilla in male moths (Steinbrecht et al. 1992). These studies also showed PBP to be synthesized in supporting cells.

The presence of PBP in pheromone-sensitive sensilla seems well established. However, few studies have investigated binding of the pheromone to PBP, and no studies to date have shown binding of an odorant, other than pheromone (Vogt and Riddiford 1981; Vogt et al. 1991a), to so-called general odorant-binding proteins which have yet to been shown to be necessary for insect olfaction. Only recently has the development of designed molecules which function as photoaffinity labels demonstrated the affinity of pheromone molecules for PBP and showed specific amino acids to which the functional groups of the pheromone may bind (Prestwich and Du this volume). Similar affinities of both the pheromone and its metabolite for PBP may indicate the involvement of PBP not only in transport of the pheromone to the receptor but also in removal of pheromone after interaction with the receptor and degradation by antennal enzymes. Detailed studies of *Antherea pernyi* showed two PBPs coded for by different genes indicating potential specificity for different pheromone components (Breer this volume). More definitive studies are needed to prove the function of OBPs in insects in general including PBP, as well as to delineate their role in receptor site interactions.

The biochemistry of pheromone reception has been one of the areas in which considerable progress has been made in the past 10 years (Breer this volume). Studies on pheromone transduction indicate the involvement of phospholipase C and of inositol triphosphate (IP_3) as the second messenger, which increases dramatically in response to low concentrations of pheromone (Breer this volume). G-proteins were hypothesized to couple phospholipase C to specific receptors in membranes of pheromone-receptive neurons. The ability of pheromone receptor neurons to resolve bursts of pheromone present in filamentous plumes was related to phosphorylation of a potential receptor protein (molecular weight 55,000) by protein kinase C which was activated by diacylglycerol, a by-product of IP_3 activation. Results of these insect studies are similar to previous results in studies of olfaction in vertebrates (Breer 1994b) and Crustacea (Michel et al. 1992; Fadool and Ache 1994). Most gratifying was the observation that these results explain, in certain instances, the nature of the electrophysiological responses of receptor neurons (Breer this volume).

One of the most sought-after but elusive goals in insect olfactory physiology has been the characterization of a putative membrane pheromone receptor. A new hypothesis presented at this time (Breer this volume) was taken from the bacterial literature (Higgings et al. 1990). For insect pheromone reception, it was hypothesized that PBPs might serve not only to bind the pheromone and its metabolites, but also to govern the orientation of the pheromone to the membrane receptor. Thus a complex of pheromone and PBP would be necessary for receptor activation (Breer this volume; Prestwich and Du this volume). This hypothesis was congruent with experimental results obtained for the moth, *Antherea polyphemus* (Ziegelberger 1995). This latter study also indicated two forms of PBP, possibly with differing functions.

Peripheral coding of pheromone components in moths has been considered to occur by the activation of specialist receptor neurons housed in trichoid sensilla on male antennae (Boeckh et al. 1965; Schneider 1969). Some coleopterous insects utilize similar specialized receptor neurons (Mustaparta et al. 1979; Dickens 1990). A hypothesis put forth for heliothine moths is that both a component of the attractant pheromone blend and an inhibitor are detected by the same receptor neuron (Mustaparta this volume). This hypothesis is similar to one put forth for the southern pine beetle, *Dendroctonus frontalis* (Dickens and Payne 1977). However, in moths, recordings were only made from one class of sensilla, thus leaving open the possibility of detection of these compounds by specialized neurons housed in other sensilla.

3. Projection Patterns

Projection of the axons of pheromone receptor neurons in moths has revealed a seemingly clear-cut compartmentalization of pheromone-specific neurons to specific glomeruli in the macroglomerular complex (MGC) (Hansson et al., 1992). Subsequent investigations showed a more complex projection pattern in which axons responsive to individual pheromone components may project to more than one MGC glomerulus (Hansson this volume). Projection patterns of axons to MGC compartments and morphologies of MGC glomeruli raise questions concerning the evolution of the MGC and other glomeruli. Knowledge gained from studies of a range of species should provide an understanding of the functional morphology of the MGC and the function of higher-order interneurons.

4. Central Nervous System

Knowledge of insect pheromones and the behaviors they elicit has progressed dramatically since the discovery of the female-produced sex attractant of the silkworm moth, *Bombyx mori* (Butenandt et al. 1959). At that time a popular theory was that each species of female moth produced a unique chemical signal which attracted conspecific males for mating. However, it was not long before multicomponent attractant pheromones were discovered in coleopterous insects (Silverstein et al. 1966; Tumlinson et al. 1969) and then in moths (e.g. Klun et al. 1973).

Since these early studies, numerous species were shown to produce and respond to multicomponent attractants, and at least two theories were put forth to explain their activity. In the *component hypothesis,* individual components were thought to be responsible for discernible units of behavior. For example, the pheromone component emitted in the greatest amount might be perceived by the insect at the greatest distance from the source and might elicit arousal and perhaps upwind flight. The remaining components were thought to modulate other aspects of mate-seeking behavior. The *blend hypothesis* considered the complete mixture of pheromone components to be necessary for all components of the behavior, from arousal to flight and upwind orientation to the source.

Since responses of peripheral receptor neurons elicited by blends of pheromone components could be explained by the effects of individual compounds (Akers and O'Connell 1988), species-specific blends had to be interpreted by higher-order neurons in the brain. Studies of these interneurons have progressed to the point where results are beginning to correlate with, and perhaps help explain, differences in behavior among moth species (Christensen this volume). In two species of noctuid moths, *Heliocoverpa zea* and *Heliothis virescens,* males have a preponderance (90% and 96%, respectively) of output interneurons in the macroglomerular complex (MGC) which are specific to one of two pheromone components, or respond indiscriminately to either component. These heliothine moths (Vickers et al. 1991) and others, for example, the cabbage looper, *Trichoplusia ni* (Mayer and McLaughlin 1993; Mayer 1995)—can be aroused, take flight, and orient to the major pheromone component alone. By contrast, the male tobacco hornworm, *Manduca sexta* (Sphingidae), which seldom takes flight unless presented with its pheromonal blend, has nearly 50% of its MGC output interneurons responsive only to the blend. Thus, whether individual components or blends are necessary for behavior seems to depend on the species being investigated.

Predicting the behavior of a whole organism from the electrophysiological responses of a relatively few identified neurons within restricted areas of the central nervous system may be dangerous because it ignores the potential roles of the thousands of neurons which make up the nervous system, each of which can have hundreds of synapses. Behavioral responses of male moths to pheromonal stimuli are further complicated by the importance of periodicity in both female emission of the signal and male responsiveness to it (Linn this volume). Early studies with biogenic amines showed that injection of octopamine into cabbage looper males increased their sensitivity to pheromone as measured by flight tunnel bioassays, while general locomotor activity was increased by serotonin (Linn and Roelofs 1986a). More recent studies of actual levels of biogenic amines in cabbage looper male moths show that decreased levels of octopamine and dopamine in thoracic ganglia correlate with photoperiodic expression of general locomotor activity, while decreased levels of octopamine in the brain were related to successful orientation of male moths to sex pheromone in a flight tunnel (Linn et al. 1994b). Increases in melatonin in nervous tissues of male moths were related to scotophase, thereby suggesting its role as a *Zeitgeber* as has been described in vertebrate studies (Reiter 1987). These studies collectively suggest that biogenic amines may modulate the responses of those interneurons within the olfactory pathways which are responsible for effecting motor responses to pheromonal stimuli. More important, however, is the need to demonstrate the effects of these compounds on identified neurons within the olfactory pathways as has been shown for motion-sensitive neurons in the visual system of the honey bee (Kloppenburg and Erber 1995).

5. Prospects and Perspectives

Progress made in our understanding of detection and translation of pheromonal messages in insects has been substantial when compared to our knowledge a decade ago (Payne et al. 1986). There remain numerous avenues for additional research and further refinement of techniques which will fuel our future understanding of this important communication system.

5.1 Quantification of Olfactory Stimuli

In most studies of the electrophysiological responses of peripheral receptor neurons and higher-order neurons, the concentration of pheromone molecules to which the animal is exposed is not known (Mayer 1993). This is also generally true of behavioral studies. Many studies simply refer to the amount of a given stimulus placed on filter paper in a glass cartridge without relating this *dose* to the *concentration* of odor molecules to which the preparation is exposed. This procedure may be acceptable if one is interested only in the relative specificity of peripheral receptor neurons. Since the hypothesis has been put forth that a receptor neuron may provide the brain with differing messages based on the apparent quality or quantity of a stimulus (Mustaparta this volume; Dickens and Payne 1977), better quantification of olfactory stimuli seems mandatory and may lead to new discoveries.

5.2. Pheromone-Binding Proteins and Membrane Receptors

Until recently it was considered likely that pheromone molecules were transported by pheromone-binding proteins (PBPs) to membranes of olfactory dendrites where interactions leading to transduction cascades occurred *between the pheromonal molecule and the membrane receptor.* Recent results (Kaissling 1994) and models presented here (Breer this volume; Prestwich and Du this volume) consider the *complex of pheromone and PBP to be necessary for interaction with the membrane receptor.* Perhaps experimental strategies based on this model may facilitate the isolation of membrane receptors for pheromones in insects.

Progress in the discovery and characterization of membrane receptors for pheromones in insects may be facilitated by results from recent studies in the nematode, *Caenorhabditis elegans* (Troemel et al. 1995), and mammals (Dulac and Axel 1995). Examination of regions around genes coding for potential olfactory signaling molecules revealed a family of genes coding for seven transmembrane proteins which were considered candidate chemosensory receptors in *C. elegans* (Troemel et al. 1995). Subsets of these genes were expressed in known chemosensory neurons with as many as four receptor genes expressed in a single neuron. The strategy used by these investigators for the discovery of candidate receptors associated with potential olfactory signaling molecules may be applied to investigations of receptor proteins in insects. A different approach was taken in the identification of putative pheromone receptors in rodents (Dulac and Axel 1995). The assumptions were made that individual receptor neurons in the vomeronasal organ (VNO), which detects pheromones in mammals, would express different receptor genes, and the expression of these genes would be exclusively in the VNO. Experimental results indicated a family of seven transmembrane receptors with only one receptor type likely to be expressed in a single neuron.

5.3. Enhancement of Pheromone Responses by Plant Volatiles

Detection by male moths of pheromonal blends emitted by calling females clearly occurs in the presence of other environmental odors. Electroantennogram measurements recorded from light brown apple moths and codling moths in the field indicated interactions between pheromonal compounds and other odors normally found in an apple orchard (Rumbo et al. 1995). Field trapping studies showed that responses to pheromones in *Heliothis virescens* (Dickens et al. 1993), *Helicoverpa zea,* and the codling moth (Light et al. 1993) are enhanced by the presence of green leaf volatiles. Green leaf volatiles and other naturally occurring plant odors enhance pheromone responses in a number of other moth species including beet armyworm, cabbage looper, variegated cutworm, oriental fruit moth, peach twigborer, artichoke plume moth, pink bollworm, European cornborer, and diamondback moth (Light et al. unpublished). Moreover, plant volatiles enhance behavioral responses to pheromone in a variety of insects (Dickens et al. 1990). In some instances where synergism is known to occur, plant volatiles excite relatively specialized neurons responsible for their detection without affecting other pheromone receptive neurons [e.g., the boll weevil *Anthonomus grandis* (Dickens 1989, 1990)]. In other instances, plant volatiles may excite specific pheromone receptive neurons [e.g., *Spodoptera exigua* (Dickens 1994)] or enhance responses of others [*Trichoplusia ni* (Mayer and Dickens unpublished)]. In any case, it is unlikely that we will gain a complete understanding of pheromone detection and integration unless the insect is evaluated within the environmental context in which it normally responds to calling females.

5.4. Practical Applications

Understanding the neural mechanisms used by insects to detect and orient to chemical signals has led to several practical applications of these chemicals in the field (Ridgway et al. 1990). Studies of olfactory receptor responses using electroantennograms and single cell recordings, especially in conjunction with chromatographic methods, have facilitated the identification of numerous chemical signals currently used in detection and control strategies (Roelofs 1984; Struble and Arn 1984; Wadhams 1984). Better characterization of receptor neurons for pheromones, a knowledge of higher-order neural processing, and effector output will further advance our understanding of pheromone disruption in the field and perhaps lead to novel control strategies. Isolation of membrane receptors for pheromones and elucidation of PBP–pheromone receptor interactions, as well as degradative processes which occur within sensilla, could facilitate the design of molecules to modulate pheromone responses.

Acknowledgments

I would like to thank the following individuals for their critical reviews: Professor R.J. O'Connell, Worcester Foundation for Biomedical Research, Shrewsbury, MA; Professor R.G. Vogt, Department of Biological Sciences, University of South Carolina, Columbia, SC; and Dr. M.S. Mayer, USDA, ARS, Insect Attractants, Behavior and Basic Biology Laboratory, Gainesville, FL.

PART III

Neuroethology of Pheromone-Mediated Responses

Organized by Thomas C. Baker

21

Odor Plumes and the Signal They Provide

John Murlis

1. Introduction

Atmospheric turbulence causes vigorous mixing of the air at a wide range of spatial and temporal scales. The largest eddies are hundreds of meters in extent and, according to the wind speed, can take minutes to pass a fixed point. They give rise to the continual changes we notice in wind direction. The smallest eddies are tens of millimeters in size and can pass in a few milliseconds. Material released from a point is entrained into the wind and carried away from its source. As it flows, in a plume, it is taken up by the turbulence. The filaments of air containing the material are stretched and distorted by the turbulence until they form the kind of pattern we see, for example, in the smoke blowing from a chimney. The fine-scale structure in an odor plume arises, similarly, through the action of atmospheric turbulence. Consequently the concentration at a stationary point fluctuates, at a frequency of some tens of cycles per second (Hz) (Jones 1983; Murlis et al. 1992).

Insect antennae have been shown to be capable of responding to changes in concentration in pheromone stimuli at similar frequencies (Kaissling, 1986), and the insect central nervous system is known to resolve input from the peripheral nervous system to at least 10 Hz (Christensen and Hildebrand, 1987b). Insect behavior is also known to be influenced by rapid fluctuations in the concentration of the odor stimuli (Kennedy et al. 1981). The importance of complex pheromone plume structure in upwind anemotaxis and anemokinesis has been demonstrated by Kramer (1986), Willis et al. (1994), Vickers and Baker (1994b), and Mafra-Neto and Cardé (1994).

Information about the fine-scale structure of odor plumes is therefore of considerable interest in the study of pheromone-mediated insect flight. This chapter summarizes the current state of knowledge about fine-scale plume structure in general and how it relates to pheromone plumes. It includes a summary of the experimental methods used to measure plume structure, along with major findings and points for future investigation.

2. How Fine-Scale Structure Arises

Turbulence in fluid flows contains considerable energy, but the energy is distributed throughout the fluid over the whole range of frequencies of motion. The usual way this is imagined is as a nested series of eddies whose sizes range from the largest turbulence scales of many hundreds of meters to the smallest, of a few tens of millimeters. Energy is injected into the largest eddies, and then it cascades down to the smallest as successively smaller eddies are set in motion by those just a fraction larger. The energy is finally lost as heat when the smallest eddies lose their "fight" against the viscosity of the fluid and collapse.

The consequence of this is that there is plenty of energy in the largest eddies to produce the continual changes in wind direction which cause a plume to meander and undulate as it leaves its source. In medium-sized eddies, sufficient energy remains to stir the filaments within the plume. At a small but finite scale, however, the turbulent energy is lost and further mixing to yet smaller scales is impossible. There is, then, a lower limit on the physical size of elements and, because medium-sized turbulence is causing the body of the plume to expand, there remain gaps between elements which grow as the plume flows downwind.

The turbulent energy cascade can be quantified by measuring the energy contained at each frequency—for example, with a series of band-pass filters—and by showing this as an energy spectrum. The close relationship between turbulence and fluctuations in concentration was shown by Mylne and Mason (1991). They measured the distribution of turbulent kinetic energy and of the "energy" of the concentration fluctuations amongst the range of frequencies in the signals: Both the turbulence and the concentration fluctuations were found to have the same spectral shape, which demonstrates the link between them.

An understanding of the turbulence in a flow therefore provides a useful insight into the structure of a plume produced in it. However, because turbulence is difficult to predict, despite considerable recent progress in mathematical modeling of turbulent flows, there remains a need for experimental work using tracers in the study of turbulent plume structure.

Very few published studies of fine-scale plume structure have been specifically directed toward the question of odor-modulated animal behavior, and the discussion below draws heavily on work done, for example, to assess the impact of accidental release of toxic chemicals. Work done in other fluids, notably water, produces results which can be related to plumes in air by using simple scaling laws.

3. Measuring Plume Structure

It has generally proved difficult to measure structure in turbulent plumes directly, and tracers have been widely used to produce more tractable physical models. This is possible because turbulence forces dominate the generation of plume structure (Jones 1983; Murlis 1986) and any material would be similarly distributed in the plume. The requirements for measuring plume structure are a combination of a tracer which is passive and conservative for realistic atmospheric lifetimes and a means of detecting it with sufficient time resolution. These requirements are not easily satisfied in practice, and published methods of plume structure measurement inevitably involve a degree of compromise.

Tracer gases—for example, sulfur hexafluoride—most easily satisfy the criteria of passivity and conservation. They have been used in trials to detect long-range transport of atmospheric pollutants such as power station emissions (Hadjitoffi and Wilson 1979), but temporal resolution at low concentrations is limited by flame ionization detectors (FIDs) to a few seconds. Improved time resolution was reported using dilute propane, with FIDs in a wind tunnel application by Fackrel (1978, 1989). In recent atmospheric trials, propylene was used with specially tuned gas detectors to produce measurements with a resolution of up to 10 Hz (Mylne 1988; Mylne and Mason 1991).

An alternative approach is to use ions as tracers (Jones 1978; Boreham and Harvey 1984). This has the advantage that ions can be detected at high temporal resolution (down to a few milliseconds), but it suffers in other respects. Unipolar ions are conservative up to hundreds of seconds flight time, which gives them a reliable range of hundreds of meters in a wind of about 1 $msec^{-1}$, and further in higher winds (Jones 1978). They are not passive, however, and are affected both by the repulsion forces between them and by image force in the ground. An analysis of these properties (Chatwin 1985), however, concludes that they are not significant in flight times of up to about 100 seconds. Bipolar ions have been used in wind tunnel work (Jones and Gulliford 1979; Boreham and Harvey 1984), but are nonconservative due to recombination and have not been widely used in atmospheric plume studies.

Ion sources are described in Jones (1978), Jones and Griffiths (1984) and Murlis and Jones (1981). For measurement of ion concentration there are two types of detector. Ion collectors (Jones 1978) are active devices, in which air is drawn at a uniform rate through a system of annular electrodes. Ions are driven on to one of the electrodes and the concentration is detected as the ground current they produce in an electrometer circuit. Ion flux probes (Boreham et al. 1985) are passive devices, in which air is driven through an electrode system by the wind, and the output current is a measure of the ion flux through the system. The measurement of flux (Murlis et al. 1989) is useful in the study of olfaction because it provides an estimate of the rate at which molecules could impinge on the peripheral sensory system. Boreham et al. (1985) showed the relationship between measurements made by these two types of instrument. They found that the signals have a similar pattern, but ion flux probes are more responsive to rapid changes in ion density at the leading and trailing edges of bursts. Results from the two types of ion detector can be reconciled by allowing for the wind speed.

In the most recent development of the ion detector system, a passive trace gas is used and ionized by a strong ultraviolet light source in the intake of a modified ion collector. Trials using butene as a tracer have confirmed the plume structures hypothesized from ion tracer measurements. The advantages of the ionized trace gas method are that it combines the advantages of a passive and conservative tracer with those of highly responsive electronic detection.

Fine-scale measurements of plume structure in water are reported by Moore et al. (1992). Using a chemical tracer and a semiconductor-based detector system, they obtained measurements of microscale turbulence-driven concentration fluctuations at high spatial and temporal resolution, at least equivalent to the level of fine-scale structure found in atmospheric plumes.

Ideally, in studies of this kind the fluctuating concentration of the active compound should be measured directly. Baker and Haynes (1987) and Murlis et al. (1989) used an electroantennogram, while Van der Pers and Minks (1993) used a single sensillum, to monitor pheromone levels in the field. Some interesting results have emerged from this work, but the interpretation of them remains an area of difficulty.

4. Summary of Findings from Plume Structure Measurements

A typical fluctuating concentration signal is shown in Fig. 21.1, taken from Mylne and Mason (1991), illustrating its intermittent nature. The signal consists of a series of brief spikes separated by longer gaps of zero concentration. In the example illustrated, the peak concentration is more than four times the third standard deviation from the mean, and individual spikes regularly extend to the third standard deviation. It follows that the analysis of the signal, as it is seen by receptors with the level of frequency response found in insect neurosystems, should focus on the strength and duration of the spikes in the signal rather than on their mean concentration. Unfortunately, signals of this kind, in which low probability events may be disproportionately important, are particularly difficult to analyze. The discussion which follows charts the possible approaches to this problem and attempts a summary of the main findings of relevance to the study of insect response to a pheromone stimulus.

4.1. Time-Dependent Signal Characteristics

Intermittency is one of the most obvious characteristics of signals derived from turbulent plumes. It is, however, notoriously difficult to define and to analyze

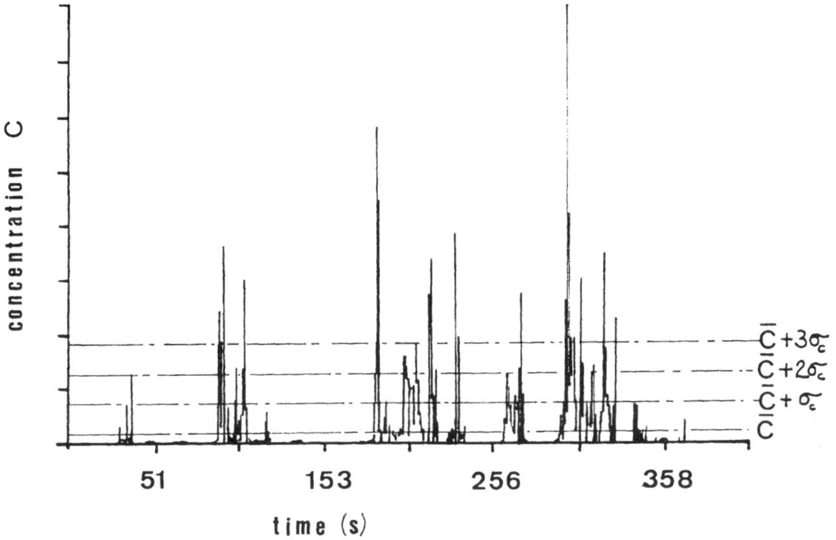

Figure 21.1. A typical tracer signal. Taken from a detector placed to the side of a plume approximately 67 m from a source of propylene mounted 2 m above ground level. The vertical scale shows the concentration. The trace gives the instantaneous value, and the levels of the mean (C) and of the standard deviations from the mean are shown by lines parallel to the horizontal axis. [From Mylne (1988).]

(Jones 1978). There are three sources of intermittency in the signal from a fixed sensor: gaps of zero concentration appearing in the main body of a plume due to imperfect mixing of the material in it, the gaps which appear due to the undulating and meandering of the plume, and the movement of the plume as a whole due to shifts in wind direction. These all depend on turbulence structure, which, itself, continually changes during the day. Jones (1983) and Mole and Jones (1994) have attempted formal treatments of intermittency. They concluded that any measure is likely to be somewhat subjective and that it would be very difficult to determine spatial trends in intermittency.

Despite this, however, intermittency, expressed as the proportion of time for which signal is present, is a useful guide to the overall "density" of the signal, even if it is not an exact measure. Intermittency varies greatly, but usually is found in the range 0.1 to 0.4. This means that in a pheromone plume the signal will be absent for about 60–90% of the time.

Treatment of intermittency in the atmospheric physics literature, for example Jones (1978), relies on long-term time averages in which the continual shifting of wind direction plays a significant part. Murlis et al. (1989), however, assumed that, in practical problems of pheromone-mediated behavior, long-term time averages were irrelevant; and to determine intermittency, they considered a series of short periods (each 90 seconds long) of particularly active signal. Estimating intermittency in this way they showed that intermittency recorded in trials in an open field fell with distance from source from about 0.4 to 0.1 between 2 m and 20 m from the source.

Another very visible characteristic of these signals is that the spikes they contain are of short duration, typically tens of milliseconds up to about half a second. Fig. 21.2, which is taken from Jones (1983), shows the frequency distributions of the durations of spikes and of the gaps between them. All distributions are heavily skewed to higher values, with modal durations considerably shorter than the mean. Murlis and Jones (1981) concluded that, for practical purposes the modal value was more useful than the average, but could find no trend with distance from source in the modal values of distributions taken from analysis of signal records 30 minutes long. Murlis et al (1989), however, found trends in typical durations of spikes and return periods in a set of 90-second records of active signal: As the source to receptor distance increased from 2.5 m to 20 m, spike duration increased by a factor of 2 but the return period increased by a factor of over 7.

The aim of much of the published work on fine-scale plume structure has been to understand the relationship with turbulence structure, and much of the experimental evidence comes from trials made in the simplest possible physical environment, where the turbulence itself is best understood: in the open atmosphere over flat, level ground. However, where the physical environment significantly affects turbulence structure, it can be expected that plume structure too would be affected (Jones and Griffiths 1984). Allen (1968) and Wang (1989) showed that trees have considerable influence on the turbulence structure in a forest, and Murlis et al. (1989) quantified the consequent impact on plume structure. In a forest they found that the duration of bursts of signal were longer in the forest and they contained more spikes on average than those measured over open ground.

4.2. Concentration Distribution in a Turbulent Plume

There are several ways of expressing the strength of the odor signal available from a turbulent plume. The most usual is concentration, but, for estimating inputs to insect sensory systems, the flux or the total amount of material contained in each spike or each burst may be of more interest. In any case the considerable contrast between the strengths of individual spikes or bursts and the overall mean strength of the signal makes for difficulty in analysis.

Jones (1983) showed that, although the mean concentration in an ion plume falls monotonically with distance from source, the relationship for the fluctuating concentration is less simple. Figure 21.3 shows that a particular value of instantaneous concentration can be found over a wide spatial range downwind from the source—although the further from the source, the less frequently it can be found.

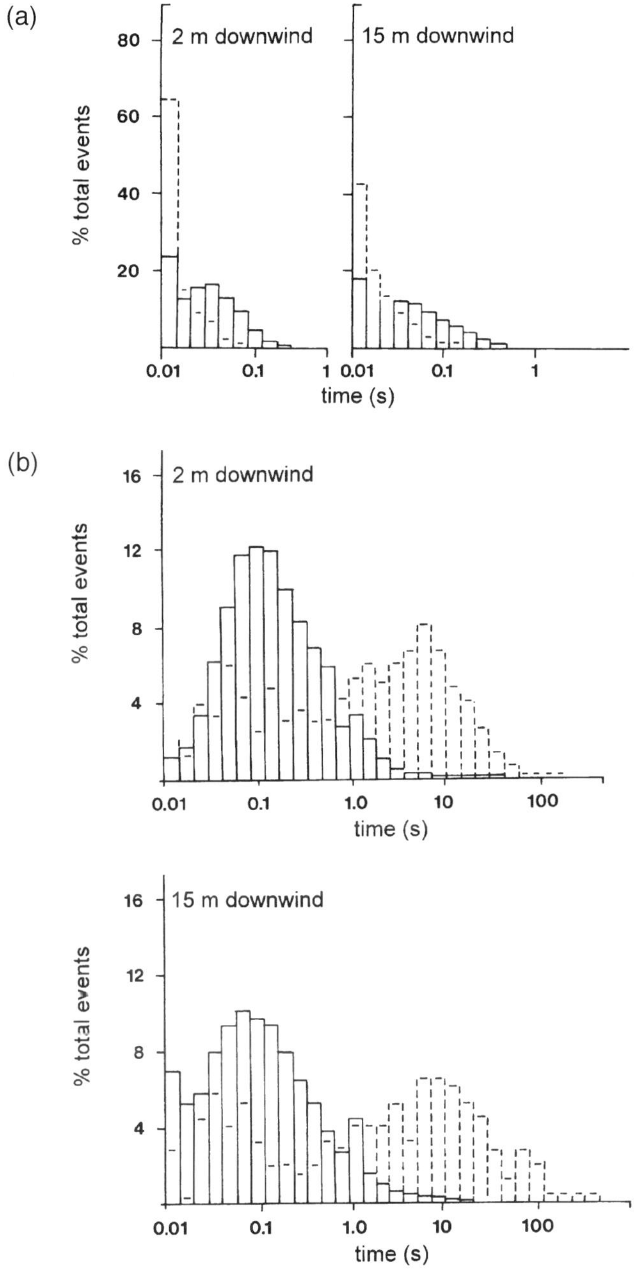

Figure 21.2. Time-dependent structure of a turbulent plume measured 2 m and 15 m downwind from an ion source mounted 1 m above ground level. Frequency distribution of burst lengths (**a**) and burst return periods (**b**) for all bursts above a 50-mV noise rejection threshold (solid lines) and for stronger bursts exceeding an arbitrary 500-nV threshold (dashed lines). [From Jones (1983).]

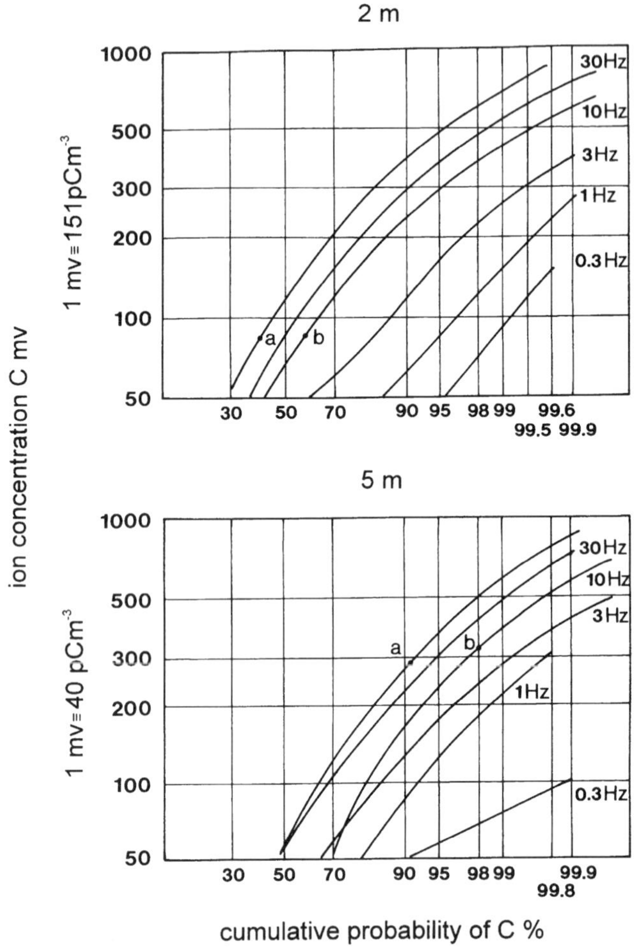

Figure 21.3. Cumulative probability functions from signals recorded 2 m and 5 m downwind from an ion source mounted 1 m above ground level. The vertical axis shows concentration and the horizontal the probability that the concentration will not be exceeded. Several lines are shown for different cutoff frequencies for the concentration fluctuations. The points marked a and b are at the same concentration in both parts of the figure. The figure shows that on occasions the same concentration can be found at 5 as at 2 m downwind of the source, though at 5 m it will be less frequent. [From Jones (1983).]

For example, the median value at 2 m from source in the Jones (1983) data set is exceeded 5% of the time 5 m from source. This is of considerable practical importance if the peak concentrations have more influence than the overall signal mean (Wright 1958), because it would make it difficult to extract reliable information about proximity to source without sampling several sequential bursts.

In order to extract information about the strengths of spikes and bursts, periods of zero signal can be omitted from the analysis, a technique known as *conditional averaging*. Mylne and Mason (1991) analyzed conditional averages of concentration in a plume of dilute propylene released 1 m above ground level, and Fig. 21.4 depicts a conditionally averaged cross section through the plume indicating the mean concentration and the intensity of the fluctuating concentration together with the intermittency. The mean concentration shows the approximately Gaussian distribution usually assumed in plume modeling (Bossert and Wilson 1963; Murlis et al. 1992). The intensity of concentration fluctuations, however, remains almost constant across the plume and the distribution of intermittency is similar to that of mean concentration, with the edges of the plume considerably more intermittent than the core. This is an important finding because it suggests that the mechanism which produces the distribution of mean concentration in plume cross sections is that filaments of the plume material become less dense at the edges, while the concentration in the filaments themselves is maintained across the plume. An insect flying through a plume might therefore experience fewer "spikes," but their strengths would not vary systematically.

The cross section through the plume body, however, is only part of the story. Meandering of the plume also affects the cross-wind distribution of the plume material. The burst return period distributions shown in Fig. 21.2 show increasing

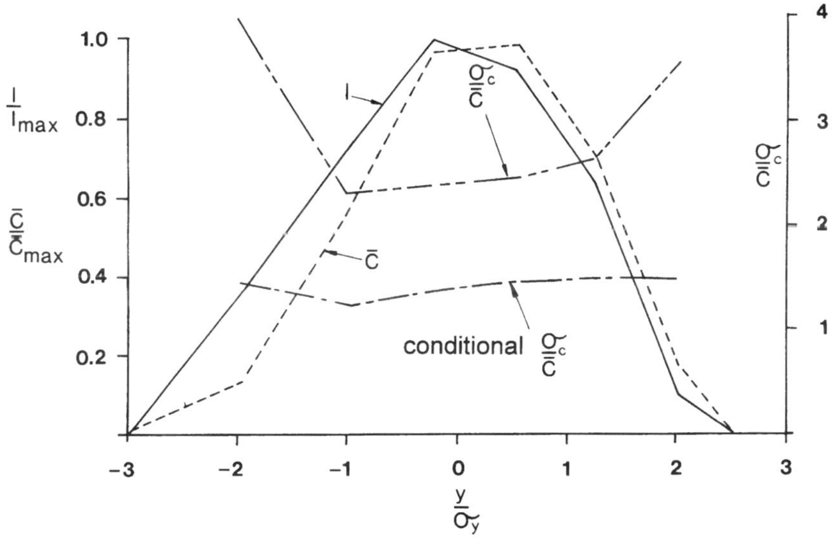

Figure 21.4. Cross sections of mean concentration (dashed line), intermittency (solid line), and conditionally sampled fluctuating concentration intensity (——— – – ———) in a turbulent plume (see Fig. 21.2 for details), taken from Mylne (1988). The conditionally sampled intensity of concentration fluctuations is also shown (——— – ———).

bimodality as the source to receptor distance increases. This can be explained as a separation between the intermittency within the plume and the effects of meandering and undulation. The typical period of the latter is a few seconds. Ride (1988) predicted that the effect of these larger-scale motions would be that the plume would spend more of its time at the boundaries of its travel than at the center of the space it occupies and that it would be more intermittent in the center than at the boundaries. This would increase the mean concentration measured toward the edges of the plume's cross-wind travel. Ride (1988) compared a mathematical model of this effect with field data from Hanna (1984). He found that the maxima of concentration appeared at the boundaries of the plume travel and were a factor of up to 2 greater than concentrations toward the center. The data Ride (1988) used came from measurements in a plume from a large, high-emission source, but for a pheromone plume, which emanates from a small source, this effect can be expected to deliver an even greater contrast between center and boundaries (Ride 1988; Murlis et al. 1989).

Murlis et al. (1989) also used conditional sampling in their analysis of flux in a plume of ions and showed that the conditionally averaged strengths of bursts, whether measured as the average of the peaks of ion flux or as the average of the total quantity of ions in bursts, fell less rapidly with increasing distance from source than did the overall signal mean strength. Mylne found a similar trend in analysis of conditionally sampled concentration. The consequence of this finding is that (a) the contrast between the signal mean and the instantaneous bursts or spikes decreases as the distance to the source decreases and (b) contrast is greatest furthest from source.

5. Conclusions and Future Work

Significant progress has been achieved recently in experimental methods for investigating plume structure, in particular the development of high-frequency detection of passive tracers. Results obtained using the most recent methods broadly confirm the findings from earlier work, including ion tracer trials. These show that the signal from a distant pheromone source can be expected to be highly intermittent with stimulus present in broken bursts some tens of milliseconds in duration occurring with a period of some hundreds of milliseconds. The peak concentrations in bursts will be a factor of 10 or more greater than the signal mean, and the contrast between the peak and the mean will be considerably more pronounced at greater source-to-receptor distances. Encounters with the plume can be expected more frequently at the boundaries of its cross-wind travel than in the center of the space it occupies.

The first priority in further work is to complete the intercomparison between results obtained with ions as a tracer and those using the new UV ionization techniques with a passive gas tracer. Once this is done, a number of extensions of the data base will be possible, including:

- Measurements of plumes in complex environments including crops, such as cotton, vineyards, orchards, and forests
- Measurement of the interaction between plumes from an array of sources
- The impact of source size on plume structure
- The influence of weather, including atmospheric stability, on plumes

It should also be possible to evaluate the measurements of pheromone levels made in the field using electroantennogram or single sensillum recordings. Simultaneous trials with passive gas detectors and single sensillum instruments (Van de Pers and Minks 1993) would be especially useful.

Although a program of work of this scale might appear ambitious, it is perfectly feasible with the more advanced of the methods currently available. The most effective means of achieving rapid progress would be to form a collaboration with an existing program of advanced atmospheric dispersion investigation. Trials of specifically entomological interest could then be added to an established work schedule with significantly reduced marginal effort.

Acknowledgments

I gratefully acknowledge the support of Professor R.T. Cardé in pursuing work on pheromone plume structure and of Dr. A.K. Minks, which enabled me to attend the Wageningen Symposium.

22

A Tentative Intercausal Nexus and Its Computer Model on Insect Orientation in Windborne Pheromone Plumes

Ernst Kramer

1. Introduction

After many years of experimental studies it seems most probable that moths have in common a fundamental algorithm which enables them to find the source of windborne pheromone plumes. This algorithm is a simple sequence of "go" (proceed upwind) and "wait" (keep position) (Kennedy 1978: "attraction" and "arrestment") which is well adapted to the inhomogeneity of natural odor plumes (Wright 1958, Murlis and Jones 1981). In this algorithm, concentration gradients of the odorants as required for tropo- or clinotactic responses are not engaged, because only the average of local concentrations over long periods of time could help to locate the source.

Although different species of moths show a variety of approach behavior, the idea arose whether all peculiarities in odor-induced anemotaxis can be combined in one and the same intercausal nexus merely by a modification of the parameters. This chapter is intended to give a brief presentation of the actual state of affairs in this direction.

The term *intercausal nexus* ("Wirkungsgefüge") is used in the sense of Mittelstaedt (1962). It denotes the input–output relations of a system which, apart from these relations, is completely unknown. This is different from a circuit-, block-, or signal-flow diagram and is completely independent of the underlying hardware. Lines in intercausal nexus just indicate variables and their direction of action but not signal-carrying pathways. Boxes represent causal connections between variables. Intercausal nexus also need to be discriminated from models. Models are always concrete solutions of nexus using distinct hardware which is actually provided by a computer. Important properties of the system may not appear explicitly within the nexus. For example, most pheromones are mixtures of compounds in set ratios, and attractivity is reduced if these ratios are incorrect.

One could be tempted to search for a neuron or a neuronal complex which continuously evaluates this ratio. The intercausal nexus of anemotactic orientation proposed herein demonstrates that such an endeavor might be in vain. There might be no distinct element of the nexus responsible for such an operation. Nevertheless, it can operate as a whole if, for example, the four elements having an olfactory input (cf. low-pass filters with functions g_1–g_4 in Fig. 22.1) receive their signal from receptors with differing specificities. In this case the system will operate successfully only if the time constants of these four elements and their weighing is adequately tuned to the ratio of pheromone components.

The intercausal nexus of this simple orientation system looks rather complex because details which initially seem to be marginal—for example, time course of receptor excitation, perception of wind direction in flight and walking, or properties of turbulent diffusion—need to be taken into account. A means to validate an intercausal nexus is to test it with a functional model. This will help to uncover relations or fuzzy definitions and stimulate new and meaningful experiments.

2. Intercausal Nexus and Computer Model

The kernel of the proposed nexus (Fig. 22.1) is designed as to meet the well known behavior of: "go" (walk upwind / fly upwind) while filaments of air with pheromone are perceived, and "wait" (walk in circles/stop upwind flight and avoid backdrift) if pheromone pulses cease. If the odor plume meanders back, resume upwind movement.

2.1. Simplified Odor Field

In contrast to natural odor plumes, the plume used here is only a two-dimensional triangle originating downwind from a point source. It is static and does not meander. These restrictions are chosen mainly because temporal processes cannot be shown in track figures without reducing their clarity. Within this triangle, odor pulses occur randomly but their mean density is constant; that is, at any place within the gray zone in Figs. 22.2–22.7, the probability of crossing an odor filament is identical and has been deliberately set. Due to the randomized occurrence of pulses, occasionally large gaps appear in the timing or spacing between pulses even though the "insect" is within the plume. In order to further reduce the complexity of resulting paths, the pulse height and pulse width is kept constant in the cases shown here. The actual wind direction at any location fluctuates stochastically around mean value (in the given examples the deviation of the mean is set to ±30°). These simplifications do not seriously affect the validity of the model. Only if the efficacy in finding a source is in question, these parameters must be considered in detail [cf. discussion: on efficacy of hitting source (Section 3.4) and Fig. 22.8].

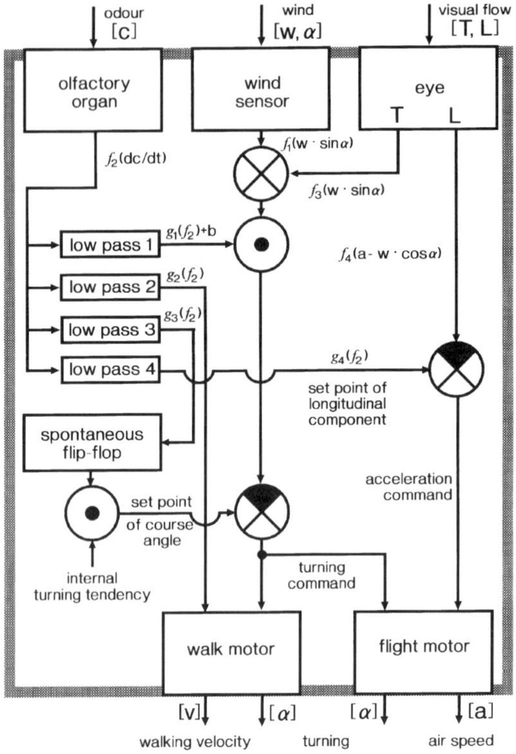

Figure 22.1. Intercausal nexus of anemotactic orientation as used for the computer simulation. Arrows symbolize variables and their direction of action, and boxes symbolize the relations between them. Crossed or dotted circles indicate either addition or multiplication, respectively. Black sectors mark a change of sign. The nexus incorporates two independent feedback loops. The significant part of the first loop runs down the middle of the figure, and that of the second loop runs along the right margin. Both loops are closed via the environment (outside the large, gray-bordered box). Loop one controls the course angle α in flight and walking. The set point of α is zero (due to upwind anemotaxis). As a decisive detail the gain is modified in this loop by the actual olfactory situation (low pass 1). The term *b* prevents a zero gain during casting flight in the absence of pheromone pulses and is possibly also a g_x-type function with a large time constant (same as g_2, determining the duration of a general arousal following a pheromone stimulation). By the action of the internal turning tendency and the spontaneous flip-flop, the set point of α can be shifted to produce menotactic orientation and "counterturning." Loop two controls the upwind advance in flight with respect to the ground. In this loop the set point is determined by the olfactory situation via low pass 4. While walking, the speed is controlled by an open loop via low pass 3. Most details of the walk section are based on behavioral and electrophysiological data of *Bombyx mori.* The flight control section is mainly devised using data on optomotor anemotaxis in *Lymantria dispar, Heliothis virescens, Adoxophyes orana, Grapholita molesta,* and others. Further explanation is given in the text.

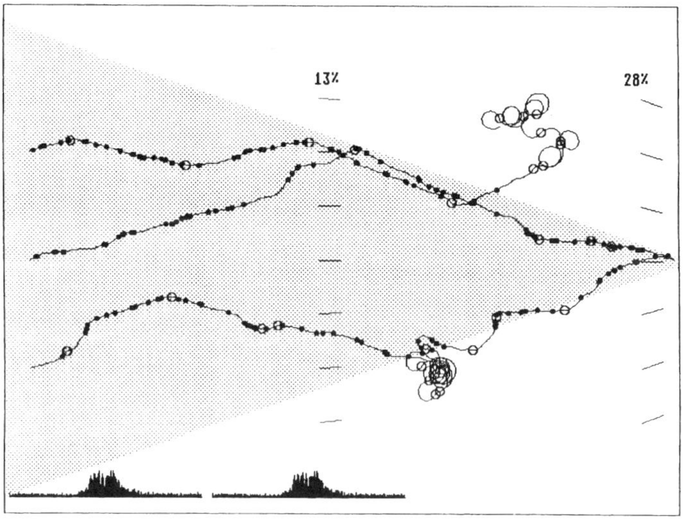

Figure 22.2. Three tracks of a modeled moth walking in a standardized odor cloud. The initial form of the nexus as designed to fit solely *Bombyx mori* data is applied. Significant is a flip-flopping internal turning tendency leading to zigzags within, and downwind pointing loops outside of, the plume. The mean flip-flop time is 4 sec. Here and in the following figures, every location in the gray zone has the same probability of containing an odor filament (pulse). Any incidence of an odor pulse during a run is marked by a black dot, and any change of the flip-flop is denoted by an empty circle. Inserted histograms (bottom) show the distribution of pertaining course angles (left) and track angles (right, differ from left only in flight). Small lines denote the assumed mean local wind vectors of only those portions of air previously having passed the odor source (in fact, vectors outside the active space are irrelevant and are only shown to illustrate the procedure). Columns are headed by the applied percentage of preservation. The shown area represents a field of about 128 × 96 cm. If not explicitly denoted, the following parameter settings are used in all track figures: time constants (τ) in the four low passes are 0.32, 6.65, 1.65, and 1.10 sec, respectively. Internal turning tendency is 180° sec^{-1}. Mean odor pulse interval is 0.33 sec and mean walking speed is chosen such that shown tracks last about 20 sec (break off after a maximum of 50 sec). Insets: distributions of course angles (α) left and track angles (right); they differ only in flight (Figs. 22.4-22.7).

2.2. *Inputs*

2.2.1. *Olfactory Input*

The olfactory part mainly refers to *Bombyx mori,* using both electrophysiological (Kaissling 1986b; Kaissling et al. 1989a; Rumbo and Kaissling 1989; Kodadová 1993) and behavioral experiments (Kramer 1975, 1986, 1992). Electrophysiological studies with flying moth species [*Grapholita molesta* and *Agrotis segetum*

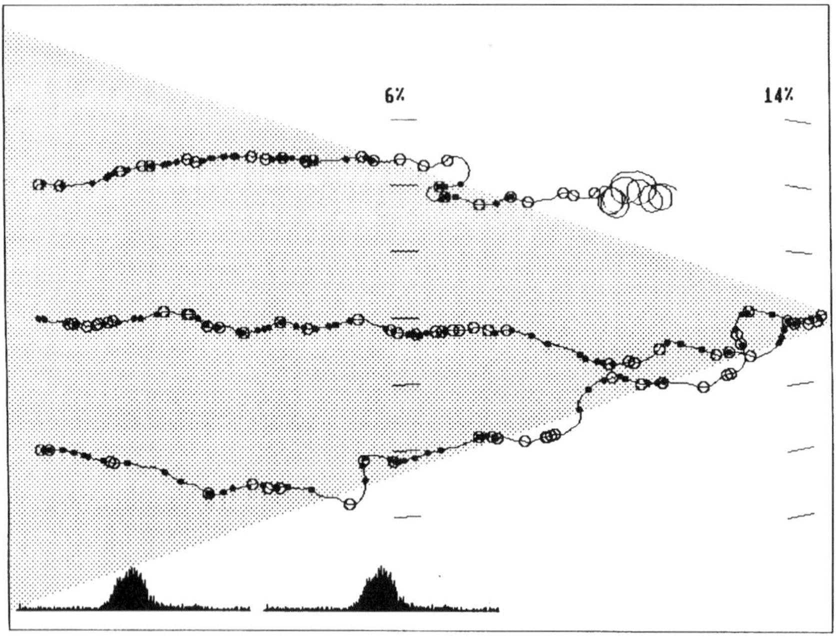

Figure 22.3. As in Fig. 22.2, except the flip-flop changes more often (nonperiodic) when the olfactory signal from low pass 3 is high (i.e., within gray zone). Tracks appear straighter within the gray zone because zigzagging is indiscernible from noise.

(Baker et al. 1989)] yielded time courses in receptor responses which correspond to those found in *Bombyx*.

We know that males of *B. mori* orient upwind only when the pheromone stimulus is pulsed (Kramer 1986, 1992) and that each pulse causes a brief anemotactic response, the intensity of this response being only little affected by the concentration within the pulses (Kramer 1986). The latter are features already found in the receptors. The first derivative of their output—$f_2(dc/dt)$—has an amplitude which, due to the receptor's wide-range dose–response relationship and phenomena of adaptation, is only weakly dependent on odor concentration (Fig. 22.9).

2.2.2. Mechanoreceptive Input

Little is known about how walking or resting moths measure wind direction. The most likely candidates for this task are the Johnston organs in the antennal bases which sense mechanical forces exerted on the antenna. If these forces are caused by wind, then, for physical reasons, the output will follow a sinusoidal curve with respect to the course angle α. The difference of the outputs of the two antennae is zero if the moth's posture is due upwind. In the nexus the output

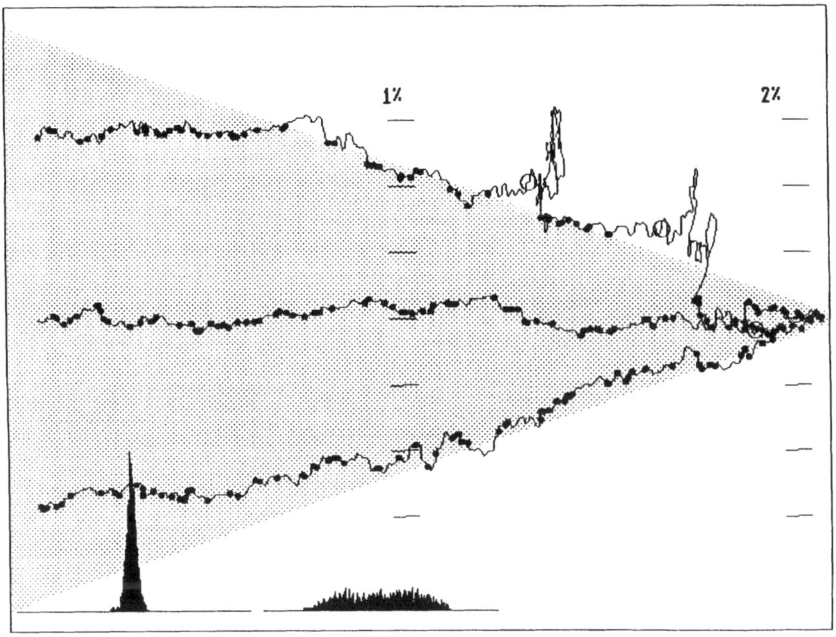

Figure 22.4. Three tracks of a modeled moth flying in the standard odor plume. The spontaneous flip-flop and the internal turning tendency have been put out of function. When the "moth" leaves the "cloud," an irregular type of casting occurs. This casting is only caused by the combination of a decreasing gain in the feedback loop controlling the course angle (*T* component) and a decreasing set point in the loop controlling upwind progression (*L* component). The model is highly efficient in finding the source if odorous filaments to only a small extent have preserved the direction of movement they had when passing over the source. The space shown with flight tracks is always 640 × 480 cm. Wind speed is set to 150 cm sec^{-1} (except in Fig. 22.7, where it is 450 cm × sec^{-1}). The longitudinal component of image flow *L* (following g_4) corresponds to 0–30 cm sec^{-1}.

of the wind sensor is therefore set to the anemotactic error signal $f_1(w \times \sin \alpha)$, where *w* is the wind speed. It is assumed that the mechanoreceptive output is not sin α itself. In fact, experiments with *B. mori* and *Lymantria dispar* have shown that the gain of the anemotactic control loop is to a certain extent proportional to the wind speed (Kramer 1975; Preiss 1980).

2.2.3. *Visual Input*

Since the early work of Kennedy (1940), we know that flying insects use optomotor information to detect wind drift and its direction. According to David (David et al. 1982; David 1986; Preiss and Kramer 1986a,b), the image flow of visual ground patterns over the moth's eye can be split into a component parallel to the longitudinal body axis (*L*) and a transverse component (*T*). If we disregard

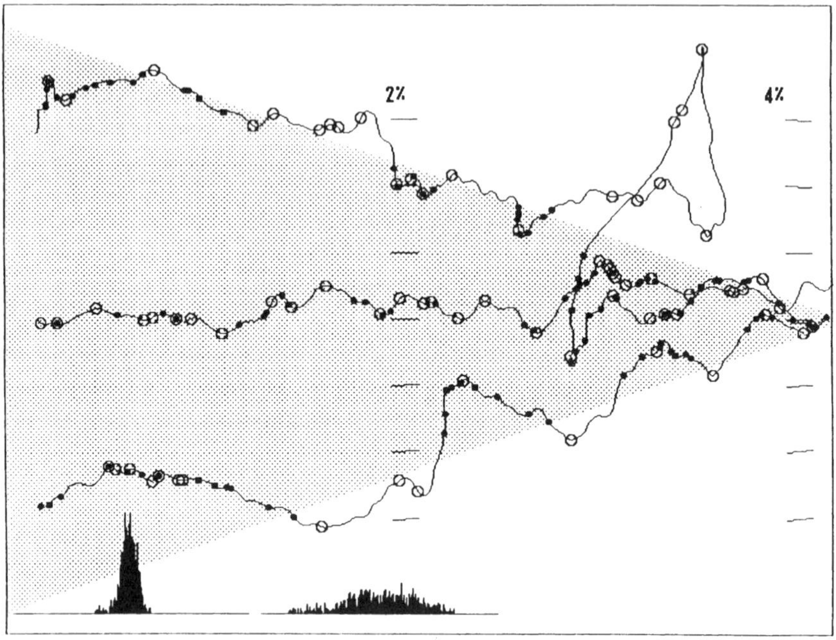

Figure 22.5. Model with internal turning tendency which stochastically changes sign (flip-flop). The probability of such a change depends on the output of low pass 3 in such a way that the mean time span between changes increases with lower output. "Counterturns" are mainly caused by the action of the flip-flop in combination with the internal turning tendency. Again, if filaments have preserved the direction of movement to some small extent, most of the "moths" arrive at the target.

the influences of flight altitude, as we do in the nexus presented here, T depends only on wind speed and course angle and is formally the same as the mechanoreceptive output, namely $f(w \times \sin \alpha)$. Therefore T is the best perceptual basis for control of the course angle α.

With the course kept upwind, $\cos \alpha$ is close to 1, whereby the longitudinal component $L = a - w \times \cos \alpha$ reduces to $a - w$ (where a is the air speed—that is, the speed of the moth with respect to the air), which is the upwind progress. As we know from wind tunnel experiments (Kennedy and Marsh 1974; Kennedy 1978), the upwind progress is the magnitude that is actually controlled rather than the air speed itself, as is the case in locusts (Gewecke 1975).

2.3. *Control of Course Angle*

2.3.1. *Wind-Related Turning Commands*

In the nexus, the outputs $f(w \times \sin \alpha)$ from the mechanoreceptive and T from the visual system may be added because the former is considered to be zero in

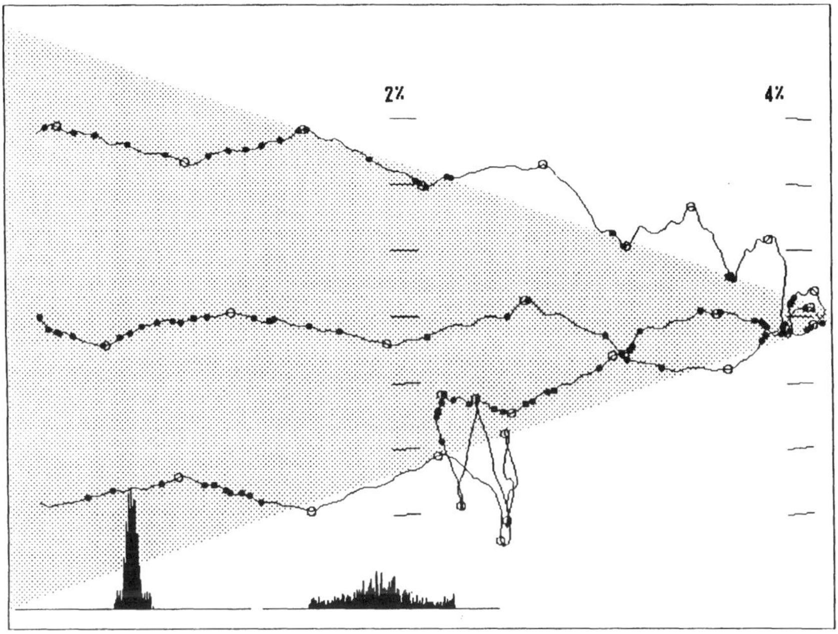

Figure 22.6. As in Figure 22.5, except for the flip-flop being coupled to an oscillator which makes the flip-flop change status regularly. In the given example a change occurs every second and the increasing amplitude of the zigzags during "casting" is only due to the compression of the ground track which has a rather constant ground speed. This effect of increasing amplitude in casting can be further enhanced if the frequency of the oscillator is arranged to be in some direct proportion to g_3.

flight (difference of left and right antenna both measuring air speed) while the latter is zero during walking.

2.3.2. Modulation by Olfactory Signals

As already demonstrated in the 1984 meeting at Oxford (Kramer 1986; Kaissling and Kramer 1990), anemotaxis in a walking silkmoth is not simply switched on or off by the pheromone plume but rather modulated by odor filaments arriving at its antennae. Each odor pulse elicits an anemotactic turning response up to stimulus frequencies of at least 3 pulses per second. In the meantime, this has been found also in flying moths (e.g., *Grapholita molesta;* see Baker and Vickers 1994; Baker and Vickers this volume), which respond to single filaments with an upwind "surge." This rapid buildup and decay of anemotaxis causes most of the characteristic features in the process of orienting; together with a likewise modulation of the set point of *L* (cf. section on internal turning tendency, Section 2.3.4), it renders possible a st·:pless transition from upwind flight to casting and

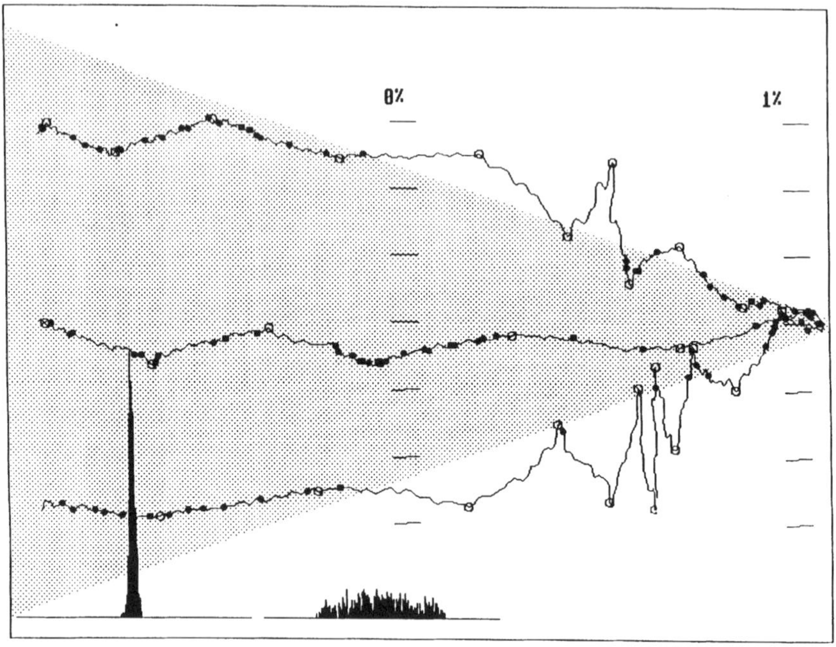

Figure 22.7. As in Figure 22.6, except that wind speed has been changed from 150 to 450 cm × sec^{-1} (without changing the average incidence of odor pulses, so that *L* has the same characteristic as before). This shows that ground tracks are hardly affected by this triplication of wind speed.

vice versa. In the nexus this is achieved by simply modifying the gain in the course-controlling loop by multiplying (weighing) the actual error signal $f(w \times \sin \alpha)$ by a function (g_1) depending on the actual olfactory situation. This signal is assumed to be proportional to the mean amplitude of every increase in the pheromone concentration (dc/dt); it is easily obtained by feeding the rectified (one way) dc/dt signal into a leaky integrator with an adjustable time constant (low-pass filter #1 in Fig. 22.1). This time constant determines the velocity by which the anemotactic response decays after every pheromone pulse or, to speak in cybernetic terms, rules the actual gain of the feedback loop which controls the course angle. The constant *b* in the output of this filter is only required in flight during casting to prevent the gain from falling to zero, which would cause a total breakdown of the anemotaxis when the olfactory stimulation ceases. In the model, *b* is constant; in fact it is rather also a function of the olfactory input but having a much larger time constant than the low-pass filters 1–4.

The modulated turning command $[(g_1(f_2) + b] \times [f_1(w \times \sin \alpha) + f_3(w \times \sin \alpha)]$ is directly used to control the angular position in flight or walk; that is, it

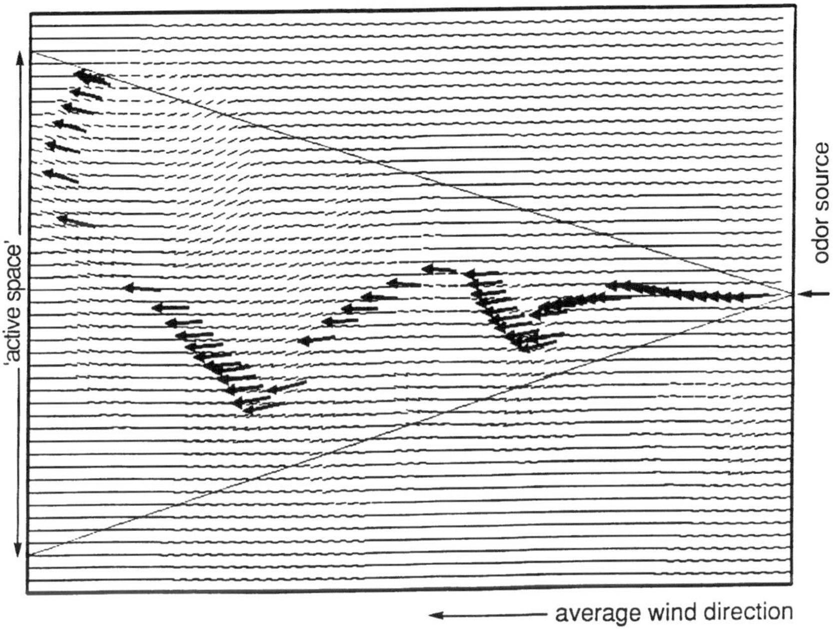

Figure 22.8. Simulated odor plume demonstrates that parts (arrows) of the odor-carrying filament travel downwind into a direction which is still related to, but not identical with, the direction they had while passing the source—that is, local wind vectors of air portions carrying pheromone point rather away from the source than into the mean wind direction. Background shows local wind vectors in a turbulent flow from the right to the left.

makes the moth turn left or right such as to minimize this command. In this simple feedback loop this is always the case when the moth heads upwind ($\alpha = 0$).

2.3.3. Remarks on Course Control

The ability to compensate for disturbances—that is, to return to a set value if it is unbalanced by external influences—is the most characteristic feature of feedback mechanisms. In our case, disturbances among others are caused by chaotic fluctuations of the wind direction. In the nexus and its computer model these disturbances are represented by adding "noise" to the actual value of α. A second source of noise accounts for the imprecision of measurement and in data processing as it occurs in the system itself. This noise is added to the wind-related turning command, and the system permanently follows these changes as precisely as the dynamics of its locomotor system allow. The course control loop is of the "integrating" type, because the locomotor system always turns as long as the actual α deviates from the actual wind direction.

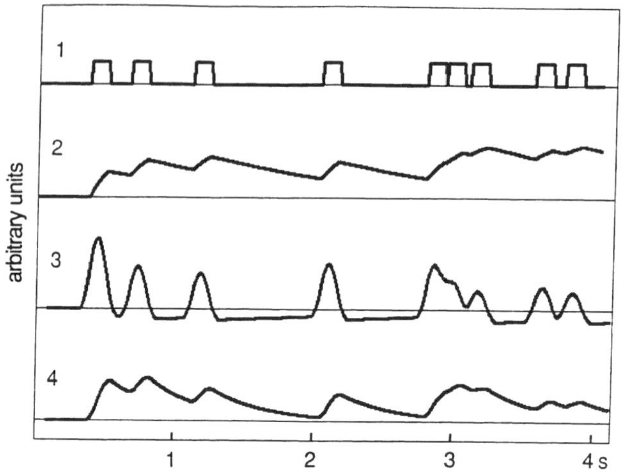

Figure 22.9. Simulated olfactory input as used in all given examples. (1) Odor pulses standardized with respect to amplitude, duration, and randomized time of occurrence (2 pulses sec^{-1}, 150-msec pulse width). (2) Pulses passed through a leaky integrator with $\tau = 2$ sec. This signal corresponds to real receptor potentials. (3) First derivative of (2) used as f_2 (dc/dt) corresponding to bursts of nerve impulses. (4) Positive values of (3) passed through low pass 1 (g_1 (f_2) with $\tau = 1$s).

The relevant properties of the nexus with respect to the control of the course angle are as follows:

(a) If there are neither pheromone pulses nor wind, the nexus "steers" a completely erratic course due to internal noise.

(b) If there is only wind, the walking version behaves like in (a). In the flight version, disturbances cause large fluctuations around the average upwind direction (casting) due to a low residual gain of the loop (in the nexus caused by the term *b*).

(c) If there is wind, any pheromone pulse elicits a brief anemotactic response (<1 second). If by chance the actual position is not due upwind, a small upwind jerk can be seen which decays corresponding to the time constant of low-pass filter #1.

(d) If pheromone pulses follow each other so rapidly that the effect of the previous is not yet faded out, the effects will add up. Up to an optimum of 3–5 pulses per second, the precision of the anemotaxis increases.

(e) If pheromone pulse frequency is so high that single pulses cannot be resolved due to the low-pass filtering by the receptor system, the effect is then the same as if there was a constant pheromone level (dc/dt = 0) as in *b*).

2.3.4. Internal Turning Tendency

Circling (which characterizes walking in *B. mori*) and anemo-menotactic behavior are not yet among the above issues. These phenomena can be explained by the assumption that there is always a small internal tendency to turn. The effect of such tendency is most evident if anemotaxis is either weak or completely absent. In the latter case it would make the insect circle, while in the former it would make the course deviate from due upwind (anemo-menotaxis). The nexus takes account of the internal turning tendency by adding a signal to the turning command which does not depend on the angular position ("set point of course angle"; see Fig. 22.1). The sign of the internal turning tendency alternates by the action of a flip-flop. In *Bombyx* these changes occur spontaneously (i.e., from unknown origin) with randomly distributed time intervals (Kramer 1975; Olberg 1983a). As a result, the "insect" either circles (no anemotaxis) with alternating direction or zigzags upwind (with anemotaxis) with varying tack lengths and alternatingly set course angles α (Fig. 22.2). This flip-flop may also be influenced by the olfactory situation such that the probability of the occurrence of sign changes is controlled via low pass #3 (Figs. 22.3 and 22.4). Furthermore, this influence may be chosen so as to generate a regular sequence of sign changes. As a consequence, the more regular zigzagging and casting patterns so often observed in wind tunnel experiments would appear (Baker 1990; Kuenen and Cardé 1994; Kuenen et al. 1994; Mafra-Neto and Cardé 1994; Willis and Baker 1987; Willis and Arbas 1991a).

2.4. Control of Locomotory Velocity

2.4.1. Walking

With respect to walking, the nexus refers to data obtained by experiments with *Bombyx mori* (Kramer 1975, 1986). Silk moths hardly move spontaneously; they only move persistently while perceiving pheromone pulses (Kramer 1992). When this perception ceases, walking activity decreases to a complete stop after a few minutes. In addition to course control, the walking speed is modulated by the pheromone pulses (Kramer 1986; Kanzaki et al. 1992). f_2 (dc/dt) is also taken to control walking speed. For this purpose it is low-pass-filtered with an adequate time constant of a few seconds (low pass #3).

2.4.2. Flight

The variable which controls speed of flight is also f_2 (dc/dt), after having been filtered (low pass #4). The controlled condition, however, is the longitudinal component of the visual flux rather than the real air speed. The result of this linkage is that, the course angle always being set to upwind (though with varying gain), the moth accelerates/decelerates until the actual value of $g_4(f_2)$ is balanced by L ($a - w \times \cos \alpha$). This leads to a velocity of upwind progress which is proportional to $g_4(f_2)$; that is, the stronger the olfactory signal, the higher the velocity of upwind progress. When $g_4(f_2)$ decreases due to ceasing pulses, upwind progress gradually drops to zero (casting, Figs. 22.4 to 22.7).

2.4.3. Remarks on Locomotory Velocity

In contrast to course control, terrestrial and aerial locomotion are completely different in the nexus because of the compensation of wind drift. The assumption is needed that in reality the transition from walk to flight is also controlled by some magnitude of the g_x (f_2) type. In the male *B. mori* moth, I see a smooth transition between walk and flight. The intensity of wing vibration increases with stimulus intensity and duration such that the moth finally lifts its legs like in flight and very few moths (small, light, starved ones?) actually do take off if temperature is high.

3. Discussion

3.1. Altitude of Flight

The described nexus restricts to the orientation in a horizontal plane because there is little known about the control of absolute or relative flight altitude (Preiss and Futschek 1985; Witzgall and Arn 1990). With respect to control of relative altitude, it has been shown that *L. dispar* responds to shrinking or expanding ground patterns by a decrease or increase of lift (Preiss and Kramer 1986a,b). There is, however, no relevant indication that this reflex is used to stabilize flight altitude in odor plumes.

Changes in size of ground pattern images are unsuited to control absolute flight altitude, which is needed when real velocities with respect to air and ground are to be extracted from the image flow. In the nexus, L and T only depend on air speed, wind speed, and course angle. Actually L and T are in inverse proportion to the flight altitude. Whether moths do weigh the image flow with respect to actual flight altitude is unknown. Yet without such weighing, the controlling variable g_4 (f_2) could also be balanced by changing the flight altitude. In this case the expectation is as follows: The lower the altitude of the odor plume, the slower the upwind progression; the moths tend to reduce their flight altitude during the approach of the pheromone source as the pheromone concentration and pulse density increase.

3.2. General Properties

Numerous runs and flights with the model have demonstrated that qualitatively all types of realistic tracks can be obtained by setting the parameters to suited values. But there are also quantitative accordances with experimental data which have not been explicitly implemented—for example, the rather constant average ground speed which was first found by Marsh et al. (1978) in *Plodia interpunctella.* Meanwhile this has also been shown in several other moth species such as in the gypsy moth (Kuenen and Cardé 1994). This results from the simultaneous operation of the *L* and *T* feedback loops. The occurrence of casting after loss of contact with the plume is likewise resulting from the parallel action of these two loops. Due to the decreasing gain in the course-controlling loop, the amplitude of fluctuations of the course angle grows, while simultaneously the decreasing set value in the *L* loop is leading to a ground track more and more compressed with respect to the mean direction of flight. The implementation of a decay in anemotaxis during gaps between pheromone pulses was initially necessary to fit experimental data on walking silkmoths (Kramer 1986). Actually the nexus does not switch between the two apparent modes of "go" and "wait" in an all-or-none manner. These two "behaviors" are only the extreme cases of a continuum of responses produced by the same intercausal nexus.

3.3. Internal Turning Tendency and "Spontaneous" Flip-Flop

From behavioral and electrophysiological experiments we know that there exist, at least in *B. mori* (Kramer 1975; Olberg 1983a; Kanzaki this volume), neuronal structures that change activity like flip-flop circuits, upon either short sensory or internal inputs. Müller and Wehner (1988) report that search excursions of desert ants *Cataglyphis* sp. consist of a series of logarithmic spirals randomly alternating between clockwise and counterclockwise rotation. An alternating internal turning command has also been suggested by Wendler and Scharstein (1974) in the corn weevil, *Calandra granaria.* It is probable that an alternating internal turning tendency is common to all insects, and its primitive effect is to prevent the animals being caught in circular orientation.

In the computer model a flip-flop behavior is simulated by a random generator which changes the sign of an internal tendency to circle in random (or regular) time intervals. The parameters of this function may be influenced by the actual odor situation in different ways—for example, that a change of sign becomes either more or less likely if the contact to the odor plume is lost. In the actual version the mean frequency is in inverse proportion to the olfactory signal [g_3 (f_2)], which means that the time between changes is greater in the absence of pheromone than in its presence. The flip-flop may also be responsible for the regular sequences of counterturnings as is often observed in moths when flown in a wind tunnel (Kennedy 1983; Baker et al. 1984).

3.4. Efficacy of Hitting Target

The described nexus would not guide moths to a quasi-point-shaped target without considering a crucial condition which makes the system converge to the source. David (1982 et al.) and Elkinton (1987 et al.) have demonstrated elegantly that series of soap bubbles when released from one point maintain, to some extent, their initial direction of wind drift for tens of meters. As a consequence, such wind pockets loaded with pheromone have vectors of movement which scatter around the direction of the source rather than around the main direction of wind. This relation diminishes when moving downwind. Nevertheless, for a moth orienting in local wind the pheromone source coincides with a virtual source of wind; this coincidence is stronger when closer to the source. Although, when developing this model, odor distribution by turbulent diffusion was also implemented (Fig. 22.8), the model for computational reasons uses a simplified odor field. Therefore the above effect does not appear per se and needs to be imitated. This is done by shifting the actual angle λ of movement of air pockets which carry pheromone somewhat away from the mean wind direction onto the direction pointing to the source δ. The amount of this shift, the "preservation" $p(=\lambda/\delta)$, is a function of the distance to the source. It is defined here to be P_0 at the source and to decrease to almost zero at large distances by following the hyperbolic relation of $p = P_0 h/(h + d)$. In this relation, h denotes the distance d from the source in which the preservation P is reduced by 50%. In numerous runs of the model, it turns out that a preservation of only a few percent suffices to guide most of the model moths to the source. The model still has too large a number of free parameters to allow for quantitative statements, which is also true for most other aspects of the model; but if this feature is totally neglected, the moths do find the source, but only by chance.

3.5. Predictive Potency

The benefit of modeling is the ability to predict specific qualities not yet observed in the original. Such predictions can stimulate new types of experiments or give a new view on previous results. In the following, some points accomplished by the model are presented:

- Moths have an "internal turning tendency" which during olfactory orientation is counterbalanced by anemotactic turning commands.
- The direction (left/right) of this turning tendency changes (spontaneously or regularly), lest the efficacy of orientation be reduced dramatically.
- The occurrence of a change of turning direction is affected by the actual olfactory input.
- Changes of turning direction will not occur upon olfactory threshold events because these do not improve the efficacy of target finding in the model.

The above four points can be tested by experiments from which all external directional cues are excluded, and the pheromone stimulus can be applied to avoid any anemotactic response. Under such conditions the internal turning command would be the only one remaining, causing the moths to circle with changing direction. In contrast to noisy zigzagging, circling is easy to observe. The model allows that these experiments can be carried out with tethered, walking, or flying moths.

Other points are as follows:

- At different wind speeds, ground tracks of flying moths are expected to be similar with respect to average ground speed and amplitude of zigzagging, whereas in tracks of walking moths the amplitude of course fluctuations is lower at higher wind speeds. This is a consequence of the error signal of course being in both cases in proportion to the wind speed ($w \times \sin \alpha$; Figs. 22.6 and 22.7).
- Unilaterally deantennized moths can be expected to walk upwind with a tendency toward the ectomized side, which differs from tropotactic responses. The orientation of flying moths, however, should only be little affected by such amputation.

Acknowledgments

I would like to thank K.-E. Kaissling for his critical remarks and his presentation of this nexus at Wageningen, and I would like to thank L. Williams for making my English readable.

23

Pheromone-Mediated Flight in Moths

Thomas C. Baker and Neil J. Vickers

Dedicated to the memory of Professor John S. Kennedy

1. Introduction

Several explanations for the existence of zigzagging upwind flight in male moths responding to sex pheromone have been given over the course of research on this topic (e.g. Kennedy 1983; Preiss and Kramer 1986b; Baker and Haynes 1987; David and Kennedy 1987; David and Birch 1989; Baker 1990; Witzgall and Arn 1990; Willis and Arbas 1991a). Although a debate about the superficial shape of the flight tracks appears to be over-reductionistic, the root of the discussion involves the mechanisms that the moths use to maneuver and reach the source of pheromone in wind. As such, it is a very important debate. The behavioral mechanisms need to be precisely understood if we are to make sense of the underlying neuronal responses at the sensory, central nervous system, and motor levels and create robust neuroethological knowledge about this powerful and agriculturally important biological process called attraction.

Kennedy and Marsh (1974) and Marsh et al. (1978) convincingly demonstrated that optomotor anemotaxis was the primary mechanism used by flying male moths to locate pheromone sources. This purely anemotactic system included the wide reversals that the moths showed during casting in clean air, which were attributed to the mechanism "reversing anemomenotaxis" (Kennedy and Marsh 1974; Marsh et al. 1978). Importantly, no one had heretofore removed pheromone and observed what males would do in response to clean air, and these results became the first recognition that both clean air and pheromone shape the flight tracks of males. Preiss and Futschek (1985) later showed that the optomotor system is more responsive to image motion when pheromone is present than when it is absent.

The studies of Baker and Kuenen (1982) and Kuenen and Baker (1983) then revealed the existence of a nonanemotactic mechanism involved in zigzagging upwind flight; they concluded that such a mechanism is present, since the zigzags of *Grapholita molesta* males persisted in zero wind when no anemotaxis was possible. Also, the zigzags widened in zero wind when the males flew into clean, still air, a pattern similar to the onset of casting flight in males out of pheromone into clean wind (Baker and Kuenen 1982; Kuenen and Baker 1983). Earlier reports of persistent, regular reversals by *Adoxophyes orana* males when flying within homogeneous clouds of pheromone were reinterpreted (Kennedy 1982) and pheromone was now hypothesized to switch on and modulate through its concentration a programmed, self-steered series of reversals (counterturns) as well as switching on optomotor anemotaxis (Kennedy et al. 1980, 1981; Kennedy 1982, 1983).

A plume created by a point source of pheromone is not a time-averaged homogeneous cloud. Turbulence within the moving air causes the plume to break up into strands of odor-laden air (filaments) interspersed with pockets of clean air where little or no odor is present as first articulated by Wright (1958) (Murlis this volume). The physical intermittency of these filaments was found to be a requirement for male moths to sustain their upwind progress; this insight emerged from the experiments of Kennedy et al. (1980, 1981), Willis and Baker (1984), and Baker et al. (1985), with combinations of homogeneous clouds of pheromone, side-corridor clouds of pheromone, point-source plumes placed within clouds, and clouds alternated with clean air. Baker et al. (1985) reported that the tracks of males engulfed in clouds of pheromone were similar in appearance to casting tracks, as if the males had just lost pheromone, even though they were engulfed by a cloud of pheromone. The persistent counterturns, as during casting, could be due to habituation, which would cause the equivalent of clean air to be experienced; when the same cloud was interrupted with swaths of clean air, upwind flight with smaller side-to-side oscillations occurred (Baker et al. 1985).

An important advance in knowledge further occurred when it was shown by the very fast reaction times of *G. molesta* males to both pheromone ON and pheromone OFF that these males were likely responding to individual filaments of pheromone and pockets of clean air (Baker and Haynes 1987). The frequencies of filaments in plumes had been measured using electroantennograms (EAGs) in the field and in the laboratory (Baker and Haynes 1989), and the intervals between filaments were shown to be over twice as long as the reaction latencies of the moths (Baker and Haynes 1987). These insights eventually led to a model (Baker 1990) that hypothesized that flight tracks of males were shaped by the reaction latencies to filaments and pockets of clean air, as well as by the frequency of filaments encountered in a particular type of plume. Filament contacts would evoke upwind surges, and pockets of clean air following pheromone would elicit the casting, counterturning program. A reiterative phasic system of fast reactions to contact and loss of pheromone was hypothesized to override an underlying tonic system, which, when allowed to be expressed by longer periods of clean air, would drive a counterturning oscillator, causing casting to be expressed.

The model (Baker 1990) emphasized for the first time that casting flight is pheromone-mediated flight, even though it is initiated and persists in clean air for a long time. This long-duration, tonic casting system in clean air was linked with the occurrence of large-scale turbulence that swings the wind and the plume to create large pockets of odorless air. There is a need for a system to persist over long periods when clean air is present if recontact with filaments is to occur. Conversely, the fast-acting phasic system of upwind surges was linked to the occurrence of small-scale turbulence that creates the plume's finely structured filaments. The need for the surge to stop so quickly in clean air (i.e., to be so phasic) was attributed to the fact that the next pocket of clean air following a filament may be a big one due to large-scale turbulence, and to plunge too far upwind in such a pocket is to move off-line from the source and also to move away from the shifted plume (Baker 1990).

After the experiments of Kennedy et al. (1980, 1981) and of Baker and Kuenen (1982; also Kuenen and Baker 1983) a continuum was now hypothesized to exist between (a) zigzagging triggered by pheromone and (b) casting triggered by clean air (Kennedy 1983). Both types of tracks were said to involve a program of counterturning and anemotaxis, with higher-frequency counterturning and more upwind steering evoked by pheromone contact. Now, however, we believe that the counterturning reversals are expressed only when there is sufficient clean air *following* pheromone relative to the response latencies of a particular species. Additionally, counterturning might be expressed when the male experiences "clean" air after pheromone as a result of habituation when flying within a homogeneous cloud (Willis and Baker 1984; Baker et al. 1985; Kennedy et al. 1980, 1981) or when flying through filaments that have excessive concentrations of pheromone and thus render antennal neurons incapable of disadapting fast enough to respond to successive filaments (Baker et al. 1988), or when flying through filaments of substandard blend quality (Witzgall and Arn 1990).

Thus we currently believe (Baker 1990) that each above-threshold filament, instead of evoking a continuum of counterturning (from casting to zigzagging depending upon the above factors), causes an upwind surge; and if repeated contacts with pheromone filaments are frequent enough, then reiterative surging will occur, resulting in straight upwind flight (Baker 1990). Flight by all moths could therefore be straight upwind if the filament quality and quantity are good enough relative to the reaction latencies of the species in question. If not, hybrid partially surging and partially casting tracks would be produced (zigzagging tracks), since clean air time interspersing filaments would allow periodic expression of the counterturning program in between surges. Although this model (Baker 1990) is superficially similar to the original idea of Kennedy and Marsh (1974; straight upwind flight in pheromone, casting in clean air), it has the distinction of incorporating the nonanemotactic component, self-steered counterturning, which previously had not been discovered (Baker and Kuenen 1982; Kuenen and Baker 1983). This model is also distinct from that of Kennedy (1983,

1986), because straight upwind flight as potentially resulting in all moths from proper pheromone filament stimulation is now included.

In a recent series of experiments by two independent pheromone research groups, this hypothesis was given substantial support (Mafra-Neto and Cardé 1994; Vickers and Baker 1994b). However, before detailing these most recent results, we would like to take some time to review the evidence for the counterturning program as an orientation mechanism, because doubts are still harbored by some groups about the existence of this mechanism (Witzgall and Arn 1990; Witzgall this volume).

2. Evidence for the Counterturning Program

2.1. Experiments in Zero Wind

Preiss and Kramer (1986b) attributed the presence of zigzags in tracks to the inability of males to steer precisely upwind when in pheromone, and they did not recognize counterturning as an element used in pheromone-mediated flight. They hypothesized that because of optomotor error in detecting transverse image drift and attaining the desired 0° (straight upwind) set-point, males steer through too much of an angle off to one side of the windline. They then attempt to compensate by reversing direction and overshoot 0° in the opposite direction. However, experiments in zero wind with *G. molesta* (Baker and Kuenen 1982; Kuenen and Baker 1983; Baker et al. 1984) and *Lymantria dispar* (David and Kennedy 1987) do not support this mechanism, but rather support the existence of counterturning. In pheromone, the zigzags of males of these two species continued *even in zero wind* and thus the Preiss and Kramer (1986b) mechanism could not be involved in the oscillations because there was no wind-induced drift to trigger any reversals (Baker and Kuenen 1982; Kuenen and Baker 1983; David and Kennedy 1987). Even more telling, when males zigzagging in a stationary pheromone plume (created by an instantaneous stopping of the wind) emerged into clean, still air, their reversals became wider and slower, with even more temporal and angular regularity than when in pheromone (Baker and Kuenen 1982; Kuenen and Baker 1983; David and Kennedy 1987). These tracks in clean zero wind, characteristic of casting flight in clean air but again occurring *without* the aid of wind-induced transverse drift, exhibited the temporally regular reversals characteristic of casting as well. Thus, pheromone exposure has also been shown to modulate the self-steered counterturning program, devoid of any possible changes in an anemotactic error function.

In another study, males took off in zigzagging flight in zero wind in pheromone, but flew with no zigzags in clean, still air (Baker et al. 1984). Thus, counterturns can *originate* in males exposed to pheromone under windless conditions without the need for wind to establish some kind of turning pattern prior to wind stoppage (Baker et al. 1984). The ability of pheromone to initiate counterturns in zero wind further establishes this self-steered system as a component distinct from anemotaxis in pheromone-mediated flight.

There is no evidence that moths that counterturn in zero wind do so with the aid of optical feedback, even though their successive turns have an angular regularity (Baker et al. 1984) in addition to a temporal regularity. If moths were using such optical feedback, perhaps it would involve a change in the *sign* of direction of rotation of their visual flow field, with males turning clockwise or anticlockwise until they receive a certain amount of rotational stimulation from their own movements and then reverse the movements. On the other hand, the regularity in the number of degrees turned may be due to the regularity of the tempo of counterturning in the two moth species in which these experiments have been performed (Baker and Kuenen 1982; Kuenen and Baker 1983; Baker et al. 1984; David and Kennedy 1987), which would fortuitously take the moth through a prescribed amount of degrees before the turn reversal generator is activated. Thus, without external feedback, such an internal program could conceivably produce an orderly succession of angular amplitudes during regular reversals in turning. When there is wind, the addition of drift may help trigger a turn back before the full number of rotational degrees has been accrued (David and Kennedy 1987) because transverse image motion may add to the degrees apparently turned through or may help trigger a reversal.

2.2. Regularity of the Turning Tempo

A program of counterturning with or without wind should be characterized by either an angular or a temporal regularity in the reversals; this regularity is quite obvious in clean air in the three species in which it has been analyzed; *G. molesta, L. dispar,* and *Amyelois transitella* (Von Keyserlingk 1984; David and Kennedy 1987; Haynes and Baker 1989). In clean wind, the temporal regularity of the reversals during casting is striking (Figs. 23.1A–C) and is evidence for a program of temporally regular oscillations that in wind is now polarized by optomotor anemotaxis to result in a spatial distribution of track angles that are centered about the windline. There was also clearly an angular and temporal regularity

Figure 23.1. The tempo of casting flight in clean wind as indicated by the lateral positions of males of three species plotted against time. (**a**) *G. molesta* male's lateral position (cm) in the wind tunnel while flying upwind in pheromone (left of clear arrow), while casting after pheromone loss (right of clear arrow), and after regaining the plume that has shifted over (right of black arrow). Note regular but diminishing tempo of turns as the exposure to clean air continues. [From Von Keyserlingk (1984).] (**b**) *A. transitella* female (top) and male (bottom) flight tracks during casting after losing odor plume (right of dashed line in both figures). [From Haynes and Baker (1989).] (**c**) *L. dispar* male's tempo of counterturning during casting flight in wind, as measured by track angle (top) and turning rate (bottom). [From David and Kennedy (1987).]

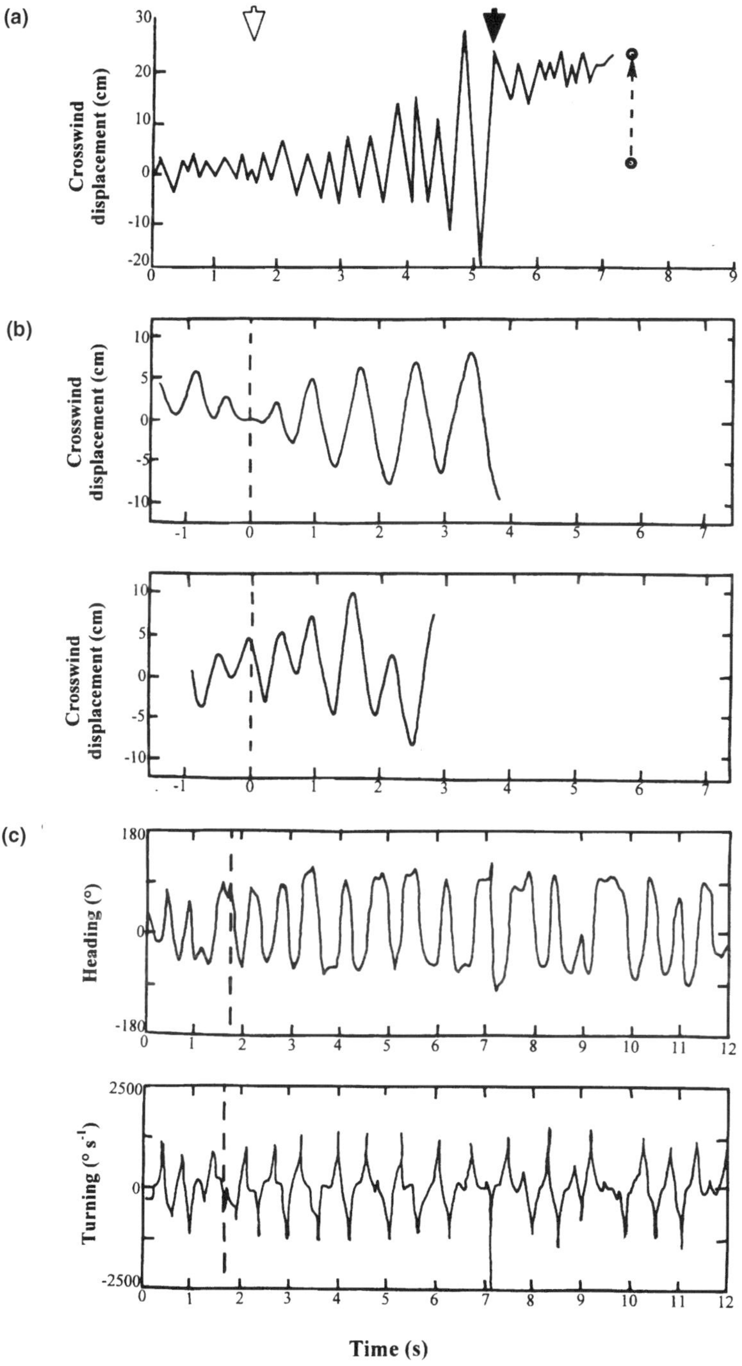

(a)
Crosswind displacement (cm)
30
20
10
0
-10
-20
0 1 2 3 4 5 6 7 8 9
(b)
Crosswind displacement (cm)
10
5
0
-5
-10
-1 0 1 2 3 4 5 6 7
Crosswind displacement (cm)
10
5
0
-5
-10
-1 0 1 2 3 4 5 6 7
(c)
Heading (°)
180
0
-180
0 1 2 3 4 5 6 7 8 9 10 11 12
Turning (° s-1)
2500
0
-2500
0 1 2 3 4 5 6 7 8 9 10 11 12
Time (s)

when moths flew in pheromone in zero wind (Baker and Kuenen 1982; Kuenen and Baker 1983; Baker et al. 1984) as well as out of pheromone during casting in zero wind (Baker and Kuenen 1982; Kuenen and Baker 1983; David and Kennedy 1987), the regularly repeating magnitude of successive angles but of opposite rotation providing evidence for a regularly timed reversal of direction. As the program of counterturns slows in tempo following the time since pheromone loss (Fig. 23.2), contact with pheromone can truncate the program and

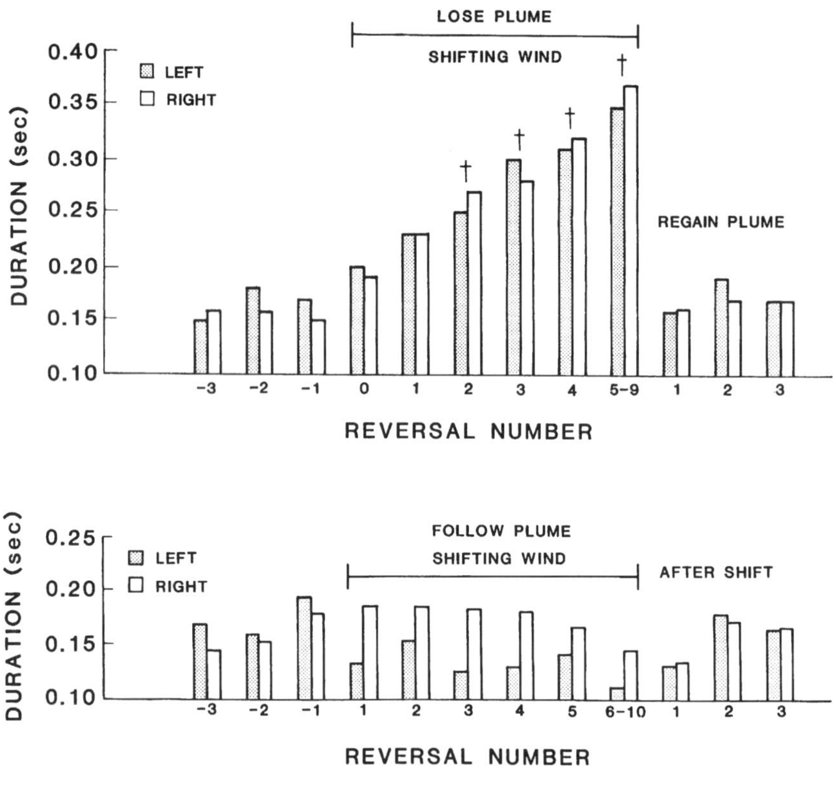

Figure 23.2. The mean inter-reversal durations of 20 *G. molesta* males that lost the plume completely and began casting (top) and of 16 males that followed the plume across and reiteratively recontacted the plume before each left-hand track leg (bottom) during an experimentally produced wind-shift of 35°. Shaded bars are during left-hand legs, while white bars are during right-hand legs, at the indicated reversal number before or after losing the plume and after regaining it (top), or before, during, and after the wind-shift (bottom). Note the regular, incremental increase in the inter-reversal duration (top) as time since pheromone contact increases, which is indicative of a retarding of the tempo of counterturning with time after pheromone contact. The tempo returns to a faster state immediately after the plume is contacted again, as well as on each leftward leg after plume contact (bottom). [From Baker and Haynes (1987).]

create shorter-duration reversals to narrow the track during an upwind surge (Baker and Haynes 1987). Exposure to clean air, beginning even on the next reversal, was shown to allow the program to begin to be expressed again and to play out with increasingly retarded reversals unless it was interrupted once more by contact with pheromone. Thus, it may be that the upwind surge is coincident with the suppression of the *expression* of casting. The results of Kanzaki et al. (1994; see also Kanzaki this volume) with walking *Bombyx mori* seem to indicate that this type of system is operating to produce reiterative straight-line rapid surges with successive pulses of pheromone that decay into increasingly greater side-to-side oscillations as the time since contact with a filament lengthens.

The reiterative interception of filaments on rightward track legs and clean air on leftward track legs revealed the rapid resetting and relaxation of the counterturning frequency, contributing to sawtooth-shaped, asymmetric tracks in males flying in an experimentally shifted plume or along a side corridor of pheromone (Baker and Haynes 1987). Thus, Baker (1990) hypothesized that it is just a question of how long it has been since the last pheromone filament was contacted and the moths' typical species-specific latency to begin casting that will determine how much casting can be expressed before a filament causes an upwind surge and resets the counterturning program. In combination with a resetting of the anemotactic system to straighter upwind, either straight-upwind or zigzagging upwind flight would occur, depending upon the above factors.

2.3. The Question of Bimodality of Track Angles

Much has been made lately of the question of bimodality or unimodality of track and course angles in the analysis of male flight tracks. Preiss and Kramer (1986b) considered such bimodality to be merely an extension of the anemotactic error system that is modulated by pheromone loss. Witzgall and Arn (1990) considered bimodality to be a function of the quality of pheromone used, with the poorer blend quality presumably increasing the imprecision of upwind steering as per the Preiss and Kramer model. Baker (1990) and Kaissling and Kramer (1990) viewed the amount of bimodality as being a function of the amount of casting that is expressed in a given track. Purely upwind, unimodal distributions would indicate that filaments of sufficient quality are being encountered by the male with such frequency that reiterative phasic upwind surges are evoked, to the exclusion of the counterturning program. Even brief episodes of casting thus could not occur in what little clean air is present between filaments.

All species of moths examined thus far in free flight begin casting when emerging into clean air from a pheromone- or host-odor plume. It is during casting that bimodality in track and course angle distributions are always and most obviously present (Fig. 23.3). Therefore, it is logical to conclude that the degree of bimodality in a distribution of track angles in moths flying upwind to pheromone should be an indication of the amount of casting that is being expressed

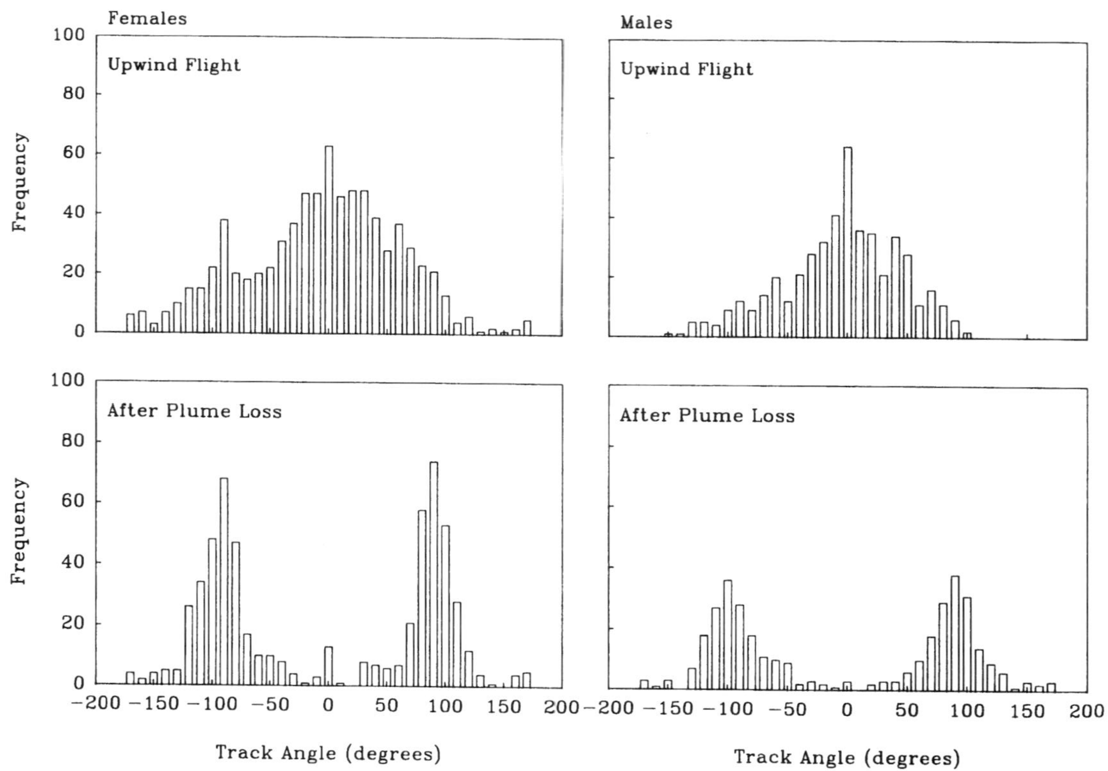

Figure 23.3. Track angles of eight female and eight male *A. transitella* moths as they flew upwind in a plume of host odor and extract from female pheromone glands, respectively (top), or engaged in casting flight following removal of the plume (bottom). [From Haynes and Baker (1989).]

in that track, or, in other words, an indication of the amount of clean air that the moth is experiencing (Fig. 23.3). Unimodal tracks of males of some species flying in pheromone plumes are not proof that a counterturning program is not used by such males, only that under the conditions used to evoke such flight, there is not enough clean air in which to observe casting as the expression of the counterturning program. As mentioned above, all moths show bimodality of course and track angles during casting. However, during flight in a pheromone plume, it is known that unimodality is dependent on the species, the wind speed (Willis and Cardé 1990; Willis and Arbas 1991a; Willis et al. 1991), and other conditions such as pheromone blend quality (Witzgall and Arn 1990), which may or may not result in bimodality of the distributions of track and course angles.

The work of Preiss and Kramer (1986b) in which removal of pheromone caused no obvious counterturning-based casting flight is the only exception to clean-air casting in moths. However, the moths in their experiments were not in free flight. Rather, they were tethered and could only rotate in the yaw plane; their ability to roll or change their pitch, both known to occur in free flight as part of normal maneuvering, was restricted. As outlined by David and Kennedy (1987), there are many problems with the tethering technique used by Preiss and Kramer (1986b) that restricted normal male maneuvers, as well as with the pheromone delivery technique, all of which could explain the unusual behavior seen in these experiments. In *Lobesia botrana,* the other species for which it is claimed that no counterturning occurs (when natural pheromone is used), casting flight in clean air has never been attempted to be evoked by removing pheromone (Witzgall and Arn 1990), and so the story is incomplete for this species.

2.4. The Question of Blend Quality

Witzgall and Arn (1990), who, similar to Preiss and Kramer (1986), questioned the need to invoke the counterturning program as a mechanism in pheromone source location, stated, "It is worth noting that most hypotheses on an internal counterturning program have been derived from experiments with synthetic compounds." This statement is incorrect, however, because in actuality the hypotheses supporting a counterturning program [or "reversing anemomenotaxis" of Kennedy and Marsh (1974)] were derived by using *clean air following pheromone* to evoke casting flight. A self-steered counterturning program, in addition to anemotaxis, is an important and most obvious underlying element to casting flight in clean wind (see above). In all species that have been examined, clean air following pheromone exposure reliably turns on casting flight, regardless of which blends of pheromone, synthetic or natural, are used.

Interestingly, the experiments of Haynes and Baker (1989), cited by Witzgall and Arn (1990) as an example of the value of using natural pheromone to produce nearly straight upwind flight in plumes, also included the evocation of casting flight in clean air following removal of these plumes of pheromone for males, or of natural host odor for females. Thus, even with optimal natural pheromone, it is only a matter of time before males will express counterturning-based casting after pheromone is removed and clean air is experienced [see also Willis and Arbas (1991a) for comparisons of male and female casting after gland-derived pheromone and host-odor loss, respectively]. It is not difficult to imagine then that even in such plumes of natural odors, the latency of the transformation of upwind flight into casting flight and the amount of time that elapses between contact with filaments will determine how frequently casting will be expressed and therefore how zigzaggy the track will be (how bimodally distributed the track angles will be) (Baker 1990). If filament contact is just frequent enough to prevent casting from being expressed (the interval between filament contacts is just shorter than the postfilament casting latency), then the missing of a filament will cause the beginning of counterturning to be expressed. Some short episodes of casting can be seen even in these flight tracks to natural odor (see Fig. 1 of Baker and Haynes 1987). The absence of filaments, such as in a large pocket of clean air due to large-scale turbulence, will allow the casting program of counterturns to be expressed even more fully.

Poorer-quality filaments will be less likely to be above threshold (less likely to cause upwind surges) than better-quality filaments. Such relationships between pheromone blend quality and upwind flight thresholds have been demonstrated repeatedly (Baker et al. 1981; Linn et al. 1984, 1986a), with off-blends being able to evoke sustained upwind flight, but at higher concentrations than optimal blends. When projected to the level of individual filaments, blend threshold differences of this type should translate into a lower *effective* frequency of filament contact by males flying in suboptimal blends and result in more clean-air time for casting to be expressed. Such is probably the case in the zigzagging tracks of male *L. botrana* (Witzgall and Arn 1990) when the (imperfect) synthetic blend was used compared to straighter tracks in response to the natural blend. More casting will be initiated and fewer upwind surges will occur in such plumes, and the tracks will thus exhibit more bimodality of course and track angles than flight in the natural plume, even though the physical structure (filament frequency) will be on average the same in both. The effective filament frequency will be greater in the optimal, natural plume of pheromone to which the male's threshold for responding to filaments is lower. It is thus not surprising that an optimal blend and emission rate, as afforded by calling females, should produce more direct and less zigzaggy upwind flight than will a synthetic blend.

Synthetic blends are not per se the cause of zigzagging flight tracks, since, as mentioned above, the interaction of filament encounter rate, moth reaction latencies, and filament blend quality and concentration all shape the track, and a synthetic blend at one filament frequency that causes zigzagging tracks can cause straight upwind flight when the filament frequency is increased to produce only successive surges (see below) (Mafra-Neto and Cardé 1994; Vickers and Baker 1994b). Conversely, the use of natural pheromone or an optimal synthetic pheromone blend will not ensure straight upwind flight if the plume structure is suboptimal.

3. Recent Advances

It is clear from past data and from recent experiments that the amount of bimodality of track or course angles in moth flight tracks is an indication of how much casting is occurring in these tracks, and the amount of casting is a measure of how frequently filaments of sufficient quality and concentration are being encountered by antennae to evoke upwind surges. This is quite apparent in the recent results of Mafra-Neto and Cardé (1994) and Vickers and Baker (1994b) in two different species of moths. The following is a summary of the findings of Vickers and Baker (1994b). A summary of Mafra-Neto and Cardé's (1994) results can be found in Cardé and Mafra-Neto (this volume).

Vickers and Baker (1994b) characterized single upwind surges of flying male *Heliothis virescens* moths in response to individual strands of pheromone generated experimentally in a wind tunnel, using a pulsing device. All males flying upwind in a 10-filament sec^{-1} pulsed plume responded to the truncation of the plume by casting. Males that intercepted one of the last filaments produced by the pulsing device prior to plume truncation began casting within an average of 0.27 sec. Of these casting males, 32% (n = 192) subsequently responded to a single filament by making a short upwind surge followed by a reversion to casting. In many cases the complete sequence of casting following plume truncation, interception of a single pulse, ensuing surge, and lapse back to casting flight was not captured entirely within the camera's field of view. Hence, only 13 (or 7%) of the responding males' tracks could be used. No casting males exhibited any upwind displacement in the control where no filament was generated.

The latency between filament interception and the surge was found to average 0.30 sec, while the surge itself had an average duration of 0.38 sec. The males covered an average of 13.4 cm in the upwind direction during this time. By calculating average track angles and average ground speeds for each 1/30th of a second, synchronized by the passage of the filament past each of the 13 responding moths, we produced an average cast–surge–cast track that then served as a template for later comparisons (Vickers and Baker 1994b). Counterturning frequencies, higher during (3.33 counterturns sec^{-1}) and lower before and after (2.38 counterturns sec^{-1}) the surge, were also integrated into the cast–surge–cast template (Vickers and Baker 1994b).

When flown upwind to pulsed plumes generated at varying frequencies, males exhibited the template for a single iteration, complete with cross-wind casting both before and after the straighter upwind surging portion, at the lowest frequency capable of sustaining successful upwind flight to the source, as hypothesized by an earlier models (Baker 1990; Kaissling and Kramer 1990). Also as predicted, when the pulse generation rate was increased and filament contact by males became more frequent, only the straightest upwind portions of the surges were reiterated, producing direct upwind flight with little cross-wind casting (Fig. 23.4).

Males were not able to fly upwind in plumes consisting of fewer than 4 filaments per second, although they did become activated and cast for short periods of time above the take-off point. This is in close agreement with our previous observational accounts detailing males' ability to fly upwind in pulsed plumes generated at various frequencies (Vickers 1992; Vickers and Baker 1992). The tracks in response to 4 filaments sec^{-1} (Fig. 23.4) were tortuous in shape with a bimodal distribution of track angle vectors around ± 90° indicative of more cross-wind casting flight. In contrast, at 10 filaments sec^{-1}, where many tracks had extended periods of nearly straight upwind flight and fewer periods of cross-wind casting, track angles were unimodally distributed about 0°. At 5 filaments sec^{-1} (Fig. 23.4) the shape of the flight tracks and the distribution of track angles were intermediate in appearance between those of the 4-filament sec^{-1} and 10-filament sec^{-1} frequencies. Since only filament frequency was varied in these experiments, the more straight upwind distribution of track angles cannot be attributed to a compensatory response to changes in wind velocity.

Many of the actual flight tracks produced in response to 4 filaments sec^{-1}, a sufficient frequency to sustain upwind flight, exhibit identifiable templates of an average moth responding to a single pulse (surge) followed by clean air (cast) strung together in succession (Fig. 23.4). Occasionally males ceased upwind progress at the higher pulse frequencies. Importantly, the initial upwind movement following interception of the pulses anew (since upwind progress was never observed in males casting in clean air) resembled the surge part of the cast–surge–

Figure 23.4. (**a**) Flight tracks (Actual) of two male *H. virescens* in response to a pulsed plume generated at 4 filaments sec^{-1}. The cast–surge–cast template (Template) (Vickers and Baker 1994b) laid end to end indicates that the upwind progress of these males is due to the repetitive evocation of single surges interspersed by casting. (**b**) Two flight tracks of males responding to a pulsed plume generated at a rate of 5 filaments sec^{-1}. The tracks are more upwind than tracks of males flying to 4 filaments sec^{-1} and less upwind than those of males responding to plumes generated at 10 filaments sec^{-1}. This is reflected by the intermediate distribution of track angle vectors. The tracks are now not obviously composed entirely of the cast–surge–cast template laid end to end (as at 4 filaments sec^{-1}), but there are instances when the template does reveal itself within the flight track. (**c**) Flight tracks of two males in response to a plume generated at a rate of 10 filaments sec^{-1}. The middle, upwindmost part of the surge is reiteratively evoked by fast filament contact resulting in straighter upwind tracks compared to 4 and 5 filaments sec^{-1}, respectively. When males did lose the train of pulses at these higher production rates (lower Actual track), the initial part of the recovery following pulse recontact resembled the first part of the cast–surge–cast template, before becoming canalized more upwind once again.

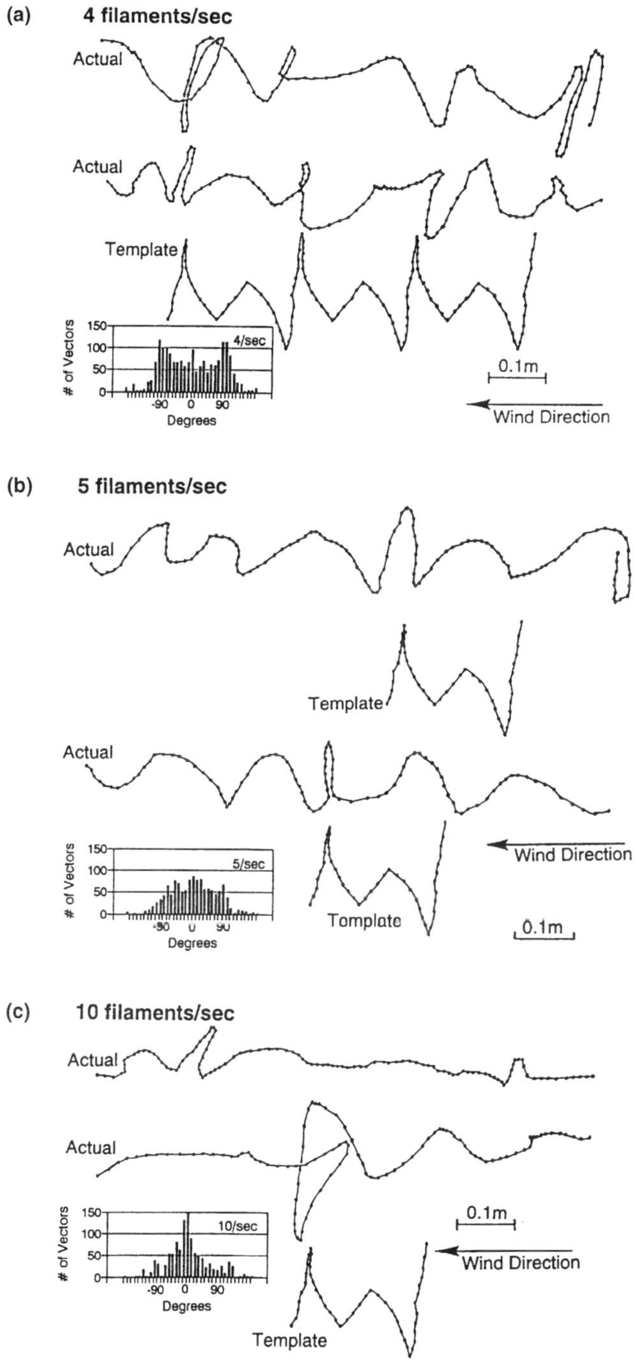

(a)
4 filaments/sec
Actual
Actual
Template
# of Vectors
150
100
50
0
4/sec
-90 0 90
Degrees
0.1m
Wind Direction
(b)
5 filaments/sec
Actual
Template
Actual
# of Vectors
150
100
50
0
5/sec
-90 0 90
Degrees
Template
Wind Direction
0.1m
(c)
10 filaments/sec
Actual
Actual
# of Vectors
150
100
50
0
10/sec
-90 0 90
Degrees
0.1m
Wind Direction
Template

cast template (Fig. 23.4). Thereafter the tracks became canalized directly upwind because faster contact with filaments resulted in the repeated evocation of only the middle, upwindmost section of the surge without time for casting to be expressed within the shorter time-span pockets of clean air.

In-flight EAGs made from males in free flight upwind in a pheromone plume further support the idea that reiterated straight surges occur when a high frequency of filaments is encountered under the usual point-source pheromone plume conditions and results in more direct upwind flight (Vickers and Baker 1994b). Intervals between successive hits with pheromone, as indicated by EAG activity, showed that males making relatively steady upwind progress with no overt casting intercepted more than 5 filaments sec^{-1} (mean interval: 0.19 sec), which was greater than for males flying crosswind (less than 4 filaments sec^{-1}; mean interval: 0.28 sec) or for antennae held stationary in the plume before flight (3.4 filaments sec^{-1}) (Vickers and Baker 1994b).

In-flight EAGs also showed that when filament contacts cease, the casting, counterturning program begins to be expressed after a latency period of 0.30 sec. Males that were flying crosswind upon encountering a filament turned their tracks upwind in an average of 0.23 sec following contact. These response latencies, this time measured in a typical pheromone plume emitted by a standard dispenser, correspond closely to the results from the single-pulse experiment confirming that the surge duration in response to pheromone was 0.38 sec before reverting to casting flight, and then casting males responded to a filament itself within 0.30 sec.

With the recent realization of the importance of measuring reaction times to odor onset and loss as being critical to understanding odor-mediated orientation by moths (Baker 1989b, 1990), it is not surprising that the reaction times to odor loss have only been estimated for a few species (Marsh et al. 1978; Baker and Haynes 1987; Baker and Vogt 1988; Haynes and Baker 1989; Willis and Arbas 1991a; Vickers 1992). Moreover, only in *G. molesta* (Baker and Haynes 1987), and *H. virescens* (Vickers 1992; Vickers and Baker 1994b) have the response

---

Figure 23.5. Flight tracks of two males viewed from above together with the simultaneously recorded electroantennogram activity of a third antenna transported by each flying male. The first male **(a)** is hit by one filament just before, and two filaments just as, it takes flight, producing an upwind trajectory. During this surge, the male is hit by three filaments (gaps between filaments are F–#1, 0.18 sec; #1–#2, 0.18 sec; and #2–#3, 0.08 sec). Filament #1 appears to have no effect on the behavior because after the average 0.23-sec latency period, now 0.26 sec later (just after #3), the moth begins to go into casting flight across the windline to the left. Filaments #2 and #3 occurred too late in the surge to prevent this cast, but 0.27 sec later (and before casting is fully established) the moth does begin another upwind surge. The male then contacts three more filaments in quick succession—filaments #4 (0.17 sec after #3), #5 (0.23 sec after #4), and #6 (0.13 sec after #5)—and these are apparently sufficiently frequent to sustain the upwind movement.

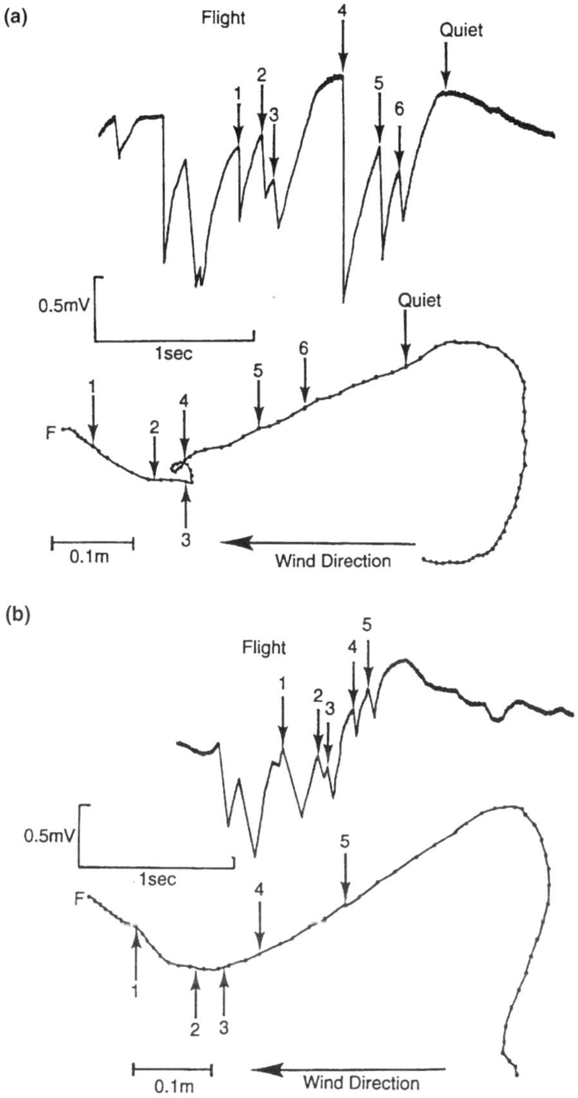

Figure 23.5. (Con't.) Following filament #6, the moth turns its track across the windline after 0.5 sec without filament contact. In **(b)** the male initiates flight in the upwind direction. Almost at take-off he receives a filament and then again shortly after taking flight (0.27 sec #1). Upwind flight appears to be sustained owing to the sequential arrival of filaments with short gaps between them (0.20 sec #1–#2; 0.07 sec #2–#3; 0.08 sec #3–#4; 0.17 sec #4–#5). After filament #5, there is a latency period of 0.43 sec without further contact with pheromone filaments before the male turns its track crosswind on a casting leg. [From Vickers and Baker (1994b).]

times (latencies and durations) to pheromone *onset* been measured. Our findings that *H. virescens* males react more slowly to clean air and hence have longer-duration surges than *G. molesta* males support the phasic–tonic model for upwind flight and explain why *H. virescens* males can have much straighter tracks than *G. molesta,* whose tracks are nearly always zigzag-shaped. *H. virescens* males will be less likely to go into casting flight and zigzag like *G. molesta* in plumes of similar structure (and filament frequency) because *H. virescens'* longer-duration surges carry them into filaments more often before casting can be fully expressed, as evidenced by their straight upwind flight tracks in the laboratory and field (Vickers 1992; Vickers and Baker 1994b). As alluded to earlier, recent results from another species, *Cadra cautella,* show relationships between track straightness and response to single filaments (Mafra-Neto, 1993; Mafra-Neto and Cardé 1994; Cardé and Mafra-Neto this volume) remarkably similar to those found in *H. virescens* (Vickers 1992; Vickers and Baker 1994b). Results from both studies support the filament-frequency-based surging–casting model of pheromone source location (Baker 1990).

They also support the existence of two mechanisms that create the surging–casting interplay: a self-steered counterturning program integrated with optomotor anemotaxis (Kennedy 1983; Kuenen and Baker 1983; Kennedy 1986).

Acknowledgments

We gratefully acknowledge the continuing support of the U.S. Department of Agriculture in this work carried out under grants 87-CRCR-1-2577, 92-37302-7636 and 93-37302-8329 (to T.C.B.).

24

Modulation of Pheromone-Mediated Flight in Male Moths

Peter Witzgall

1. Pheromone-Mediated Optomotor Anemotaxis

Male moths find sources of sex pheromone by flying into the scented wind. The upwind flight response is triggered immediately upon perception of pheromone, up to many meters downwind from a source releasing nanogram amounts per hour. This behavior has fascinated naturalists for centuries and has inspired studies on the chemistry of the female signal, the male receptor physiology, and the underlying orientation mechanism. The strength of the male response has even encouraged speculations and research toward the use of pheromones to control harmful insects.

In flight, the moth cannot sense the wind directly, but visually detects the wind direction through its displacement over the ground. While flying due upwind, the image flow is in line with the body axis; during lateral excursions, the male experiences visual drift (Kennedy 1940; Kennedy and Marsh 1974; Marsh et al. 1978; David et al. 1983; Cardé 1984; David 1986; Preiss and Kramer 1986a,b).

The upwind orientation behavior has been termed "zigzagging," due to a meandering flight path with irregular excursions across the wind line. This sequence of track reversals is thought to be the result of visually steered upwind flight (optomotor anemotaxis) integrated with an internal program of counterturns (Kennedy et al. 1980, 1981; Baker and Kuenen 1982; Kuenen and Baker 1983; Kennedy 1983, 1986; Baker et al. 1984; Baker 1990). However, experiments with tethered moths in a flight simulator have failed to confirm self-steered counterturning: In regard to "the male's inability to fly precisely upwind," optomotor anemotaxis alone suffices to explain the observed flight pattern (Preiss and Kramer 1986a,b).

The anemotactic response is modulated by pheromone; most studies on male orientation behavior have relied on synthetic chemicals, released from rubber septa or filter paper. Yet, there is evidence that male flight behavior changes with the composition of the pheromone blend (Palaniswamy et al. 1983; Willis and Baker 1987, 1988; Witzgall 1990) and that the behavior toward synthetic pheromone and calling females is not the same (Vetter and Baker 1983, 1984; Sanders 1985b, 1986; Baker et al. 1991). Male *Lobesia botrana* are capable of flying almost straight upwind toward calling females; convoluted flights are only obtained with synthetic pheromone and obviously result from an inadequate stimulus (Witzgall and Arn 1990, 1991). This has now been confirmed with two other tortricid species, *Grapholita molesta* and *Eupoecilia ambiguella* (P. Witzgall et al., in preparation).

2. Flight Recording in the Wind Tunnel

The flight section of the wind tunnel was 165 × 63 × 63 cm. The position of the moths was recorded on-line by two CCD cameras (25 frames sec^{-1}) and an image processor, at a distance of 100 to 50 cm from the pheromone source. The live camera images and the horizontal and vertical projections of the insect's track were displayed simultaneously on video monitors to facilitate visual observations. The spatial structure of the flight tracks was calculated off-line (Witzgall and Arn 1990, 1991).

Males of the European grape moth, *Lobesia botrana,* and the Oriental fruit moth, *Grapholita molesta* (Lepid., Tortricidae), were flown to conspecific females and to optimized or incomplete synthetic blends (*L. botrana,* ternary blend: 1 ng E7, Z9-12Ac, 0.2 ng E7, Z9-12OH, 0.05 ng Z9-12Ac; main compound alone: 1 ng E7, Z9-12Ac; *G. molesta,* ternary blend: 1 ng Z8-12Ac, 0.2 ng Z8-12OH, 0.06 ng E8-12Ac; binary blend: 1 ng Z8-12Ac, 0.06 ng E8-12Ac) (Linn and Roelofs 1983; Arn et al. 1988).

A calling female or a filter paper with synthetic pheromone was confined in a glass bulb (2 cm ID), leading into a glass capillary (1 cm long, 2 mm ID) (see Witzgall and Frérot 1989). Air from a gas cylinder left the capillary at 30 cm sec^{-1}. The glass bulb was behind a perforated metal sheet (63 × 63 cm, pore size 3 mm), and the capillary outlet was fitted into a pore.

3. Geometry of Flight Trajectories

3.1. Track and Course

The male's *track angle* and *ground speed* result from its *course angle* to the wind, its *air speed,* and the *wind speed* (Fig. 24.1; Marsh et al. 1978). The ratio of air speed to wind speed determines accordingly the track angle, as a function of the course angle. For flights of *L. botrana* males to a calling female and to an optimized synthetic blend, at a wind speed of 30 cm sec^{-1}, this ratio was 2.3 and 1.5, respectively (Fig. 24.1). A reduced air speed during flights to synthetic

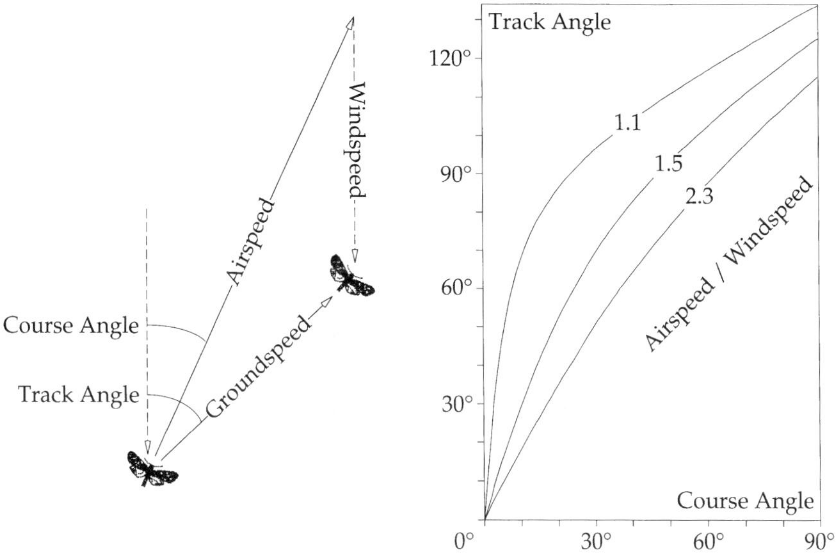

Figure 24.1. Triangle of velocities [after Marsh et al. (1978)] and relation between course and track angle, at three different ratios of air speed to wind speed [after Preiss and Kramer (1986a)]. Mean ratios were calculated from flights of *L. botrana* ($N = 60$) to a calling female at a wind speed of 30 cm sec^{-1} (2.3) and to synthetic blend (see Section 2), at a windspeed of 30 cm sec^{-1} (1.5) and 60 cm sec^{-1} (1.1).

pheromone, instead of flights to calling females, enhances the effect of wind-induced drift; a given course angle results in a larger track angle.

Upon loss of contact with the pheromone plume, net upwind progression is reduced to almost zero (Marsh et al. 1978) and the optomotor control of flight direction and flight altitude is greatly diminished (Preiss and Futschek 1985); the male is thus "casting" in wide loops transverse to the wind line (Kennedy and Marsh 1974; Marsh et al. 1978; Preiss and Kramer 1986a; Kuenen and Cardé 1994). Changes in flight direction may be steered by an internal counterturn generator or by visual stimuli.

Track angle and ground speed can be measured from flight track recordings; course and thrust can only be measured from tethered insects, flying in optically simulated wind (Preiss and Kramer 1986b). It should not be overlooked that course angles computed from flight tracks, using the triangle of velocities (Fig. 24.1), reflect the sum of both the set-point of the male's course and course corrections. Compensatory flight maneuvers result from inaccuracies in detecting the wind direction (Preiss and Kramer 1986a,b), as well as from latencies in the response to visual drift and airborne odor (Kaissling and Kramer 1990). Therefore, a bimodal frequency distribution of course angles calculated from flight tracks

Figure 24.2. Bicycle tracks on asphalt covered with snow. The cyclists' inability to ride straight results in a meandering track. Lateral deviations are followed by course corrections, as shown by divergent tracks of front and hind wheel. (View picture by turning 90° counterclockwise.)

does not necessarily derive from a counterturning program (Figs. 24.2–24.4; Marsh et al. 1978; Willis and Baker 1987; Willis et al. 1991).

The angular and temporal resolution for detection of and response to wind-induced drift have not yet been measured; but they cannot be neglected, especially in turbulent wind. In Fig. 24.3, the moths' position is plotted at intervals of 0.04 sec; the duration of "interturn reversals," in the horizontal projection, is up to 0.36 sec. The response time of antennal receptors to a pulsed pheromone stimulus is about 0.1 sec in *Antheraea polyphemus* (Kaissling 1986b; Rumbo and Kaissling 1989). The latency for generation of a wing-fanning or turning response to an odor pulse is 0.2 to 0.3 sec in walking *Bombyx mori* (Kaissling and Priesner 1970; Kramer 1986), and flying males of four different species reacted to loss of the pheromone stimulus within 0.15 to 0.5 sec (Baker and Haynes 1987; Baker and Vogt 1988; Haynes and Baker 1989; Willis and Arbas 1991a).

3.2. Spatial Structure of Upwind Flights

Analysis of moth flight trajectories has mostly been restricted to the horizontal plane, despite both vertical and horizontal deviations from the plume axis (Murlis and Bettany 1977; David et al. 1983; Baker 1989b). Three-dimensional reconstruction of flight tracks of *L. botrana, G. molesta* (Fig. 24.3), and *E. ambiguella* shows that lateral excursions occur concurrently with changes in height (Witzgall and Arn 1990, 1991; Witzgall et al., in preparation).

Recordings from one camera image a distorted two-dimensional view of the moth's flight maneuvers, but not their spatial extension. What appears as zigzag-

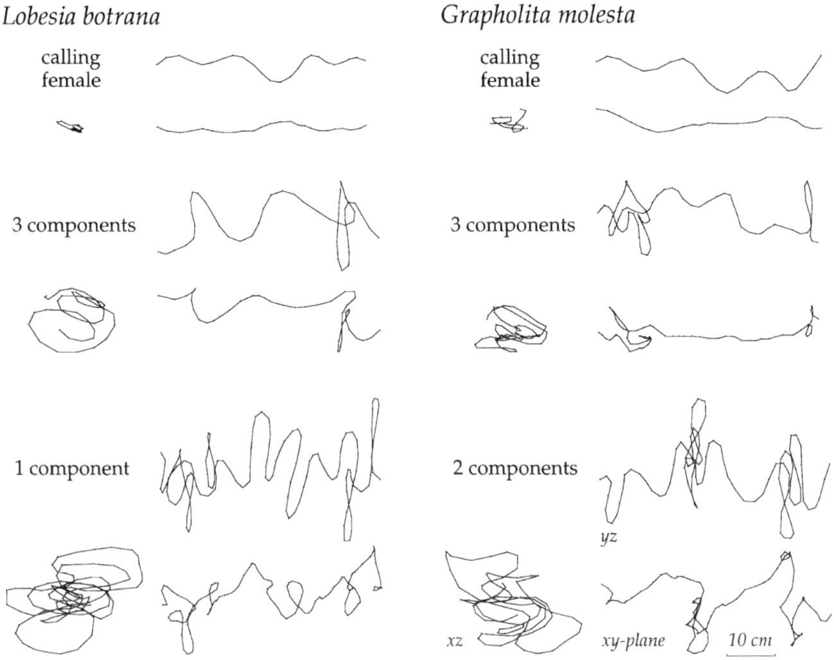

Figure 24.3. Typical flight tracks of *L. botrana* and *G. molesta* males to a calling conspecific female and to optimized and incomplete synthetic pheromone blends (see Section 2). Each track is shown in three projections (top: *zy*-plane, horizontal plane; bottom right: *xy*-plane, view from side; bottom left: *xz*-plane, view from source).

ging, seen from above, is actually a smooth, looping path in space. A "counterturn" or "cross-wind track reversal" has been defined as a switchover from one side of the wind line to the other (Marsh et al. 1978). Accordingly, the second flight track in Fig. 24.3 (*L. botrana* to three-component blend) shows seven counterturns and eight counterturn apices in the horizontal projection.

Three-dimensional reconstruction of this track shows that the male has been continuously moving anticlockwise, except during the last two frames. It did not switch from one side of the wind line to the other, but rather looped around the plume axis. Therefore, the orientation or the temporal regularity of zigzags, observed with one camera from above (Marsh et al. 1978; Baker et al. 1984; David and Kennedy 1987; Willis and Cardé 1990; Willis and Arbas 1991a; Willis and Baker 1994), does not give evidence of a counterturn oscillator.

4. Modulation of Flight Behavior

4.1. Wind Speed

Male moths maintain a rather constant ground speed over a wide range of wind speeds. At higher wind speeds, rather straight flights have been observed; the

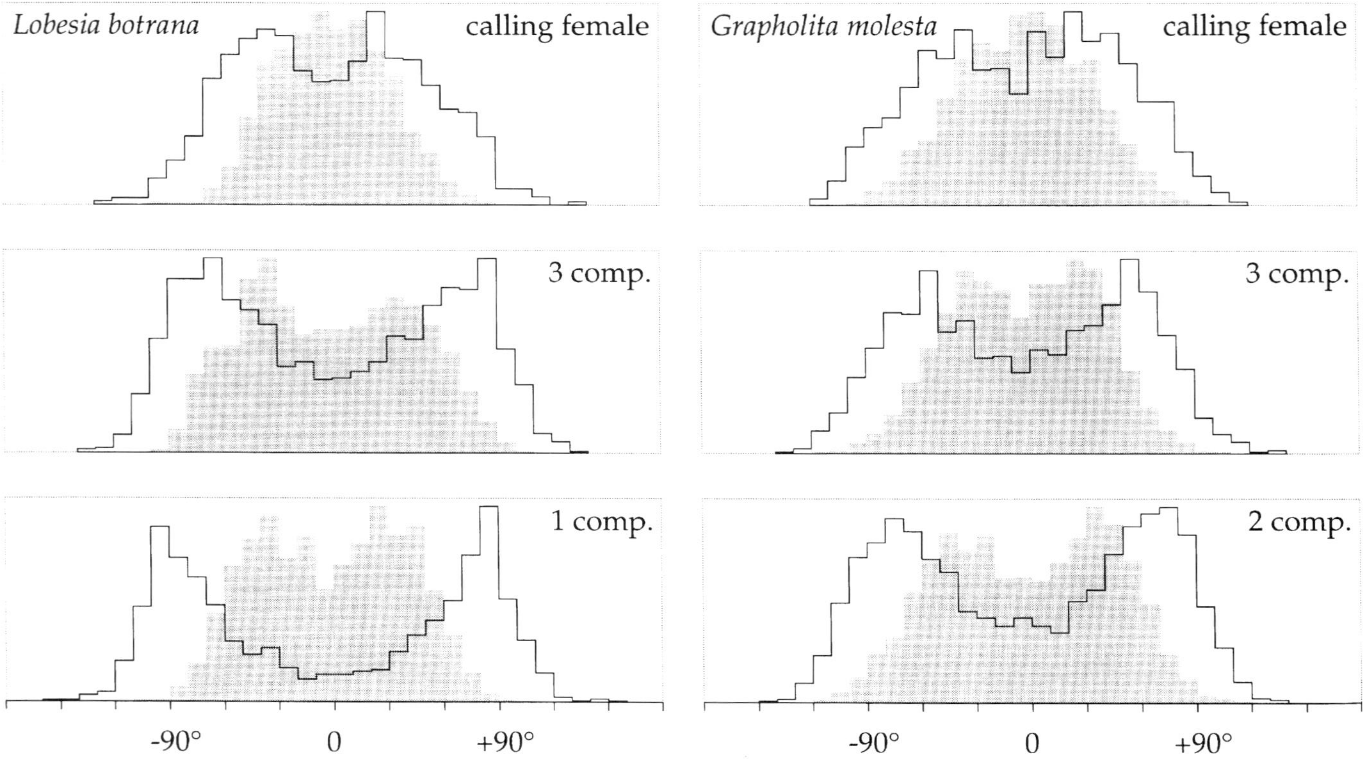

Figure 24.4. Frequency distribution of track angles (empty) and course angles (shaded) in *L. botrana* and *G. molesta* flights ($N = 60$) to calling female and to optimized and incomplete synthetic pheromone blends (source composition as in Fig. 24.3).

course angles are strongly diminished and their calculated frequency distribution is focused around due upwind (Marsh et al. 1978; Murlis et al. 1982; Sanders 1985; Willis and Cardé 1990; Willis and Arbas 1991a). For flights of *Lymantria dispar* at a wind speed of 1.5 m sec^{-1} (Willis and Cardé 1990), *Manduca sexta* at 2.0 m sec^{-1} (Willis and Arbas 1991a) and *L. botrana* at 0.6 m sec^{-1} (Witzgall et al., in preparation), the ratio between mean air speed and wind speed was 1.1. At this ratio, a course angle of 5° results in a track angle of 45° (Fig. 24.1).

Accordingly, at high wind speeds, active steering across the wind line is either negligibly small or absent. This does not lend support to the idea of an internal counterturning program. Rather, more direct flights in stronger wind indicate that the uncertainty for detection of the wind direction is reduced (Preiss and Kramer 1986a).

On the other hand, in "zero wind" the males fly along an irregular, undulating track (Kuenen and Baker 1983; Baker et al. 1984; David and Kennedy 1987; Willis and Cardé 1990). This indicates that there might be an internal program of counterturns, since in the absence of wind the male's track and course are identical (Fig. 24.1). The degree of self-steered counterturning would accordingly be regulated by wind speed and would be most pronounced during wind lulls.

On the contrary, optomotor anemotaxis is no longer operational without wind and the males may not be able to head straight toward the source, as they do in strong wind. In the absence of wind-induced drift, stimulation by pheromone may enhance visual orientation to landmarks, leading to a behavioral pattern different from anemotactic flight. Investigations of odor-modulated flight in dead calm also meet experimental difficulties, because the males do not initiate orientation flights without wind. Immediately after shutting off the wind, the air movement is not strictly zero, but omnidirectional at speeds of <2 cm sec^{-1} (Willis and Cardé 1990). The males may even retain visual information, which they may use for location of the source after the wind has been stopped (Baker et al. 1984). It is therefore unclear whether orientation flights in the absence of wind can provide evidence for counterturning during upwind flight.

4.2. Temporal Structure of the Odor Stimulus

Pheromone emanates from a small source as a filament, tangled and stretched by turbulences. The flying male perceives this filament as a series of pulses varying in concentration, at irregular intervals (Murlis and Jones 1981; Kramer 1986; Murlis this volume). The filamentous structure of the "odor plume" remains intact downwind from the source, since orientation to pheromone is not suppressed by a nearby source of attraction antagonist (Priesner and Witzgall 1984; Witzgall and Priesner 1991; Liu and Haynes 1992, 1993a; Rumbo et al. 1993). Continuous, uniform stimulation with pheromone is not sufficient to initiate and maintain upwind-oriented flight (Kennedy et al. 1980, 1981; Willis and Baker 1984; Baker et al. 1985; Kramer 1992).

The modulation of upwind flight by the frequency of encounters with the pheromone filament has been demonstrated with pulsed odor stimuli. Successful source location is elicited at frequencies of 2–10 pulses per second (Vickers and Baker 1992), as indicated by the peak frequency of ion plumes (Murlis and Jones 1981) and the temporal resolution of odor pulses by peripheral and central olfactory neurons (Kaissling 1986b; Christensen and Hildebrand 1988; Rumbo and Kaissling 1989; Almaas et al. 1991). At higher wind speeds (Section 4.1), the plume is more elongated (Sanders 1985b), and field electroantennogram recordings showed more frequent and sharper antennal responses (Baker and Haynes 1989).

Male silk moths, *Bombyx mori,* respond to a single pheromone pulse by a rapid increase of walking speed and a sudden upwind turn; their course is straightest at the highest pulse frequency (Kramer 1986). This has now been confirmed in a flight tunnel. Male *Cadra cautella* fly straight upwind in response to fast pulses of pheromone and zigzag at lower pulse frequencies. During casting flight, males respond to a single pulse by an upwind surge (Mafra-Neto and Cardé 1994; Cardé and Mafra-Neto this volume).

Upon loss of contact with pheromone, the transition from upwind to casting flight takes place within 0.15 to 0.5 sec in four different species (Section 3.2). In *B. mori,* the rate of change of walking direction is highest at low pulse frequencies (Kramer 1986). This indicates that males zigzagging under a suboptimal stimulus frequency may repeatedly switch from upwind to casting flight in between filament encounters (Baker 1990; Baker and Vickers this volume). This "dual response model" proposes separate neural pathways: A phasic response to single pulses elicits upwind turns and an underlying tonic response is driving casting flight. However, it is still unknown whether the moth changes flight direction according to a counterturn oscillator or in response to external stimuli.

4.3. Chemical Composition of the Odor Stimulus

Stimulation of the males with an incomplete pheromone blend decreases the rate of successful upwind flights and landings at the source (e.g. Linn et al. 1986a) and generates flights with wide excursions and reduced upwind progression (Fig. 24.3; Palaniswamy et al. 1983; Sanders 1985b). Male *L. botrana, G. molesta,* and *E. ambiguella* zigzag in response to synthetic pheromone, but fly rather straight upwind toward calling females (Figs. 24.3 and 24.4) (Witzgall and Arn 1990, 1991; Witzgall et al., in preparation). In *Amyelois transitella,* direct upwind flights have been observed in males orienting to calling females and in females orienting to host odor (Haynes and Baker 1989). This implies that convoluted flight tracks with a bimodal distribution of course and track angles are an artifact caused by an inadequate chemical stimulus.

According to studies with tethered *L. dispar* (Preiss and Futschek 1985; Preiss and Kramer 1986a,b), males orient upwind by compensatory reactions to the longitudinal, transverse, and centrifugal/centripetal image flow components: They keep upwind progression at a small positive value, they minimize drift across the wind line, and they stabilize flight altitude at the height of the source. These optomotor responses are independently controlled, and they are enhanced by the perception of *pheromone*. The behavioral evidence is strengthened by recordings from descending interneurons, showing that "... pheromone amplifies some neural pathways mediating male optomotor responses, especially the directionally selective responses to the transverse movement of the image" (Olberg and Willis 1990). Projection neurons responding specifically to a blend of the main compound and synergists or antagonists, and not to single compounds, have been identified in the antennal lobe of two noctuids (Christensen et al. 1991a; Hansson et al. 1994).

The female gland secretion usually contains compounds which do not seem to augment male attraction in field trapping or wind tunnel tests (see Arn et al. 1992, 1995). However, a great variability is inherent to bioassays based on countings of attracted males; this approach may not allow us to entirely distinguish behaviorally active from redundant components (e.g., Linn et al. 1984). The straightness of the tracks shown in Fig. 24.3 reflects the completeness of the synthetic blend; the differences between calling females and optimized synthetic blends may accordingly be due to missing components.

Flight behavior also depends on the proportions of minor pheromone components within the blend. Off-blends containing too large amounts of synergists have an antagonistic effect on male attraction and modify upwind flight in *G. molesta, Cacoecimorpha pronubana,* and *L. botrana* (Linn and Roelofs 1983; Willis and Baker 1987, 1988; Witzgall 1990; Witzgall et al., in preparation). Likewise, antagonists obstruct attraction and deteriorate the flight behavior of the responding males (Liu and Haynes 1992, 1993a). In *L. dispar,* compensatory lift reactions are blocked by a blend of pheromone and an attraction antagonist (Preiss and Kramer 1983; Preiss and Futschek 1985).

Apart from analytical problems to identify all pheromone components released by the female, it is quite difficult to mimic known female blends with synthetic compounds formulated on filter papers or rubber septa. This may not at all be possible in species using, for example, dienic compounds (Witzgall et al. 1993). Even gland extracts do not necessarily equal the calling female, quantitatively and qualitatively. In three *Helicoverpa* species, the female gland contains a strong attraction antagonist for conspecific males, which is converted into a pheromone component on the gland surface (cf. Teal and Tumlinson 1986).

When investigating the male orientation mechanism, calling females should be used as pheromone source; if not, they must be included as a control. The chemical stimulus modulates male orientation behavior, and discrepancies between natural and synthetic pheromone should be taken into account.

5. Conclusion and Outlook

There is irrefutable evidence that males orient upwind to pheromone sources by optomotor anemotaxis. In contrast, unequivocal proof of an internal counterturn generator is still lacking and cannot be obtained from wind tunnel studies alone. The observed flight behavior, including zigzagging flight under suboptimal stimulus conditions, can also be explained by a modulation of the optomotor responses. Nearly straight upwind flights with a unimodal distribution of course angles occur at high wind speeds (Section 4.1), at a high frequency of pheromone pulses (Section 4.2), and in response to females (Section 4.3).

Flight tracks illustrate differences in the chemical composition of pheromone blends. A quantitative evaluation of male and female upwind flight behavior to pheromone or host odors can be used as a sensitive bioassay for the identification of behaviorally active compounds. Recording techniques will have to be improved to assess orientation behaviors also in the field, especially in order to advance the mating disruption method.

Upwind orientation also serves as a model for the study of neurobiological mechanisms that generate behavioral responses to an olfactory signal (Hildebrand this volume). However, not only the orientation behavior as elicited by calling females, but notably its modulation by chemical signals of different composition, may provide cues to trace the sensory processing of the pheromone stimulus. One potential approach to link neurophysiological and behavioral studies is the comparative analysis of synergism and antagonism (Mustaparta this volume).

Attraction synergists and antagonists function to amplify the behavioral response to the main pheromone component and to facilitate communication within phylogenetic groups with only a few, structurally related compounds. Closely related species perceive both the respective synergists and antagonists via specialized antennal receptors (Priesner 1979a, 1986; Almaas et al. 1991), whereas differences in the processing of pheromone synergists and antagonists in the antennal lobe are becoming apparent (Christensen this volume; Hansson this volume)—yet these differences cannot explain opposite behavioral effects.

It is still unclear where "feature detectors" differentiating attractive from repellent signals could be situated, if they exist at all. To this purpose, tethered moths flying in optically simulated wind should be used for simultaneous electrophysiological recordings from flight muscles (Willis and Arbas this volume) or from interneurons in the ventral nerve cord, connecting the brain and thoracic motor centers. The validity of flight measurements with tethered moths can be verified by calculating flight tracks from the recorded motor output and comparing these to free flight.

Acknowledgments

I am grateful to Drs. H. Arn and S. Rauscher, Wädenswil, H. A. Beyer, Courgenay, and M. Bengtsson, Alnarp, for their collaboration and to R. Preiss, Hamburg, for helpful comments on the manuscript.

25

Mechanisms of Flight of Male Moths to Pheromone

Ring T. Cardé and Agenor Mafra-Neto

1. Introduction

The renowned ability of male moths to discover the location of a pheromone-emitting female by following her aerial trail of pheromone to its source appears to be a basal characteristic of moths. Female-emitted male attractants occur amongst the oldest lineages, such as the more primitive genera of Hepialidae (Wagner and Rosovsky 1991). Among the higher moths or Ditrysia, which comprise 98% of moth species, it is also the male which conventionally is lured to the female; the exceptions to this rule are almost certainly secondarily derived. Another indicator of a common lineage for male attraction to female ditrysians is that the location of pheromone production appears confined almost exclusively to abdominal segments VII–X and typically is situated in segments VIII and IX (Percy-Cunningham and MacDonald 1987). The chemistry of the attractants as well appears to have remained remarkably undiversified, comprising mainly acetate, alcohol, or aldehyde terminal moieties on evenly numbered chains 12–18 carbons long, mostly with one or two positions of unsaturation, even in long diverged lineages such as the Noctuoidea, the Bombycoidea, and the Pyraloidea (Arn et al. 1992). The combined evidence thus supports a female-to-male chemical signaling system as a basal condition of moths (see also Löfstedt and Kozlov this volume).

The ability of the females of some moth species to advertise their location over distances perhaps up to hundreds of meters likely accounts for the capacity of such species to persist at densities which would preclude mate location using other communication modalities. Successful mate finding by pheromone at very low population densities may in part be one ultimate evolutionary factor in the proliferation of moths to some 120,000 described species, despite their seemingly undiversified larval life style as consumers of plants.

Rau and Rau (1929) provided perhaps the most extensively documented case of long-distance mate finding by luring male silkworm moths released several kilometers from caged virgin females held on a second-story roof of their home in St. Louis, Missouri. C.H. Fernald also demonstrated evident location of distant females in the gypsy moth *(Lymantria dispar)* in New England: a few males released several kilometers away located females after several hours (Forbush and Fernald 1896). These and other similar cases show that males potentially might detect and orient toward such distant chemical cues, but unfortunately this method cannot disentangle the relative contributions of ranging flight which might bring a male into contact with a pheromone plume and of pheromone-dictated maneuvers which close the distance between pheromone sender and responder.

2. Précis of Mechanisms Used by Flying Moths to Orient to a Source of Pheromone in Wind

2.1. *Optomotor Anemotaxis*

The precise way in which flying moths integrate pheromonal and environmental cues to achieve upwind displacement has been subject to continuing controversy (Arbas et al. 1993; Willis and Arbas this volume). Determining wind direction once an organism is airborne is not a simple task. The eminent ant biologist W.M. Wheeler wrote in 1899 that a flying organism orients itself to the wind by assuming a "position in which the pressure exerted on its surface is symmetrically distributed." However, once an organism is airborne it is not possible to use mechanoreceptors to detect the direction of wind flow; mechanoreception is only useful in estimating air speed. Instead, wind direction is judged entirely by sensing the organism's displacement relative to its visual surround. J.S. Kennedy first verified this mechanism, termed *optomotor anemotaxis,* with mosquitoes (Kennedy 1940).

The actual path that a male moth takes in following a pheromone plume is usually characterized as an upwind zigzag. Most flight tracks are recorded in wind tunnels and analyzed for movement in the horizontal plane, on the assumption that vertical displacement in the zigzag path is minimal (cf. Murlis et al. 1982; Witzgall and Arn 1990, 1991; Witzgall this volume). Visual inputs typically are segregated into longitudinal and transverse components, the former reflecting the organism's image flow front to rear and the latter side to side (Marsh et al. 1978). The image flow actually experienced by a moth during zigzagging maneuvers is presumably much more complex than evidenced by this method of analysis. Gypsy moths roll during the apex of a turn and therefore the visual surround in turn would rotate, unless the moth turned his head to maintain a fixed relationship with the ground (Zanen 1993). *Manduca sexta* moths accelerate along the leg of a turn and slow at the turn's apex, suggesting that acquisition of visual information to decipher course, wind drift, and velocity must adjust somehow for these perturbations (Willis and Arbas 1991a; Willis and Arbas this volume).

2.2. *Why Do Moths Zigzag?*

2.2.1. *Counterturning Set by an Internal Motor Program*

Wright (1958) proposed that the zigzag course is dictated by a central nervous system "program." Kennedy (1977) elaborated this hypothesis and further suggested that each reversal in the zigzag might reflect exiting the plume and turning back toward the direction in which the plume was last sensed. Such a pattern of turns relative to plume's long axis clearly is occurring when male moths zigzag along a narrow ribbon plume in wind tunnel trials. Kennedy's subsequent wind tunnel experiments, however, in which males zigzagged along the boundary between a wide corridor of pheromone and clean air, executing turn reversals both in clean air and *within the cloud* of pheromone, led to the conclusion (reviewed by Kennedy 1983) that reaching the plume's margin was not a requirement for a counterturn.

Behavioral evidence favoring an underlying motor program governing counterturns has accumulated in several moth species. Baker et al. (1984) documented very similar counterturn frequencies in male *Grapholita molesta,* the oriental fruit moth, in zero and continuous wind. In the gypsy moth, the metronome-like timing of zigzags (ca. 4 reversals sec^{-1}) is seen in a variety of conditions in the wind tunnel: flying to various concentrations of pheromone, in two ambient temperatures (Charlton et al. 1993a), in still air and in several wind velocities (Willis and Cardé 1990), and with males of different sizes (Kuenen and Cardé 1993). Free-flying males in the forest also show about the same turning rate at distances from 2.5 to 20 m from a point source of pheromone (Willis et al. 1991). Such invariant turn rates are consistent with the output of a putative turn generator (David and Kennedy 1987). Casting, considered explicitly in Section 2.3, also appears governed by the same putative turn generator.

2.2.2. *Counterturning as a Mechanism of Course Correction*

The view that counterturning is governed by an internally set motor program has been challenged by Preiss and Kramer (1986a,b) with two lines of evidence. Using an elegant flight simulator, they have shown that tethered male gypsy moths in a defined visual surround performed alternating right and left turns similar to those that would be expected in free flight. They concluded that the turns were reiterative corrections of course, caused by the male's inability to employ optomotor cues effectively enough to head straight upwind. Second, a computer simulation of the male's maneuvers, given an assumption of an inability to head directly upwind, also yields a zigzag output (Preiss and Kramer 1986a,b). We have termed this the *flight imprecision* model (Mafra-Neto and Cardé 1995b).

The counterturning evoked by a navigational correction system versus a counterturn generator would appear difficult to reconcile solely from an examination of behavioral output. The metronomic features of counterturning might be explained by (a) a time lag for the moth to appraise the direction of drift before a counterturn and (b) the limitations of how rapidly this maneuver can be executed. There are two sorts of evidence, however, that are inconsistent with the flight imprecision model. First, male moths can zigzag along a pheromone plume in zero wind, which we consider in Section 25.2.5. Second, zigzagging would be difficult to interpret as a corrective navigational maneuver if males were capable of straight upwind flight along a pheromone plume. Nearly straight line flight upwind has been documented by Haynes and Baker (1989) in the pyralid *Amyelois transitella* and by Witzgall and Arn (1990) and Witzgall (this volume) in the tortricids *Lobesia botrana* and *G. molesta.* In both species, flight aimed nearly straight upwind occurred when a male flew to a natural source of pheromone. Segments of straight flight along the plume have been documented in the field with the gypsy moth (David et al. 1983). Recent wind tunnel trials with a pyralid (Mafra-Neto and Cardé 1994) and a noctuid (Vickers and Baker 1994b; Baker and Vickers this volume) also have shown suppression of counterturning and essentially straight upwind flight, which we consider in Section 25.3.

2.2.3. *Counterturning Influenced by Blend Quality*

Willis and Baker (1988) noted that the zigzag path of male *G. molesta* widened when the *(E)*-acetate component of the pheromone blend was presented in a lower-than-natural ratio. Witzgall and Arn (1990, 1991) and Witzgall (this volume) compared flight tracks of *L. botrana* and *G. molesta* males flying along identically-sized plumes from calling females and synthetic pheromone. Males flew straight or nearly straight upwind in pheromone plumes from females and would be expected to do so in synthetic pheromone plumes that fully duplicate the composition of the female's blend and match her rate of pheromone emission. Instead, in *G. molesta* and *L. botrana,* synthetic imitations of the natural blend yielded paths that were of zigzag form if considered in the horizontal plane, with both horizontal and vertical deviations from the plume's central axis in 3-D. Witzgall and Arn (1990, 1991) suggested that the extent of deviations from the plume's central axis might be the consequence of using a synthetic pheromone source that did not precisely match the female's effluvium, and they noted that most studies showing moth zigzag flight used synthetic sources of pheromone. Murlis et al. (1982), for example, found in field observations in Crete that the flight track of male *Spodoptera littoralis* following a plume of synthetic pheromone also had an undulating feature.

Witzgall (this volume) raises the additional issue of whether the analyses of flight tracks only in the horizontal plane have camouflaged our view of when counterturns occur. In a 2-D analysis, counterturning reversals are interpreted as a zigzag, but vertical undulations are undetected. Clearly more 3-D analyses need to be performed. The extent to which moths undulate and zigzag needs to be verified under a spectrum of visual conditions. For example, the visual cues available to moths in the Witzgall and Arn (1991) wind tunnel were not as prominent as in the wind tunnels used in other studies (e.g., Kennedy and Marsh 1974; Willis and Baker 1988; Charlton et al. 1993a). A visually depauperate setting makes it difficult for moths to maintain a relatively steady height and might contribute to vertical movement (Vickers and Baker 1994a).

2.3. Loss of the Pheromone Plume and the Significance of Casting

Another strategy crucial to finding a pheromone emitter is *casting:* zigzagging without upwind displacement, with progressively wider lateral excursions. Casting occurs following the loss of a plume, and this reaction is readily evoked in the wind tunnel by removal of the pheromone source (Kennedy and Marsh 1974). In the field, a flying male can lose the plume when the wind shifts direction or when a male exits the plume (David et al. 1983). Because the direction upwind and along the plume's long axis are aligned only infrequently (Elkinton et al. 1987; Brady et al. 1989), males flying upwind undoubtedly fly out of a pheromone plume with some regularity. Casting is thus interpreted as a strategy that maximizes the opportunity of a male regaining contact with a plume that has been lost (David et al. 1983). In the gypsy moth, the rate of turn reversals in casting (Kuenen and Cardé 1994) and upwind zigzagging along a plume are similar enough to suggest that both maneuvers are governed by the same putative turn generator.

2.4. Structure of the Pheromone Plume and Its Significance to Orientation

The successful location of a point source of pheromone in wind is a task complicated by the signal's spatial distribution which itself is dictated by turbulent diffusion. Pheromone is transported in eddies ranging in scale from less than a centimeter to many meters (Murlis et al. 1992; Murlis this volume). The resulting signal appears in filaments which are interspersed with relatively clean air (Wright 1958). The probability of encountering pheromone filaments at a fixed sampling point declines with distance from the pheromone's source, although packets of pheromone can harbor relatively undiluted pheromone many meters from the source. Thus the signal available to flying males is highly intermittent, and its temporal and structural characteristics vary with distance to the source.

An early field test demonstrating the effects of plume structure on attraction was provided by the trapping tests of Lewis and Macaulay (1976). These trials showed that the structure of a plume emanating from a trap had a marked influence on the success of male orientation.

A definitive verification of the influence of the plume's structure on orientation stems from wind tunnel trials comparing the flight tracks of males flying in homogeneous clouds of pheromone with tracks of males flying in clouds "pulsed" with pheromone and clean air (Baker et al. 1985). Following their release into a standard stimulus of a narrow, ribbon-like plume, males presented with the heterogeneous, pulsed cloud of pheromone progressed upwind in typical zigzag fashion; males encountering a homogeneous cloud quickly changed to casting. In other wind tunnel tests, males presented with a narrow plume within a more dilute, homogeneous cloud of pheromone were able to zigzag upwind along the superimposed plume (Kennedy et al. 1980, 1981). Together these innovative manipulations of plume structure verified that the principle that changes signal strength are requisite for continued upwind progress, possibly because with a continual, even stimulation, sensory adaptation can cause the signal to "fuse" (Baker et al. 1989).

An explicit demonstration of the influence of plume structure in the field was provided in the analyses of gypsy moth flight to narrow plumes from a point source of pheromone versus plumes issuing from cylinders of 0.2-m diameter and therefore of at least this initial size. Willis et al. (1994) found that tracks of males flying along narrow plumes were slower and narrower and aimed more to the crosswind (i.e., zigzaggy) than to plumes generated from visible or invisible cylinders.

The importance of plume structure is evidenced in Baker's (1990) phasic–tonic model of response. The counterturning program is a tonic response maintained by current or recent pheromone contact. The phasic portion is determined moment-to-moment by contact with filaments of pheromone which induce "reiterative upwind surges." These reactions, depending on their latency and the frequency of encounters with filaments, produce either (a) a flight track with zigzags across the wind line or (b) a straightened-out upwind flight. Early reports on casting *Adoxophyes orana* males narrowing their reversals and surging upwind when reaching the leading edge of a cloud of pheromone (e.g., Kennedy et al. 1980, 1981; Kennedy 1983), along with more recent reports on casting male moths surging upwind after they intercept a single, brief pulse of pheromone (Mafra-Neto and Cardé 1994; Vickers and Baker 1994b), lend support to the idea that these flight maneuvers are determined moment-to-moment by their interaction with the fine-scale structure of the plume. New evidence on the importance of fine-scale plume structure will be considered in Section 25.3.

2.5. Orientation in Still Air

In still air, or under wind or visual conditions below the threshold for the optomotor response, males may rely on the spatial distribution of pheromone concentration to orient to the pheromone's source. Such sequential sampling of the pheromone (longitudinal klinotaxis), coupled with a self-steered series of counterturns, appears to account for the ability of male moths to continue flying for approximately a meter along a narrow plume in zero wind (Kuenen and Baker 1983), as seen with *Pectinophora gossypiella* (Farkas and Shorey 1972), *G.*

molesta (Kuenen and Baker 1983) and *L. dispar* (Willis and Cardé 1990). Under the wind tunnel conditions of the above experiments, males initially oriented to the pheromone plume in wind, which was then abruptly stopped, leaving the pheromone plume temporarily suspended in the tunnel. It is plausible that the visual cues used by males before the wind ceased aided the males in continuing their progression along the plume in still air. The importance of longitudinal klinotaxis under field conditions remains to be established.

2.6. Summary of Explanations of Counterturning

The zigzag path of a male moth orienting along a pheromone plume in wind has been explained by three systems. All rely on optomotor anemotaxis to polarize the direction of flight, but the counterturning reversals are viewed either as the output of a counterturning program (Kennedy 1983), reiterative course corrections (Preiss and Kramer 1986a,b), or the consequence of emitting an incomplete synthetic copy of the natural pheromone (Witzgall and Arn 1990). The similarity in the timing of the counterturns seen in upwind flight along the plume in wind, in zero wind, and in casting following the short-term loss of pheromone supports the existence of a counterturning program. The ability of moths to fly under some circumstances an essentially straight upwind path argues against zigzagging being a maneuver to correct for an imprecise upwind course heading.

3. Effects of Plume Structure on the Maneuvers of *Cadra cautella*

3.1. Pheromone Blend and Its Influence on Attraction

Our selection of *Cadra cautella,* the almond moth, as an experimental subject originally stemmed from our interest in seeing how well this species conformed to the *blend hypothesis,* which holds that the entire natural blend of pheromone has the lowest threshold for all behaviors (Linn et al. 1986a). The pheromone blend of *C. cautella* is a mixture of *(Z,E)*-9,12-tetradecadienyl (Brady et al. 1971; Kuwahara et al. 1971) and *(Z)*-9-tetradecenyl acetates (Brady 1973). The first component alone is attractive and the second, itself unattractive, augments attraction when combined with the first. Our wind tunnel experiments (Mafra-Neto 1993) found that at a wide range of tested concentrations the “initial” responses, activation and taking flight, were evoked in roughly equal proportions by either the first component or both components. The thresholds for “late” responses in the sequence of attraction, including locking on to the plume, upwind flight and close approach to the pheromone-bearing dispenser, however, were lowest for the natural blend of the two components. If the responses prior to locking on to the plume are considered typical of initial response in the field, this outcome does not conform to the blend hypothesis. However, no combination of dose or of one or two components tested evoked source location in more than half the males assayed, a surprising outcome if the pheromone blend was correctly characterized and the assay conditions optimized. Our tests were conducted in a laminar-flow wind tunnel (Mafra-Neto 1993) and pheromone was dispensed from a small point source, yielding a narrow, continuous ribbon of pheromone.

3.2. Evidence That Fine-Scale Plume Structure Modifies Flight Tracks

In some wind tunnel assays, stray *C. cautella* males from a previous treatment joined the test male in flight along the plume. Analysis of the video tapes resulting from these encounters showed that the male in the upwind position always maintained a zigzag path, while the male in the downwind locale shuttled between zigzagging and brief dashes straight upwind. These observations led to tests in which two males were released simultaneously in a narrow ribbon plume consisting of pheromone and a visible tracer of "smoke" from a titanium tetrachloride source. Visualizing the plume's structure allowed us to verify that the upwind male deformed the ribbon plume as he traversed it, creating a turbulent pulse. When the downwind male encountered a turbulent pulse, he briefly surged upwind (Fig. 25.1) (Mafra-Neto and Cardé 1995a). These reactions implicated the plume's fine-scale structure as influential in governing a male's orientation to pheromone.

3.3. Effects of Turbulent Structure

Other than tests with corridors of pheromone and pulsed clouds, there has been little systematic study of how plume structure modulates flight maneuvers (but

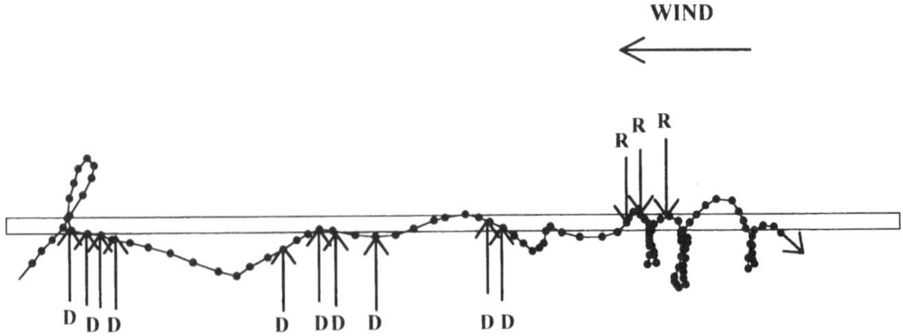

Figure 25.1. Representative flight track of a male *Cadra cautella* flying upwind along a thin ribbon plume of pheromone mixed with a visual marker ($TiCl_4$). The arrows indicate where the male contacted the plume; **R** indicates that the plume is an undisturbed ribbon and **D** indicates a disturbed region of the plume caused by the wake of a second upwind male flying through the plume. The time-averaged position of the pheromone and smoke plume is indicated by the rectangle. Dots indicate the position of the moth at 0.03-sec intervals. [After Mafra-Neto and Cardé (1995a).]

see references in Section 2.4). We tested if the plume's overall structure altered the flight path. A wide plume with a turbulent structure was generated simply by positioning a small deflector immediately upwind of the pheromone dispenser. Instead of the zigzaggy flights seen along continuous narrow ribbon plumes, males in turbulent plumes flew straighter and more rapidly upwind, with some flights taking essentially a straight-line path to the pheromone source (Fig. 25.2) (Mafra-Neto and Cardé 1994, 1995a). These disparate trends caused by plume structure were little altered by changes in pheromone concentration (Mafra-Neto 1993).

These findings led us to manipulate the plume's fine-scale structure. A mechanical stimulus generator delivered pheromone-laden pulses at frequencies up to 5 Hz into the wind tunnel through a pipette. This device allowed us to vary the frequency of pulses and, by the volume injected, the overall dimensions of a continuous plume or train of pulses. Pulses at 5 Hz yielded fast, straight-line flights or fast flights with shallow zigzags. Slowly pulsed (0.63 Hz) plumes produced zigzag flights characterized by a slow progression upwind with wide lateral excursions (Fig. 25.3) (Mafra-Neto and Cardé 1994, 1995b).

3.4. Effect of a Single Pulse of Pheromone

Both the observations of two males simultaneously flying along a plume and the reactions of males flying in plumes with 0.63-Hz pulses suggested that contact with individual filaments of pheromone dictated the male's maneuvers. This was tested explicitly by presenting males with a rapidly pulsed plume, turning off the pheromone source, and introducing a brief (0.25- or 0.02-sec duration) pulse of pheromone (Fig. 25.4). Males initially flew upwind along rapidly pulsed plumes, casted after the pheromone was withdrawn, surged briefly straight or nearly straight upwind following contact with the pulse, and then returned to casting (Mafra-Neto and Cardé 1994, 1996). The degree of crosswind suppression and duration of the upwind surge was dictated by the duration, or size, of the pheromone pulse intercepted (Mafra-Neto and Cardé 1996). Interception of 0.25-sec pulses at winds of 50 cm sec^{-1} induced males (after a latency of 0.21 ± 0.07 sec) to adopt a straighter upwind flight, for 0.72 ± 0.10 sec. At a net upwind speed of 22.12 ± 3.15 cm sec^{-1}, this response translated into 18.33 ± 4.40 cm of net upwind displacement in the surge (Fig. 25.4). During the last third of the surge, upwind displacement started to subside and crosswind displacement increased, marking a transition to casting. Interception of shorter 0.02-sec pheromone pulses resulted in briefer upwind surges (0.57 ± 0.13 sec) and shorter upwind displacements (11.43 ± 2.34 cm) than evoked by longer 0.25-sec pulses (Fig. 25.4). The net upwind speed for short pulses was 15.87 ± 3.22 cm sec^{-1}, significantly lower than for longer pulses. Interception of shorter pulses also elicited an incomplete suppression of counterturning, which resulted in upwind surges that were more sinusoidal than those in response to the interception of larger pulses. In *C.*

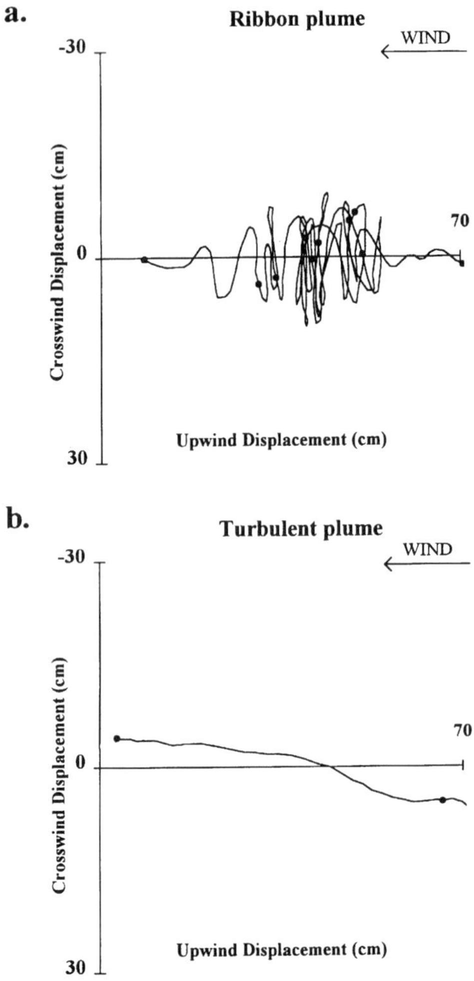

Figure 25.2. Representative flight tracks of *Cadra cautella* males flying upwind along plumes of differing sizes; both plumes are from a dispenser with a dose of one female equivalent of pheromone. Note the wide zigzags and slow speed of the male flying along a ribbon plume **(a)** and the lack of counterturning and increased speed of the male flying along the wide turbulent plume **(b)**. Similar trends are observed in tracks of males flying along ribbon and turbulent plumes issuing from dispensers of 10 and 100 female equivalents. Dots indicate a moth's position every second. [After Mafra-Neto and Cardé (1994).]

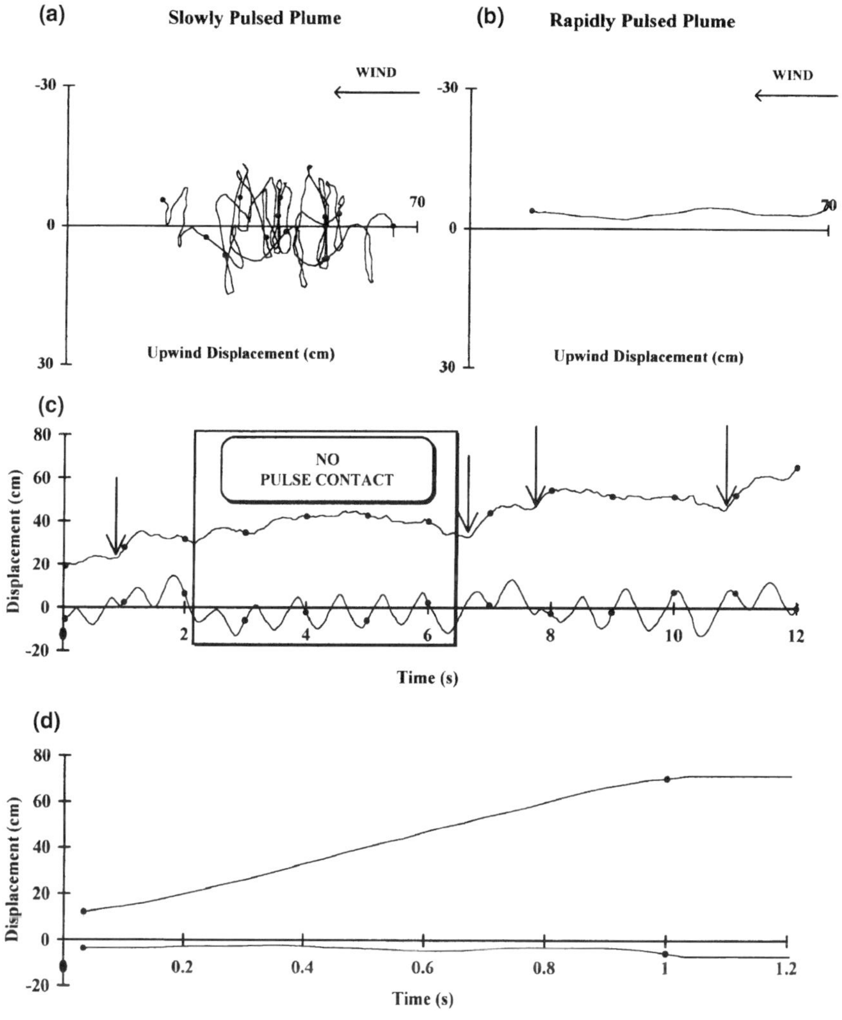

Figure 25.3. Representative flight tracks of *Cadra cautella* males flying to two structurally different pheromone plumes. **(a)** Graph showing the track of a male flying to a slowly pulsed plume, comprised of 0.1-sec pulses generated every 1.5 sec; **(b)** Graph showing a flight track of a male flying along a rapidly pulsed plume, comprised of 0.1-sec pulses generated every 0.2 sec. **(c, d)** The upper trace depicts the longitudinal component and the lower trace depicts the transverse component of the tracks. The *x* axis indicates time in seconds and the *y* axis indicates the distance traveled in centimeters. In graph **c**, when the moth surged upwind, counterturning was interrupted. In graph **d**, the male flew rapidly upwind without counterturning. Dots represent the position of the moth every second. [After Mafra-Neto and Cardé (1966).]

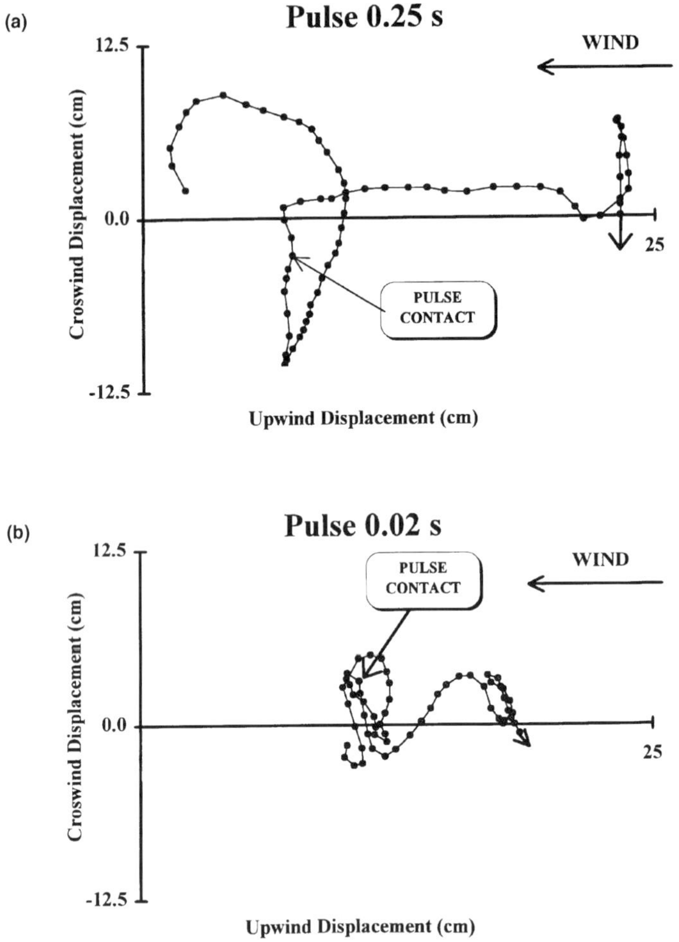

Figure 25.4. Reaction of casting *Cadra cautella* males to the interception of a single pulse of pheromone. Males flying along a rapidly pulsed (5 Hz) plume casted following the removal of the plume. Casting males were then challenged with a single puff of pheromone. **(a)** Graph showing a flight track of a male intercepting a 0.25-sec pulse. **(b)** Graph showing a male intercepting a 0.02-sec pulse. After a brief delay, males surged upwind and then returned to casting. The straightness and duration of the surge were correlated with the duration of the pheromone pulse. If the pheromone pulse was short, the suppression of the crosswind displacement was incomplete. The brief upwind surge waned and the male returned to casting. For simplicity, only a portion of the long casting flight is shown. Dots represent the moth's position every 0.3 sec. [After Mafra-Neto and Cardé (1996).]

cautella there appears to be a common mechanism promoting such upwind dashes following a single encounter with a filament of pheromone or sustained upwind flight with shallow counterturns or a due upwind heading along either wide turbulent plumes or plumes containing a train of pulses. In *C. cautella* such upwind dashes following a single encounter with a filament of pheromone help us to understand how structural characteristics of pheromone plumes mediate sustained upwind flight. If males surge to single pulses, then plumes that ensure a rapid sequence of pulses (of optimal duration for the blend and concentration of pheromone) will promote the sustained expression of the upwind surge, resulting in nearly straight upwind flights; deviations from "optimal" stimulation result in the expression of casting flight interspersed among the surges.

Straight upwind flight and a suppression of counterturning are explained in *C. cautella* by the following five conditions (Mafra-Neto and Cardé 1994, 1995b, 1996): (1) The tempo of encountering filaments of pheromone is rapid enough to prevent the initiation of counterturning. (2) During a refractory period, clean air or contact with pheromone does not affect response. (3) The pheromone filament is at a concentration suitable for the maneuver, but below a level promoting counterturning. (4) The duration of contact with the pheromone pulse is sufficiently long to ensure optimal transmission of the signal and sufficiently brief to ensure its replacement by "clean" air before arrival of the next filament. (5) The plume's lateral dimensions are sufficient to ensure that the male encounters another filament of pheromone before the time interval specified in the first condition. In combination with optomotor anemotaxis, these conditions seem to explain straight upwind flight and counterturning in *C. cautella.*

In 1990 Baker suggested that the upwind course of a moth was dictated by moment-to-moment contact with pheromone. Loss of odor would trigger toward crosswind flight while contact with odor would promote a change in steering angle toward upwind. A suitable rate of encountering pheromone filaments would tend to increase the rate of counterturning, resulting in shallower zigzags or heading more upwind (Baker 1990). In independent experiments, Vickers and Baker (1994b; Baker and Vickers this volume) have shown that straight upwind flight in *Heliothis virescens* also is dictated by the frequency of encounters with individual filaments. The parallel findings that *Cadra* and *Heliothis* are capable of essentially straight upwind flight suggest that this control system may be widespread among moths. The quality of the stimulus (interaction among blend, concentration and duration) will determine how effectively and for how long counterturning is suppressed after interception of a single pulse (Mafra-Neto and Cardé 1996), whereas the time lag between a male encountering a filament of pheromone and turning upwind, and between loss of pheromone and turning crosswind, will determine the frequency of shift between the upwind surges and casting flight in turbulent plumes (Baker 1990; Vickers and Baker 1994b; Baker and Vickers this volume; Mafra-Neto and Cardé 1994, 1996).

The behavioral reaction time for the interception of a pheromone filament has been measured for a few moths, and it is surprisingly similar among taxa (Baker and Vickers 1994): It takes 0.2–0.3 sec for *C. cautella* to turn upwind, 0.2–0.5 sec for *H. virescens* (Vickers and Baker 1994b), and 0.2–0.3 sec for walking *Bombyx mori* (Kramer 1986). Casting *C. cautella* and *H. virescens* moths did not increase their ground speeds significantly during a surge upwind in response to a single pulse, whereas walking *B. mori* increased its ground speed. The temporal characteristics of the response to filaments in a plume presumably vary somewhat among moth groups and different environmental situations that alter the plume's structure. Together these factors may account for the somewhat different paths that moths take when flying along pheromone plumes, especially in the field. Even within a species, differences in moth size can influence moth velocity and the pattern of zigzag (Kuenen and Cardé 1994).

In plumes of manipulated structural complexity, *C. cautella* males produced flight tracks that reflected either the full cast–surge–cast template of response to interception of a single-pulse or only certain portions of this model, depending on the structure of the pheromone plume encountered. Flight along plumes in which contact with pheromone was frequent (due to size or frequency of the pulses) resulted in relatively straight flight tracks that appeared to be a string of upwind surges. Males intercepting pulses at high frequency increased their average ground speed substantially, probably because a new surge was elicited before the last surge waned. Because the flight prior to the new surge was already aligned upwind, such males achieved an increased ground speed. Flight along plumes in which contact was less frequent generated tracks reflecting the complete cast–surge–cast response. In these tracks, average ground speed was maintained at a relatively constant level, probably because the male reverted to casting flight before a new pulse of pheromone was intercepted. Such casting males ceased counterturning before surging upwind.

Flying males modify their maneuvers according to the current structural characteristics of the plume. The plasticity of moth upwind flight to pheromone sources appears to be the result of a constant modulation of single-pulse responses to the characteristics of the newly intercepted filaments. The behavior the male was performing when the new pulse was intercepted becomes the baseline for the expression of the new single-pulse response. If the male is counterturning, the interception of a new pulse will result in a change in course, whereas if the male is already flying straight upwind, the male may accelerate. Thus, single pulse reactions seem to be modified by the male's recent history of pheromone interception instead of simply being the expression of a rigid, single-pulse template. Interactions among the single pulse responses provide plasticity to pheromone-mediated flight, which may allow moths to perform source location efficiently under variable environmental conditions.

Although most flight tracks of *C. cautella* to its two-component blend were nearly straight upwind at a wide range of concentrations, similarly straight tracks also may be induced by using a particular concentration of the main component alone or the two pheromone components plus an inhibitor (Mafra-Neto 1993). Thus straight flights are obtained at a wide range of concentrations using the complete blend, or at a particular concentration using the "wrong" blend. If males perform upwind surges in response to interception of the pheromone at a wide range of concentrations, then most pulse interceptions in a turbulent plume of the right blend will promote the surge and result in a straight upwind path. But in plumes of the wrong blend, the male has to intercept filaments at an optimal concentration (and perhaps for the right amount of time). Thus most pulse interceptions in a turbulent plume of the wrong blend will promote an incomplete surge and result in a more zigzaggy upwind path.

5. Conclusion

Optomotor anemotaxis polarizes the flight of male moths in wind. The finding in two long-diverged lineages, the phycitid *C. cautella* and the noctuid *H. virescens* (Baker and Vickers this volume), of straight upwind flight in plumes providing a suitable rate of pheromone filament encounter is inconsistent with the flight imprecision model of Preiss and Kramer (1986a,b). A due or nearly due upwind wind heading indicates that these moths can accurately set their course due upwind; zigzags are a consequence of filament encounter rate and a presumed counterturning program and are not a necessity of upwind navigation. *C. cautella* and *H. virescens* also do not conform to the predictions of the Witzgall and Arn (1990) model, which attributes zigzags as being a consequence of an incomplete pheromone blend. Straight upwind courses occurred at suitable rates of filament encounter. In *C. cautella* nearly straight upwind flights were obtained for several blends, but at different concentrations. Although *C. cautella* males flew nearly straight upwind to off-blend and to complete blend sources, they only flew straightened-out paths to off-blend sources at a narrow range of doses, whereas straightened-out paths to complete blends occurred at a wide range of doses. This difference in response to complete and incomplete blends might help to explain why flight tracks along turbulent plumes of incomplete blend are more tortuous than tracks along plumes of complete blends.

Our view of how flying male moths locate a pheromone source in wind has been established mainly by wind tunnel manipulations of plume structure and the completeness of the pheromone blend. Although detailed experiments have been performed on relatively few moth species, these exemplars represent a spectrum of moth lineages, suggesting that our efforts so far to understand the principles underlying moth flight to pheromone will have wide applicability to all moth groups. We should not, however, be surprised to uncover differences among moth groups in their solution to the problem of locating a female by following her pheromone plume, given that most of these lineages diverged tens of millions of years ago, and the selective pressure to locate a female quickly remains paramount.

Acknowledgment

We thank M. Willis for his helpful review. Our research was supported by grants from the National Research Initiative of the USDA.

26

Pheromone Processing in the Lateral Accessory Lobes of the Moth Brain: Flip-Flopping Signals Related to Zigzagging Upwind Walking

Ryohei Kanzaki

1. Introduction

Male silkworm moths, *Bombyx mori,* in response to the sex-attractant pheromone released by conspecific females, walk upwind in a zigzag path with fluttering their wings, in what is called the "mating dance" (Kramer 1975, 1986; Obara 1979; Olberg 1983a; Kaissling and Kramer 1990; Kanzaki and Shibuya 1992b; Kanzaki et al. 1992). Even though they vibrate their wings at 40–50 Hz (Kanzaki and Shibuya 1986b), they are too heavy to fly. Zigzag walking is also initiated by a synthetic primary pheromone component *(E, Z)*-10,12-hexadecadienol or bombykol) (Butenandt et al. 1959).

As proposed in other flying moth species (Kennedy 1983; Mafra-Neto and Cardé 1994), including tobacco hornworm moth, *Manduca sexta* (Willis and Arbas 1991; Arbas et al. 1993), we have suggested that the *Bombyx* walking toward the pheromone source is controlled by self-generated zigzag turning programs which are triggered by intermittent pheromonal stimulation to the antennae (Kanzaki et al. 1992). We also suggested that these programs start in response to individual filaments (odor-laden strands) of pheromone that comprise the plume and are modulated by the frequency at which males encounter pheromone filaments in the plume (Kanzaki et al. 1992).

Olfactory receptor neurons in trichodeal hairs on the antennae transmit information about pheromone via encoded firing patterns to the brain through the antennal nerve (Kaissling and Kramer 1990). In the brains of moths, olfactory information passes through several stages of synaptic processing involving the first-order olfactory center, the antennal lobes (ALs) (Olberg 1983b; Kanzaki and Shibuya 1986a), and higher centers in the protocerebrum (PC) (Light 1986; Kanzaki and Shibuya 1992a). After local synaptic processing in the ALs and PC, the integrated multimodal information is relayed by PC neurons which have an axon descending

(DNs) in the ventral nerve cord (VNC) to thoracic motor circuitry to affect the behavior (Olberg 1983a; Kanzaki and Shibuya 1986b; Kanzaki et al. 1994).

Our research is providing an understanding of how information about pheromones is processed in the olfactory neural system of the brain and how this information participates in generating and controlling the pheromone-modulated maneuvers during locomotion.

2. Zigzagging Walking Toward the Pheromones in *Bombyx mori*

Bombyx mori males were tested in a wind tunnel using pulsed pheromone plumes produced by switching the clean air stream to the stimulant cartridge, which was controlled by a three-way solenoid valve (Kanzaki et al. 1992). The pulsed plumes were introduced into the constant air stream of the wind tunnel (50 cm sec^{-1}).

A single pulse stimulation (100 msec) with pheromone to either antenna elicited a long-lasting zigzag turn (<4 sec). After zigzagging, the pattern shifted to "looping" (turns of more than 360°) (Fig. 26.1) (Kanzaki and Shibuya 1992b).

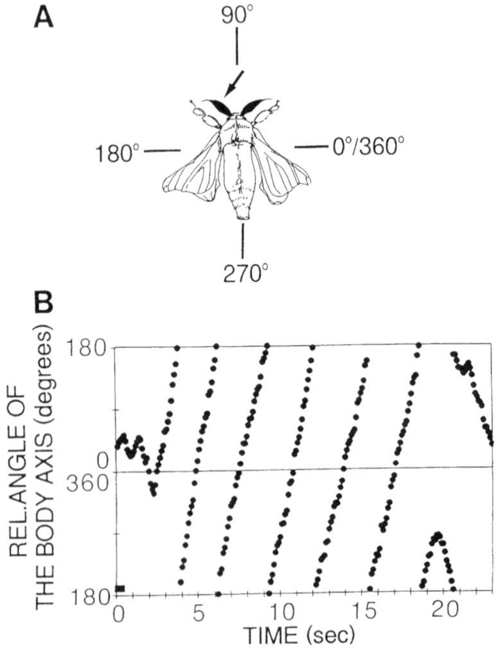

Figure 26.1. Example of *Bombyx mori* male walking tracks in response to a single pulsed pheromonal stimulation. **(A)** Single pulsed stimulation (100 msec) with pheromone extract was applied to the left antenna (arrow). **(B)** Zigzagging walking pattern was represented as changes of the body axis as in part A. Plots of the relative angles of the body axis against time when single pulse of pheromone was applied to the antenna, as shown in part A. Bold line in bottom left corner in part B indicates the stimulation period.

The direction of the first turn of the zigzag was strongly correlated with which antenna was stimulated (Kanzaki et al. 1992). When repetitive pulsed stimulation was applied, a zigzag pattern similar to that triggered by a single pulse was set with each stimulation. The interval of time and the angle of the zigzag increased gradually turn by turn but were reduced with each successive stimulation ('reset') (Kanzaki et al. 1992). Similar results were obtained even when both compound eyes of a male were painted black to prevent visual information. *B. mori* males walk straighter upwind to fast-pulsed or continuous plumes than to slowly pulsed plumes (Fig. 26.2) (Kramer 1975, 1986; Kanzaki et al. 1992). There may be two control mechanisms during upwind walking toward the pheromones: (1) self-generated zigzag turning programs triggered by a pulsed pheromonal stimulation and (2) "reset" mechanisms of the programs (Kanzaki et al. 1992). After the "rest" mechanisms of the programs, the increment of the stimulation frequency eventually elicited only small turns repeatedly. Therefore, the walking direction appears to be more and more straight. The walking pattern gradually became more complex as the stimulation frequency was decreased (Fig. 26.2).

Even a single pulse of pheromone (100 msec) applied to either antenna elicited the zigzagging for approximately 4 sec followed by looping (Fig. 26.1). This suggests that some neurons in the central nervous system preserve this neural information and show a long-lasting (tonic) activity.

We have found that in *B. mori* as well as in *M. sexta,* many olfactory PC neurons (some restricted to the PC) and some DNs innervate a particular neuropil area in the protocerebrum, the lateral accessory lobes (LALs) (Kanzaki and Shibuya 1986b, 1992a; Kanzaki et al. 1991a,b, 1992). These PC neurons, especially the DNs, show a long-lasting excitation activity and state-dependent or flip-flopping responses (see below) to repetitive pheromonal stimulation (Kanzaki and Shibuya 1986b, 1992a; Kanzaki et al. 1991a,b, 1994). This is of particular interest because the information the DNs carry represents the integrated output of the brain circuits that can act on thoracic motor circuitry to cause olfactory-modulated behavior. The LALs appear to be important for processing olfactory

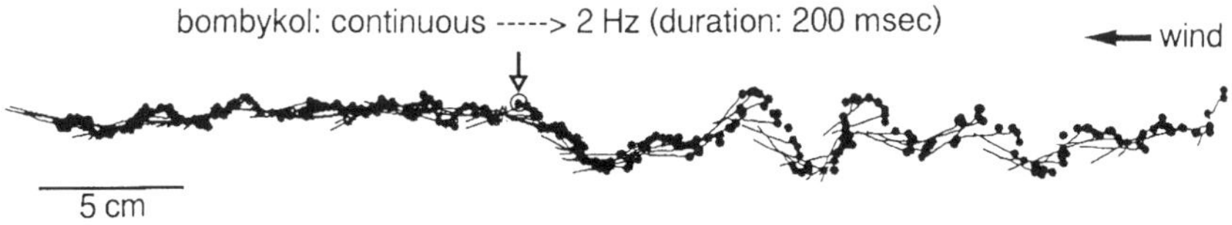

Figure 26.2. Plots of the *Bombyx mori* male walking tracks, as viewed from above. Black circles represent the head position of the moth at 0.1-sec intervals. Bars indicate the body axis. Wind was from the right (0.5 m sec^{-1}). Pheromonal stimulation (100 ng bombykol) was introduced in the air stream of the wind tunnel continuously (C) or at frequencies of 2 Hz (duration: 200 msec). The stimulation was switched from C to 2 Hz at the arrow.

information in the higher olfactory centers in the PC (Kanzaki and Shibuya 1992a; Kanzaki et al. 1991a,b, 1994).

It is reported that some DNs of *B. mori* males show a characteristic state-dependent activity which has two distinct firing frequencies, high and low. Switching back and forth between the two states occurs with each repetitive pheromonal stimulation (Olberg 1983a; Kanzaki et al. 1994) (see Fig. 26.4). The activity pattern is called a "flip-flop" one after the basic element of "memory" in electronic circuit operation (Meyer et al. 1968). In *M. sexta,* the state-dependent activity pattern, similar to the flip-flopping activity pattern in response to the pheromonal stimulation, was characterized (Kanzaki et al. 1991b).

We have studied the details of the flip-flopping DNs to provide a basis for understanding the neural system required to produce flip-flopping activity patterns, which may be an important signal in controlling the pheromone-modulated zigzagging walking of *B. mori* males (Olberg 1983a; Kanzaki et al. 1992, 1994).

3. Lateral Accessory Lobes

3.1. Morphology

The spheroidal neuropil structure of the lateral accessory lobe (LAL, 150 μm × 100 μm) exist on each side of the PC along the stalk of the mushroom body, lateral to the central body (Homberg et al. 1991; Strauss and Heisenberg 1993) and posterior to the ALs (Fig. 26.3) (Kanzaki and Shibuya 1992a; Kanzaki et al. 1994). Each LAL is linked with each other through the commissure of the LALs by protocerebral bilateral neurons (PBNs) (Kanzaki and Shibuya 1992a). Some PC neurons provide the anatomical substrate for linking the LAL to the lateral PC, where projection neurons from the ALs terminate (Kanzaki et al. 1991a). No antennal lobe projection neurons terminate in the LALs (Kanzaki and Shibuya 1986a). The LALs are also linked to adjacent neuropils. Thus, the LALs are interposed in the pathways of olfactory information flow from the ALs through the lateral PC to DNs. Similar functional pathways were also found in *M. sexta* (Kanzaki et al. 1991a,b).

3.2. Protocerebral Bilateral Neurons Innervating the LALs

Intracellular recording and staining of PC neurons revealed that some neurons innervated both LALs with heavily branched arborizations linked by a single major neurite which crossed the midline of the brain through the commissure of LALs (protocerebral bilateral neurons or PBNs) (Fig. 26.5A) (Kanzaki and Shibuya 1992a). Branches ipsilateral to the cell body exhibited smooth profiles, and those contralateral to the cell body had numerous varicosities. Although it must be verified with electron microscopy, it is possible that the polarized information flows from the side ipsilateral to the cell body to the contralateral side, suggesting

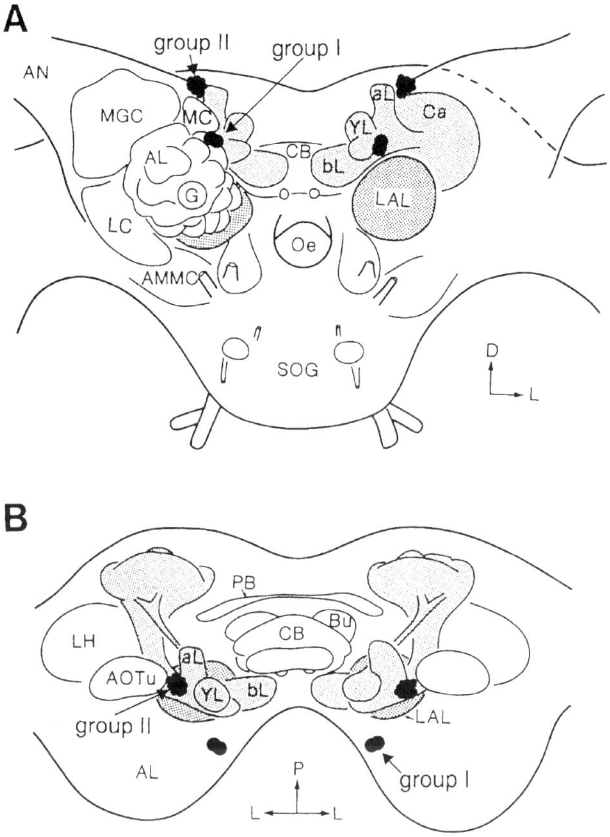

Figure 26.3. Schematic representation of the neuropils in the midbrain of male moths. **(A)** Dorsal view. **(B)** Frontal view. The lateral accessory lobes (LALs) exist on both sides of the protocerebrum and are lateral to the central body (CB) and posterior to the antennal lobe (AL). Two groups of descending interneurons (DNs) which have dendritic arborizations in the LAL were found: (1) DNs which have a cell body in the cluster ventral to the medial cell cluster (MC) of the AL neurons (group I) and (2) DNs which have a cell body adjacent to the α-lobe (aL) of the mushroom body (group II). Other abbreviations: AMMC, antennal mechanosensory and motor center; AN, antennal nerve; AOTu, anterior optic tubercle; Bu, buttress; bL, β-lobe; Ca, calyces of the mushroom body; D, dorsal; G, ordinary glomerulus; L, lateral; LC, lateral cell cluster; LH, lateral horn; MGC, macroglomerular complex; Oe, oesophagus; P, posterior; PB, protocerebral bridge; SOG, suboesophageal ganglion; YL, Y-lobe.

that the PBNs may be important for comparing olfactory information between the LALs.

Some of these PBNs typically showed a characteristic, long-lasting excitation (LLE) which outlasted the stimulation period in response to bombykol. In the case of a stimulation with 500 ng bombykol applied to the antenna for 1 sec, firing levels >70% of the peak frequency lasted more than 20 sec (Kanzaki and Shibuya 1992a). Such LLE-type responses have not yet been encountered in the local interneurons or projection neurons of the first-order olfactory center, the antennal lobes of *B. mori* (Olberg 1983b; Kanzaki and Shibuya 1986a). The LLE may be a response arising at the level of the PC. Similar robust LLE responses were induced in bilateral neurons of *M. sexta* which link with each LAL (Kanzaki et al. 1991a). The PBNs of *M. sexta* also showed the LLE in response to female pheromone extract or a mixture of an individual component of the pheromone (Tumlinson et al. 1989)—that is *(E, Z)*-10,12-hexadecadeinal, the primary component of the pheromone (bombykal), and *(E, Z)*-11,13-pentadecadienal, the pheromone mimic that selectively stimulates the *(E, E, Z)*-10,12,14-hexadecadeinal receptors in the antenna (Kaissling et al. 1989b). Individual components of the pheromone mixture did not elicit remarkable LLE-type responses in *M. sexta* (Kanzaki et al. 1991a). In *M. sexta,* such LLE-type responses have not yet been recorded in the antennal lobe neurons (Christensen and Hildebrand 1987a,b; Homberg 1987; Kanzaki et al. 1989).

3.3. Descending Interneurons Innervating LALs

Cobalt staining of the neurons within a connective, or within a small bundle split from a connective, revealed the morphology of descending interneurons (DNs). The DNs which have a projection in the LAL were consistently stained when a small bundle of axons split from a dorsal part of the connective was filled with cobalt (Kanzaki et al. 1994). These DNs fall into two classes according to the position of their cell bodies in the brain (Fig. 26.3). DNs in the first group have a cell body in the cluster ventral to the medial cell cluster (MC) of the AL neurons (group I, Fig. 26.3). Two to three cell bodies usually make up the cluster. They have widely spread aborizations throughout the LAL ipsilateral to the cell body and the neuropil ventral to the LAL. The major neurite projects to the contralateral LAL through the commissure of the LALs and descends the contralateral connective (Fig. 26.5A). DNs in the second group have a cell body adjacent to the α-lobe of the mushroom body (group II, Fig. 26.3). Ten to fifteen cell bodies usually make up the cluster. The major neurite projects to the ipsilateral LAL, where it has sparse arborizations and descends the ipsilateral connective (Fig. 26.5A). Cell bodies of each group exist in bilateral clusters in the brain (Kanzaki et al. 1994).

4. Flip-Flopping Descending Interneurons

4.1. Two Types of Flip-Flopping Activity Patterns (FF and ff)

Since group I and group II DNs were selectively stained by cobalt backfilling from the dorsal part of the connective, multiple suction recordings were made from the region in order to determine their physiological properties (Fig. 26.4,

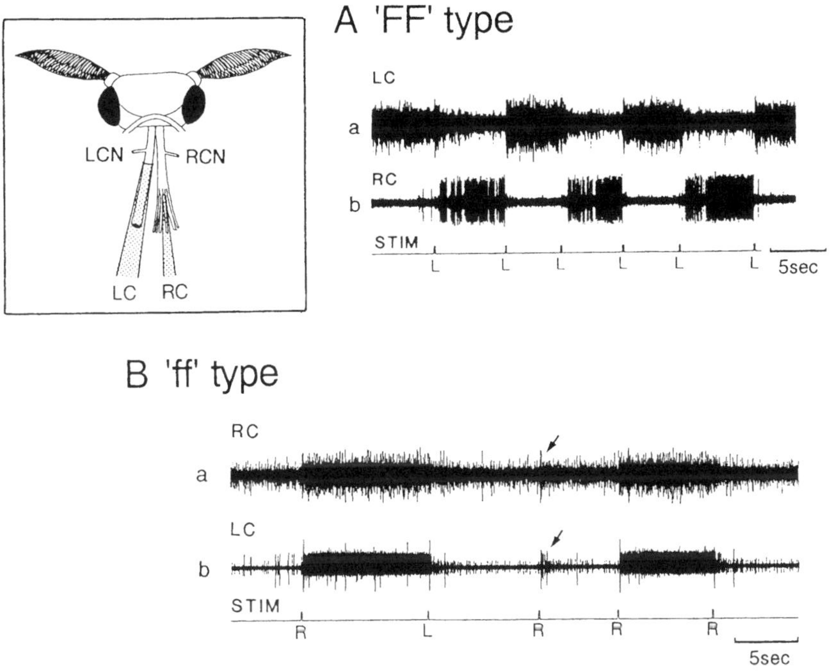

Figure 26.4. Two types of flip-flopping state-transition patterns (FF and ff). **(A)** The flip-flopping activity pattern recorded from a small bundle split from a right connective (RC) had an antiphasic relationship to some of the largest impulses recorded from the whole left connective (LC). The flip-flopping activity pattern in (b) is classified as FF type according to the criteria (see text). **(B)** The flip-flopping activity pattern recorded from a small bundle split from a left connective (LC) had a synchronized relationship to some of the largest impulses recorded from the whole right connective (RC). The flip-flopping activity pattern in (b) is classified as ff type. STIM indicates the sequential pheromonal stimulation applied to the left (L) or right (R) antenna individually (concentration: 10 ng bombykol in part A and 1 ng in part B; duration: 100 msec). For explanation of arrows, see text. **Inset:** Schematic representation of the experimental preparation. LCN, left cervical nerve; RCN, right cervical nerve.

inset). As shown in Fig. 26.4Ab, flip-flopping activity patterns were usually recorded from a small bundle of the dorsal part of the connective. The diameter of the electrode was less than 10 μm. Even when a whole left or a whole right connective was picked up in a suction electrode (Fig. 26.4, inset), some of the largest impulses showed a typical flip-flopping activity pattern in response to brief sequential pheromonal stimulation (Fig. 26.4Aa). Cross sections of the connective indicate that axons with a large diameter (>4 μm) are restricted to the dorsal region of each connective (Kanzaki et al. 1994). This suggests that the axon diameter of flip-flopping DNs is larger than that of other DNs, and the axons of these interneurons occupy the dorsal part of the connective.

The flip-flopping activity has been defined by Olberg's criteria; that is, a state transition is at least a doubling or a halving of the prestimulus firing frequency which persists for at least 5 sec after the end of the stimulus (Olberg 1983a). Pure air stimulation did not elicit a state transition. Although a state transition was hardly elicited under dark conditions (<1 lux), typical state transition was observed under light conditions (approximately 3500–7000 lux) (Kanzaki et al. 1994).

The flip-flopping activity patterns that consistently accompanied some of the largest impulses recorded from a whole right connective (RC) and a whole left connective (LC) were antiphasic (Kanzaki et al. 1994). In order to clarify the flip-flop activity patterns, one suction electrode was applied to a whole left or a whole right connective, and an other electrode was applied to a small bundle split from the other connective (e.g., Fig. 26.4, inset). As illustrated in Fig. 26.4A, some flip-flop activity patterns recorded from a small bundle split from a connective (Fig. 26.4Ab) showed an antiphasic relationship to some of the largest impulses recorded from the other whole connective (Fig. 26.4Aa). On the other hand, other flip-flop activity patterns recorded from a small bundle split from a connective (Fig. 26.4Bb) exhibited a synchronized relationship to some of the largest impulses recorded from the other whole connective (Fig. 26.4Ba). Even when a flip-flopping activity pattern was not elicited in response to the stimulation, synchronized phasic firing was recorded between the RC and the LC (arrows in Fig. 26.4Ba,b). We classified the flip-flop activity patterns which showed an antiphasic relationship as FF type (Fig. 26.4Ab) and those which showed a synchronized relationship with respect to some of the largest impulses recorded from the other whole connective as ff type. In some examples, FF and ff responses were recorded simultaneously by a single suction electrode (Kanzaki et al. 1994). Thus, each connective contains FF and ff types of flip-flopping DNs (Fig. 26.5). The action of the flip-flopping activity pattern descending the VNC of *B. mori* is similar to that of an electronic "trigger flipflop," which has one input (trigger pulse) and two antiphasic long-lasting outputs (Q and $\overline{Q}$). Such a flip-flop state-transition processing may be involved in some neural systems of the insect brain.

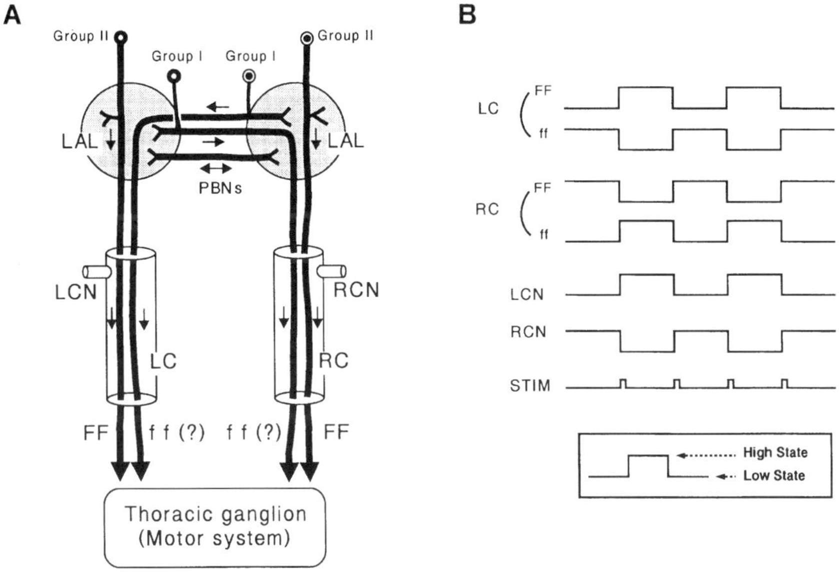

Figure 26.5. **(A)** Schematic summary of LAL olfactory pathways in the brain of male *Bombyx mori* moths. The two lightly shaded circles represent the left and right lateral accessory lobes (LALs) in the protocerebrum. Lines show group I and group II DNs and protocerebral bilateral neurons (PBNs). **(B)** Schematic representation of the activity correlations between the flip-flopping descending interneurons (DNs) in the VNC and some neck motor neurons (NMNs) in the cervical nerve (CN). Activity patterns are represented as a timing chart. The activity of some NMNs is synchronized to a flip-flopping activity pattern classified as FF but not to ff, when recordings were made from the same side of the connective and the CN. STIM indicates the sequential pulsed pheromonal stimulation applied to left or right antenna. LC, left connective; LCN, left cervical nerve; RC, right connective; RCN, right cervical nerve. *Note:* Although in some cases the state transition occurred spontaneously without pheromonal stimulation, correlations in activity patterns between the flip-flopping DNs and NMNs were maintained consistently (not shown in the schema).

4.2. Morphology of Flip-Flopping Descending Interneurons

After recording flip-flopping activity from a small bundle split from a connective, the neurons within the bundle were stained with cobalt by passing current through the recording suction electrode. Five examples of neurons, which exhibited FF-type flip-flopping activity patterns, were stained successfully. In all cases, group II DNs (Figs. 26.3 and 26.5A) were stained. In four out of five examples, a few other groups of DNs which had cell bodies ventral to the calyces of the mushroom body or in the suboesophageal ganglion (SOG) were also stained. In each example the morphology of these stained DNs differed. In one case, only group II DNs were stained in the brain. This suggests that the stained group II DNs showed FF-type activities. By the staining methods used, we have not yet succeeded in staining a bundle showing only ff activity patterns. We have characterized intracellularly DNs which were morphologically similar to group I DNs in *B. mori* (Kanzaki and Shibuya 1986b). These DNs exhibited an LLE which in response to a pulsed pheromonal stimulation outlasted the stimulation period by more than several seconds (Kanzaki and Shibuya 1986b). Since sequential pulsed stimulation was not applied in that preparation, we do not know whether the cell exhibits a flip-flopping activity pattern. In *M. sexta,* DNs morphologically similar to group I, which have arbors in the LALs and an axon descending a connective contralateral to the cell body, were characterized intracellularly (Kanzaki et al. 1991b). These neurons exhibited state-dependent responses, in that stimuli applied when a neuron was in a state of low-frequency firing elicited accelerated firing for several tens of seconds. However, the clear reduction of firing elicited by subsequent pheromonal stimuli spontaneously reverts to the state of high-frequency firing after several seconds (Kanzaki et al. 1991b). It is possible that group I DNs of *B. mori* show ff activity patterns (Fig. 26.5A).

4.3. Hypothesis of the Functional Links Required for Flip-Flopping Activity Patterns

Figure 26.5A illustrates schematically the functional links between LALs by DNs and PBNs of male *B. mori* moths. Morphologically similar DNs and PBNs, except for group-II-like DNs, have been characterized in *M. sexta* (Kanzaki et al. 1991a,b). Although group-II-like DNs have not yet been morphologically characterized in *M. sexta,* group-I- and group-II-like DNs were found in a related sphingid, the sweet potato hornworm moth, *Agrius convolvuli* (Kanzaki unpublished). Similar functional links in the LAL system may exist in *B. mori, M. sexta,* and *A. convolvuli.*

Physiological and morphological studies on DNs of *B. mori* lead us to the following hypothesis of the functional links required for flipflop activity patterns. That is, through one LAL, group II DNs may conduct an FF signal to the connective ipsilateral to the LAL while group I DNs conduct an ff signal to the connective contralateral to the LAL. Moreover, mirror symmetrical signal transmission would occur through the other LAL.

When the FF-type and ff-type flip-flopping activities were simultaneously recorded from left (LC) and right connectives (RC) respectively or vice versa, the activity patterns were synchronized as shown in Fig. 26.5B. Thus, FF DNs in the LC and ff DNs in the RC, or vice versa, carried a synchronized signal pattern. These results suggest that FF DNs in the LC and ff DNs in the RC may receive a similar activity input (e.g., flip-flopping activity patterns) in the same neuropil, maybe in the left LAL or vice versa. Since it is strongly suggested that group II DNs show FF activities and their arborizations were restricted to the LAL, group II DNs may receive the FF signal only in the LAL. On the other hand, when FF activities were simultaneously recorded from the left and right connectives, each of the FF activities had an antiphasic relationship (Fig. 26.5B). Therefore, flip-flopping activities which had an antiphasic relationship were transmitted to each FF DN through left and right LALs. These FF activities may be transmitted to group II DNs, and ff activities may be transmitted to group I DNs (Fig. 26.5A).

Each LAL is linked with each other by PBNs, some of which showed a long-lasting excitation (Kanzaki and Shibuya 1992a). In these functional links between LALs, PBNs or other bilateral PC neurons which connect the two LALs might have reciprocal inhibitory connections to produce antiphasic flip-flopping activities (Kanzaki et al. 1994). When the LAL commissure linking each LAL was microsurgically dissected between the left and right LALs, the *B. mori* male showed only straight-line walking or one way looping (Kanzaki unpublished). This result is consistent with the idea that the LALs might have reciprocal inhibitory connections through the commissure of the LALs. In *M. sexta,* each LAL is also linked to a central complex by some protocerebral neurons which showed γ-aminobutylic acid (GABA)-like immunoreactivity (Homberg et al. 1987).

4.4. The Role of the Flip-Flopping Descending Interneurons

The flip-flopping state-transition signals carried by DNs in the connectives, which link the brain and the thoracic motor system, are thought to be important instructions for pheromone-directed turns in the upwind approach of the male *B. mori* to the female (Olberg 1983a; Kanzaki et al. 1994). We have studied the direct physiological correlations between the flip-flopping activity patterns and the motor system involved in this behavior.

Bombyx mori males perform a variety of sophisticated and synchronized head movements when they show zigzagging turning in response to the pheromonal stimulation applied to the antenna. The head movements were recorded by a high-speed video system during the zigzagging walking elicited by the pheromone (Kanzaki and Mishima 1996). As observed in flies (Land 1974; Hengstenberg et al. 1986; Milde et al. 1987; Strausfeld et al. 1987), when the male *B. mori* turns to the right, its head shows sidewise movement to its right, and vice versa (Kanzaki and Mishima 1996). In flies, the morphological organization of the neck motor system [i.e., neck motor neurons (NMNs) and the neck muscles they supply] was well studied by Strausfeld and collaborators (Milde et al. 1987; Strausfeld et al. 1987). It is reported in flies that the neck sidewise movement is regulated by oblique horizontal muscles (OH1, OH2) supplied by the NMNs (CNM2) in the cervical nerve (CN) and others.

Figures 26.5B and 26.6 represent the correlations in activity patterns between the flip-flopping DNs and some NMNs in the CN of *B. mori.* We found that the

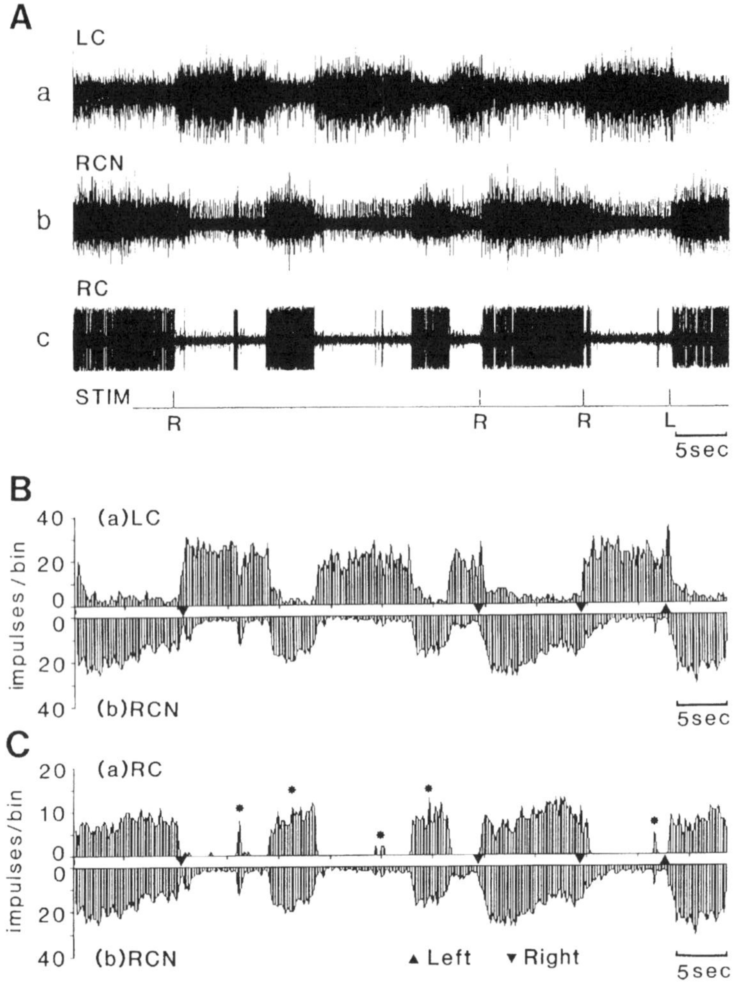

Figure 26.6. Correlations of activity patterns between flip-flopping descending interneurons in each connective and some neck motor neurons in a cervical nerve. **(A)** Flip-flopping activity patterns recorded from a whole left connective (LC) and a small bundle of the dorsal part of right connective (RC) are shown in (a) and (c), respectively, in response to sequential pheromonal stimulation (concentration: 10 ng bombykol; duration: 100 msec). Stimulation (STIM) was individually applied to the right antenna (R) or left antenna (L). Pheromonal responses of neck motor neurons in the right cervical nerve (RCN) were simultaneously recorded (b). **(B, C)** Impulse-frequency histograms (0.5-sec bins) of part A are illustrated. The frequencies are indicated on the *y*-axis with opposite direction between LC and RCN **(B)** and between RC and RCN **(C).** Even when the state transition occurred spontaneously (*), synchronized activity pattern between the RC and RCN is consistently maintained. Triangles (▲, ▼) indicate the stimulation applied to the left (L) or right antenna (R), respectively. [From Kanzaki and Mishima (1996).]

neck motor system of male *B. mori* is controlled by pheromone, showing strong correlation with the flip-flopping state-transition signal produced in the brain. In other words, the flip-flopping activity pattern classified as FF and the activity of some NMNs are consistently synchronized when recorded from the same side of the connective and the CN. Thus, the FF-type flip-flopping activity shown by some DNs (i.e., group II DNs) is strongly related to the NMNs' activity, suggesting that the state-transition signal descending the VNC to the thoracic motor system is related to the turning during pheromone-mediated walking. Although each CN contains about eleven NMNs (Mishima and Kanzaki 1996), we do not know yet which NMNs elicit the flip-flopping state-transition activities.

In many cases, spontaneous state transition of flip-flopping activity occurred repetitively after a single pulsed stimulation. Even when the flip-flopping DNs showed repetitive state transitions, some NMNs contained in the CN consistently exhibited a synchronized activity pattern (Fig. 26.6C, Kanzaki and Mishima 1996). In response to a single pulse of pheromone, *B. mori* males exhibit a zigzag turning (Fig. 26.1B). It seems that the repetitive state transitions may be the instructions for the zigzag turns of *B. mori* males. How the repetitive transition of activity patterns is produced in the LAL olfactory system and how the information participates in generating the zigzagging behavior is the subject of ongoing research.

Acknowledgments

This research was supported in part by a Grant-in-Aid for General Scientific Research from the Ministry of Education, Science and Culture of Japan (No. 05640761), a grant from the University of Tsukuba Research Projects, and a grant from Chemical Materials Research and Development Foundation.

27

Active Behavior and Reflexive Responses: Another Perspective on Odor-Modulated Locomotion

*Mark A. Willis and Edmund A. Arbas**

1. Our Initial Approach

That male moths orient their flight into the wind upon sensing wind-borne pheromone from a conspecific female, and modulate their subsequent flight performance with respect to pheromone and wind stimuli, is generally accepted by most that study this behavior (see Baker and Vickers this volume; Kramer, Cardé and Mafra-Neto this volume; and Witzgall this volume). The behavioral and physiological mechanisms that enable a male moth to accomplish this complex feat are still a matter for considerable (and lively) debate. As we initiated our approach to this behavior from an explicitly neuroethological perspective (see Arbas this volume) using the moth *Manduca sexta* as our model, we decided that, rather than assume all male moths locate pheromone sources using the same mechanisms, we would start "from scratch." This approach has forced us to examine our data in new and different ways and, at times, to come to conclusions and propose mechanisms that are at odds with some of the currently accepted views of the control of pheromone-modulated flight in moths (Arbas and Willis 1994, Willis and Arbas 1994). However, careful review of the work of earlier researchers in this field, especially that of J.S. Kennedy, has revealed hints of many of the interpretations that we have come to with regard to our own data. In order to proceed in the development of our own ideas regarding the generation and control of this biologically and agriculturally interesting behavior, it has been critical to clearly understand the existing hypotheses.

The one behavioral mechanism that is universally agreed upon is odor-modulated optomotor anemotaxis (Kennedy 1940, 1977). That is, upon contacting an

*Ed Arbas died unexpectedly June 18, 1995. His knowledge and ideas contributed significantly to the content of this chapter.

attractive odor in flight a moth will turn into the wind and then control its velocity and steering by using visual flow field cues to determine the direction and speed of the wind in which it is flying. The sinuous look of these flight tracks arises from the sequential left–right counterturns that often occur with remarkable temporal regularity (Willis and Baker 1987; Baker and Haynes 1987; Baker and Vogt 1988; Willis and Cardé 1990; Willis and Arbas 1991a). The temporal regularity of these counterturns has led many to propose some sort of internal counterturn generator as an explanation (Wright 1958; Kennedy and Marsh 1974; Baker et al. 1984). Together with optomotor anemotaxis, the idea of an internal mechanism to generate temporally regular counterturns is perhaps the most widely accepted behavioral mechanism involved in the odor-modulated locomotion of moths (Kramer 1975; Kennedy 1983, 1986).

1.1. Current Models for the Control of Pheromone-Modulated Flight

A number of models have been proposed to account for the sensory-motor mechanisms underlying this behavior (Marsh et al. 1978, Preiss and Kramer 1986a,b; Kaissling and Kramer 1990; Baker 1990; Mafra-Neto and Cardé 1994; Kramer, this volume). The most parsimonious of these (Preiss and Kramer 1986a) does not invoke an internal timing mechanism for the counterturning pattern mentioned above, but, instead, relies exclusively upon odor-modulated optomotor anemotaxis. According to this hypothesis, a flying male moth turns into the wind in response to contact with a pheromone plume, but his inability to orient precisely into the wind, together with a tendency to turn which is proportional to the component of visual feedback indicating lateral slip (transverse image flow), results in a continuous series of visually driven compensatory turns left and right across the wind line. Having a turning response that is proportional to transverse (T) image flow results in a faster and stronger turning response to the increased lateral image flow caused by the increased drift in higher wind velocities. Progress toward the source is thought to be sustained by the visual-motor system maintaining a small constant velocity of positive (i.e., front-to-back) longitudinal (L) image flow over the eyes, with the final result being the sinuous track flown by the male moths. An implicit assumption of this model is that the moths would fly directly upwind to the source if they could. A recent modification of this model includes an odor-independent internal turning tendency that interacts with the optomotor anemotactic system, enabling the moths to maintain a set course at an angle to the wind direction (Kaissling and Kramer 1990).

Another recent hypothesis (Baker 1990) attempts to explain the variations of zigzagging flight in different species of moths, using a combination of behavioral and neurophysiological data. Pheromone contact is thought to cause the male moth to orient his flight into the wind using optomotor feedback. Upwind progress toward the pheromone source is then thought to be mediated by the interaction of two flight control systems simultaneously activated by contact with pheromones. The first, a tonic drive system, is thought to be responsible for providing a long-lasting (i.e., several seconds) tonic drive to an internal counterturning generator. The second, a phasic drive system, causes the moth to steer directly upwind. Immediately upon pheromone contact, the phasic system is thought to override or suppress the simultaneously activated tonic system, resulting in an upwind "surge" (*sensu* Baker 1990) toward the pheromone source. This suppression wanes with some relatively rapid time course (<1 sec) upon entering clean air, allowing the program of counterturning to emerge. If the gap between pheromone contacts is long enough, upwind progress ceases, and counterturning continues in a period of casting flight. Differences between species that generate straighter (Baker and Vogt 1988; Haynes and Baker 1989; Vickers and Baker 1994b) or more sinuous (Baker et al. 1984; Willis and Cardé 1990; Willis and Arbas 1991a) flight tracks would be explained by differences in their abilities to respond to the fine structure of the pheromone plume and the time constants of suppression and recovery of the counterturn system. Thus, the structure of the flight track would be determined by the pattern of contacts with pheromones or clean air within the plume (caused by both intrinsic plume structure, and the moth's maneuvering), together with the response latency of the individual and the particular species in question. Indeed, recent evidence shows that an alternating casting/surging pattern can be elicited by experimentally generated intermittent strands of pheromone (Mafra-Neto and Cardé 1994; Vickers and Baker 1994b) and that such a mechanism may contribute to the sinuosity of the flight tracks, or lack thereof, in plumes with different levels of turbulence (Mafra-Neto and Cardé 1994).

In these previous two models, the patterning of the moth's behavior was posited to arise primarily from reactions to the changing conditions of the environment; either the slip of visual flow fields (Preiss and Kramer 1986a,b) or the fine structure of odor concentrations within the pheromone plume (Baker 1990; Vickers and Baker 1994b). A different view, formalized by J.S. Kennedy and his colleagues (Kennedy and Marsh 1974; Marsh et al. 1978; Kennedy et al. 1981; Kennedy 1983; 1986; Ludlow 1984; David 1986), proposed (although not always explicitly) that the pattern of zigzagging upwind flight may be generated actively and be adapted to acquire information about the wind and pheromonal signal. In this model, the onset of pheromonal stimulation is thought to activate a hypothetical internal counterturn generator and bias optomotor anemotaxis toward flight in the upwind direction. The ground speed (velocity along the flight track) and track angle (orientation of the movement vectors with respect to due upwind) are postulated to be maintained at internally set, "preferred" values while moths perform pheromone-modulated upwind flight (Marsh et al. 1978). The preferred ground speed and track angles would be maintained by monitoring the velocity of L and T visual slip across the eyes (Ludlow 1984; David 1986) and by making compensatory adjustments in air speed (magnitude of thrust) and course angle (direction of thrust) in response to deviations from the preferred values of slip velocity caused by environmental perturbations and motor errors (Marsh et al.

1978). The set points of the optomotor control system would also be modulated by the encounters with pheromone filaments in the turbulent plume, in addition to information on air flow around the moths sensed directly via mechanoreceptive inputs and indirectly via proprioceptive inputs (Marsh et al. 1978). Thus, in this model, the structure of the flight track is the result of internally regulated parameters—for example, set points for steering and ground speed control, and the pattern of counterturning, combined with and modulated by moment-to-moment exteroceptive and proprioceptive information. Therefore each maneuver would be *"action-and-reaction in one"* (Kennedy 1986).

The debate over the control of pheromone-modulated flight is a reflection of an ongoing argument about one of the fundamental questions in the neural control of behavior: the relative roles of peripheral sensory input and centrally patterned activity in the organization of behavior. A review of Kennedy's later work on odor-modulated flight (Kennedy 1983, 1986) reveals that he apparently avoided this false dichotomy, explicitly discussing the behavior in question as a dynamic interaction of ongoing centrally organized behavior that is shaped and modulated by peripheral inputs. In addition, throughout the development of these ideas the work of Kennedy and his co-workers has explicitly included and discussed the role of modalities other than vision and olfaction (i.e., mechanoreceptive and proprioceptive inputs) in the shaping of odor-modulated flight (Marsh et al. 1978).

1.2. A Neuroethological Perspective

As has been mentioned elsewhere in this volume (see Arbas this volume), a neuroethological approach to understanding behavior and its neural control has more to do with the philosophy of the researcher than the techniques used or the aspect of the system being studied. It is our belief that the fullest understanding of a behavior and its underlying neural control will derive from applying new knowledge from one level of organization and experimentation to our understanding and thoughts about all other levels. These levels may span a very broad range of issues: from the interaction of the animal with its physical environment (e.g., fluid dynamics of the interaction of olfactory sensilla with airborne pheromone), to the integration of sensory inputs from many modalities (e.g., the effect of olfactory input on the weighting of visual information), the successful incorporation of this processed sensory input to ongoing motor behavior, and the probable feedback from the active motor circuits into the integrative and sensory processing centers of the brain.

2. Averaged Behavior Versus Finer-Grained Analysis of Movements

A typical approach to experiments and data analysis on pheromone-modulated flight in male moths has been to make measurements (usually each 33 msec) from the flight tracks representing the responses of individual males, average them across the entire flight track, then lump these with the averages of other individuals responding to the same treatment (i.e., generate grand means) (Fig. 27.1A). Initial studies of the flight responses of *M. sexta* to pheromones presented in different wind speeds (Willis and Arbas 1991a) showed results consistent with those from other species of moth (Marsh et al. 1978; Willis and Cardé 1990). Data analyzed in this way indicated that the moths maintained relatively constant ground speeds and track angles by altering the direction and magnitude of their thrust to compensate for the effects of increased wind speeds (Fig. 27.1B,C). The most stereotyped component of pheromone-modulated flight in these experiments was the time interval between sequential counterturns (i.e., interturn durations, Fig. 27.1D). On average, the interturn durations were not affected by flight in different wind speeds, but did decrease significantly as the moths approached the source (Fig. 27.1D). This analysis of our data showed behavior that was consistent with other moth species, and thus could be explained with previously proposed hypothetical mechanisms (Marsh et al. 1978). However, since the nervous system generates and controls behavior on a milliseconds time scale, relying on averaged data for our analysis cannot provide the temporal resolution necessary to understand the neural bases of this behavior.

Currently, and for the foreseeable future, our focus will be to examine the responses of individual animals at the finest level of resolution (spatial and temporal) available to us. The first of these analyses has recently been completed (Willis and Arbas unpublished), and the results indicate that the control of this behavior may be considerably more complex than previously thought (Marsh et al. 1978; Ludlow 1984; David 1986).

2.1. Modulation of Ground Speed

By plotting the movement parameters (e.g., ground speed, track angle, etc.) of individual moths for each sequential 33-msec movement vector during their flight to the odor source, a pattern of responses very different from that observed from averages emerges. Individual *M. sexta* males flying upwind in a plume of pheromones did not maintain a constant ground speed. Rather, the ground speeds of most moths was modulated in a predictable pattern over a range from approximately 0 cm sec^{-1} to 115 cm sec^{-1}. We observed two distinct patterns in the moths

Figure 27.1. **(A)** Illustration of the pheromone-modulated flight track of an *M. sexta* male. The dots on the track represent the position of the moth every 33 msec. The figure of the moth on the track illustrates the effect of wind-induced drift; the moth directs its thrust along the course angle, but due to the influence of wind the resulting movement is in the direction of the track angle. **Inset:** The triangle of velocities as described by Marsh et al. (1978), whereby one can calculate approximations of the air speed, course angle, and drift angle using the measured values of ground speed, track angle, and wind speed and direction. **(B)** Plot of the grand means of 10 individuals flying upwind in a

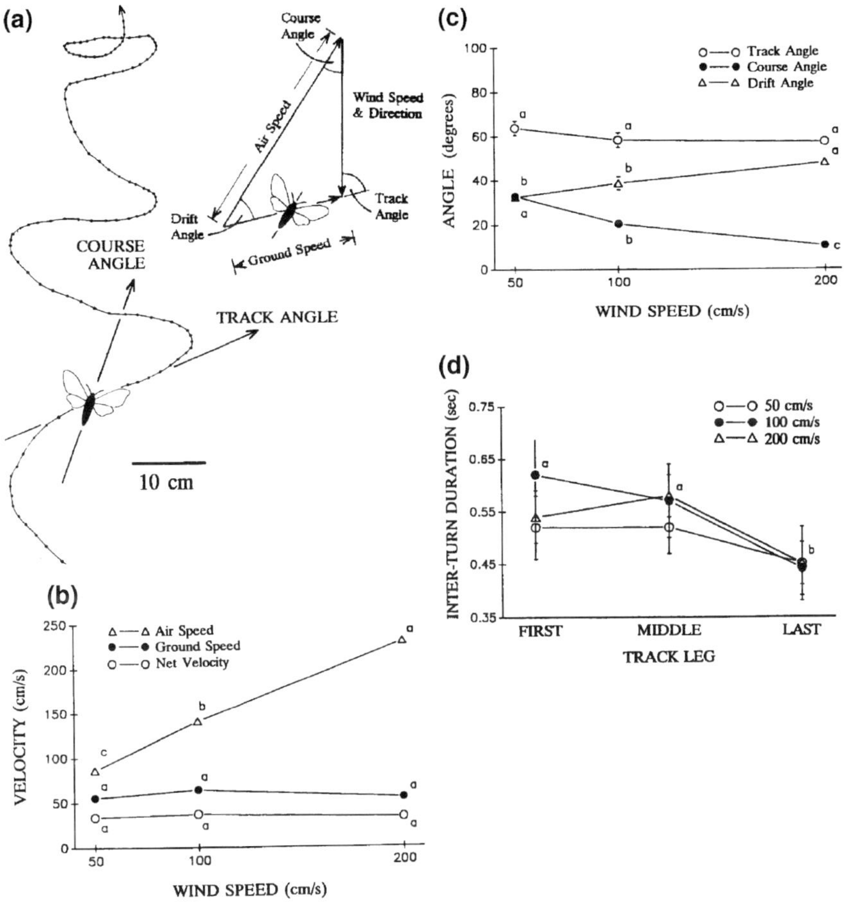

Figure 27.1. (Con't.) pheromone plume at three different wind velocities. The 95% confidence intervals are so small that they are obscured by the symbols. This plot shows the mean air speed increasing to compensate for the increased wind velocity, while the mean ground speed and net velocity remain the same. **(C)** Plot of the grand means of the angular parameters measured and calculated from 10 moths flying in three different wind speeds. Grand means are all plotted with 95% confidence intervals. This plot shows the mean course angle decreasing (steering more directly into the wind) to compensate for the increase in wind speed, while the resulting mean track angle shows no significant change. **(D)** Plot of the grand means of the interturn durations (time between the apices of sequential counterturns) from flight to pheromones at three different wind speeds measured at three different distances from the odor source. This plot shows that there is no significant effect of wind speed on the mean interturn durations, but that the time between turns decreases (i.e., counterturning frequency increases) as the moths approach nearer to the odor source. In panels B, C, and D, means with no letters in common are significantly different according to a one-way ANOVA and Duncan's New Multiple Range test.

modulation of ground speed: (1) In over 80% of the moths the ground speeds reached maximum values approximately midpoint during the straight legs between counterturns, and minimum values at or near the apices of the counterturns; (2) in about 50% of the moths the peak ground speed decreased as the animal approached the odor source (Fig. 27.2B illustrates an individual whose ground speed was modulated in both ways). Thus, *M. sexta* (and we suspect other moths) do not maintain constant ground speeds during odor-modulated flight, but instead most individuals show a predictable modulation within a range of ground speeds.

Is this modulation of ground speed active, or is it a purely physical result of a relatively constant thrust being directed more into the wind without a compensatory increase? An experiment performed in parallel with the fine-scaled behavioral analysis suggests that at least part of this modulation in ground speed may be actively generated (Willis and Arbas 1991b, 1994). In this study the motor patterns from the indirect flight muscles that generate wing beating were recorded while the moths were flying up a pheromone plume in a laboratory wind tunnel. These results show that the burst frequency of these flight muscles decreases (i.e., the wing beat frequency decreases) during some, but not all, counterturns in many of the moths. Thus, part of the decrease in ground speed during some of the counterturns appears to be the result of modulation of motor output during maneuvering, and not just the result of pure aerodynamics.

2.2. *Modulation of Track Angles*

Similar patterns were observed to underlie the constant mean track angles in our previous analyses (Fig. 27.2). During flight along certain straight legs in the flight tracks (~70%), the moths maintained their track angles clustered tightly (coefficient of variation <20%) around the mean value for that particular straight leg (Fig. 27.2D—last three straight legs). Along other straight legs (~30%) the track angles were not maintained with such precision (Fig. 27.2D—first four straight legs). During counterturns the track angles swept rapidly through a large range as the moths switched the orientation of their tracks from one side of the wind to the other (Fig. 27.2A,D). There was no correlation between the mean orientation of a given straight leg and whether it was maintained in a more or less precise orientation by the moths (Willis and Arbas unpublished). The grand mean values for track angles and ground speed in this study matched those from the earlier study (Willis and Arbas 1991a), but the real structure of the behavior underlying the means is considerably more complex than previously supposed.

2.3. *What Does It All Mean?*

What is the significance of this newly appreciated complexity in the structure of pheromone-modulated flight? These new results are important because many of the hypothetical mechanisms proposed to control this behavior are the result of an

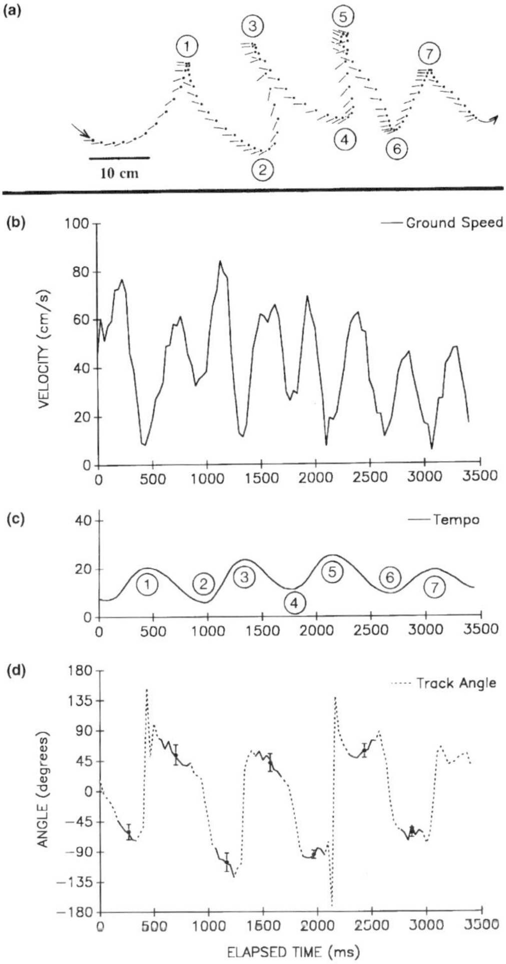

Figure 27.2. Example of ground speed modulation and track angle distributions during upwind flight in a pheromone plume. **(A)** Track of an *M. sexta* male flying upwind in pheromones. Dots represent the moth's position every 33 msec. Lines associated with the dots represent the orientation of the longitudinal body axis. Circled numbers next to counterturns in the track correspond to labels on the apices of turns in panel C. **(B)** Plot of the ground speed for each 33-msec movement·vector along the track in panel A plotted against time. Plotted on the same horizontal axis, but with an arbitrary *Y*-scale, is the side-to-side component of the track. Circled numbers in this plot correspond to the same numbers in panel A above, allowing easy identification of the apices of the counterturns. **(D)** Plot of the angular orientation of the individual movement vectors (track angle) measured every 33 msec (dashed line) along the flight track in panel A plotted against time. Mean (filled circles) and standard deviations (error bars) are depicted for the track angles distributed along each straight leg in this performance.

understanding of the behavior based upon the performances of many individuals averaged together, and thus may be of little relevance to the individual moth swooping back and forth across the wind every 500 msec. The initial analysis of flight tracks of moths challenged to fly in pheromone plumes at different wind speeds revealed that the moths apparently compensated by changing both their course angles and air speed to maintain "virtually the same set of track angles and ground speeds" (Marsh et al. 1978). The inference that Marsh et al. drew from this was that the moth could "define the triangle of velocities, and hence the track angle, using air speed, rate of image movement, and angle of image movement (i.e. two sides and the enclosed angle)" and, by maintaining some of these variables (i.e., ground speed and track angle) at constant values could make compensatory changes in the others. The origins of this idea may have come from earlier observations (Kennedy 1951) where it was suggested that flying locusts maintain air speed and the rate and direction of image movement over the ventral ommatidia at "preferred" values to maintain tracks along the windline. Indeed, population averages of ground speeds and track angles measured from *M. sexta* flying at different wind speeds in our own previous studies seem to substantiate these findings (Willis and Arbas 1991a). These ideas taken together, and reinforced by averaged flight performance data from a variety of studies, have been widely taken to mean that these insects regulate their flight to achieve a *constant, preferred velocity* of image flow over the ommatidia of the compound eyes which then enables them to maintain a constant, preferred ground speed and track angle (Cardé and Hagaman 1979; Kuenen and Baker 1982a; Cardé 1984; Willis and Baker 1987; Willis and Cardé 1990; Willis and Arbas 1991a).

Our fine-scaled analysis of individual flight tracks revealed that *M. sexta* (and we suspect many other moth species) do not maintain a constant preferred ground speed and, therefore, do not regulate flight motor output to achieve a *constant velocity of retinal slip* (e.g., Kuenen and Baker 1982a; Preiss 1992; Preiss and Gewecke 1991). Although it is clear from numerous studies that flight behavior of insects can be entrained to moving visual patterns (e.g., Kennedy and Marsh 1974; David 1982), *M. sexta* flying freely over a stationary ground pattern showed no stabilization of velocity to a constant value. What parameters of visual space the moths are using as feedback to control their velocity and steering flight is not clear. It is, however, universally hypothesized that the moths use optomotor feedback to control their ground speed and track angle by decomposing the visual flow fields impinging on the ventral surfaces of the compound eyes into their L and T components (Ludlow 1984; David 1986). It is known that there are neurons in the visual systems of flying insects (Collett and Blest 1966; Milde 1989a,b; Suder et al. 1990; Hausen and Egelhaaf 1989) that are sensitive to wide-field movements of the visual panorama, and the responses of these neurons typically break flow fields into perpendicular vectors (i.e., T and L image flow). However, it is clear from recent research on the visual systems of flying insects that it is possible that many important parameters of the visual surround are being extracted, in parallel, from the inputs to the compound eyes (Farina et al. 1995; Milde 1993; O'Carroll 1993; Ronacher and Wehner 1995; Srinivasan 1992; Srinivasan et al. 1993). At any given moment the moth may be using any, all, or a specific combination of them as a feedback to link its sensory-motor performance to the environment.

What benefit might a moth derive from actively stabilizing the orientation of some straight legs at a single value? This question has been posed, and a possible explanation has been offered (Cardé 1984; Baker 1985; Kennedy 1986). By maintaining a constant orientation of the track angle between turns the moth would expose itself to wind-induced drift and thus might acquire information on the wind direction and speed (i.e., "feeling for" a wind-induced error sign; Kennedy 1986). The behavior of the moths in our sample lends some support to this hypothesis [e.g., the straight legs with more precisely maintained track angles (Fig. 27.2A,D)]. However, straight legs with low variation in track angles are often separated by straight legs with highly variable distributions of track angles (Willis and Arbas unpublished). If a moth computes its course angle relative to due upwind, decomposing the triangle of velocities (Marsh et al. 1978), and steers off of the wind line to capture wind-induced off-axis image flow (Kennedy 1986), our observations suggest that it may not require this information continuously. Rather, this information may be updated as necessary on a straight leg by straight leg basis. The implication is that the tightly controlled straight legs may represent sampling periods during which the moth is updating its information on the relation of its course to the upwind direction. Alternatively, precise stabilization of the straight leg may not always be necessary for the acquisition of wind information, and other maneuvers or sensory mechanisms may suffice (Willis and Arbas unpublished).

Implicit in this hypothesis is the idea that the pheromone-modulated flight of moths is under several layers of control. The first level is a flight stabilization system similar to that elucidated in locust, by Rowell and his co-workers, and dubbed "the autopilot" (Rowell 1988). This system generates compensatory responses to external perturbations, stabilizing the flight performance. Other higher-order flight control systems would rely on exteroceptive (visual, olfactory, mechanoreceptive, and proprioceptive) and internal information (counterturn timing, possibly memory of previously performed maneuvers, or minimally the sign of the last turn) to determine the set points that the autopilot would then maintain. The combination of these systems acting in concert could then generate the complex pattern that we recognize as the zigzagging upwind flight track.

2.4. *Stereotyped Nature of Interturn Intervals*

When analyzed and compared at the level of grand means of 10 individuals per treatment, the time interval between sequential counterturns is one of the most consistent parameters of odor-modulated flight in *M. sexta* (Willis and Arbas

1991a); it is, on average, 500–550 msec from the apex of one turn to the apex of the next counterturn. The interval between turns was modulated as the males approached the odor source (shorter interturn interval closer to the source), but there were no statistically significant differences related to the wind speeds. Remarkably, many of the individual interturn durations that comprise the performances of single moths fall precisely on, or very close to, the 500-msec mean value (Willis and Arbas unpublished). The results of our recent study revealed approximately 50% of the individuals analyzed (N = 11) had mean interturn durations with coefficients of variation of <15% (Willis and Arbas unpublished). For the moths analyzed in this study, the individual with the most variable performance had interturn durations ranging from 833 to 466 msec, while the moth with the least variable performance had interturn durations ranging from 533 to 433 msec. These results lend further support to the idea of an internal timer triggering and underlying these temporally stereotyped counterturns, and that this internal timer is modulated continuously during upwind flight in the plume (Kennedy 1983).

It is clear from our work with *M. sexta* and previous work (Marsh et al., 1978; Willis and Arbas 1991a; Willis et al., 1991; Willis and Baker 1994) that both the frequency of counterturning and the ground speed are modulated as the moths fly up the pheromone plume to the source. A possible mechanism underlying this modulation of flight behavior was initially proposed to explain the ability of moths to continue flight up a pheromone plume during periods of zero wind (Baker and Kuenen 1982; Kuenen and Baker 1983), and later extended to encompass pheromone-modulated flight in all wind conditions (Kennedy 1983; Kuenen and Baker 1983). This mechanism was termed *longitudinal chemotaxis,* or steering with respect to a difference in chemical concentration sampled over time, along the path of the animal (Kennedy 1983). A hypothetical internal mechanism underlying the proposed behavioral mechanism was "short-term averaging" (Kennedy 1983) of the characteristics of the pheromone plume. All observations and experiments performed using *M. sexta* are consistent with this idea of a continuously modulated, ongoing process. However, results of recent experiments indicate that male moths are also capable of responding to individual encounters with pheromone or clean air (Mafra-Neto and Cardé 1994; Vickers and Baker 1994b; Cardé and Mafra-Neto, and Baker and Vickers this volume).

In these latter experiments (Mafra-Neto and Cardé 1994; Vickers and Baker 1994b) the plume was abruptly truncated by removing the source, resulting in the performance of counterturning in place across the wind, a behavior known as *casting flight* (Kennedy and Marsh 1974). Once the moths were casting in clean air, a device was used to deliver regular puffs of pheromone at different frequencies, from 1 to 10 Hz. Each laboratory studied different moth species (Mafra-Neto and Cardé: *Cadra cautella;* Vickers and Baker: *Heliothis virescens*) but observed similar behavioral responses; the moths' movements were oriented more directly upwind during their encounters with pheromones, and the moths returned to casting flight during encounters with clean air. As the emission rate of pheromone puffs increased (and the time spent in clean air between puffs decreased), the moths' flight was oriented more directly upwind and little casting flight was observed. These data lend support to the idea that reiterative switching from upwind surges to crosswind casting, as proposed by Baker (1990), forms the sinuous upwind flight tracks.

In view of the clear ability of some moths to react to fine-scale changes in onset and loss of the pheromonal stimulus and the apparent alternation between casting and surging flight (Mafra-Neto and Cardé 1994; Vickers and Baker 1994b), the consistency of the interturn durations, regular patterns of ground speed modulation observed from most moths, and the seeming ability to tightly control track angles along certain straight legs observed from *M. sexta* males are puzzling. These characteristics suggest a continuously modulated ongoing process of upwind flight with counterturning, rather than a chain of reiterative reflexive reactions to momentary conditions alternating between surges (with counterturning suppressed) and casting. Recent studies of the responses of *L. dispar* males to loss of the pheromone plume in a wind tunnel (Kuenen and Cardé 1994) support the idea of some ongoing process underlying the behavior, as well. In an experiment designed to study the effects of pheromone source concentration on the persistence of casting flight after pheromone loss, Kuenen and Cardé (1994) demonstrated that up to 40% of *L. dispar* males continue to fly along a modified counterturning track for more than 1.5 sec and up to a meter upwind after their last possible encounter with pheromones. In this case, upwind progress in the absence of pheromonal stimuli might be explained by the persistence of counterturning flight, possibly maintained by a short-term motor memory such as that discovered by Möhl (1988) in locusts, but not on the basis of a rapidly extinguishing surge triggered by the final pheromone encounter.

3. Odor-Modulated Zigzagging as an Active Sampling Strategy (or Everything Old is New Again)

It is our feeling based on an intimate knowledge of our own results, and an extensive review of the literature, that the interpretation that best fits the data from *M. sexta* is that zigzagging upwind flight is an active behavior adapted for continuously sampling the information necessary for successfully locating resources important for survival. Although this hypothesis may not be stated as simply elsewhere, many of the hypotheses and interpretations of earlier workers strongly imply a similar conclusion.

It has been suggested previously that the adaptive values of zigzagging upwind while maintaining a constant ground speed and specific preferred angles between counterturns are as follows: (1) Repetitive counterturning across the wind increases the amount of air sampled for pheromonal cues (Kennedy 1983), and (2) flying upwind at a set angle to the wind enhances the moth's ability to acquire information on wind velocity and direction from the sensory inputs generated by actively exposing itself to wind-induced drift (Cardé 1984; Baker 1985; Kennedy 1986). Although these responses are integrated smoothly into the behavior that we observe, for the sake of discussion we will treat the anemotactic component and counterturning separately.

3.1. Active Behavior Generates Information for Flight Control

Moths likely use information from the many sensory modalities available to them to control the maneuvers that comprise their zigzagging flight tracks. This information could be acquired "passively" by sensory monitoring of reafferent and exafferent input during the performance of maneuvers organized principally as reactions to the instantaneous sensory surround (Preiss and Kramer 1986a,b; Baker 1990). Additionally (or perhaps alternatively), sensory information may be acquired and accumulated during the performance of maneuvers, or sequences of maneuvers, organized principally to scan actively for the relevant environmental parameters (Marsh et al. 1978; Heisenberg and Wolf 1984; Kennedy 1983, 1986; Wolf and Heisenberg 1990, 1991; Wolf et al. 1992). This idea finds parallels in the visual systems of many arthropods and other animals, including humans, where scanning movements of the retinae (Land 1969, 1992), eyes (Land et al. 1990), or whole animal (Collett 1978; Sobel 1990) enable the visual system to extract features from the environment otherwise not available to it (see also Buchner 1984; Schwind 1989; Warren and Wertheim 1990).

The idea that the performance of specific maneuvers may be necessary for the acquisition of information used in flight control and guidance is supported by recent observations of a number of nest building hymenopterans (bees: Lehrer 1991, 1993; solitary-nesting wasps: Zeil 1993a,b; socially nesting wasps: Collett and Lehrer 1993). In all of these cases an individual insect flies a stereotyped pattern (a very moth-like counterturning track in most species) upon leaving the hive or nest. It has been proposed that embedded in this pattern may be specific visual sampling points, during which information on the orientation and location of the next entrance are acquired (Collett 1992; Collett and Lehrer 1993). In some cases, it has been proposed that the points of visual information acquisition might be the counterturns (at the ends of straight legs flown at an angle to the target), during which the head and longitudinal axis of the insect are aimed at the entrance to the nest. In contrast, Zeil (1993a,b) suggested that the wasps he studied might be memorizing elements of parallax fields surrounding major landmarks associated with the nest, and thus might be acquiring these cues during the sweeping interturn straight legs.

An elegant series of experiments on the optomotor control of flight in *Drosophila melanogaster* concluded that actively generated fluctuations in torque result in the continuous recalibration of the optomotor control system (Heisenberg and

Wolf 1984; Wolf and Heisenberg 1990). Continual minor fluctuations in torque (even during extended periods of stable flight), typically alternating in direction, have been interpreted as the optomotor control system actively "feeling for" input [in the same sense that Kennedy (1986) proposed that males flying at an angle to the wind are "feeling for" wind-induced slip of the visual surround]. In these closed-loop tethered flight experiments it was discovered that these small periodic fluctuations in torque allow the flies to compute a "correlation coefficient" between some aspect of their motor performance [the first time derivative of torque, called *pretorque* by Heisenberg and Wolf (1988; Wolf and Heisenberg 1990)] and the immediately consequent visual input (angular acceleration of the visual surround). If there is a correlation between the motor signal and the visual feedback indicating sufficiently coincident movement, then the control system calculates a "coupling coefficient" (i.e., to what extent does the fly's motor system affect the visual feedback?) which allows the system to determine the direction of movement and to set the internal gain that will enable the fly to compensate for any asymmetric aspect of its flight system and achieve stable orientation (Wolf and Heisenberg 1990). This optomotor control system enables the tethered fly to discriminate between self-generated movement and externally generated movement in <50 msec. This type of periodic sampling via self-generated movement has been interpreted as being crucial to the ongoing stability of this visually controlled behavior (Wolf and Heisenberg 1990). Based on similar closed-loop tethered flight studies, a similar form of "self-calibration" has been proposed to underlie the visual flight stabilization system in locusts (Möhl 1989).

Our own experiments with tethered flying *M. sexta* males suggest that internally programmed counterturning may serve a related function for flight control. In this experiment, male *M. sexta* were tethered to a wooden dowel 30 min prior to lights-off on their LD cycle, and they experienced lights-off at the expected time. Over half of these tethered males initiated flight activity at the expected time, starting approximately 30 min after lights-off. Approximately 25% of the moths that activated to tethered flight spontaneously executed bouts of 3–8 counterturns (Baker et al. 1994). These counterturns were easily interpreted as such, because these maneuvers involved coordinated movements of the entire body; the head turned into the attempted maneuver, the abdomen simultaneously ruddered in the same direction as the head, and the wings developed an asymmetric posture in a manner consistent with a directed turn. Interestingly, these coordinated counterturns alternated back and forth approximately every 500 msec, similar to the timing of counterturning observed during odor-modulated free flight to a pheromone source (Willis and Arbas 1991a). The critical observation for this experiment is that these naive animals had received no visual flow field feedback and no wind stimuli and had not been exposed to pheromone during the course of the experiment (or prior to it). Thus, the results of this study demonstrate that the central nervous system of the moth is capable of spontaneously organizing and executing a series of temporally regular counterturns, without being activated by pheromones. Since moths were able to generate counterturns without prior exposure to pheromones, we suggest that neuronal circuits responsible for this pattern are not obligatorily configured, or activated, by olfactory stimuli. Counterturning may be called up in contexts other than strictly odor-modulated flight. Counterturns may contribute to active maneuvers used to "capture" useful afferent information—for example, visual slip due to wind-induced drift, or other information about wind velocity, direction, or the olfactory stimulus. Active maneuvers may also contribute to flight control and orientation by providing reafference as in the *Drosophila* example above (Wolf and Heisenberg 1990). Spontaneous counterturns, or even individual turns, may be activated during free flight of *M. sexta* as a means of generating reafference which permits the calibration of internal sensory-motor gains. Indeed, the observation that brief bouts of counterturns were exhibited at the initiation of tethered flight after fictive take-off fits with this interpretation. Additionally, the observation that brief sets of counterturns often occur near the end of flight bouts prompts the interpretation that failure to receive appropriate reafference upon attempted maneuvers signals abnormal conditions to the moth and triggers the cessation of flight.

3.2. Acquiring Olfactory Information: Interaction of a Turbulent Chemical Signal and Active Behavior

Counterturning observed during odor-modulated flight ranges from the relatively broad counterturning in place that characterizes casting flight after loss of odor, to the narrower more frequent counterturns that are a major component of upwind zigzagging flight (Kennedy 1983; Willis and Arbas 1991a; Kuenen and Cardé 1994). Kennedy (1983) proposed that "The common function of the counterturning, whatever its mode and scale, seems to be the regaining of contact with the elusive wind-borne odor." Thus, even during upwind flight in the plume, counterturning has been seen as a strategy for maintaining contact with, and continuous sampling of, the odor plume. The act of turning back and forth across the wind, and modulating the velocity along the straight leg between those turns as the plume is traversed, will drive the array of olfactory sensilla on the antennae through the odor plume, often achieving peak velocities near the plume axis.

It is known from both mathematical and physical modeling studies that the transfer of odor molecules from fluid media to a sensillum can be significantly effected by changes in velocity (of either the fluid or the sensillum) (Koehl 1995a,b). Measurements from flow visualization studies of a moth antenna (Vogel 1983), and odorant concentrations around the olfactory organs of lobsters (Moore et al. 1991) demonstrate the impact of velocity changes on the detection of odor signals. Increases in flow velocity (or velocity through the fluid) resulted in a greater proportion of the flow passing through the antenna (Vogel 1983), along with an increase in the peak concentration sensed and a decrease in odor peak duration (Moore et al. 1991).

These measurements and observations make it abundantly clear that the characteristics of the odor signal to which the moth is responding are not determined solely by the turbulence of the airstream or the wake generated by the structure from which the female moth is releasing pheromone. The olfactory information that the moths are responding to is actually an interaction of the fluid dynamics of the physical environment with their behavior. Thus, the modulations in velocity that we have measured almost certainly have an impact on the olfactory input the moth receives during upwind flight to the source.

4. Conclusion: From Fluid Dynamics to the Virtual Wind Tunnel

In our effort to approach this system in an explicitly neuroethological context we have incorporated knowledge and ideas from such apparently disparate fields as biological fluid dynamics and the neurobiology of pattern-generating circuits. The recent development of a behavioral-level computer simulation of pheromone-modulated flight (Belanger and Arbas 1995) has enabled us to compare directly the performance of (a) simulated moths operating under a more reflexive control system (Baker 1990) and (b) those operating under a continuously modulated counterturning program (Kennedy 1983). The results of simulation studies suggest that the success of the particular behavioral strategy is critically dependent on the performance limits of the animal and the (simulated) environmental conditions encountered (Belanger and Arbas 1995). Using the mechanisms proposed in the literature, combined with all of the pertinent empirical measurements available to us, it was found that even the most successful simulated behavioral strategy only got the simulated moth to the pheromone source 50% of the time (starting in the plume 2 m away from the source). This result indicates that we have much yet to learn before we understand how moths actually locate a distant unseen source of odor. By continuing to approach this complex behavioral question from as many viewpoints as possible, we hope to gain more insights into the many possible ways that moths may utilize olfactory and other information available to them to locate distant resources critical for survival.

Acknowledgments

We would like to thank Drs. Reg Chapman, Jim Belanger, and Ring Cardé for suggestions and comments on earlier versions of this manuscript. We also express our thanks to Drs. R.T. Cardé and A.K. Minks, the local organizers, and the other participants for a high-quality symposium, full of stimulating conversations and interesting ideas.

28

Neuroethological Study of Pheromone-Modulated Responses

Edmund A. Arbas

1. A Neuroethologist's Approach to the Study of Insect Behavior

Attempts to define the principal aims of neuroethology have appeared, with subtle variations, on numerous occasions (e.g., Hoyle 1970, 1984; Ewert 1980; Fentress 1980, 1991; Bullock 1983; Camhi 1984; Ingle and Crews 1985). It is a fair distillation of these to say that this research enterprise aims to elucidate neural mechanisms underlying naturally evolved behavior in animals, including man. At first glance, a statement so broad seems to provide little guidance. It seems to overlap the aims of psychology, behavioral biology, and the many branches of the neurosciences. How is this discipline different? The principal difference in the neuroethological approach does not lie in the research tools employed or in the organisms studied, but in the psyche of the investigator. Drawing its roots from comparative ethology, the neuroethological approach stresses the comparative analysis of behavior referred to the natural environment in which the organism has evolved. There is emphasis on a phylogenetic perspective. Drawing historically also from neurophysiology, the neuroethological approach stresses experimental laboratory research aimed at revealing proximate mechanisms of behavior in terms of sensory inputs, integration by neuronal networks, and neuromuscular outputs. The most rigorous methods of phylogenetic analysis, of behavioral biology, and of neurophysiology are to be employed, and the interpretations of results obtained are to be synthetic. The benefits flow in many directions. The approach forces experimenters working at a behavioral level to abandon mere description of phenomenology and to focus on underlying mechanisms. For a neurophysiologist, neuronal structures or activity patterns cannot be divorced from their role in generating an adapted act, the behavior of the animal. For both behavioral and physiological laboratory studies, the analysis must be tied as much as possible to the conditions encountered by the animal interacting with its natural environment.

How, in the context of (a) the First International Symposium on Insect Pheromones and (b) a session on "Pheromone-Mediated Responses," does the neuroethological perspective contribute to the framing of the questions that we are analyzing, and the hypotheses that we are testing? Does this perspective help us to interpret recent experimental advances and to identify future directions?

It is useful first to set the global perspective. The behavior of insects in response to stimulation by pheromones is of deep scientific interest and practical importance, and a valuable subject to study. This statement is unlikely to draw criticism from any of the symposium's participants, but the question of what the scientific interest and practical importance specifically are would motivate much discussion. Obviously, the pheromonal communication system which brings male and female insects together for mating has evolved a crucial role in the life cycles of these insects and provides a marvelous opportunity for specific, biorational intervention where they are agricultural pests. Aspects of this perspective were explored in Part IV of this volume: "Use of Pheromones in Direct Control." From a broader perspective, pheromone-modulated behavior is just a special case of odor-modulated behavior. Understanding how males find signaling females by tracking the odor (pheromone) plume may contribute a great deal to the understanding of how females locate host plants by tracking their odors. The converse is also true. Indeed, where examined specifically, many features of the odor-modulated behavior of females resemble closely the pheromone-modulated behavior of males (e.g., Haynes and Baker 1989; Willis and Arbas 1991a). Notably, when females were experimentally equipped with male antennae and pheromone receptors by transsexual transplantation (Schneidermann et al. 1986), they were able to orient and fly up a plume of pheromones with normal (i.e., male) behavior. Remembering the natural environment, in which the pheromonal odors are received against the backdrop of a complex mixture of plant volatiles, it might be profitable to explore further the interactions of plant and pheromonal odors in influencing the behavior of males.

From a broader biological perspective still, odor-modulated flight of insects is but one example of odor-guided locomotion in a fluid medium. Whether a sea bird orients to the smell of a dead fish that it might like to eat, a salmon orients to a plume of its home-stream odor in an off-shore ocean current, or a male moth flies toward a calling female, the problem of tracking an odor borne by the same fluid medium (air or water) in which the organism is locomoting is very similar. Interestingly, birds, fish, and moths all approach their attractive odor sources from down-current by locomoting in a zigzagging pattern (reviewed in Arbas et al. 1993). Hence, algorithms employed to locate odor sources (i.e., the behavior) may convergently share characteristics in all of these diverse organisms.

From the perspective of neurobiology, the ability of male insects to detect and discriminate pheromonal odors in a complex olfactory environment and to guide their locomotion with respect to a plume of odor embodies numerous problems of signal discrimination, feature detection, multimodal sensory integration, and motor control, all areas being hotly pursued in a number of vertebrate and invertebrate model systems. Many of the problems encountered and solved by the orienting moth are common to those experienced during autonomous locomotion by robots (e.g., Cliff et al. 1994). As such, the odor-guided locomotion of moths provides an excellent model in which to study these fundamental aspects of sensation, integration, and control, with possible benefits being derived for many other fields of inquiry and application.

2. A Neuroethological Perspective on the Pheromonal Communication System

The pheromonal communication system of insects shares the same components as any communication system: a sender (the female moth); the signal and its transmission line (the pheromone in its plume); and a receiver (the male moth) (see Sebeok 1977). Each component of the system has its own properties and contributes to the dynamics and successful functioning of the entire system. The female determines the onset of the communication when she begins to release pheromones during calling behavior. Endocrine factors determining the production of pheromones by females were discussed in an early session of the symposium with particularly spirited discussions on the role of pheromone biosynthesis activating neuropeptide (PBAN). Yet, other factors also play a role in determining the onset and timing of this behavior. A circadian rhythm of activity activates calling behavior in the first several hours of scotophase (e.g., Sasaki and Riddiford 1984; Itagaki and Conner 1988) and also plays a permissive role in males, in that they are entrained to nearly the same cycle and, thus, are typically active and responsive at the times females begin to call. This important aspect of coordination between circadian rhythm and pheromone sensitivity has received considerable attention (e.g., Linn and Roelofs 1986, 1992; Linn et al. 1992) and was represented in the sessions at the meeting (Linn this volume) and deserves further research. In many of the examples considered, the signal from sender to receiver consisted of presence or absence of pheromones (i.e., the mere release of the chemical blend). An interesting variation on this theme is represented by species which may superimpose a higher-order temporal code on the signal by pulsing pheromones (i.e., arctiid moths—Cardé and Roelofs 1973; Conner et al. 1980; Itagaki and Conner 1987). Such signaling may aid species recognition or enhance the signal-to-noise ratio in a noisy transmission environment, or even influence the details of the behavior of the approaching males.

The noisy transmission channel remains largely mysterious despite considerable study (Murlis this volume). Odor is released by the signaling female into an air current. Turbulence mixes odor-laden and clean air nonuniformly so that the flying male passing through the normal plume receives discontinuous stimulation by the odor. The details of how the pheromonal signal is modified within the turbulent plume are important since there is potential information for the receiver in these details. Filaments of odor-laden air near to the source are likely to be more compact and of higher concentration than those further from the source which have expanded by diffusion since being formed. The rate at which pheromonal filaments will be encountered will depend on the wind velocity, rates of vortex shedding from the pheromone source (e.g., plume structure will be different if the female calls from a shrub versus a tree trunk), the fluid dynamics of sensory hairs (Koehl 1995), and, importantly, the air speed of the male moth itself.

Do male moths pay attention to these features? Clearly, presence of the appropriate blend of odor has profound effects. For example, contact with pheromones activates a complex of behaviors in the responsive male. If inactive, he initiates wingfanning and, following warmup, takes off into flight. If already flying, he begins to orient with respect to the wind employing visual cues from the wind-induced drift (i.e., optomotor anemotaxis), and possibly other modalities, and directs his locomotion upwind toward the pheromone source. Loss of contact with the pheromones following the initial encounter activates a pattern of casting flight in which the moth ceases upwind progress, and it executes a series of counterturns with ever-widening horizontal and vertical excursions across the wind line. If the odor is re-encountered, upwind progress begins again. Recent experiments have demonstrated unequivocally that the pattern of encounters with odor and clean air can profoundly influence the behavior. By changing the structure of the plume using different-sized odor sources to produce different kinds of wakes, Mafra-Neto and Cardé (1994; Cardé and Mafra-Neto this volume) showed changes in the sinuosity of flight tracks in *Cadra cautella*. Next, using a controlled pheromone emitter to produce discrete puffs of pheromones in an ongoing wind, Mafra-Neto and Cardé (1994) and Vickers and Baker (1994) (see Cardé and Mafra-Neto, this volume; Baker and Vickers in this volume) demonstrated the clear behavioral transitions between (a) casting in clean air and (b) upwind "surges" following encounters with odor. By manipulating the frequency of odor puffs emitted, the upwind flight of *Heliothis virescens* could be entrained to produce widely sinuous tracks where a greater proportion of casting flight was being exhibited with infrequent encounters with odor, or to produce straighter tracks where a greater proportion of surging flight was exhibited largely by the pattern of encounters with odor. In other species, such as *Manduca sexta*, where the timing of counterturns produced during zigzagging flight seems to be highly stereotyped in many types of plume (i.e., whether measured from females flying in plant odor, or males flying in pheromones at different wind speeds, or in different visual environments) (Willis and Arbas 1991a; Willis and Arbas this volume), other mechanisms, such as an ongoing pattern of self-steered turns, may dominate in shaping the structure of the flight tracks. Even there, however, evidence exists that the turning pattern may be modulated by other features of the odor plume, possibly such as pheromone concentration, as counterturning accelerates at close proximity to the source. Experiments manipulating the quality of odor blends in yet other species, such as *Lobesia botrana*, indicate that the character of locomotor performance, as well as success in source localization, is profoundly affected by the composition of the pheromone blend (Witzgall this volume). Hence, multiple signals within the plume (e.g., pattern of odor encounters, blend composition, and concentration) may be integrated with self-steered behavior such as counterturning (see also below) to produce successful source localization.

These various studies have examined the pheromone-modulated behavior of many different species of Lepidoptera, and many commonalities in mechanism are emerging. However, explicitly comparative studies have not, as yet, been performed. Behavioral or physiological traits have not been mapped onto a cladogram, and similarities and differences have not been evaluated on the basis of phylogenetic relationships. Such studies will be important because although these insects may share many mechanisms in common (e.g., all species tested exhibit casting flight upon odor loss), they may well be employed in subtly, or not so subtly, different ways to achieve source localization. Need for such caution is not without precedent, because precisely analogous differences have been discovered in the mechanisms for visual flight control in different species of fly (e.g., Heisenberg and Wolf 1979).

Manipulations of the odor plumes and analysis of flight tracks have permitted workers to infer a number of behavioral mechanisms which have motivated physiological studies. These latter studies, involving neuroanatomy and electrophysiology, have led to significant headway toward the identification of underlying physiological mechanisms. Finally, the development of computational approaches will permit the testing of our understanding at behavioral and physiological levels through simulation studies (e.g., Kramer this volume).

3. Behavioral Mechanisms

3.1. Optomotor Anemotaxis

Past studies employing rotating floor patterns in wind tunnels (Kennedy 1940), tethered moths flying under open-loop conditions in an optical stimulator apparatus (Preiss and Kramer 1986a,b; Arbas et al. 1993; Milde 1993; Wendler et al. 1993), or moths flying freely over different striped patterns all have established firmly the importance of the visual modality in stabilization and regulation of flight. It has been demonstrated repeatedly that behavior can be entrained by manipulations of visual stimuli: Moths can be made to fly in place over moving floor patterns, or they can be made to track stripes moving in patterns that simulate pitch, roll, or yaw with attempted corrective maneuvers when tethered under open-loop conditions. Precisely how optomotor stabilization of flight is achieved and how it is overridden during active maneuvering is a very large question in insect neurobiology, with significant advances toward the answer having been made in the Lepidoptera (e.g., Ibbotson et al. 1991; Olberg and Willis 1990; Milde 1993; Wendler et al. 1993), but with the greatest detailed knowledge of underlying mechanisms still being in flies (e.g., Egelhaaf and Borst, 1993; Strausfeld and Gronenberg 1990; Gronenberg and Strausfeld 1990; Gilbert and Strausfeld 1992), and in locusts (e.g., Hensler and Rowell 1990; Bääder et al. 1992; Rowell 1993). Only one of the studies in Lepidoptera explicitly studied the interaction of stimuli of olfactory and visual modalities combined (Olberg and Willis 1990).

3.2. Counterturning

The existence and function(s) of counterturning continue to be subjects of intense discussion (e.g., see Baker and Vickers this volume; Cardé and Mafra-Neto, this volume; Witzgall this volume). Although some flight tracks recorded in any species may be relatively straight, while others may be relatively more sinuous, all flight tracks in pheromones contain some measure of side-to-side movement. Are such movements active turns, or merely uncompensated deviations from a straight line due to an imprecise motor system or unpredicted environmental perturbation? Baker and Vickers (this volume) and Cardé and Mafra-Neto (this volume) cogently argue several lines of evidence supporting the view that counterturning is, in fact, an expression of active, self-steered behavior.

1. Loss of pheromone contact initiates casting flight which comprises a series of counterturns, back and forth across the wind line, in the absence of upwind progress. This pattern of behavior has been observed in every species where tested. Since, when elicited in a wind tunnel, wind velocity is controlled, casting cannot be ascribed to an unpredictable environmental perturbation such as a wind gust. Since it is activated by loss of odor, it is unlikely to be a manifestation of a noisy motor system. On the other hand, since the turns are performed in wind, they might still be considered to be organized by wind-induced drift and an optomotor control system where the set point has been altered from a value near 0° (i.e., due upwind) to some other value (e.g., 90°—across the wind line).
2. Moths have been observed to perform casting flight after exposure to pheromones and subsequent loss, even in zero wind (Baker and Kuenen 1982; Kuenen and Baker 1983; David and Kennedy 1987). In another case, counterturns had no particular orientation to compass direction (i.e., were not oriented to the former wind line or any other landmark), but still showed temporally regular alternation (Baker et al. 1984). Since they were performed in zero wind, they could not be initiated or organized by wind-induced drift. It might still be argued, in this experiment, that the turns were organized by an optomotor servo-mechanism which oscillates when operating at high gain (i.e., the corrective reaction

overshoots zero and causes another error signal in the opposite direction), based upon the visual slip produced by the moth's own movements. Interestingly, it could be argued that this temporally regular counterturning observed in zero wind was unlikely to have arisen from the instantaneous contacts with pheromone filaments either.

Recent experiments in our laboratory (Baker et al. 1994; and unpublished data) indicate that counterturns can be organized by moths under open-loop conditions in which there exist neither visual feedback cues from wind-induced drift nor those from the moth's own movements. Male moths, *M. sexta*, were tethered with zero rotational degrees of freedom and the spontaneous behavior exhibited during the active period of the circadian cycle was observed *in the absence of wind, pheromones, or moving visual stimulation*. Most moths initiated wingbeating and bouts of tethered flight within a short time after the onset of scotophase, their normal activity period. Nearly all of the moths exhibiting flight made postural adjustments indicating attempted turns. Approximately 25% of these not only made attempted turns, but also organized them into regular bouts of side to side alternations. Among these, the mean period of attempted turning was 560 msec, a value remarkably similar to the period of stereotyped counterturning exhibited in unrestrained, pheromone-modulated flight in a plume (Willis and Arbas 1991a; Willis and Arbas this volume). These observations indicate that male *M. sexta* are able to organize a pattern of counterturns endogenously and without reference to any phasic exteroceptive sensory input such as apparent slip of the visual panorama, turbulent wind, or even intermittent contact with pheromone. Interestingly, motor patterns recorded from flight muscles during these spontaneous turns (Baker et al. unpublished data) are also remarkably similar to those recorded during free flight (Willis and Arbas unpublished data).

4. Physiological Mechanisms

Recent electrophysiological studies in freely behaving, tethered, and reduced preparations have provided intriguing new information. Techniques developed for recording from wing muscles during free flight have revealed highly flexible patterns of phase and frequency modulation in the firing of key motoneurons known to underlie the maneuvers generated during pheromone-modulated flight (Willis and Arbas 1991a; Arbas et al. 1993; Willis and Arbas this volume). Similar motor patterns are revealed in the corresponding muscles when recording from tethered moths producing spontaneous, self-generated counterturns under open-loop conditions (Baker et al. 1994), as well as during optically stimulated maneuvers in more reduced preparations (Arbas et al. 1993; Wood and Arbas 1994). Recordings with long, flexible leads from a cyclopean donor antenna mounted on the head of a flying moth have permitted monitoring of pheromone encounters (through EAG responses) during flight in pheromone plumes and correlations of these encounters with changes in behavior (Vickers and Baker 1994b).

Intracellular recordings from CNS continue to reveal new properties of neurons and circuits in primary olfactory neuropil (Christensen and Hildebrand 1987b; Kanzaki et al. 1989) and higher brain centers (Kanzaki et al. 1991a,b). Antennal lobe neurons have been characterized which behave as "pheromone generalists," in responding equally well when any of several individual components or the complete pheromone blend stimulates the antenna, or "pheromone specialists," which respond to some specific components and not to others, or show excitatory responses to some components and inhibitory responses to others and also show mixed excitatory/inhibitory responses when stimulated with blends (Christensen and Hildebrand 1987b, 1988; Christensen et al. 1989a; see also Christensen this volume and Hildebrand this volume). Importantly, intracellular recordings during the delivery of pulsatile stimuli indicate that male moths can resolve intermittent pheromonal signals at high frequencies in the CNS (Christensen and Hildebrand 1988; Christensen et al. 1989a). Recordings from AL projection neurons showed that fusion frequencies for pulsed inputs vary significantly in different classes of AL neurons. The pheromone-specialist neurons are able to follow pulsed inputs up to 10 Hz (Christensen and Hildebrand 1988). Other AL projection neurons follow only lower frequencies of pulsatile stimulation and then, at some threshold, fuse their firing into a single burst. At present, the meaning of these responses to olfactory processing and behavior can only be speculated upon. A rigorous neuroethological approach dictates that the activity of neurons and circuits be linked in a causal relationship to the specific behavior being studied.

Recordings from neurons in higher centers followed by anatomical characterization have tracked olfactory processing pathways from antennal lobes to lateral protocerebrum, with substantial inputs to the calyces of the mushroom bodies (Kanzaki et al. 1991a; Christensen and Hildebrand 1987a), structures implicated in olfactory learning and memory in other insects (e.g., Erber et al. 1987) and which may contribute to some of the computations made by the moth brain during oriented zigzagging flight. Recordings from neurons in the lateral accessory lobes, bilateral structures associated with the central complex, revealed a novel long-lasting excitation (LLE) in local and descending neurons. This response, elicited in descending neurons only by pheromone blend stimuli, showed state-dependent changes to repetitive pheromonal stimuli applied to the antennae. An initial puff of pheromones elicited an increased rate of firing characteristic of LLE. A subsequent identical puff of pheromones elicited a different response. Rather than increasing firing, or showing saturation, the rate of firing declined. Following the decline, the firing rate then gradually recovered to the higher level of the previously elicited LLE over the next several seconds, where another similar stimulus again caused a decline in firing rate (Kanzaki et al. 1991b). It appears that the normally intermittent odor filaments of a natural plume would cause a significant state-dependent modulation or suppression of the firing of such neurons. Loss of odor would cause a restoration of LLE that would wane over some longer time course without additional pheromonal stimuli. Although such neuronal responses show tantalizing similarity to aspects of the whole insect's behavior, no direct linkage to behavior has yet been made.

Long-lasting increases in firing elicited by pheromone components or visual stimuli have been described in a class of axons termed "flip-flopping" interneurons in male *B. mori* (Olberg 1983a). These neurons exhibited conditional responses, in that stimuli applied when a neuron was in a state of low-frequency firing elicited accelerated firing, while identical stimuli applied when the neuron was in a state of high-frequency firing caused decelerated firing (hence the term *flip-flop*). The conditional responses of *M. sexta* descending neurons differ from the flip-flopping of descending neurons in *B. mori* in several ways. In *B. mori*, long-lasting increases in firing could be activated in flip-flopping neurons by the primary component of the pheromone blend, (*E*,*Z*)-10,12-hexadecadienol (*E*10,*Z*12-16:OH or bombykol), which also activates the entire sequence of zig-zagging walking by which the flightless males approach signaling females (Olberg 1983). In *M. sexta*, individual pheromone components failed to elicit LLE, but blends of the major pheromone components or female extract were effective (Kanzaki et al. 1991b). In *B. mori*, both high and low states of firing elicited by pheromone components were stable and lasted as long as 4 min. In *M. sexta*, LLE elicited by pheromone blends can last several tens of seconds, but the conditional reduction of firing elicited by subsequent pheromone-blend stimuli spontaneously reverts to the state of high-frequency firing after several seconds (Kanzaki et al. 1991b).

Continued study of the *B. mori* descending neurons has revealed additional complexity in the flip-flopping system of neurons (Kanzaki et al. 1994; Kanzaki this volume). Not only have activity patterns, flipping and flopping reciprocally, in the connectives on the two sides of the moth been recorded, as described earlier by Olberg (1983), but bursts of activity have been observed to alternate in different units within each ipsilateral connective. Staining of neurons by backfilling with cobalt dye after recording has revealed two groups of descending neurons within the brain that appear to be responsible for generating these patterns of activity (Kanzaki et al. 1994; Kanzaki this volume). Further characterization of this system may reveal the pattern-generating mechanism, hypothesized by Kanzaki et al. (1994) to lie in the linkage between the bilateral lateral accessory lobes and, possibly, the role of the descending activity in shaping motor patterns. Preliminary studies indicate that activity patterns of neck motoneurons are closely correlated with flip-flopping activity recorded from the connectives. Thus, there is indication that this activity of descending neurons may control head movements associated with zigzag walking in pheromones.

In response to the question posed at the outset, it is clear that the neuroethological perspective figures prominently in current research on pheromone-modulated responses and is woven intricately into the fabric of our thinking on the subject.

It provides a framework for research, aids in identifying new directions, and provides a particularly useful umbrella under which to bring together workers of diverse disciplines, in symposia such as this one, for discussion, debate, and cross-fertilization of ideas. The field is healthy, progress is accelerating, and there is every reason for optimism that the Second International Symposium on Insect Pheromones will mark significant new advances toward the understanding of pheromone-mediated responses.

PART IV

Use of Pheromones in Direct Control

Organized by Albert K. Minks and Ring T. Cardé

29

Mechanisms of Mating Disruption in Moths

C.J. Sanders

1. Introduction

Insect control (population regulation) by the use of synthetic semiochemicals to disrupt normal mating behavior is now operational for a number of species of Lepidoptera. But, although there have been successes, there have also been failures. Moreover, in some instances a previously successful operation has subsequently failed. Also, there is always the possibility that the technology can be improved, or that it can be made to work more cheaply. In order to improve efficacy and to be able to fix it when it is broken, we have to understand how it works. Unfortunately, the mechanisms of how disruption works are still largely unknown. Because the technique involves complex behavioral interactions between insects in a communication mode (olfaction) to which humans have little inherent affinity, understanding disruption is not easy. First we need to understand what we are trying to disrupt—that is, the mating behavior of the insects. Then we have to know the characteristics of the formulation used to disseminate the disruptant chemicals and how the synthetic material behaves in the airspace, and finally we have to understand the behavior and population dynamics of the target insect that affects its susceptibility to mating disruption. This chapter will therefore be divided into four sections: first, a brief discussion of mating behavior in moths; second, a review of the possible mechanisms by which mating behavior of moths may be disrupted; third, a discussion of the mode of action of different disruption formulations; fourth, a review of some of the factors which affect the success of mating disruption as a control technique.

2. Mating Behavior

Mating behavior has been thoroughly studied under laboratory conditions for a number of moths [e.g., gypsy moth (Charlton and Cardé 1990a), oriental fruit moth (Baker and Cardé 1979a), artichoke plume moth (Haynes and Birch 1984b), spruce budworm (Sanders and Lucuik 1992)]. The process begins with the release of sex pheromone by the female moth, a behavior which is regulated by hormones (see Section 1) and modulated by photoperiod, temperature, and the age of the female (maturation). When a receptive male receives a sufficiently high concentration of the pheromone, he takes flight (with or without wing-fanning, depending on the pheromone concentration) and "locks-on" to the pheromone plume. Under field conditions, because males may become active before the females start to call, it is presumed that many males are already flying when they first perceive and lock-on to the pheromone plume. Once he is locked-on to the plume, the male proceeds upwind and arrives in the vicinity of the calling female moth. He then engages in what can best be described as searching behavior, until he locates the female. There is then a brief courtship which, if successful, results in the female accepting the male, and the male attempting copulation.

Although upwind flight behavior is triggered by olfaction, it is also dependent upon vision. A flying moth perceives its movements by visual reference to its surroundings; in some species, vision may play additional roles. For example, Charlton and Cardé (1990b) showed that male gypsy moths orient to the boles of trees. Close-range courtship involves olfaction, vision, and mechanoreception, to varying degrees, depending upon the species. The female-emitted pheromone acts as a primer in the courtship behavior of many, if not all, species (Grant 1987). In many species, pheromones are released also by the male moths, which may allow the female to identify the male and possibly indicate his fitness. In addition, male pheromones could act as "aphrodisiacs." Vision plays a crucial role in the courtship behavior of some species. In the Sesiidae, males orientate to females and copulate while still on the wing (Barry and Nielsen 1984). Male oriental fruit moths use vision to orient to the female as they waft their pheromone toward her with their wings. Mechanoreception is also critical in mate recognition and in the positioning of the male at the time of copulation. Critical cues lie in the scales of the insects, although whether this is entirely mechanoreception or partly chemoreception is not entirely clear (Grant 1987; Charlton and Cardé 1989).

Pheromones may have other effects on conspecific behavior. Male-emitted pheromones could have a "mate-guarding" role by having a repellent effect on other males. Female-emitted pheromones could relay messages to other females, causing them to aggregate or to disperse. For instance, it has been reported that the pheromone emitted by female spruce budworm is perceived by other female spruce budworms and that high concentrations make the females more active (Palaniswamy and Seabrook 1978). For this reason it has been speculated that it may trigger dispersal, which is an important feature of spruce budworm population dynamics. However, firm evidence for this is lacking.

Olfactory cues involving semiochemicals are therefore implicated in at least three stages in the interactions between male and female moths that lead to copulation: (i) when the female signals she is ready to mate by releasing pheromone, (ii) when the male signals his proximity to the female by releasing a pheromone, and (iii) when contact with the female signals to the male that he has located the "right thing." Each of these stages presents the applied entomologist with an opportunity to interfere with the normal semiochemical communication, but it is important to remember that vision and mechanoreception may also play a role and that the effectiveness of mating disruption as a control technique will depend on the interactions of all the different factors.

3. Disruption of Orientation

Because close-range courtship may involve vision and/or mechanoreception as well as olfaction, attempts to disrupt the mating process may be less effective at this stage than during upwind flight to the pheromone source, which is modulated almost entirely by olfaction. Consequently, applied entomologists have focused on disrupting upwind flight. Mating disruption should therefore be more accurately described as disruption of orientation. The processes by which a male uses the female-emitted pheromone to locate the female moth have been described in detail in the previous section. The salient points are as follows. In a uniform, unstructured cloud of pheromone, the sensory system of the moth soon becomes fatigued and ceases to respond to the pheromone (Kennedy et al. 1981; Willis and Baker 1984). However, it is now well established that odor "plumes" emitted from point sources have a distinct structure of puffs, eddies, or filaments caused by turbulence (Murlis and Jones 1981), and this structure provides the moth with a sequence of "on/off" signals which are essential for the orientation process. The antennal receptors of male moths can differentiate up to five pulses of pheromone per second (Rumbo and Kaissling 1989; Baker and Haynes 1989), and moths respond behaviorally to changes within a few tenths of a second of them occurring (see Baker and Vickers this volume). In response to pheromone the male orientates upwind, with wider crosswind excursions in response to lacunae of lower pheromone concentration between the eddies. Note that the lacunae between the high concentrations do not need to be free from pheromone; the olfactory receptors are tuned to specific compounds and can discriminate small changes in odor concentration (e.g., Mayer 1993). Even when there is a background of pheromone (Kennedy et al. 1981; Kuenen and Baker 1982b) or of pheromone antagonists (Sanders 1981; Liu and Haynes 1992), male moths can proceed upwind in response to a superimposed pheromone plume. Therefore, rather than describing the signals as "on/off," it is perhaps more appropriate to refer to them as phasic or "more/less" signals. As a result of upwind surges in response to pheromone and crosswind casting in the absence of pheromone, males exhibit the characteristic oscillating side-to-side flight which we call locking-on and which is characteristic of males orientating to a pheromone plume. This behavior is well suited to conditions in the field, where changes in the direction of wind eddies produce a very erratic odor trail. Crosswind casting optimizes the chances of relocating the plume (David et al. 1982, 1983), although Kuenen and Cardé (1994) have recently questioned the complete dependency of males on crosswind casting for relocating plumes. Those males that do start crosswind casting when they lose the odor plume show extraordinary persistence; male spruce budworm will continue casting in the wind tunnel for 5 min or more after losing a pheromone plume.

3.1. Mechanisms

The possible mechanisms of mating disruption have been reviewed by Bartell (1982) and Cardé (1990) and can be categorized as follows: (i) sensory fatigue, (ii) competition between natural and synthetic sources, (iii) camouflage of the natural plume, (iv) imbalance of sensory input, (v) antagonists to attraction and pheromone mimics, and (vi) combinations of pheromone and insecticide ("lure and kill," "activate and kill").

3.2.1. Sensory Fatigue

Sensory fatigue includes two processes, adaptation and habituation. Adaptation is a reduction in the firing rate of the chemoreceptors; habituation is a reduction in response to the incoming signals that takes place in the black box of the insect's central nervous system. Both are dependent on pheromone concentration, but habituation is also affected by other nerve inputs and by the physiological state of the insect—it has to be in the right mood. Adaptation and habituation can be analyzed by the neurophysiologist, but to a behaviorist working with the whole animal, the outward effect of the two is the same: Response is reduced. The term "sensory fatigue" is therefore useful to cover both processes.

Sensory fatigue caused by exposure to pheromone has been demonstrated in many species. Insects that have been pre-exposed to pheromone show reduced response levels when subsequently exposed to pheromone, and the evidence suggests that effects are most pronounced when the insects are pre-exposed to pulses of pheromone, rather than to a continuous stimulus (Bartell and Lawrence 1973: Kuenen and Baker 1981), possibly because of the phasic-tonic nature of the response (Kaissling 1986b). Because of this, it may be possible to achieve sensory fatigue more efficiently with fluctuating concentrations of pheromone than with a constant high concentration. If a flying moth comes into contact with high pheromone concentration, he can become sensorially fatigued, and as a result he starts crosswind excursions just as if there were no pheromone present (Kuenen and Baker 1982b). This can easily be demonstrated in a wind tunnel, where a male that is approaching a high concentration pheromone source will cease upwind flight and start crosswind casting in ever-wider excursions, finally hitting or landing on the side or roof of the tunnel. In the field it is presumed that these males would fly away and disappear. However, the question of how males behave in the field when they have become sensorially fatigued has not been studied in any detail. They may disperse out of the plot, or they may remain inactive. Preliminary studies carried out with mark and release of oriental fruit moth in peach orchards in Australia suggested that males do not disperse, but remain in the orchards (R.A. Vickers, E. Rumbo, and C.J. Sanders, unpublished information). Much could be learned from operational programs if they were coupled with mark and release experiments.

3.2.2. Competition Between Natural and Synthetic Sources

Conceptually, this is the most straightforward of the processes and is sometimes referred to as "false-trail following." The principle here is that males lock-on to a plume from one of the dispensers and follow that instead of the plume from a calling female. The success of this mechanism (in the absence of sensory fatigue) depends on three factors: (i) Each dispenser should be as potent or more potent than a calling female moth, (ii) there must be enough dispensers to make the probability of a male finding a female instead of a synthetic source very low, and (iii) the males must spend sufficient time going to, or at, the dispensers so that they never find the calling female moths. The first point implies that the complete pheromone blend must be known, and that it must be released from each source at or above the rate released by a female moth. This in turn implies that the rate that pheromone is released by individual female moths must also be known. Average figures will not be adequate, because the dispensers must outcompete even the most potent females. If not, the system will not only be less effective, but it will also select for the most potent females, which could jeopardize the technique. The second point, the numbers of dispensers required, is a question of probabilities, a trade-off between the numbers of dispensers released and the numbers of female moths present. Efficacy will also depend on the density of the moth population. This would be a logical reason for believing that the technique will be more effective at low densities, and this is supported by field evidence (e.g. Webb et al. 1990; Howell et al. 1992).

The term "trail-following" is misleading. A trail is an anthropocentric concept, implying goal orientation. We humans recognize a trail because we know of or can see our destination. Insects do not have that perception; they are reacting only to the stimuli in their immediate surroundings. In a sense they are flying blind. A useful concept is to think of the eddies of pheromone as stepping stones. It is as if the moths are moving from stepping stone to stepping stone across a river; they know which way to head because of the current of the river, but they have no idea of where they are going to land on the far bank. This is very evident in wind-tunnel observations. When presented with multiple pheromone sources in a wind tunnel, the males rarely remain locked-on to a plume from one source. They react from eddy to eddy, so that while constantly moving upwind, their path often moves from plume to plume and they appear to lock-on to several plumes before they arrive at one source of pheromone. Even if male moths are disrupted by the sources of synthetic pheromone (in the sense that most of them fly to a pheromone source other than a calling female), if left for any length of time they may eventually find a calling female. We have found this to be true for spruce budworm male moths in a wind tunnel. Even after male spruce budworms were completely disorientated on first exposure to multiple sources of synthetic pheromone, when left in the tunnel they remained active and eventually located the target females. This raises the question of just how effective the tactic of "competing sources" is by itself in preventing mating; possibly it causes more activity, which increases the probability of a female being located. Efficacy could be enhanced if the dispensers themselves caused prolonged courtship activity, which might be achieved by adding close-range olfactory cues such as extracts from moth scales. Alternatively, the tactic can be highly effective if the dispensers are laced with insecticide (or are placed in traps) to kill the attracted moths. For such "lure-and-kill" techniques it is again of paramount importance to ensure that the lures are as competitive with the calling female moths as is possible.

3.2.3. Camouflage of the Natural Plumes

If the density of release points is sufficiently high, or if the releasers are well upwind, it is assumed that a uniform pheromone fog will be created downwind which will mask or camouflage a pheromone plume released by a calling female. A uniform fog is fairly easy to obtain in a wind tunnel, and it has been shown with several species that male moths flying in a structureless pheromone fog cease upwind flight and engage in wide crosswind excursions with no upwind progress as if they have become sensorially fatigued (Kennedy et al. 1981; Willis and Baker 1984). If a structured plume is superimposed upon the fog, then the males respond to it and, provided that the overall concentration is not too high, they will proceed upwind to the source. In the field it is assumed that males will fly around in a disoriented fashion unless they come close to a female and can perceive her plume against the background fog. The higher the concentration of the fog, the closer a male will have to come to a female before he detects her pheromone. The effectiveness of this technique would therefore appear to be a function of the number of wild females and the concentration of the fog. However, it is questionable if it is possible to achieve "uniform permeation" in the field. Eddies and vortices caused by the air passing over and through the vegetation will create pockets of high and low pheromone concentration, even if the vegetation is adsorbing and releasing pheromone. This may provide enough contrast in concentration to allow the moths to orientate. If true, the situation as far as the perception of the moth is concerned will not be very different from that created by "female-equivalent" dispensers as discussed in the previous scenario. There will be stepping stones of pheromone wherever a male moth turns, although in this case the stepping stones will lead nowhere unless the male moth happens on a series of stepping stones that lead to a calling female.

3.2.4. Sensory Imbalance

In some cases, and many would even say in most cases, the complete, female-produced pheromone blends have not been identified. In other cases the complete blend is difficult to obtain, expensive, and/or unstable. Therefore considerable attention has been given to finding out how important it is to use the complete blend for mating disruption.

Clearly, if the goal is to lure males to a pheromone source in order to kill them, then it is important to be as competitive as possible and the natural blend will almost certainly produce the best results. But, as suggested above, it may not be necessary to lure the males all the way to the lure. Efficacy may depend on the males following "stepping stones" of odor long enough to cause sensory fatigue. It then becomes important to know how crucial it is to use the natural blend and what the consequences are of not using it. Laboratory studies have shown that at pheromone concentrations close to those released by the virgin female moths, the olfactory receptors are tuned to specific pheromone components (e.g., Mayer 1993), which suggests that the natural blend will be the most effective. Minks and Cardé (1988) reviewed the available evidence from field trials and concluded that in most cases evidence supports the conclusion that the natural blend is the most efficacious. However, there are several examples of effective control with off-blends, and my impression from the evidence presented by Minks and Cardé (1988) is that we do not really know.

If the pheromone plume is not too dissimilar from the naturally produced blend, some males will lock-on to it, which will result in some disorientation. This alone could cause some reduction in mating, and it is probable that sensory fatigue will also occur. But male moths possess a highly effective sensory system for differentiating a plume of the correct pheromone blend from a background of competing blends. Disruption will occur if the off-blend is superimposed completely upon the natural blend, which can be achieved by releasing the two from the same point. However, males can differentiate between two sources that are separated by only a few centimeters, and thus they will proceed to the correct source (Liu and Haynes 1993b; Rumbo et al. 1993). Pheromone receptors on the antennae are present in approximately the same proportions as the molecules in the natural pheromone blend. In a bicomponent blend, most receptors respond to the major pheromone component, while only a few respond to the minor component. If the synthetic material contains an abnormal amount of the minor component, then initially the insect will perceive an abnormal blend and will behave abnormally. In the field this can be seen as reduced catches in traps baited with off-blends; in the wind tunnel it can be seen as males failing to respond, or as males that take flight but fail to reach the pheromone source. However, in our studies with the spruce budworm we have noticed a peculiar behavior in the wind tunnel that suggests that sensory fatigue is also playing a role. The natural pheromone of the spruce budworm is a 95:5 blend of (*E*) and (*Z*)-11-tetradecenal.

Males are activated by a 50:50 blend but will not fly upwind and quickly veer off to one side of the tunnel out of the plume. However, these males can be made to readily fly upwind to the source of the 50:50 blend by first exposing them to the natural blend from calling, virgin female moths. For example, if a synthetic lure is placed at the upwind end of the tunnel and a caged female is placed 10–20 cm downwind, males released at the downwind end of the tunnel will lock-on to the resulting pheromone plume. If the female is then quickly removed, the male reverts to crosswind casting but eventually moves upwind to the source. Exposure to the natural, female-produced blend causes the male to maintain crosswind casting for several seconds after the female has been removed. This keeps him exposed to the off-blend long enough for the receptors that are tuned to the minor component to become adapted. However, this evidently occurs before the receptors that respond to the major component have become adapted to the lower ratio of the major component. After a while the insect will "perceive" the 50:50 blend as closer to the natural 95:5 blend and will act accordingly. So disruption will occur because of sensory fatigue. Indeed, in the field, where the males cannot escape the 50:50 blend by veering off to the side, we have found that disorientation is as effective with the 50:50 blend as with the 95:5 blend, although not so in the tunnel. If this theory is correct, then disruption of mating behavior with incomplete blends, such as has been achieved with apple leafrollers in the Netherlands (van Deventer et al. 1994) and British Columbia (Judd and McBrien 1994), is caused by sensory fatigue and not false-trail-following.

3.2.5. Antagonists and Pheromone Mimics

This technique depends upon the synthetic compounds having sufficient similarity in structure to the natural blend for them to bind to the pheromone receptor sites and cause responses similar to those evoked by the true pheromone, or for them to remain bound to the receptor sites so that there are insufficient free sites to bind with the true pheromone to cause a response. Successful control using this mechanism is rare (Minks and Cardé 1988), but the principle may be effective in mating disruption of a few species—for example, the navel orangeworm (Curtis et al. 1987) and, more recently, the pea moth (Bengtsson et al. 1994).

3.3. Formulations and Field Efficacy

In any given field situation, it is probable that several of the above mechanisms will be in operation. Which these are, and the relative importance of each, will depend upon the synthetic pheromone blend and the type of formulation used. Formulations can be assigned to one of three categories: (i) microdispersibles, (ii) female-equivalent point sources, and (iii) reservoirs. With a microdispersible formulation, such as microcapsules and beads, the individual particles release minute quantities of pheromone; and even if they form clumps, which they often do, they will be releasing far less than a calling female from a single point.

"Female-equivalent" formulations, such as Hercon chips, Scentry fibers, and similar formulations, have release rates from the individual particles close to those from calling females. Reservoir-type formulations, such as BASF pouches, Hercon tape, Scentry coils, and Shin-Etsu ropes, represent the other end of the spectrum of release rates from the microdispersible formulations. Individual reservoir dispensers release pheromone at rates of up to several micrograms per hour, 100–1000 times the release rate of a calling female moth.

The manner in which the various types of formulation affect insect behavior depends not only on the release rates from the individual dispensers, but also on the concentration of pheromone in the airspace, on how it is adsorbed and re-released from foliage, and on the structure of the pheromone cloud. Techniques have recently been developed for measuring the concentration of ambient pheromone compounds (see Färbert et al. this volume), and these techniques will play an important role in elucidating the mechanisms of disruption.

3.3.1. Microdispersibles

The merit of microdispersible formulations is that they can be applied by conventional spray equipment from the ground or from the air, which reduces application costs and makes them suitable for treating large areas. The downside is that, so far, such formulations do not maintain constant release rates over long periods, although improved formulations are continually being developed. Because the individual particles are releasing pheromone at rates below that of individual calling females, males will not be lured to the pheromone sources. It is often assumed that this will result in a pheromone fog with very little structure and that the mechanism of disruption will be a combination of camouflage of the female pheromone plumes and sensory fatigue. However, it is questionable whether "uniform permeation" or a pheromone fog can be achieved under field conditions. Vegetation causes turbulence, which in turn will cause eddies of different pheromone concentration that may be sufficiently distinct to allow upwind orientation. The end result is that microdispersible formulations may evoke similar in-flight responses in the male moths to those evoked by the female-equivalent formulations. However, the male moths will not be able to locate the individual sources, which precludes the use of additional chemicals either to kill the moths or to induce them to remain near the source.

3.3.2. Female-Equivalent Sources

In female-equivalent formulations, such as plastic flakes or chopped fibers, the individual particles release pheromone at about the same rate as calling virgin female moths; consequently each particle produces a pheromone plume that competes with the calling female. They are not as easy to disseminate as micro-dispersible formulations, but custom-designed equipment is available for spraying most of them, including aircraft-mounted sprayers, although the addition of a sticker is required if it is necessary to have the particles adhere to the foliage. These formulations have the advantage that release rates are easier to control and can be maintained over longer periods than with the microdispersible formulations. Both wind-tunnel experiments and field observations indicate that the principal mechanism by which these formulations operate is to lure male moths away from the females.

The effect of the density of competing sources can be clearly demonstrated in a wind tunnel. For instance, with two calling spruce budworm female moths (target females) as a single pheromone source in a wind tunnel, more than 98% of males locked-on and flew to the females. When the density was raised to 8 female moths/m^3, 14.1% of males flew to the target females and 83.1% went to the competing females. At a density of 35 competing females/m^3 the figures were 5.9% and 84.7%, and at a density of 69 females/m^3 the figures were 1.1% and 96.7%. Clearly the effect is dependent on the density of the disruptant sources. Moreover, there was no evidence of sensory fatigue—even with the array of 69 female moths/m^3, 97.8% of the males were able to lock-on and fly upwind to one or other of the females. Similar results can be obtained in the field. Observations have been made on the oriental fruit moth in peach orchards at Cobram, NSW, Australia (R.A. Vickers, E. Rumbo, C.J. Sanders, unpublished information). Three concentrations of pheromone dispensers were deployed, and by far the highest level of activity was recorded in the treatment with lures that released pheromone at a similar rate to virgin females. The application rate was 729 lures in a 3- × 3-m cubic arena (9/m^3). Male oriental fruit moths readily flew into the arena, and most visited several of the dispensers. Some spent several minutes at one dispenser, all the while displaying their hair pencils with no indication of sensory fatigue. Evidently these lures evoked heightened activity; and this suggests the possibility that where treated areas are surrounded by untreated areas, male moths may well be lured into the treated plot from the untreated areas.

These observations can be summarized as follows. Even at high densities of "female-equivalent" dispensers there is still sufficient structure for males to orientate without sensory fatigue (even if the plumes meld downwind to create a more "uniform permeation," the plume created by a calling female would still provide significant increases in concentration for males to orientate). The effectiveness of competition alone, as mentioned before, depends on the combination of lure potency (in terms of female-equivalence) and numbers of lures. Because the males have to operate in an environment where they are continually being exposed to pheromone, it is assumed that sensory fatigue also plays a role in disruption, and several papers have demonstrated reduced responses after preexposure to synthetic pheromone (see Liu and Haynes 1993b). However, the relative effects of competition and sensory fatigue are not easily resolved. Disruption of flight orientation of male spruce budworm to calling female moths increases as both number and concentration of pheromone dispensers increased, but sensory fatigue was evident only at higher concentrations, where males were already being "lured" away from the calling female (Sanders 1985a). Prolonged pre-exposure to lower concentrations showed no effects of sensory fatigue; as soon as males perceived the female pheromone they locked-on successfully and maintained upwind-oriented flight for sustained periods. Because sensory fatigue was not evident at concentrations below those where most males were lured away from the female-emitted pheromone plume by the synthetic pheromone sources, it is concluded that sensory fatigue only supplements the other mechanisms. However, without the effects of sensory fatigue, prolonged activity by the male moths would eventually bring them into contact with the females and mating would result.

Another question is just how discrete the plumes are. Pheromone will be adsorbed and re-released from the vegetation (Wall and Perry 1983; Noldus et al. 1991). This will create background "noise," which in turn will tend to blur the differences in concentration between the eddies of the pheromone plumes. Carried to extremes this would create a uniform fog of pheromone, similar to that created by a microdispersible formulation.

3.3.3. Reservoirs

Reservoir-type formulations differ from "female-equivalent" formulations in release rate, with the individual dispensers releasing pheromone at rates many times that of a calling female. Their advantage is that they are generally long-lasting, and fewer of them are needed to achieve the same overall release rate, which makes them cheaper to deploy by hand. Their disadvantage is that they cannot be disseminated effectively from aircraft. Their mode of operation is probably as follows. At some distance from each pheromone source, the concentration is low enough to allow the males to "lock-on" and to orientate upwind. Some males of some species are able to orientate upwind to reach the sources of pheromone, but in many cases it is probably that the increasingly high pheromone concentration will cause sensory fatigue and that the moths will veer off to one side long before they reach the source. The reservoir types of formulations therefore combine the effects of false-trail-following and sensory fatigue to best advantage.

4. Attributes of the Target Species

The efficacy of mating disruption is dependent upon the population density, the food preferences of the target species, the degree of isolation of the treated areas, and other attributes of the species, such as characteristics of mating behavior (which have already been discussed), the release rates of pheromones by the female moths, and the responsiveness of the males, which includes both the range of concentrations to which the males respond and the length of time during the day that they respond.

4.1. Population Density

If mating disruption depends upon male moths being lured away from the females by competing synthetic sources, then clearly population density will be an important factor: The more insects, the greater the probability that they will encounter each other by chance. Sensory fatigue as a mechanism causing disruption will be less dependent upon density, but probabilities will still favor some mating at high densities. Experiments on caged populations of spruce budworm confirmed these conclusions (Schmidt and Seabrook 1979). Although there have been few definitive studies under field conditions, Webb et al. (1990) and Howell et al. (1992) concluded from field trials on gypsy moth and codling moth that mating disruption is less effective as density increases.

4.2. Food Preferences and Isolation

Mating disruption becomes far less effective if the target area is surrounded by untreated areas of host species or if the target pest occurs on other vegetation surrounding the treated crop. Female moths of most pest species, especially among the Microlepidoptera, are capable of dispersal over at least a few tens of meters. In some species, such as the spruce budworm, dispersal can occur over hundreds of kilometers, which renders attempts to control populations in patches of forest adjacent to untreated areas ineffectual. Other species are polyphagous, feeding on "weed" species surrounding the crop. To be effective, mating disruption must account for these potential invaders.

4.3. Female Potency

Competing pheromone sources, to be effective, must outcompete the calling moths. Similarly, male moths that are sensorially fatigued will respond when exposed to a higher concentration of pheromone. It is therefore crucial to know the potency of the female moths. Generalities are invalid, because pheromone release rates vary widely among species and among individuals within the same species (e.g., Lacey and Sanders 1992). Synthetic pheromone sources must outcompete the most potent of the females, and it is therefore important to know the maximum release rate of individual females and not just average rates. Techniques are now available which make this relatively simple to determine (Tóth and Buser 1992; Lacey and Sanders 1992), and these measurements should be a consideration in all evaluations of mating disruption trials.

4.4. Male Responsiveness

There is a lower threshold of pheromone concentration below which males will not respond. At high concentrations there is a threshold above which males become sensorially fatigued (as demonstrated by arrestment of upwind flight). Experiments to determine these two thresholds have been carried out for several species (Fig. 29.1) and show that the breadth of the response window varies

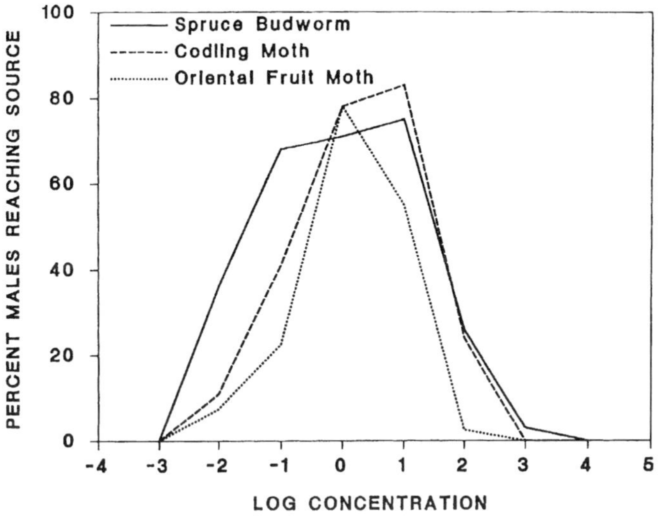

Figure 29.1. Numbers of male moths of three species responding to synthetic pheromone in wind tunnels, showing lower threshold where no males respond and upper threshold where male upwind flight is arrested. Spruce budworm data are from Sanders (1990), oriental fruit moth data are from C.J. Sanders unpublished data, and codling moth data are from McDonough et al. (1993).

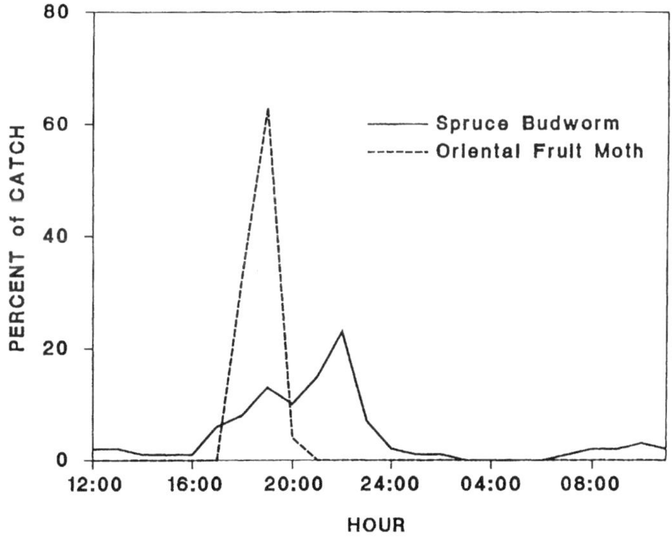

Figure 29.2. Diel periodicity of catches of male oriental fruit moth and spruce budworm in traps baited with synthetic pheromone. Oriental fruit moth data are from Rothschild and Minks (1974), and spruce budworm data are from Sanders (1971).

among species. Logically those species with a narrow window should be more amenable to control by synthetic pheromone than those with wide windows.

The length of time during the diel cycle that males respond to pheromone also varies with the species (Fig. 29.2). Logically, mating disruption will be more effective in those species where the males start responding to the synthetic source long before the females begin to call.

5. Conclusion

Based on the available evidence and the above reasoning, it can be hypothesized that mating disruption is caused by a combination of false-trail-following and sensory fatigue. Formulations such as high release rate dispensers, which combine the two attributes, appear, in general, to be the most effective, and they are even more effective so if the synthetic material is identical to the natural pheromone blend. Species which are monophagous, occurring in isolated patches of host type, with relatively low rates of pheromone release, narrow windows of male response, and a brief period of daily activity, such as the oriental fruit moth, will probably be the most vulnerable to the technique.

30

Measuring Pheromone Concentrations in Cotton Fields with the EAG Method

Petra Färbert, Uwe T. Koch, Andreas Färbert, and Robert T. Staten

1. Introduction

Mating disruption has proved to be an important method of pest control for many cropping systems. However, there seems to remain a large potential for improvement and optimization of the method. For example, there are different types of formulations available, but there is very little data about their actual performance in the field: What level of pheromone concentration can each formulation maintain, and for how long? Apart from the performance of the pheromone sources, other questions arise. How homogeneous is pheromone distribution? How far does pheromone drift into adjacent fields? Do the plants' leaves interact with pheromone?

To answer these questions, a portable, fast-reading pheromone measurement system would be very desirable. Standard gas chromatography–mass spectrometry (GC–MS) techniques permit measurements using adsorbent pheromone collectors, but they need sampling times of 5–24 hours (e.g., Flint et al. 1993). These systems therefore cannot resolve pheromone density changes caused by wind or temperature variation. However, this problem can be solved by using the insect antenna as a detector, generating electroantennograms (EAGs) from pheromone responses. We have developed and applied a portable EAG pheromone measurement system to assess pheromone density distributions in pheromone-treated cotton fields in Phoenix, Arizona (USA).

2. Methods

The EAG pheromone measurement system utilizes the antennae of male insects capable of detecting sex pheromones. If a sensitive direct current (dc) amplifier is connected to the ends of an excised antenna, a dc voltage drop can be measured each time pheromone-containing air is blown over the antenna. The voltage drop A in the EAG is related to the pheromone density x in a logarithmic dose–response function in such a way that A is increased by a constant amount a each time the pheromone density x is increased tenfold, as long as x is above the detection threshold x_0. For quantitative measurements, the antenna is mounted inside an airtight chamber (Fig. 30.1). A constant flow of air across the antenna is maintained by a suction pump. For calibration, the air intake flows through a charcoal filter removing potential stimulating air components (Sauer et al. 1992). The calibration signals are generated using glass syringes containing a mixture of oil and pheromone in a small vial. Inside these syringes, a constant equilibrium mixture of air, oil vapor, and pheromone vapor is established. The pheromone density in the syringes is dependent on the dilution of pheromone in the oil (Kafka 1970; Sauer 1989). When the piston of a syringe is moved by a step motor drive, a reproducible amount of pheromone is added to the main airstream, generating an EAG. By applying pheromone stimuli from syringes with different

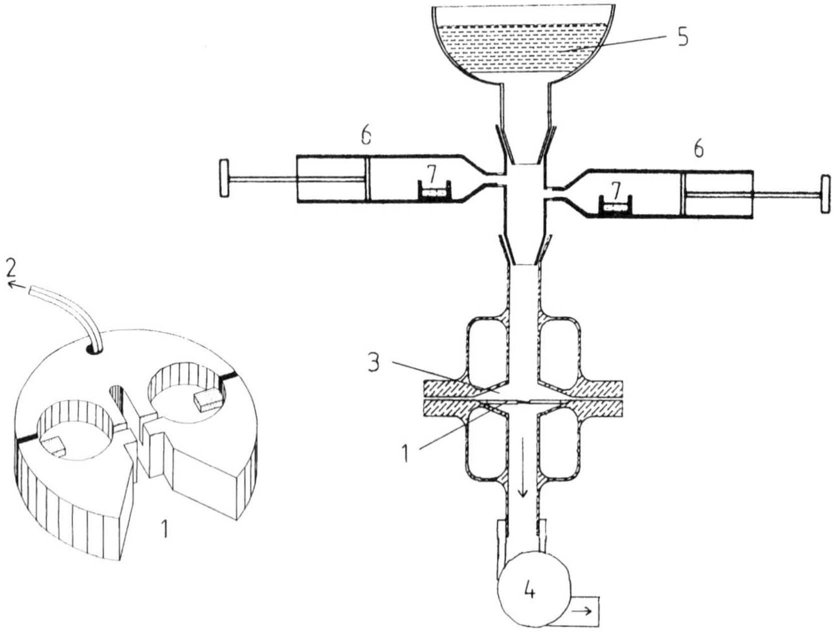

Figure 30.1. Schematic diagram of the EAG measurement system. The antenna is mounted in the antenna holder (1), bridging the gap between two pools of saline. Electrodes immersed in the saline bath make connection (2) to the amplifier. The antenna holder is mounted in an airtight chamber (3) in which a constant airflow is generated by a suction pump (4). The filter (5) is in place during calibration and removed to measure ambient air. Three syringes (6, only two are shown) contain a vial (7) with a pheromone–oil mixture to generate calibration pulses.

pheromone contents (Termer 1992), different points of the dose–response function can be obtained (Fig. 30.2), and the constants *a* (sensitivity) and x_0 (threshold) can be determined.

Pheromone in the ambient air can be measured by removing the charcoal filter, generating a rise in the EAG signal similar to a step function. The height of the step can be used to determine the ambient air pheromone density using the dose–response function established in the calibration. This simple protocol fails, however, if the antenna reacts to ambient concentrations of other stimuli such as plant odors. In this case, the step function recorded while removing the filter contains the sum of responses to pheromone and nonpheromone stimuli, and the relative contribution of each component cannot be deduced from the signal. In a modified measuring protocol (Milli 1990; Sauer 1991), further calibration pulses are added to the airstream while the charcoal filter is removed. If there is no pheromone in the outside air, the calibration pulses added will produce an addi-

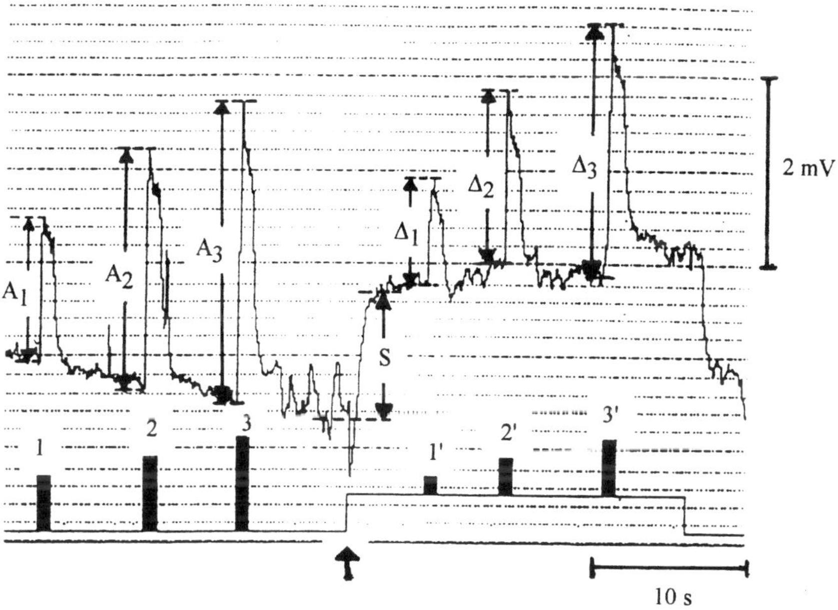

Figure 30.2. Original recording of one measurement cycle. **Upper trace**: EAG recording. **Lower trace**: Recording of syringe and filter activity. Black vertical bars indicate the activation of syringes 1, 2, and 3, respectively, and bar height indicates syringe number. *Left part*: Antenna calibration. One pulse out of each syringe generates a graded EAG amplitude A_1, A_2, A_3, permitting the construction of a dose response function. *Right part*: Opening of the filter (arrow) generates step responses in the EAG. Superimposed syringe pulses 1′, 2′, 3′ produce strong additional EAG response amplitudes Δ_1, Δ_2, Δ_3 indicating little pheromone in the ambient air.

tional response similar to a standard calibration response riding on top of the step function plateau (Fig. 30.2). However, if there is a substantial amount of pheromone present in the ambient air, the response to the calibration pulse added to the incoming air will be negligible, provided that the pheromone density outside is high compared to the calibration (Fig. 30.3). Thus, ambient air pheromone density can be calculated from the height of responses to additional calibration pulses added to the incoming air. The portable EAG system consists of the following functional components:

(a) A measurement chamber containing the antenna mounted in a perspex holder

(b) Calibration sources: three syringes containing pheromone diluted in oil (Silicon-Öl Typ 3, Merck No. 12525), dilution 10^{-6}, 10^{-5}, 10^{-4} ml/ml; syringe step motors and drive electronics

(c) A system to remove and put on the charcoal filter with motor and drive electronics

(d) A miniature air pump and a flow monitor to generate a continuous airstream over the antenna

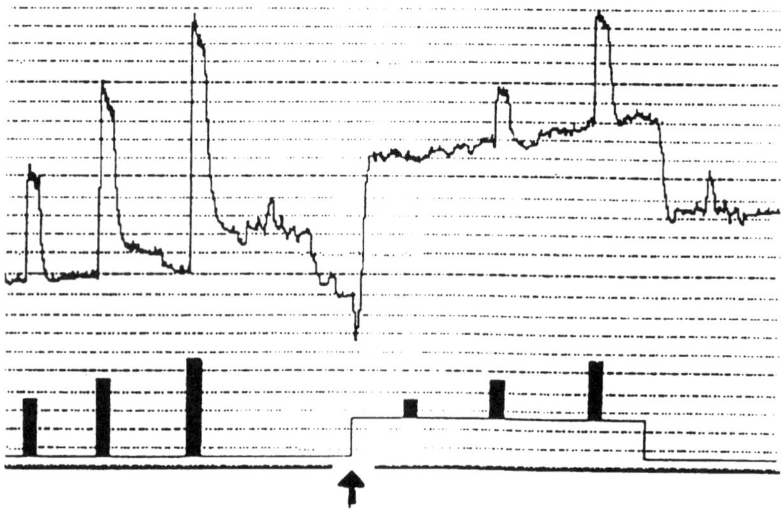

Figure 30.3. Measurement cycle in a pheromone-treated plot. Traces are as in Fig. 30.2. *Left part*: Calibration sequence as in Fig. 30.2. *Right part*: Opening the filter (arrow) again generates a response in form of a step function in the EAG. The superimposed syringe pulses generate additional EAG amplitudes only for syringes 2 and 3, and responses are much reduced compared to the calibration. Ambient pheromone concentration is higher than that caused by activation of syringe 1.

(e) An EAG preamplifier

(f) A thermistor wind probe

These components are mounted together on a measuring probe which is connected to the control unit by a cable up to 10 m long. The control unit consists of the EAG amplifier, the wind probe amplifier, the timers controlling the calibration syringes and the charcoal filter movements, and the associated power supply. The output of the analog amplifiers and a monitor trace recording syringe and filter activity is converted to digital signals using a 12-bit DT 2814 (data translation) A-D converter. Using the program "Easylab" (Stemmer Elektronik) the data are stored on the hard disk of a booksize 386 SX-type computer with LCD screen. The control unit and computer are mounted in a small rack. The measurement system is powered by a 12-V 30-Ah car battery driving a voltage converter to generate 220 V ac. The battery and the control unit including the voltage converter are also mounted on the portable rack. Four handles and four adjustable legs can transform the rack into a small table that can stand on uneven ground. In order to measure pheromone at different heights, an aluminum pole is used to position the probe at heights between 0 and 2.5 m. When preparing a measurement, the antenna of a male pink bollworm *Pectinophora gossypiella* (3–4 days old) is cut near the pedicellus. The antenna is positioned in such a way that it will bridge the gap between the two pools of saline in the antennae holder. To improve contact between antenna and saline, the tip and the base of the antenna are cut with miniature scissors while immersed in the saline. The antennal holder is then mounted on the probe, and measurements can start. Each measurement cycle (as shown in Figs. 30.2 and 30.3) comprises three syringe pulses for calibration, and three pulses superimposed to the step function response while the filter is open, which yields up to three pheromone density readings per cycle. Usually, three measurement cycles (50 seconds each) are taken at each measurement location and height, so that up to nine readings can be averaged to yield a pheromone density measurement reading with a small standard error. It turned out to be essential to repeat the calibration at each measurement cycle, since the properties of the antenna (a and x_0) can vary. Usually, an antenna provides satisfactory readings for 30 minutes; exceptionally good antennae lasted up to 100 minutes.

For evaluation, the individual EAG amplitudes are read from the display screen while replaying the recordings on the computer. These readings are stored in an Excel database together with data on the measurement location, temperature, humidity, averaged wind velocity, and dilution factor of the syringes used. Programs are applied to calculate a, x_0, and the pheromone relative density. The final pheromone density results including standard errors as well as the other parameters were stored in an Excel file to produce diagrams or to perform further analysis.

3. Results

3.1. Comparison of Different Formulations

We have measured pheromone concentrations in cotton fields as generated by different kinds of formulations, such as rope dispensers, sprayable microcapsules (beads), and fibers. These measurements were conducted in the middle of treated areas at least 0.5 ha in size. The time of measurement ranged from 21:00 to 1:00 hours in order to cover wind, temperature, and humidity values present near the time when the pink bollworms fly. The pheromone density values are all given in relative units, defined by the characteristics of the calibration syringes. A relative density of 10^{-6} therefore is the pheromone density existing inside a calibration syringe containing 10^6 parts of oil and 1 part of pheromone.

The performance of the rope dispensers was measured over 3 months in 1992 (Fig. 30.4). On the day of application, the first density values were 2.5×10^{-6} at

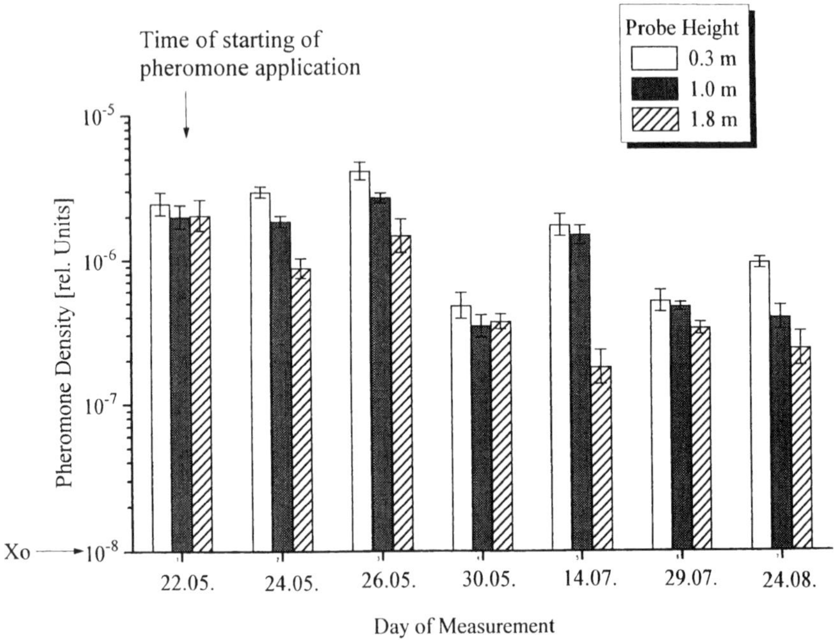

Figure 30.4. Pheromone densities in a cotton field treated with the rope formulation. Over a period of 3 months, the pheromone density remains approximately at the same level, with the exception of May 30, when strong winds reduced the pheromone concentration. The pheromone density (y-axis) is given in relative units. In this and in the following figures the position of the x-axis represents the average level of the detection threshold x_0 of the measurement system. Measurements were taken at three different probe heights (0.3, 1.0, and 1.8 m). All measurements shown in this and all following figures were made with antennae of male *Pectinophora gossypiella.*

0.3-m height. The pheromone density increased slightly to 4×10^{-6} in the following 4 days and then gradually declined until 3 months later, when the density was still 1.5×10^{-6}.

In general, the highest pheromone density was found near the ground, decreasing toward 1- and 1.8-m probe heights. The values measured on May 30 were an exception. On this day, there were strong winds in the field with velocities of more than 5 m sec^{-1} which reduced the overall pheromone density. Due to increased turbulence, the pheromone was transported up into higher air zones, thus explaining why there was little difference among the pheromone densities at the three heights.

Similar data obtained in a field treated with a sprayable microcapsule formulation ("beads") are shown in Fig. 30.5. In the evening of the day of the first application, pheromone density reached up to 4×10^{-6}, which is somewhat higher than the density reached by the rope dispensers on the day of application. However, only two days later, the density had dropped to 3×10^{-7} (at 0.3-m height) and even further to 1×10^{-7} in the next two days. A new treatment was applied on May 28. But two days later, the pheromone density had risen only moderately to 1.5×10^{-7}. After a third application on June 30, pheromone density was again

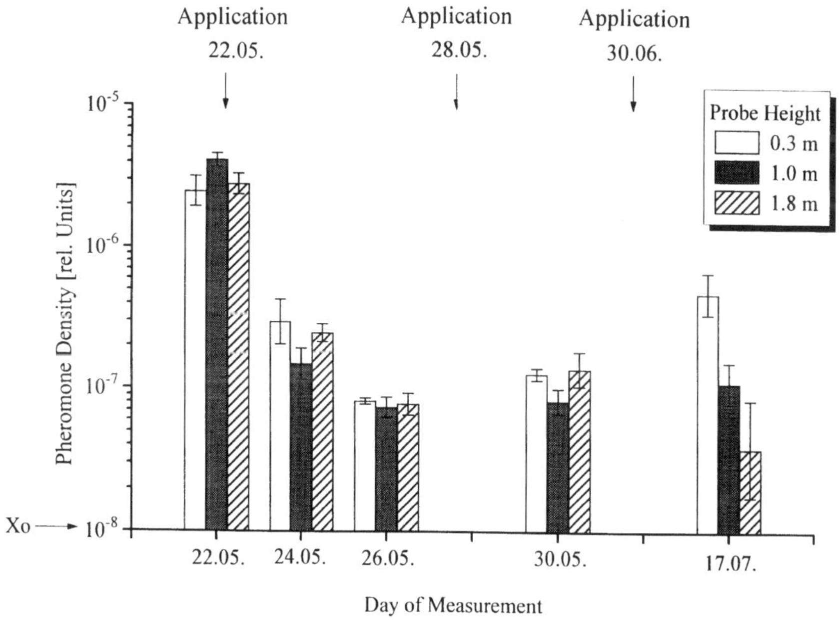

Figure 30.5. Pheromone density in a cotton field treated with a sprayable microcapsule formulation. Note that on the day of pheromone application, pheromone density exceeds that of the rope formulation. Later values were on a much lower level despite renewed treatments.

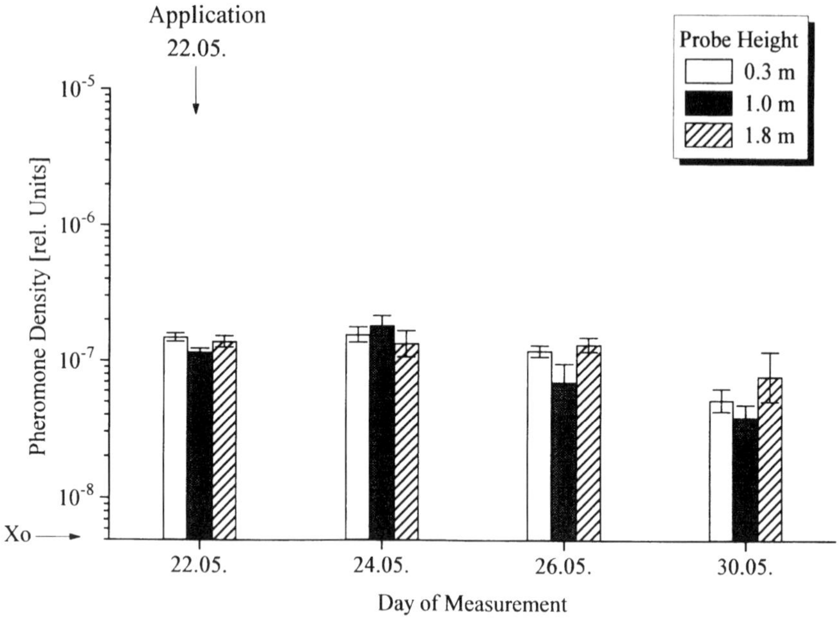

Figure 30.6. Pheromone densities in a cotton field treated with a fiber formulation. Density values were about 10 times lower than those obtained with the rope formulation.

measured on July 17. This resulted in considerably higher values of up to 5×10^{-7} at 0.3 m.

Results of measurements taken in a field treated with a fiber formulation are presented in Fig. 30.6. Starting off with a density of 1.5×10^{-7} on the day of application, the pheromone levels declined gradually to values ranging from 4 to 7×10^{-8}, 8 days later.

3.2. Spatial Distribution

Results of a series of measurements taken to investigate how far pheromone drifts into adjacent untreated cotton fields are shown in Fig. 30.7. Measurements were conducted in a 300- × 200-m cotton field. In one corner, an area of 50 × 50 m was treated with rope dispensers. Plant height was 40 cm. The wind blew with an average speed of 0.5 m sec^{-1} over the treated area into the untreated part of the field. The measurement position plotted on the horizontal axis has its origin at the border between the treated and untreated part of the field. Inside the treated area the average pheromone density is in the usual 10^{-6} range found in other rope-treated fields. In the untreated area, pheromone signals (up to 9×10^{-7}) were measured 80 m from the treated area. In contrast, on a dirt road only 10 m upwind from the treated area (measurement # 9), pheromone density was below the antenna's threshold (x_0).

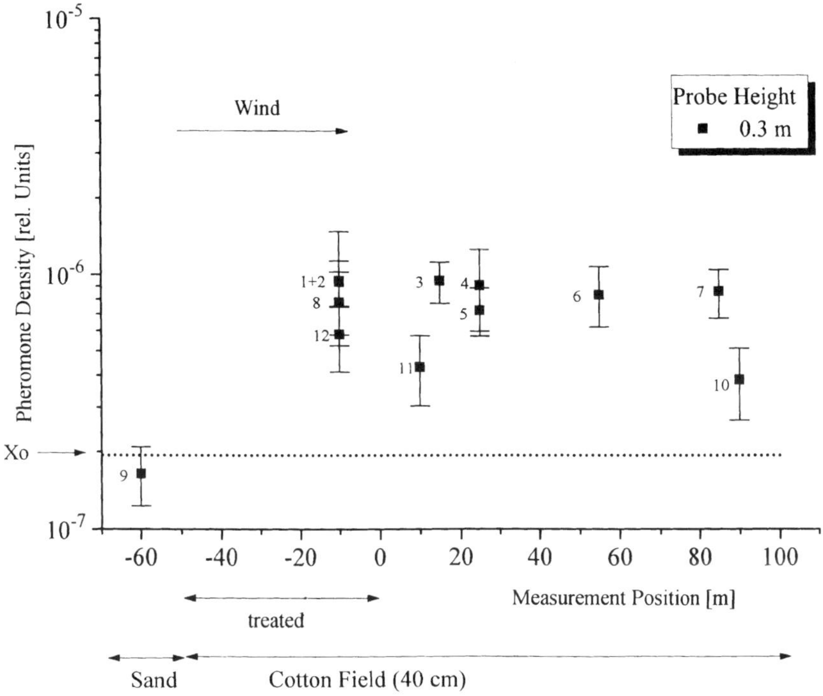

Figure 30.7. Pheromone transport into adjacent untreated fields. On the horizontal axis, the distance from the border between a treated (*left*) and untreated (*right*) cotton field is plotted. Measurement points are numbered to indicate the time sequence in which the measurements were taken. Note that pheromone can be measured at a distance of up to 90 m downwind from a treated field.

A new way of dispenser application has been proposed and evaluated in vineyards on *Lobesia botrana* by Charmillot (et al. 1995). Instead of distributing the dispensers evenly in the field, all dispensers were concentrated on the border, generating a "dispenser belt" (translation of Charmillot's *ceinture*). We also set up such a dispenser belt treatment around an isolated 200- × 100-m cotton field. The rope dispensers were placed at a density of 2 per meter in the borders of the field. In Fig. 30.8, the results are shown of pheromone density measurements on a pathway parallel to the wind and in perpendicular direction to the line of the belt. As expected, the highest densities (5×10^{-6}) were found near the site of application. Substantial amounts of pheromone (up to 4×10^{-7}), however, were measured 100 m downwind from the belt. In these measurements, the density was also determined at heights of 1.0 and 1.8 m. Pheromone density was reduced at higher elevations, but 100 m downwind from the belt the density was about 1.5×10^{-7}, far above the detection threshold of 3×10^{-8}.

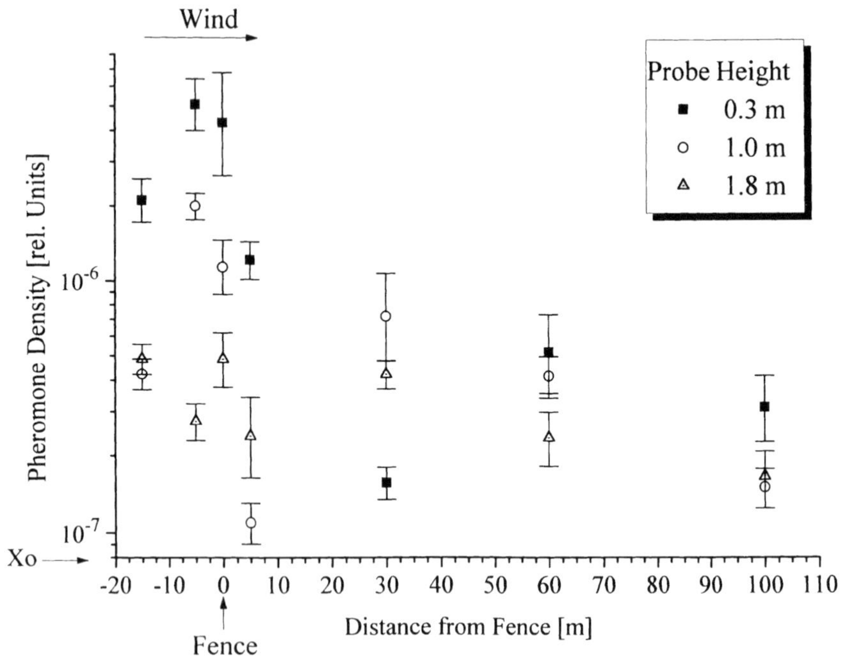

Figure 30.8. Pheromone distribution in a field where all dispensers were concentrated on a belt surrounding the field. Near the contiguous row of dispensers, pheromone density is very high; it drops off gradually at larger distances. Note that at a probe height of 1.8 m, the pheromone density is in general much lower than at 0.3 m.

4. Discussion

4.1. EAG Method

The present results show the usefulness of the EAG measurement system. The experiments measuring pheromone transport into adjacent fields and the distribution downwind from pheromone sources require a measurement system with a response time in the order of minutes, because wind variability results in rapid density changes in that time scale. These short response times cannot be achieved with the "classical" method of sampling pheromones from ambient air into adsorbent material for GC analysis.

4.2. Dispenser Comparison

Our system can be used to compare the actual performance of dispensers in the field. These results can help to decide how often treatments have to be renewed and permit optimal dispenser selection comparing dispenser performance (pheromone output, lifetime) and cost (material and application).

4.3. Pheromone Distribution

Our results show that under conditions of weak (<0.5 m sec^{-1}), steady winds, which may be found during the flight time of *Pectinophora gossypiella*, behaviorally relevant concentrations of pheromone can be transported over distances of hundreds of meters. Sauer et al. (1992) reported that at the border of a pheromone-treated vineyard, pheromone density decreases sharply within a distance of 10 m. This may be the case if the wind from an untreated area is blowing into the treated area. This is consistent with our results, since point #9 in Fig. 30.7 shows a sharp decline of pheromone density at the upwind edge of the treated area. Unfortunately, Sauer et al. did not indicate wind direction and velocity in their tests.

The large distances over which pheromones are transported suggests two important consequences:

1. If a pheromone is used as mating disruptant, males may be attracted from the fields downwind of the pheromone-treated area. This may explain why in some mating disruption tests, damage levels are often highest at the borders of the treated areas.
2. Untreated controls should not be influenced by long-distance pheromone transport. Control areas should be situated upwind from the treated area (considering the predominant wind direction during flight), or be more than 200 m from the pheromone-treated area.

The pheromone distribution in the belt experiments shows that within at least 100 m, an effect on the behavior of the males is to be expected. Charmillot (et al. 1995) concluded that good control inside the belted area can be achieved. In one of Charmillot's dispenser-belt experiments in vineyards, pheromone density profiles were measured with our EAG system. Those results are in good agreement with the data of Fig. 30.8.

Finally we note that further improvements of the device are in progress, so that absolute values of pheromone density will soon be feasible.

Acknowledgments

We acknowledge the support from the USDA NRI to Prof. Ring Cardé. We thank Andrea Termer for additional measurements in 1992, Winfried Galm and Ulf Andrick for programming and technical support, and PD Dr. Claus Pfau for contributing to development of the measurement system. Additional support was provided by the Bundesministerium für Forschung und Technologie, via a research contract with BASF AG. We also thank Dr. J.J. de Kramer for helpful discussions and Ms. Sybille Watt for typing the manuscript. This work would not have been possible without the constant support given by Prof U. Bässler, University of Kaiserslautern.

31

Measuring Pheromone Dispersion in the Field with the Single Sensillum Recording Technique

J.N.C. van der Pers and A.K. Minks

1. Introduction

Pheromones are active at extremely low concentrations. Under natural conditions, airborne pheromones are embedded in a background of many other volatiles. Direct measurement of pheromone components in the field is not possible with currently available physicochemical detection techniques, because of the limited sensitivity and the lack of specificity of these techniques. However, sensitive and selective pheromone receptors are present on the antennae of many insect species. With the help of electrophysiological recording techniques, these receptors can be used as biological detectors. Two of these techniques are well established: electroantennography (EAG) and single sensillum recording (SSR). Because of the rather complex instrumentation, both methods were employed exclusively until recently in laboratory settings. The EAG technique was first used in the field by Baker and Haynes (1989) in a stationary setup, in which the antennal preparation was freely exposed to the air. In 1990 Koch described a transportable EAG apparatus, in which the antenna was conditioned by a continuous filtered airflow interrupted by exposures to the surrounding air and reference sources. This device was used to measure relative pheromone concentrations (Milli and de Kramer 1989; Karg et al. 1990; Sauer et al. 1990; Sauer 1991; Milli and Dickler 1992; Sauer et al. 1992) under various field conditions (Bengtsson et al. 1994; Karg et al. 1994) and in wind tunnels (Färbert et al. this volume). The method used by Baker and Haynes (1989) allows registration of fast fluctuations (<1 sec) of pheromone concentrations, but the antenna is not shielded from air turbulence and nonpheromone volatiles. In the device designed by Koch (1990) the airflow over the preparation is kept constant and the effect of nonpheromone volatiles is canceled by a special measuring protocol and the use of reference sources. As a consequence of this method, the device produces data about pheromone concentrations and concentration gradients in space, but fast fluctuations in concentrations cannot be recorded.

From wind tunnel and field experiments (Baker et al. 1985; Baker 1986), evidence has arisen that the discontinuous character of pheromone plumes is necessary for sustaining upwind flight. The dynamic character of odor plumes was well demonstrated by the experiments of Murlis and Jones (1981), Murlis (1986), and Murlis et al. (1989), in which ionized air and appropriate detectors were used to simulate pheromone plumes under various field conditions. The study of real pheromone plumes and the dynamic effects introduced by air movements require a measuring device able to perform continuous recordings with both a high temporal resolution and a high specificity to pheromone components. The required specificity is not obtained with EAG recordings, in which pheromone responses and responses elicited by environmental odors are found to be dependent from each other (Rumbo et al. 1995).

The required specificity and sensitivity can be met by the SSR technique. Pheromone receptor cells respond to changes in concentration within a few milliseconds, and stimulus concentration fluctuations of up to 5 Hz can be followed (Kaissling 1986b; Rumbo and Kaissling 1989; Marion-Poll and Tobin 1992). Moreover, pheromone receptor cells are extremely selective, making the detector indifferent to nonpheromone background volatiles and other modalities like temperature, moisture, and air-speed fluctuations. On the other hand, SSR produces the response from only one or few receptor cells out of a population of many thousands, whereas an EAG response is generally assumed to reflect the activity of a large number of receptor cells. The main obstacle of making the SSR technique suitable for use in the field is the rather complicated arrangement of the required instrumentation. Therefore, a special device was developed in which all essential components for SSR recording were combined in a compact apparatus. The device is truly portable and can easily be moved around in the field.

The primary objective of using SSR in the field is to collect information about *dynamic* aspects of pheromone dispersion as a function of source intensity and source arrangement, distance from the source, air movements, and influence of vegetation. Information about instantaneous and average pheromone flux over the antennal preparation can be derived from SSR recordings using simultaneous air-speed recordings. For that purpose the portable SSR device is equipped with a miniature air-velocity sensor positioned as close as possible to the antennal preparation. Calibration of the response was attempted in the laboratory by coupled gas chromatography–SSR (GC–SSR), in which the response to a precisely defined stimulus can be measured.

Recordings obtained with the portable SSR device from several species of moths and under different field conditions are presented. Potentials, future developments, and limitations of the technique are discussed. The information may provide a basis for the development of more effective techniques of mating disruption.

2. Materials and Methods

2.1. Insect Material

In the preliminary experiments, antennae from males of the apple glasswing sesiid moth *Aegeria* (=*Synanthedon*) *myopaeformis* were used mainly because of its easily accessible pheromone sensilla. These males were collected from bait traps in the field. Further experiments were performed on antennae of male moths of several species of tortricids: *Adoxophyes orana*, *Pandemis heparana*, and *Archips podana*, which were available from laboratory cultures. In one laboratory experiment, male moths of *Spodoptera exigua* from a laboratory culture were used.

2.2. Chemicals

In the experiments with *A. myopaeformis* a hexane solution (1 mg/ml) of synthetic (*Z*, *Z*)-3,13-octadecadienyl acetate, described as the main pheromone component of this species (Arn et al. 1992), was used. In the field, recordings were made from antennae of *A. orana* male moths by administering polyethylene dispensers loaded with 1 mg of the two major sex pheromone components of this species, (*Z*)-9-tetradecenyl acetate (Z9-14:Ac) and (*Z*)-11-tetradecenyl acetate (Z11-14:Ac) in a 9:1 ratio (Arn et al., 1992). Experiments on *P. heparana* and *A. podana* were conducted using polyethylene dispensers containing 5 mg Z11-14:Ac, which is one of the main pheromone components of both species (Arn et al. 1992). Recordings in an open field were performed using filter paper strips loaded with either 1 mg Z11-14:Ac or E11-14:Ac dispensed in 5-ml disposable syringes, which served as remotely controlled odor sources. In some of the experiments, one of the odor sources was combined with a smoke plume generated by smoke tubes (100 × 8 mm; plume diameter at orifice 2 mm; Dräger, Germany). For coupled GC–SSR, a hexane solution of synthetic Z11-14:Ac (0.2 ng/μl) in hexane was used. In the laboratory experiments with *S. exigua*, 1 ml of a hexane solution of 10 μg Z9,E12-14:Ac, the main pheromone component of this species (Arn et al. 1992), was applied on a filter paper (30 × 50 mm). After evaporation of the hexane the filter paper was placed in a glass jar.

2.3. Portable SSR Equipment

Preliminary experiments using a prototype of the SSR apparatus were performed in the field and in the laboratory (Van der Pers and Minks 1993). Based on the experiments conducted with the prototype, an improved version was built and special software was developed to analyze the recordings.

The recording module (Fig. 31.1), measuring about 20 × 20 × 20 cm, consists of a central aluminum column (1) with two joystick micromanipulators (2, 3) (Narishige, MN-151) mounted on both sides at an angle of 90° with respect to

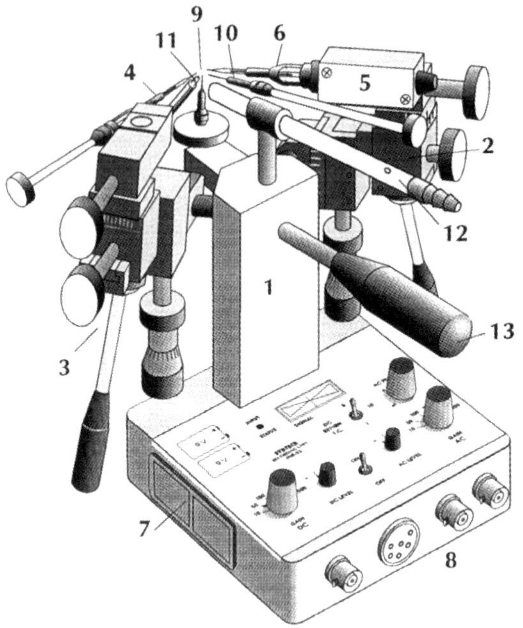

Figure 31.1. Portable SSR recording module. 1, Central column; 2, right micromanipulator; 3, left micromanipulator; 4, indifferent electrode holder; 5, probe containing preamplifier; 6, different electrode holder; 7, battery compartment; 8, signal receptacles; 9, holder with vertical micro-knife; 10, horizontal cutting micro-tool; 11, micro-thermistor air-speed sensor; 12, flow tube for application of filtered air to the preparation; 13, carrying handle. Probe cable and grounding wires are not shown for sake of clarity. Height of instrument is about 22 cm.

each other. The left-hand micromanipulator carries the pipette holder for the indifferent electrode (4). The right-hand micromanipulator holds a miniature probe (5) containing a high-impedance input amplifier (Analog Devices type AD546). The holder for the different electrode (6) is mounted directly onto the probe. To minimize leakage current and the influence of electrode capacitance, the body of the electrode holder is connected to the inverting input of the amplifier. The probe is connected via a cable to the aluminum base of the instrument containing the signal conditioning circuits, gain control switches, and battery compartment (7). The base contains separate circuits for DC (bandwidth 0.1–40 Hz) and AC (bandwidth 200–2500 Hz) signal amplification with adjustable gain (10–1000). The DC signal is monitored by an analog signal indicator meter. The DC amplified signal output is directly available from a BNC receptacle (8) and also converted to a frequency-modulated signal provided by a voltage-to-frequency converter. The aim of the signal conversion is to enable recording of the DC signal on a conventional audio tape recorder. Recordings can be made using tungsten electrodes or by means of glass capillary pipettes combined with silver wire.

In our experiments the apparatus was accommodated for the tip recording method (Van der Pers and Den Otter 1978) using tungsten instead of glass micro-knifes to cut the tip of a sensillum and bring it into contact with the recording pipette electrode. For this purpose a holder for a vertical micro-knife (9) was placed on the central aluminum column, and a horizontal tungsten-cutting micro-tool (10) was mounted in an adjustable clamp on the right-hand micromanipulator. To make the antennal preparation, the unit was placed under a stereo dissecting microscope, which was mounted on a universal stand and placed on a collapsible camping table. After the contact with a sensillum was established, the quality of the signal was tested by application of a short puff from a pheromone dispenser. A tube (inside diameter 7 mm) could be mounted on the central column and directed to the antennal preparation to flush the preparation with filtered air from a miniature air pump. A handle attached to the central column enables the operator to carry the unit with one hand, leaving the other hand free for adjustment of controls and operation of a microphone for voice comments.

A micro thermistor (11) (Fenwal Electronics, type BC32L1) mounted between two wires (diameter 0.8 mm, distance 1.5 mm) on a stainless steel shaft (diameter 6 mm) and positioned within short distance (2 mm) from the recording site was used as air velocity sensor. The sensor was connected via a cable to the signal conditioning circuit in the base compartment. The output of the sensor was directly available as a voltage proportional to the air speed over the sensor and as a modulated frequency. The latter signal could be recorded on one channel of a conventional audio tape recorder. All electronic circuits were designed for low-power consumption from two 9-V alkaline batteries housed in a compartment in the base of the unit. The batteries lasted for several hours of continuous recording.

During recording from a stationary position all signals (AC, DC, air-speed sensor, and voice comments) were stored on a four-channel cassette instrumentation recorder (Kyowa, model RTP-50A) and monitored on a dual-channel oscilloscope (Hameg, model HM-502). A separate audio amplifier with loudspeaker was used to monitor the AC signal acoustically. Power for the recorder, oscilloscope, and audio amplifier was provided by standard 220-V 50-Hz power outlets present in the experimental orchard, or from a 12-V car battery via a static sinus converter. During mobile recording, a stereo audio cassette tape recorder (Sony TCD-5M) was used to store the AC signal on one channel and the air-speed signal on the other channel. A switch allowed the operator to change from recording air speed to voice comments via a microphone.

2.4. Signal Processing

The examination and analysis of continuous SSR signals of relative long duration with traditional methods—oscilloscope and high-speed paper recorder—is very laborious, if not impossible. Conventional analog-to-digital computer interfaces are not suitable either, because they generally cannot handle signal trains of more than 10 sec. Therefore, we converted the spike frequency in our first SSR signals to a voltage to enable recording on a conventional chart recorder (Van der Pers and Minks 1993). However, this method does not permit reliable conversion of small spike amplitudes and does not work well for low frequencies. In 1994 a computer interface (Syntech IDAC-01) became available with associated software, Autospike™ (Syntech, Hilversum, The Netherlands), specially developed for electrophysiological signals of long duration. All recordings were analyzed using this system. The Autospike program samples the spike signal at a rate of 50,000 samples sec^{-1} with a resolution of 16 bits. Spikes are detected and classified in amplitude classes with a maximum resolution of 127 classes. Stimulus indications and comments are stored together with the data.

During replay of the recordings in the laboratory, signals were inspected on a dual-channel oscilloscope and the spike signal was monitored by a loudspeaker. Subsequently, suitable sections of recordings were processed and recorded using the Autospike program and the data were stored on disks.

2.5. Experiments in the Field

In all field experiments the sensitivity of the receptor cells and the quality of the recorded signal were evaluated by application of a test stimulus from an artificial pheromone source, a pheromone-loaded polythene dispenser, or a pheromone-impregnated rubber septum.

One series of experiments was performed in apple orchards treated with pheromone dispensers for mating disruption. Recordings were made under two conditions: (1) during exposure of the preparation to ambient air and (2) when the preparation was shielded by means of a plexiglass hood or by flushing with filtered air.

Another series of experiments was conducted in an open terrain in which the preparation was placed at distances of 7 or 10 m downwind from an array of 5–10 artificial pheromone sources. The pheromone sources consisted of a disposable 5-ml plastic syringe loaded with a filter paper onto which 1 mg of a pheromone component was pipetted. The plunger of the syringe was removed and replaced by a connector, on which a tube was mounted leading to a membrane pump via an activated carbon filter. Pump, filter, and syringe, with the needle pointing upward, were mounted on a vertical stick that could be fixed into the ground. The supply of pheromone from the syringes could be switched on and off from a distance.

In a third set of experiments the recording module was held in one hand and moved toward a pheromone source while walking. In one experiment the odor source was combined with a smoke tube (Dräger, Germany) in order to enable the operator to visually follow the plume.

2.6. Recordings in the Laboratory

To study the relation between air velocity and the receptor response in ambient air with a rather homogeneous pheromone level, field conditions were simulated in the laboratory. After contact with a pheromone receptor was established, the recording module was covered by a plexiglass hood, and pheromone-containing air was blown into the hood. A remotely controlled electric fan was also placed under the hood to create air turbulence on command. Recordings were made in still air with the fan switched off, alternated by periods of turbulent air with the fan running.

A series of coupled GC–SSR experiments measured the response of pheromone receptor cells of *A. orana* to precisely defined amounts of Z11-14:Ac. A Chrompack CP9000 gas chromatograph equipped with a heated effluent transfer line (200°C) and a CPC-5 WCOT column (10 m; 0.24 mm i.d.) was used. The injected quantity of Z11-14:Ac was split 2:1 by an adjustable micro-needle splitter valve (SGE, system OSS-2). The main fraction was conducted to the FID. The smaller fraction was transported to the preparation and mixed with an airstream (activated charcoal-filtered and humidified), flowing at 1 $msec^{-1}$ through a PTFE coated tube with an inner diameter of 7 mm. The antennal preparation was positioned at a distance of 2 mm from the outlet of the tube. After contact was made with the receptor and spontaneous activity was recorded from one or two cells, a short puff of Z11-14:Ac was administered to check the sensitivity of the cell. Subsequently, samples of 1 µl hexane containing 0.1–10 ng of Z11-14:Ac were injected. The responses of the cells to the eluting Z11-14:Ac were measured as the increase in number of spikes during the elution period compared to spontaneous activity. A fresh antennal preparation was used for each experiment.

3. Results and Discussion

3.1. Recordings in the Field

In most recordings a constant background firing of one, two, or three cells could be recorded. As in laboratory experiments, the firing frequency of the cells associated with spontaneous activity is rather different among individual antennae and receptor cells. The relatively long sensilla trichodea on the antennae of male moths contain the pheromone-sensitive receptor cells. The response of these cells was tested by administering a short puff from a pheromone dispenser or pheromone-loaded syringe to the preparation. The main pheromone component usually activates a receptor cell with a large spike amplitude (Fig. 31.2). The firing frequency is in some preparations lower than 1–2 spikes sec^{-1}, and other preparations fire at a steady rate of 3–7 spikes sec^{-1}. Some cells generated shorter or longer bursts of activity at irregular intervals.

In several experiments with *A. orana* and *A. podana* conducted in an apple

Figure 31.2. Typical field response of male *Archips podana* pheromone receptor to stimulation with pheromone dispenser containing Z11-14:Ac. The dispenser was positioned 50 cm upwind of the antennal preparation.

orchard treated for mating disruption, the preparations were shielded during 10–15 sec from ambient air movements by means of the plexiglass hood. The recordings showed an apparent decline in firing rate immediately after placement of the shield, and air movement around the preparations was suppressed almost completely (Fig. 31.3A). One can also note that the firing rate is to some degree positively correlated with the air speed. In this situation a rather high average wind speed (up to 4 m sec^{-1} combined with a dense arrangement of pheromone dispensers created a fairly homogeneous distribution of pheromone. Pheromone receptors respond proportionally to the pheromone flux rather than to concentration (Kaissling and Kramer 1990). That is why we could observe that in a homogeneous pheromone cloud the receptor responses fluctuate with wind speed.

Pheromone receptors generally produce spikes, even in the absence of a stimulus. This spontaneous activity differs between preparations, but is generally a few spikes sec^{-1} or less. In the recordings made from male pheromone receptors of *A. myopaeformis* in an orchard where pheromone traps were suspended in the trees, the firing rate of the cells was relatively low and showed a regular pattern due to low wind speeds. In this experiment a continuous recording over 260 sec was made. After 90 sec of recording, the antenna was shielded from ambient air for 80 sec by flushing the antenna with a charcoal-filtered stream of air followed by another unshielded period of 90 sec (Fig. 31.3B). The firing frequency of two types of receptor cells, one with a large and one with a small spike amplitude, was higher during exposure to ambient air, indicating that both cell types responded to pheromone molecules. The firing frequency of the large-amplitude cells changed from 3.6 spikes sec^{-1} in the first period to 2.3 spikes sec^{-1} in the shielded period, after which it recovered to 3.2 spikes sec^{-1}. For the small-amplitude cells, firing frequencies during the three periods were, respectively, 2.9, 1.1, and 2.8 spikes sec^{-1}. At these low average firing rates the decrease in spike frequency during shielding was observed because of the relatively long recording periods. The firing pattern was quite regular without distinct bursts of activity, indicating a uniform distribution of the pheromone molecules. At this low spike frequency the effect would not have been recognized in a conventional SSR experiment using a stimulus period of one or a few seconds. However, the intact male moth possessing several tens of thousands of receptors is certainly able to detect variations in firing frequency of less than one spike per second in a very short

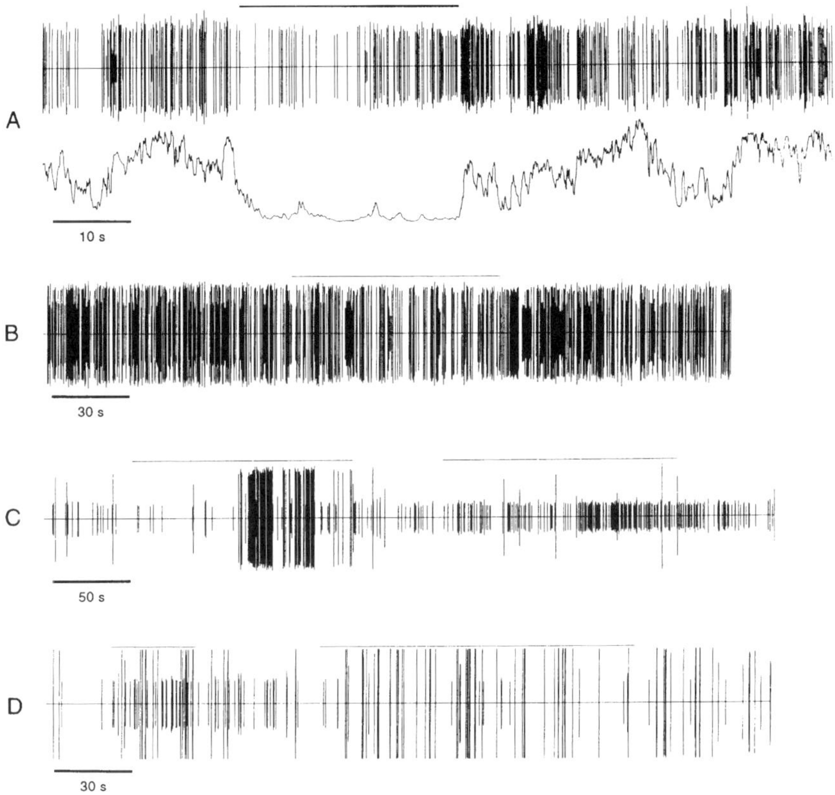

Figure 31.3. **(A)** Recording of pheromone receptor cell activity of male *Adoxophyes orana* (upper trace) and air velocity (lower trace) in an apple orchard supplied with dispensers for mating disruption. Horizontal line above upper trace: preparation shielded by means of plexiglass hood. **(B)** Recording of pheromone receptor cell activity of male *Aegeria myopaeformis* in an orchard where pheromone traps were suspended in the trees. Horizontal line above recording indicates period during which the preparation was flushed with charcoal-filtered air. **(C)** Recording of pheromone receptor cell activity of *Pandemis heparana* in an open field. Horizontal lines above recording indicate release of Z11-14:Ac and E11-14:Ac, respectively, from remotely controlled pheromone sources placed 10 m from the preparation. **(D)** Recording of pheromone receptor cell activity of *Pandemis heparana* in an open field. Horizontal lines above recording indicate release of respectively E11-14:Ac and Z11-14:Ac from remotely controlled pheromone sources placed 7 m from the preparation.

sampling period. In our experimental situation, in which the response of only one individual cell is recorded, a much longer sampling time is required to detect differences in pheromone concentration.

A completely different reaction pattern is observed during experiments in open terrain where pheromone sources were placed 10 m from the antennal preparation Figure 31.3C shows a recording from antennal receptor cells of a male *P. hepa-rana*. Low spontaneous activity of two cells was recorded before the remotely controlled pheromone sources were activated. Different pheromone components were detected during recording from sensilla-containing receptor cells specifically sensitive to one pheromone component. The cell with the large amplitude, which is known to be activated by Z11-14:Ac, started firing approximately 50 sec after the five sources loaded with Z11-14:Ac were switched on. The small-amplitude cell, responding to E11-14:Ac, slightly increased its firing rate immediately after the source loaded with E11-14:Ac was activated, but the strongest responses occurred after more than 1 min. In these recordings the firing patterns of the receptor cells were very irregular, and we could observe that the time between the onset of pheromone release and the reaction of the cells can be considerable. During this experiment, wind speed was low and its direction fluctuated. Initially the pheromone sources were placed 10 m upwind from the preparation. However, the actual position of the sources turned out to be only occasionally upwind. This meandering character of ambient air movements in open terrain may explain the long delay between onset of pheromone source and receptor response observed in this experiment. In another recording, in which the pheromone sources were placed closer to the preparation (7 m), cells started responding within a few seconds of pheromone release (Fig. 31.3D).

When pheromone sources were placed in the apple foliage, where the average wind velocity typically is low (< 0.2 m sec^{-1}) the response pattern of receptor cells were not correlated with the periods of pheromone release from remotely controlled pheromone sources. This was demonstrated in a long-lasting recording from antennal pheromone receptor cells of *A. podana* (Fig. 31.4A). Five pheromone sources were placed at distances between 5 and 7 m from the recording unit between apple trees. During recording, the pheromone sources were switched on and off for 1 min. No relation between the activity of the receptor cell (only one cell in the sensillum was firing) and the release periods of pheromone was noted. The pheromone plumes from the sources are dispersed by air turbulence and the dense vegetation between sources and recording site. However, we observed a gradual increase in spike frequency from 0.7 spikes sec^{-1} in the first 2 min to 2.3 spikes sec^{-1} in the last 2 min. This increase is probably due to (a) a steady rise in pheromone concentration in the ambient air near the pheromone sources and (b) the preparation during the course of the experiment.

Several recordings were made in which the operator approached a pheromone source while carrying the recording apparatus. Only at short distances from the source, about 20–50 cm, did the pheromone plume occasionally contact the

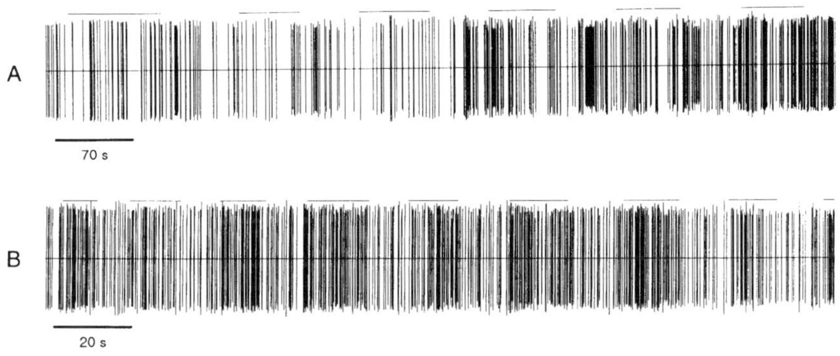

Figure 31.4. **(A)** Recording of pheromone receptor cell activity of male *Archips podana* in apple orchard. Five pheromone sources containing Z11-14:Ac were placed 5–7 m from the antennal preparation between apple trees. Horizontal lines above registration indicate periods of pheromone release. **(B)** Response pattern of Z11-14:Ac-sensitive receptor cell on male antenna of *Spodoptera exigua* recorded in the laboratory during continuous exposure to air containing Z11-14:Ac. Horizontal lines above recording indicate running periods of ventilator inducing air turbulence.

antennal preparation, as measured by the responses. When the pheromone source was combined with a smoke source and the operator followed the plume, antennal responses were recorded each time the smoke crossed the recording module. The probability of a pheromone plume emitted by a single source hitting the antenna at a distance is obviously very low, as was also observed by Baker and Haynes (1989). For this reason we used arrangements of multiple (5–10) pheromone sources.

3.2. Recordings in the Laboratory

The influence of air movements on the receptor response to pheromone molecules in a homogeneous distribution was demonstrated by the laboratory experiment, in which the preparation was placed under a plexiglass hood with a fan to create turbulence. The space within the plexiglass hood was filled with air loaded with Z11-14:Ac from a pheromone source. Figure 31.4B shows the response of a receptor cell sensitive to Z11-14:Ac on a male antenna of *S. exigua* during periods of still air alternated by periods of turbulent air. In turbulent air the firing rate of the cell was substantially higher than in periods of still air. The data from the field as well as those from the laboratory experiments demonstrate that the response to pheromone outdoors is dependent on both the concentration and the velocity of the air. This means that for estimating the average pheromone concentration in a given environment both the firing frequency of the receptors and the air speed are essential.

Figure 31.5 shows the results of 20 coupled GC–SSR experiments on antennal pheromone receptors of male *A. orana* moths. The amounts of Z11-14:Ac reaching the antennal preparation were estimated to range from 50 pg to 3.3 ng. During the elution of 50 pg Z11-14:Ac, average spike frequencies from 2.8 to 7.9 spikes sec^{-1} were recorded. Even rather high amounts of 2 ng evoke an average of only 20 spikes sec^{-1}. This relatively low sensitivity was also found in the first coupled GC–SSR experiments (van der Pers and Löfstedt 1983) showing a much higher sensitivity of pheromone receptors of *Agrotis segetum* than those of *A. orana*. The result of our GC–SSR measurements also show a considerable difference in individual recordings, which made it impossible to construct a reliable dose–response curve. In earlier SSR measurements of *A. orana* pheromone receptors (Den Otter 1977), considerably less variation in responses was found. Unfortunately, because of the observed variation and the low sensitivity of the pheromone receptors of *A. orana*, calibration of field data by comparison to GC–SSR measurements is not reliable. Calibration of the responses probably is only possible by using a reference stimulus during outdoor recordings, as applied in EAG recordings in the field (Färbert et al. this volume; Karg et al. 1990, 1992, 1994; Koch 1990; Milli and de Kramer 1989; Milli and Dickler 1992; Sauer et al. 1992). However, the references used so far only allow relative calibrations. To measure absolute pheromone concentrations in the field a new calibration technique has to be developed. This is necessary for both the SSR and the EAG method.

With the portable device described in this study we were able to conduct SSR measurements in the field. The preparation in the field of an antennal sensillum

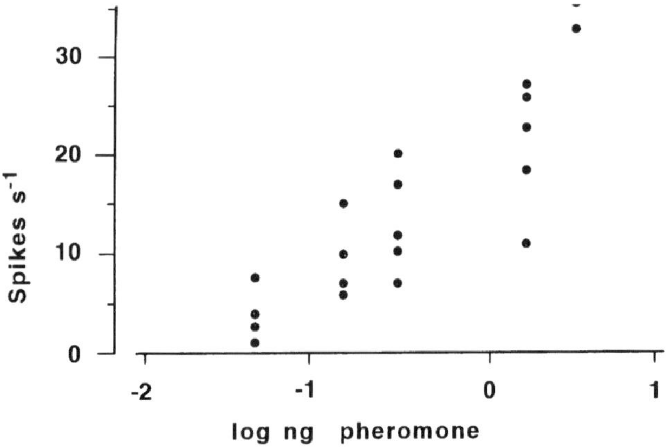

Figure 31.5. Response frequencies of pheromone receptor cells of male *Adoxophyes orana* moths during stimulation with distinct amounts of Z11-14:Ac ranging from 50 pg to 3.3 ng, recorded by coupled gas chromatography–SSR. Black dots represent spike frequency values from individual recordings.

for tip recording did not present special difficulties. However, the major obstacles to collecting sufficient data for analysis are the highly fluctuating and unpredictable conditions outdoors. In laboratory SSR configurations, care is taken to flush the antennal preparation with a well-conditioned and constant flow of filtered, humidified air. In the field, wind speed can vary within seconds between zero and several meters per second. Large variation in wind speed between experiments as well as during experiments is possible. Moreover, the instantaneous wind direction may also be highly variable, even if the prevailing wind direction is fairly constant. Due to the widely fluctuating parameters in the field it is difficult or even impossible to repeat experiments. Nevertheless, the observed fluctuations form the natural background in which pheromone communication takes place. For a better understanding of orientation disruption it is necessary to continue to collect data in the natural situation.

Acknowledgments

The authors thank L.M.H. Blommers and P. van Deventer for supplying insects and field support, and they thank S. Voerman for the supply of chemicals.

32

Mating Disruption of the Codling Moth

Albert K. Minks

1. Introduction

The codling moth, *Cydia pomonella*, is the most important pest in pome fruit orchards in Europe, North and South America, South Africa, and Australia (Chapman 1973). The codling moth assumed key pest status where control with broad-spectrum insecticides has reduced the beneficials and created secondary phytophagous mite problems. Mating disruption using sex pheromones could contribute—alone, or together with insect pathogens—to the development of an integrated control system against codling moths, because of its species-specific and environmentally safe action.

The major component of the sex pheromone of the codling moth, *C. pomonella*, was characterized as (*E,E*)-8,10-dodecadien-1-ol (E8,E10-12:OH, also called codlemone) (Roelofs et al. 1971; Beroza et al. 1974). Einhorn et al. (1984) and Arn et al. (1985) subsequently identified a number of additional components present in the gland as well as in the effluvia, of which dodecanol (12:OH) and tetradecanol (14:OH) were the most significant ones in quantitative as well in behavioral respect. These two compounds, together with E8,E10-12:OH, are used as standard mixture in some of the most widely used commercial formulations, such as the "twin dispensers" (RAK 3+4, BASF Inc.) and the "twisted ropes" (Isomate-C, Biocontrol Inc.). In small plots in an apple orchard, McDonough et al. (1994) found that an equilibrium mixture of the four isomers of E8,E10-12:OH (61% *EE*, 14% *ZE*, 20% *EZ*, and 5% *ZZ*) produced better disruption of trap catches than the standard. However, until now this finding has not been confirmed at a larger, more practical scale and consequently has not been adopted in commercial dispenser types.

The identification of E8,E10-12:OH by Roelofs et al. (1971) marked the beginning of intensive research on mating disruption, continuing up to the present.

All these efforts illustrate the great economic significance of the codling moth and the urgent need to find selective and environmentally friendly control methods. Unfortunately we must conclude that so far reports with consistently positive experiences with mating disruption of the codling moth are scarce and that the development of this technique for widespread commercial use is a painstaking process. Everywhere where tests have been conducted—in the United States, Canada, Western Europe, Australia, and New Zealand—results often showed too much variability: Most tests were successful, but some of them were not. Recently, however, several reports have appeared with encouraging, more consistent results. They will be discussed hereafter.

2. First Experiences

Two recent reviews summarize the state of affairs until 1988. In one review Vickers and Rothschild (1991) provide a complete account of the tests done with formulations that are either sprayable (such as microcapsules and chopped hollow fibers) or dispensers (such as hollow fiber tapes, rubber tubing, and laminates). They concluded that mating disruption of the codling moth is only feasible if initial numbers are low and if the pheromone blend used closely resembles that of the female moth.

In the second review, Charmillot (1990) postulated that mating disruption of the codling moth is only possible under certain stringent conditions: (a) orchards should be isolated and not smaller than 3 ha, (b) the distance to external infestation sources should be at least 100 m and the borders of the treated orchard should be protected by a double set of dispensers, (c) when comparing plots with mating disruption and with insecticide treatments a buffer zone of 40–50 m should be maintained, (d) the borders of the treated orchard should not have an irregular form, (e) the treated orchard should be homogeneously planted with small trees not more than 3 m high, and (f) initial populations should not have more than 2–3 diapausing larvae per tree, so that the tolerance threshold is not exceeded during the first year of application.

These many stipulations make it almost inconceivable that mating disruption of the codling moth would be technically or commercially feasible.

3. Recent Developments in Europe

In Switzerland (Mani et al. 1984; Charmillot and Bloesch 1987) such positive results were obtained initially with the Hercon plastic laminate releasing E8,E10-12:OH as active ingredient that in 1987 this formulation was registered provisionally for use as a mating disruptant. Unfortunately this permit was canceled 3 years later because further results were inconsistent.

In 1987, a twin-dispenser formulation was introduced: a plastic matrix containing two ampullae, one filled with the codling moth pheromone mixture and the other with the major leafroller compound (*Z*)-11-tetradecenyl acetate, present in the pheromone bouquet in most tortricine species. This new formulation, which is designed for hand application, stimulated efficacy testing in many European countries. In most tests, effects of the mating disruption treatment are assessed not only by means of reduced catches in pheromone traps, but also by measuring the copulation rate of tethered female moths, by counting infested fruit in the summer and at harvest, and by determining the number of larvae in overwintering treebands and in the spring on the trees.

Tests conducted in Germany, Switzerland, Austria, France, Belgium, and The Netherlands yielded good control in orchards with a low initial population of codling moths. The application rate used in these tests ranged from 100 to 200 g E8,E10-12:OH per hectare, depending on the number of generations of codling moth present in the area in question (Neumann et al. 1990). Three years of testing of the twin dispensers in three Dutch apple orchards (total area of 6 ha) at a density of 500 dispensers per hectare (in 1989 220 mg E8,E10-12:OH per dispenser, in 1990 370 mg, and in 1991 310 mg). Van Deventer et al. (1992) confirmed that codling moth fruit damage could be kept between 1% and 2%, which means that economically acceptable control is possible in orchards with a low population density. In more recent tests in the same orchards it became clear that a decrease of the dispenser density from the recommended 500 to 300 per hectare affected the orientation disruption to traps, resulting in higher catches of male codling moths (Van Deventer et al., 1994). Mani and Schwaller (1992) reported that in many orchards in Central Switzerland the mating disruption trials failed, mostly but not always because of high populations. In some years, particularly in 1987 and 1991, failure was due to inferior quality of the dispensers. In the lower Rhône Valley in France, results in 1988–1991 also were variable and not fully satisfactory (Audemard et al. 1992). But in the Styra region in Austria, mating disruption was reported to be a great success and therefore adopted as the standard treatment on 400 ha or 10% of the total acreage of the apple orchards in that region (U. Neumann personal communication). At the research station at Laimburg in South Tyrol (N. Italy), tests of the twin dispensers yielded good control in an orchard of 20 ha during 1989–1991 (Boscheri et al. 1992). From 1991 a serious outbreak of resistance of the codling moth against diflubenzuron (Dimilin) developed in the region, creating the prospect that most of the orchards could no longer be treated along integrated principles. Many commercial growers were "forced" to adopt mating disruption more rapidly than was anticipated, because this was the only available method for control. Moreover, they were encouraged by the positive results described by Boscheri et al. (1992) and by Waldner (1991, 1993a). Mating disruption started with an area of 110 ha in 1991, increasing to 600 ha in 1992, to more than 1200 ha in 1993, and to an acreage of approximately 2400 ha in 1994 (Waldner 1994a). In lightly infested orchards (<1% fruit damage), only mating disruption was used; but in orchards with infestations of >1% damage, mating disruption was applied in combination with one to three insecticidal sprays (Waldner 1993a). Early observations indicate positive results of this integrated method in most orchards in the region (W. Waldner personal communication).

One problem of testing the twin-dispenser formulation was great variation from year to year in initial quantities of the active ingredient and the emission characteristics, which made comparison of the results between years problematic (Charmillot and Pasquier 1992b). In the Lake Geneva region in Switzerland it was found in 1988 with an initial loading of 200 mg E8,E10-12:Ac and 50 mg 12:Ac that the half-life was 43 days. In 1989 it was 220 mg E8,E10-12:Ac and 50 mg 14:Ac with a half-life of 75 days, in 1990 and 1991 the half-life was approximately 200 days with a much heavier filling of 584 and 395 mg, respectively, per ampulla.

The twin-dispenser formulation has been registered in Germany, Austria, and Belgium, whereas registration is not required in Italy. Beside trials with the twin dispensers, tests with positive results were reported with the "Isagro" formulation, developed by the Italian company Donegani, existing of platelets of resin-treated filter paper impregnated with pheromone (Ioriatti and Dalla Serra 1992) and a polymeric device developed by TNO in The Netherlands (Batllori et al. 1992).

4. Recent Developments in North America

After some early attempts to evaluate mating disruption in field conditions in the eastern (Cardé et al. 1978) and western United States (Moffitt and Westigard 1984), tests were started in Washington State, in 1986 to determine the efficacy of the twisted-rope formulation. The ropes were 20-cm-long hollow polyethylene strips, sealed at both ends and filled with a mixture of E8,E10-12:OH, 12:OH, and 14:OH (in a ratio 10:5:1) as active ingredients. Two applications in the season, each at a rate of 1000 ropes per hectare (224 g pheromone mixture per hectare), reduced 67–92% of the codling moth infestation in various orchards between 1986 and 1991 (Howell et al. 1992). This was confirmed by other investigators who reported that treatments of 1000 ropes per hectare yielded economically acceptable fruit injury levels of at most 0.5% in the majority of the orchards. Occasional negative results were attributed to either extremely high codling moth pressure from adjacent orchards or to a too low application rate of dispensers (Gut and Brunner 1992). That the initial population is of crucial importance for the success of mating disruption of codling moth was confirmed by the same authors (Gut and Brunner 1994) in tests in pear orchards in central Washington in 1992 and 1993. In blocks with low population pressure, pheromone applications of 1000 or 626 rope dispensers per hectare provided good control (<0.1% injury) equivalent to conventionally treated blocks. At moderate levels of codling moth, fruit injury was 1.6% and 3.6% at two sites in pheromone treated blocks. Treatment of these blocks in 1993 with more dispensers (1625 respectively 2000 per hectare) provided significantly better control (<0.5% fruit injury).

Encouraging results have also been obtained in tests conducted from 1987 to 1992 using rope dispensers on ±150 ha of apple and pear orchards in British Columbia, Canada (Judd and McBrien 1994).

In Virginia, three orchards of 2 ha were each treated with 1000 dispensers per hectare, causing an almost 100% reduction of male catch in pheromone traps (Pfeiffer et al. 1993). Fruit damage in treated blocks ranged from 0% to 4.7%, compared with 0% to 1.1% in a plot with conventional chemical control and 20% to 58% in abandoned blocks. Excellent control of the codling moth was also obtained in an isolated pear orchard of 1 ha in the Sacramento valley in California with an application of 1050 rope dispensers (Barnes et al. 1992). They described four characteristics for a reliable trial on codling moth disruption, which are usually difficult to realize: (a) isolation of the treated orchard from immigrating gravid females, (b) a known background of moth population, (c) lack of interference by other control methods and (d) continuity year after year in the same location.

The rope formulation was registered for commercial use in the USA in 1991 at a rate of 1000 dispensers per hectare, two applications per season. The total acreage in use for rope dispensers in the United States expanded rapidly from 800 ha in 1991 to 2750 ha in 1992 to almost 4000 ha in 1993 (Gut and Brunner 1994), with the largest part of the treated acreage in Washington and in California. However, in recent years it was reported that at the registered rate, results have been mixed and supplementary spraying appeared to be unavoidable in selected parts of some orchards. Howell (1992) concluded that the emission rate of pheromone was insufficient in cool periods and recommended application of more dispensers per hectare, or a redesign of the dispenser, so that more pheromone would be emitted at lower temperatures.

5. Concluding Remarks

Despite considerable field evaluations, routine, large-scale commercial use of the mating disruption technique against the codling moth is not yet possible, although progress is encouraging. So far the required technical reliability of the technique has not been achieved. Serious limitations are the absolutely necessary low initial populations in the area to be treated and a high degree of isolation, conditions which are difficult to realize in most fruit growing regions. It also appears that the chances of successful mating disruption are higher in regions with moderate temperature regimes and with only one or two codling moth generations per season. In this respect it is important that the producers of the available formulations guarantee that the pheromone emission rates of their products are sufficient even on cool nights.

Acknowledgment

I thank P.-J. Charmillot for his comments on this chapter.

33

Mating Disruption in European Vineyards

Heinrich Arn and Friedrich Louis

1. Introduction

The tortricids *Lobesia botrana* and *Eupoecilia ambiguella* are the most important pests of European vineyards. Larvae feed on the grape berries, reducing their yield and rendering them susceptible to gray mold, *Botrytis cinerea*. Generally, two generations are observed, but in *Lobesia botrana* under favorable conditions there may be even three or four (Bovey 1979; Hillebrand et al. 1990). The two species coexist over a wide area; *E. ambiguella*, with a requirement for high humidity, is absent from the dry regions of the South, while *L. botrana*, which needs a warmer climate, has rarely been reported from the cooler regions of the North. The local importance of the two species has recently been monitored in Austria (Höbaus 1988). Another tortricid, *Argyrotaenia pulchellana*, occasionally causes damage similar to the species mentioned, while *Sparganothis pilleriana*, a leaf feeder, was a devastating pest during the last century and has retained local importance in warm climates.

The use of pheromones to control pest insects was proposed by Götz (1941) in his pioneering work on sex attraction in *E. ambiguella* and *L. botrana*. It was not until 30 years later that the chemical structures of pheromones of vineyard pests became known, first of *L. botrana* (Roelofs et al. 1973; Buser et al. 1974) and later *E. ambiguella* (Arn et al. 1976; Saglio et al. 1977a), *S. pilleriana* (Saglio et al. 1977b; Guerin et al. 1986), *A. pulchellana* (Frérot et al. 1984), and the geometrid *Peribatodes rhomboidaria* (Buser et al. 1985). In most of these species, additional components were later identified in the female secretion and the attractant blends were modified accordingly; for reviews see Arn et al. (1988, 1992). Traps baited with sex attractants are widely used for insect detection and monitoring.

Apart from a brief report on *A. pulchellana* (Brechbühler and Meyer 1992), mating disruption in vineyards has been directed against *E. ambiguella* and *L.*

botrana. The active ingredients for the two species are (*Z*)-9-dodecenyl acetate and (*E*)-7,(*Z*)-9-dodecadienyl acetate, respectively. An earlier review on mating disruption in the two species was given by Descoins (1990).

2. Control of *Eupoecilia ambiguella* by Mating Disruption

Mating disruption of *E. ambiguella* was demonstrated to be feasible in field tests in Switzerland (Rauscher and Arn 1979; Arn et al. 1981) and was then developed on a large scale in Germany (Winkelmann-Vogt 1986; Vogt et al. 1993). The technique has received registration by BASF in Germany, Switzerland, and Austria. Since then, the area treated has been on the rise. In 1993, an estimated 14,000 of a total of 100,000 ha of German vineyards, situated in Baden-Württemberg, Bayern, Hessen, and Rheinland-Pfalz, were treated with pheromones. In six eastern cantons of Switzerland, where *E. ambiguella* is often the only tortricid pest found on grapes, mating disruption has become the most important method for pest control, accounting for 30% of the 2000 ha of vineyards. In Austria, pheromones have been used on 300 ha of grapes located in Niederösterreich.

Dispensers are placed each second or third row, 500 per hectare, at a rate of 170 g of active ingredient per hectare. BASF recommends that vineyards should be at least 1 ha in size if isolated and 10 ha if located within a larger grape-growing area. Inclusion of an additional 30 to 50-m strip around the area to be protected is also recommended. Depending on the case, this area may also be treated conventionally. As vineyards are often fragmented, treatments are mostly in the hands of grower organizations which also monitor further developments throughout the season. If the damage gets out of hand, an insecticide treatment is recommended. Such measures, however, have rarely been reported to be necessary for *E. ambiguella*.

3. Mating Disruption in *Lobesia botrana*

First field tests demonstrating feasibility of mating disruption in *L. botrana* were carried out in the Bordeaux area (Roehrich et al. 1979), but large-scale production of pheromone was not achieved until 10 years later. Numerous reports on field tests using BASF dispensers leave no doubt that mating disruption can be effective (Neumann et al. 1993), but that it also sometimes fails unexpectedly (Arias et al. 1992; Perez Marin 1992; Borgo 1992; Charmillot and Pasquier 1992a; Schmid and Raboud 1992; Brechbühler and Meyer 1992). On the other hand, detailed monitoring of pest populations in the course of integrated production in recent years has shown that pest control in general is not always as effective as previously assumed (Lorenz and Louis 1993; Louis 1994).

High population densities have been deemed responsible for failures of mating disruption (Neumann 1992). Another hypothesis states that population pressure leads to dispersal (Neumann et al. 1993). As Schmitz (1992) demonstrated, doubling the population density by releasing *L. botrana* into a disruption plot indeed led to a twofold increase in attack; more surprising was the observation that the attack was five times higher when the insects were released from one central point than when they were evenly distributed throughout the plot. Based on population sampling with field cages, Feldhege et al. (1995) specify a critical population density for *L. botrana* with four females and four males per 10 m^2.

Recent recommendations, resulting from five years of experience in the Bordeaux area, are to knock down the first generation of *L. botrana* with an insecticide and to continue with pheromone for the current and the following season (Stockel et al. 1994). Evidently, a close watch on effectiveness is necessary to be able to intervene with an insecticide if the situation begins to get out of hand. Local outbreaks have mostly been observed at the borders of the pheromone-treated area.

With the registration of mating disruption of *L. botrana* in Germany in 1994, twin ampoules with separate chambers for the *E. ambiguella* and *L. botrana* pheromones are used. The amount of (*E*)-7,(*Z*)-9-dodecadienyl acetate (E7Z9-12:Ac) is 120 g per hectare in 500 ampoules.

In spite of these achievements, many questions connected with mating disruption in vineyards are still open. Improvements can still be expected in the controlled-release formulation. Apart from occasional determinations of weight loss or pheromone remaining in the dispensers, little quantitative information on airborne pheromone during these tests is available. The measurements of airborne pheromone (Sauer et al. 1992; Färbert et al. this volume; Van der Pers and Minks this volume) and of actual release rates of dispensers (Pop et al. 1993) is expected to provide more accurate information on pheromone concentration and dispersal in the vineyard.

4. Effects of Mating Disruption on Vineyard Fauna

Simultaneously with the introduction of mating disruption against *E. ambiguella*, a program was initiated in Neustadt an der Weinstraße to investigate possible effects of the technique on beneficial arthropods. In laboratory studies, no negative effect of high doses of (*Z*)-9-dodecenyl acetate (Z9-12:Ac) on the predatory mite, *Typhlodromus pyri* (Hoos 1989), or the lacewing, *Chrysoperla carnea*, could be found. On the other hand, a slight increase of mortality was found in the European earwig, *Forficula auricularia*, and the wolf spider, *Lycosa agrestis*: In the latter case, an increased level of intraspecific aggression was observed (Schirra et al. 1991).

In the field, 3 years of mating disruption against *E. ambiguella* led to a continuous decrease in the population of the European red mite, apparently caused by a recovery of *Typhlodromus pyri* (Fig. 33.1). Sampling of the arthropod fauna of vineyards revealed an effect of the pheromone treatment on spider populations on grapevine trunks (Louis and Schirra 1992). More spider species and, on average, twice as many individuals were found in pheromone-treated plots than

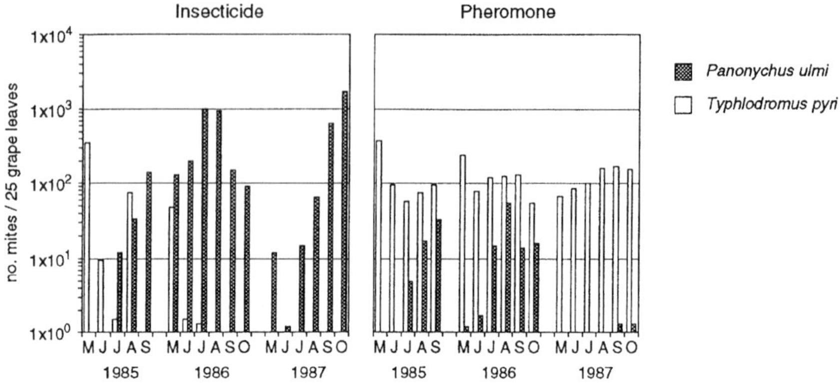

Figure 33.1. Population densities of the European red mite, *Panonychus ulmi*, and its predator, *Typhlodromus pyri*, over three growing seasons of *Eupoecilia ambiguella* control with deltamethrine (**left**) and Z-9-dodecenyl acetate (**right**). [From Hoos (1989).]

in plots treated with parathion. The most striking effect was observed for the jumping spiders *Dictyna uncinata*, *Synageles venator*, and *Marpissa muscosa*. The only pest occurring in the pheromone-treated vineyards was *S. pilleriana*, but its populations remained well below the economic threshold (Schirra et al. 1991).

5. Economic Considerations

Mating disruption campaigns require slightly more personnel than conventional chemical or even biological control techniques. For many users, this inconvenience is more than offset by the fact that once the dispensers are deployed, no further treatment schedules need to be observed. Considering the cost of material, however, mating disruption is the most expensive pest control technique in viticulture. In 1995, mating disruption of *E. ambiguella* and *L. botrana* cost DM 545 per hectare for the material alone, which is two to four times the price of a conventional insecticide. Some regional governments in Germany and Switzerland have paid subsidies to growers using pheromones. The state of Rheinland-Pfalz, where most of the German wine is grown, contributed DM 180 per hectare in 1995.

Mating disruption could be made cheaper by reducing the cost of the pheromone or improving the economy of its use. Compared with early experiments, application rates have been increased, perhaps beyond those necessary. Dispensers used for disruption in general emit most of the material during the hot hours of the day when the insects are not responsive; efforts to develop release devices with a low temperature gradient or which emit the chemical only in the dark would be worthwhile. Saving cost on the chemical itself could be expected through the use of an impure product or an analog; bioassays for rapid optimization in this field still await development.

6. Notes on the Mechanisms of Disruption

"Confusing" the males by evaporating relatively large amounts of pheromone was a relatively simple concept 30 years ago (Babson 1963; Wright 1963). What is now called "mating disruption" appears to be the result of various mechanisms (Cardé 1990; Sanders this volume). For the practical-minded, the main difference between what is called "competition" or "false-trail-following" and other mechanisms (e.g., sensory fatigue or camouflage) lies in the composition of the active ingredient: While the first case requires an attractive product (e.g., the blend produced by the female), the second conceivably allows the use of off-blends, single components, or impure materials. As long as we do not know the importance of each mechanism, it is difficult to predict which chemical is best suited to disrupt mating.

E. ambiguella females produce Z9-12:Ac at high isomeric purity, and proportions of more than 0.1% *E* isomer reduce the attractivity of a synthetic product (Arn et al. 1986). Technical Z9-12:Ac contains a few percent of *E* isomer and is not attractive to *E. ambiguella* males. We know that mating disruption operates very well with such a product; therefore, false-trail-following cannot be a critical factor in this species. In *Cydia nigricana*, a blend of pheromone plus attraction inhibitors has been suggested to be a more efficient disruptant than pheromone alone (Bengtsson et al. 1994).

In *L. botrana*, isomeric impurities do not seem to affect attractancy (Ideses et al. 1982); and synthetic E7Z9-12:Ac, which typically contains 10–20% *E,E* and some of the other two geometric isomers, is very attractive to males. Schmitz (1992) observed that BASF dispensers made for mating disruption attract male moths in the field. This raises the possibility that occasional failures in disruption of this species could have been caused by an attraction of males into the treated area. Wind-tunnel experiments show that *L. botrana* and *E. ambiguella* males pre-exposed to pheromone respond less to calling females; this gives support to the idea that mechanisms other than false-trail-following play a role (Arn 1979; Arn and Rauscher 1987). On the other hand, Z9-12:Ac and (*E*)-7,(*Z*)-9-dodecadien-1-ol, components produced by *Lobesia botrana* females, augment male attraction when added to the main component, and some unknown factors improve the straightness of the flight tracks (Witzgall and Arn 1990; Witzgall this volume). These chemicals might be of potential interest in mating disruption.

7. Outlook

The acceptance of mating disruption by grape growers in certain areas of Germany, Austria, and Switzerland is a success story similar to that reported for the oriental fruit moth or the pink bollworm. The technique has become popular not only due to a growing concern about the environment, but also because of t value of the end product. In these regions, local wine is precious; people drink it with measure, love to talk about it, and are willing to pay a good price for it. So wine produced with pheromones and other "green" plant protection agents such as *Bacillus thuringiensis* can share a quite important specialty market.

As long as mating disruption is limited to *E. ambiguella*, it cannot grow much beyond the acreage treated now, especially since *L. botrana* has recently spread into vineyards in which it was not previously reported (Louis 1994). With the availability of a pheromone formulation against *L. botrana*, however, mating disruption has a potential use in the prestigious wine production areas further south—for example, the "appellation contrôlée" wines of France. Its limitations with respect to population density and perhaps other factors need to be established. In many areas, the cost may be prohibitive, at least as long as conventional insecticides remain effective and safe. That this can dramatically change with the buildup of resistance has been exemplified in the German-speaking Südtirol province of Northern Italy, where mating disruption of the codling moth is the order of the day (Waldner 1993b, 1994b).

34

Successful Area-Wide Program to Control Pink Bollworm by Mating Disruption

Robert T. Staten, Osama El-Lissy, and Larry Antilla

1. Introduction

This review deals with the pragmatic aspects of pheromone use for the control of the pink bollworm (PBW), *Pectinophora gossypiella*. We will provide a brief historical summary of the principal pheromone formulations used in the United States, defining them within two categories based largely on amounts of active ingredient used. We will summarize results from two area-wide management programs. Within these programs, population suppression has been maximized. It has exceeded the degree of suppression obtained when pheromone has been used on limited numbers of fields within a large geographically defined growing area. We will then briefly speculate on long-term implication of these programs and where they fit within the overall theory and practice of integrated pest management.

2. Historical Summary of Formulation Use

Pheromone formulations used in suppression of PBW within the United States and throughout the world can be broadly classified in two categories. Low-rate formulations require multiple applications to suppress more than one generation per growing season. These formulations use small dosages of active ingredient for each application. Total annual application of pheromone seldom exceeds 20 g/ha. High-rate formulations are normally applied once early in the season and impact at least two generations within a year. This single application normally contains 78 g/h. The impact of a high-rate system restricts population development through the majority of the season.

2.1. Low Rate Formulations

Low rate formulations represent those in longest usage and with the most diversity of formulation types. These formulations in PBW require repeated applications of material at rates from 0.75 to 10 g of a synthetic pheromone per hectare. Effectiveness, as measured by drastic reduction of male capture in sentinel, pheromone-baited traps, usually lasts 8–16 days before an additional application of pheromone or another control measure is needed. These first-generation systems required specialized equipment for application of either laminated plastic chips or hollow fibers with an appropriate adhesive to attach the formulation to the cotton canopy. These formulations yielded about 8000–20,000 point sources per hectare. The principal formulation of this type (as of 1994) is a hollow fiber produced by Ecogen and is used with a polybutene adhesive (Brooks et al. 1979). A similar formulation of laminated layers of plastic cut into small 0.16-cm^2 flakes was available in the United States from Hercon until 1990 (Kydonieus and Beroza 1978). These two systems were often used with a minute amount of insecticide in their adhesive (U.S. Patent 4671010, Staten and Conlee). Under this concept patent, this usage is termed "attract and kill." The extent of use of these two systems was reviewed by Baker et al. (1990).

These two earliest registered systems have been followed by sprayable formulations using the same general levels of active ingredient (AI). These systems can be applied using conventional ground or aircraft spray equipment. Some of the formulations tested by our Phoenix Methods Development Lab have been plagued by weather problems. Those with larger particle or capsule size are easily dislodged from plant surfaces by wind or rain. Those contemplating use of sprayable formulations would be well advised to check these limitations under local climate conditions. All systems are similar in that repeated applications are needed for continuous population suppression.

Management of a pest population with low rate pheromone formulations is much more demanding than using conventional insecticides. Timing is extremely important, because no low rate formulation as it is commercially applied suppresses mating for more than 8–16 days. The actual time frame in which disruption is effective is modulated by temperature and climatic conditions. Thus, monitoring of pest populations via the level of captures in pheromone traps is essential and, from a Pest Control Advisor's perspective, is labor-intensive. The primary mode of action is not to kill the insect, but to keep a male from finding a female and mating with it. Thus, the higher the density of a pest population, the more likely a male is to encounter a female. Field evidence shows that the efficacy of these formulations with PBW is inversely density-dependent. Because of this relationship, these materials are most effective against the low-density spring moth population from diapausing larvae and the first spring generation of moths. Overwintering larvae are subject to mortality because they are at an energy and/or nutritional deficit. The next generation is found only in flower buds before the plant fruiting cycle has provided PBW with its optimum host form, the boll (Brazzel and Martin 1957). Unfortunately, growers typically are reluctant to expend resources before populations have reached high and damaging levels.

Low rate systems were used in an early area-wide program in the Imperial Valley of California in 1981 (Baker et al. 1990). This program had considerable impact, but it was not continued in 1982. An important lesson to be learned from this effort was that for an area-wide program to supplant a conventional insecticide program, it had to provide enhanced control.

2.2. High-Rate Systems

In 1984, the use of long-life, high-rate AI formulations marked a radical departure in pheromone control of the PBW (Flint et al. 1985). The most common high-rate formulation is the PBW Rope, sold in the United States by Mitsubishi. The PBW Rope was a departure from all other commercial disruption systems used for this insect in several respects:

1. Only one application with a high level of AI (78 g/h) is applied.
2. Only 1000 dispensers per hectare are hand-placed in the field.
3. Mating disruption is effective for at least 50–60 days.
4. It is, from our observations, more efficacious than low rate systems against higher level populations.
5. There is only one decision period. It must be applied early in the season when cotton is small and just before the first hostable flower buds are present.
6. Cost of the treatment is perceived as high.

This system was first tested in small field plots in Arizona in 1984 by Flint and in large-scale field tests in 1985 in the Imperial Valley of California and the Mexicali Valley of Baja California, Mexico (Staten et al. 1987). During the mid-1980s, pesticide usage in the Imperial Valley was extensive. All growers were using a long season cotton growing system, which had maximum host potential for the PBW. Population pressure was extreme, and suppression was critical. Under these conditions, the PBW Rope was successful (Table 34.1).

3. Coachella Valley Trial

After this first year of replicated field testing, the Phoenix Methods Lab initiated an area-wide test in a small (350–470 ha) semi-isolated area of Southern California, the Coachella Valley. This effort represented the first integration of sterile insect technology with pheromone disruption for control of an established area-wide population (Staten et al. 1993). In the Coachella Valley, cotton is a secondary

Table 34.1. Comparison of boll damage and insecticide usage in Imperial and Mexicali Valley fields treated in 1985 with high-rate PBW Rope, low-rate pheromones, and only conventional insecticides

	Larvae/100 bolls		
Treatment	August	September	Number of insecticide treatments
		Imperial Valley[a]	
Insecticide only	0.85[a]	0.88[a]	11.4[a]
Low-rate pheromone	0.90[a]	2.1[a]	10.4[a]
PBW Rope	0.32[a]	0.39[a]	6.6[b]
		Mexicali Valley[b]	
Insecticide	1.72[a]	1.55[b]	4.9[a]
PBW Rope	0.7[b]	0.72[b]	2.9[b]

Source: Staten et al. (1987).

[a]There were seven PBW Rope-treated fields, eight conventional-insecticide-treated fields, and eight low-rate pheromone-treated fields (NoMate PBW or Disrupt). Low-rate pheromone-treated fields were fields in which local Pest Control advisors used one to four applications of the low-rate pheromone systems in either a fiber form or a flake form. Means in same column having no letters in common are significantly different according to ANOVA followed by Duncan's multiple range test $p = 0.01$).

[b]There were 16 pBW Rope-treated fields and 14 conventional insecticide-treated fields. Means in same column having no letters in common are significantly different according to Student's t test ($p = 0.01$). Within this test, each plot was a full field. Although the Imperial Valley of California is separated from the Mexicali Valley by only a fence, the economic structure of cotton growing is completely different. The rope was successful in both systems, in that conventional pesticides were eliminated in an amount at least equivalent in value to the cost of the rope. Population suppression was equal to or better than a conventional program.

crop to dates, grapes, citrus, and vegetables. The cotton growing season there is virtually identical to the adjacent Imperial Valley, approximately 40 miles to the southeast, but these two areas are separated by a brackish water lake, the Salton Sea. Records were obtained on pesticide use in 1985 in all fields before we began work in 1986. In this trial, a high-rate formulation (PBW Rope) was used in all fields in 1986 without sterile insects. In 1987–1989 this formulation was only used in fields in which a low ratio of sterile PBW to native PBW was apparent at first available flower bud. As in all our large-scale field tests, when a Pest Control Advisor deemed it appropriate, a developing PBW larval population or other insect pest was treated with insecticides. The key to this and all other pheromone strategies for the PBW is to hold populations *below* a level which is a normal economic threshold for conventional insecticides. Table 34.2 shows the pheromone and sterile insect treatment strategies and the required conventional pesticide usage before we started our project in 1985 and in subsequent years of the area-wide trial.

Conventional insecticides were reduced drastically from a grower system in which conventional treatments average from 7.3 per field in 1985 to no treatments

Table 34.2. Annual summary of control strategies and conventional insecticide use patterns for premanagement and area-wide management systems in the Coachella Valley

Year	Treatment system	Fields treated with pheromone	Mean number of conventional insecticides/field	Number of fields treated with insecticides	Total conventional insecticide treatments
1985	Insecticide only (pretrial data)		7.3	56/57	414
1986	Pheromone and insecticides	31/31	1.8	17/31	56
1987	Sterile insects, pheromone and insecticides	17/21	1.0	7/27	28
1988	Sterile insects and pheromone	4/31	0	9/31	0
1989	Sterile insects, pheromone and insecticides	3/23	1.9	18/23	44

in 1988 (Table 34.2). Even in 1986 and 1987, the fields requiring treatments typically were those closest to the Imperial Valley. Insecticide treatments began after moths were collected in traps placed in the 40 miles between the Imperial Valley and the Coachella Valley. The PBW moth becomes more mobile as the season progresses, with dispersal flights occurring after irrigation of cotton is stopped (Stern and Savacherean 1978). This occurred in 1989 when the adjacent Imperial Valley stopped irrigation and applied a defoliant to cotton a full month earlier than the Coachella Valley. Within a few days of the start of defoliation in the Imperial Valley, massive moth flights were found in the desert between the two valleys. This appeared to be a key factor in the rise in population at the end of the season. In all fields, population levels were the lowest of any areas in the Colorado River Basin. Until late 1989, PBW populations appeared to be a constant downward cycle in our test area. The most important features of the Coachella Trial were that it dealt with all fields in the area and demonstrated that the high-rate pheromone system alone or in combination with sterile insect technology could regulate the PBW over a large area. The trial also represented the beginning of a new approach: integration of sterile insect and pheromone systems in area-wide management. In 1994, a trial of this nature was initiated in a large project in the Imperial Valley. It will be necessary to balance and refine the two technologies based on costs.

4. Parker Management Trial

4.1. History and Control Strategy

In late 1989, key personnel of the Arizona Cotton Research and Protection Council considered improving management of the PBW. A very successful boll

Table 34.3. Boll infestation data from the Coachella Valley Pink Bollworm Management Study, showing total larvae per sample and number of bolls sampled

	1986		1987		1988		1989	
Sample week	Bolls sampled	Total larvae	Bolls sampled	Total larvae	Bolls sampled	Total larvae	Bolls sampled	Total larvae
1	50	3			370	0		
2	500	3	500	0	726	0	560	0
3	960	2	1700	0	1200	0	720	0
4	1180	0.9	2400	7	1600	1	960	0
5	1520	1	2400	33	2000	0	1520	0
6	1520	3	2500	34	2160	0	1600	1
7	1600	1	2700	1	2320	0	1760	0
8	1520	1	2700	6	2480	5	1564	18
9	1600	1	2700	5	2480	5	1860	39
10	1760	43	1350	25	2480	2	1960	39
11	2200	138	1350	41	2480	0	1240	47

weevil eradication program was in its final stages. Based on successful strategies in previous projects, a strategy for management with pheromones on an area-wide basis was developed. This grower organization elected to set up a match-funded program in which the cotton grower organization state-wide would initially provide one-half the cost of project management and treatment. As the project turned out to be successful, local growers have taken a progressively larger share of the costs. By 1994, all but supervisory costs were assumed by the local grower community. The USDA provided technical assistance. The Parker Valley (Fig. 34.1), in the north central area of the Colorado River Desert cotton growing region, was chosen because at 11,250 ha it was of significant size. The PBW was a major problem, and pesticide usage was extensive.

The relatively simple goals of the Parker Project set out in 1990 became more stringent with each year. They were as follows: (a) in 1990, maintenance of the

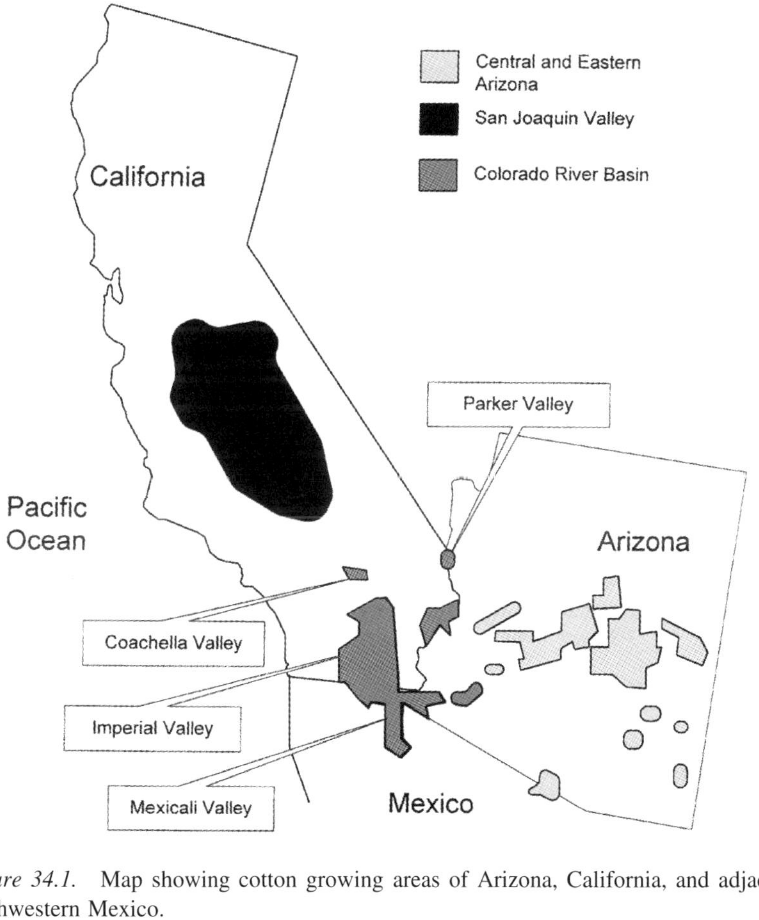

Figure 34.1. Map showing cotton growing areas of Arizona, California, and adjacent northwestern Mexico.

PBW at low enough levels to avoid nonprogram-applied insecticides in more than 10% of the fields before August 15, (b) in 1992, to maintain populations at levels below this same threshold until September 1, and (c) in 1993, to completely control all populations throughout the growing season without the need for grower-applied insecticide. The treatment strategy for use by the program was not exclusively pheromone, but one in which pheromones were used in virtually all treatments. During the first years of the program, a dual application (pheromone with an insecticide overspray) was used for the first treatment and at times of hazard or perceived risk. Dual applications were not employed after 1993. Insecticides without pheromone were used very sparingly—for example, when a rope treatment was thought to be in jeopardy. No insecticides without pheromone were applied for the PBW after 1992 (see Table 34.4).

Although it is beyond the scope of this chapter to detail the rationale for each treatment, the PBW Rope generally was applied in the first cotton to have flower buds so long as the stand and plant height at the six-leaf stage was uniform enough for good rope distribution. It was also applied selectively to fields with highest spring trap capture and fields which were near human activity. All other low-rate pheromone treatments used the hollow fiber formulation.

4.2. Parker Program Results

Results are expressed as boll infestation data in larvae per 100 bolls. Table 34.5 gives boll infestation data for 45 evaluation fields selected randomly from the yearly average of 650 fields. Prior to the program in 1989, virtually no pheromone as a disruptant was used by growers. In spite of intensive chemical treatments, larval populations were usually above 25/100 bolls (weeks of August 13, 20, and 27 and September 3). In subsequent years, there was an incremental decline in boll infestations. Of particular significance is the fact that no larvae were found during the entire season in 1993 from a season-long sample total of 22,852 bolls.

Trap data show similar declines. Critical for comparison are the catches in late April and early May from the overwhelming generation before pheromone is applied. Lower captures from this generation are found in each succeeding year (Fig. 34.2). In the first and second years of this program, 1990 and 1991, no pheromone applications were made after July 30. In subsequent years, phero-

Table 34.4. Treatments used in Parker Valley from 1990 to 1993

	Hectares treated			
Treatment system	1990	1991	1992	1993
Rope (high-rate system)	4,788	5,070	3,864	2,756
PBW fiber	12,339	12,628	24,914	16,588
Fiber with insecticide overspray	13,346	11,088	10,113	1,089
Insecticide	0	340	2,456	0

Table 34.5. Summary of boll data from Parker Valley showing total larvae per 100 bolls and number of bolls sampled 1989 through 1993[a]

	1989		1990		1991		1992		1993	
*Week[a] of	Total bolls sampled	Larvae/ 100 bolls	Total bolls sampled	Larvae/ 100 bolls	Total bolls sampled	Larvae/ 100 bolls	Total bolls sampled	Larvae/ 100 bolls	Total bolls sampled	Larvae/ 100 bolls
7/09	—	—	2910	0.3	—	—	2147	0.19	1837	0
7/16	—	—	2885	0.6	3332	0.6	3161	0.09	—	—
7/23	2680	3.6	3123	1.4	3378	0.03	3259	0.95	—	—
7/30	2758	7.7	3037	2.7	3320	0.09	3440	0.61	4800	0
8/06	3107	17.9	3167	1.5	3591	0.03	2287	0	3049	0
8/13	3080	25.9	3074	5.9	3614	0.03	3063	2.45	3280	0
8/27	3033	36.4	3537	20.6	3661	1.6	2519	1.19	3143	0
9/03	3067	34.5	2932	10.9	3499	1.9	2930	1.60	2973	0
9/10	3095	21.6	3338	10.4	3725	3.7	2797	1.78	2806	0
9/17	3027	28.4	3627	33.3	3585	6.6	—	—	964	0

[a]All data are from 45 fields randomly selected from ±650 fields before season started.

[b]After 1989 all sample weeks were adjusted by heat units for climatically similar weeks.

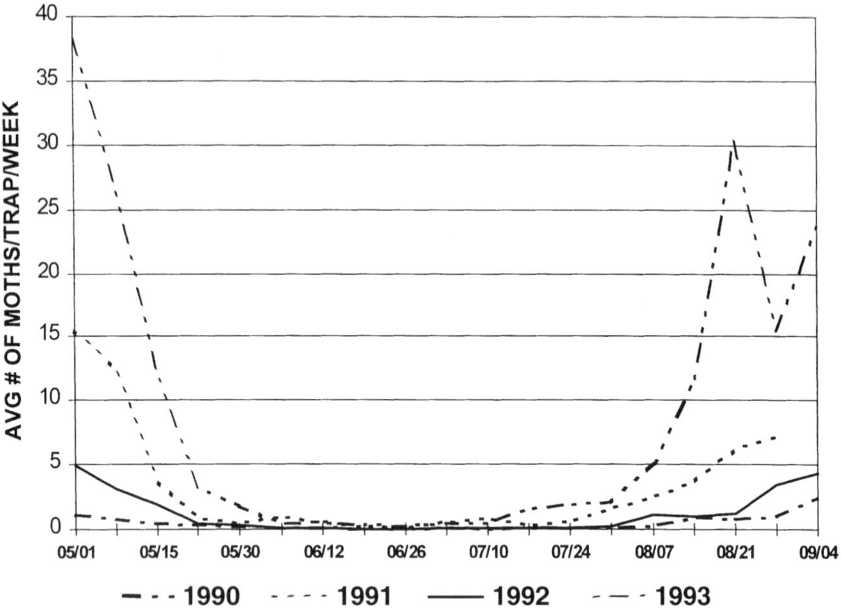

Figure 34.2. Moth capture data from all program monitoring traps in Parker pink bollworm program over a 4-year period. [Modified from Brooks et al. (1979).]

mone was applied until September 1. Each year moth capture in September declined. It is important to realize that to properly manage this program, there is one trap for every 4 ha, giving a total of 2200–2700 traps. These figures are, therefore, from a very large sample. There is a high degree of certainty that this overall population decline is real.

Because these programs are designed to deal with an entire growing area of cotton, an experimental check is not possible. The two programs discussed are constructed with the hypothesis that pheromone systems or pheromone and sterile insect systems use soft technology which can be damaged by immigration. We use the term *soft technology* because control measures do not actually have significant mortality components and result in reduction of total pesticide usage. We expect improved performance on larger areas of contiguous treatment. The PBW is particularly mobile late in the season and as it emerges from overwintering sites in the spring. Our only comparisons are from other survey projects. Figure 34.3 illustrates population trends from a state-wide trapping program conducted each spring by The Arizona Cotton Research and Protection Council. This survey is run in all nonprogram areas throughout the period that moth emergence is expected from overwintering populations. It represents data from one trap on every fourth field throughout Arizona, or 1400 to 2115 traps each year. Figure

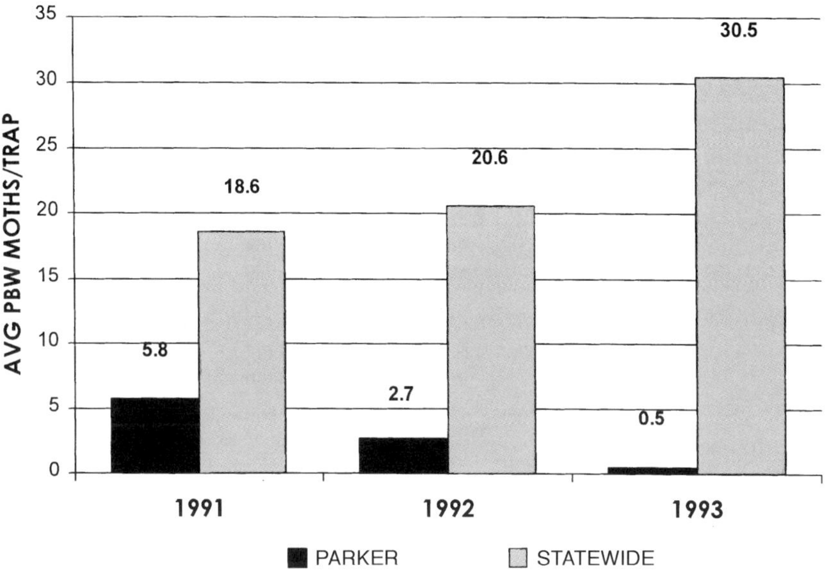

Figure 34.3. Comparison of state-wide spring trapping program date with spring tapping data from Parker for April 4–May 5.

34.3 illustrates an escalating population each spring from April 4 to May 5 in all nonprogram areas combined from 1991 to 1993. The opposite is true of Parker.

An accurate boll survey is a very laborious task. In 1992, the program management personnel of Arizona conducted boll surveys in 42 fields in Central Arizona. These data are compared to data collected in the same fashion in Parker in Table 34.6. In 1990, program personnel also collected data in the adjacent Palo Verde

Table 34.6. Boll infestation data from Central Arizona compared to the data from the Parker Program in 1992

	Parker (pheromone IPM)			Central Arizona (conventional treatments)		
Date	Bolls	Larvae	Larvae/100 bolls	Bolls	Larvae	Larvae/100 bolls
7/09	2147	4	0.19	3300	37	1.12
7/16	3161	3	0.09	3400	27	0.79
7/23	3259	31	0.95	2600	59	2.27
7/30	3440	21	0.61	2700	209	2.27
8/06	3287	0	0.00	3300	321	9.73
8/13	3461	0	0.00	3400	337	9.91
8/30	3063	75	2.45	2700	248	9.19
8/27	2519	30	1.19	2800	401	14.32
9/03	2930	47	1.60	—	—	—
9/10	2797	50	1.78	—	—	—

Table 34.7. Boll infestation in the Palo Verde Valley and adjacent Parker Valley in 1991

	Parker (Pheromone IPM)		Palo Verde Valley (Conventional treatments)	
Date	Bolls	Larvae/100 bolls	Bolls	Larvae/100 bolls
7/9	—	—	—	—
7/16	3332	0.06	1204	0.09
7/23	3378	0.03	1138	0.08
7/30	3320	0.09	920	1.74
8/6	3591	0.03	1144	3.1
8/13	3614	0.03	1223	1.1
8/27	3661	1.6	1142	4.5
9/3	3499	1.9	1150	12.9
9/10	3725	3.7	1204	20.8
9/17	3585	6.6	1178	33.3

Valley of California using Parker sampling procedures. They sampled 15 randomly selected fields in Palo Verde to compare infestation with randomly selected fields in Parker. The differences in Table 34.7 are likewise considerable. In the scientific world, as in a courtroom, these data would be considered anecdotal, or circumstantial. Our trials were not run as replicated or controlled experiments. The variables affecting Parker were not the same as those affecting the adjacent Palo Verde Valley. Diversity is even greater comparing Parker with the rest of Arizona. Numerical differences in boll infestation are, however, great and numerically high. These differences occurred even though virtually all growers applied control measures for the PBW in the areas compared to Parker.

5. Discussion

In our view, not universally shared by all of our colleagues, use of pheromones has been a major advance in integrated pest management for cotton insect control. The rest of our discussion will also be controversial. Throughout the area of severe general infestation, pest management has previously been conducted with a field-by-field mentality. Profits are continuously limited. Control costs are relatively high year after year. It is an attempt by growers to match their insect control with other resource utilization. Farmers vary their resource utilization drastically, depending on a wide array of variables: water use and quality, land ownership, availability of equipment, and the price of cotton. Some growers will tolerate significant populations on leased or rented ground. However, PBW populations are not restricted to single fields in their influence. Where field-by-field management occurs, population reservoirs are created. The overall general equilibrium position remains high.

The first significant departure from this in the U.S. deserts was in an area-wide pheromone project in the Imperial Valley in 1981. In an area-wide paradigm, using soft technology that can be highly effective, costs from year one to subsequent maintenance years must be drastically lowered if it is to be accepted by the cotton grower. In the Imperial Valley, this was not true. This has proven to be true in Parker. Total average program expenditures per hectare of cotton for PBW were $106.00 in 1990, $106.00 in 1991, $136.00 in 1992, and $56.00 in 1993. In the first year, only the first three generations were targeted by the program. By 1993, PBW was controlled season-long. So long as pheromone disruption remains effective, a new general equilibrium position is manageable. If eradication is not pursued through more intensive resources allocation, this new equilibrium position may be maintained with lower resource input than previously used with conventional insecticide. Only if the program is abandoned would an old population dynamic return. We have taken the original PBW disruption paradigm of Brooks et al. (1979) and expanded this model to illustrate this phenomenon on a multiyear area-wide basis (Fig. 34.4). Although the population curves are hypothetical, they in fact reflect what occurred in Parker. Our third year representing Parker in 1992 and 1993 represent, at minimum, a new general equilibrium position over a protracted period of time. Our contention is

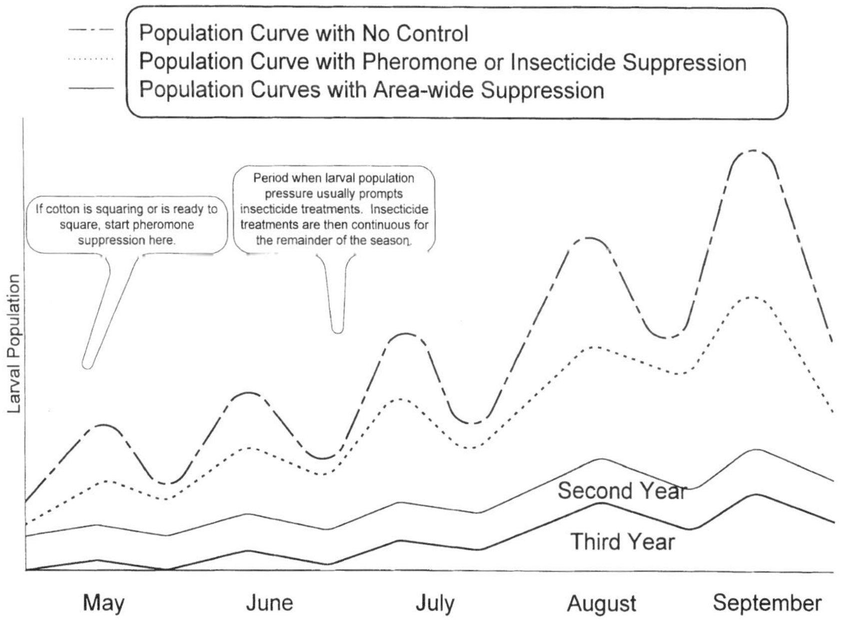

Figure 34.4. Hypothetical population dynamics in fields on a field-by-field basis versus an area-wide management approach. [Modified from Brooks et al. (1979).]

simply that without changes in technology and/or a commitment to eradication at a short-term escalating cost, area-wide management with carefully run mandatory control of the PBW is the best alternative for the irrigated southwestern areas of the United States and northwestern Mexico. The only deleterious effect we can see from this scenario is the elevation of the risk of resistance. No case of resistance to pheromones has been established. Our program will make an effort to evaluate this potential and develop a monitoring system.

35

Integrating Pheromones into Vegetable Crop Production

John T. Trumble

1. Introduction

Vegetable growers historically have been cautious about adopting the use of new technologies. Their crops are frequently of high value and short duration and require exceptionally high cosmetic standards. As a result, unless there is a clear economic benefit, most producers are slow to incorporate new procedures; the adoption of pheromone-based technology for insect monitoring has been no exception. However, the use of pheromones in vegetables has been further inhibited by a series of recurring difficulties. Initially the user is faced with decisions regarding trap selection and placement, confusion over best pheromone dose for longevity, and a host of conflicting claims by industry representatives. Then there is the revelation that trap catches, more often than not, do not reflect larval populations in the field. Weather, stage of crop growth, and horticultural practices produce variable impacts on the reliability of pheromone trap data. Nontarget pests sometimes confound the trap collection information and even occasionally represent a health threat. Given these problems, the real surprise is that pheromones have now begun to achieve widespread use in commercial vegetable production for such diverse practices as documenting flight phenology, setting economic injury levels, and monitoring for pesticide resistance.

In a few cases, the mating-disruption technique has been employed successfully for control of vegetable crop pests. The use of this technique for control of *Keiferia lycopersicella*, the tomato pinworm, is probably the best documented. When integrated with the release of biological control agents and the judicious use of insect pathogens and pesticides with low mammalian toxicity, the mating disruption technique was shown to provide a significant reduction in *K. lycopersicella* populations and a clear economic benefit to tomato growers in Mexico. The experimental design, population data, and economic analyses of the project will be presented. The approach has also been widely used in California during the past 2 years.

There are many reasons for the relatively rapid acceptance of the mating disruption technique in tomatoes. Development of pesticide resistance and related loss of most effective chemicals are certainly important. Similarly, incorporating pheromones into a comprehensive integrated pest management program and demonstrating compatibility with biocontrol agents such as leafminer parasites, pinworm parasites and *Trichogramma* spp. was also of value. However, the single most important factor in grower acceptance was the evidence from a detailed cost–benefit analysis showing greater net profits from the program using pheromones as compared to the standard pesticide treatment approach.

This chapter is organized into two parts. The first part, Section 2, provides a brief history of the difficulties experienced by researchers and industry personnel in implementing pheromones for use as monitoring tools in vegetable crops. The second part, Sections 3 and 4, will focus on the use of pheromones specifically for monitoring and mating disruption of *K. lycopersicella*, the tomato pinworm.

2. Barriers to Acceptance of Pheromone Technology

Vegetable growers historically have been cautious about adopting the use of new technologies. Their crops are frequently of high value and short duration and require exceptionally high cosmetic standards. As a result, many growers suffer from what has been termed the "sleep at night syndrome" (terminology after B. Cartwright). That is, growers or their pest control advisors (PCAs) feel that if a pesticide has been sprayed, then they have done their jobs. If a catastrophic loss subsequently occurs, the loss was due to an "act of nature" and could therefore not be ascribed to their growing practices. In some developing countries, growers have the perception that implementation of IPM practices such as scouting or pheromone use are simply beyond their economic means or capabilities (Pollard 1991). Ironically, as pesticide resistance develops, this often means that growers spend more for less effective pesticide-based procedures.

The concerns just described have been alleviated to some extent by recent successes (see other chapters in this part) as well as the realization that pesticide availability is rapidly becoming limited. Expansion of the United State's urban population has led to an increase in the political influence of environmental activists, labor unions, and consumer groups which are increasingly adopting anti-pesticide agendas and sponsoring ballot initiatives designed to restrict pesticide use (Trumble 1990). Furthermore, fear of ground water pollution and concerns regarding cancer have stimulated a surge in legislation which further limit pesticides (Trumble 1989). These problems are exacerbated by the continuing trend for reduction in registration of new pesticides due to increased testing requirements and registration costs (Georghiou 1990). Thus, as growers become sensitized to the need to adopt new approaches to pest management, the opportunity exists for implementing the use of pheromones and other relatively new pest management technologies into commercial farming.

Nonetheless, even a brief discussion with growers regarding acceptance of any new arthropod control system will reveal several valid reasons for caution. New technologies are often flawed and require further study and modification before they become economically viable. Many growers have expensive pieces of equipment which promised to revolutionize pest control (electrostatic sprayers, tractor mounted insect vacuums, etc.) that are now rusting behind the barn. Juvenile hormones were promised to provide long-term control of most insect pests because these compounds were required by the insects for normal life processes. Despite the best guesses of researchers, resistance developed fairly rapidly to the "first-generation" products (Georghiou and Lagunes-Tejeda 1991), and many growers were caught by surprise when the products failed. Even with pheromones, much of the potential usefulness of the technology is only now being realized, some 20–30 years after the initial expectations following discovery (McNeil 1991).

2.1 Impediments to Application in Commercial Agriculture

Like any new technology, pheromone use in vegetables suffered from incipient problems in application. One recurring difficulty has been the selection of the right trap, pheromone dispenser, dose, and field placement. Although this problem appears to be easily solved by simple comparative studies designed to measure best capture rates, grower acceptance of pheromone traps for monitoring is often based on ease of use and cost rather than effectiveness. As a result, species that can be collected in the inexpensive wing traps, delta traps, and bucket traps have been monitored more readily than those requiring omnidirectional traps or large cone traps (Sharma et al. 1973; Struble 1983; Mitchell et al. 1989). Over the past 10 years, numerous papers have been published on selection and efficiency of different trap designs and pheromone dosages (Trumble and Baker 1984; Gray et al. 1991; Valles et al. 1991; Jansson et al. 1992; see also the references in these articles). This diversity of trap designs and pheromone delivery systems, coupled with conflicting claims from manufacturers and less-than-ideal transfer of information from researchers to grower, may cause some PCAs or growers to avoid adopting an apparently confusing technology.

Researchers also have addressed related problems of trap saturation, pheromone longevity, convenience, placement, and nontarget captures (Proshold et al. 1986; Drapek et al. 1990; Campbell et al. 1992; Derrick et al. 1992). Even though most of these problems can be easily resolved through research, implementation of the modified methodology still can be difficult. Growers and PCAs are often reluctant to set up traps at specific heights if this makes finding the traps difficult or if poles or special stakes are required. The need for placing traps above the height of the canopy or between rows appears particularly onerous, because this can interfere with routine crop maintenance or pesticide application. Although some traps have proven quite efficient at collecting males attracted to pheromones, the lack of commercial availability, along with the time constraints on PCAs which limit their ability to manufacture the traps, has hampered implementation (Jansson et al. 1992). The inconvenience of sorting through related, nontarget moth species not only increases monitoring time, but makes identification difficult and may distort reports of trap captures (Adams et al. 1989; Weber and Ferro 1991). In some cases the attractancy of traps to nontarget insects may be more than just a nuisance. For example, cone-orifice traps designed and baited to catch *Spodoptera exigua* in tomato fields actually caught large numbers of bumble bees (*Bombus*), much to the distress of the people checking the traps (Trumble and Baker 1984; J. Trumble unpublished). Similar collections of stinging Hymenoptera have since been reported for other trap designs (Adams et al. 1989).

Perhaps the single most important factor which has limited the use of pheromone trapping for monitoring pests has been a lack of correlation between trap collections and egg or larval densities in the field; coefficients of determination in the range of 25–40% are common (Faccioli et al. 1993; Latheef et al. 1993; Witz et al. 1992). In some cases this relationship has been nonsignificant (Wiesenborn et al. 1988, *Helicoverpa zea* on sweet corn), while others have found no consistency in relationships (Campbell et al. 1992, *H. zea* on tomato). Efforts to improve correlations between *H. zea* trap catches and egg numbers in sweet corn have been almost heroic; researchers have evaluated a series of 1- to 2-hr time periods throughout the crepuscular/dark period of flight and found that selected time periods provided better relationships (up to $R^2 = 51\%$ for moths caught between between 4 and 5 A.M.) (Latheef et al. 1993). Unfortunately, while of considerable interest to researchers, the potential for implementation of this approach by PCAs and growers may be limited to those suffering from insomnia. In similar studies, researchers have found that relating trap collections to egg counts between minus 1 and +12 days following trapping could also improve the predictive value of relationships [Latheef et al. (1991) found correlation coefficients (r) ranging from 0.33 to 0.84; see also Weber and Ferro (1991)]. However this unpredictability of the delay variable makes implementation difficult if not impossible, and Weber and Ferro (1991) concluded that the pheromone traps might be most effective for timing egg searches.

A number of key factors probably have substantial influence on this relationship through effects on oviposition behavior, larval survival, or trap efficiency. For example, the stage of crop growth and availability of alternate hosts is of considerable importance for *H. zea* ovipositing in sweet corn (Wiesenborn et al. 1988). The presence of pesticide residues, the crop accession or variety grown, and so on, could also reasonably be expected to impact oviposition behavior or subsequent larval feeding behavior (Penman et al. 1988; Hoy et al. 1990). Pesticide residues in particular have been documented to change pest distributions (Trumble

1985), which would affect the relationship between trap catches and larval or egg population estimates. In some moths with wide geographic distributions, local regional "dialects" have been found which would affect responses to specific pheromone blends (Löfstedt et al. 1986b). Furthermore, a variety of weather conditions have been implicated as key factors influencing trap catches of males for many different systems (Dent and Pawar 1988; Elkinton and Cardé 1988; Jönsson and Anderbrant 1993; Pitcairn et al. 1990; Royer and McNeil 1993a). Because variations in weather conditions between years, crops, and locations are likely, and given that these factors can have differential effects on mortality and oviposition behavior, anticipating stable, predictive relationships between pheromone trap catches and larval or egg densities may be naive. In fact, the real surprise is that some pheromone-based monitoring programs have been remarkably successful in vegetables. This success is due, in part, to the tenacity of researchers who have overcome these obstacles to document reliable, efficient, and species-specific monitoring techniques employing pheromones.

3. Examples from the Literature

Although there are many creative ways that pheromones have been used for monitoring pest populations in vegetables, three basic approaches are most common. These approaches include employing pheromones to (1) document flight phenology, (2) set economic injury levels and (3) monitor for pesticide resistance.

3.1. Documenting Flight Phenology

Information on flight phenology of pest insects is typically used in two ways. In many cases, the occurrence of a peak in flight activity can serve as a signal to initiate sampling procedures for eggs or larvae. This approach is currently suggested for *H. zea* infesting sweet corn in Massachusetts, USA (Weber and Ferro 1991) and strawberries in California, USA (Wiesenborn et al. 1988), as well as for the potato tuberworm, *Pthorimaea operculella*, infesting potatoes in Peru (Raman 1988).

A second common use for information on flight phenology is the documentation of migration patterns and overwintering habitats. Using pheromone trap collection data, Mitchell et al. (1989) were able to provide evidence that *Spodoptera frugiperda* migrations from endemic areas in the Caribbean to the U.S. mainland were not a critical component in annual reinfestations. Wiesenborn et al. (1988) used trap collections of *H. zea* in conjunction with weather data to suggest that the rare infestations of California strawberries were due to migrations resulting from the confluence of mild winters and a coastal counterclockwise low-pressure system called the "Catalina Eddy." Pheromone traps also were successfully used in conjunction with weather patterns to prove that *S. exigua* was overwintering in California, rather than migrating northward from Texas or Mexico (Trumble and Baker 1984).

3.2. Setting Economic Injury Levels

P. operculella is the target of one of the most geographically widespread control programs utilizing pheromones. This small gelechiid moth causes damage when larvae mine the leaves, stems, or exposed fruit of potatoes. Damage may occur both before and after harvest. Unlike most other systems, a useful predictive relationship often exists between pheromone trap catches and larval density in the foliage or percent fruit damage. This relationship has been used to establish threshold levels for treatment in Australia (Valencia 1981), India (Lal 1989), Peru (Raman 1982, 1988), and the United States (Shelton and Wyman 1979). Unfortunately, the authors note that changing the potato cultivar, the selection of pesticides used, irrigation schedules or practices, time of harvest, or even the soil type in which the potatoes are grown can dramatically alter the correlative relationship and render the technique less useful.

3.3 Monitoring for Pesticide Resistance

Riedl et al. (1985) and Haynes et al. (1986) pioneered the concept of using pheromone traps to monitor pesticide resistance in pest insects. The first work required that insects collected in traps be transported back to the laboratory for standard bioassays based on topical applications. The second eliminated the need for topical applications by incorporating the pesticide directly into the sticky material in the trap. The procedure of Haynes et al. (1986) was further developed by Brewer and Trumble (1989, 1991) to monitor the occurrence of pyrethroid resistance in *S. exigua* in tomatoes. This required the development of a considerable amount of background information including, a discriminating dose of fenvalerate in the sticker of pheromone traps (that is, a dose which will kill nearly all susceptible adults but few of the resistant moths) and information on the relationship between adult and larval resistance.

A sequential sampling plan based on the sequential probability ratio test then was constructed (Brewer and Trumble 1991). The technique allows populations to be rapidly categorized as highly resistant, moderately resistant, or susceptible with a high degree of reliability. This approach has yielded several substantial benefits including the following: (1) determination that probably field failure of the pesticide could be made prior to application, (2) sample sizes, and thus cost and effort, could be minimized, and (3) the rapidity of the test provided more timely information on resistance levels than the use of a topical application method. Such techniques may become increasingly important as the availability of pesticides declines. However, this approach has some substantial drawbacks such as the need for a large amount of developmental effort and the demonstration of a predictive relationship between insecticide resistance levels in the adult stage and larval stage of the pest species.

4. Monitoring and Mating Disruption of the Tomato Pinworm, *K. lycopersicella*

K. lycopersicella has been recognized as an important pest of tomatoes throughout much of North and Central America since its initial discovery in the Imperial Valley in California in 1923 (Oatman 1970; Charlton et al. 1991). Although some photosynthetic potential may be lost when larvae mine and fold the leaves, most economic damage occurs when larvae penetrate the fruit and become contaminants (Lin and Trumble 1983, 1985). Until the development of pheromones for this pest, control efforts had focused primarily on pesticides, because biological control agents, while present, did not provide an adequate level of control for commercial production (Oatman 1970; Oatman et al. 1979). Fortunately, following developmental studies in the 1970s and early 1980s, a commercial pheromone product became available. This product was a 96:4 *E:Z*-4-tridecenyl acetate blend (Jenkins et al. 1990). The development, refinement, and some applications of the pheromone have been described in detail by Jenkins et al. (1990) and Charlton et al. (1991) and will not be repeated here.

4.1. Initial Studies Using Pheromones for Controlling K. lycopersicella

Beginning in the late 1970s through the 1980s, several attempts were made to use pheromones for control of *K. lycopersicella* in fresh market tomatoes. Results from the first studies were inconclusive (Jenkins et al. 1990). In the first replicated study, moth captures were reduced in plots of tomatoes in Florida that were treated with 40 g a.i./ha of pheromone, but no damaged fruit were found in either treatment. Unreplicated mating disruption studies in Mexico were encouraging, with a reduction in average percent fruit damage from 5.4% to 1.2%. However, because of the experimental design, statistical significance could not be determined.

A large-scale, 2-year study by Van Steenwyk and Oatman (1983) in southern California produced inconsistent results. In the first year, although significant reductions were seen in collections of adult moths in pheromone traps between treatments, there were no differences in foliage infestations or fruit infestations regardless of the level of pheromones applied (2.5, 10.0, or 40.0 g/ha). In a second year where only the 10.0 g/ha rate was used, significant reductions were reported in moth captures and for both larvae and mining injury in the foliage. However, differences in total fruit damage were not significant ($p < 0.05$ level), probably due to very low levels of fruit damage: 0.8% fruit damaged in the control versus 0.06% damage in the pheromone treatment. Unfortunately, the results of this study are confounded by the application of some unspecified pesticides for *S. exigua* control during both years that the authors conceded probably impacted *K. lycopersicella*.

In a related study, Van Steenwyk et al. (1983) developed a treatment threshold using pheromone traps. A pest density level of 10 moths/trap/night was suggested for initiating pesticide applications in California, based on a correlation coefficient of $r = 0.76$ between trap captures and fruit damage 2 weeks later. However, this relationship is only weakly predictive, and it could not be validated at other locations in California (Toscano et al. 1987). Levels of 10 or more moths/trap/night in Sinaloa are common (Alvarado-Rodriguez and Rivera-Rubio 1990; Trumble and Alvarado-Rodriguez 1993), and the threshold has not been adopted in Mexico. High levels of moth catches in Sinaloa are probably due to the presence of alternate hosts, abandoned fields, and volunteer tomatoes (Alvarado-Rodriguez and Rivera-Rubio 1990) as well as relatively high mean temperatures which do not fall below the thermal flight threshold of 11°C (Lin and Trumble 1985).

Jiménez et al. (1988) reported that the mating disruption technique was highly effective for *K. lycopersicella* suppression in cherry tomato production in northern California. In a 2-year study, peak fruit infestations ranged from 14% to 65% in plots treated 12 to 16 times with the broad-spectrum pesticides azinphosmethyl, methomyl, diazinon, or naled, whereas plots treated with pheromones only or pheromones and one application of *Bacillus thuringiensis* had infestations of only 1–5%. Unfortunately, no statistical analyses were conducted and significant differences between treatments could not be reported. Interestingly, the data suggested that some pesticide resistance may have developed to methomyl and/or diazinon because a field treated 16 times had 65% damage, whereas the completely untreated field had only 12% peak damage.

4.2. Mating Disruption in Commercial Tomatoes in Sinaloa, Mexico

Studies on mating disruption of *K. lycopersicella* conducted in Mexico by Trumble and Alvarado-Rodriguez (1993) were initiated largely as a result of concern over development of pesticide resistance. In the 1970s and 1980s, *K. lycopersicella* developed resistance to a wide variety of pesticides in Mexico, and became a major pest of tomatoes (Brewer et al. 1993). Repeated applications of broad-spectrum insecticides applied to control the pest resulted in two major problems; residue levels of some exported fruit from Mexico were above U.S. tolerances (Food and Drug Administration 1979), and the pesticides were causing outbreaks of other pest species such as the leafminer, *Liriomyza sativae*. Throughout the 1980s, Mexico's 1 billion dollar tomato industry was in jeopardy; large populations of this pest were causing considerable economic hardship and some fields had to be abandoned prior to harvest (Alvarado-Rodriguez and Rivera-Rubio 1990).

By the late 1980s, it was evident that continuing with a control program for *K. lycopersicella* based solely on intensive pesticide use would probably result in rapid resistance development to any new pesticide and additional economic losses. At the request of the tomato growers in the State of Sinaloa, a series of experiments were designed and conducted to determine the feasibility of implementing an integrated control program based on mating disruption, parasite releases, an insect pathogen (*B. thuringiensis*), and abamectin, a pesticide with potential for *K. lycopersicella* control which was previously unused in Mexico and had low contact toxicity to mammals. The intent was to develop a program before abamectin became commercially available and thus could be used in a fashion that would maximize control while minimizing resistance development.

From the outset, several criteria for the program were agreed upon. First, because tomato production in Sinaloa is exceptionally labor-intensive, any control strategies chosen had to minimize human exposure to chemicals with high mammalian toxicity. Second, environmentally benign techniques would be given priority, because most of the runoff from this region feeds into the Gulf of California, which supports environmentally sensitive sport and commercial fisheries as well as vacation resorts. Finally, the program had to be economically viable. Without a clear indication that profits could be increased or at least maintained, there was little hope that the program would be commercially implemented. However, pheromone-based control alone could not provide a complete solution to the insect problems on tomatoes.

Sequential plantings in Sinaloa assure that migrations of *K. lycopersicella* from nearby abandoned or harvested fields could be anticipated. Developing (1) a larval treatment threshold (a density which would trigger controls other than mating disruption) and (2) a control technique which would not disrupt the biological control strategies for the other insect pests therefore was critical to the potential success of the program. Nearly concurrent studies by Wiesenborn et al. (1990) on the economics of insecticide treatment programs for tomatoes in California provided larval thresholds for treatment which could be modified for use in Sinaloa. Other IPM studies in California indicated that abamectin could provide substantial suppression of *K. lycopersicella* without causing significant increases in populations of *Liriomyza* species or killing adults of key hymenopterous leafminer parasites (Trumble 1990).

Minimal activity by abamectin against a complex of hymenopterous parasites attacking *L. sativae* suggested that abamectin would probably have little impact on release of the hymenopterous parasite, *Trichogramma pretiosum*, which was released for *H. zea* suppression. Release rates for this parasite were based on previous work by Elizondo-Alapisco and Alvarado-Rodriguez (1988). Similarly, *B. thuringiensis*, a pathogen specific to insects, was employed for control of *S. exigua* since this material has little or no demonstrated effect on parasites (Trumble 1990).

The experimental design and the economics of the IPM program have been reported (Trumble and Alvarado-Rodriguez 1993) and will not be repeated in detail here. However, additional results have been analyzed and a brief overview of the project will be necessary in order to interpret them. Following a series of preliminary studies to adjust thresholds and refine survey techniques, comparisons of insect populations, fruit damage, control costs, and net profits were made between a conventional insecticide treatment program (20–35 applications of two pesticides per application), an IPM program (mating disruption, *T. pretiosum* releases, as needed applications of *B. thuringiensis* or abamectin, etc.), and an untreated control. Each treatment was replicated (0.5 ha planting/treatment) in each of three valleys (Culiacan, Guasave, and Los Mochis) and in each of three seasons (fall, winter, and spring plantings, except for Culiacan, which was not planted in the spring). Thus, each treatment was evaluated on a total of eight 0.5-ha plantings. All plots were at least 1 km from each other and from commercial tomato fields. Mixtures of pheromones in hollow fibers and a flowable polybutene sticker (Ecogen, Inc.) were applied monthly at 13 g a.i./ha. Pheromones were applied at approximately 2-m intervals along every other row.

Data on *K. lycopersicella* were collected for males/trap/night (m/t/n), eggs/leaf, and larvae/plant. Adults were monitored at each plot with three wing-style traps. Larvae were counted on 10 randomly selected plants from each of row 1, row 5, and row 10 and from the center of the plots for 40 whole plant counts per week. Eggs were counted on 30 randomly selected leaves from the same field locations. Eggs were brought back to the laboratory and evaluated for percent parasitism. Data were pooled across locations and compared within plantings for each treatment using ANOVA and Fisher's Protected Least Significant Difference Test.

The mean number of m/t/n from the fall planting in Culiacan (Fig. 35.1A), the winter planting in Guasave (Fig. 35.1B), and the spring planting Los Mochis (Fig. 35.1C) show that populations of adults were generally high in all three growing seasons with most moths collected in the latter half of each season. The untreated control plots typically had the highest numbers collected, followed by the commercial treatments and then the IPM treatments. Although the corresponding larval counts per plant provided in Fig. 35.2 (A, B, and C) did show peak populations toward the end of the season, the population density increased by an order of magnitude with each consecutive planting. Thus, correlations between adult collections in traps and larval counts would not be reliable across plantings. Somewhat better correlations may have occurred within plantings, but predictive relationships were not evident and no effort was made to pursue adult trapping as a threshold treatment mechanism.

Information on the mean number of eggs per leaf and the mean percent parasitization of those eggs did show some interesting trends (Fig. 35.3). First, the egg density and percent egg parasitism increased in consecutive seasons as did the larval density. This may have occurred in part due to the greater availability of hosts. Second, there were significant treatment effects: Egg densities were different between the control and commercial treatments in the spring planting. Although the percentage of parasitized eggs was highest in the spring planting,

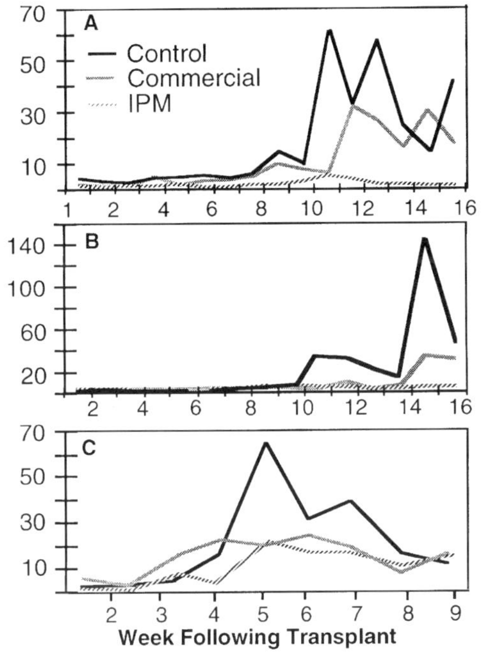

Figure 35.1. Mean number of *K. lycopersicella* males per pheromone trap per night in a fall planting in Culiacan (**A**), a winter planting in Guasave (**B**), and a spring planting in Los Mochis (**C**) valleys in the State of Sinaloa, Mexico. Comparisons are made between collections from tomato plantings treated with a commercial insecticide program, an IPM program, and an untreated control.

and there was a trend across plantings for higher parasitism in the IPM treatment, the commercial treatment was only significantly different from the IPM treatment in the winter planting. Nonetheless, such differences suggest that programs attempting to correlate egg counts with adult trap catches or larval densities should focus on viable eggs, because variable parasitism rates between treatments would add another confounding factor to the analysis.

In the fall planting, when *K. lycopersicella* larval populations were low, there were no differences in mean percent fruit damage between the commercial and IPM treatments (Fig. 35.4A). However, input costs of pesticides and pheromones were lower in the IPM program and net profits (value of the crop minus all production/harvest and sales costs) were correspondingly higher (Fig. 35.4B). The fruit damage in the winter and spring plantings was substantially higher than during the fall plantings, with commercial plots averaging significantly more damage (range 75–90%) than the IPM plots (range 33–35%). As a result, at the typical carton values of US $5 to $9, our commercial plots lost money (up to $1000/ha) in the winter and spring. However, even with these losses, a grower

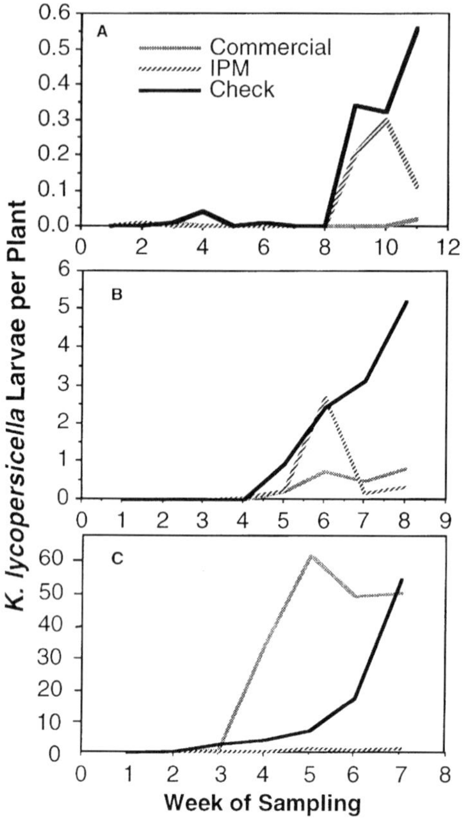

Figure 35.2. Mean *K. lycopersicella* larvae per plant in a fall planting in Culiacan (**A**), a winter planting in Guasave (**B**), and a spring planting in Los Mochis (**C**) valleys in the State of Sinaloa, Mexico. Comparisons are made between collections from tomato plantings treated with a commercial insecticide program, an IPM program, and an untreated control.

using the commercial program on 1000 ha and selling fruit at $9/carton would generate a net profit for the year (fall profits minus winter and spring losses) in excess of 1.5 million dollars. For the IPM approach at the same carton value profits were made in every planting, and the total net profit for the year exceeded 5 million dollars. Thus, a substantial economic incentive exists for using the IPM program.

The use of this IPM program meets the desires of the consumers for reduced pesticide usage. Furthermore, reduction in the potential for pesticide resistance development through the introduction of biological control agents as well as use of less pesticide allows growers a more sustainable option for long-term production of tomatoes. Finally, the substantial reduction in use of pesticides with

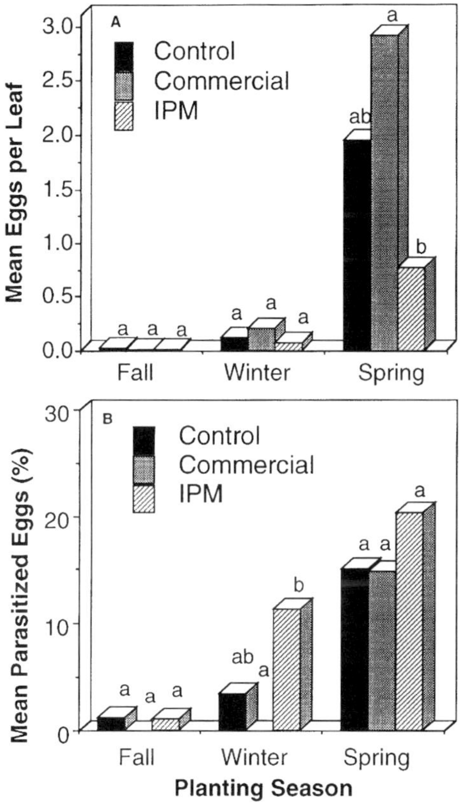

Figure 35.3. Mean eggs per leaf (**A**) and mean percent parasitized eggs (**B**) in fall, winter, and spring plantings in Sinaloa, Mexico. Letters above bars indicate significant differences within plantings at the $p < 0.05$ level, ANOVA, Fisher's Protected Least Significant Difference Test. Percent data were transformed by the arcsin square root transformation prior to analysis.

high levels of mammalian toxicity will help to minimize any potential human health concerns.

A caveat is in order. Given the extensive populations of *K. lycopersicella* presently in Sinaloa, along with the population isolation generated by growing tomatoes in geographically separated valleys, the possibility exists that behavioral adaptations or even physiological changes may occur which will reduce the effectiveness of the current pheromone formulation. Should a physiological resistance occur that results in change in pheromone blend, a simple modification in chemical composition might be adequate to deal with the problem. More than one blend may be required if the population is not homogeneous. However, development of a behavioral resistance (i.e., the adults move out of the field

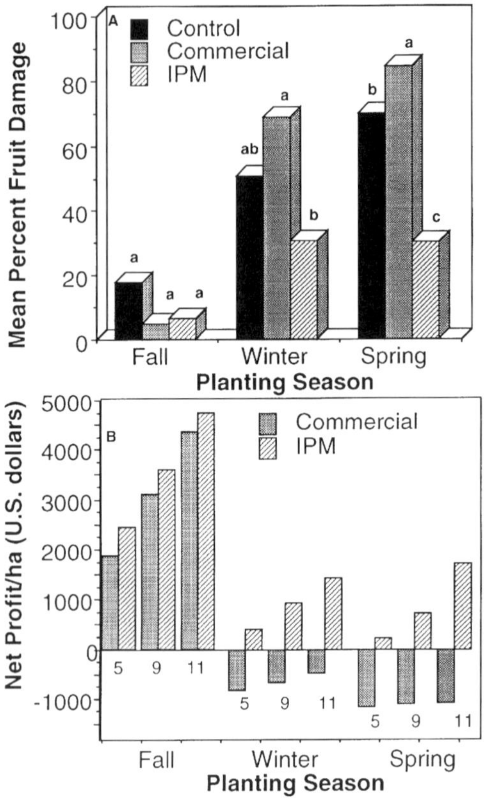

Figure 35.4. Mean percent fruit damage (**A**) and net profit information (**B**) for control, commercial, and IPM plantings in three seasons in Sinaloa, Mexico. Letters above bars indicate significant differences within plantings at the $p < 0.05$ level, ANOVA, Fisher's Protected Least Significant Difference Test. Percent data were transformed by the arcsin square root transformation prior to analysis. Numbers directly below bars in part B indicate prices for a standard carton of tomatoes.

to mate, etc.) may require additional biological studies to determine possible application strategies to alleviate the problem.

Acknowledgments

Dr. Benito Alvarado-Rodriguez deserves credit for the data collected from fresh market tomatoes in Mexico. His input into the development of the IPM program, along with his subsequent efforts at extending this knowledge, was crucial to the success of the program. The reviews of W. Carson, K. White, and M. Berdegue improved this chapter.

36

Mating Disruption of the Lightbrown Apple Moth: Portable Electroantennogram Equipment and Other Aspects

D.M. Suckling and G. Karg

1. Introduction

The New Zealand export apple industry produces very high quality fruit sold in over 60 countries worldwide. Apples must meet stringent quarantine and residue standards, with a zero tolerance of pests and damage in exported fruit. Insecticides are applied to control the key pests, which include the lightbrown apple moth (*Epiphyas postvittana*), the codling moth (*Cydia pomonella*), and four related native leafrollers, although only the lightbrown apple moth is abundant in all orchard districts (Wearing et al. 1991). Increasing consumer concern over food safety, particularly pesticide residues, has spurred the search for alternatives to broadspectrum insecticides (Wearing et al. 1993), although the development of integrated pest management for New Zealand apples has been underway for some time (Collyer and van Geldermalsen 1975). Insecticide applications also result in selection for resistance, with consequent potential for control failures and problems in meeting quarantine standards. This chapter will review aspects of field applications of lightbrown apple moth pheromone in New Zealand, including some new results from electroantennogram measurements of instantaneous changes in pheromone concentrations in treated orchards.

2. Pheromones in Insecticide Resistance Management

Insecticide resistance was detected in the lightbrown apple moth during the early 1980s, and the distribution of the resistant phenotype was mapped using a pheromone-based monitoring system, combined with topical application of the insecticide (Suckling et al. 1985, 1987). Pheromone trapping was also used as a sampling tool for monitoring the level of a detoxication enzyme, which was found to be elevated in insecticide-resistant moths (Armstrong and Suckling

1988). The resistance was very confined in distribution, but there was good evidence of dispersal, which exceeded 400 m in some feral males (Suckling et al. 1990a). The edge of the resistance zone was also sharply defined over several hundred meters by the frequency of resistance in offspring from feral females, as well as from crosses between tethered virgin susceptible females and feral males (Suckling and Khoo 1993). No barriers to gene flow were detected by electrophoresis.

One response to the insecticide resistance problem was to investigate pheromone blends suitable for mating disruption, using rubber septa and polyethylene rope dispensers (Shin Etsu) (Suckling and Clearwater 1990). Some disruption was demonstrated with major or minor pheromone components alone [respectively, 100 μg (*E*)-11-tetradecenyl acetate (*E*11-14:Ac) and 5 μg of (*E,E*)-9,11-tetradecadien-1-yl acetate (*E*9,*E*11-14:Ac)] (Bellas et al. 1983), although the best results were achieved with both components presented together in a blend. However, disruption was achieved with a nonattractive blend, suggesting that false trails may not be necessary for disruption.

Mating disruption with pheromone dispensers proved to be an effective input: It supplemented a failing insecticide program and managed the resistant population, with complete prevention of trap catch in the treated area for close to 12 months (Fig. 36.1) (Suckling et al. 1990b). Disruption of trap catch was accompanied by a greater than 200-fold reduction in damage in the first year of pheromone application, from 5.4% damage in 1985–1986 to 0.02% damage in 1986–1987. The incidence of damage underwent a further drop to 0.0072% damage by

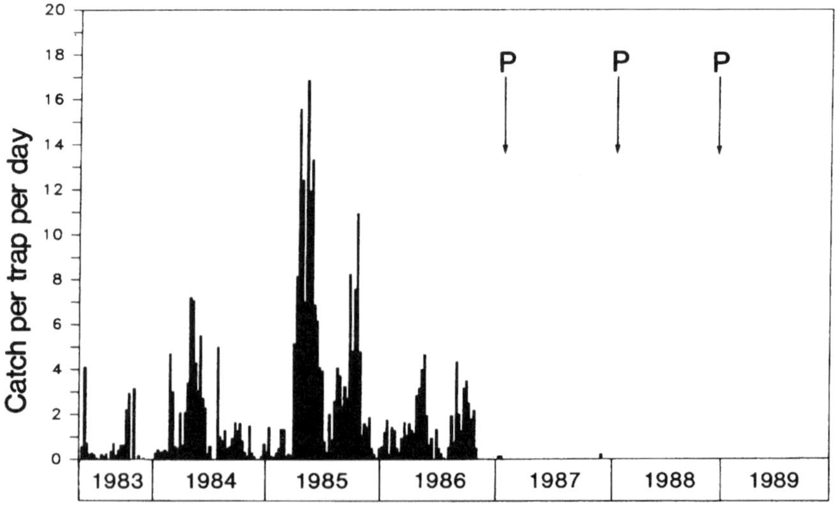

Figure 36.1. Pheromone trap catches of *Epiphyas postvittana* at Moutere Bluffs, New Zealand before and after deployment of mating disruption to control insecticide resistant insects. [Reproduced with permission from Suckling et al. (1990b).]

1988–1989, despite a concurrent reduction in the number of insecticide applications from 11 to 7. Disruption has continued to be applied annually for resistance management at this site, and export standards have been met consistently, since the inception of the combined pheromone and insecticide program. Mating disruption can offer a very powerful solution for insecticide resistance problems, but adoption of disruption as part of the pest management program, in advance of control problems, would be a more proactive approach to resistance management.

3. Wider Evaluation of Mating Disruption

This initial success at resistance management led to other trials of disruption at sites not affected by insecticide resistance, with the aim of providing an alternative control tactic to insecticides (Suckling and Shaw 1990, 1991, 1995). Trials in Australia have also commenced recently (Williams 1993). Four years of disruption trials (1990–1994) in New Zealand have involved >320 ha of treated orchards. Equivalent insect control to the standard insecticide program was achieved, with fewer applications of organophosphates near harvest along with reduced residues on fruit (Suckling and Shaw 1995). Blocks treated with higher rates of pheromone dispensers (2000 per ha) had significantly lower fruit damage at harvest than the standard rate (1000 per ha), suggesting that there is scope for improving the efficacy of disruption. Dispensers in current use contain the two-component pheromone blend: 54.9 mg *E*11-14:Ac and 2.8 mg *E*9,*E*11-14:Ac, with 19.1 mg *Z*11-14:Ac, an inhibitor for trap catch in this species (Rumbo et al. 1993), present as an impurity. This inhibitor reduces attractancy when presented with the pheromone blend, but low catches to these dispensers can occur in untreated orchards (Suckling unpublished data). The inadvertent change by the manufacturer, from the presence of a small to a significant quantity of *Z*11-14:Ac present with the pheromone in polyethylene dispensers, apparently did not affect the efficacy of disruption (Suckling and Shaw 1992), although the two blends have never been compared directly.

We have predominantly assessed the efficacy of disruption in field trials using catch in pheromone traps, mating frequency of tethered virgin female moths, and the incidence of fruit damage at harvest. Mating frequency was significantly correlated with trap catch (Fig. 36.2), although it was easier to disrupt trap catch than mating (note the *y* intercept of zero catch corresponded to 23% mating). The incidence of mating in the presence of dispensers was very low, with complete disruption of tethered females on 68% of occasions with release rates from 1 to 5 mg per ha per h, 84% of occasions with 8–12 mg per ha per h and 100% of occasions with > 12 mg ha^{-1} hr^{-1} (Suckling and Shaw 1992). Traps and tethered females provide useful information about levels of disruption over time, but are not very efficient at providing information on (a) changes in pheromone concentration in the field and (b) the immediate impact of operational factors,

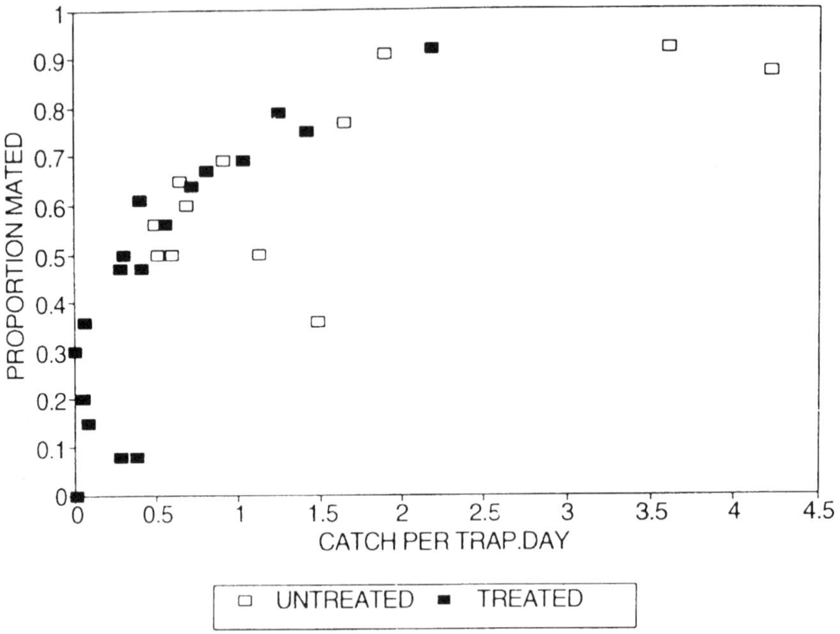

Figure 36.2. Relationship between weekly pheromone trap catch of males and mating frequency of nearby tethered female lightbrown apple moths, 50 m from pheromone-treated or untreated orchard blocks. [Reproduced with permission from Suckling and Shaw (1992), copyrighted by the Entomological Society of America.]

such as dispenser formulation, placement, or density of sources. Pheromone inputs from polyethylene dispensers into orchards can now be accurately predicted using a model of release rates (Bradley et al. 1995). The model is based on meniscus length and the temperature at the time of interest (e.g., dusk). Estimation of atmospheric concentrations and determination of a mass balance will require much more information.

4. Electroantennogram Recordings in Orchards—Mean or Variance?

The spatial and temporal distribution of pheromone in orchards or other crops treated with disruption is not well understood, but one promising technique for providing this information is the portable electroantennogram (EAG) (Sauer et al. 1992; Bengtsson et al. 1994; Suckling et al. 1994). This approach can offer information on seasonal or daily changes in the mean level of pheromone present, the effect of sampling (or insect) location within the crop, or other factors pertinent to the mechanism and efficacy of disruption. The technique is difficult to calibrate in absolute terms (Rumbo et al. 1995), but it can offer comparative information on the impact of operational factors. Regular calibrations of antennae are needed to overcome changes in antennal responsiveness. In our work (e.g., Karg et al. 1994, Suckling et al. 1994), we have normalized the EAG peak height resulting from stimulus with an orchard air sample containing pheromone, through division by the response to calibration pulses of known pheromone loading. This results in a unitless EAG, corrected for antennal responsiveness, which we have used for relative measurements of mean pheromone concentration.

One problem seen with EAGs of lightbrown apple moth recorded in orchards was the high level of background response to environmental odors, which limited the resolution available for converting EAG recordings to relative concentration levels (Suckling et al. 1994). Other species such as *Lobesia botrana* (Sauer et al. 1992) and *Cydia nigricana* (Bengtsson et al. 1994) appear to have much more sensitive EAG responses to pheromone and a lower response to other odorants. A lack of independence of EAG responses to pheromone and environmental odors was also encountered with *C. pomonella*, and consequently Rumbo et al. (1995) noted the problems with this general approach in some detail.

In terms of insect behavior and the way in which mating disruption works, it is possible that fluctuations in pheromone concentrations may be just as important as the mean concentration. These fluctuations will be largely the result of air entering the orchard from above the canopy (Raupach 1988), bringing with it parcels of air which have a much lower atmospheric concentration of pheromone. In a preliminary field experiment, we used portable EAG equipment (Suckling et al. 1994) to record time series of fluctuations in atmospheric pheromone in a treated apple orchard in New Zealand. We mounted a 3-D sonic anemometer beside the moth antennal preparation to record fluctuations in both the horizontal and vertical components of the instantaneous wind vector. We observed periodic updrafts and downdrafts in the orchard, and these appeared to be negatively correlated with EAG signals, as expected. Figure 36.3 shows a 3-min time series of the EAG and the vertical wind speed, recorded at 10 Hz. A 10-sec moving average was applied to the EAG signal in an attempt to remove some of the drift. As expected, there was a negative correlation between vertical wind and the EAG response. Clearly, the orchard air was being periodically (~10–20 sec) flushed of pheromone by clean air entering from above the canopy. At the same time, there was little evidence of any correlation between horizontal wind fluctuations and the EAG signal. This was because our measurements were made at a distance of more than 100 m from the edge of a treated block, where advective edge effects would be small.

We have used this equipment to record the instantaneous changes in antennal response to pheromone with the response of the same antenna in an untreated orchard block, more than 500 m away. The apple orchard block was treated with 750 polyethylene rope point sources of pheromone per hectare. Antennal response to the air samples were recorded for 3 min, with calibrations before and after each recording. The signals were then divided by the responses to the calibration

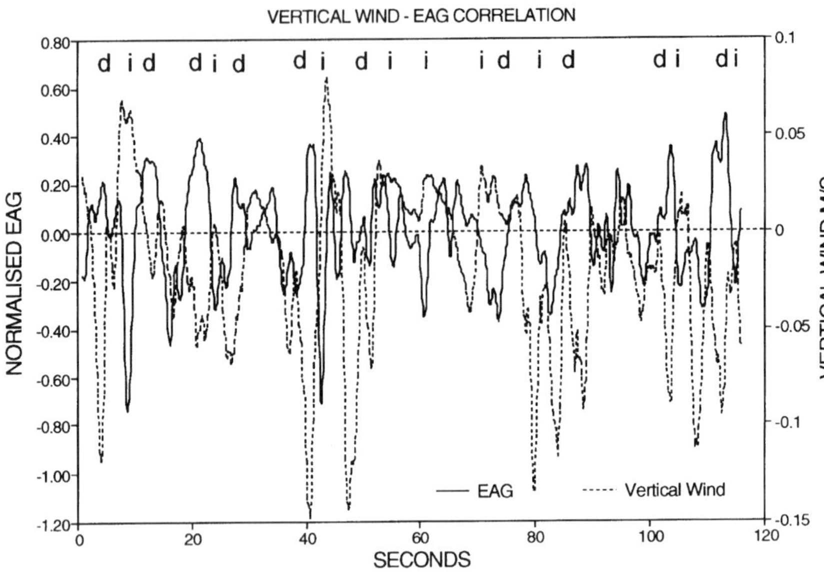

Figure 36.3. Fluctuation in male lightbrown apple moth antennal response to pheromone in relation to vertical wind speed in a New Zealand apple orchard treated with Shin Etsu rope dispensers (750/ha). Electroantennogram response goes down with increasing concentration [shown as (i) for increasing, (d) for decreasing concentration] and is at times inversely correlated with vertical wind.

pulses (Suckling et al. 1994), producing traces which can be compared between treatments. One statistic we have used for comparing these recordings between treatments is the mean-squared error. Table 36.1 indicates that the mean-squared errors of normalized EAG signals contain information about the variance in pheromone plume, although some variance due to environmental odours is present

Table 36.1. Mean-squared error of normalized (unitless) adult male Epiphyas postvittana *moth electroantennogram responses of 3 minutes' duration, recorded in pheromone-treated and untreated apple orchards, using the same antenna on April 8, 1993*[a]

Untreated	Pheromone (750 points/ha)
0.0006	0.0033
0.0003	0.0021
0.0007	0.0021
0.0012	—

[a]See text for statistical analysis.

in the signals recorded in untreated areas. The variances recorded with the same antenna were significantly different between treated and untreated areas ($t = 4.31$, $p < 0.01$). The variance in antennal responses in the untreated block is probably partly due to the numerous sources of plant volatiles (e.g., apple leaves). Instantaneous changes in concentration of plant or other environmental volatiles are clearly detectable by antennae, since the baseline response in charcoal-filtered air is, by contrast, very stable (Rumbo et al. 1995).

The mean-squared error of fluctuations in EAG responses over time does not discriminate between major events such as (a) concentrated filaments of pheromone reaching the antenna (Murlis and Jones 1981; Murlis et al. 1992) and (b) minor fluctuations. A more refined analysis would improve documentation of the micrometerological events, such as downdrafts (Fig. 36.3).

5. Problems with Field EAGs

Insect species may vary in their suitability for this technique, due to factors such as a higher background response to environmental volatiles, which has the effect of reducing the resolution available for pheromone detection (Suckling et al. 1994). Apart from the lack of specificity of response, the portable EAG equipment has several other problems, because data processing is time-consuming, and it can be hard to interpret the impact of external effects such as wind, without very sophisticated equipment. The antennal lag time for recovery from a stimulus (typically 1–2 sec) causes the results from time series to be biased toward increased concentrations, because new filaments are detected before the baseline has been covered. Antennae vary in sensitivity to pheromone and other odors, and antennal response in untreated air needs to be considered. However, despite these problems, EAGs recorded in the field can teach us about perception of pheromones by insects and may help to optimize dispenser deployment. The development of a portable sensillogram apparatus (Van der Pers and Minks 1993) may also obviate some of these difficulties, particularly lack of specificity.

6. Electroantennogram Measurement of Pheromone in Leaves

The EAG apparatus is a very sensitive detector of pheromone and has other potential applications—for example, in the investigation of the role of apple foliage as a buffer for the pheromone released from dispensers. Karg et al. (1994) used EAGs to show that absorption and release of pheromone by apple leaves can take place. The assumption was made that the response to pheromone being released from the leaf surface should produce an EAG that was above the baseline response of the antennae to the background leaf volatiles. Hence a normalized pretreatment measurement (on the same leaf) was subtracted to give the differential EAG amplitude, due to the response to pheromone. In one experiment, apple leaves were picked and the EAG elicited by each leaf was measured to determine the background antennal response to the green leaf volatiles. The petioles were then placed in a vase of water, and the leaves were held 15 cm downwind from two Shin Etsu rope pheromone dispensers mounted on the outlet side of a small fan (wind speed of 1.5 m/sec at the leaf), simulating field conditions of leaves in pheromone-treated environments. Single leaves were exposed in this way for varying time intervals from 3 sec to 30 min and were then put in a clean glass "cold trap." The cold trap was connected to the EAG apparatus, and EAG measurements of the pheromone release rate were taken using 1-sec air pulses into the cold trap within 30 sec of removal from the treated airstream. These EAGs were normalized to the related calibration pulses.

Leaves exposed to an airstream containing pheromone reached a constant level of pheromone release within 3 min (Fig. 36.4). The rate of uptake of pheromone in an airstream by apple leaves was therefore very rapid. These measurements are an indirect assessment of the process of pheromone uptake, given that our EAG results reflect the rate of release of pheromone from the leaf within 30 sec of removal of the leaf from the pheromone airstream. In addition, we do not know the degree of saturation of the leaves with pheromone, only that the leaf is able to release pheromone in proportion to the length of exposure. In these experiments, we took the release as a measure of the uptake of pheromone by the treated leaves.

The behavioral significance of this level of pheromone was indicated by catch of moths to leaves used as lures in sticky traps. Karg et al. (1994) recorded catch of moths to leaves which had been treated with pheromone, for up to three nights after treatment.

In another experiment, Karg et al. (1994) reported the dissemination of phero-

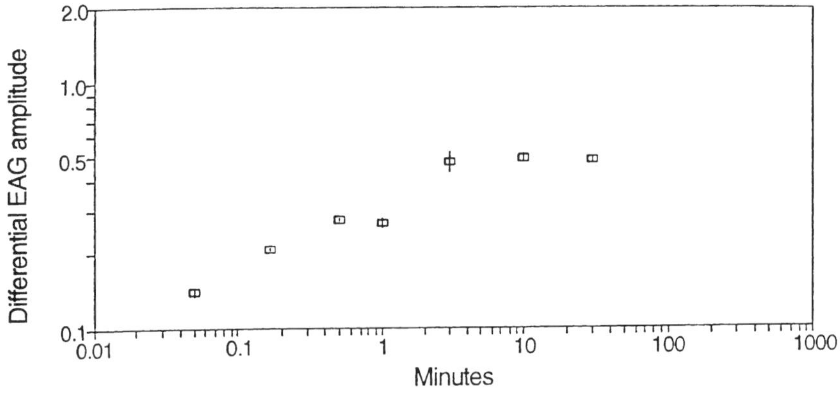

Figure 36.4. Changes in differential EAG amplitude (±SE) over time, following increasing periods of exposure of leaves to a steady airstream containing pheromone of *Epiphyas postvittana*. The rate of uptake of pheromone by single apple leaf was very rapid, with the asymptote reached after about 3 min. [Reproduced with permission from Karg et al. (1994), copyrighted by Plenum Press.]

mone released from dispensers, into nearby apple foliage in the field. Two weeks before the EAG measurements commenced, two dispensers were placed in an apple tree grown on a horizontal canopy system, with upright shoots from the base of lateral branches providing a uniform array of leaves in all dimensions. Leaves of similar size (~20 cm^2) were picked on *x, y*, and *z* axes from the two central dispensers, and immediately placed individually in a glass cold trap. The differential EAG amplitude was measured, in which the normalized EAG response elicited by an untreated leaf was subtracted from the normalized EAG measured for a leaf post-treatment. An isosphere (Fig. 36.5) showing the location of equal response around the two rope dispensers (indicated by a differential EAG amplitude of 0.2) was generated by mathematical interpolation of the measured date

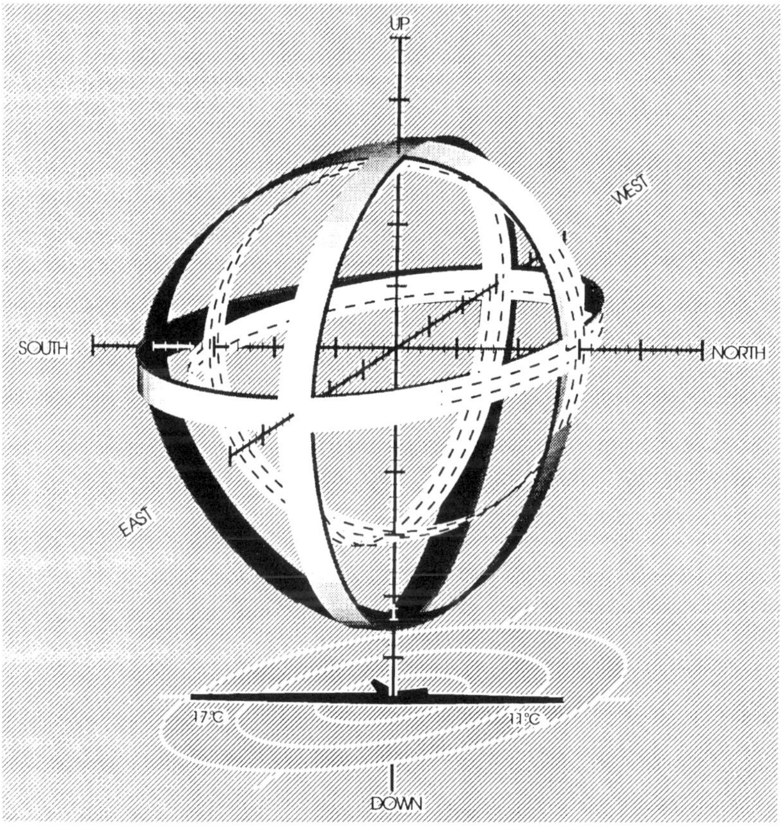

Figure 36.5. Predicted isosphere (cm) of differential EAG response of normalised value of 0.2 surrounding two *Epiphyas postvittana* pheromone dispensers placed at the center of a volume of apple leaves. White bands represent a perfect sphere at 15-cm radius, while gray bands represent the derived EAG responses. Prevailing wind direction and temperature are shown below. [Reproduced with permission from Karg et al. (1994), copyrighted by Plenum Press.]

based on the inverse square root of distance, the Pythagorean function expected to explain changes in concentration with distance.

Leaves harvested from 5 cm under the dispensers showed the highest pheromone release rate. Leaves downwind of the dispensers also had higher releases of pheromone. Pheromone molecules, once released from the dispensers, presumably dispersed downward. Pheromone release rates are greater at warm temperatures, so the spatial pattern is probably more strongly influenced by warmer winds. Temperature may also affect the capture and release of pheromone by the leaves. The significance of the role of leaves on disruption requires further evaluation, although it is clear from these experiments that biologically significant amounts of pheromone can be trapped and released by leaves, and it seems probable that this effect will have an impact on disruption.

For many moths a fluctuating, intermittent pheromone concentration is a prerequisite for a sustained upwind flight (Kennedy et al. 1981; Baker et al. 1985) and males do not fly upwind in a uniform pheromone cloud. Upwind flight, once evoked through an intermittent pheromone signal, can cease before the source is reached, turning into stationary flight (arrestment) if the pheromone concentration reaches the threshold level (Kennedy 1978). Our results to date suggest that in a pheromone-treated orchard, disruption does not operate through the maintenance of a uniform pheromone cloud. However, it appears that a flying male moth might not perceive the signal structure required for sustained upwind flight to a female.

Acknowledgments

We would like to thank Drs. Steve Green and Keith McNaughton (HortResearch) for sampling wind within the orchard canopy with sonic anemometers, and we would also like to thank Peter Shaw and Stephen Bradley for technical assistance. The research was funded by the N.Z. Foundation for Research Science and Technology, with assistance from the N.Z. Apple and Pear Marketing Board and the N.Z. Lottery Grants Board.

37

Disruption of Semiochemical-Mediated Aggregation in Bark Beetles

John H. Borden

1. Introduction

Bark beetles (Coleoptera Scolytidae) are major worldwide pests, from the comfortable country towns of southern England (Greig and Gibbs 1983) to the remote eastern Himalayas in Bhutan (Schmutzenhofer 1988). They cause immense losses of killed timber—for example, 5×10^6 m^3 of Norway spruce killed by the spruce bark beetle, *Ips typographus,* in Norway during the 1970s (Bakke 1989) and an estimated 96×10^6 m^3 of lodgepole pine killed by the mountain pine beetle, *Dendroctonus ponderosae,* in British Columbia from 1972 to 1985 (Borden 1990). It is therefore not surprising that much research has been done on developing strategies and tactics for the management of bark beetles. In the latter three decades of this century, most of this research has involved the use of semiochemicals (Borden 1993).

2. Semiochemical-Mediated Decision-Making Pathway by Host-Seeking Beetles

Ideally one might envision research on a subject arising from an insightful examination of natural processes. For host selection by bark beetles, this examination would then reveal the beetles' strategy for survival and could lead to avenues of potential practical exploitation by mankind. Had such a process occurred prior to any attempts to manipulate bark beetles with semiochemicals, applied research might have been guided by the hypothetical decision-tree for host-seeking bark beetles in Fig. 37.1.

This pathway includes four critical choices leading to the rejection of unfavorable potential hosts, but in nature a truncated pathway could involve a single choice, a binary or ternary choice pathway, or even a take-it-or-leave-it decision

Table 37.1. Summary of results from field experiments and operational trials in which disruption of response by scolytid beetles to attractive semiochemicals was achieved through an antiaggregative effect.[a]

Species and common name, if any	Semiochemicals[b]	Biological activity	Practical evaluation
Conophthorus coniperda (white pine cone beetle)	(*E*)-7-Methyl-1,6-dioxaspiro[4.5]decan (AnP)	At a release rate of 1.0 mg/24 hr greatly reduced response to sex pheromone-baited traps (de Groot 1991).	
Conophthorus resinosae (red pine cone beetle)	(*E*)-7-Methyl-1,6-dioxaspiro[4,5]decane (AnP)	Response to sex pheromone-baited traps almost completely inhibited by this spiroacetal released at 1.0 mg/24 hr (Pierce et al. 1995).	
Dendroctonus adjunctus (roundheaded pine beetle)	Verbenone	Verbenone released at an estimated rate of 12 mg/24 hr caused a 93% reduction in response to sticky traps enclosing newly infested ponderosa pine bolts (Livingston et al. 1983).	
Dendroctonus brevicomis (western pine beetle)	Verbenone (AnP) *trans*-Verbenol (MP,S) Ipsdienol (AnP,S) Estragole (K)	At release rates of 4 and 40 mg/24 hr, verbenone reduced catches in attractant-baited traps. *trans*-Verbenol attractive at low release rates and inhibitory at higher rates; did not increase inhibitory effect in combination with verbenone (Bedard et al. 1980). When released from 48 stations at 4 mg/24 hr/station, verbenone had no effect on capture of beetles at centrally located attractive bolt (Tilden et al. 1981). Response to attractant baited traps was inhibited by ipsdienol (Byers 1982). Verbenone and ipsdienol inhibit response in a synergistic manner (Paine and Hanlon 1991), reducing response to attractant-baited traps by 93% when released at 1.2 and 0.4 mg/24 hr, respectively (Bertram and Paine 1994b). Estragole at 0.5 mg/24 hr caused a 60% reduction in response to attractant-baited traps (Hobson 1995).	Verbenone inhibited attack on treated ponderosa pines at a release rate of 196 mg/24 hr/tree in 1 of 2 experiments, but had no effect at 20 mg/24 hr/tree (Bedard et al. 1980). On attractant-baited ponderosa pines, ipsdienol and verbenone released at 2 and 100 mg/24 hr/tree, respectively, reduced attack density by 88% and 73% after 4 and 7 days, respectively, lowering the density below the trees' lethal threshold (Bertram and Paine 1994a).

Dendroctonus frontalis (southern pine beetle)	Verbenone (AnP) *endo*-Brevicomin (AnP) *exo*-Brevicomin (S) Hexan-1-ol (K) Hexanal (K) Estragole (K)	Verbenone caused up to 86% reduction in response to attractant-baited traps (Renwick and Vité 1969; Payne et al. 1978), and was most effective if used as 30–50% (+) enantiomer (Salom et al. 1992). (1*S*,5*R*,7*R*)-(−)-*endo*-brevicomin caused pronounced inhibition of response to attractant-baited traps (Vité et al. 1985), confirming earlier results with the racemic compound alone or in combination with (±)-*exo*-brevicomin (Vité and Renwick 1971; Payne et al. 1977, 1978). *exo*-Brevicomin alone inhibited response to attractant-baited traps in one experiment (Vité and Renwick 1971), but not in another (Payne et al. 1978). When released at 0.3 ml/24 hr, hexanol and hexanal caused 32 and 46% reduction, respectively, in responses to attractant-baited traps; the combination of hexanol with hexanal did not enhance inhibition (Dickens et al. 1992). Estragole released at ≥160 mg/24 hr caused reduction in response to attractant-baited traps at a level comparable to that caused by verbenone (Hayes et al. 1994).	*exo*- and *endo*-Brevicomin applied together halved attack density on treated trees (Richerson and Payne 1979); when deployed with verbenone, caused pronounced reduction in landing rates and attack density on treated trees (Richerson and Payne 1979; Watterson et al. 1981) and induced competitive displacement of *D. frontalis* by *Ips avulsus* (Payne and Richerson 1985). Verbenone was effective in controlling spot infestations when trees at advancing front were treated; and freshly attacked trees were felled, causing olfactory and visual disruption (Payne and Billings 1989; Payne et al. 1992).
Dendroctonus ponderosae (mountain pine beetle)	Verbenone (AnP) *exo*-Brevicomin (MP) Frontalin (MP) Ipsdienol (AnP,S) Estragole (K) (*E*-2-Hexen-1-ol (*Z*)-3-Hexen-1-ol	(−)-Verbenone, but not its antipode, released at 0.07 mg/24 hr reduced catches in attractant-baited traps by 78% and 86% in ponderosae and lodgepole pine forests, respectively (Ryker and Yandell 1983), and reduced attack density on attractant-baited lodgepole pines to well below lethal threshold (Borden and Lindgren 1988). Reduction in response to attractant-baited traps raised to 96% by increasing the release rate to 5 mg/24 hr	Grid deployment of verbenone release devices induced dispersal of beetles from infested lodgepole pine stands and excluded beetles from uninfested areas (Amman et al. 1989, 1991; Lindgren et al. 1989a). Aerial application of verbenone impregnated plastic pellets significantly reduced infestation levels in lodgepole pine stands (Shea et al. 1992). Efficacy of treatments was inconsistent between years, geographic areas or tree species

(Continued)

Table 37.1. *Continued*

Species and common name, if any	Semiochemicals[b]	Biological activity	Practical evaluation
Dendroctonus ponderosae Con't.		(Schmitz and McGregor 1990). *exo*-Brevicomin and frontalin increased response to attractant-baited traps at release rate of 0.05 mg/24 hr and inhibited response at 5.0 mg/24 hr (Borden et al. 1987). *exo*-Brevicomin active in inducing attack on lodgepole pines (Borden et al. 1990), but inhibited attack on white pines (Pitman et al. 1978). (±)- or (*S*)-(+)-Ipsdienol released at 0.5 mg/24 hr caused up to 90% reduction in response to attractant-baited traps but had no effect on attack density on attractant-baited trees (Hunt and Borden 1988). At release rates of 0.5 ml/24 hr (Hobson 1995) and 160 mg/24 hr (Hayes and Strom 1994), estragole caused respective 70% and 77% reductions in response to attractant-baited traps. A blend of (*E*)-2-hexen-1-o1 and (*Z*)-3-hexen-1-o1, each released at 3.8 mg/24 hr almost completely inhibited response to attractant-baited traps (Wilson 1995).	(Bentz et al. 1989; Lister et al. 1990; Gibson et al. 1991; Shea et al. 1992). Combination of grid treatments with verbenone and attractive tree baits in adjacent areas was better than either tactic alone in altering attack distribution (Lindgren and Borden 1993).
Dendroctonus pseudotsugae (Douglas-fir beetle)	3,2-MCH (seudenone) (MP) 3,3-MCH (AnP) Verbenone (MP) 6-Methylhept-5-en-2-one (AnP) 3-Methylcyclohexanone Pulegone Piperitone Menthone	At release rates ≤ 0.05 mg/24 hr, 3,2-MCH was attractive in combination with other attractants, repellent at ≥ 0.5 mg/24 hr (Rudinsky (1973). 3,3-MCH released at 2 mg/24 hr caused 77% reduction in response to attractant-baited traps (Rudinsky and Ryker 1979). At a release rate of 0.3 mg/24 hr, verbenone repressed catches in attractant-baited traps, but at 10% and 0.0001% dilutions, attraction was enhanced (Rudinsky et al. 1974a). When released at 5 mg/24 hr,	Field tests with 3,2-MCH released from vials or granular formulations were of variable, but promising, effectiveness in preventing or reducing attack on trees and logs (Furniss et al. 1972, 1974, 1977; Rudinsky et al. 1974b; St. Clair et al. 1977). Large-scale field trials with broadcast 3,2-MCH-impregnated granules indicated operational readiness of disruption tactic (Furniss et al. 1981, 1982; McGregor et al. 1984). In combination with aggregation pheromone, 3,2-MCH reduced the attack

		6-methylhept-5-en-2-one caused a four-fold decrease in response to attractant-baited traps (Ryker et al. 1979). Reductions of 61–90% in responses to female-infested logs and attractant-baited traps were caused by 3-methylcyclohexanone, pulegone, piperitone, and menthone (Rudinsky et al. 1979).	density on standing trees, causing death of the attacking beetles while allowing the trees to survive (Hedden and Pitman 1978).
Dendroctonus rufipennis (spruce beetle)	3,2-MCH (seudenone) (AnP) Estragole (K)	When released from vials, 3,2-MCH caused almost complete repression of response to female-infested logs or synthetic attractants in sticky traps in Idaho and Alaska (Kline et al. 1974; Furniss et al. 1976). Estragole released at 0.1 mg/24 hr reduced response to frontalin-baited traps by 82% (Werner 1995).	Attack on felled spruce logs strongly suppressed by 3,2-MCH released from vials or bagged granules (Rudinsky et al. 1974b), or bubble caps releasing 3,2-MCH at 9–45 mg/24 hr per log (Lindgren et al. 1989b). Placement of three vials containing 3,2-MCH on stumps reduced attack density by only 50% (Dyer and Hall 1977).
Dendroctonus simplex (eastern larch beetle)	3,2-MCH (seudenone) Frontalin (S) Estragole (K)	Response to attractant-baited traps was reduced by 92% and 86% when open vials containing 3,2-MCH and frontalin, respectively, were placed in the traps (Werner et al. 1981). Estragole released at 0.1 mg/24 hr reduced response to seudenol-baited traps by 73% (Werner 1995).	
Dryocoetes affaber	*exo*-Brevicomin (MP) (–)-*endo*-Brevicomin	In a 1:2 ratio of (+) enantiomers, *exo*-brevicomin was synergistic with (–)-*endo*-brevicomin in eliciting attraction to traps, but at a 9:1 ratio at a release rate of 0.2 mg/24 hr *exo*-brevicomin was inhibitory. When added to the optimal 1:2 blend of (+) enantiomers, (–)-*endo*-brevicomin at 0.2 mg/24 hr caused 74% and 62% reductions in response to traps by males and females, respectively (Camacho et al. 1994).	

(Continued)

Table 37.1. Continued

Species and common name, if any	Semiochemicals[b]	Biological activity	Practical evaluation
Dryocoetes confusus (western balsam bark beetle)	*endo*-Brevicomin (MP) (−)-*exo*-Brevicomin	In a 9:1 ratio of (+) enantiomers, *endo*-brevicomin was synergistic with (+)-*exo*-brevicomin in eliciting attraction to traps, but at a 1:1 ratio at a release rate of 0.2 mg/24 hr *endo*-brevicomin was inhibitory (Stock et al. 1990; Camacho et al. 1993). When added to the optimal 9:1 blend of (+) enantiomers, (−)-*exo*-brevicomin at 0.2 mg/24 hr caused a 68% decline in response to traps (Camacho et al. 1993).	
Gnathotrichus retusus	(*R*)-(−)-sulcatol (*S*)	(*R*)-(−)-sulcatol caused up to 93% reduction in response to traps baited with an equal amount of (S)-(+)-sulcatol (Borden et al. 1980, 1981).	
Ips accuminatus	Amitinol (S)	Response to attractant-baited traps was significantly reduced by presence of amitinol (Kohnle et al. (1988b).	
Ips calligraphus	(*R*)-*cis*-Verbenol (S) (*S*)-(+)-Ipsdienol (S)	At a release rate of 0.1 mg/hr (*R*)-*cis*-verbenol caused a 76% reduction in response to pheromone-baited traps (Vité et al. 1976b). (*S*)-(+)-Ipsdienol at a release rate of 2 mg/hr halved the response to pheromone-baited traps (Vité et al. 1978).	
Ips duplicatus	Ipsenol (S)	Attraction to ipsdienol was strongly inhibited when ipsenol was released from sticky traps at 0.3 or 0.5 mg/24 hr (Schlyter et al. 1987).	
Ips erosus	Amitinol (S)	Response to attractant-baited traps was significantly reduced by presence of amitinol (Kohnle et al. 1988b).	

Ips latidens	*cis*-Verbenol (S) (*S*)-(+)-Ipsdienol (S) verbenone (S)	At a release rate of 3–6 mg/24 hr, both enantiomers of *cis*-verbenol caused an 88% decrease in catches in attractant-baited traps (Miller et al. 1991). At 0.1 mg/24 hr (*S*)-(+)-ipsdienol, but not the antipode, had an even greater effect (Miller and Borden 1992). Verbenone was inhibitory in a dose-dependent manner (Miller et al. 1994a).	
Ips paraconfusus (California five-spined *Ips*)	(*R*)-(−)-Ipsdienol (*S*) Verbenone (*S*)	(*R*)-(−)-ipsdienol released at 2 mg/24 hr reduced response to traps baited with male-infested logs by 91% (Light and Birch 1979). At a release rate of 4 mg/24 hr verbenone caused a 92% reduction in response to attractant-baited traps (Byers and Wood 1980). (±)-Ipsdienol or verbenone released alone at 10 mg/24 hr caused respective 91% and 87% reductions in catches in pheromone-baited traps, but no interaction in impact occurred between them (Paine and Hanlon 1991).	Verbenone and ipsdienol released at 10 and 0.2 mg/24 hr, respectively, used to repel beetles from high-risk stand of Torrey pines in combination with mass trapping to pheromone-baited traps in successful "push–pull" operation (Berson 1992; Shea and Neustein 1995).
Ips pini (pine engraver)	Ipsenol (S) (*S*)-(+)-Ipsdienol (S) (*E*)-Myrcenol (MP) Verbenone (S)	Ipsenol released at 1.0 mg/24 hr reduced response to traps baited with male-infested logs a mean of 80% in four experiments in California (Birch and Wood 1975). In similar subsequent experiments (±)- and (*S*)-(−)-ipsenol, but not (*R*)-(+)-ipsenol, caused 83–98% reduction in response (Birch et al. 1977). In Idaho, ipsenol released at rates up to 2 mg/24 hr caused a maximum reduction of 68% in response to traps baited with male-infested logs (Furniss and Livingston 1979). In California (*S*)-(+)-ipsdienol inhibited response to male-infested logs or to traps baited	When ipsenol was released at 1.0 mg/24 hr from each of four release points around a 3-m-long log baited with male beetles, attack density was reduced by 99% (Birch and Light 1977). Verbenone and ipsenol almost completely inhibited attack on groups of three felled lodgepole pines when released from five devices per log at rates of 50 and 1.5 mg/24 hr/log, respectively (Borden et al. 1992). Application of impregnated plastic beads releasing verbenone and ipsenol at 5.0 and 1.0 mg/m^2/24 hr in 15- × 15-m plots provided almost complete protection of six felled trees

(Continued)

Table 37.1. Continued

Species and common name, if any	Semiochemicals[b]	Biological activity	Practical evaluation
Ips pini Con't.		with (*R*)-(−)-ipsdienol (Birch et al. 1980); but in other parts of the geographic range, both enantiomers are synergistically attractive (Lanier et al. 1980; Miller 1990). (*E*)-Myrcenol reduced responses to attractant-baited traps in dose-dependent manner, but induced attack on logs (Miller et al. 1990). Verbenone reduced catches to attractant-baited traps in a dose-dependent fashion (Miller et al. 1995). Verbenone and ipsenol acted synergistically in reducing attack on ipsdienol-baited logs (Borden et al. 1992).	within the plots for 10 weeks when application was made 2 weeks before the first attack (Miller et al. 1994; Devlin and Borden 1994).
Ips schmutzenhoferi	Verbenone Ipsdienol (MP)	Verbenone inhibited response to attractant-baited traps by 86%. Ipsdienol appeared to enhance attraction at low ratio with attractive compounds and to inhibit response at ratios ≥ 10 times higher (Kohnle et al. 1988a).	
Ips sexdentatus	Ipsenol (S)	When released at 8 mg/hr, ipsenol caused an 87% reduction in response to ipsdienol-baited traps (Vité et al. 1974).	
Ips typographus (spruce bark beetle)	Verbenone (AnP) Ipsenol (AnP,S) Ipsdienol (MP)	Verbenone alone, ipsenol alone, or both together at release rates of 1.0 mg/24 hr, strongly reduced catches in attractant-baited traps (Bakke 1981; Schlyter et al. 1989). Ipsdienol caused an increase in response to attractants at release rates of ≤ 0.04 mg/24 hr, but a decrease at 0.6 mg/24 hr (Schlyter et al. 1987).	When population was at a moderate (but not high) level, verbenone and ipsenol released from laminated dispensers at 16–32 and 4–8 or 36–72 and 9–18 mg/24 hr/felled tree, respectively, caused a reduction in numbers of trees attacked and 83% and 98% reduction in attack density. Subsequent experiments in two different years using laminated flakes with 1200 and 600 mg of verbenone and ipsenol, respectively, per pair of logs produced similar results (Bakke 1987).

Leperisnus varius (ash bark beetle)	(*E*)-7-Methyl-1,6-dioxaspiro[4.5]decane (AnP)	Response was inhibited by 98% when this spiroketal was released from attractant-baited traps (Kohnle 1985).
Pityokteines curvidens	Ipsdienol (S)	Both enantiomers of ipsdienol caused a complete inhibition of response when released at 1 mg/hr in pheromone-baited traps (Harring 1978).
Pityokteines vorontzovi	(*S*)-(+)-Ipsdienol (S)	When released at 1 mg/hr in 30-min tests, (*S*)-(+)-ipsdienol caused a complete inhibition of response to pheromone-baited traps (Harring 1978).
Poligraphus poligraphus	(+)-Terpinene-4-ol	Inhibited response by 87% when released from attractant-baited traps (Kohnle et al. 1985).
Scolytus multistriatus (smaller European elm bark beetle)	α-Multistriatin (MP)	Enhanced response to attractant-baited traps at release rate of 5–10 μg/day, but inhibited response at 57 or 365 μg/day (Blight et al. 1983).
Scolytus scolytus (large elm bark beetle)	α-Multistriatin (AnP,S)	Response to attractant-baited traps reduced in dose-dependent manner at 5–10, 57, and 365 μg/day, respectively (Blight et al. 1983).
Xylocleptes bispinus	1-octen-3-ol (K)	Inhibited response by 83% when released from pheromone-baited traps (Klimetzek et al. 1989).

[a]References were selected for convincing and (usually) statistically analyzed data on materials capable of causing considerably more than 50% repellency.

[b]Symbols in parentheses following common names of semiochemicals refer to functional classification as follows: AnP, antiaggregation pheromone; MP, multifunctional pheromone; K, kairomone; S, synomone. Where no symbols are given, the biological role of the semiochemical is unknown or obscure. IUPAC names for semiochemicals listed under their common name are as follows:

amitinol: *trans*-2-methyl-6-methylen-3,7-octadien-2-ol
exo-brevicomin: *exo*-7-ethyl-5-methyl-6,8-dioxabicyclo[3.2.1]octane
endo-brevicomin: *endo*-7-ethyl-5-methyl-6,8-dioxabicyclo[3.2.1]octane
estragole (4-allylanisole, methyl chavecol): 1-methoxy-4-(2-propenyl)benzene
frontalin: 1,5-dimethyl-6,8-dioxabicyclo[3.2.1]octane
ipsenol: 2-methyl-6-methylene-7-octen-4-ol
ipsdienol: 2-methyl-6-methylene-2,7-octadien-4-ol
3,2-MCH: 3-methylcyclohex-2-en-1-one
3,3-MCH: 3-methylcyclohex-3-en-1-one
menthone: 5-methyl-6-(1-methylethyl)cyclohexanone
(*E*)-myrcenol: 2-methyl-6-methylene-2,7-octadien-3-ol
α-multistriatin: 2,4-dimethyl-5-ethyl-6,8-dioxabicyclo[3.2.1]octane
piperitone: 3-methyl-6-(1-methylethyl)cyclohex-2-enone
pulegone: 2-isopropylidene-5-methlcyclohexanone
sulcatol: 6-methylhept-5-en-2-ol
terpinen-4-ol: 4-methyl-1-(1-methylethyl)cyclohex-3-en-1-ol
cis-verbenol: *cis*-4,6,6-trimethylbicyclo[3.1.1]hept-3-en-2-ol
trans-verbenol: *trans*-4,6,6-trimethylbicyclo[3.1.1]hept-3-en-2-ol
verbenone: 4,6,6-trimethylbicyclo[3.1.1]hept-3-en-2-one

(Thorsteinson 1960) following direct arrival at a host suitable for attack. The most fundamental choice (Fig. 37.1, no. 1), is to reject a tree or stand of an inappropriate host species. Next in the hierarchy of decisions is to reject any tree of the correct host species that is unsuitable for attack (Fig. 37.1, no. 2). A mistake in either of the above choices can be fatal (Raffa et al. 1993). Finally, a host-seeking beetle must be able to evaluate potential hosts with enough sensitivity that it can select a susceptible tree with enough bark real estate in which to raise a brood. It has a good chance of attacking this tree successfully. A mistake leading to attack of a host tree already colonized by either hetero- or conspecific beetles (Fig. 37.1, nos. 3 and 4, respectively) can lead to an extreme competitive disadvantage through exploitative competition by the established beetles for a limited phloem resource (Berryman 1982; Rankin and Borden 1991).

Successful navigation of the decision-making pathway characteristically leads to a mass attack mediated by a blend of beetle-produced aggregation pheromones and host tree kairomones (Borden 1985). Several operational tactics utilizing these attractants to mass trap beetles or to induce them to attack trap trees have been developed (Borden 1993). A few courageous researchers (Vité et al. 1976a; Richerson et al. 1980; Tilden et al. 1981) have explored the potential for using these attractant semiochemicals as disruptants, as is now done routinely for Lepidoptera. However, incomplete efficacy, or concern as to the potential of the treatment to induce attack (e.g., along the boundaries of treated areas), has precluded development of attractant-based disruption for bark beetles in the forest environment.

In the course of exploring the bioactivity of attractive semiochemicals for bark beetles, researchers have discovered a total of 25 compounds that have proven antiaggregative activity in the field against beetles in 29 species (Table 37.1). The majority of these compounds can be classified functionally into three groups: antiaggregation pheromones, multifunctional pheromones (attractive at low concentrations and repellent at high concentrations), and interspecific synomones that cause mutual or unilateral repellency involved in resource partitioning and reproductive isolation between species. Because species-specificity in pheromone blends in the Scolytidae often is imparted by structural or optical isomerism of pheromones (Borden 1985), a surprising degree of disruption effect occurs when a beetle encounters the antipode of the bioactive isomer (Table 37.1). In many cases these antipodes function as pheromones in other species.

3. Exploitation of Multifunctional and Antiaggregation Pheromones

Very early in the history of research on bark beetle pheromones, Rudinsky (1968, 1969) disclosed the startling occurrence of a "pheromone mask" in the Douglas-fir beetle, *Dendroctonus pseudotsugae.* The mask was produced if stridulating males were placed outside the galleries of attractive females. If the females under

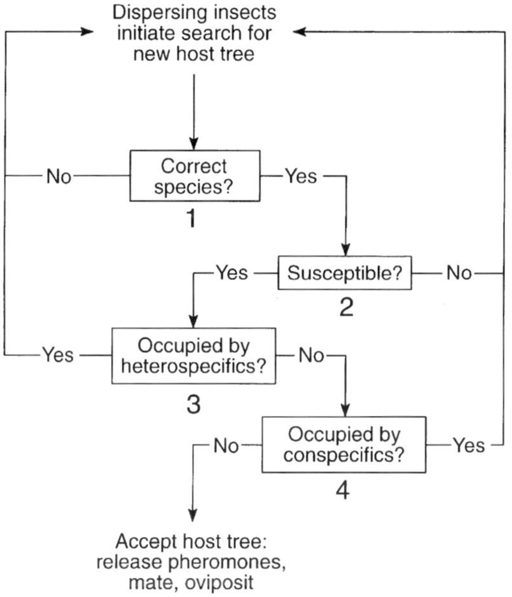

Figure 37.1. Hypothetical decision-tree, in which bark beetles must make four critical choices that lead to rejection of non-hosts or unfavorable hosts, prior to a take-it-or-leave-it choice as to whether or not to attack a suitable host.

the bark were killed with a hammer, the mask was lifted, indicating that the females produced the mask in response to a male sonic signal. This pheromone was soon identified as 3,2-methylcyclohex-2-en-1-one (3,2-MCH or seudenone) (Kinzer et al. 1971), and further research demonstrated its multifunctionality (Rudinsky 1973).

Any possibility of exploring the role of diverse potential disruptants through a deliberate, experimental examination of the critical decisions involved in host selection (Fig. 37.1) was lost in the excitement of the moment. The discovery of the antiaggregative properties of 3,2-MCH in the Douglas-fir beetle stimulated an ambitious search for other disruptants for this beetle, as well as investigations into their bioactivity and practical potential as pest management agents (Table 37.1). Another multifunctional pheromone, verbenone, was soon discovered (Rudinsky et al. 1974a), as were two antiaggregation pheromones, 3,3-MCH (Libbey et al. 1976) and 6-methylhept-5-en-2-one (Ryker et al. 1979). Four parapheromones (analogs of 3,2-MCH of no known significance in nature) were also shown to have antiaggregative properties (Rudinsky et al. 1979). Because the spruce beetle, *Dendroctonus rufipennis,* is closely related to the Douglas-fir beetle, opportunistic researchers hypothesized and soon demonstrated, that 3,2-MCH could disrupt aggregation in that species as well (Table 37.1).

Keeping pace with research that demonstrated the antiaggregative capacity of 3,2-MCH against Douglas-fir and spruce beetles was research on the practical evaluation of this promising new semiochemical. Various means of controlled release of 3,2-MCH from point sources or broadcast pheromone-impregnated granules were tried and proven to be very effective in deterring attack on single or multiple host logs. Ultimately an operational formulation of pheromone-laden plastic beads was developed for broadcast delivery from a hopper slung below a helicopter. This technology was proven to be fully efficacious in preventing attack by the Douglas-fir beetle on large areas of felled trees, as would occur after a violent windstorm (Furniss et al. 1981, 1982). In preventing attack for a single season, the phloem of downed trees would be allowed to deteriorate through natural processes to a state unsuitable for bark beetle attack. Beetle populations could not then develop in this host material to levels that would threaten adjacent standing trees the next year (McGregor et al. 1984). Despite the demonstrated efficacy of 3,2-MCH, it has yet to be registered for use as a pesticide in the United States. Its minimal potential for use against the Douglas-fir beetle in Canada would not justify registration, and applied alone it is not efficaceous enough to be recommended for use against the spruce beetle.

In applying the antiaggregant 3,2-MCH against Douglas-fir and spruce beetles, researchers were using a simple subversive variant (Fig. 37.2) of the decision-tree presented in Fig. 37.1. Within a treated area, dispersing beetles are given no choice but to take it or leave it, because the clear semiochemical message is given that the available host material is fully colonized by conspecifics. The intention is that the beetles' decision will be to continue dispersing.

The expectation of this simple model became the driving force behind numerous studies on the bioactivity and practical evaluation of verbenone, which acts as an antiaggregation pheromone for several *Dendroctonus* spp. (Borden 1985). Unlike 3,2-MCH, however, verbenone has not produced consistent results, as is

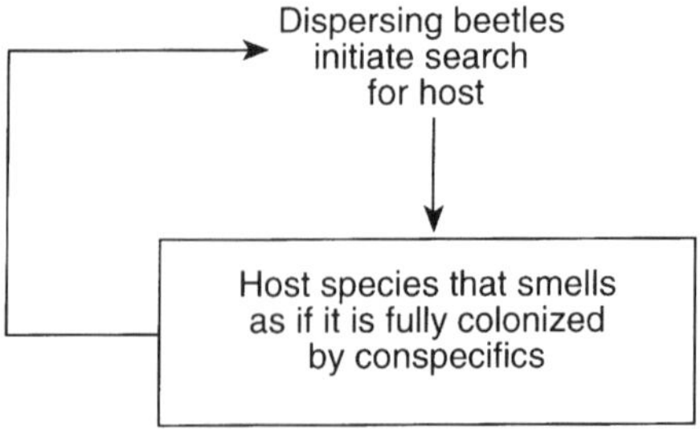

Figure 37.2. Desired decision-making pathway, in which a single multifunctional or antiaggregation pheromone is used to create the false semiochemical message that a tree or stand is fully colonized by conspecifics, leaving the target bark beetles no choice but to resume dispersal.

most depressingly evident for the mountain pine beetle, *Dendroctonus ponderosae* (Table 37.1).

Initial applied research with verbenone release devices deployed in a grid pattern in large blocks of lodgepole pines, *Pinus contorta* var. *latifolia,* clearly showed that it can cause populations of mountain pine beetles to disperse or not to aggregate (Amman et al. 1989, 1991; Lindgren et al. 1989a). However, in other experiments, similar treatments with verbenone caused a significant reduction in infestation in only one of two tests in lodgepole pine (Gibson et al. 1991), and in no case was there any effect in stands of ponderosa pine, *Pinus ponderosa* (Bentz et al. 1989; Lister et al. 1990). An encouragingly low ratio of newly attacked lodgepole pines in 1988 to trees previously attacked in 1987 in plots treated with an aerially applied granular formulation of verbenone compared to a much higher ratio in untreated control plots gave cause for optimism (Shea et al. 1992). This optimism was further justified by a threefold higher number of unsuccessful attacks in verbenone-treated than in control plots. However, the optimism was tempered by the fact that a similar experiment conducted the following year failed to yield similar results (Shea et al. 1992).

Among the explanations for poor or inconsistent efficacy of verbenone are technological problems with controlled release devices, changing stand microclimate between years, high temperatures in ponderosa pine stands, and selection for beetles refractory to verbenone as an infestation proceeds (Amman 1994). One other factor is the photodecomposition of verbenone to an inactive compound chrysanthenone (Kostyk et al. 1993). Such a phenomenon might limit the active range of verbenone to a local area around an attacked tree, preventing widespread disruption of orientation behavior from occurring in nature. In practice it might cause rapid degradation of verbenone in the open-grown sunny and hot conditions that are characteristic of many ponderosa pine stands.

4. Integrated Tactics

Failure to achieve a state of operational readiness with verbenone used alone against any *Dendroctonus* spp. (Table 37.1) does suggest that a reconsideration of the critical decision-making pathway for host-seeking bark beetles (Fig. 37.1) would be profitable. Dickens et al. (1992) found that the green leaf volatiles, hexan-1-ol and hexanal, characteristic of herbaceous and woody angiosperms, caused a moderate but significant repellency against the southern pine beetle, *Dendroctonus frontalis.* Two other green leaf volatiles, (*E*)-2-hexen-1-o1 and (*Z*)-3-hexen-1-o1, were much more strongly repellent to *D. ponderosae* (Wilson 1995). Would these and other constituents of non-host species be useful antiaggregative disruptants, when combined with multifunctional or antiaggregation pheromones? The deterrent effect of unknown volatiles from aspen, *Populus tremula,* on the European species, *Tomicus piniperda* and *Hylurgops palliatus,* suggests that this might be the case (Schroeder 1992). Similar bioactivity appears to reside in estragole (4-allylanisole or methyl chavecol), a compound that is found in high levels in physiologically weakened pines (Cobb et al. 1972) and in the volatiles captured from galleries occupied by both sexes (particularly female) of spruce beetles boring in logs of Engelmann spruce, *Picea engelmannii* (Gries et al. 1988). Its ability to deter response by spruce, larch, mountain pine, western pine and southern pine beetles to attractant-baited traps (Hayes et al. 1994; Hayes and Strom 1994; Hobson 1995; Werner 1995) (Table 37.1) suggests that it may have considerable potential as an operational disruptant. A quick glance at Table 37.1 reveals a plethora of mutually or unilaterally repellent synomones that might have practical potential in an integrated disruption program.

Incorporation of repellent synomones, as well as host and non-host kairomones, in a blend with a multifunctional or antiaggregation pheromone would lead to a still simple, if idealistic, model that describes a potentially highly disruptive effect (Fig. 37.3). In practice it would only be necessary to incorporate enough components to impart consistent results at the desired level of efficacy. If components involving more than one of the four critical choice phenomena (Fig. 37.1) were unknown or unavailable, it might be possible to combine agents involved in *one* choice (e.g., a potent blend of non-host kairomones) or to integrate a semiochemical-based tactic with some other tactic (e.g., a silvicultural manipulation of a forest stand). In the past few years the direction of research has shifted markedly toward the integration of tactics, with considerable early success.

In two cases, the antiaggregation pheromone, verbenone, has been combined successfully with a multifunctional semiochemical. For the European spruce bark beetle, *Ips typographus,* the additional semiochemical is ipsenol, which is produced by *I. typographus* males and has an antiaggregative function (Bakke 1981; Birgersson et al. 1984). However, it is also produced by sympatric *Ips* spp. (Borden 1985) and most probably also functions as a mutually repellent synomone in nature. The western pine beetle, *Dendroctonus brevicomis,* can

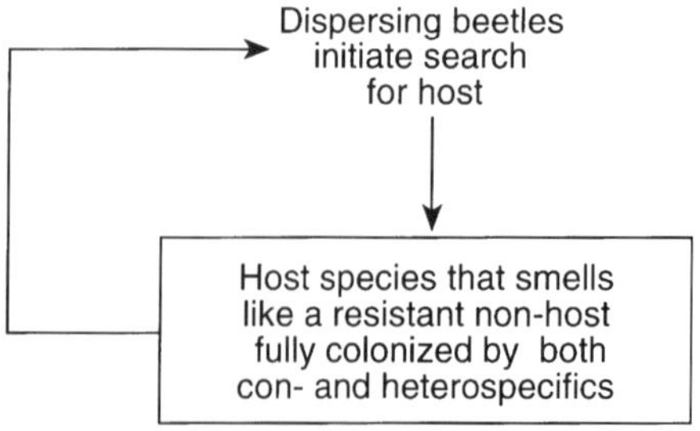

Figure 37.3. Desired decision-making pathway for bark beetles similar to that in Fig. 37.2, but conveying the integrated false semiochemical message that an unsuitable tree or stand is fully colonized by both con- and heterospecifics.

convert the host monoterpene myrcene to (+)-ipsdienol, which has an antiaggregative effect (Byers 1982). (+)-Ipsdienol is also one of three aggregation pheromones produced by the California fivespined ips, *Ips paraconfucus* (Borden 1985), and most probably also functions as a synomone in partitioning the phloem resource when it and the western pine beetle attack the same host tree, as they usually do.

In both *I. typographus* (Bakke 1981) and *D. brevicomis* (Paine and Hanlon 1991; Bertram and Paine 1994b) the blend of components had a significantly greater antiaggregative effect in field trapping experiments than either compound alone. When applied as a broadcast formulation of semiochemical-impregnated flakes on felled logs of Norway spruce, *Picea abies,* the blend of verbenone and ipsenol caused striking decreases in attack density by *I. typographus,* except where populations were very high (Bakke 1987). As for 3,2-MCH on the Douglas-fir beetle (McGregor et al. 1984), the treatment should be useful in preventing a buildup of large populations in windthrown timber or slash that could then threaten to kill living trees.

The challenge of developing an operational treatment using an antiaggregant blend for the western pine beetle is daunting, because this species attacks and kills large standing trees and rarely infests downed timber. The treatment thus must be effective in three-dimensional space. In a 12-replicate experiment, the boles of ponderosa pines were treated at 1-m intervals up to 5 m above ground with devices releasing ipsdienol and verbenone, respectively, at 2 and 100 mg per 24 hr per tree. The treated trees and 12 control trees were then baited with an attractant formulation, and the arrival and attack of western pine beetles was monitored. The inhibitor treatment caused reductions in attack density of 88% and 73% within 4 and 7 days after treatment, respectively. The mean attack density after 7 days on treated trees was 34 per m^2 of bark surface, below the hypothesized threshold for death of the trees (Bertram and Paine 1994a). In an operational setting without the attractive baits, it is thus probable that this treatment would be highly effective in protecting high value trees or small stands from attack.

As for *I. typographus,* populations of the pine engraver *Ips pini* can build up in felled timber and slash. Because intensive silviculture can leave copious amounts of vulnerable slash on the ground, *I. pini* is becoming a silvicultural pest (Livingston 1979). There are no known antiaggregation pheromones for this species, but two repellent synomones, ipsenol and verbenone, are respectively produced by a sympatric *Ips* species, *Ips latidens,* and by the mountain pine beetle (Miller et al. 1991; Borden et al. 1992). Used together to create the unnatural message that suitable lodgepole pine logs were inhabited by two heterospecifics, the antiaggregant blend released from individual dispensers stapled to the logs reduced attack by *I. pini* to a negligible level (Borden et al. 1992). Taking the tactic to a state of operational readiness, Devlin and Borden (1994) used semiochemical-impregnated plastic beads in a simulated aerial application.

When the application was made 2 weeks before the first attack by *I. pini* in the spring, this treatment provided 10 weeks of essentially complete protection from attack of slash in 225-m^2 thinning plots.

Blends of antiaggregants have also been used with moderate effectiveness against the southern pine beetle (Table 37.1). However, the greatest success with this species has been to integrate the use of antiaggregation pheromone, verbenone, with a silvicultural treatment in spot infestations (Payne and Billings 1989; Payne et al. 1992). During outbreaks, these spots are highly dynamic entities with the head of the spot advancing in a directed manner, and with trees rapidly changing in status from uninfested to newly infested, brood-producing, and dead, as the spot rolls through the forest. The dynamics are predictable and have been effectively modeled (Stephen and Lih 1985).

An integrated strategy devised for treating these spot infestations is now in a state of operational readiness. In this strategy, verbenone is applied to uninfested trees to disrupt attack at the head of the infestation, and newly infested trees behind the head are felled to disrupt visual orientation of beetles emerging from the brood-producing trees. When evaluated in comparison to pretreatment spot growth, and in comparison to post-treatment growth predicted by the model in the absence of treatment, this integrated bait and cut tactic was highly effective in eliminating spot infestations. It was also superior to using either the pheromone or visual disruptant treatment alone. The fate of beetles affected by the disruptant tactic is unknown; but prolonged survival during dispersal in high summer temperatures is doubtful, and there was no evidence in any instance of new spot infestations developing (Payne et al. 1992).

Because a semiochemical disruptant for one species may be an attractant for another (particularly if the disruptant is a synomone), it is possible that a disruptant treatment for a primary pest species may induce attack by a secondary species, causing the former to be competitively displaced (Borden 1992). This, in fact, occurred when southern pines treated with a blend of *exo*- and *endo*-brevicomin for the southern pine beetle were heavily colonized by *Ips avulsus* (Payne and Richerson 1985). Judicious selection of disruptants may enhance such effects in the future. In addition, Bakke (1987) hypothesized that clerid predators (*Thanasimus* spp.) would respond to the ipsenol in the integrated disruptant treatment for *I. typographus,* accentuating the treatment effect by consuming the few beetles that were not affected by the disruptant. Because entomophagous insects can respond to any signal indicating the presence of potential hosts, including natural antiaggregants (Borden 1985), enhanced (but not detected) predation may have already occurred in past disruption treatments.

The uncertain fate of behaviorally disrupted beetles could potentially be addressed by introducing attractive semiochemicals (Fig. 37.1) into the treatment. In this ultimate tactic for disrupting aggregation (Fig. 37.4), two types of semiochemicals would be employed: disruptants to push beetles away from the resource to be protected, and attractants to pull them away. This push–pull maneuver

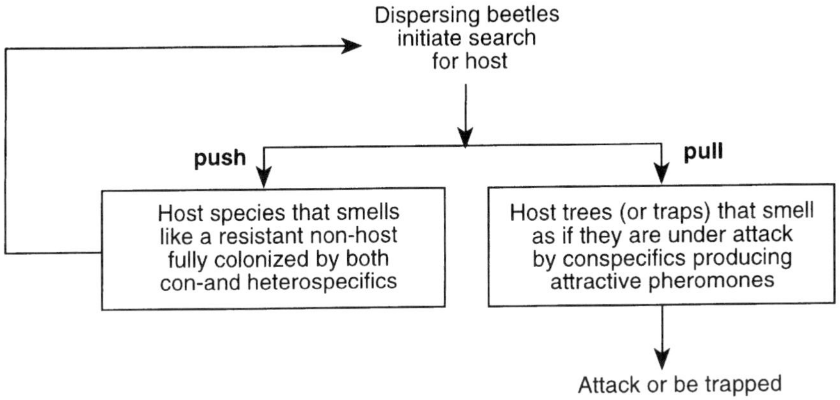

Figure 37.4. Optimal decision-making pathway integrating all possible negative semiochemical stimuli in Fig. 37.1 to "push" the target bark beetles away from a tree or stand at risk of attack, coupled with a lethal alternative, a source of attractive, semiochemical-baited trees or traps that "pull" them away.

incorporates the attractive and deterrent elements used by Hedden and Pitman (1978) in the attack density regulation tactic developed for single trees. In practice it more closely resembles the stimulo-deterrent diversion technique proposed for onion maggots, *Delia antiqua* (Miller and Cowles 1990).

In a seven-replicate test of the push–pull tactic, Lindgren and Borden (1993) applied verbenone release devices at 10-m centers in 50 × 150-m panels in lodgepole pine stands infested by the mountain pine beetle. These rectangular areas simulated a wildlife corridor or riparian zone which might require protection from further attack. In identical flanking panels, three large pines at 50-m centers were baited with attractive semiochemicals to create three centers of attraction on each side of the disruptant panel. The ratio of newly attacked trees in the flanking panels over those in the middle panel was used as the primary means of assessment. The integrated tactic introduced a consistency not found in the use of verbenone alone. In each of seven replicates, the ratio of newly attacked trees in the baited flanking blocks over those in the central block was significantly higher than random, and the normalized (by area) mean ratio was 4.9 compared to 3.6 for the verbenone treatment alone, and 1.7 and 1.6 for baits alone and untreated controls, respectively. A large-scale trial is required, however, before this tactic is judged to be operational.

The push–pull tactic was also used heroically in an operational IPM program implemented in southern California to save a rare stand of Torrey pines, *Pinus torreyana,* from devastation by *Ips paraconfusus* (Shea and Neustein 1995). Two synomones were used: verbenone produced by western pine beetles (Renwick 1967), and (−)-ipsdienol [formulated as (±)-ipsdienol] produced by pine engravers (Birch et al. 1980). Separate dispensers of each were placed 30 m inside the green uninfested portion of the stand at 10 centers along a 270-m transect. A counterpart line of 10 groups of three multiple-funnel traps was placed along a similar transect 30 m inside an adjacent area of beetle-killed pines. This program was implemented for 86 weeks from May 1991 to September 1993. A total of 330,717 beetles were captured, and not a single Torrey pine died in the reserve after August 1992.

5. Prognosis

The future for disrupting aggregation by bark beetles appears bright. A vigorous research effort is ongoing, particularly on developing new integrated tactics. Promising initial results with hexanal, hexan-1-o1, and estragole suggest that these and other compounds from host and non-host trees will soon be used operationally in blends with beetle produced pheromones and synomones. Concurrently, demands on wild and urban forests are increasing. A single tree vulnerable to attack by bark beetles on the grounds of a million dollar house in the wilderness may have enormous monetary value, justifying even very expensive tactics to protect it. Of less monetary but higher social value are non-timber uses of forests in areas set aside for recreation, water production, and the protection of natural flora and fauna. High investments will be made in the future for silvicultural treatments to maximize the productivity and quality of stands grown for a harvest of timber. As they have for centuries as part of nature's recycling strategy, bark beetles will threaten to kill trees and destroy forests. In the quest to moderate the course of natural events, we will increasingly seek to disrupt aggregations of bark beetles to protect our trees and forests.

38

First Steps in the Use of Aphid Sex Pheromones

John A. Pickett, Lester J. Wadhams, and
Christine M. Woodcock

1. Introduction

In 1977, two pioneers in the field of aphid chemical ecology stated that "At this point, methods by which aphid sex pheromones could be used to inhibit mating involve sheer conjecture, since synthetic pheromone is unavailable for testing" (Nault and Montgomery 1977). After the first identification of an aphid sex pheromone (Dawson et al. 1987), this testing has now proceeded to the extent that commercially funded programs are developing these pheromones for agricultural and horticultural use. This chapter charts earlier investigations into aphid semiochemicals, the identification of the sex pheromones, and recent developments in aphid chemical ecology (Pickett et al. 1992) in the context of aphid control.

2. The Need for Aphid Control

Aphids represent the main insect pests on many agricultural and horticultural crops in temperate and northern climates. They cause direct feeding damage and also transmit virus diseases to crop plants; for example, the peach-potato aphid, *Myzus persicae,* transmits beet yellows virus in the semipersistent mode and potato virus Y in the nonpersistent mode. Control is usually by systemic aphicides, although these have been superseded by more selective direct-acting agents such as the carbamate pirimicarb. Resistance has developed to the organophosphorus compounds and is now beginning to cause problems with pirimicarb (Moores et al. 1994). New agents with greatly enhanced selectivity such as imidacloprid have been developed, but resistance will evolve if such compounds are overused or used inappropriately (Denholm and Rowland 1992). Other approaches to aphid control should be explored to deal with legal requirements for reduced pesticide inputs and also to provide alternatives where resistance to conventional pesticides develops. Novel control methods based on use of semiochemicals should also include exploitation of pathogens, predators and parasitoids that naturally regulate aphid populations on crop plants.

3. Early Studies on Aphid Semiochemicals

Hille Ris Lambers initially attempted, but without success, to use the aphid alarm pheromone, (*E*)-β-farnesene, which causes dispersal of aphids when attacked by predators, to reduce landing of alate aphids on seed potatoes (Hille Ris Lambers and Schepers 1978). Following suggestions by Nault (1973), attempts were made to combine the alarm pheromone activity with other control agents against the cotton aphid, *Aphis gossypii,* on glasshouse ornamental crops, particularly chrysanthemums. Although the alarm pheromone gave only a small increase in mobility of *A. gossypii,* this was sufficient to increase pick-up of the pathogen *Verticillium lecanii* and thereby cause acceptable mortality (Pickett et al. 1986). Similar results were obtained in the field using contact pesticides such as permethrin. In both cases, electrostatic spraying systems were employed to present the alarm pheromone to the aphids in the most efficaceous way. (*E*)-β-Farnesene is now available commercially and there have been further attempts to use this material in the field, including higher levels of treatment where some direct toxicity can be observed (Ester et al. 1993), presumably via a hormonal effect noted previously (Mauchamp and Pickett 1987).

Antifeedants have been successfully employed in the field against transmission of virus by aphids. The antifeedant drimane (−)-polygodial, extracted from the water-pepper *Polygonum hydropiper* (Polygonaceae), was used against transmission of barley yellow dwarf virus and gave results statistically similar to those obtained with the broad-spectrum pyrethroid cypermethrin (Dawson et al. 1986). However, three applications of the antifeedant were required, compared to one application of the pyrethroid. It is hoped that recent studies on the mechanisms by which polygodial inhibits aphid feeding, and thereby virus transmission, will lead to further practical developments (Powell et al. 1993).

4. "Push-Pull" Against Aphids

Despite initial success with the alarm pheromone and antifeedants, it was considered that a more sophisticated strategy for using aphid semiochemicals should be developed in line with other work (Miller and Cowles 1990). A push–pull or stimulo-deterrent diversionary strategy against aphids has been proposed: Treatments on the harvestable crop would include inhibitors of kairomones acting as attractants for aphids, antifeedants where appropriate, and attractants for aphid predators and parasitoids. In addition, a sacrificial or trap crop would be utilized with kairomonal attraction enhanced by choice of cultivar or application of attractant formulations. On the sacrificial crop, selective control agents such as fungal pathogens would be employed.

5. Aphid Sex Pheromones

Depending on species and climate, many aphids feed and reproduce parthenogenetically on the secondary or summer host, where the main pest impact is observed. In the autumn, presexual female morphs, the gynoparae, migrate to the primary or winter host. Here, true sexual females, the oviparae, are produced and on maturity these release a sex pheromone which is used by the males to locate mates. The sex pheromones are detected principally by olfactory cells in the secondary rhinaria on the male antenna, and electrophysiological studies played an important role in the identification of the active components. Using tungsten microelectrodes (Boeckh 1962), single cell recordings (SCRs) can be obtained from individual olfactory sensilla; these SCRs, combined with high-resolution capillary column gas chromatography (GC) (Wadhams 1990), were used to locate physiologically active peaks in volatiles released by the females. For many species of pest aphids, the sex pheromones were found to comprise one or both of the cyclopentanoids nepetalactone and nepetalactol (A and B, respectively, in Fig. 38.1) in different ratios (Pickett et al. 1992). It was interesting to observe that there are two different cell types in the secondary rhinaria responding separately to the two compounds, and dose–response data showed high specificity between the two cell types for the respective compounds (Dawson et al. 1990). Electroantennogram (EAG) studies on males of the black bean aphid, *Aphis fabae,* have shown that the whole antenna is 1000–10,000 times more sensitive to the sex pheromone components than to green leaf volatiles such as (*E*)-2-hexenal (Hardie et al. 1994d).

6. Field Work with the Aphid Sex Pheromone

The damson-hop aphid, *Phorodon humuli,* was selected for initial field work, in collaboration with C. Campbell at Horticulture Research International, East

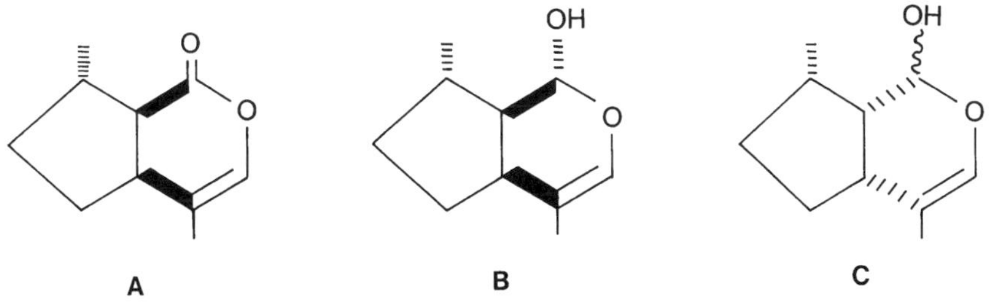

Figure 38.1. Structure of aphid sex pheromone components.

Malling, United Kingdom, and provided the first demonstration of long-range attraction of flying aphids (Campbell et al. 1990; Pickett et al. 1992), which had previously been disputed (Steffan 1987). Hops, *Humulus lupulus* (Cannabiaceae), are the summer host for this species and *Prunus* spp. provide the winter host. Combined GC–SCR on the males, using an extract of volatiles from calling oviparae, showed that the sex pheromone for *P. humuli* comprised two diastereoisomers of a nepetalactol with a different stereochemistry (C, Fig. 38.1) from that observed for most other aphid species (Campbell et al. 1990). In an autumn field trial, six Petri dish water traps containing this compound caught over 3000 male *P. humuli,* whereas a suction trap, sampling over 500 m^3 of air per hour, caught less than 400. In these trials, male aphids were observed to orient toward the pheromone traps in surprisingly strong winds. This phenomenon is being quantified in collaboration with J. Hardie's group at Imperial College, Silwood Park, United Kingdom, using a portable wind tunnel in the field to generate a discrete plume. Preliminary trials indicate that male behavior can be influenced by a sex pheromone lure several meters away from the source. Although mating disruption and trap-out are being considered commercially as control methods, collaborative efforts between Rothamsted and East Malling involve use of the pheromone to attract male *P. humuli* into traps containing a particular strain of *Verticillium lecanii* effective at the relatively low field temperatures occurring during the autumn migration. The study has now been extended to other species—for example, the blackberry-cereal aphid, *Sitobion fragariae* (Hardie et al. 1992). In Korea, catches of over 1000 aphids per day per water trap have been obtained for a number of species, including *Tuberocephalus momonis, Lachnus tropicales,* and *Aphis citricola* (K.S. Boo, Seoul National University, Suwon, South Korea, personal communication).

It was considered likely that volatiles from the primary or winter host might also have an influence on the attraction of males. In initial trials, an extract of *Prunus* sp. bark synergized the activity of the sex pheromone for *P. humuli* (Campbell et al. 1990) and a similar situation was found for the bird-cherry-oat aphid, *Rhopalosiphum padi,* with volatiles from the bird-cherry, *Prunus padus* (J. Pettersson, personal communication; Hardie et al. 1994c). However, it appeared that host volatiles did not have a role in the attraction of male *Cryptomyzus galeopsidis* to conspecific females (Guldemond et al. 1993).

During field trials on the sex pheromone components, it was observed that certain aphid parasitoids were also caught in the traps containing the nepetalactone A (Hardie et al. 1991, 1994a). EAG studies, initially on *Praon volucre,* but now including many other parasitoid species such as *Aphidius* spp. and *Diaeretiella rapae,* showed that there was a high inherent sensitivity to this component, with a lower response to related compounds, including nepetalactone isomers and nepetalactols (Hardie et al. 1993). It therefore appeared that an "innate" response was possible with these parasitoids. Indeed, in initial field trials, increased parasitism on the grain aphid, *Sitobion avenae,* by *P. volucre* was observed in nepetalactone-treated plots (Lilley et al. 1994; C.P.M. Tripathi, personal communication). In these trials, *P. volucre* was the main parasitoid attracted. However, for certain crops, it would be preferable to attract other aphid parasitoids, such as *Aphidius ervi,* against the pea aphid, *Acyrthosiphon pisum. A pisum* employs both the nepetalactone A and the nepetalactol B as its sex pheromone, in a 1:1 ratio (Pickett et al. 1992). By using lures with this formulation in a field pot trial, it was found that *A. ervi* parasitized *A. pisum* to over three times the extent (C.P.M. Tripathi, personal communication). This work has now formed the basis of a program in which set-aside strips are used to build up aphid populations so that parasitoid searching behavior can be stimulated in cereal crops by means of aphid sex pheromones. A semiochemical released by hyperparasitoids causing dispersal of aphid parasitoids has recently been identified (Höller et al. 1994), and this knowledge could also lead to ways of improving parasitoid aggregation within the crop canopy for aphid control.

7. Sex Pheromone Biosynthesis

Development of sex pheromones for attracting aphids and possibly disrupting mating, and also as attractants for aphid parasitoids, is creating a need for substantial quantities of these compounds for field testing. Synthesis is relatively difficult on a large scale, and a new program has been established to explore the possibilities of producing these pheromones by genetically transformed fermentation organisms. During the search for the identity of aphid sex pheromones, very small amounts of the open-chain compound citronellol have been found. It has been suggested that, in plants, citronellol could be the biosynthetic precursor of nepetalactones (Grandi et al. 1984), and this is being investigated in aphids by isotopic labelling studies. However, a more convenient route to the biochemistry is to use plants in the family Labiatae (= Lamiaceae), such as *Nepeta* spp. which biosynthesize nepetalactones with the necessary stereochemistry (Eisenbraun et al. 1980). By synthesizing the labeled precursors, early steps in the biosynthesis of the *P. humuli* sex pheromone have been studied in compounds ranging from geraniol through to those oxidized at the *trans*-8-methyl. A partial clone has been obtained for the gene that encodes the cytochrome P450 in the initial oxidative step, and the enzyme effecting the next oxidation to the aldehyde groups is now characterized (Hallahan et al. 1992; Hallahan et al. 1994). In collaboration with M. Rhodes at the Institute of Food Research, attempts are being made to establish tissue cultures from *Nepeta* spp. as another route to providing fermentor production of the sex pheromone. Currently, the most convenient method for producing these compounds is to cultivate *Nepeta* spp. plants in the field and extract the material by steam distillation, followed by chromatographic purification and, where necessary, reduction to the lactols (Dawson et al. 1989).

8. Other Aphid Semiochemicals

In the spring, alates are produced on the primary host which migrate back to the herbaceous summer host. Coupled GC–SCR on the primary rhinaria on the fifth and sixth antennal segments of these spring migrants have been used in the identification of attractant compounds from summer hosts—for example, for *P. humuli* (Campbell et al. 1993). However, other cells located in the primary rhinaria respond to volatiles from non-host plants. These compounds may be employed by aphids to avoid inappropriate hosts and thus could provide candidate repellents. Indeed, one compound identified in this way for *A. fabae* is (1*R*,5*S*)-myrtenal, which in the linear track olfactometer acts as a repellent and also inhibits attraction to host volatiles (Hardie et al. 1994b). J. Pettersson (personal communication) had suggested that, while autumn migrants may find volatiles from the primary host attractive, the spring migrants may find such compounds repellent. By GC–SCR on spring migrants of *R. padi,* using volatiles from the winter host, *Prunus padus,* methyl salicylate was identified as having strong electrophysiological activity and was found to reduce attraction to cereal leaves (Pettersson et al. 1994). In Sweden, in a field trial on barley, two formulations of methyl salicylate reduced colonization by *R. padi* by approximately 50%. Models predict that this level would be sufficient to allow parasitism to bring the aphid population below the acceptable damage level (Wiktelius and Pettersson 1985). *R. padi* is not found at high population densities in the United Kingdom in the summer, but further trials using methyl salicylate on wheat demonstrated similar reductions in populations of *S. avenae* and the rose-grain aphid, *Metopolophium dirhodum* (Pettersson et al. 1994). The combined use of aphid sex pheromones (to attract parasitoids) and repellent compounds, initially methyl salicylate, is now being explored.

9. Conclusion

The search is on for other non-host plant chemicals that will act as repellents for pest aphids. The identification of sex pheromones has been extended to a wide range of aphid pests and could include more, where there is a need or scientific interest. The use of these semiochemicals is being attempted in a wide European collaboration from southern Italy to Sweden, and attempts are being made to exploit some aspects by biotechnology.

Acknowledgments

We gratefully acknowledge permission from our valued colleagues to quote much material as personal communications. The work was in part supported by the United Kingdom Ministry of Agriculture, Fisheries and Food. IACR receives grant-aided support from the Biotechnology and Biological Sciences Research Council of the United Kingdom.

39

Putting Pheromones to Work: Paths Forward for Direct Control

Tristram D. Wyatt

1. Introduction

Insect pheromones offer exciting examples of pure science applied to human benefit. Even before the first pheromones had been identified, the potential for practical exploitation was recognized—a potential that has now been realized in many successful schemes using pheromones for direct control of insect pests. Although most uses have been in agriculture and forestry, manipulation of vector insects with behavior-modifying chemicals looks increasingly possible in medical entomology—for example, via the sex pheromones of sandflies (e.g., Hamilton et al. 1994) and oviposition attractants of mosquitoes (e.g., Beehler et al. 1994; Mordue et al. 1992). There still remains the question why the application of pheromones has not been greater.

Each pheromone symposium since the 1960s has emphasized the potential for pest control, but even "potential has a shelf-life" (Atwood 1990). However, we should remember that it is still less than 40 years since the first chemical identification of an insect pheromone and that we have made massive strides since then. Far from being a time for pessimism, I will argue that the basis for optimism is strong and there should be a big push for support. Compared with conventional pesticides, pheromone research has been only modestly funded. The rationale for increasing funding on pheromone research includes the negligible environmental impact of pheromones and increasing evidence of their effectiveness in pest control. Nonetheless, pheromones represent only 2% of the dollar sales of insecticides worldwide (1985 data in Jutsum 1988), despite great efforts on the part of researchers, companies, and extension workers. How far can slow deployment be blamed on unanswered questions in the scientific understanding of pheromone communication as opposed to practical difficulties in implementation? I will return to each of these topics.

Much of the above could have been written any time since 1960, so what makes now a good time? In part this is due to recent significant advances in pheromone research, discussed in this volume, particularly in our understanding of orientation mechanisms (Baker and Vickers this volume, Cardé and Mafra-Neto this volume) and in our ability to measure pheromone concentrations in the field (see below). Developments in neurobiological understanding are also promising. However, equally important are changing attitudes to pest control, by both growers and the public.

2. Current Success of Pheromones

We have come a long way since Rothschild (1981) said that "pheromone researchers . . . will have to accept that it may be some years before disruption is accepted as a viable control method for any more than 1 or 2 species." Cardé and Minks (1995) review some of the major successes of direct control of moth pests by pheromones in field crops and orchards: commercially viable formulations as effective as, or better than, the conventional insecticides they replace. Borden (1993) reviews a similar picture of success in the use of pheromones for control or monitoring of forest pests, both beetles and Lepidoptera. Chapters in this part of the volume describe progress within these and other systems.

2.1. Using Pheromones: Monitoring and Control

The focus of this section is on using pheromones for direct control of insect pests, but a major use of pheromones continues to be for monitoring pest populations both of crop and orchard pests (Wall 1989), stored products (Burkholder 1990), and forestry pests (Borden 1993). Pheromone-based monitoring provides one of the most effective survey methods for detecting pest species. Trumble (this volume) describes some of the persistent problems with monitoring as a quantitative decision tool—in particular the low correlations between adult male trap catches and caterpillar levels in the crop. However, more sophisticated schemes incorporating decision thresholds have been tried with some success, and significant reductions in pesticide use can result (Wall 1989).

Pheromones have been used for direct control of insect pest populations in three main ways: mass trapping, "lure and kill" (including "push–pull" mechanisms), and mating disruption. Much of what follows emphasizes control of moth pests by mating disruption through use of synthetic sex pheromones released in the field to permeate the crop or provide numerous point sources of synthetic pheromone. In pest moths (among the most important of agricultural pests), because it is the larvae which do the damage, the aim of pheromone intervention is to prevent fertilization of eggs by preventing mates meeting. However, direct control in other systems, especially with bark beetles (Borden 1993; Borden this volume), does not involve interference with sex pheromones but, instead, involves manipulation of aggregation pheromones and the process of tree colonization, including exploitation of sometimes complex species interactions mediated by semiochemicals. There are many other potential uses of semiochemicals for control (Tumlinson 1988), including marking pheromones such as oviposition marking pheromones in both flies (Prokopy 1981) and moths (e.g., Schoonhoven 1990), oviposition attractants, larval interactions mediated by pheromones, kairomones which modify the behavior of natural enemies (e.g., Pickett et al. 1992; Pickett et al. this volume), and direct control by nonpheromone attractants such as male-lures for tephritid fruit flies for lure and kill (Cunningham et al. 1990). Reviews of pheromone applications to particular crops can be found in Ridgway et al. (1990). Jutsum and Gordon (1989a) and Howse et al. (1996) review applied pheromone use in agriculture more generally.

2.2. How Have Research Agendas Changed?

This book comes some 20 years after the classic volume edited by Birch (1974). Since then the use of pheromones as control agents has been reviewed regularly (e.g., Shorey and McKelvey 1977; Mitchell 1981; Nordlund et al. 1981; Kydonieus and Beroza 1982; Campion 1984; Wall 1984; Jutsum and Gordon 1989a; Ridgway et al. 1990; Jones 1994). It is sobering to see how little research agendas and concerns have changed since then. Many of the points raised by Birch and successors are still concerns: the mechanism of mating disruption (involving sensory biology and behavior) and the need to understand the population biology (including migration and other characteristics) of the pest species. Other issues have included the difficulty of evaluating the effectiveness of pheromone trials and practical questions of registration. The major question now must be how to move research and applications forward.

Where initially the challenge was chemical—the enormous quantities of material originally needed for analysis are legendary—with the integration of gas chromatography—coupled mass spectrometry (GC–MS) with gas chromatography–electroantennogram/single sensillum recording (GC–EAG/SCR) and the increasing sophistication of chemical analysis, chemical identification is rarely the rate-limiting step. Instead, the persistent problem is the complexity of insect behavior, and in particular our ignorance of behavior in the field. Whereas pheromones can be used for monitoring purposes within a few years of discovery, mating disruption takes much greater understanding.

3. Progress in Understanding Mating Disruption: Successes and Problems

Since the effectiveness of direct control with pheromones has been demonstrated in many systems, one could argue that we do not need to understand the mechanisms. It may seem a luxury, but as Sanders (this volume) points out, to improve methods and understand how to change them when they fail, we need to know how they work. Such an understanding would also help in predicting likely resistance mechanisms (although insects are likely to surprise us). Much of the discussion that follows is devoted to mating disruption in moths, but many of the methodological problems are relevant to all pheromone-mediated control systems.

3.1. Poor Understanding of the Mechanism(s) of Mating Disruption

In most species, males find the female by flying upwind in response to her pheromone. To prevent successful mating, a number of stages in the mate location and courtship sequence could be interrupted, but long-range orientation may be the link most vulnerable to interference from synthetic pheromone because short-range orientation and courtship may involve other sensory cues (Sanders this volume). Mating disruption describes the result not the mechanism(s) by which this is achieved.

Sanders (this volume) and Cardé (1990) review the supposed mechanisms, much as Bartell (1982) proposed: (1) sensory adaptation (at the peripheral receptor level) or habituation in the CNS (lumped as "sensory fatigue" in Sanders), (2) false-trail-following (competition between natural and synthetic sources), (3) camouflage of natural plumes by ubiquitous high levels of synthetic pheromone, (4) imbalance in sensory input by massive release of a partial pheromone blend, and (5) the effects of antagonists to attraction and pheromone mimics.

As Sanders explains, field results combined with laboratory wind-tunnel tests and electrophysiological recordings can sometimes suggest the main mechanisms which might be involved in particular situations but given the variety of proposed mechanisms, the real breakthrough will be to design critical field and laboratory experiments to distinguish the hypotheses. However, even in one species, many mechanisms may be acting, perhaps synergistically (Cardé 1990).

Below I suggest that collaboration of behavioral and field scientists with neurobiologists might lead to greater understanding.

3.1.1. Orientation Mechanisms and Mating Disruption

The unifying paradigm is the growing understanding of how males track upwind toward an odor source (Baker 1990; Baker and Vickers this volume; Cardé and Mafra-Neto this volume; Mafra-Neto and Cardé 1994; Vickers and Baker 1994b), building on the work of the last 25 years. The link between male behavior and the filamentous structure of odor plumes (Murlis et al. 1992) has been clarified. The key features seem to be the male's brief surge upwind in response to each antennal contact with filaments of pheromone-laden air and casting on its loss.

The male's response to pheromone "filaments" helps explain why the natural blend is often the most effective for mating disruption (Minks and Cardé 1988). Vickers and Baker's (1992) wind-tunnel experiments show that the individual components of multicomponent blends (the norm) need to be in the same puff for optimum up-wind flight, in line with other behavioral work showing the importance of the full blend at all distances from the female (Linn and Roelofs 1989). Nonetheless, some mating disruption systems have used off-blends successfully (see Sanders this volume; Arn and Louis this volume).

Sanders (1981) observed that in field trials, "inhibitors" were only effective if released very close to the female (see also Cardé 1990); he suggested that this was due to the structure of plumes, as above: Only if superimposed on the pheromone, in the same filaments, would inhibitory compounds work. Recent laboratory work supports this (Liu and Haynes 1992; Rumbo et al. 1993). Although inhibitors may offer control in some cases (e.g., Bengstsson et al. 1994), the prospect is not generally promising.

3.1.2. Primer Effects in Mating Disruption

Most behavioral work on mechanisms has understandably focused on the immediate (releaser) effects of synthetic pheromones during the males' search for the female. Much of the effect could be due to the longer-term (primer) effects of permeation of the field with pheromone during the rest of the day (or even the period of activity immediately preceding flight). Despite pioneering work by Bartell (review 1982) and later work (e.g., Linn and Roelofs 1981; Liu and Haynes 1993b), the effects of pre-exposure of different types and duration are not clear. There is a pressing need to expand the timescale and range of experiments to investigate the more subtle primer effects.

Primer effects also reinforce the need to know where resting animals are. For example, during the day, pink bollworm male adults go down to the base of the plants into the soil and may avoid exposure to the pheromone in the air (Flint et al. 1975).

3.1.3. Wider Taxonomic Variations: Generalizing to Other Insect Orders

While our understanding of moth orientation increases, an underlying question remains about how far these inflight mechanisms apply to other insect groups such as beetles, sawflies, and now aphids (Pickett et al. 1992; Pickett et al. this volume) which also use long-distance pheromone communication (Cardé 1990). All moths, through common ancestry, might well share the same mechanisms, but other taxa could have evolved orientation behavior independently. This is important because moth orientation models inform much of the theoretical discussion of disruption. The same question may apply to other pure and applied aspects of mating disruption.

3.1.4. Tracking Individual Moths in the Field

Currently a major handicap is that it is not usually possible to track individual moths in the field, except for short distances close to the pheromone source in experiments (e.g., Willis et al. 1994). An unanswered question is the behavior of males before first contact with the pheromone. Could radiotracking yet come within reach for the larger moths, as miniaturization of electronics continues? This technique would also help answer what the animals do during the day.

3.2. Pheromone Concentrations in the Field and the Structure of Plumes

One of the major unknowns in pheromone research is now partly solved: New portable electroantennogram (EAG) and single sensillum recording (SCR) equipment mean that we can now measure instantaneous concentrations of pheromones in the air (Färbert et al this volume; Van de Pers and Minks this volume; Suckling and Karg this volume). As recently as 1990, Arn wrote that the invention of a "sniffer" to monitor concentrations continuously in the field would be "a golden key to pheromone applications" (Arn 1990). Earlier techniques could only average over a longer period by adsorption on a pheromone collector because air concentrations (at pg/liter) are too low for instantaneous measurement by GC–MS. The new systems exploit the sensitivity and specificity of the moth's antenna to its own pheromone and combine this with a mechanism for calibration of the EAG or SCR responses. Whether EAG or SCR is the best technique may depend on the question.

The new techniques will help in two ways—first, by allowing the fine structure of pheromone plumes to be measured directly rather than by proxy of ionized air techniques (Van de Pers and Minks this volume). Already these studies confirm earlier results on plume fine structure (Murlis et al. 1992; Murlis this volume). Second, rather than speculating, we can now investigate the concentrations (and the spatial and temporal patterns) of airborne pheromone which different mating disruption strategies produce (e.g., Färbert et al. this volume). Similarly, the effects of adsorption of pheromones onto leaves and the effects of the canopy on air movements can be investigated (Suckling and Karg this volume). We can now start to measure the "active spaces" of plumes downwind of lures in traps to study trap interference in a new way.

3.2.1. Wind-Tunnel Concentrations

The new technologies for pheromone air concentration measurement now mean that laboratory studies can be made more realistic. Until now, pheromone concentrations have usually been given as dose on a source rather than the achieved air concentration.

3.3. Complete Blends and Why They Matter

While pheromones for large numbers of insect species have been identified, it is likely that many earlier identifications are incomplete (Tumlinson 1988; Linn and Roelofs 1989). As Borden (1993) points out, mating disruption projects abandoned in the past might have been using incomplete blends. In some of these trials, real females outcompeted synthetic sources—implying that suboptimal pheromones were being used (e.g., Sanders 1984). It would be worth going back to many of these with a fresh identification of the blend. The role of host plant volatiles as synergists with insect pheromones is also being increasingly recognized for some moths and bark beetles (e.g., Borden this volume; Klein 1981).

Producing effective mating disruption schemes is further complicated by the existence of geographical races using different blends (isomers, enantiomers, ratios) for example bark beetles (Birch 1984; Lanier 1990) and moths such as the European corn borer (*Ostrina nubilalis*) and turnip moth (*Agrotis segetum*) (Löfstedt 1993).

3.4. Unsolved Practical Problems for Field Trials

3.4.1. Controls and the Large-Scale Needed to Show an Effect

The design of field trials still presents a major practical problem (e.g., Birch 1974; Sanders 1989). Control sites are needed, but flying insects (whether males or fertilized females) and an airborne pheromone do not respect field boundaries. No field is an island. It can be difficult to find matching control plots far enough away to be unaffected by the pheromone treatments on the experimental plots, but close enough to offer similar conditions. As Minks noted (tongue-in-cheek) at the symposium, pheromone trials in a small country like Holland might need experimental controls in Germany! The dilemma of field trials is that small-scale plots might not work, but large-scale experiments, to reduce edge effects, need blends and application rates largely already determined. This reinforces the need to be able to do realistic tests of these factors before field trials. A new review of the design of small plot experiments along the lines of Roelofs and Novak (1981), as well as an up-to-date review of general trial procedures along the lines of Rothschild (1981), would be timely.

3.4.2. Problems with Assessing Success of Mating Disruption

In common with IPM trials generally, measuring the effectiveness of mating disruption trials can be difficult and has concerned researchers from the beginning. For example, Minks (1977) argued that monitoring trap catches is not enough. In some cases, it is possible to disrupt trapping but still have mating occur in the plots (e.g., Suckling and Shaw 1992). The difficulty of sampling females is unsolved for most species. Economic endpoints (including percentage damage) are ultimately called for. However, although essential, such measures do not help an understanding of *how* mating disruption succeeds or fails in reducing pest populations (as distinct from the mechanism of interference with male moth behavior). Instead, for this, population studies of the pest are needed (below).

3.4.3. Reproducibility of Trials

Field trials are not controlled environments, and variable results between trials and locations can make assessment of efficacy difficult. As Minks (this volume) says of tests of mating disruption of codling moth, most tests were successful but some were not. Borden (this volume) describes similar variability in the success of trials of verbenone to manipulate the mountain pine beetle, *Dendroctonus ponderosae.* The reasons for failure are diverse and we should not dismiss a mating disruption project on basis of one failure [Arn and Louis (this volume) remind us that conventional control can also fail some years]. The failure of otherwise often successful schemes emphasizes our need to better understand what is going on in the field.

3.5. Delivery of Pheromones in the Field

McLaughlin et al. (1981), for example, argued that a lack of reliable pheromone delivery systems was a major problem. Since then, many new systems have been developed (Jutsum and Gordon 1989a; Leonhardt 1990), but providing constant release (zero-order) systems remains a problem. Quality control can also be patchy.

Sanders (this volume) discusses the likely effects of different formulations on mating disruption. In some ways, only now does it make sense to make things more sophisticated because now we can measure the concentrations produced (above). A new challenge for dispenser technology comes from our recognition of the need for release blends close to the natural one (Tumlinson 1988).

4. The Way Forward: Pure Questions, Applying the Answers

Many of the fundamental questions about the mechanisms of mating disruption and the reasons for success or failure in the field need the combined skills of both pure and applied biologists. Traditionally, many of the leading research scientists working on "pure" pheromone topics have also been very active in applied research. Two links are rare, however: first, neurobiology with mating disruption, and second, population studies of pests with mating disruption field trials.

A consistent theme in reviews over the last 25 years is the need for a deeper understanding of the processes underlying mating disruption instead of continued field trials contributing relatively little to wider questions (e.g., Wall 1984). I fear that reviews 10 years from now will be saying the same thing. What we need now is to bring together the different areas of research. Collaborative research needs encouragement from funding agencies, but pure research would also benefit from the greater understanding that could result. These collaborations do already occur successfully, but the link to population studies is perhaps the weakest: too applied for university researchers, too fundamental for extension workers.

4.1. *Clues to Mechanisms of Mating Disruption from Neurobiology and Sensory Physiology*

Insect pheromone communication systems have provided remarkably productive models for studying peripheral receptor neurones and their integration into higher neuronal levels, ultimately leading to motor outputs and behavior (Part 2, this volume). Neurobiology has naturally tended to be laboratory-based, and neurobiologists have been perhaps less active in the applied aspects than other pheromone researchers. However, greater collaboration would benefit both sides. Neurobiological research could help in two ways: first, by a direct contribution to the study of mating disruption itself; and second, by providing ways of predicting the likely effects of blends.

A major shortcoming of much mating disruption work is our poor understanding of the physiological mechanisms of the observed effects. For example, what *is* "sensory fatigue" (e.g., see Sanders this volume)? It is a useful shorthand but does not help explain the mechanism (is it adaptation at the peripheral level or habituation in the CNS? and at what level in the CNS?). As Kennedy (1992) argued, this kind of shorthand discourages us from asking the next level of questions. Together with behavioral biologists, these are the questions neurobiologists could help answer by laboratory studies (e.g., Baker et al. 1989). Neurobiologists could also talk with field biologists to help design better field experiments: Applied pheromone researchers are not able to keep up with the latest in neurobiological and behavioral neuroscience.

Second, in the distant future, neurobiology may be able to help interpret EAG and SCR responses which show that an antenna can detect a compound, at the peripheral receptor, but usually give no indication of the behavioral response that would be given by the animal. At the moment we are a very long way from being able to make such predictions of behavior (Part III, this volume). Patterns are starting to emerge in the projections of different receptor neurons into macroglomeruli compartments in the brain (Hansson this volume), but differences between species are not encouraging and the techniques are currently too difficult to use as a tool for other ends.

4.2. *Clues to Mechanisms of Mating Disruption from Behavioral Ecology, Genetics and an Evolutionary Approach*

The significant advances in our understanding of the genetics of pheromone production by female moths and its reception by, and behavior of, males (reviews Löfstedt 1990, 1993; Butlin this volume; Haynes this volume) show the strength of an approach which brings together chemists, biochemists, behavioral biologists, geneticists, and sensory physiologists. The benefits for evolutionary biologists include unique opportunities for investigating natural selection at the genotype level in nature (Löfstedt 1993; Phelan 1992), and the benefits for the applied biologist include more effective IPM methods and a greater understanding of the potential for resistance.

Many of the questions about the strength of mate selection that interest behavioral ecologists (together with population biologists and geneticists; e.g., see Butlin this volume) are similar to those requiring answers for mating disruption studies. For example, among the questions Löfstedt (1993) asks are: How many female moths (individuals and species) are simultaneously calling in a given habitat? What is the variance in male and female mating success, respectively? How long does it take before a male or female mates? Although the focus of Löfstedt's work is speciation, there is much common ground here for pure and applied biologists. Among the techniques that might be employed to answer these questions, DNA fingerprinting, used more routinely in other ecological specialities, could prove valuable to determine paternity within egg clusters.

Despite the central role of mate finding behavior and courtship underlying mating disruption, our knowledge of these processes, in even the common pest species, is extremely limited (McNeil 1991). In part, this is because most moths fly at night. Nonetheless, this type of more basic behavioral work is desperately needed. Paradoxically we may know more about the more unusual systems such as the saltmarsh moth, *Estigmene acrea* (Arctiidae) where leks of calling males seem to attract females but later in the evening solitary females call (Willis and Birch 1982). We need much more study of what goes on in seemingly conventional moth species. It may seem rather like "me-too"-type studies, but given the variation in behavior that is already apparent these are needed. A greater knowledge of pest behavior might make for fewer surprises later when we start to interfere with mating. Similarly, host selection by beetles needs further study, in addition to interspecific interactions including habitat partition (Borden this volume).

4.3. Pest Biology and Population Density Effects

"Research on population dynamics of pest insects hand-in-hand with control methods is one of the most needed areas of applied research" (Birch 1974) but one that happens rarely. The problem is that the life-table assessment, including egg counts, larval counts, and harvest counts needed, are very time-consuming and thus expensive to carry out. The information on population dynamics could also reveal the contribution of beneficials (e.g., predatory mites; see Arn and Louis this volume), increased in numbers by a reduction in spraying, to the overall reduction of the pest population. If significant, this would have implications for spraying for other pests.

High pest population densities are often blamed for failures of mating disruption. Among the ideas suggested are that competition between artificial and real females will be less effective with increasing population density. Alternatively, male searching strategy at higher densities (or indeed, permeation with pheromone) may change toward other cues such as visual orientation [although more recent evidence in the species commonly used as the example of this, *Lymantria dispar,* suggests that male behavior does not change (see Willis et al. 1994)]. The reality is that we don't know.

For most species a key unknown population parameter is migration (both short and long) and whether females migrate before or after mating (McNeil et al. this volume). Species in which fertilized females migrate may be poor candidates for mating disruption (Sanders this volume; Trumble this volume).

4.4. Resistance to Pheromone Control

Although clear resistance to direct control by pheromones has not yet been reported, pheromones are unlikely to be immune. The existence of geographical races and sibling species differing in pheromone blend should be evidence enough to worry us *a priori.*

An understanding of the potential routes for resistance is one of the major contributions a collaborative approach, involving genetics, biochemistry, sensory, and field biology, could make. Discussing the potential for resistance, McNeil (1992) reviews the evidence for interindividual variation in signal variation and reception (and the genetic basis for these) and records examples of such variation in the age females start pheromone release, pheromone titer, emission rate, proportion of different components present, and variation in the sensory sensitivities and behavior of males. Resistance to a pheromone could entail using other chemical cues, already produced and perceived but not currently part of the pheromone, in place of the original components (Pickett 1992).

Laboratory selection experiments show differing potential for directed change by selection (references in McNeil 1992; Löfstedt 1990). There have been few field studies. Haynes et al. (1984) found no change in emission rates or blend in the pink bollworm *Pectinophora gossypiella* after 3–5 years of mating disruption, although a later increase, perhaps coincidental, in emission rate was noted (Haynes and Baker 1988). That mutations changing pheromone blend can occur was found serendipitously in a laboratory culture of *Trichoplusia ni* and was tracked down to a step in the biosynthesis pathway (Jurenka et al. 1994b; Haynes this volume). Selection of males in the culture led to a broadening of their response spectrum to include the new blend.

It may be that behavioral changes (Haynes et al. 1984; McNeil 1992; Sanders 1989) in features such as mating sites (e.g., "conventional encounter" sites), other mating cues, or selection for dispersal of mated females will prove to be among the most important sources of resistance. Parthenogenetic strains could in theory develop, but there is no evidence of this (see Sanders 1989).

Given the selection pressures we may be generating in successful mating disruption programs, should we already be considering strategies for minimizing development of resistance, analogous to those proposed for *Bacillus thuringiensis* (see Tabashnik 1994)?

5. Putting Pheromones to Work—Progress and Problems with Applications

5.1. Barriers to Take Up

While most of this chapter has been about science, it will not surprise the reader that, in common with much of IPM (Dent 1991), the most serious challenges to implementation of pheromones in pest control are probably political and economic, cogently reviewed by Silverstein (1990). Moving from the science, or even research field trials, to take-up by farmers is the most difficult step (Dent 1991). Each review over the last 25 years has identified the problem which, as the characteristics of pheromones remain the same, have changed little. Here I will focus on some of the principal ones and suggest some ways forward.

5.2. Who Pays

The problems of commercial development have been rehearsed many times. It was recognized early that it is unrealistic to expect the large agrochemical companies to develop commercial pheromone technologies for other than a very few major crops, notwithstanding the significant involvement of some large companies (Jutsum and Gordon 1989b). The barriers to commercial interest mostly concern the problem of recouping high development costs. They include small markets (and thus relatively high unit cost of active ingredients), formulation difficulties, marketing and technology transfer difficulties, problems with patentability (Silverstein 1990) to which Arn (1990) added the difficulties presented by increasing complexity of pheromones, in both chemistry and biology, and, for pesticide companies, competition with their own conventional products.

There are strong parallels with biological control which shares many of the same barriers to commercial development (Jutsum 1988; Van Lenteren and Woets 1988). The opportunities for recovering development costs commercially may be limited, but government or similar public sector investment can be more than justified by the public benefits: to individual farmers, to consumers, to the environment, and (for Third World countries) in savings of foreign exchange. It is always easy to argue for more funds, but here is a case where public investment would pay dividends: Not all "public goods" come from choices dictated by return on capital narrowly defined by company interest. This comment may seem overpolitical but reflects commercial realities.

Most companies involved in commercialization of pheromones, which tend to be small, do not have the resources for the basic research needed in developing the technology. Much of the research and extension work will need to be government-supported. However, as Jones (1994) points out, it is ironic that at the same time as politicians are legislating for reductions in conventional pesticide use (below), they seem most keen to reduce the worldwide capacity for scientific research into alternative pest control methods at government and academic institutions. The trends in many western countries toward commercialization and privatization of government research (e.g., HortResearch in New Zealand, CSIRO in Australia, and GHRI and NRI in the United Kingdom) do not bode well. Even today, current investment is small. Van Lenteren and Woets (1988) estimate that the amounts spent on biological control, largely in the public sector, represent about only 1% of investment by chemical companies in pesticide research. Levels of investment in pheromone research are likely to be on the same order.

Campaigning for increased investment will need readily accessible (and up-to-date) information, including successes (e.g., Cardé and Minks 1995), current usage, costs, and benefits [like Klassen et al. (1982) but updated regularly]. Without political lobbying, it is unlikely that we will have progress with pheromone adoption. A major handicap is the lack of powerful equivalents to chemical industry organizations, although with the formation of the American Semiochemical Association and European Semiochemical Association (Jones 1994) we perhaps have a good beginning.

Changing consumer attitudes toward pesticide use (e.g., Coats 1994; Trumble this volume) are already having effects on the climate for alternatives to conventional pest control, and pheromones are likely to benefit. Other incentives for the development of mating disruption and other IPM techniques come from growing insecticide resistance and the rising costs of developing new pesticides.

5.3. Extension, Persuasion, and Take-up

Gaining acceptance for pheromone-based pest control, like adoption of other new IPM methods, requires considerable effort. Trumble's (this volume) description of the conservatism of vegetable growers and their reasonable suspicion of new pest control technologies (not all have matched their promise) is typical. Adoption will be more likely if it is clear to farmers that there *must* be change, in response (for example) to pesticide resistance problems (e.g., Suckling and Karg this volume; Staten et al this volume; Trumble this volume). Getting acceptance of ultimately better but less certain pest control methods is difficult. Schemes requiring much greater sophistication on the part of farmers will be less successful than those incorporating a more sympathetic appreciation of farmers' perceptions and constraints. Extension workers need to match the pesticide company sales effort at the farm level. The chances of adoption will be improved if mating disruption techniques can be seen to be cost-effective in schemes involving farmers [best of all from the start (Dent 1991)].

Some of the most effective uses of pheromones for mating disruption have involved region-wide schemes (e.g., Staten et al. this volume; Campion 1984; Jones 1994). Once implemented, involvement of the government can ensure adoption across the whole region, but getting the initial decision can be more difficult than dealing with individual farmers.

Investment in implementation, the key to adoption of pheromone technology, is often skimped, or not properly addressed at all. Unfortunately, this crucial activity is one of the most dependent on public funds.

Pheromones could be very appropriate technology for developing countries, but Silverstein (1990) reports excellent critiques of many current approaches and emphasizes the need for ingenuity and simplicity in the integration of pheromones into pest control systems.

5.4. Government Influences Through Legislation

Registration requirements can have a profound effect on pheromone adoption. Weatherston and Minks (1995) suggest that the time and cost required to complete the registration process provide the major hurdle to introduction of semiochemical products into the market place. These authors review the pattern of current regulation in 21 countries. They acknowledge that the special nature of semiochemicals has been recognized by many national regulatory bodies and the Food and Agricultural Organization of the United Nations but argue that more could be done to harmonize and expediate pheromone registration. Among their proposals is the suggestion that structure–activity data should be used for environmental and health-risk assessment of pheromones, in the first instance for lepidopteran pheromones, to avoid the need to test each fairly similar pheromone for toxicological and other properties. Adoption of a 90-day evaluation deadline would also help. They propose developing this approach first with the U.S. Environmental Protection Agency (EPA), but they hope it might be accepted by other national regulatory authorities in time.

Legislation restricting pesticide use will also be important in the future. For example, Trumble (this volume) describes how increasingly restrictive legislation at the state level has dramatically changed use patterns in California (see Weatherston and Minks 1995 for other initiatives). Carrots can also be used: Arn and Louis (this volume) report that some regional governments in Germany and Switzerland offer subsidies to vineyard growers toward the costs of pheromones.

5.5. Selecting Good Target Species for Mating Disruption

Rothschild (1981) feared that there would be few "universal truths" that can be applied to mating disruption, which would mean that each pest species (and crop system?) might have to be studied individually. To make the task manageable, we need to search for patterns, good predictors of success (or likely failure) in pest species and agricultural systems, resting on basic studies on field behavior, pest biology, and mating disruption experience. Testable suggestions for such characteristics have been made (e.g., Koehler et al. 1977; Rothschild 1981; Sanders 1989 this volume). This challenge should be taken up.

5.5.1. Pest Complexes

A major strength of pheromones is their effectiveness as part of IPM schemes, because of their compatibility with biological control agents and other beneficials. Trumble (this volume) describes use of pheromones in IPM of tomato pests in Mexico. Pheromones fit well in the virtuous spiral—for example, in greenhouse IPM (Van Lenteren and Woets 1988), where the use of one biological control agent encourages (or requires) moves from conventional pesticides for other pests.

5.6. Chemical Advances

Costs of active ingredients have been a limiting factor in some mating disruption schemes. Developments in synthetic chemistry—in particular, chiral synthesis (driven by the pharmaceutical industry after thalidomide)—may bring down costs. The use of biotechnology for pheromone synthesis could produce biologically correct pheromones cheaply (Lindgren 1990; Pickett 1992).

6. Conclusions

Butenandt could have had no idea in 1950, contemplating 500,000 moths to identify the silkworm pheromone, bombykol, that today such a task could be attempted with a single female. Similarly, the idea of measuring pheromone concentrations in the wind would have seemed a fantasy. Problems with mating disruption remain, but in many cases we now have the tools to answer them. Finding the answers needed to make pheromones an even more effective weapon in the direct control of insect pests will be of mutual benefit to both pure and applied biologists.

Acknowledgments

I would like to thank J. and V. Wyatt, M.C. Birch, D. Dent, H. Fadamiro, and J. Pickett for helpful suggestions on the manuscript, and I am grateful to G. P. Thomas for continued support.

V

Evolution of Pheromone Communication

Organized by Christer Löfstedt

40

Pheromone and Morphology-Based Phylogenies in New Zealand Tortricid Moths

John S. Dugdale

1. Introduction

Systematics try to construct phylogenies—that is, routes of relationship and descent. We have, up until recently, based our phylogenetic reconstructions on morphology, because that is the character field for which we have the greatest amount of comparative information covering all the putative taxa. Where new character fields have become available, we have often regarded these as a form of independent check. There are two dangers inherent in this concept of independence: (a) the inclination to see only what supports other morphology-based conclusions and (b) the (probable) illusion of character independence (Kluge and Wolf 1993). The first danger is often compounded by a lack of appreciation of the confidence limits that are placed on information provided by a new character field, such as the pheromone systems and processes explored at this symposium. The second is influenced by our knowledge—or rather, lack of knowledge—of the genetic (transcription) control affecting the character states we choose to employ. For both dangers, as Kluge and Wolf (1993, p. 196) point out, "the limits of phytogenetic systematics will be clarified considerably when cladists understand how their knowledge claims are made." It is better (i.e., logically secure), they argue, to accommodate what they call the whole evidence.

For systematics, accommodating the "whole (i.e., available) evidence"—which can include character fields in behavior, host choice, and specific mate recognition—also implies a wide coverage of all of these character fields for taxa in and around the chosen group and its outgroup(s). Systematists attempt to recognize patterns; biochemists and neurophysiologists attempt to define processes. Too often, for too long, and quite illogically, these two research groups have tended to be separate. Yet patterns cannot be interpreted in isolation from processes, and processes need their universality tested.

Systematists have always concentrated on aspects of mating systems; originally, genitalia structure was recognized as a field rich in characters and therefore a source of speculation as to form and function, evolutionary directions, and specificity (e.g., Hammond 1981). Mating systems involving biosynthetic pathways for pheromone production are of great interest to systematists who want to test the predictability of our classifications.

As a morphological systematist, I was fortunate to join a team investigating the structure and biosynthesis of female sex pheromones (and male responses) in some orchard leafrollers in New Zealand. Foster et al. (this volume) recapitulate some of the work and its conclusions. Much relevant information is summarized by Foster et al. (1991b) and Dugdale (1990).

Tortricidae is a large family (>5000 species) of Lepidoptera, instantly recognizable but with subgroupings poorly defined morphologically. There is an uneasy consensus (Horak and Brown 1991) on the definition of subgroups, based largely on an analysis of postulated underived/derived morphological character states (Horak 1984). One conclusion of the latter study is that two important orchard leafroller genera in New Zealand, *Ctenopseustis* and *Planotortrix,* are relatively underived (plesiomorphous) archipine tortricids (Horak and Brown 1991). These two genera have unusual attractant pheromones among Tortricidae (Roelofs and Brown 1982; Arn et al. 1992). The series of comprehensive catalogues of pheromones by the above authors allow morphologically rooted systematists to look for patterns that may assist phylogenetic interpretation of this morphologically rather homogeneous family. There is also clear information on both sex pheromone blends and of the biochemical pathways of many of the active constituents used by *Ctenopseustis* and *Planotortrix* species (Foster et al. 1991b).

Section 2 discusses problems inherent in the interpretation of morphology and biosynthetic pathway character fields, Section 3 lists morphological and biosynthetic characters in *Planotortrix* and allies and discusses morphology versus pheromone-based phylogenies in *Planotortrix,* and Section 4 indicates some conclusions of possible general relevance and suggests where systematists may assist pheromone research.

2. Problems Inherent in Interpretation of Morphology and Biosynthetic Pathway Character Fields

There are four problem areas in phytogenetic interpretation of both morphological and biosynthetic pathway characters.

1. *Polarity.* Which state is underived (primitive, plesiomorphous), and which state is derived (advanced, apomorphous)?

2. *Homoplasy*. Has a state common to two or more taxa arisen independently as a response to an extrinsic factor (convergence), or as a result of genetic relationship (strict homoplasy, including reversals)?
3. *Resolution power*. Has the character field been studied comparatively for most or all of the members of a monophyletic group?
4. *Appropriateness of interpretation*. Morphologically, are we dealing with homologous structures? Biochemically, can we regard a biochemical pathway as we do a structure?

2.1. Polarity

In morphology interpretations, while polarity is clear for some characters (e.g., those involving reduction, or loss), it is not so for others. In the latter instances the soft option of assigning polarity by virtue of the association of that state with another character showing a clear polarity is involved. Essentially, they are the same pitfalls as in assuming guilt by association. This is one area where cladists can make explicit the basis of their knowledge claim. Levels of interpretation vary between character states, and there is a reliance on Dollo's rule (i.e., once lost, an organ cannot be regained) and on the concept of character reversal. Assigning polarity is basic to a phylogenetic reconstruction because monophyletic groups are recognized by shared derived character states (synapomorphies); groups defined on shared primitive character states (symplesiomorphies) are paraphyletic or uninformative because shared primitive character states are no argument for close relationship.

Currently the polarity of a given biosynthetic pathway system may be impossible to assess, except by invoking "guilt by association." While the concurrence of putatively plesiomorphous morphological characters and a certain desaturase system is easily shown, there is as yet no chemical or gene-specific evidence for relationship. At present we have the situation whereby the Δ9- and Δ10-desaturase systems characterizing *Ctenopseustis* and *Planotortrix* are regarded as less advanced than the Δ11-desaturase system occurring in most Tortricidae (Roelofs and Bjostad 1984) and, indeed, most ditrysian moth families (Löfstedt 1991). In short, a Δ11-desaturase system appears to be a ground-plan feature of most of the Ditrysia examined so far.

2.2. Homoplasy

Homoplasy brings similar problems. In morphology, states due to convergence (similar solutions to common extrinsic problems) or to genetic relationship (including "reversals") are distinguished by comparison with other characters regarded as independent—for example, those not thought to be affected by that extrinsic factor or that field of genetic influence. There is a robust argumentation plan for morphological characters, but is there one for biosynthetic pathways?

2.3. Resolution Power

Resolution power for phylogenetic reconstruction is greater for morphology-based studies than for biosynthetic pathways studies, because of the ease of gathering a multiplicity of morphological data for all members of the group under study. The resolution power of morphology-based systems can extend to the population (e.g., see Higgins 1975). Even so, current reconstructed phylogenies for major groups in Lepidoptera show a number of unresolved higher taxa (Minet 1991; pp. 80–81, Figs. 54 and 55). Unresolved taxa in well-argued cladograms can be evidence that, over time, most of the related forms (sister groups) have been lost and the terminal entity is all that remains (e.g., the tree *Ginkgo biloba* or the moth family Tortricidae). The abundance of evidence is not always welcomed by some workers. As Gauld and Mound (1982, p. 74) note, for certain groups, the richer the database "—in terms of structural characters, distributional, host organism or other biological information—the greater can be the problems of classification."

In contrast, the resolution powers of biosynthetic pathway systems for phylogenetic reconstruction are limited. Until we learn what is appropriate to ask, they can only provide a coarse resolution in terms of relationships between taxa (e.g., as in Fig. 40.1B). For differentiation between species, biosynthetic pathways can be critical (see below); but usually, most differentiation at the species level is assessed on biosynthetic products and the ratios or blends that are produced (e.g., Roelofs and Brown 1982; Arn et al. 1986). Because a product can be produced by more than one biosynthetic pathway and because of the variety of such pathways, it has long been recognized that biosynthetic pathways present a potentially useful phylogenetic character field (Roelofs and Brown 1982; Roelofs and Bjostad 1984; Foster and Dugdale 1988). What is needed is evidence of evolutionary direction.

2.4. Appropriateness of Interpretation

Appropriateness of interpretation is a major problem in the use of biochemical characters. Is it appropriate for us to regard a biosynthetic pathway as we do a

---

Figure 40.1. Reconstructed phylogenies for *Planotortrix.* (**a**) Reconstruction based on morphological synapomorphies (shared derived characters). (**b**) Reconstruction based on shared biosynthetic pathways (based on Table 40.1). (**C**) Combined reconstruction. Dashed lines: alternative hypothesis with *octo* and *octoides* as sister group of all other species (but this hypothesis lacks support from morphological synapomorphies). Abbreviations: fm, plant family monophagous; gm, plant genus monophagous; p, polyphagous. In part A, note the lack of convincing apomorphies for *excessana* and *octo,* which are grouped on plesiomorphies; the reconstruction for these four species appears paraphyletic. In part B, note the lack of species resolution (e.g., no distinction between *excessana* and *puffinis*).

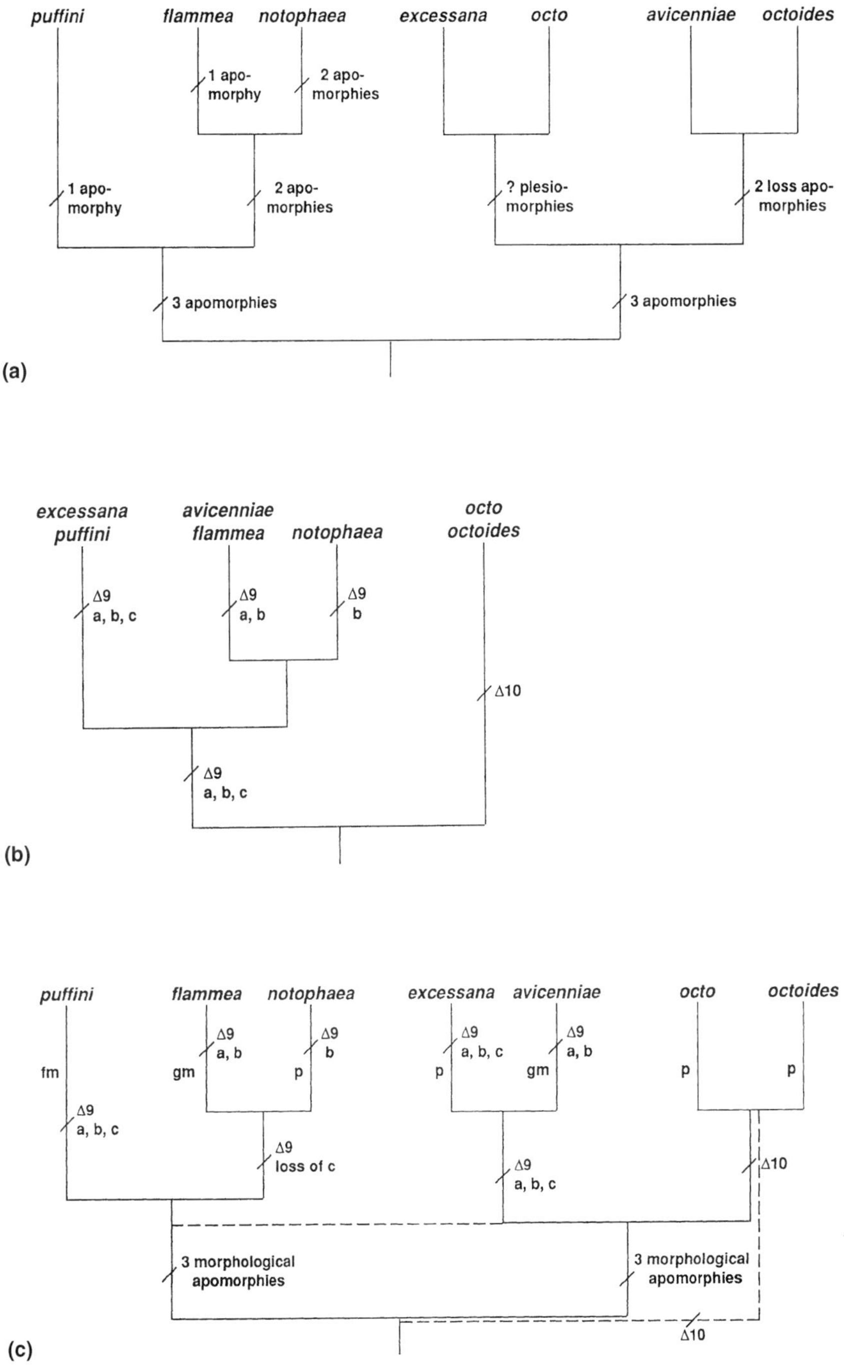

puffini
flammea
notophaea
excessana
octo
avicenniae
octoides
1 apo-morphy
2 apo-morphies
1 apo-morphy
2 apo-morphies
? plesio-morphies
2 loss apo-morphies
3 apomorphies
3 apomorphies
(a)
excessana puffini
avicenniae flammea
notophaea
octo octoides
Δ9 a, b, c
Δ9 a, b
Δ9 b
Δ10
Δ9 a, b, c
(b)
puffini
flammea
notophaea
excessana
avicenniae
octo
octoides
fm
gm
p
p
gm
p
p
Δ9 a, b
Δ9 b
Δ9 a, b, c
Δ9 a, b
Δ9 a, b, c
Δ9 loss of c
Δ9 a, b, c
Δ10
3 morphological apomorphies
3 morphological apomorphies
Δ10
(c)

structure? Once lost, can it reappear? Can we regard a Δ9-, Δ10-, and Δ11-desaturase array as a transformation series, and can we confidently assume that one of these desaturase systems is less advanced than the others? Can we legitimately think like that, or are morphological systematists asking questions as meaningless or inappropriate as "What is virtue's color?" Thus concepts of polarity, reversals, and parallelisms due to "underlying apomorphy" (Horak 1984, p. 38) may be highly suspect. The concept of monophyly [a group defined on the possession by all its members of one or more derived (apomorphous) character states] may be inappropriate at worst, or currently untestable at best, when applied to biochemical characters.

3. Morphological and Biosynthetic Pathway Characters in *Planotortrix*

In New Zealand, the two endemic tortricid genera *Ctenopseustis* and *Planotortrix* include species of economic importance to horticulture (Wearing et al. 1991). These have been the subject of intense investigation of their female sex pheromone systems, because both genera share an unusual variety of biosynthetic pathways, none of which appear to show evidence of Δ11-desaturase involvement (e.g., Foster and Dugdale 1988). Both genera are regarded as being relatively rich in plesiomorphies (primitive characters) (Horak 1984). Because information on biosynthetic pathways is available for more *Planotortrix* species than for *Ctenopseustis,* the former genus is discussed here.

Planotortrix is currently regarded as being a monophyletic group with seven species (Dugdale 1990). The species can be defined morphologically, enzymatically, and pheromonally (Foster et al. 1991b), and biosynthetic pathways are either known or inferred for each species (Foster and Dugdale 1988). Because *Planotortrix* species use one, two, or three of four possible biosynthetic pathways, comparing relationships based on shared morphological apomorphies (presumed derived characters) with the occurrence of pathways for each species may provide a clue as to underived versus derived biosynthetic pathways (if it is appropriate to do so, see the preceding section).

The four biosynthetic pathways known or inferred for *Planotortrix* species (Foster and Dugdale 1988) are as follows:

1. ex oleate via Δ9-desaturation → Z5-14:Ac
2. ex palmitoleate via Δ9-desaturation → Z7-14:Ac
3. ex myristoleate via Δ9-desaturation → Z9-14:Ac
4. ex palmitate via Δ10-desaturation and β-oxidation → Z8-14:Ac

Known or inferred presence of these pathways for each species is given in Table 40.1.

Morphologically, *Planotortrix* can be divided (Fig. 40.1A) into two groups:

Table 40.1. Known or inferred occurrence within Planotortrix *species of four biosynthetic pathways.*

Biosynthetic[a] pathway	*Planotortrix* species[b]						
	aviceniae	*excessana*	*flammea*	*notophea*	*octo*	*octoides*	*puffini*
a (Δ9):	(+)	+	–	–	–	(+)	(+)
b (Δ9):	(+)	+	(+)	(+)	–	–	(+)
c (Δ9):	–	+	–	–	–	–	(+)
d (Δ10):	–	–	–	–	+	+	–

[a]Foster and Dugdale (1988).

[b]Dugdale (1990): +, confirmed; (+), inferred.

the *notophaea* group, with a modified aedeagus apex, a narrowed uncus apex, and the cestum sclerotized only laterally; and the *excessana* group, with the male costal fold "ceiling" lacking broad scales, the vesica with a spinulose-capped basal process, and the uncus apex broad. In appearance, adults of the *notophaea* group are characteristically somewhat slender-winged, and those of the *excessana* group are broad-winged. Within the *notophaea* group, morphological synapomorphies (shared derived states) define the arrangement [*notophaea* & *flammea*] + [*puffinis*]. Within the structurally homogeneous *excessana* group, the pair *octoides* & *avicenniae* is defined on reduction synapomorphies, and the pair *excessana* & *octo* is defined only on symplesiomorphies; the morphological evidence available has led to a paraphyletic or uninformative arrangement.

On the type of biosynthetic pathways, *Planotortrix* is divided (Fig. 40.1B) into those with Δ9-desaturase systems (i.e. *flammea* & *avicenniae* & *notophaea* and *excessana* & *puffini*) and those with a Δ10-desaturase system (i.e., *octo* & *octoides*). Within the Δ9-desaturase-defined group, either there has been loss of some pathways—from those with pathways a, b, c (Table 40.1) (*puffini, excessana*) to those with pathways a, b (*avicenniae, flammea*) to one pathway, b (*notophaea*—or, alternatively, there has been addition (i.e., from b → a, b, → a, b, c). If, and only if, it is appropriate to regard these pathways as states of a character, then the direction (either toward loss of states or to addition) can be inferred by comparison with morphological data. Figure 40.1B indicates coarseness of resolution of the biosynthetic data at the species level.

Because the biosynthetic pathway grouping based on a Δ10-desaturase system (*octo–octoides*) is singular, direction can be inferred only in relation to the Δ9-desaturase-defined group. Thus if the presence of Δ10-desaturase in two members of a monophyletic group can be regarded as a synapomorphy, then *octo* and *octoides* together are either (a) the sister group of the other *Planotortrix* species or (b) the sister group of *excessana* and *avicenniae*. Biosynthetic pathway information has partially resolved a paraphyletic and thus phylogenetically unacceptable hypothesis.

Combining the morphological and biosynthetic pathway synapomorphies (Fig.

40.1C), for the group *puffini* & *flammea* & *notophaea,* both morphological and pathway synapomorphies agree, with *puffini* structurally and biosynthetically separate from *flammea* and *notophaea.* The assumption here is that there has been progressive loss of Δ9-desaturase pathways. The assumption that possession of three Δ9-desaturase systems is (relatively) underived is supported by their presence in *puffini* and *excessana* (Table 40.1; Fig. 40.1B). This assumption rests on the presumed validity of treating biosynthetic pathways like morphological structures.

Morphological studies have failed to provide a resolved grouping in the *excessana* group, where *excessana* & *octo* is defined on symplesiomorphies, and *avicenniae* & *octoides* is defined on doubtfully homologous and possibly convergent reduction of two structures. The biosynthetic pathway data show that *excessana* & *avicenniae* is part of the Δ9-desaturase group and that the Δ10-desaturase pair *octo* & *octoides* is a basic dichotomy. Combining the data sets suggests that the *excessana* group and the *puffini* group morphologically represent a basic dichotomy and that the Δ10-desaturase arisal is a later event within the *excessana* group.

The assumption that two Δ10-desaturase defined species are sister species may be false; on morphological grounds, a hypothesis that the Δ10-desaturase system has twice arisen independently within the *excessana* group can be postulated. (Fig. 40.1C, dashed line).

Thus, a combination of morphological and biosynthetic pathway data has provided a testable hypothesis for one species group in *Planotortrix* (where the species are morphologically clearly demarcated). If, and only if, use of biosynthetic pathway data in this way is valid, then the combination has provided an equivocal answer regarding the other species group (where the members are not clearly demarcated morphologically) and clearly defines what needs testing. In cladistic terms, species relationships in one group (*puffini* et al.) appear resolved, but those in the other group are unresolved. Biosynthetic pathway information here has been useful in defining a problem.

4. General Conclusions and Possible Future Perspectives

(a) *Usefulness of monophyletic species groups:* The (morphologically) monophyletic tortricid genera *Ctenopseustis* and *Planotortrix* (as summarized by Foster et al. 1991b), and some of their allies, have a range of biosynthetic pathways. Analysis of such groups, where either every species or all but one species have been studied, can provide robust information for studies on evolutionary direction within a group and provide comparisons between groups. One question regarding the evolution of tortricid pheromones is whether the Δ11-desaturase system is underived or derived (see Horak et al. 1988). The presence in some Australian and two New Zealand genera of Δ9- and Δ10-desaturase systems (Horak et al. 1988; Foster and Dugdale 1988), along with the apparently relatively underived morphological character states of some of these genera (Horak 1984), might be considered as evidence of the derived or advanced nature of Δ11-desaturase systems. There is a coincidence between possession of certain underived states and Δ9- & Δ10-desaturase systems (Table 40.2), but it is not unequivocal. In *Planotortrix,* one presumed monophyletic group provides an indication that for biosynthetic pathways, evolution has proceeded by simplification—that is, reduction in the number of biosynthetic pathways. Also in *Planotortrix,* there is, in one species group, an indication that the Δ10-desaturase system has arisen either once or possibly twice.

(b) *Validity of interpretation:* Biochemical data can provide some support for phylogenetic studies. Currently, they can indicate problem areas, but cannot resolve them (see *Planotortrix* section). While both morphologically and biochemically based analyses have similar problems, the major stumbling block is interpretation of biochemical information in an evolutionary/phylogenetic context. Is it valid to treat the Δ9-desaturase systems (listed in Table 40.1) in *Planotortrix* as forming an elaboration transformation series (e.g., a → a, b → a, b, c) or as a reduction/loss series (e.g., a, b, c → b → a)? Is it also valid to propose a phylogenetic sequence (e.g., Δ9- → Δ10- → Δ11-desaturase systems)? If so, what are the bases for the validity of such an interpretation?

(c) Studies on pheromone biochemistry and pheromone-mediated behaviour force systematists to think of the organisms as not only living, but also reproducing, entities. Systematists will try to interpret data in

Table 40.2. Pathways, male genitalia, and host specificity in some New Zealand and Australian harmologoid Archipini

Genus	Desaturase system	Male mating structures	Host-plant specificity[c]
Ctenopseustis	Δ10 ,Δ9[a]	Complete	P(4),M(2)
Planotortrix	Δ10,Δ9[a]	Complete	P(4),M(3)
Leucotenes	Δ11[a]	Socii reduced	M(1)
Apoctena	Δ11[a]	Complete or socii reduced	M(7),P(4)
Pyrgotis	Δ11[a]	Socii reduced	M(7),P(2)
Ecclitica	Δ11[a]	Socii reduced, transtilla modified	M(4)
Epitymbia	Δ9[b]	Socii, transtilla reduced modified	n.e.
Meritastis	Δ9[b]	Socii, transtilla reduced	n.e.
Capnoptycha	Δ10[b]	Socii, valvae reduced	n.e.
Isochorista	Δ9[b]	Complete	n.e.

[a]Foster and Dugdale (1988, pp. 228–229).

[b]Horak et al. (1988, p. 1169)

[c]P(*n*), polyphagous (number of species); M(*n*), monophagous (number of species); n.e., not estimated.

a phylogenetic way, and papers given at this conference show that biochemists and physiologists find this useful.

Systematists can therefore, indicate, or even provide, candidate groups for study. The orchard leafroller work in New Zealand covered more than the orchard-infesting species and indeed intentionally covered three monophyletic groups. This was done because there were no recorded studies of a multispecies monophyletic group except for those on *Yponomeuta* (Menken et al. 1992). There are still far too few studies of this sort. For Tortricidae, any useful generalizations of the evolution of pheromone biosynthetic pathways must await such studies. Based on an analysis of underived and derived character states in Tortricidae (Horak 1984), genera postulated as relatively underived could be compared with those postulated as more derived, or genera known or inferred to have a variety of biosynthetic pathways could be investigated. Such studies would provide more comprehensively based hypotheses about biosynthetic pathway and pheromone evolution.

Systematists can also suggest candidate groups that are critical to phylogenetic understanding within an insect order (e.g., Hepialoidea in Lepidoptera) or which may be used to test the limits of the universality of the processes uncovered by biochemists and neurophysiologists. In view of the work reported at this symposium based on the sphingid moth *Manduca sexta,* it would be desirable to extend this to members of the Bombycoidea with which, on morphological interpretations, sphingids are now aligned. Such an extension would test the validity, or at least the predictive values, of morphology-based phylogenies.

To a systematist/morphologist, this symposium has been most exciting because of the potential new character fields presented. Systematists have an expanding opportunity to collaborate in studying the processes on which we base interpretation of our patterns.

Acknowledgments

I am grateful to Manaaki Whenua—Landcare Research New Zealand, Dr. S. C. Löfstedt, and Dr. A. Minks for assistance with travel funding, to Dr. Löfstedt for hospitality in Lund, and to Mr. Tymone Duval for formatting the tables. I am particularly grateful to Dr. R. Cardé, Dr. C. Löfstedt, and Dr. A. Minks for critical comments on drafts of this chapter.

41

A Phylogenetic Analysis of Pheromone Communication in Primitive Moths

Christer Löfstedt and Mikhail Kozlov

1. Introduction

A robust phylogeny exists for Lepidoptera, but the phylogenetic arrangement of the archaic lineages, as well as several lineages within the Ditrysia, has not been fully resolved (Kristensen and Skalski 1996). Sex pheromones have been chemically identified from more than 400 lepidopteran species, sex attractants have been reported for many more (Arn et al. 1992), and the pheromone communication systems of many moths have been investigated in great detail (Baker 1989c), making the Lepidoptera an interesting target group for a phylogenetic analysis of the communication systems. Can pheromone characters provide increased resolution in the reconstruction of the lepidopteran phylogenetic tree? How have pheromones evolved within an insect order, from archaic to more advanced character states? To what extent can the design of pheromone communication systems be considered adaptive?

The answer to such questions will provide important information from both general and specific perspectives and in order to address them the evolutionary trends of the important characteristics of pheromone systems should be elucidated, taking into account both their adaptive value and within-group variation. In the present chapter we restrict our analysis to a few morphological conditions (presence and position of pheromone glands) and related physiological/behavioral traits (presence and composition of long-distance pheromone, male copulatory behavior). The recent phylogeny of lower lepidopteran lineages will be used for the reconstruction of character evolution; and the order Trichoptera (caddisflies), which is the sister group of Lepidoptera (butterflies and moths), will be used as an outgroup (= reference group) for the analysis of lepidopteran pheromone systems. The advantages of using caddisflies are their common origin with lepidopterans and their lower diversification (about 10,000 species, i.e., about 7%

of the Lepidoptera). It should be mentioned, however, that pheromones of caddisflies are much less studied than those of lepidopterans, and phylogenetic relationships among the main lineages are still debatable (Ross 1967; Wiggins and Wichard 1989; Weaver 1992a,b). The present chapter aims to (1) summarize the recent knowledge of pheromone communication systems in Amphiesmenoptera (Table 4.1), with an emphasis on the limited but interesting data available on primitive moths, (2) discuss evolutionary trends in both morphological and physiological traits related to pheromone communication, as well as significance of these changes for the phyletic evolution of caddisflies and lepidopterans, and (3) identify the taxonomic groups that have to be researched before the main evolutionary events can be attributed to a certain clade/grade.

2. Pheromone Communication in Trichoptera

Mate-finding in caddisflies occurs by either (a) male searching for receptive females or (b) females attracted to male swarms. Swarming in this order is not only aerial, but includes behaviors performed on terrestrial substrates and the water surface (Solem 1984). Although information on courtship is sparse, Solem (1986) reported that males of *Agrypnia obsoleta* Hagen may perform a swarming-courtship dance around the female, and Erman (1984) found that a male *Parthina linea* Denning, when encountering conspecifics while walking, may flare its wings, vibrate them rapidly, and often simultaneously curl its abdomen up dorsally, extending the claspers.

Evidence of female sex pheromones has been reported for a large number of trichopteran species in several different families, including the Sericostomatidae, Limnephilidae, Rhyacophilidae, and Hydropsychidae (Wood and Resh 1984; Resh and Wood 1985; Solem 1985; Solem and Petersson 1987; Löfstedt et al. 1994b). In all cases the pheromones appear to be produced in the pair of exocrine glands opening on the Vth abdominal sternite. These glands are widespread in Trichoptera and often occur in both sexes (Nielsen 1980).

In *Rhyacophila fasciata* Hagen, females produce a mixture of heptan-2-one, nonan-2-one, heptan-2-ol, and nonan-2-ol, which attracts conspecific males (Löfstedt et al. 1994b). *R. fasciata* males possess the same glands but produce different compounds (hexanoic and octanoic acid) of unknown biological significance. The same chemicals were identified from *R. nubila* Zetterstedt, but to date there are no behavioral data to support their pheromonal function (Ansteeg and Dettner 1991; C. Löfstedt and E. Petersson unpublished). Duffield (1981) identified nonan-2-ol from the glands of the nearctic caddisfly *R. fuscula* Walker. In *Hydropsyche angustipennis* Curtis, both males and females produce a mixture of ketones and alcohols, with nonan-2-one being the major constituent, but the titer of the identified compounds was much higher in males than in females. Both males and females were attracted to a synthetic mixture of the compounds identified

Table 41.1. Characters used in phylogenetic analysis

No. Character	TRICHOPTERA	Micropterigoidea	Agatiphagoidea	Heterobathmioidea	Eriocranioidea	Acanthopteroctetoidea	Lophocoronoidea	Neopseustoidea	Mnesarchaeoidea	Hepialoidea	Nepticuloidea	Incurvarioidea	Palaephatoidea	Tischerioidea	DITRYSIA
	Taxa[a]														
1 Long-distance pheromone present	+	?	?	?	+	?	?	?	+	+	+	+	?	+	+
2 Female-produced pheromone	+	?	?	?	+	?	?	?	+	±	+	+	?	+	+
3 Sternum V glands	±	±	+	+	+	+	+	–	–	–	–	–	–	–	–
4 Short-chain pheromones	+	?	?	?	+	?	?	?	?	?	+	–	?	?	–
5 Male wing-fanning	?	+	?	?	+	?	?	?	–	+	+	+	?	+	+
6 Terminal glands	–	–	–	–	–	?	?	?	–	+	?	+	?	+	+
7 Long-chain pheromones	–	?	?	?	–	?	?	?	?	?	–	+	?	?	+

[a]+, character present; –, character absent; ±, both character states recorded (absent/present); ?, uncertain character state.

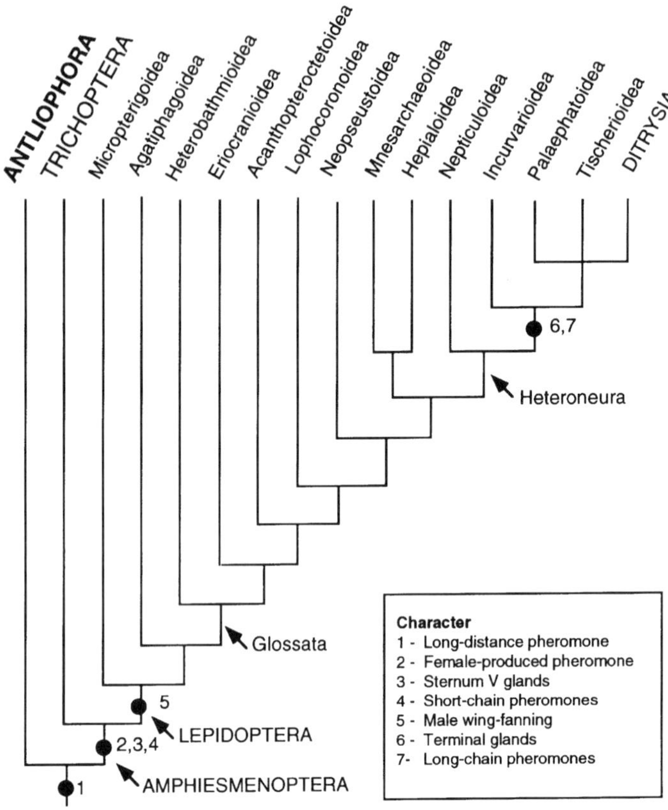

Figure 41.1 Cladogram of the basic lineages of the panorpoid insects and of Lepidoptera. Sequence of lineages after Kristensen (1984a) and Kristensen and Skalski (1996). Character enumeration (Arabic numerals) is the same as in text and Table 41.1. Filled circles are apomorphic character states. Resolution of the Heteroneura is based on characters 6 and 7, as discussed in the present chapter.

from the gland extracts, suggesting that the pheromone functions as an aggregation pheromone (Löfstedt et al. 1994b).

Gland secretions without proven pheromonal activity have been identified from a number of additional species. Duffield et al. (1977) reported the occurrence of *p*-cresol, indole, and skatole in the sternum V glands of the limnephilid *Pycnopsyche scrabipennis* Rambur. Ansteeg and Dettner (1991) found male *Polycentropus flavomaculatus* Pict. to produce 2-methyl-butanoic acid and several aromatic compounds, whereas sternum V glands of conspecific females contained only trace amounts of a smaller selection of compounds. Female *Phrygaena grandis* L. produced 1-pentanol, 1-octanol, 3-methyl-2-heptanone, and phenylethanol.

3. Pheromone Communication in Lepidoptera

3.1 Micropterigoidea

The single family Micropterigidae comprises about 130 mostly day-active moth species. The genus *Micropterix* contains most of them. [Species numbers given here and in the following overview of Lepidoptera refers to figures given in Scoble (1992).] The other genera are informally treated as the *Sabatinca* group (Kristensen and Nielsen 1979).

Adults of both *Micropterix* Hübner and *Sabatinca* Walker have been observed to aggregate on flowers/inflorescences of the adult host plants (Kozlov 1982, 1985, and unpublished; Pellmyr et al. 1990) or on reproductive fern fronds (J. Dugdale and G. Gibbs, personal communication). This aggregation is probably related to both adult feeding and mate-finding. For *Sabatinca,* moth attraction by synthetic blends with components of floral volatiles has been demonstrated (Pellmyr et al. 1990). Precopulatory behavior has been described in detail for *Micropterix maschukella* Alphéraky (Kozlov 1982), whose males demonstrate the typical lepidopterous "dance" with wing fanning. However, results of pilot experiments conducted by Pirngruber (1944) suggested absence of pheromone communication in *Micropterix calthella* L. Abdominal sternal V glands were not found in *Micropterix* (Kristensen 1984b), and these species were supposed to rely primarily on visual cues for species recognition (Kozlov 1985).

Sternum V glands were reported in species belonging to the *Sabatinca* group (Philpott 1924; Razowski 1975; Kristensen 1984b), but their chemical contents have not been investigated and no pheromonal activity has been demonstrated. When two New Zealand species, *Sabatinca* (*Palaeomicra*) *chalcophanes* Meyrick and *Sabatinca* (s. str.) *demissa* Philpott, were investigated with respect to the potential existence of a pheromone communication system, all the obtained results (behavioral observations, screening of potential pheromone candidates, examination of antennal responses to extracts) were negative and as a consequence we suggest that these species lack a long-distance pheromone communication system (M.V. Kozlov, E.L. Zvereva, S. Foster, J. Zhu, and C. Löfstedt unpublished results).

3.2. Agathiphagoidea

A single family with two species is included in this superfamily. Males of *Agathiphaga vitiensis* Dumbleton possess glands on the Vth abdominal sternite (Davis 1975). The gland is a long twisted tube, different from the saccular structure encountered in other Lepidoptera and Trichoptera (Kristensen 1984b). No information on the mate-finding behavior of this family is known to us.

3.3. Heterobathmioidea

This superfamily consists of a single small genus in a single family. In both sexes of *Heterobathmia pseuderiocrania* Kristensen et Nielsen, the anterolateral area of sternum V bears a pair of conical, apically tongue-shaped tubular processes with a scaly texture, with the openings of the sternal glands located on the subapical areas of these processes (Kristensen and Nielsen 1979). No behavioral data are available.

3.4. Eriocranioidea

The single family Eriocraniidae contains less than 30 species. Male attraction by conspecific females of these day-active moths was reported for three species (Kozlov 1985, 1988; Wagner and Rosovsky 1991), and the courtship behavior of *Eriocrania cicatricella* Zetterstedt has been described in some detail (Zhu et al. 1995). Virgin females do not demonstrate any special pose when "calling." Males fly around the female, land 10–15 cm away from the female, and then run to her displaying a typical wing fanning behavior. If the population density was high, up to five males "dancing" around a female could be observed.

In most species in the family Eriocraniidae, both sexes possess a pair of glands on the Vth sternite (Davis 1978), although in the genus *Eriocrania* Zeller we only found the glands in the females (Zhu et al. 1995). An electroantennogram assay of different body parts showed that pheromone production in *E. cicatricella* is associated with the sternum V glands.

The active compounds were identified as (*Z*)-4-hepten-2-one, (2*R*)-heptan-2-ol and (2*R*)-(*Z*)-4-hepten-2-ol. In field tests, a subtractive assay confirmed that the unsaturated alcohol is a major pheromone component. A bait containing the two alcohols with (*S*)-configuration was attractive to male *E. sparrmannella* Bosc (Zhu et al. 1995). Subsequent work on *E. semipurpurella* Stephens and *E. sangii* Wood has demonstrated (*Z*)(-6-nonanen-2-ol in different enantiomeric composition as the pheromone of these species (Kozlov et al. 1996). We conclude that in the Eriocraniidae sex pheromones are probably of general importance for mate-finding, and species-specificity is based on permutations of alcohols and ketones which have not been described in ditrysian Lepidoptera.

3.5 Acanthopteroctetoidea

This superfamily contains only two genera with a few species. No information on their mating biology is available to us. Abdominal sternites IV and V are unmodified (Scoble 1992). Males of *Acanthopteroctetes unifascia* possess a pair of glandular-like structures on the third abdominal pleuron. Females of *A. unifascia* were not yet discovered and thus nothing is known about the existence of the presumed glands in the opposite sex. Furthermore, these structures are missing in other species of *Acanthopteroctetes,* and their exact function remains unknown

(Davis 1978). Males of *Catapterix crimeaea* Zagulajev et Sinev (the family Catapterigidae recently synonymized with Acanthopteroctetidae, N.P. Kristensen personal communication) possess a pair of oval transparent areas on the IInd abdominal sternite, with a duct opening in the middle of each, whereas sternites IV and V are unmodified (Zagulajev and Sinev 1988).

3.6 Lophocoronoidea

A single genus with three species is included. Sternum V glands are well developed in both sexes (E.S. Nielsen and N.P. Kristensen personal communication), but no behavioral or chemical information is available.

3.7 Neopseustoidea

The superfamily contains a single family with 10 species, of which the biology is essentially unknown. No specialized organs have been observed on the Vth sternite of either sex in Neopseustoidea, although paired, circular, membranous areas are present near the caudal margin on the IVth sternite of the female. These structures appear as thin, membranous "windows" in the otherwise pigmented integument, similar to those observed in Eriocraniidae and Nepticulidae, but have no external openings or internal ducts (Davis 1975).

3.8 Mnesarchaeoidea

The New Zealand family Mnesarchaeidae with one genus currently forms a separate superfamily, Mnesarchaeoidea. In *Mnesarchaea acuta* Philpott and *M. loxoscia* Philpott we showed that males were attracted to caged females. The sex pheromone appears to be released by female thoracal structures as extracts of these produce significant electroantennogram responses from males, but the active components are not yet identified (M.V. Kozlov, E. Zvereva, S. Foster, J. Zhu, and C. Löfstedt unpublished data).

3.9. Hepialoidea

The superfamily Hepialoidea is diverse and speciose (>600 species). It includes one major family Hepialidae with about 500 species and four minor ones. Female attraction of males was observed in several hepialid genera (*Fraus* Walker, *Gazoryctra* Hübner, *Triodia* Hübner, *Korscheltellus* Börner, and *Pharmacis* Hübner) and was considered an ancestral state for the family as a whole (Wagner and Rosovsky 1991 and references therein). More recently, Dugdale (1994) reported on female attraction of males among both an advanced (*Wiseana* Viette) and a more archaic (*Aoraia* Dumbleton) New Zealand genus of hepialids which supports this conclusion. Field observations of *K. gracilis* Grote indicate that females release a pheromone while fanning their wings (Wagner and Rosovsky

1991). In some *Aoraia* species, females are brachypterous. After handling such newly emerging species, males can be attracted to ones hand (Dugdale 1994). Recent observations suggest that *Dumbletonius,* unlike its close relatives, has females that fly to a raised site and then disseminate an attractant (Dugdale 1994).

In other genera (*Hepialus* F., *Phymatopus* Wall., *Sthenopis* Packard, and *Zenophassus* Tind.) sex roles are reversed, with males attracting females. However, because we are addressing family-level phylogeny, we will not discuss this secondary situation further (for details, see Wagner and Rosovsky 1991).

The site of pheromone production seems to be variable within Hepialoidea. Females of some New Zealand species emit pheromones from the VIIIth abdominal tergum. *Dumbletonius* females have an eversible integumental area dorsally, hidden under tergum 8 at rest and with a distinctive tuft of long hair-like scales. Other related genera have a less hypertrophied reversible area (Dugdale 1994). The sex pheromone of *K. gracilis* is evidently released from the female's hind wings (Kuenen et al. 1994), whereas the male-produced pheromones of several species are released from androconial brushes on metatibia.

3.10. Nepticuloidea

Nepticuloidea includes two families, Nepticulidae and Opostegidae. Nepticulidae has the highest number of species, around 600, of all monotrysian families. No female calling behavior prior to mating was observed in *Stigmella kozlovi* Puplesis, *S. malella* Stainton, and *Ectoedemia* sp., but the males demonstrated typical wing-fanning behavior (Kozlov 1985; M. Tóth personal communication).

As in the Eriocraniidae and Neopseustidae, the females in Nepticulidae possess membranous "windows" in the otherwise pigmented integument of the IVth sternite. Internally the fenestrae are normally associated with the globular reservoirs of the sternum V glands. However, in this family they are situated close together, reduced in size, and possibly vestigial, and they may be absent in some species (Davis 1975). Males of some species also have morphological structures such as anal tuft of hairs (Wilkinson and Scoble 1979) and paired spots on abdominal sternums V, VI, and VII (Newton and Wilkinson 1982) considered as adaptations to pheromone release, but no evidence has been reported of a male pheromone actually being produced.

Diunsaturated methylcarbinols, (*S*)-(*E*)-6,8-nonadien-2-ol and the corresponding (*Z*) isomer, in a 10 : 3 ratio, were identified as a female-produced sex pheromone in *S. malella.* This blend attracted large numbers of conspecific males, and chemically similar baits attracted other nepticuliid species (Tóth et al. 1995). Short-chain alcohols and ketones also attracted numerous specimens of *Stigmella* sp. in trapping experiments with *Eriocrania cicatricella, E. semipurpurella,* and *E. sangii* (M.V. Kozlov unpublished).

In Opostegidae the presence of sex pheromone in *Opostega salaciella* Treitschke is suggested as Bankes (1890) reported that a female attracted a cluster of males. However, paired fenestrae are absent on sternum IV, as they are in all Monotrysia except Nepticulidae (Davis 1989).

3.11. *Incurvarioidea*

The superfamily Incurvarioidea includes a large number of monotrysian moths, which are organized into six families (Nielsen and Davis 1985). In the family Prodoxidae, female calling behavior has been observed in *Greya politella* Walsingham (Davis et al. 1992a) and *Lampronia capitella* Clerck (J.-W. Zhu personal communication). In the latter species the ovipositor is extruded during calling, suggesting that the pheromone-producing gland is located at the terminal abdominal segments. Males of *Greya* show a typical "dance" with wing-fanning (Davis et al. 1992a). Extracts of the *L. capitella* ovipositor elicit significant electroantennogram responses when tested on male antennae, and gas chromatographic analyses with a male antenna as an electroantennographic detector revealed three active gland constituents, which have not yet been positively identified (C. Löfstedt, W. Francke, M.V. Kozlov, J. Zhu, J. Löfqvist, and B.G. Kovalev unpublished data). According to retention times and preliminary mass spectrometric data, the structures should be long-chain pheromone components, similar to those identified in Ditrysia.

Most observed representatives of the family Adelidae are diurnal. They probably lack a pheromone communication system, and mate-finding instead seems to depend on visual cues. Lack of evidence for a pheromone communication system holds for species having male swarming behavior [*Nemophora amatella* Staudinger (= *amurensis* Alphéraky), *N. ascoldella* Milliére, *N. schrenckii* Bremer, *N. degeerella* L., and *N. staudingerella* Christoph], as well as the only nonswarming species studied (*N. karafutonis* Matsumura, erroneously cited as *N. schrenckii* Bremer by Kozlov 1985), in which males search individually for females. In all studied species the male precopulatory "dance" includes the typical lepidopteran wing-fanning (Kozlov 1985).

3.12. *Palaephatoidea*

This recently discovered superfamily includes a single family with six genera. Neither sex possesses abdominal sternum V glands or fenestrae on sternum IV (Nielsen 1987). No behavioral data are available and no pheromones have been identified.

3.13. *Tischerioidea*

Tischerioidea includes a single family with one genus. Female calling behavior, with the ovipositor extended and with a pronounced diel periodicity, has been observed in *Tischeria ekebladella* Bjerkander. Abdominal tip extracts show pheromonal activity, suggesting that the pheromone is produced in a terminal gland (M. Tóth personal communication), similar to what has been observed in higher moths.

3.14. Ditrysia

Mate-finding and mating in ditrysian Lepidoptera, with the exception of butterflies, are typically pheromone-mediated behaviors. Sex pheromones are released by the females from glands located on the intersegmental membrane between the 8th and 9th abdominal segments (Percy-Cunningham and MacDonald 1987). The females assume a more or less pronounced calling posture in which the pheromone gland is exposed, and the female-released sex pheromone evokes a typical series of behaviors in pheromone-stimulated males. Males that have been stimulated by the pheromone respond by flying upwind. In some species the sexes will mate immediately when a male has located a calling female, whereas in other species a more or less elaborated close-range behavior follows on the long-distance attraction. The attracted males land close to the calling female and initiate courtship which includes typical male wing-fanning behavior and in some cases the release of male pheromones (Birch et al. 1990). Pheromone chemistry of ditrysian Lepidoptera is well studied. The majority of the identified pheromone compounds are long-chain (12–18 carbon atoms) unbranched acetates, alcohols, and aldehydes with one or two double bonds. A second major group of pheromone components, consisting of unsaturated hydrocarbons or their epoxy derivatives, occurs mainly within the superfamilies Noctuoidea and Geometroidea (Löfstedt 1991; Arn et al. 1992).

4. Reconstruction of the Ancestral Character State and Character Evolution

4.1. Presence of Long-Distance Pheromones (Character 1)

The superorder Amphiesmenoptera (Trichoptera + Lepidoptera) is considered a sistergroup of the superorder Antliophora (Mecoptera + Siphonaptera + Diptera) (Kristensen 1981). These two superorders are comparable in diversity and morphological progress, but very different with respect to chemical communication. The presence of long-distance sex pheromones is established in all orders mentioned above, except in the wingless Siphonaptera, but it is only in the Lepidoptera that the use of long-distance sex pheromones for mate recognition seems to be the norm.

The order Mecoptera is recently included in the superorder Antliophora, but scorpionflies are generally believed to have many traits that resemble the ancestor of the Amphiesmenoptera (Rasnitsyn 1980; Kristensen 1981). Therefore data on pheromone communication in scorpionflies would be of special interest for our discussion on the appearance of pheromone communication in caddisflies and lepidopterans. Unfortunately the available information is restricted to evidence of a male-produced long-distance sex pheromone in *Panorpa japonica* Thunberg, released by a gland situated within a large genital complex (Thornhill 1992). This observation may suggest that presence of a long-distance pheromone (*Character 1*), which certainly is an ancestral state for Lepidoptera, may be considered a synapomorphy (similarity between groups in an advanced trait evolved by their common ancestor) of all panorpoid orders. However, considering the diversity of pheromone communication systems within the Antliophora, we are inclined to conclude that these are not homologous, but are better interpreted as examples of convergent evolution (homoplasy), the outcome of strong selection for chemical communication systems in insects.

Abdominal glands in some caddisflies are known to produce substances that are repellent or toxic to invertebrates (Duffield et al. 1977; Ansteeg and Dettner 1991). Thus, one must ascertain whether, in some groups, defensive secretions have evolved from pheromones or if the pheromone is derived from defensive secretions. In our opinion, the current knowledge does not allow us to reject either alternative. Pheromones may well have evolved from glands producing defensive secretions in a common amphiesmenopteran ancestor. There is a strong selection, especially in males, to locate members of the opposite sex for mating. Regardless of their original function, the sternum V gland secretions may be exploited as cues for mate-finding. The defensive function of these secretions may have been lost repeatedly, because maintenance of "expensive" defense chemicals is profitable only under high predation pressure, which may not be a major force for selection in ephemeral small swarming insects such as adult caddisflies that have no known specialized predators. Once established, a pheromone system should facilitate mating in most derived species. Its communicative function requires only a small amount of compound to be released, and the glands producing the signal may very well be lost in the nonsignaling sex (See *Characters 2* and *3* below). It is interesting to note that in no taxa where sex pheromones are produced by terminal glands (*Character 6,* below), the sternum V glands are retained serving a possible defensive function. The evolution of sex pheromones from compounds with defensive function has been suggested also by Leal (this volume) in the case of scarab beetles. On the general level, however, the evolution of pheromones from defensive secretions is only one of several possibilities.

4.2. Female Sex Pheromones (Character 2)

The release of long-distance pheromones by the female (*Character 2*) was suggested by Wagner and Rosovsky (1991) as an ancestral character state of Lepidoptera. Their suggestion was based on a cladistic analysis of mating behaviors in Hepialidae, in which they found that the plesiomorphic condition is for females to release the long-distance attractant. This interpretation is supported by our current analysis, suggesting that only a few groups of Lepidoptera have shifted to mate-finding systems based on male-released pheromones (see also Krasnoff this volume). The data available for caddisflies are scanty, but nevertheless point in the same direction. Rhyacophilidae, generally considered an archaic trichopteran family, possess female-produced sex pheromones as do several other caddisfly families. Thus, we prefer to consider female sex pheromones as a possible synapomorphy of the whole clade Amphiesmenoptera rather than a trait evolved independently in Lepidoptera and several trichopteran lineages.

If a long-distance pheromone originally attracted both sexes and served as an aggregation pheromone, or attracted one sex only as a true sex pheromone, can only be guessed. Some data suggest that aggregation is a common type of behavior in both caddisflies and archaic moths (see review above), and the primary presence of abdominal glands in both sexes suggests that their secretions may have facilitated cross-attraction/aggregation in an ancestral group of Amphiesmenoptera. On the other hand, female and male Trichoptera of several species have the same glands but produce different compounds (Ansteeg and Dettner 1991; Löfstedt et al. 1994b), which suggests a sex-specific derived role of the secretions.

4.3. Sternum V Glands (Character 3)

While in Mecoptera the pheromone-releasing structures are located in the male genital complex (Thornhill 1992), pheromone production in Amphiesmenoptera is primarily related to paired abdominal glands with ducts opening on sternum V (*Character 2*). This amphiesmenopteran autapomorphy (= an advanced trait typical only for the group under consideration) was discussed by Kristensen (1984b). However, these glands have been secondarily reduced on numerous occasions within both orders. Among Lepidoptera, Lophocoronidae is the most advanced family possessing these glands in their typical form (E.S. Nielsen and N.P. Kristensen personal communication), whereas they are lacking in all higher taxa.

The data on presence/absence of these glands in caddisflies are primarily based on research of Nielsen (1980), whereas results concerning lepidopterans were summarized by Davis (1975). The sternal abdominal glands are not easy to observe under a dissection microscope, and therefore the number of groups where they are really absent could be less than presently believed. For example, Ansteeg and Dettner (1991) discovered glands in females of two limnephilid species when Nielsen (1980) had concluded that in females they were absent.

4.4. Short-Chain Pheromone Components (Character 4)

Pheromone communication in the investigated caddisflies and archaic moths of the Eriocraniidae and Nepticulidae utilize short-chain components—that is, ketones and secondary alcohols with less than 10 carbon atoms. We consider this character as a possible groundplan autapomorphy of Amphiesmenoptera. In the

Trichoptera and eriocraniid moths, production of these compounds is associated with the sternum V glands (*Character 3*), suggesting that *Characters 3* and *4* are linked. However, this is not necessarily the case as the nepticuliid *Stigmella malella,* which has no known sternum V glands, produces the same type of pheromone components from a presently unidentified site.

4.5. Male Wing-Fanning (Character 5)

The relatively long-lasting (a few seconds to minutes) precopulatory "dance" of males with rapid wing-fluttering (*Character 5*) is characteristic for Lepidoptera. Among ditrysian moths it is a common response to sex pheromones, and a similar behavior was reported for eriocraniid moths. As a matter of fact, this type of behavior was observed in most Lepidopteran families where it was looked for, including the most primitive family Micropterigidae with no identified sex pheromones so far. Thus ancestrally, this behavior was not necessarily a response to pheromones. Whether the wing-vibration reported for *P. linea* (Erman 1984) occurs in more trichopteran species and if it is homologous to the wing-fanning described in Lepidoptera remain to be investigated.

4.6. Female Pheromone Glands at the Terminal Abdominal Segments (Character 6)

Female ditrysian moths have pheromone glands on the intersegmental membrane between the terminal abdominal segments and display a more or less typical calling behavior, extending the terminal segments, thereby facilitating the emission of the pheromone. Such female calling and/or electrophysiological activity of ovipositor extracts has recently been described also in two incurvarioid moths (Davis et al. 1992a; J.-W. Zhu personal communication), as well as in the tisheriid *Tischeria ekebladella* (M. Tóth personal communication), and we suggest that *Character 6* is a synapomorphy of the more advanced heteroneuran lineages. Circumstantial evidence for post-tergum VIII pheromone glands in (homoneurous) Hepialoidea may appear to contradict this interpretation (see Hepialoidea section, above); but since neither Mnesarchaeoidea nor even many genera of Hepialidae possess these glands, we suggest that their appearance in some genera of ghost moths represent a result of parallel evolution. If the homology of terminal glands in Hepialidae and Ditrysia is demonstrated, however, *Character 6* should be considered a synapomorphy of Neolepidoptera, a clade including Exoporia (Mnesarchaeoidea and Hepialoidea) and all heteroneuran moths.

4.7. Long-Chain Pheromone Components (Character 7)

All sex pheromones positively identified from non-ditrysian moths are chemically distinct from ditrysian pheromone components. Having no information on the biosynthesis of non-ditrysian pheromones, we make the somewhat arbitrary distinction between short-chain (<10) and long-chain (≥10) pheromone components based on the number of carbon atoms in their carbon skeletons. The long-chain pheromone components typical of Ditrysia appear to be a relatively late development but not necessarily an autapomorphy of Ditrysia. Preliminary information on the sex pheromone of *L. capitella* suggests that this incurvarioid moth has pheromone components similar to those identified in ditrysian moths (C. Löfstedt, M.V. Kozlov, J. Zhu, J. Löfqvist, B.G. Kovalev, and W. Francke unpublished data). Thus it is more likely that the occurrence of long-chain pheromone components is a synapomorphy of the more advanced heteroneuran lineages. Based on this character and *Character 6* we group Incurvarioidea, Palaephatoidea, Tischerioidea, and Ditrysia in one clade, whereas Nepticuloidea with short-chain pheromone components and no evident terminal pheromone gland should be regarded as more primitive. A more primitive position for Nepticuloidea was suggested by Minet (1984), who separated Nepticuloidea from the infraorder Eulepidoptera, which he reestablished to comprise Incurvariina, Tisheriidae, and Ditrysia.

5. Increasing the Resolution of the Phylogenetic Analysis

Investigation of pheromone communication in a few critical taxa (Table 41.1) could (a) clarify the evolution of pheromone communication in the basal lineages of Amphiesmenoptera, (b) provide the necessary information to resolve some ambiguities concerning the lepidopteran phylogeny among Heteroneura, and (c) give the background to allow a deeper understanding of how and why the diversity and the patterns in ditrysian moth pheromone communication evolved.

Knowledge about pheromone communication in Trichoptera is limited to a few taxa. Pheromone-mediated behaviors should be studied and pheromones should be identified in the major trichopteran families. Such studies could also contribute to the reconstruction of an improved phylogeny of Trichoptera.

Pheromone communication, including mating behavior and the chemistry of the secretions of the sternum V glands, should be studied in Micropterigidae (especially in *Sabatinca* Walker), the sister group of all more advanced Lepidoptera. Similarly, Agathiphagidae, another family with many archaic traits, should also be subject to pheromone studies.

The Exoporia (i.e., Hepialoidea and Mnesarchaeoidea) assume an important place for further research on pheromone communication systems. The hypothesis on the homology of the terminal glands in Hepialoidea and Ditrysia should be tested. If Exoporia and Ditrysia have homologous terminal pheromone glands as suggested by circumstantial evidence, one should reconsider the position of Exoporia within the classification of Lepidoptera. This group is significant in that the similarities with Ditrysia in internal female genital structure (Dugdale 1974) are currently regarded as parallelisms (Common 1975; Nielsen and Common 1991, "tentative cladogram," p. 833). An alternative explanation would be that Exoporia is a sister group of Ditrysia. Identification of female-produced pheromones and location of their production sites in Hepialoidea and Mnesarchaeoidea could contribute to the testing of this hypothesis and clarify the relationships between Exoporia and Nepticuloidea.

Location of the pheromone glands in Nepticulidae and Opostegidae would be of special significance because Nepticulidae lack the sternum V glands present in most lower taxa, but produce pheromone components similar to those identified from Eriocraniidae and Trichoptera. Further investigation of nepticulioid moths may thus reveal how and when pheromone biosynthesis in terminal abdominal glands replaced pheromone production in the sternum V glands.

The divergence of the heteroneuran taxa Nepticuloidea, Incurvarioidea, Palaephatoidea, and Tischerioidea is not resolved based on traditional morphological characters, and several more or less conflicting phylogenies have been suggested (Minet 1984; Nielsen 1989; Scoble 1992; N.P. Kristensen and Skalski 1996). The ovipositor gland and the type of pheromone components present in ditrysian moths may have evolved already in connection with the evolution of monotrysian Heteroneura. As already mentioned, short-chain pheromone components in Nepticulidae support a more primitive position of Nepticuloidea relative to the other superfamilies included in Heteroneura. Terminal pheromone glands definitely evolved before divergence between Ditrysia and other heteroneuran lineages. Presence of long-chain pheromones in almost all ditrysian moths investigated so far, as well as our preliminary information on pheromone chemistry in Incurvarioidea, suggests that also the ditrysian pheromone chemistry evolved earlier in the evolution of Heteroneura. These hypothesis should be tested by investigation of pheromone communication in the heteroneuran taxa mentioned.

The biosynthetic pathways producing the alcohols and ketones identified from Trichoptera, as well as Nepticulidae and Eriocraniidae, are unknown. One possibility, suggested to us by W. Francke (personal communication), is that (*Z*)-4-hepten-2-ol and (*Z*)-6-nonen-2-ol are produced by $\Delta 11$ desaturation of tetradecanoic acid, followed by chain-shortening (β-oxidation) and eventually decarboxylation of the β-keto acids produced as chain-shortening intermediates. This hypothesis, if not rejected, would bridge the dichotomy between pheromone components identified in archaic and more advanced moths. β-Oxidation in mitochondria is complete, whereas β-oxidation in ditrysian moth pheromone biosynthesis is limited to a few cycles. The pathway proposed for the pheromones produced in the sternum V glands could represent an intermediary stage. However, we want to emphasize that this pathway is hypothetical and awaits experimental evidence.

6. The Adaptive Value of Pheromones

The adaptive value of female-produced moth sex pheromones was recently challenged by the evolutionary biologist G.C. Williams (1992, pp. 111–113). Williams suggested that *"female sex pheromones usually do not exist or are no more common than female initiation of courtship by singing or by waving ornamental fins or feathers."* Williams developed his provocative standpoint further: *"Male moths do indeed respond to chemicals from reproductively ready females . . . My dispute is with the claim that these compounds are signals. . . . As argued by Burghardt (1970), Otte (1974), and Lewis (1984), a stimulus should be called a signal only if it is normally adaptive for the sender and produced by machinery attributable to selection for the proposed communicative function"*, which according to Williams is not the case for female moth sex pheromones, because he found evidence only for *"extreme development of male abilities to detect female odors, not of any female machinery for releasing them."* However, his theoretically plausible standpoint has no empirical support.

Our phylogenetic analysis demonstrates that a female sex pheromone belongs to the groundplan of Lepidoptera, and probably also to that of the superorder Amphiesmenoptera. It is widely recognized, not only by Williams, that sexual selection is stronger on males than on females and that male moths would generally be the ones searching for females (see e.g., Löfstedt 1990 and Phelan 1992 for references to data and for further discussion). This would be compatible with our suggestion that the communicative function of sex pheromones evolved from compounds originally used for defense in ancestral moths and caddisflies, without suggesting that the adaptation is *"entirely male"* as proposed by Williams (see also Phelan 1997 for discussion of the Williams pheromone fallacy).

Evolution of a communication system is better understood as a coevolution between the sender and the responder (Löfstedt 1993 and references therein). Stimuli need matching receptors to eventually function as signals. In caddisflies and moths, pheromone receptors (and the accompanying biochemical machinery) may have evolved from the insects' ability to detect various plant compounds because the short-chain pheromone components used by caddisflies and primitive Lepidoptera are similar to many frequently encountered plant volatiles (Visser 1986). Once established, the sex pheromone communication system has subsequently evolved toward a higher level of sophistication as can be seen in more advanced moths. In the heteroneuran moths, specialized terminal glands with no other known function than pheromone production have evolved in the females, accompanied by evolution of more or less unique long-chain pheromone components. Both biosynthesis of the pheromone and the female calling show a diel periodicity; and intriguing details of the hormonal control of pheromone production, including action of a pheromone biosynthesis activating peptide named PBAN, have been revealed (Raina this volume).

Since Lepidoptera is one of the most successful orders of insects, as evaluated by the diversity at both species (>150,000) and family (>110) levels (Kristensen 1996), the question arises whether the evolutionary success of lepidopterans was at least partly determined by their ability to use pheromones for mate recognition. The appearance of pheromone communication could be considered as an adaptive trait for the whole clad, not restricted to the lower hierarchical levels such as genera and species where the role of pheromones in mate recognition is well established. To test the hypothesis that evolution of pheromone communication with a certain type of pheromone components has led to adaptive radiation, it is necessary to compare the number of species in sister groups with and without these pheromones. Trichoptera and primitive Lepidoptera share the same type of pheromone components and the same pheromone-producing gland. Trichoptera and non-ditrysian moths comprise comparable numbers of species, 10,000 and 2,200 species respectively, whereas the ditrysian moths consist of approximately 150,000 species (Brusca and Brusca 1990; Kristensen 1996). It is thus possible that the evolution of the ditrysian type of pheromone created new possibilities for specific mate-recognition signals and gave rise to adaptive radiation. Long-distance pheromones were lost in butterflies (Hesperioidea + Papilionoidea), but in these day-active lepidopterans the pheromone-based mate-finding system may have been replaced by an equally successful visual one. A rigorous test of the hypothesis involves comparison of species numbers in taxonomic groups possessing the ditrysian pheromone type, with sister groups showing character reversals to the ancestral state.

Acknowledgment

Knowledge about pheromone communication in Trichoptera and in primitive moths, of critical value for the current analysis, was gained in collaborative projects involving several co-workers, especially Drs. J.-W. Zhu, W. Francke, and E. Petersson. Drs. M. Tóth and P.L. Phelan kindly provided unpublished information and valuable suggestions. Comments by Siana LaForest and by Drs. J. Dugdale, N.P. Kristensen, J. McNeil, E.J. van Nieukerken, and O. Pellmyr were greatly appreciated. This work was supported by grants from the Swedish Natural Science Research Council, The Royal Swedish Academy of Science, and the Bank of Sweden Tercentenary Foundation.

42

Evolution of Male Lepidopteran Pheromones: A Phylogenetic Perspective

Stuart B. Krasnoff

1. Introduction

Male-specific scent structures (hereinafter referred to as MSSs) intrigued early lepidopterists (Müller 1877; Longstaff 1912) and continue to fascinate those who study their diverse morphologies and the role they play in mating behavior. Several recent reviews comprehensively catalogue the occurrence of these structures, the compounds they produce, the relatively small number of these compounds with proven pheromonal activity, the responses the pheromones induce in females, and the theories of their evolutionary origin (Boppré 1984; Birch and Hefetz 1987; Fitzpatrick and McNeil 1988; Birch et al. 1990; Phelan 1992). In this chapter I consider historical patterns along with the evolutionary processes that shaped male lepidopteran pheromone systems, I critically examine some of the methods that have been used to test scenarios for the evolution of MSSs, and, finally, I advocate the use of rigorous phylogenetic approaches to test these scenarios.

2. Evolutionary Contexts

2.1. History of the Ideas

Most workers have converged on two major selective forces, sexual selection and reproductive isolation (or species recognition), to explain the evolution and proliferation of MSSs. Phelan (1992) noted Müller's early speculation that MSSs evolved via sexual selection (Müller 1878). Although species-specificity of MSSs was also noted in the early literature (Eltringham 1915), Brower and co-workers (Brower et al. 1965; Brower and Jones 1965) were the first to speculate that species-specific male scents produced by MSSs might play a role in species isolation.

The species recognition/reproductive isolation function of male lepidopteran pheromones was first explored in interspecific mating experiments by Taylor

(1973) in butterflies and by Grant et al. (1975) in moths. The role of sexual selection in the origin of MSSs and associated pheromones was discussed theoretically by Baker and Cardé (1979a) in their study of courtship of the oriental fruit moth. Phelan and Baker (1986) subsequently demonstrated that sexual selection occurs in the courtship of a moth, *Ephestia elutella,* manifested by female preference for larger males. They also showed a positive correlation of male size to pheromone titer, establishing the possibility that females chose larger males based on their pheromone output.

2.2. Reproductive Isolation Through Sexual Selection (Phelan and Baker 1987)

Phelan and Baker (1987) united the two principal selective contexts for viewing the evolution of male lepidopteran pheromones, reproductive isolation and sexual selection, invoking Baker and Cardé's (1979a) application of Fisher's "runaway" sexual selection model (Fisher 1958). They referred to the notion that females with a heritable preference for a species-specific male courtship display might gain an adaptive advantage by avoiding heterospecific mating. An advantage of this type is deemed necessary to initiate the episode of rapid directional selection thought to account for exaggerated male-specific traits used by females as instruments of choice in courtship.

The argument for the reproductive isolating function of male pheromones was set against the background of the allopatric speciation model which describes a process that begins with the geographic division of a population into two populations that diverge genetically in allopatry (Dobzhansky 1937; Brooks and McLennan 1991). When contact is reestablished, matings between members of the two deems result in fitness penalties for the participants. If such matings have no issue or produce inviable or infertile offspring, then there is no gene flow, speciation is essentially complete, and ethological barriers to mating between the two species may evolve in one or both species by selection against vain expenditures of gametes and/or mating effort. The term "reproductive character displacement" has recently been restricted to describe this "interspecific" process (Butlin 1987). If interdeem matings result in hybrids of reduced fitness, then there is gene flow, and ethological traits that foster positive assortative mating may arise under certain restricted conditions by the process of "reinforcement" (Butlin 1987). In this case the fitness cost of not mating assortatively (in the positive sense) is considered capable of driving the evolution of reproductive isolating mechanisms that complete the speciation process.

Paterson (1993) challenged this model of speciation and the "adaptive" view of the evolution of ethological isolating mechanisms, arguing that geographic speciation is completed in allopatry, via natural selection, with the evolution of "specific-mate recognition systems" (SMRSs) adapted to the unique environmental conditions encountered by the incipient species. However, should ethological traits serve to keep potentially introgressing sister-species discrete if, subsequent to speciation, they should come into contact, they do so only as an incidental effect of their evolution in allopatry. West-Eberhard (1983) proposed a similar "incidental" model but, in contrast to Paterson, favored sexual selection as the force driving the rapid and divergent evolution of species-specific mating signals and associated traits.

Thus Phelan and Baker (1987) divide the alternative view of the evolution of ethological isolation into adaptive and incidental models, and subdivide each as driven by either natural or sexual selection. They present evidence that male lepidopteran pheromones evolved as ethological isolating mechanisms in the final step of the speciation process with sexual selection rather than natural selection as the driving evolutionary force in this context.

Phelan and Baker (1987) argued that if male pheromones arose by selection against mating mistakes between members of sympatric populations of incipient species that are incompletely isolated, then the presence of MSSs should be associated with a higher likelihood of fitness-depressing heterospecific mating encounters between members of closely related extant species. Conversely the absence of MSSs should correlate with a low or absent likelihood of such a mating encounter. The incidental model would predict no such relationship.

The principal evidence supporting this argument is a set of 2×2 chi-squared contingency tables, each representing a different moth taxon and MSS and each tabulating species with and without MSSs in one direction and tabulating species with and without host–plant overlap with a congeneric species in the other direction. The rationale for this approach is that interspecific mating mistakes can only occur between closely related sympatric species, and such sympatry can be "indexed" by the sharing of a host plant with a congeneric species. Conversely, absence of MSSs should correlate with absence of host overlap.

Phelan and Baker (1987) found significant correlations in all cases between the occurrence of MSSs and host overlap, which they interpreted as verifying the prediction of the adaptive model of the origins of MSSs and male pheromones in a context of reproductive isolation. However, the validity of this approach rests on several questionable assumptions. First, there is an implicit assumption that the presence of MSSs is a valid index of the presence of a unique (autapomorphic) species-specific male pheromone system capable of isolating each species from a closely related species. Just as a plesiomorphic female pheromone gland has been reused to carry different highly specific signals in different, often closely related species, it is certainly plausible that various MSSs have been recycled once or many times throughout the phyletic history of a lineage to carry uniquely derived pheromonal signals. However, knowledge of pheromones as well as knowledge of the structures thought to produce and release them is necessary to support such an assumption. Given that MSSs in extant species may be nonfunctional (see below), the numbers in the MSS present cells of the tables could be inflated relative to the numbers in the MSS absent cells.

A second assumption is that any pair of congeneric species were potential players in a reproductive isolation scenario with each other. This greatly oversimplifies the process of identifying legitimate participants in a reproductive isolation scenario (Maddison and Maddison 1992). The allopatric speciation model upon which Phelan and Baker (1987) based their argument is specific in this regard. For an ethological isolation mechanism to function in the speciation process, reproductive isolation must have taken place between the products of a speciation event (i.e. sister species) whose status as such is best determined by cladistic analysis. Pairs of "congeneric species" will certainly include pairs of species that are not sister species.

Evolution of isolating mechanisms between non-sister species is conceivable by either reproductive character displacement or reinforcement (both *sensu* Butlin 1987; see above). Of course, in the latter case hybridization would have to be possible between non-sister species. Assuming that speciation is largely a dichotomous process, evidence of interspecific hybridization between one species and more than one other close relative indicates that this is possible (Kilias et al. 1989). However, exclusive mating signal characters observed in non-sister pairs cannot have evolved as reproductive isolating mechanisms in the strict sense because separate episodes of speciation produced those species. For mutually exclusive mating signals to be held up as reproductive isolating mechanisms, it must be shown that an interaction could have occurred either between one of the extant species and an ancestor of the other extant species identifiable in a cladogram as a sister group of the first, or between two ancestors (sister groups) of both extant species that have a sister group relationship (Fig. 42.1).

The most serious weakness in this approach is its dependence on a third problematic assumption—that the numbers of species accumulated into the cells of the contingency tables, although purporting to represent an equivalent number of independent evolutionary origins of MSSs, undoubtedly do not, and thus are inflated. No adaptive argument can be made regarding the evolutionary origin of a trait during the speciation process that produced two extant species if phylogeny indicates that the trait evolved in a common ancestor and is simply conserved in extant species. In other words, if an MSS evolved very early in the phyletic history of a lineage, then its presence in extant species cannot be validly used in an argument pertaining to its origin without making another assumption, that the evolutionary forces supporting the maintenance of the structure are the same as those that drove its origin (Coddington 1988). Although Phelan (1992) remarked on the need to distinguish between these two categories of selective contexts, this distinction was not taken into account in the interpretation of the results of the test originally presented by Phelan and Baker (1987).

Furthermore, even if the correlations discovered by Phelan and Baker (1987) are used to argue that MSSs have been maintained in species exposed to interspecific sexual challenge, but were lost or never evolved in species not so challenged, the argument would still be hampered by the problem that affected the argument as originally worded. Because the occurrences of loss, recorded as absences of

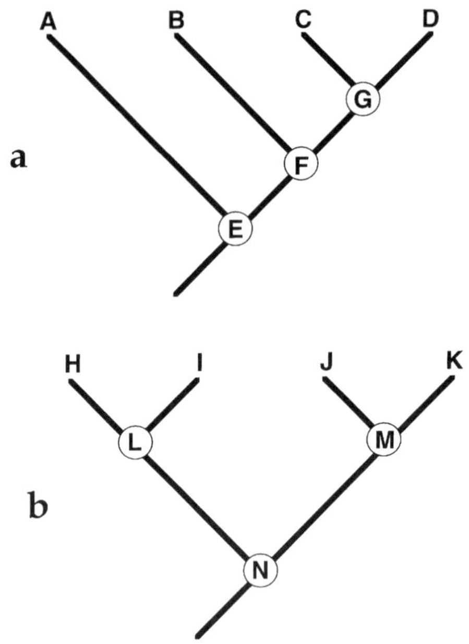

Figure 42.1. Reproductive isolation in the strict sense can only occur between sister species. In part **a** those pairs would be C and D, B and G (the ancestor of C and D), and A and F (the ancestor of B, C, and D), and part **b** the pairs would be H and I, J and K, L and M. Thus observing potential species isolating traits in extant species does not ensure that those traits evolved in extant species. Observations of potential species isolating functions of pheromones and related courtship behaviors between non-sister species pairs (e.g. C and B, D and B, B and A, C and A, D and A) could be the result of either (1) the inheritance of features that evolved in a reproductive isolating context (*sensu strictu*) between an extant species and its sister group (e.g., differences between B and C could have been inherited from features that evolved during interactions between B and G), (2) interactions between two ancestors of extant species such as L and M, or (3) as post-speciation ethological isolating mechanisms.

MSSs in each contingency table, may be phylogenetically nonindependent, these could also be inflated in the samples as presented.

To further illustrate this problem I examined a source of data used for one of the contingency tables presented in the Phelan and Baker (1987) study, the yponomeutid fauna of Japan monographed by Moriuti (1977). Many species of yponomeutids and their sister group, the argyresthiids, have abdominal coremata, the presence or absence of which, in most cases, does not vary below the generic level. For example, all 18 species of the genus *Yponomeuta* possess coremata. If the genus is monophyletic, a character analysis is simple: The coremata originated in the ancestral species that gave rise to the genus, or they originated in some more basal ancestor.

This latter possibility appears likely. Moriuti presented a partially resolved phylogenetic hypothesis and classification for the genera of the Yponomeutidae. Mapping the occurrence of coremata onto Moriuti's tree and carrying out a character analysis to determine the most parsimonious interpretation of the evolution of coremata in this group produces ambiguous results in certain branches with unresolved polytomies, but the general picture that emerges is consistent with the hypothesis that the coremata arose early in the history of the group, no later than in the common ancestor of the Yponomeutidae and the sister family Argyresthiidae and were lost numerous times in more terminal taxa (Figure 42.2).

This perspective is given to illustrate the point that sampling the Yponomeutidae at the level of species greatly overestimates the number of independent origins of the coremata in the group. If, after cladistic treatment, preferably excluding the corematal character in the analysis, the coremata are shown to have arisen at or below the root of the group, then the occurrence of coremata in all 44 species accumulated into the cells on the "androconia present" side of the contingency table would be attributable to a single origin of an MSS in that ancestral branch and the sample of 44 would be reduced to one. This would invalidate the chi-squared approach (see also Felsenstein 1985; Donoghue 1989; Maddison and Maddison 1992).

Following this line of reasoning, the species in the sample in which the coremata are absent may also be nonindependent with regard to their possession of this character state (coremata absent)—that is, if they share this character state by common descent with other species in the sample. For example, Moriuti (1977) assigned "coremata absent" as the character state for the subfamily that includes the genus *Prays*. Thus the absence of the coremata in all eight species of *Prays* must be due to the absence or loss of coremata in the common ancestor of all species in the subfamily or a more basal ancestor, not to losses in the individual species. It should be noted that secondary losses must be distinguished by phylogenetic analysis from primitive absences; they are not equivalent (i.e., homologous) and cannot be used in the same way to make deductions about the evolutionary origins of a character of which they represent the absence (Coddington 1988).

Rigorous analysis of the role of lepidopteran MSSs in reproductive isolation requires evidence from sister-species pairs (preferably many such pairs), sympatric in distribution, and with autapomorphic MSSs or, preferably, male pheromones occurring in one or both species. Many MSSs that are currently treated as single characters may, on further scrutiny, be broken down into multicharacter complexes such that many autapomorphies may be recognized that would provide a database more amenable to a rigorous test of adaptive versus incidental models of reproductive isolation.

In the second half of their argument, Phelan and Baker (1987) offer several lines of evidence to support sexual selection rather than natural selection as the driving force in the evolution of MSSs. First cited is the lack of examples of reproductive character displacement. If natural selection were the selective force

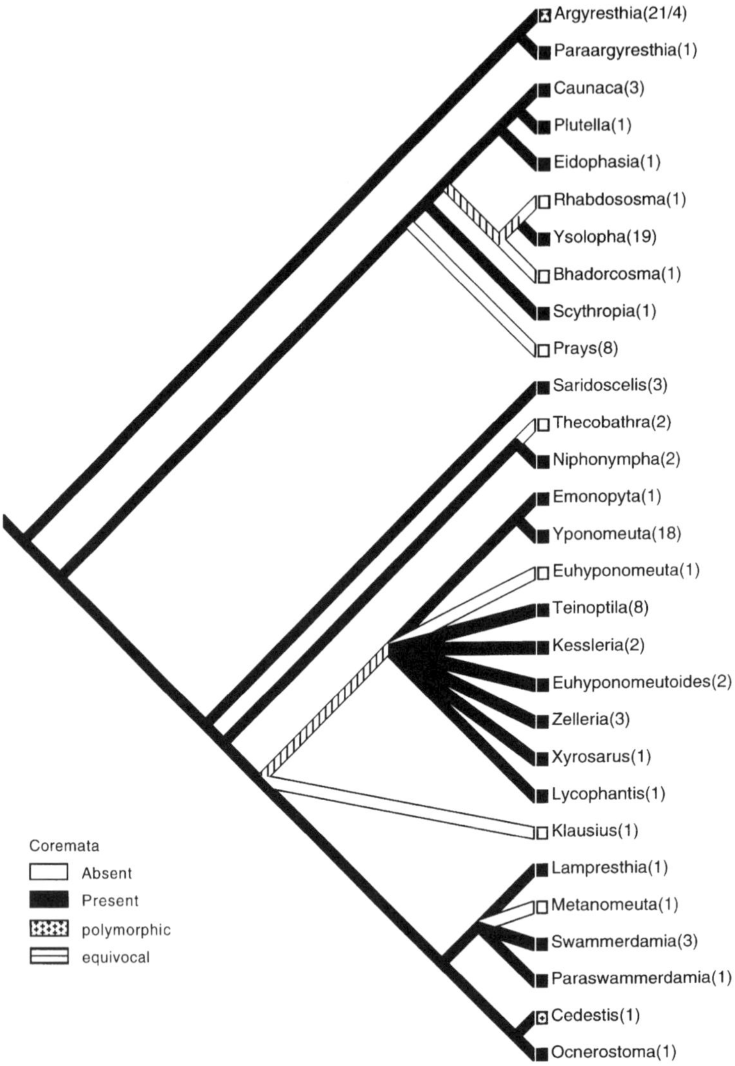

Figure 42.2. Character analysis using MacClade (Maddison and Maddison 1992) showing the most parsimonious reconstruction of the evolution of the coremata as a two-state character (present/absent) mapped on Moriuti's (1977) phylogenetic tree for genera of the Yponomeutidae, *sensu latu.* Figures in parentheses next to generic names refer to number of species treated by Moriuti. Two genera have species with coremata and without coremata. Numbers indicate that 21 species of *Argyresthia* have coremata, while four do not. The single Japanese species of *Cedestis* included lacks coremata.

at work, then we should observe intraspecific geographic variation in male pheromones or related structures with greater signal divergence evident in zones of overlap where closely related species compete for the sexual communication channel. However, Phelan and Baker argue that the homogeneity of MSSs and male lepidopteran pheromones throughout the range of species in which they are found argues for their rapid spread by runaway sexual selection (Fisher 1958).

To further support the role of sexual selection in the evolution of MSSs as ethological isolating mechanisms, Phelan and Baker (1987) also cite the "striking morphological diversity" of MSSs and state that these structures "typically appear sporadically within a [taxon] and their relative complexity bears little relation to lepidopteran phylogeny." There is no dispute that MSSs have evolved many times independently within the Lepidoptera. This is offered as evidence of the kind of rapid and divergent evolution that has been claimed as a footprint of sexual selection (West-Eberhard 1983; Eberhard 1985). However, the widespread assertion, attributed to McColl (1969), that MSSs are "ephemeral in appearance" (Phelan 1992) and that their distribution in the Lepidoptera or in a particular lineage is "taxonomically untidy" (Phelan 1992) is dubious without reference to a particular phylogenetic hypothesis. Such statements imply that by some objective measure, MSSs have been determined to be highly homoplastic.

The "striking" diversity of MSSs in the Lepidoptera, when viewed from a phylogenetic perspective, may in fact look more like variation on a surprisingly small number of themes. Although a proliferation of autapomorphic MSSs have been identified in some lineages (e.g., in some dioptine notodontids—J.S. Miller personal communication), most species in which MSSs occur share them with other closely related species, often many others. With due respect to the adaptationist view that MSSs would persist only in species subject to selective pressures for their maintenance, the occurrence of homologous MSSs in a large number of extant species may be, in large part, the legacy of origins early in the history of the lineages in which they occur.

One of the numerous examples that could be used to illustrate this point is the androconial organ associated with the pleated sacculus on the genitalia of many notodontid species. In this case there is good phylogenetic evidence indicating that a widespread MSS evolved early in the phyletic history of a lineage and was lost many times (Miller 1991). A treatment of exemplar species from 50 notodontid genera reveals a terminal clade of 22 genera, in which the saccular scent organ is conserved in 15 genera and absent in 7. Character analysis indicates that this MSS was present in the basal branch of the clade and was lost five times (Figure 42.3).

The phylogenetic record may show that MSSs have been lost far more often than they have originated. So, far from solely supporting the view that sexual selection must have governed their origin, this record also begs the question of why, if runaway sexual selection is supposed to be followed by stabilizing

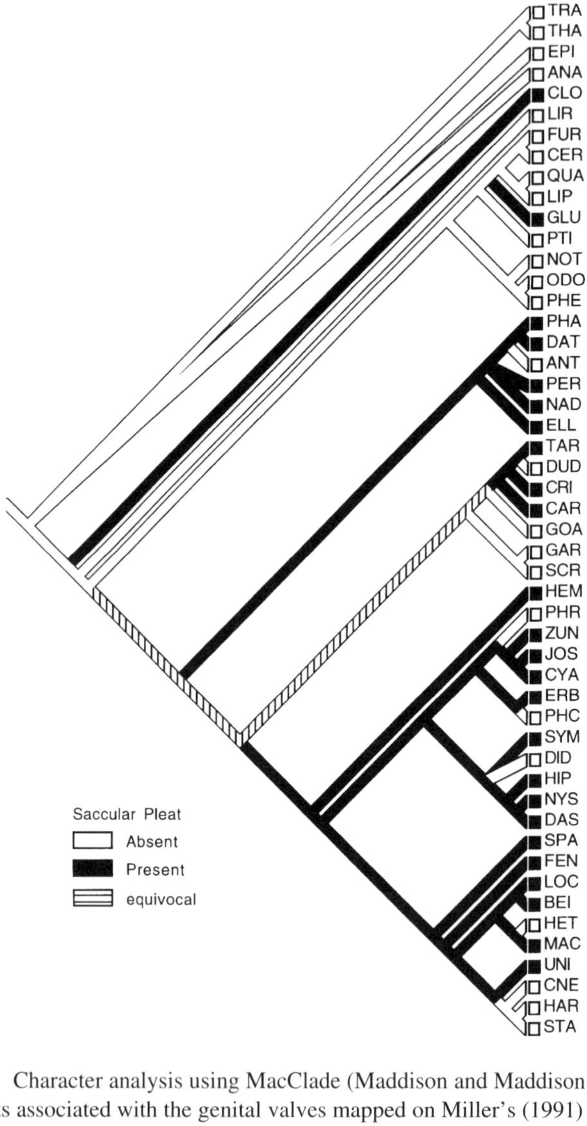

Figure 42.3. Character analysis using MacClade (Maddison and Maddison 1992) on the saccular pleats associated with the genital valves mapped on Miller's (1991) phylogenetic tree of selected genera of the Notodontidae. For simplicity the four character states defined by Miller (absent, present but "not highly modified," "very faint," and huge and curls inward) were reduced to two state character states (present/absent). Generic abbreviations are as follows: TRA, *Traumatocampa;* THA, *Thaumetopoea;* EPI, *Epicoma;* ANA, *Anaphe;* CLO, *Clostera;* LIR, *Lirimiris;* FUR, *Furcula;* CER, *Cerura;* QUA, *Quadricalcarifera;* LIP, *Liparopsis;* GLU, *Gluphisia;* PTI, *Ptilophora;* NOT, *Notodonta;* ODO, *Odontosia;* PHE, *Pheosia;* PHA, *Phalera;* DAT, *Datana;* ANT, *Antheua;* PER, *Peridea;* NAD, *Nadata;* ELL, *Ellida;* TAR, *Tarsolepis;* DUD, *Dudusa;* CRI, *Crinodes;* CAR, *Cargida;*

selection for the maintenance of the traits that arose by the runaway process, should they be so frequently lost.

3. Phylogenetic Approaches to the Evolution of Male Lepidopteran Pheromones

3.1. Two Examples

In the last 20 years, as the center of gravity in systematics has shifted toward cladistics, phylogenetic approaches to evolutionary questions that were formerly the sole province of behavioral and evolutionary ecologists have proliferated (Wenzel 1992; Miles and Dunham 1993; Miller and Wenzel 1994), and comparative methods have undergone a renaissance (Brooks and McLennan 1991; Gittleman and Luh 1992). The principal reviewers of MSSs and male lepidopteran pheromones have acknowledged the need to go better armed with phylogenies when confronting key questions about their evolution.

The application of phylogeny in this area is, so far, rare, but there are at least two significant exceptions. Wagner and Rosovsky (1991) used a morphological phylogeny of selected hepialid genera to address basic questions about the evolution of pheromone communication in the family. Their character analysis indicates that male pheromonal calling, an unusual behavior exhibited by a few widely scattered genera, represents a derived condition that evolved only once in the group from ancestors in which female calling characterized the mating system.

Vane-Wright and co-workers (Vane-Wright et al. 1992; Schulz et al. 1993; Vane-Wright and Boppré 1993) used a cladistic approach to examine the evolution of male pheromones in a species complex of danaine butterflies in which constraints on the divergence of visual signals due to mimicry may have increased selection pressure favoring the evolution of species-specific courtship pheromones. The importance of this study lies in (a) the construction and testing of models of character evolution consistent with various solutions to the species-recognition problem (see Vane-Wright and Boppré, 1993) and (b) the careful consideration of the debate over whether or not to use the chemical/pheromone characters in deriving the phylogeny used to deduce their evolutionary history (Vane-Wright et al. 1992; see also Kluge 1989).

Vane-Wright et al. (1992) delineate several modes of pheromone transforma-

---

Figure 42.3 (Con't.) GOA, *Goacampa;* GAR, *Gargetta;* SCR, *Scrancia;* HEM, *Hemiceras;* PHR, *Phryganidia;* ZUN, *Zunacetha;* JOS, *Josia;* CYA, *Cyanotricha;* ERB, *Erbessa;* PHC, *Phaeochlaena;* SYM, *Symmerista;* DID, *Didugua;* HIP, *Hippia;* NYS, *Nystalea;* DAS, *Dasylophia;* SPA, *Spatalia;* FEN, *Fentonia;* LOC, *Lochmaeus;* BEI, *Schizura beidermani;* HET, *Heterocampa;* MAC, *Macrurocampa;* UNI, *Schizura unicornis;* CNE, *Cnethodonta;* HAR, *Harpyia;* STA, *Stauropus.*

tion consistent with a putative species-recognition function, including (1) addition of unique compounds with conservation of ancestral compounds, (2) addition of new compounds with loss of some ancestral compounds, (3) progressive loss of ancestral compounds from a complex blend. The total phylogenetic evidence supports an additive/subtractive model in which new volatiles are added and/or old ones lost at each speciation event, resulting in unique blends of compounds for each species.

They recognize that a paucity of behavioral data limits the macroevolutionary interpretation of their extensive chemical survey of androconial volatiles from 10 African danaine species. In fact, they convey a sense of despair that the behavioral role of androconial scents in the courtship of these butterfly species will ever be adequately understood (Schulz et al. 1993). Despite this concern, and the acknowledged limitations of the data set (it contains no nonoverlapping species) in shedding light on the issue of incidental versus adaptive models of the evolution of male pheromone specificity in this group, this study stands as an exemplary use of phylogeny to explore the role of male lepidopteran pheromones in speciation.

3.2. A General Phylogenetic Test for Species Recognition Function

Brooks and McLennan (1991) describe a cladistic approach for specifically testing the predictions of the adaptive and incidental models of the evolution of prezygotic reproductive isolation that may be profitably applied to male lepidopteran pheromones. Required for this test are (1) a cladogram for a relevant study group, preferably derived from characters independent of those being analyzed, (2) characters indicative of postzygotic reproductive isolation between relevant species pairs on the cladogram, and (3) characters indicative of prezygotic reproductive isolation such as ethological data bearing on the success or failure of interspecific mating attempts. Depending on how the critical characters map on the cladogram, this method may provide a definitive answer to the question of whether prezygotic isolating mechanisms evolved before postzygotic isolating mechanisms (supporting the incidental model) or after (supporting the adaptive model). Chemical and/or behavioral data pertinent to the species-specific nature of male pheromones have been generated in several studies of moth courtship (Grant et al. 1975; Hendrikse 1986; Phelan and Baker 1990a,b).

4. Scenarios Amenable to Phylogenetic Tests

4.1. Sensory Exploitation

Observing that androconial compounds from several lepidopteran lineages exhibit low species-specificity, Phelan (1992) suggest that male pheromones may have evolved through manipulation of the female host-location system. This notion traces to the work of Edgar et al. (1974) on interactions of danaid butterflies with plants containing pyrrolizidine alkaloids (PAs) and has been invoked several times since as an explanation for the evolution of the PA-derived male courtship pheromones of the danaid butterflies and the arctiid moths (Baker 1989a; Birch et al. 1990). The observation that most compounds identified from MSSs are either known to be derived from plant compounds or are identical to plant-produced compounds has led to speculation that plant chemistry has been exploited many times in the evolution of various male pheromone systems either directly (as a source of precursors) or indirectly to provide templates for endogenously biosynthesized male hormones (Baker 1989a; Birch et al. 1990).

The idea that male pheromones evolved as keys to fit locks consisting of preexisting female olfactory sensitivities to host-plant odors, possibly with an attendant repertoire of exploitable behavioral responses involved in host-location, recognition, or oviposition, is appealing, especially in light of evidence that such scenarios have been played out in mating-signal systems in animals that use other sensory modalities (Basolo 1990; Ryan et al. 1990; Proctor 1992), and because there is, at least in theory, a robust phylogenetic approach for testing this hypothesis.

The "sensory exploitation" hypothesis, most thoroughly explored in courtship songs of frogs by Ryan and co-workers (Ryan and Keddy 1992), has been proposed as an alternative to Fisherian female-choice-based sexual selection (Fisher 1958) for the evolution of mating signal communication systems. The Fisherian model specifies that the male signal trait and the female preference for the trait coevolve, albeit rapidly. If, however, it can be demonstrated that a female "preference" for a male signal trait exists in species in which the male signal trait does not occur and phylogenetic evidence indicates that the female preference is ancestral and the male signaling trait derived (i.e., that the female "preference" evolved in an ancestor of the lineage in which the male trait in question evolved), then the hypothesis of coevolution of trait and preference is falsified and the "sensory exploitation hypothesis" is supported.

This hypothesis has yet to be tested for an olfactory signal system, and male lepidopteran pheromones would seem an ideal arena for doing so. Requirements are (1) a lineage with species in which male pheromones are both present and absent, (2) an experimental approach for discerning "female preference" for an olfactory signal not produced by conspecific males (e.g., an electrophysiological and/or behavioral response to a host-plant compound that may be a male pheromone in related species), and (3) phylogenetic evidence supporting the ancestral nature of the female preference and with the male lepidopteran pheromone derived at least once and preferably many times. The moth family Arctiidae offers a potential opportunity for such a study, because it includes many species that feed, either as adults or larvae, on PA-containing plants and at least nine species with MSSs known to produce PA-derived volatiles (Culvenor and Edgar 1972; Conner et al. 1981; Schneider et al. 1982; Krasnoff et al. 1987; Krasnoff and

Dussourd 1988; Krasnoff and Roelofs 1989). As noted above, pheromonal function has been demonstrated for the PA-derived corematal compounds of several of these arctiids. Behavioral and/or electrophysiological responses to these compounds can be readily demonstrated for several species (Conner et al. 1981; Wunderer et al. 1986; Krasnoff et al. 1987; Grant et al. 1989). Moreover, PA-plant feeding and PA-sequestration have been demonstrated for species that lack MSSs (Benn et al. 1979; Rothschild et al. 1979). The sum total suggests that an arctiid lineage could be found that would be suitable for testing the sensory exploitation hypothesis.

4.2. Male Pheromones: "Use Them or Lose Them"?

As noted above, a conspicuous feature of the macroevolutionary record of MSSs is how frequently they have been lost. For many moth species there is compelling circumstantial evidence of the reduction or complete loss of male pheromones, particularly where vestigialized scent organs are present (Birch 1972; Grant 1978; Ferguson 1985; Krasnoff and Roelofs 1990).

How does loss of an MSS occur? The old theory stating that "an organ not in constant use tends to atrophy and decay" comes to mind; that is, if the function disappears the adaptation eventually disappears (Phelan 1992). However, few selectionists have taken sufficient interest in this question to recognize it as a possible wellspring of intriguing adaptive scenarios.

4.2.1. Losses Due to Host Shifts

If a species with a plant-derived male pheromone (see above) and very narrow host range, perhaps consisting of a single plant or a group of plants with similar secondary chemistry, gives rise to a daughter species with a host shift away from plants with the ancestral secondary chemistry, there will be strong and immediate selection against the maintenance of the ancestral male pheromone because females that do not accept males lacking the pheromone will not mate. Evidence that this has occurred more generally in the Lepidoptera than we currently recognize[1] may be lurking in the phylogenetic record. Two types of phylogenetic patterns would be consistent with an ancestral biosynthetic relationship between host-plant chemicals and a male pheromone: (1) hypotheses indicating that host-plant shifts are highly correlated with the loss of an MSS in many sister groups of a lineage and/or (2) concentrations of host shifts (host diversification) in branches in which the loss of an MSS occurs at the root. The latter case would lend itself to testing by the "concentrated-changes" correlation method proposed by Maddison (1990) (see also Maddison and Maddison 1992).

[1]Apart from the PA-exploiting arctiid moths and danaid butterflies, only the oriental fruit moth has been shown to be capable of biosynthesizing its male pheromone from host-plant precursors (Löfstedt et al. 1989b, 1990).

4.2.2. Losses Associated with Founder-Effect Speciation

Kaneshiro describes a process by which species-specific repertoires of courtship behavior tend to evolve toward reduced complexity following speciation by the founder effect (Kaneshiro 1983, 1988; Kaneshiro and Boake 1987; Kaneshiro and Giddings 1987). He argues that in the sparse density conditions that founder populations experience, the pressure for a female to mate at the first opportunity outweighs the benefits of choosiness which in the ancestral population may have been advantageous in the context of intersexual selection: Choosy females may fail to mate and so may not pass on their heritable tendencies toward choosiness. Thus a putative sequela of founder-effect speciation is an adaptive simplification of primitively complex courtship sequences. The direction of evolution is therefore predictable in lineages in which this type of speciation is the rule: Complex courtship will be the ancestral condition, whereas simple courtships, lacking identifiable behavioral elements found in closely related species, will be derived.

Kaneshiro finds support for this scenario in the Hawaiian drosophilid fauna in which closely related species from adjacent islands exhibit what he refers to as "asymmetric mating isolation" (Kaneshiro 1983). In experimental interspecific matings, females from one species reject males of the other; but in the reciprocal crosses, females accept the heterospecific male. Analysis of courtship behavior shows that females from the species with the more complex courtship reject males from the species with the less complex courtship. Conversely, females from a species with the less complex courtship will accept males from the species with the more complex courtship. The telling corollary is that in species pairs in which this relationship obtains, the species with the less complex courtship comes from the island of more recent geological origin and thus may be deemed to be derived relative to an "ancestral" species originating on an older island.

Although Kaneshiro's hypothesis has generated criticism on several fronts (Ehrman and Wasserman 1987), phylogenetic evidence is considered to support it, at least in part (DeSalle and Templeton 1987). The hypothesis seems germane to the evolution of MSSs in that it speaks directly to the issue of loss of courtship characteristics and, as noted above, the loss of MSSs is conspicuous in the phylogenetic record. I view the Kaneshiro model as compelling for a hypothetical case in which a foundress female moth deriving from a species with a narrow host range and a plant-derived male pheromone disperses into an isolated refugium in which the preferred host plant does not occur. To survive, any brood produced would have to accept an available substitute for the preferred food. Under these circumstances, any larvae that live to adulthood will do so by shifting to a new host. Thus, speciation by founder event might be accompanied by a host shift. If the new host plant does not provide the precursors for the male pheromone, the ultimate selection pressure would be exerted against excessively choosy females. As described above for the more general case, no males will have the ancestral pheromone. Females who do not accept these odorless males or males with the "wrong" odor will not mate. Thus loss of the male pheromone would be a likely result.

Although it may be difficult to find groups that lend themselves to testing this hypothesis, it is appealing in its plausibility and could be corroborated by a fully resolved phylogeny for a group in which losses of MSSs are concentrated on branches that can be shown to have resulted from dispersal into a new geographic area rather than from vicariance (Brooks and McLennan 1991). We might also look for other types of phylogenetic correlations—for example, a higher frequency of cladistic evidence supporting this scenario in lineages with good dispersal traits (strong fliers like the true armyworms, *Pseudaletia unipuncta, Leucania* spp.) compared with that found in lineages of weaker fliers (e.g., lymantriids, arctiine arctiids).

5. Conclusion

I have highlighted some of the advantages of adopting a more rigorous phylogenetic perspective on the evolution of lepidopteran male pheromones and associated scent structures. Certainly male pheromones can continue to serve as model systems in studies of important evolutionary processes, especially species recognition and speciation. Studies cited (Wagner and Rosovsky 1991; Vane-Wright et al. 1992) demonstrate that application of phylogenetic evidence will help us sort among the scenarios that have been proposed to explain the origin and frequent loss of MSSs and male pheromones in the Lepidoptera.

Phelan (1992) noted that the huge fund of information on pheromone-based mating systems is underutilized by evolutionary biologists in other fields. Conversely, as shown here, some of the history-based approaches used in studies of the evolution of mating systems in other animal groups have yet to be exploited by pheromone biologists. This chasm can be bridged. We will reap great rewards when pheromone biology, chemistry, and phylogeny are all combined on the same laboratory bench. Much hope for future progress in this area lies in the interest taken by systematists in groups for which there is a wealth of data on courtship and male pheromone systems and, conversely, in the interest taken by pheromone biologists in groups whose phylogeny has been rigorously studied.

Acknowledgments

Many of the ideas relating to problematic correlation methods were developed convergently with Dave Wagner, whose direction to recent literature and comments on the manuscript are greatly appreciated. Thanks go to Ring Cardé, John Dugdale, Stephen Foster, John Franclemont, Kathie Hodge, Christer Löfstedt, Jim Liebherr, Charlie Linn, Steph Menken, James S. Miller, Albert Minks, Wendell Roelofs, and Sandrine Ulenberg for exchange of ideas and/or editorial comments.

43

Evolution of Sex Pheromone Communication in Plant-Feeding Scarab Beetles

Walter Soares Leal

1. Introduction

Throughout the world, herbivorous scarab beetles (Coleoptera: Scarabaeidae) include various economically important pests in agriculture, horticulture, and forestry. Both larvae and adults damage plants by feeding on underground and aerial parts, respectively. Despite this group's economic importance as agricultural pests and the high demand for safer agrochemicals, our knowledge of their pheromone chemistry was very limited until recently. This was mainly due to the lack of uniform and consistent bioassays, which precludes the identification of these semiochemicals. This barrier has been overcome by coupling gas chromatography–electroantennographic detection (GC–EAD) (Leal et al. 1992a) or gas chromatography–behavior bioassay (GC–BB) (Leal et al. 1992b) with simplified laboratory bioassays or field tests. The use of these techniques has allowed our group to identify the sex pheromones of some scarab beetles of economic importance in Japan. Although only a handful of sex pheromones have been identified to date, our findings suggest possible scenarios for the evolution of pheromone communication in this group.

2. Classification and Phylogeny of Herbivorous Scarab Beetles

The superfamily Scarabaeoidea (= Lamellicornia) contains eight families: Lucanidae, Passalidae, Trogidae, Ceratocanthidae, Pleocomidae, Geotrupidae, Diphyllostomatidae, and Scarabaeidae (Parker 1982) (Fig. 43.1). The family Scarabaeidae encompasses about 2000 genera and 25,000 species, usually placed in about 20 subfamilies, four of which are the most notorious because their members are agricultural pests. These are the monophyletic plant-feeding (pleurostict) groups Rutelinae, Melolonthinae, Cetoniinae, and Dynastinae (Howden 1982) (Fig. 43.2),

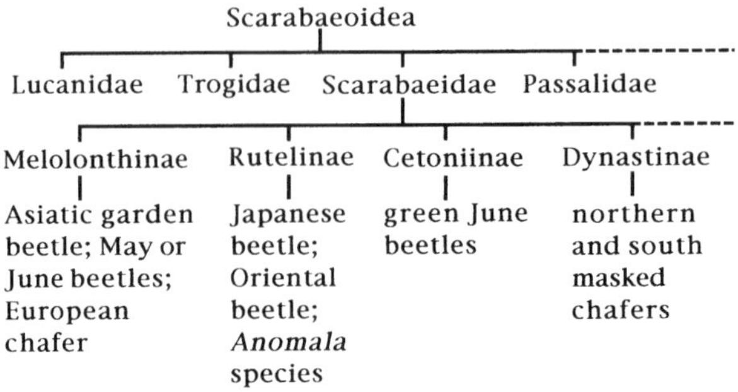

Figure 43.1 Classification of scarab beetles.

whose larvae are soil-dwelling white grubs that feed on living roots, some being destructive to crops.

Our discussion on the possible scenarios for the evolution of pheromone communication in herbivorous scarabs will be limited to the Rutelinae and Melolonthinae because these are the only groups from which sex pheromones have been identified to date.

3. Sex Pheromones of Melolonthines and Its Possible Evolution from Defensive Substances

The first sex pheromone of a scarab beetle was identified from *Costelytra zealandica,* a grass grub from New Zealand. In the flight season of 1968, it was observed that a commercially available glue, Pliobond, was very attractive to male beetles

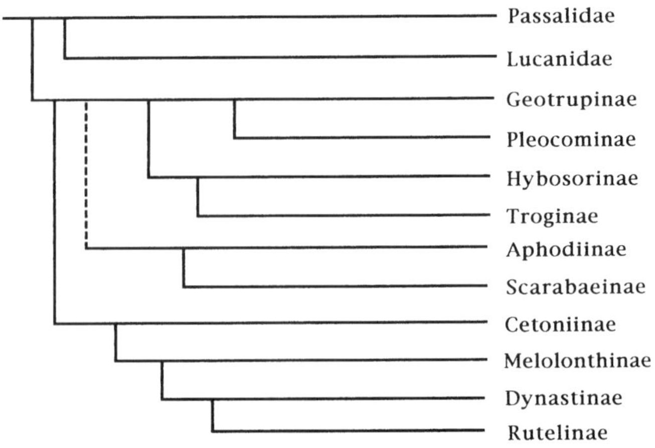

Figure 43.2 Phylogeny of some Scarabaeoidea, according to Howden (1982).

(Osborn and Hoyt 1969). Several ingredients of the glue and related materials have been tested separately and it was found that the phenol-formaldehyde resin, Durez 12687, a Pliabond ingredient, was very attractive to male beetles (Osborne and Hoyt 1970). Several known reaction intermediates, as well as phenol itself (about 2% phenol remains unpolymerized in the final resin), were field tested alone and after polymerization with formaldehyde. In these tests, phenol was the only active pure compound found. In the following year, the natural sex pheromone was isolated from an extract obtained by washing abdomens of 10-day-old virgin female beetles with ether for 2–3 min and was identified as phenol (1, Table 43.1) (Henzell and Lowe 1970). More recently, the major sex pheromone constituent of the large black chafer *Holotrichia parallela* has been isolated from the gland extracts and identified as L-isoleucine methyl ester (2, Table 43.1) (Leal et al. 1992c).

Further investigation of the gland constituents revealed that, although not attractive per se, (*R*)-(–)-linalool (3, Table 43.1) enhances the attractiveness of the major constituent (Leal et al. 1993c). Traces of L-valine methyl ester were found in the pheromone gland of some female beetles, but this compound was inactive. Some Melolonthinae species display a calling behavior by extruding an abdominal gland in a fashion similar to the calling of moths (Leal et al. 1993c). Such is the case with *Phyllophaga cuyabana* (Fig. 43.3), a soybean pest in Brazil, and *H. consanguinea,* native to India. Examination of the chemistry of these

Table 43.1. Sex pheromones of melolonthinae

Costelytra zealandica	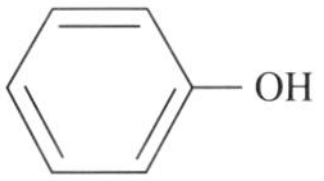 (1)
Holotrichia parallela	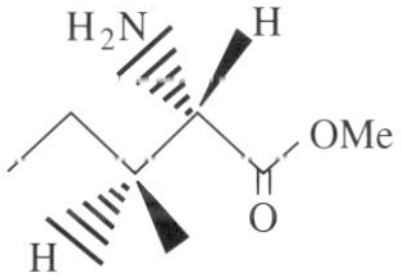 (2) 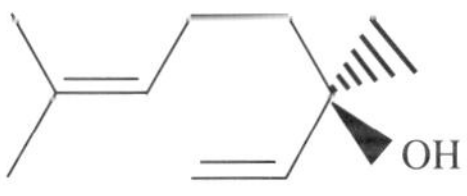 (3)
Holotrichia consanguinea	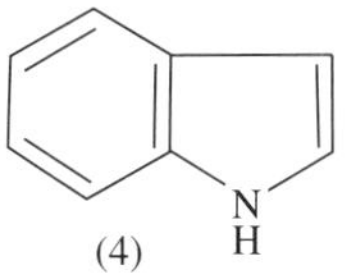 (4) 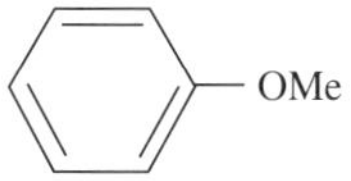 (5)
Phyllophaga cuyabana	(1) 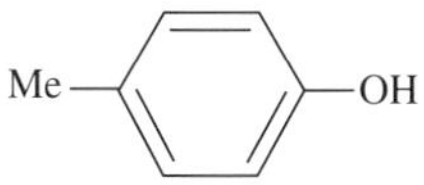 (6)

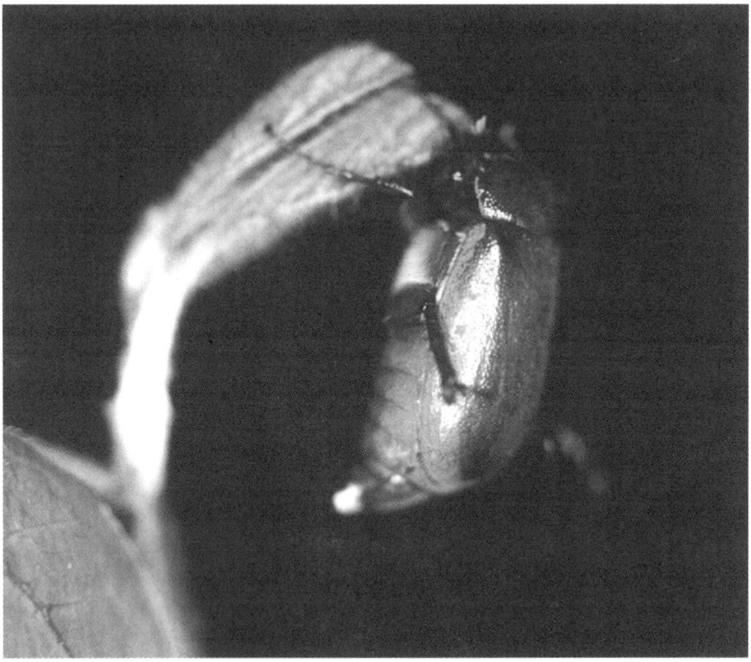

Figure 43.3 Calling position of *Phyllophaga cuyabana* with the abdominal gland exposed.

glands and field tests with natural extracts suggested that the latter species uses indole (4, Table 43.1) and anisole (5, Table 43.1) and that the former utilizes phenol (1, Table 43.1) and *p*-cresol (6, Table 43.1). However, biological activities of these authentic chemicals have yet to be tested in the field. Preliminary field tests with individual components revealed that *H. consanguinea* males were attracted only to anisole, the major compound in the pheromone gland (Leal and Yadava unpublished data).

Despite the fact that only a few species have been investigated to date, the results point to the use of phenolic, amino acid, and terpenoid compounds. Although at first sight these compounds seem to be biosynthetically unrelated, with few exceptions, they are likely to have amino acids as precursors. The deaminating L-tyrosine phenol-lyase is known to transform L-tyrosine (7, Fig. 43.4) into phenol (1, Table 43.1), and this compound has been reported to be produced by a symbiotic bacteria present in the colleterial gland of female *C. zealandica* (Hoyt et al. 1971). The huge amount of indole (4, Table 43.1) found in the glands of *H. consanguinea* could be produced by deamination of L-tryptophan (8) with tryptophanase (see Fig. 43.4). The fact that the *H. parallela* pheromone gland possesses high amounts of the two sex pheromone constituents in both photophase and scotophase (Leal et al. 1993c) led us to investigate the possible role of isoleucine methyl ester and linalool as defensive substances by assaying them against insect and plant pathogens. The antimicrobial activity of

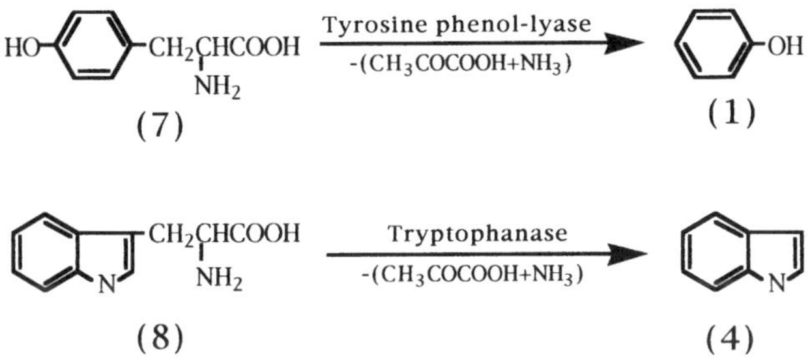

Figure 43.4 Deaminating enzymes probably involved in sex pheromone production of some scarab beetles.

the semiochemicals was tested by a modified paper disk method where the chemical was transferred to the center of Petri dishes inoculated with the microorganisms or by the action of their vapor in a closed vial (Leal and Shirata 1992a). The former test showed that both L-isoleucine methyl ester and (*R*)-(−)-linalool were very active against plant pathogens (both bacteria and fungi), whereas linalyl acetate (used for comparison) was inactive against bacteria and slightly active against some fungi (Table 43.2). Additionally, the vapor-phase test demonstrated that both (*R*)-(−)-linalool and L-isoleucine methyl ester were antimicrobial, although the former was more active (Leal and Shirata 1992b) (Table 43.3). Although the other melolonthine compounds have not been tested, they are likely

Table 43.2. Diameter (mm) of growth inhibition of plant pathogens

Microorganism (origin)	Tested compound		
	L-isoleucine methyl ester	Linalool	Linalyl acetate
Bacteria			
Erwinia cartovora (shoot soft rot)	11	10	0
Pseudomonas cichorii (bacterial rot)	10	3	0
Pseudomonas syringae (bacterial blight)	9	3	0
Pseudomonas tolaasii (bacterial blotch)	6	2	1
Fungi			
Diaporthe nomurai ("dogare" blight)	9	4	3
Fusarium solani (twig blight)	9	7	0
Cochliobolus miyabeanus (helminthosporium leaf spot)	6	3	1
Colletotrichum morifolium (anthracnose)	9	10	5

Table 43.3. Antimicrobial activity in the vapor-phase test against Beauveria *and* Metharizium *spp.*

Insect pathogen (origin)	Amount (μg)	Tested compound L-isoleucine methyl ester	Linalool	Linalyl acetate
B. brogniartii	100	–	+	–
(Anomala costata)	500	+	+	±
B. amorpha	100	–	+	–
(Anomala cuprea)	500	±	+	±
B. brogniartii	100	–	+	–
(Mimela testaceipes)	500	+	+	+
B. bassiana	100	–	+	–
(Anomala cuprea)	500	+	+	±
M. anisoplae	100	–	+	–
(Monochamus altanatus)	500	+	+	±
M. anisoplae	100	–	+	–
(Anomala cuprea)	500	+	+	+

to have antimicrobial activity, especially phenol and indole. Therefore, I suggest that these semiochemicals have evolved from a primary role of defensive substances into sex pheromones. Investigation on the effect of stereochemical configuration on the sex pheromonal activity of isoleucine methyl ester and linalool further support this hypothesis. Chirality plays a pivotal role in the activity of other scarab sex pheromones (see discussion below), and the fact that contamination with small amounts of the non-natural enantiomers has an inhibitory effect has been well documented in the literature (Tumlinson et al. 1977; Leal and Mochizuki 1993). In that regard, *H. parallela* is an interesting case study because both the major and minor sex pheromone constituents are chiral compounds (Table 43.1). It has been demonstrated, however, that in field tests the attractancy of the major constituent, L-isoleucine methyl ester, was not affected by the addition of the non-natural stereoisomers (Matsuyama et al. 1994). Although the natural pheromone is methyl (2*S*,3*S*)-2-amino-3-methylpentanoate (L-isoleucine methyl ester), field capture of male *H. parallela* with a lure composed of this stereoisomer in combination with (*R*)-(–)-linalool was as attractive as a mixture of the four isomers, L-, D-, L-allo-, and D-allo-isoleucine methyl ester, along with (*R*)-(–)-linalool. Similarly, there was no inhibitory effect due to the non-natural enantiomer, (*S*)-(+)-linalool. The binary mixture L-isoleucine methyl ester plus (*R*)-(–)-linalool was a better lure than the major component alone, but the addition of (*S*)-(+)-linalool (which is not found in the glands of *H. parallela*) did not decrease the activity of the lure.

4. Sex Pheromone of Rutelines, Possible Fatty Acid Derivatives

Sex pheromones have been identified for the following Rutelinae species (Table 43.4): *P. japonica* (Tumlinson et al. 1977), *A. rufocuprea* (Tamaki et al. 1985),

Table 43.4. Sex pheromones of Rutelinae

Popillia japonica	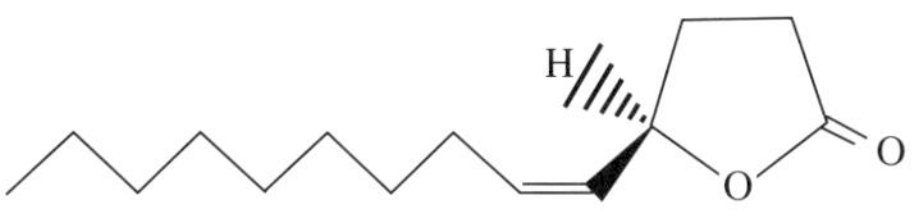 (9)
Anomala rufocuprea	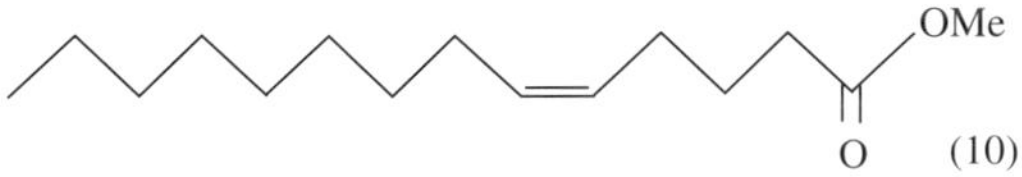 (10)
Anomala cuprea	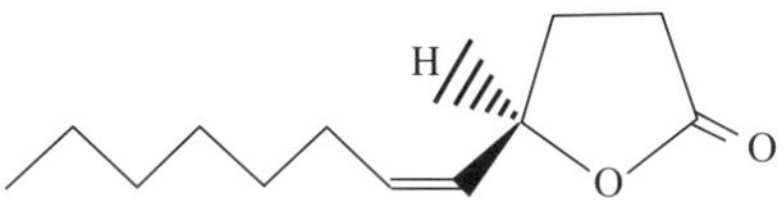 (11) + (9)
Anomala schonfeldti	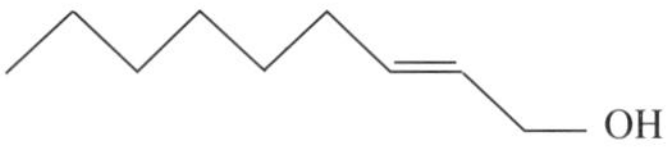 (12)
Anomala daimiana	(12) + (11)
Exomala orientalis	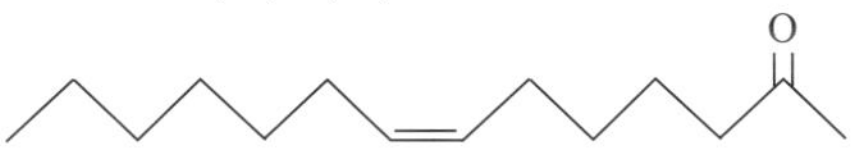 (13)
Anomala octiescostata	(11) + (9)
Anomala albopilosa sakishimana	(11) + (9)
Anomala japonica	(10) + (11) +  (14) + 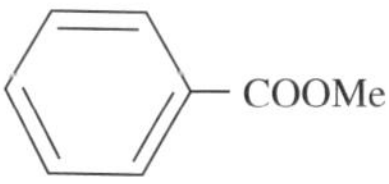 (15)

A. cuprea (Leal 1991; Leal et al. 1993a), *A. schonfeldti* (Leal et al. 1992b), *A. daimiana* (Leal et al. 1993b), *Exomala orientalis* (Leal 1993; Leal et al. 1994a), *A. octiescostata* (Leal et al. 1994b), *A. albopilosa sakishimana* (Leal et al. 1994c), and *A. a. albopilosa* (Leal et al. unpublished data). Most of these pheromones seem likely to be derived from fatty acids. Japonilure (9, Table 43.4) and methyl 5-(*Z*)-tetradecenoate (10, Table 43.4) are likely to be derived from oleic acid (16, Fig. 43.5). Two cycles of β-oxidation would give rise to (17, Fig. 43.5) that by esterification would yield (10). Allylic oxidation of (17) would produce (18, Fig. 43.5), which would undergo cyclization to yield japonilure (Fig. 43.5). Similarly, buibuilactone (10) would be derived from palmitoleic acid. 2-(*E*)-

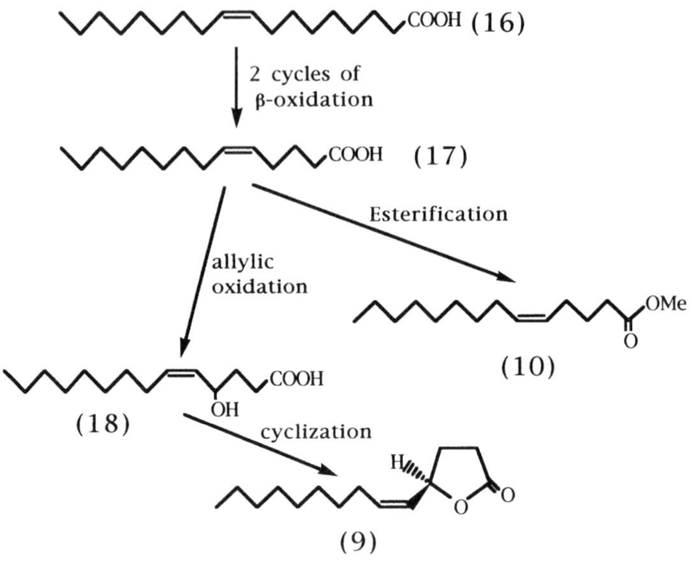

Figure 43.5 Synthesis of japonilure.

nonenol (12, Table 43.4) and 2-(*E*)-nonenal (14, Table 43.4) are likely to be produced from linolenic acid by the same biosynthetic pathway of plants.

5. Species Isolation and Pheromone Chemistry

Although three species of the genus *Anomala,* namely *A. cuprea, A. octiescostata,* and *A. albopilosa sakishimana,* have been shown to utilize the same pheromone blend (Table 43.4), there is no cross-attraction among these native Japanese species because they are isolated either geographically or seasonally. In addition, *A. albopilosa sakishimana* seems to have a more complex aggregation behavior in which plant volatiles seem to play a part (Leal et al. 1994c).

Both *A. octiescostata* and *A. cuprea* occur in the mainland of the Japanese archipelago, but the former is active in early spring, whereas the flight season of the latter is in summer. On the other hand, *A. albopilosa sakishimana* is native to Miyakojima Island in the South of Japan. Therefore, there was probably no selective pressure for the evolution of sex pheromone systems with different constituents.

It has been observed that the Japanese beetle is attracted to, and attempts to mate with, a large brown scarab, *Pelidnota* sp., native to the United States, and this pheromonal "mix-up" has been attributed to the fact that *P. japonica* is an exotic species in the United States and there was never selection for divergent pheromones (D. Whitman personal communication).

On the other hand, *A. cuprea, A. schonfeldti,* and *A. daimiana* co-occur in the field in Japan, but cross-attraction has not been observed thus far. This species isolation can be at least partially explained by the use of divergent pheromone mixtures. *A. cuprea* is not attracted to *A. daimiana* because this species does not possess japonilure (an essential minor component for *A. cuprea* attraction). That *A. cuprea* is not attractive to *A. daimiana* might be due to the fact that the former does not produce 2-(*E*)-nonenol, the major constituent of the sex pheromone system of the latter. There is no cross-attraction between *A. cuprea* and *A. schonfeldti* because these two species do not have common constituents. On the other hand, *A. daimiana* is not attracted to *A. schonfeldti* probably due to the lack of the minor component, buibuilactone. However, it is not clear why *A. schonfeldti* is not attracted to *A. daimiana* when the former needs only 2-(*E*)-nonenol. A possibility is that buibuilactone, the minor component of *A. daimiana,* is antagonistic to the action of 2-(*E*)-nonenol. Such an antagonistic effect has been observed for other scarab beetles. Captures of *A. rufocuprea* in traps baited with japonilure and methyl 5-(*Z*)-tetradecenoate have been demonstrated to decrease dramatically, although the use of the binary lure had no effect on catches of the Japanese beetle (M. Ono personal communication).

Recently, we found that *A. a. albopilosa,* a species similar to *A. cuprea,* utilizes a more complex sex pheromone blend (Leal et al. unpublished data) containing some of the presumably fatty acid derivatives (10), (11), (14), and methyl benzoate. Possibly, this species has evolved to use a more complex pheromone blend to maintain species-specificity.

6. Conclusions

The study of scarab beetle sex pheromones has suggested some hypotheses about the evolution of chemical communication in this group. Because our present knowledge on pheromone chemistry is still scanty, these hypotheses are speculative, but I hope they will form the basis for further research in this field, especially those aimed at gaining better understanding of the relationship between ecological parameters and the evolution of chemical communication.

Acknowledgments

I would like to thank Prof. Wendell L. Roelofs (Cornell University at Geneva, New York, USA) and Prof. Kenneth F. Haynes (University of Kentucky, Lexington, Kentucky, USA) for their critical revision of the manuscript.

44

A Genetic Study on Pheromonal Communication in Two *Ctenopseustis* Moths

Stephen P. Foster, Susan J. Muggleston, Christer Löfstedt, and Bill Hansson

1. Introduction

Most evolutionary processes, particularly those involved in speciation, are difficult to observe in nature because, in many cases, they probably occur over a time period that is not conducive to direct observation, and also because one usually knows little about the process of the change to be able to recognize that it is actually occurring. Thus, most evolutionary processes are usually studied by identifying closely related taxa, characterizing their differences and their genetic bases and attempting to construct some plausible explanation as to the process of change from one taxon to another.

Sex pheromones are an integral part of the mating systems of many species of insects. These chemicals usually elicit behaviors involved in orientation to, and recognition of, conspecific mates (Shorey 1973). Hence, it is clear that significant changes in sex pheromone production and release, or in the perception and processing of these chemicals, must profoundly influence the potential of individuals to mate. For insects using sex pheromones, the evolution of sex pheromone systems is likely to be a key process involved in speciation (White et al. 1995).

Löfstedt (1990) pointed out that any understanding of the evolution of pheromone systems must involve some understanding of the genetic basis of their control. Only in the last decade or so has the complex task of characterizing and understanding some of the physiological processes in sex pheromone systems been undertaken. Despite impressive progress (as reported elsewhere in this volume), we have little idea of how many processes, and therefore how many genes, are involved in sex pheromone-based mating systems of insects. There are, however, approaches that can lead to such an understanding without full characterization of all the processes involved in the system. One approach is to interbreed closely related taxa and observe the changes in the sex pheromone system of the resulting crosses. This approach, rather than investigating all the genes involved, focuses on the genetic factors and allelic effects involved in the control of pheromone differences between the taxa. However, this approach is somewhat limited in that it can be used only in those cases in which the two taxa are sufficiently closely related so that they are able to interbreed and produce viable offspring.

To date, this technique has been used successfully only for the study of pheromone systems of two closely related "strains" of the European corn borer moth, *Ostrinia nubilalis.* One of these strains uses predominantly (*Z*)-11-tetradecenyl acetate (Z11-14:AC) and a small amount of the opposite geometric isomer, (*E*)-11-tetradecenyl acetate (E11-14:AC), as its pheromone (Klun et al. 1973), and the other uses mainly E11-14:AC and a small amount of Z11-14:AC (Kochansky et al. 1975b). Through a series of crosses and backcrosses, differences in three characters of the pheromone systems of the taxa were found in (i) production of the two pheromone components by females, (ii) peripheral perception of the two components by males, and (iii) behavioral responses of males to ratios of the two pheromone components. These differences fitted a simple model in which pheromone production of the two strains differed at a single autosomal locus with two alleles, pheromone perception differed at a single autosomal locus with two alleles, and the behavioral responses differed at a single sex-linked locus (Roelofs et al. 1987 and references therein). A subsequent study found that the two autosomal loci were not linked (Löfstedt et al. 1989a). To date, studies on other species have given less clear-cut results, precluding any generalization on the genetic control in pheromone communication systems (see Löfstedt 1993 and references therein).

During study of the sex pheromones of several New Zealand tortricid moths, we found that certain morphologically indistinguishable populations of what were supposedly the same species use distinct sex pheromones and are therefore probably best considered sibling species (Foster et al. 1991b; Dugdale 1990). Two of these sibling species, *Ctenopseustis obliquana* and *C. herana,* have similar pheromone systems. The former species uses a mixture of (*Z*)-8-tetradecenyl acetate (Z8-14:AC) and (*Z*)-5-tetradecenyl acetate (Z5-14:AC) in a ratio of approximately 80:20 (Young et al. 1985), whereas the latter species uses only Z5-14:AC (Foster and Roelofs 1987). The close relationship of these two species, apparently differing predominantly in their sex pheromone systems, offered an opportunity for studying the genetic differences and evolution of their respective pheromone systems. When we started this study, *C. herana* was not described as a distinct species from *C. obliquana.* In an earlier report (Hansson et al. 1989), we described *C. herana* as *C. "ropeana,"* which was, at that time, thought to be the appropriate holotype for the population we were investigating, but which is now recognized as a synonym for *C. obliquana* (Dugdale 1990).

2. Methods and Materials

2.1. Insects

The *C. obliquana* were from a laboratory colony, maintained at HortResearch, and were originally collected from Auckland, New Zealand. The *C. herana* were collected as larvae from Lincoln (near Christchurch). For the F1s and the parents, the insects were allowed to mate in groups of three females and five males (of the appropriate species) in small clear plastic containers lined with a removable polyethylene strip. For the F2s and backcrosses, the insects were mated and reared as families, resulting from single females (originally placed with three males). A 10% honey solution soaked onto a cotton wick was provided for the insects to feed on. The insects were maintained at 18 ± 1°C under a 16:8 (L:D) photoperiod.

Eggs laid on the plastic sheets were removed from the containers and placed in humid conditions until the larvae hatched. The larvae were fed on a synthetic diet (Singh 1974) under the same light and temperature conditions as the adults. The two sexes were separated as pupae and maintained thereafter in separate incubators. The adults were used either in the various experiments or for the various crosses and colony maintenance as above. In this report, all crosses are given in the order females × males.

2.2. Sex Pheromone Analyses

The techniques and analyses used were similar to those used in the study on *O. nubilalis* (Roelofs et al. 1987). Briefly, three types of analyses were conducted. The electrophysiological responses of males were recorded from the SW1 olfactory sensilla on male antennae (for a detailed description of the method, see Hansson et al. 1989). For each of the parent species—F1s, F2s and backcrosses—recordings were made from 10 insects. Three sensilla from one antenna were tested for each insect.

The behavioral responses of 2 to 5-day-old males were tested in a 1.6-m-long wind tunnel (see Muggleston and Foster 1989) during the second and third hours of the scotophase. The tunnel was illuminated to a light level of 1.0 lux by 6- × 15-watt voltage-regulated tungsten lamps. Individual males for each of the types were flown to one of four different blends loaded on a red rubber septum (A.H. Thomas Co., Philadelphia, PA, USA): (1) 10 μg Z5-14:AC (the blend for *C. herana;* Foster and Roelofs 1987), (2) 7 μg Z5-14:AC and 3 μg Z8-14:AC, (3) 3 μg Z5-14:AC and 7 μg Z8-14:AC (the blend for *C. obliquana;* Young et al. 1985), and (4) 1 μg Z5-14:AC and 9 μg Z8-14:AC. The pheromone source was placed in the middle of a 10- × 10-cm horizontal, aluminum surface, placed on a 12-cm-high wire stand. The flight responses of males were recorded as "commenced flight upwind" or "landed on the horizontal surface." At least 50 males of each type were tested to each blend.

The pheromone gland content of 2- to 4-day-old female moths was analyzed by capillary gas chromatography. Two gas chromatographs were used for the analyses: a Pye Unicam 4500 gas chromatograph with a bonded phase Carbowax 20M column (50 m × 0.32 mm i.d.; Quadrex Corp., New Haven, CT, USA), programmed from 100°C to 200°C at 4°C/min after an initial delay of 1 min and a Varian 3500 gas chromatograph with a SP2340 column (30 m × 0.25 mm i.d.; Supelco Inc., Bellefonte, Philadelphia, PA, USA), programmed from 60°C to 140°C at 8°C/min (initial delay of 1 min), then to 180°C at 2°C/min. Tridecanyl acetate (13:AC) was added to the extracts as an internal standard.

3. Results

3.1. Electrophysiology

The electrophysiological results have been reported previously (Hansson et al. 1989), and they are summarized in Fig. 44.1. The parent *obliquana* olfactory sensilla had a large-spike amplitude cell that responded strongly to Z8-14:AC, along with a small-spike amplitude cell that responded weakly to Z5-14:AC. Conversely, the parent *herana* olfactory sensilla had a large-spike amplitude cell that responded strongly to Z5-14:AC, along with a small-spike amplitude cell that responded weakly to Z8-14:AC. The responses of the reciprocal F1 crosses resembled those of the parent *herana,* but with greater variability; some of the moths possessed atypical sensilla which did not respond like sensilla of either of the parents. Individuals with these sensilla were also observed in the F2s and most of the backcrosses. The *obliquana* × *herana* F2s showed responses similar to the F1s; that is, males showed a response similar to the *herana* parents, although atypically responding cells were also observed. None of these males showed an *obliquana*-type response. In the other F2 cross (*herana* × *obliquana* × *herana* × *obliquana*), most males had either *obliquana*-type or atypical sensilla, although one male had a mixture of *herana*-type and atypical sensilla.

Males of the *obliquana* × *herana* paternal backcross (i.e., with *C. herana* males) all had sensilla of the *herana*-type, whereas in the *herana* × *obliquana* paternal backcross (i.e., with *C. obliquana* males) there were no males with *herana*-type sensilla. Males of this backcross were *obliquana* phenotypes or phenotypes with a mixture of *obliquana* and atypical sensilla (with one male having all atypical sensilla). The maternal backcross of the *obliquana* × *herana* males (i.e., with *C. obliquana* females) yielded males with *obliquana*-type, *herana*-type, atypical, or a mixture of atypical and *obliquana*-type sensilla. Most (eight) of the males of the *herana* × *obliquana* maternal backcross (i.e., with *C. herana* females) were of a *C. herana* phenotype; the other two males had a mixture of *herana*-type and atypical sensilla.

Through a principal component analysis (see Hansson et al. 1989), the response pattern of the various crosses could be ascribed to a single-sex linked locus on

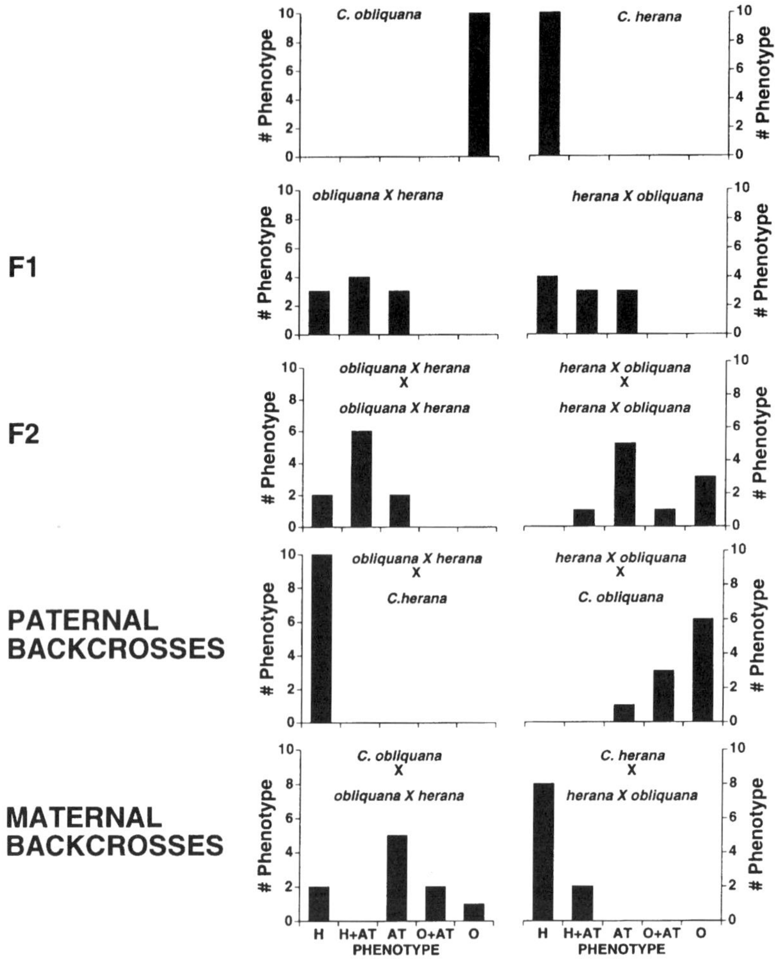

Figure 44.1. The electrophysiological responses of sensilla on male antennae of *Ctenopseustis obliquana, C. herana,* and various crosses. Responses are shown as number of males of a particular phenotype; H, had sensilla that responded similarly to those of *C. herana;* O, had sensilla that responded similarly to those of *C. obliquana;* AT, had sensilla that responded atypically (i.e., unlike those of the parents); H + AT, had sensilla that responded similarly to *C. herana* and sensilla that responded atypically; O + AT, had sensilla that responded similarly to *C. obliquana* and sensilla that responded atypically. (Data from Hansson et al. 1989.)

the Z-chromosome, with *C. herana*-type dominance. However, the considerable variability in the responses of the hybrids could not be explained by such a simple model of inheritance and suggested that other genes may be involved in the electrophysiological differences between the two species.

3.2. Male Behavior

C. obliquana males flew upwind and landed to all three blends containing Z8-14:AC, with the greatest numbers responding to the two blends with the highest amounts of Z8-14:AC; no *C. obliquana* males landed in response to the Z5-14:AC alone (Fig. 44.2). In contrast, the greatest number of *C. herana* males flew to the blend of Z5-14:AC alone; no male *C. herana* completed flight to the two blends with the highest amounts of Z8-14:AC, and only a small number of males completed flight to the other blend containing this chemical [but in a significantly, ($p < 0.05$) lower proportion than to the blend of Z5-14:AC alone]. Males of both F1 crosses showed their greatest response to the Z5-14:AC alone, although males of the *obliquana* × *herana* cross also responded well to the 30:70 blend. Thus the F1 males responded similarly, although with a broader response profile (i.e., they also responded to blends with higher proportions of Z8-14:AC), to the *C. herana* males (i.e., there is apparent *C. herana* dominance in the inheritance of this character).

Males of the two F2 crosses had response patterns different from each other. Given that the hybrids showed a predominantly *C. herana*-type response, the pattern of inheritance for both F2 crosses appeared to fit a sex-linked, rather than an autosomal, model (for which the response patterns would be expected to be the same, consisting of a 1:2:1 mixture of *C. obliquana*:hybrids:*C. herana*). This fit was more apparent for the *herana* × *obliquana* F2 males, which responded predominantly to the blends that *C. obliquana* males preferred (i.e., to blends higher in Z8-14:AC). Only roughly 25% of these males would be expected to show *obliquana*-type responses if the inheritance was autosomal (compared to 50% for sex-linked inheritance).

The pattern of responses of males from both paternal backcrosses was that expected from sex-linked inheritance (i.e., males of the *herana* × *obliquana* paternal backcross had a response pattern similar to that of *C. obliquana* males, and males of the *obliquana* × *herana* paternal backcross had a response pattern similar to that of *C. herana* males). Males from both maternal backcrosses had response patterns that fit a 1:1 mixture of hybrids and the respective parents. Such patterns are expected regardless of whether the inheritance is autosomal or sex-linked.

In summary, the pattern of male behavioral responses in the crosses appeared to fit a model similar to that found for male antennal responses; that is, the differences in responses largely appear to fit a model of inheritance through a single sex-linked locus with *C. herana*-type dominance. As for the electrophysiological responses, however, there was considerable variability in the patterns of responses, which suggests that other genes may also play a role.

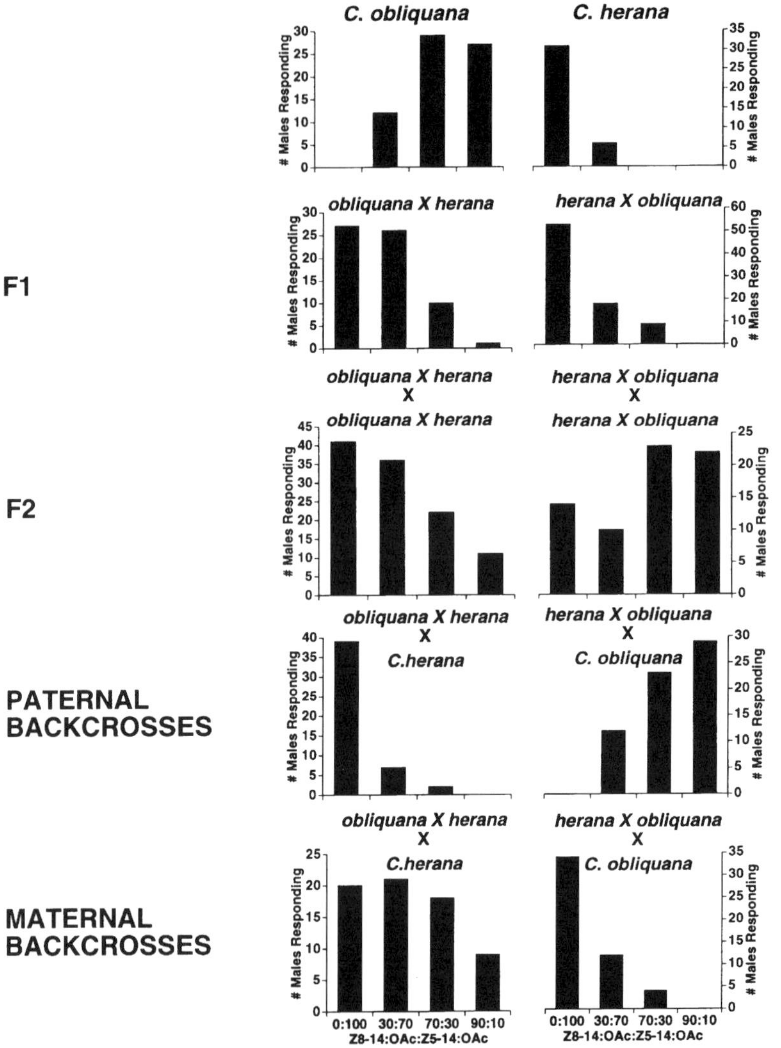

Figure 44.2. The behavioral responses (completed flights) of male *Ctenopseustis obliquana, C. herana,* and various crosses to four different blends of (Z)-8-tetradecenyl acetate (Z8-14:AC) and (Z)-5-tetradecenyl acetate (Z5-14:AC).

3.3. Female Production

There is a clear distinction between the sex pheromone gland content of females of *C. obliquana* and *C. herana: C. obliquana* females produce a mixture of Z8-14:AC and Z5-14:AC in a ratio of approximately 80:20, while *C. herana* females produce only Z5-14:AC (Fig. 44.3). Females of both F1 crosses had pheromone gland contents more similar to that of *C. obliquana* females (i.e., they produced

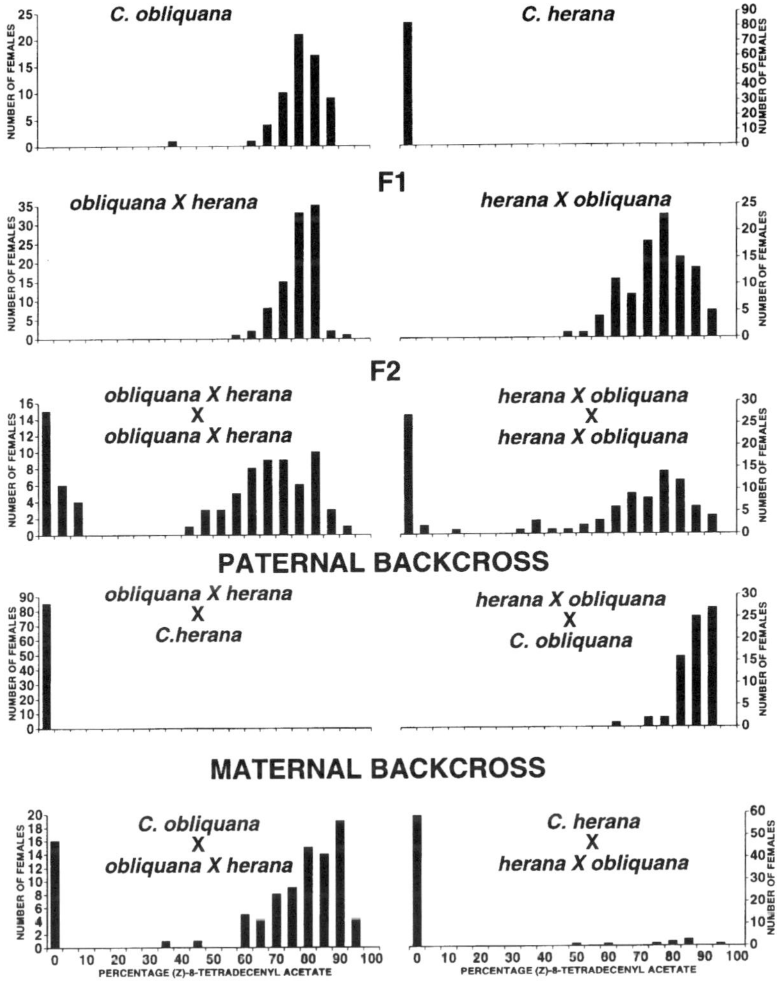

Figure 44.3. The production of pheromone [percentage of (Z)-8-tetradecenyl acetate in the blend] by females of *Ctenopseustis obliquana, C. herana,* and various crosses.

a mixture of Z8-14:AC and Z5-14:AC) than to *C. herana* females. However, in the *herana* × *obliquana* cross there was a much broader range of ratios of the two components than in the parent *obliquana* females. Both F2 crosses were comprised of females that produced *obliquana*-type and females that produced *herana*-type pheromones, in a proportion of approximately 3:1. The range of ratios of the two components in both F2 crosses was also broader than that observed in the *obliquana* parents. Given that the hybrids produced an *obliquana*-type pheromone, the ratios of phenotypes appeared to fit a pattern of a single autosomal gene with two alleles (i.e., a ratio of 1:2:1, *obliquana*:hybrid:*herana*).

Females from the paternal backcrosses, *obliquana* × *herana* (with *C. herana* males) and *herana* × *obliquana* (with *C. obliquana* males), were exclusively of the phenotypes *C. herana* and *C. obliquana,* respectively. Although the results from the latter backcross would not differentiate between autosomal or sex-linked inheritance (because the hybrids are of *C. obliquana* phenotype), the results from the former backcross suggest that the inheritance was sex-linked. The maternal backcross of the *herana* × *obliquana* F1 to the (*C. herana* females) would be expected to yield an equal mixture of *C. obliquana* and *C. herana* phenotypes for a model of inheritance of a single gene, regardless of whether the gene was on an autosome or a sex chromosome. However, in this backcross there was a predominance of *C. herana* over *C. obliquana* phenotypes. In contrast, the other maternal backcross of *obliquana* × *herana* (with *C. obliquana* females) would be expected to yield all *C. obliquana* phenotypes if the difference in inheritance was on an autosome, or a mixture of 1:1 *C. obliquana* and *C. herana* phenotypes if the difference in inheritance was sex-linked. Again the pattern did not fit perfectly either model, with a predominance, but not complete, of *C. obliquana* over *C. herana* phenotypes. All the *obliquana* phenotypes in the backcrosses showed a range of ratios distinct (either a broader range, or a narrower range, higher in Z8-14:AC content) from that of the *obliquana* parents.

The various crosses showed a pattern for the difference in pheromone production between females of *C. obliquana* and *C. herana* that did not fit easily a simple single-gene inheritance model. In particular, certain crosses (such as the F2s) indicated that a single autosomal gene model may account for the differences in pheromone production between *C. obliquana* and *C. herana,* whereas other crosses (such as the *obliquana* × *herana* paternal backcross) suggested that the factor may be sex-linked. Still others (such as the maternal backcrosses) showed a pattern that did not appear to fit either model. The broadening of the range of ratios of the two components in most of the crosses suggests that other genetic factors are involved in the differences between the two species.

4. Discussion

The two species *C. obliquana* and *C. herana* are closely related (Dugdale 1990), and their pheromone systems are similar (Clearwater et al. 1991). Females of both species produce a common component, and males of both species have antennal receptor neurons for both Z8-14:AC and Z5-14:AC.

The differences between these two species in pheromone perception and behavioral responses of males can be ascribed largely to a simple model of single gene inheritance, although variability in the data suggests that other genes are involved. The patterns of inheritance in the crosses suggest that these single genetic factors both reside on the sex chromosome and must therefore be linked. This finding contrasts with that found in the two *O. nubilalis* strains, where the differences in electrophysiological and behavioral responses appeared to be on an autosome and a sex chromosome, respectively, and cannot be linked (Roelofs et al. 1987). It is worthwhile emphasizing that in both cases (i.e., for *Ostrinia* and *Ctenopseustis* species), the difference in the behavioral response character fits or largely fits a single sex-linked gene model (see Löfstedt 1993 for a discussion of this).

The data for the production of sex pheromone by females of the two species are more difficult to interpret. These data are not clear-cut in favor of either an autosomal or sex-linked model for inheritance of this character and suggest that the difference between the species is due to the involvement of more than one gene, especially because the F1 hybrids and other crosses show a broader range of ratios of Z8-14:AC and Z5-14:AC than do the parent *C. obliquana.* The biosynthesis of the chemical Z8-14:AC (i.e., the apparent difference between the two species) involves a number of key enzymes, including those that play a role in Δ10-desaturation, two-carbon chain shortening, reduction, and acetylation. These processes, with the notable exception of Δ10 desaturation, also take place during the biosynthesis of Z5-14:AC (see Fig. 44.4). Although Δ10 desaturation involves a unique (to *C. obliquana*) Δ10-desaturase (Löfstedt and Roelofs 1985; Foster and Roelofs 1988) (as well as possibly other unique enzymes), it is possible that one or more of the enzymes involved in the other processes are unique to the production of Z8-14:AC. For example, the respective reductase enzyme systems involved in pheromone production in the two strains of *O. nubilalis* appear to be different,

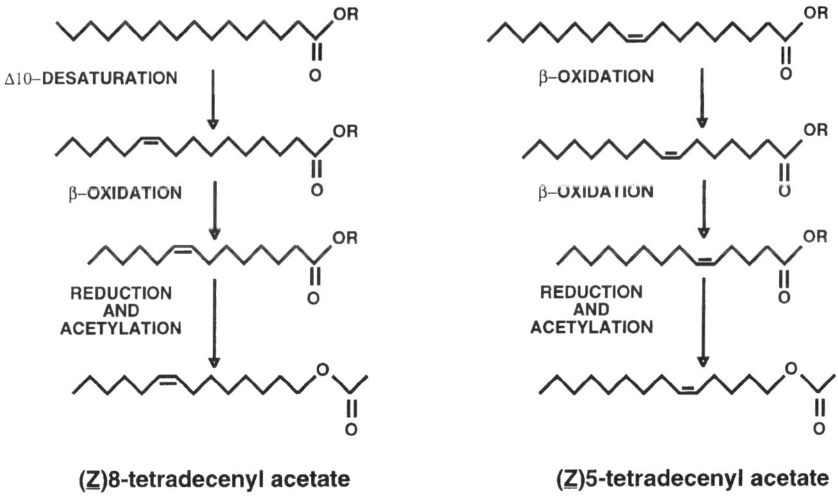

Figure 44.4. Proposed biosynthetic routes to the two chemicals, (*Z*)-8-tetradecenyl acetate and (*Z*)-5-tetradecenyl acetate, of the sex pheromones of the sibling species *Ctenopseustis obliquana,* and *C. herana.*

with the reductase system of the E-strain more selective for the (*E*)-11-tetradecenoate [compared to the (*Z*)-11-] precursor, and the reductase system of the Z-strain being more selective for the (*Z*)-11-tetradecenoate precursor (Zhu 1995). Thus, in the case of the two *Ctenopseustis* species, provided that the genes encoding for the Δ10-desaturase and the other gene(s) (such as those involved in fatty acyl reduction) were not linked, multiple gene differences between the two species would be observed for pheromone production.

These genetic data on the two *Ctenopseustis* species emphasize the complexity of pheromone systems and how limited our understanding is of what actually constitutes these systems and how these systems may have evolved. Attempts to understand the evolution of sex pheromone systems have for the most part been restricted to using data concerning the endpoints of the systems (i.e., the actual pheromone chemicals). Because relatively little is known about how pheromone systems can change, such attempts are usually constrained by fitting the pheromonal pattern to an existing pattern usually derived through morphological methods (see, for example, Roelofs and Brown 1982; Foster and Dugdale 1988). Patterns in changes of morphological characters may not necessarily reflect the changes in pheromone systems. Furthermore, because the diversity of pheromonal chemicals in the Lepidoptera is not great, their use for deriving evolutionary relationships between groups is somewhat limited. For example, Δ11-unsaturated tetradecenyl compounds are found in species in a number of different families of Lepidoptera (see Löfstedt 1991; Arn et al. 1992), yet it is absurd to suggest, based on common sex pheromone components, that species in different families are closely related.

To gain a better understanding of the evolution of insect sex pheromone systems, it will be necessary to understand more fully the actual processes that lead to the endpoints (pheromone chemicals, perception of the chemicals, and the neurophysiological processing of pheromonal inputs), such as those we have analyzed in this study. A greater understanding of pheromone systems will allow the molecular bases for these systems to be unraveled and will allow differences at the genetic level between closely related and relatively unrelated species to be determined. With this knowledge, along with the development of more accurate and complete phylogenies, it should be possible to determine the mechanisms involved in the evolution of insect sex pheromone systems.

Acknowledgments

The help and suggestions of Drs. W.L. Roelofs and B.O. Bengtsson during this study are gratefully acknowledged. This work was supported by the New Zealand Foundation for Research, Science and Technology (contract CO6328), the Swedish Research Councils NFR and SJFR, and the Bank of Sweden Tercentenary Foundation.

45

Genetics of Pheromone Communication in the Cabbage Looper Moth, *Trichoplusia ni*

Kenneth F. Haynes

1. Introduction

The process of speciation and the related diversification of communication systems are not well understood (e.g., Dobzhansky 1951; Mayr 1970; Paterson 1978; Kaneshiro 1980; Butlin 1989; Phelan 1992; Löfstedt 1993). Studies of chemical communication systems have a great deal of potential to clarify how these evolutionary processes occur. An understanding of the evolution of pheromone communication has been, and will continue to be, aided by several complementary approaches. Comparison of pheromone communication systems of closely related species, as well as broader studies relating phylogeny to the composition of pheromone blends, have given insights into how diversification of pheromone blends may have occurred (e.g., Roelofs and Comeau 1969; Tumlinson et al. 1974; Cardé et al. 1977; Klun et al. 1979; Roelofs and Brown 1982; Löfstedt et al. 1991). Similarly, studies of geographical variation in the pheromone blends within a species have provided insights into the process of speciation (e.g., Klun and Cooperators 1975; Arn et al. 1983; Löfstedt et al. 1986a; Haynes and Baker 1988; Hansson et al. 1990; Thomson et al. 1991; Tóth et al. 1992). Hybridization between species, or between subpopulations demonstrating pheromonal polymorphisms, have elucidated the nature of genetic changes that have occurred (Hansson et al. 1987; Roelofs et al. 1987; Löfstedt et al. 1989a). In addition, heritability determinations and selection experiments have demonstrated the potential for evolutionary change in a few pheromone communication systems (Collins and Cardé 1985; Roelofs et al. 1986; Collins and Cardé 1989a,b; Collins et al. 1990). Theoretical population genetics studies have suggested conditions under which a new pheromone blend could become established (see Butlin and Trickett this volume). Studies of the cabbage looper moth, *Trichoplusia ni*, have begun to contribute to an understanding of the evolution of chemical communication by following several of these approaches. In particular, an investigation of the impact of a single gene mutation that affects the pheromone blend of this species has allowed a unique opportunity to study the evolution of pheromone communication.

The pheromone communication system of the cabbage looper moth has been studied rather extensively for over 30 years. H. H. Shorey, L. K. Gaston, and their colleagues focused on this species as a model for understanding basic characteristics of pheromone biology and mating disruption (e.g., Shorey et al. 1964, 1972). The most abundant pheromone component, (*Z*)-7-dodecenyl acetate (Z7-12:Ac), was identified by Berger (1966). Subsequently, Bjostad et al. (1980) identified dodecyl acetate (12:Ac) as a second component. Studies of the biosynthetic pathways for pheromone components led to the discovery of four additional components, (*Z*)-5-dodecenyl acetate (Z5-12:Ac), 11-dodecenyl acetate (11-12:Ac), (*Z*)-7-tetradecenyl acetate (Z7-14:Ac), and (*Z*)-9-tetradecenyl acetate (Bjostad and Roelofs 1983; Bjostad et al. 1984). In a wind-tunnel test, the full blend of six components was more effective in stimulating upwind orientation and source contact than either the major component alone or any combination of the major component with any two other components (Linn et al. 1984). Deletion of any component other than Z7-12:Ac from the full blend did not reduce attraction. Linn et al.'s (1984) study indicated that with the exception of the major component, the pheromone blend of the cabbage looper moth is highly redundant. There is another unexpected dimension to redundancy in this species pheromone communication system. Landolt and Heath (1989) discovered that a male-produced pheromone released from the hairpencils was effective in attracting females. This pheromone was comprised of *d*-linalool, *m*-cresol and *p*-cresol, and its release was stimulated by exposure of males to Z7-12:Ac (the major pheromone component produced by the female) or Z7-12:Ac plus cabbage odor (Heath et al. 1992). Despite 30 years of extensive study, the cabbage looper moth still is yielding fascinating results in areas ranging from redundancy in its communication system (Linn et al. 1984) to details of its biosynthetic pathways (Bjostad and Roelofs 1983; Bjostad et al. 1984). Results from our genetic studies of the cabbage looper moth should contribute to a better understanding of the evolution of chemical communication systems.

2. Interpopulational Variation in Pheromone Blend

The pheromone blend of *T. ni* showed little variation in three populations from the United States (Riverside, California; Gainesville, Florida; and Lexington, Kentucky) (Haynes and Hunt 1990a). A laboratory colony that originated from Riverside California over 100 generations prior to the study differed significantly from the three field populations in the blend ratios of several minor components (Fig. 45.1). Because the pheromone blend ratios of the laboratory population differed to the greatest extent from the California population from which it was

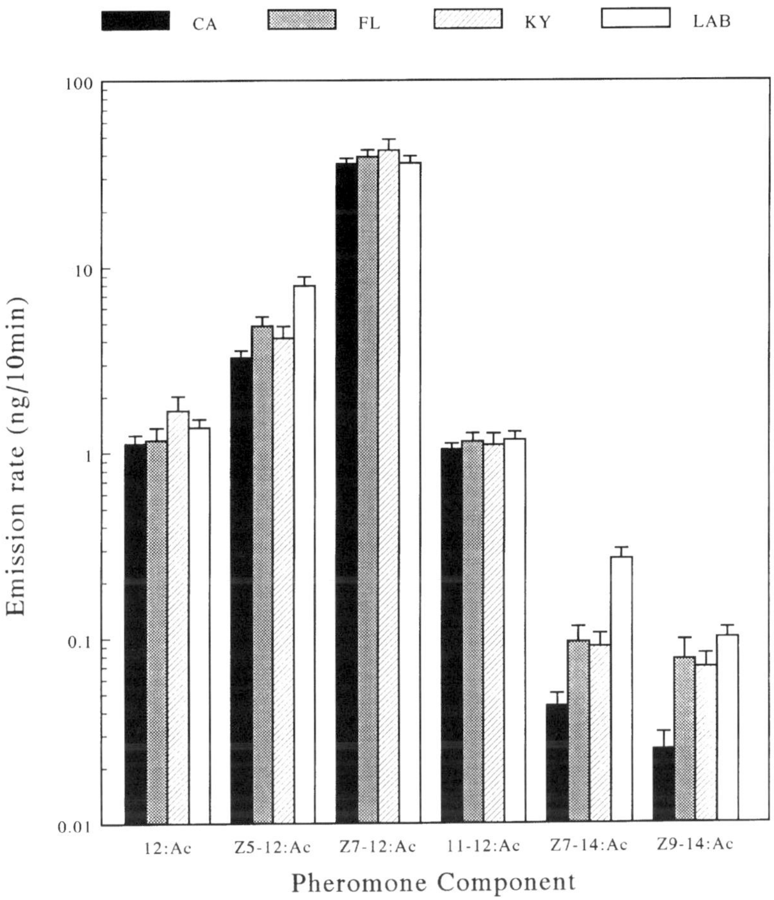

Figure 45.1. Comparison of the quantities of emitted pheromone components from four populations of *Trichoplusia ni*. The laboratory population was found to differ significantly from the field populations in the relative proportions of several minor pheromone components. Note the log scale of the *y*-axis, which was necessary to illustrate the abundance of all of the pheromone components together in the same graph. For CA (Riverside, California), FL (Gainesville, Florida), KY (Lexington, Kentucky), and LAB (laboratory colony), 58, 35, 35, and 50 individuals were analyzed, respectively [Modified from Haynes and Hunt (1990a).]

founded, we conducted selected crosses between these two populations to gain an understanding of genetic aspects of these differences (Hunt et al. 1990). Results from F1 and F2 crosses and backcrosses suggested that the differences involved one or more autosomal genes and one or more sex-linked genes. Despite these relatively subtle differences in the pheromone blend, a blend of pheromone components similar to that found in females from the field was more effective than a laboratory blend in trapping male *T. ni*. These results suggest that stabilizing selection imposed by the communication system itself might serve to limit diversification. In laboratory colonies, this stabilizing selection may be relaxed because of (a) the proximity of males to females and (b) the large number of insects in a small space. Under these conditions, discrimination among the pheromone blends of individual females may not be possible or necessary. Thus genetic drift could cause the small but significant deviation in the pheromone blend that was observed. To test this hypothesis adequately would require many independent laboratory colonies. If the deviation from the field-blend is a random event, then one might expect that females from these laboratory colonies would not all have the same pheromone blend. In other words, the variation amongst independent populations in the laboratory would be greater than variation amongst independent populations from the field.

3. Mutation Affects Pheromone Blend

During the course of studying pheromone blends from these populations, an abnormal pheromone phenotype within the laboratory colony was discovered. In contrast to the relatively subtle differences that have an underlying polygenic basis, the impact of this single gene mutation was very dramatic (Haynes and Hunt 1990b). Females that express this gene have a 20-fold increase in Z9-14:Ac compared to normal females (Fig. 45.2). In addition, they have 30-fold less Z5-12:Ac. Initially, a female that possessed this abnormal phenotype was identified and mated to an arbitrarily selected male from the mass laboratory colony. Her female offspring showed a 1:1 ratio of normal pheromone blend and abnormal blends. Subsequently a colony was established in which all females expressed the abnormal phenotype. When appropriate crosses were conducted (F1, F2, and backcrosses), it became clear that the abnormal phenotype was caused by a recessive autosomal allele at one locus (Fig. 45.3). Thus we had been fortunate to mate the original female to a male that was heterozygous at this locus. All F1 females had a normal pheromone blend. A 3:1 ratio of normal to mutant individuals was found in the F2, and a 1:1 ratio of normal to mutant females was found in the backcross to the pure mutant line.

Jurenka et al. (1994b) found that the differences in the pheromone blend in these mutant females could not be accounted for by changes in specificity of Δ11 desaturation, reduction, or acetylation. Rather, their radiolabel studies pointed to a specific effect on the chain-shortening step (Fig. 45.4). An effect on the number of cycles of chain-shortening can account for both the reduced level of Z5-12:Ac and the increased quantity of Z9-14:Ac. Z5-12:Ac is three cycles of chain-shortening "downstream" of the Δ11-desaturated fatty acyl precursor (Z11-18:CoA), whereas Z9-14:Ac is only one cycle downstream of its corresponding precursor (Z11-16:CoA).

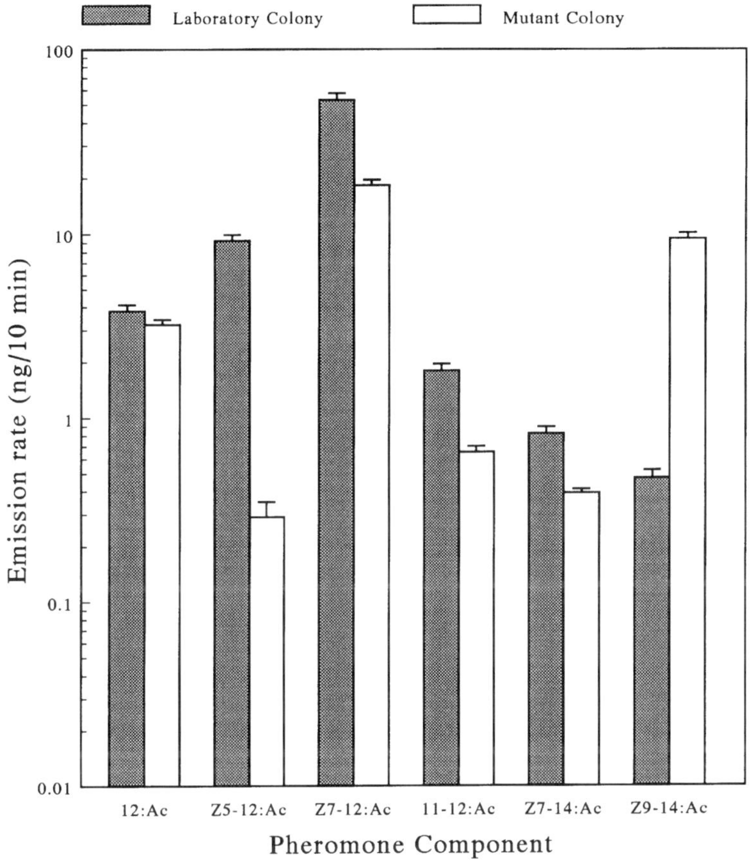

Figure 45.2. Comparison of the quantities of emitted pheromone components from laboratory and mutant colonies of *Trichoplusia ni*. Females from mutant colonies have 20-fold more Z9-14:Ac and 30-fold less Z5-12:Ac than normal females. Note the log scale on the *y*-axis. For the laboratory colony and mutant colony 68 and 125 females were analyzed, respectively. [Modified from Haynes and Hunt (1990b).]

Females that expressed the abnormal phenotype were not as effective in attracting male *T. ni* as were normal females (Haynes and Hunt 1990b). In fact the mutant females were more effective in attracting male *Agrotis ipsilon* than conspecific males. The selective disadvantage of this pheromone characteristic is obvious. One would not expect this characteristic to persist in a large population. In fact it is difficult to explain how it has persisted in our laboratory colony, unless males do not discriminate between normal and mutant females when the density of females is high.

Initially, males that carried the mutant gene responded preferentially to the

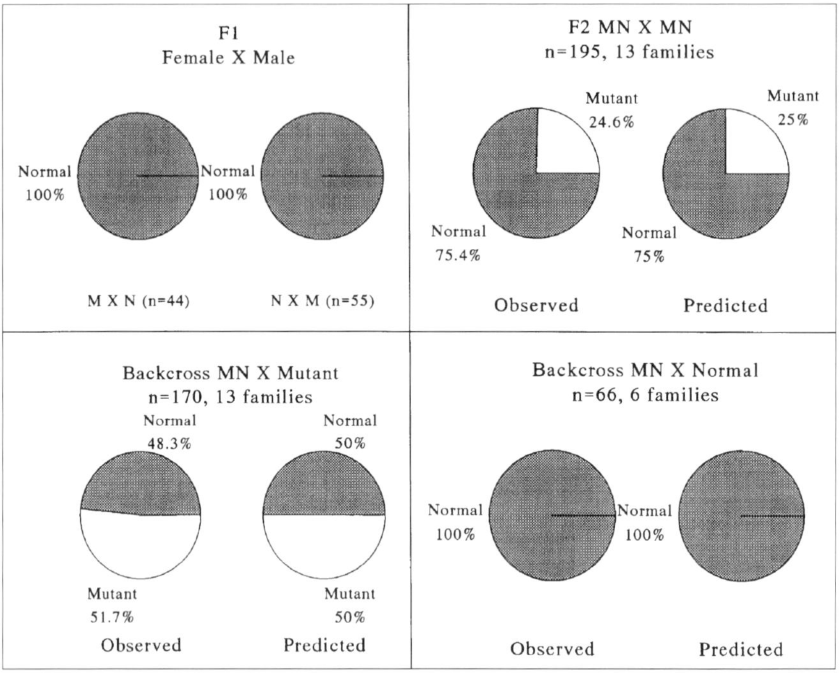

Figure 45.3. Crosses were evaluated to determine the genetic basis of the mutant phenotype of *Trichoplusia ni*. The proportion of normal individuals (shaded areas) and mutant (unshaded areas) are indicated in these pie charts. For F2 and backcrosses, predictions of the hypothesis that the mutation is a single recessive autosomal gene are shown on the right. All F1 crosses between normal and mutant moths resulted in female offspring with a normal pheromone blend. F2 crosses resulted in offspring that are one-fourth mutant and three-fourths normal. Offspring of the backcrosses to the mutant line were half normal and half mutant. Backcrosses to the normal line were all normal. These results were consistent with the mutant pheromone blend being controlled by a single recessive autosomal gene. [Modified from Haynes and Hunt (1990b).]

normal pheromone blend compared to the mutant pheromone blend in a wind-tunnel assay (Liu and Haynes 1994). This indicated that there was no genetic correlation (neither linkage nor pleiotropy) involved in the signal and response. These mutant and normal males had similar proportions of receptors sensitive to each of the pheromone components (Todd et al. 1992). However, over 49 generations in colonies in which all females expressed the gene for the abnormal pheromone blend, the response of males changed (Fig. 45.5) (Liu and Haynes 1994). At the end of this period, males from two mutant colonies responded well (~80% source contact) to the mutant pheromone blend. However, their response to the normal pheromone blend remained at high levels. Thus there was no

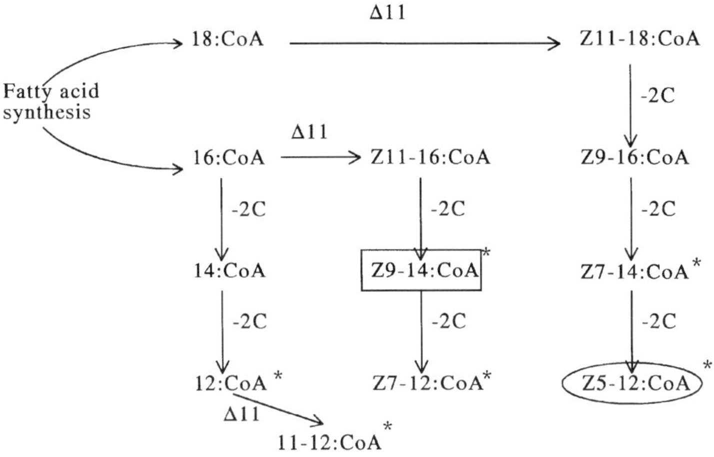

Figure 45.4. The biosynthetic pathway that produces the six pheromone components of the cabbage looper. Fatty acyl CoA precursors marked with an asterisk (*) are reduced and acetylated to form the six components. The box indicates a pheromone precursor for Z9-14:Ac, which is significantly more abundant in mutant females than in normal females. In contrast, the oval indicates a pheromone component that is significantly less abundant in mutant females than in normal females. Chain-shortening differs significantly between mutant and normal females. Δ11 = Δ11-desaturation, –2C = chain-shortening. [Modified from Jurenka et al. (1994).]

evidence for a decline (or tradeoff) in the responsiveness to the normal blend as a result of the improved response to the mutant blend. A possible explanation for this change is that in the mating colonies in which all females were now of the abnormal type, there was selection pressure on males to respond to this pheromone blend. Those males that could respond would have a mating advantage over those males that would discriminate against females with the abnormal phenotype. One could speculate that this type of evolutionary change could play a role in the process of speciation under special ecological circumstances in which the advantage of being discriminating in mating is outweighed by the advantage of having a greater pool of potential mates. These circumstances might arise when there is little cost to making a mating mistake, such as an environment in which there are few or no other species that share the same pheromone components. In addition, colonization of a new environment by a mated female whose female offspring deviate from the norm may lead to selection favoring less discrimination. These two conditions were inadvertently simulated when we established pure colonies consisting of females of the mutant phenotype. It would be useful to take a step back and examine the relative abundance of sensory receptors for the individual pheromone components, now that there is evidence

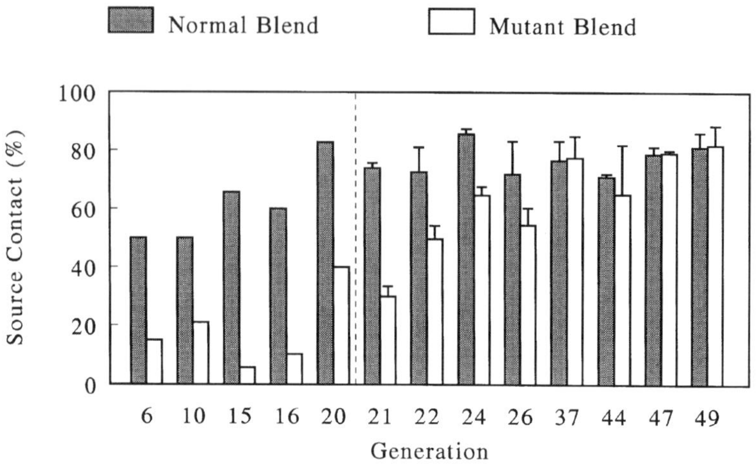

Figure 45.5. Evolutionary changes in the responsiveness of mutant males to the mutant pheromone blend. Initially the response of males from mutant colonies to the mutant pheromone blend was poor. However, after 49 generations their response to the mutant blend was equivalent to their response to the normal blend. The dashed line indicates that at the 20th generation the colony was split into two isolated colonies. [Modified from Liu and Haynes (1994).]

for an evolutionary change in the behavioral response. In *Agrotis segetum* there is evidence that geographical variation in the pheromone blend is paralleled by changes in the relative abundance of receptors (Hansson et al. 1990; Löfstedt et al. 1986a).

Phelan (1992) suggested the "asymmetric tracking" hypothesis to account for changes in a communication system. According to this hypothesis, because females are a limiting resource to males, the male's sensory system and behavioral response specificity should adapt to changes in the females' pheromone blend. Changes in male response specificity in response to changes in the female's pheromone blend should be more likely than the converse. This prediction agrees with our observation that the male's behavioral response specificity in the mutant line has changed in response to the female's pheromone blend. In our laboratory colonies of mutants, there is apparently a cost in terms of reproductive success in maintaining the status quo, which is a preferential response to the normal pheromone blend. However, it appears that the male's behavioral response specificity has broadened, suggesting that at least in the laboratory colony there is no cost to maintaining the good responsiveness to the normal pheromone blend.

Kaneshiro (1980) proposed a speciation hypothesis to explain the tremendous diversity of species of Hawaiian *Drosophila*. Kaneshiro's hypothesis has some parallels to the changes that were observed in our colonies of mutant *T. ni*.

According to this hypothesis, selection should favor less discriminating females after a new colonization of an area because of the limited population size. Thus in these populations, females might be expected to be less selective than in well-established populations. This result could explain the pattern of limited mate selectivity in "derived" species relative to "ancestral" species. This is a similar concept to the broadening of the behavioral response specificity that was observed in colonies of mutant *T.ni*, with the obvious difference that the change in discrimination has occurred in males rather than females. Löfstedt et al. (1986a) suggested that a similar stage of asymmetrical reproductive isolation might have occurred as an intermediate step in speciation in the small ermine moth, *Yponomeuta rorellus*, in which there has been a major shift from unsaturated (Δ11-unsaturated acetates) to saturated major pheromone components (tetradecyl acetate).

4. Conclusions

The implications of these studies of the cabbage looper moth on questions about speciation or the potential for development of resistance to pheromones are important. These studies have documented that there is polygenic variation in the pheromone blend when field and laboratory populations are compared. In addition, the impact of a single gene change on the pheromone blend has been documented. Isolation of pure colonies of these mutants apparently results in selection that leads to a broadening of the behavioral response specificity of males, clearly indicating that there was heritable variation in the males' specificity. Thus, we have a model system that has all of the components necessary for studying the process of speciation. However, to date, none of the studies have been conducted in a way to simulate realistic natural conditions. One should expect that in a large field population a rare mutant individual would have minimal mating success. However, a recessive gene may be maintained in a population for some time. More realistically one needs to consider the special ecological circumstances that might improve the chances of establishment of an isolated population. A founder effect, where a mated female with an abnormal pheromone blend colonizes an isolated area, could affect the subsequent evolution of the male's behavioral response specificity. In order to complete the process of speciation in a system like mutant *T. ni*, evolution of behavioral response specificity to the new pheromone blend is essential. In the future, it would be useful to test the hypothesis that this specificity could result from selection against hybridization in areas of secondary overlap. In addition, the impact of overlap with other species utilizing common pheromone components on the response specificity should be investigated. The mutant pheromone phenotype provides an important opportunity to study various evolutionary scenarios involved in the process of speciation.

Acknowledgments

Dr. Randy Hunt and Dr. Yongbiao Liu helped to conduct much of the research reported in this chapter. I appreciate the comments of R. E. Hunt and A. J. Moore on a previous draft of this chapter. In part, this chapter summarizes work supported by the Cooperative States Research Service, U. S. Department of Agriculture under agreements 87-CRCR-1-24100, 89-37153-4354, 91-37302-5875, and 94-37302-0613. This paper (no. 94-7-125) presents results of experiments conducted in connection with projects of the Kentucky Agricultural Experiment Station.

46

Effects of Interspecific Hybridization Between *Heliothis virescens* and *Heliothis subflexa* on the Sex Pheromone Communication System

Peter E. A. Teal and James H. Tumlinson

1. Introduction

Among vagile, closely related species which share common ranges, reproductive isolation is often effected by mechanisms which act prior to copulation (Mayr 1970). Numerous studies have demonstrated the significance of sex pheromones for the maintenance of premating reproductive isolation among insect species. Generally, lepidopterous species rely on differences in the chemical components or the component ratios of the sex pheromones to promote reproductive isolation (Roelofs and Cardé 1974), although among certain groups, behavior associated with sex pheromones appears to be of primary importance (Teal et al. 1978).

The species-specific nature of the pheromone communication systems of insects and, in particular, of Lepidoptera has been exploited to develop species-specific methods of population monitoring and control. However, few studies have been devoted to defining the genetic mechanisms that regulate the specificity of the pheromone communication system. This stems, in great part, from the fact that the majority of the economically important species have developed highly effective pre- and postmating mechanisms of reproductive isolation between closely related species. Therefore, the potential to study the effects of interspecific hybridization on the pheromone communication systems rarely exists.

One pair of species that can be induced to hybridize under no-choice laboratory conditions is *Heliothis virescens* and *Heliothis subflexa* (Laster 1972; Proshold and LaChance 1974). These two closely related species share similar ranges and habitats and have intersecting reproductive periods. However, they have never been found to hybridize under natural conditions. Females of the two species produce different blends of pheromone components (Heath et al. 1991; Table 46.1), and, as indicated in Fig. 46.1, males do not exhibit upwind-oriented flight when exposed to extracts of the female pheromone glands of the other species

Table 46.1. Sex pheromone components and ratios identified from Heliothis virescens *and* H. subflexa[a]

	Compound	*H. virescens*	*H. subflexa*
A	Tetradecanal	3.1	nd[b]
B	(Z)-9-Tetradecenal	4.1	nd
C	Hexadecenal	10.8	2.7
D	(Z)-7- and (Z)-9-Hexadecenal[c]	1.8	37.0
F	(Z)-11-Hexadecenal	100.0	100.0
G	(Z)-9-Hexadecan-1-ol	nd	1.9
H	(Z)-11-Hexadecen-1-ol	6.0	17.3
I	(Z)-7-Hexadecen-1-ol acetate	nd	3.7
J	(Z)-9-Hexadecen-1-ol acetate	nd	6.2
K	(Z)-11-Hexadecen-1-ol acetate	nd	28.4

[a]Data are those obtained by Heath et al. (1991) at the peak of pheromone production for each species.

[b]nd, not detected.

[c](Z)-7-hexadecenal and (Z)-9-hexadecenal were not resolved when chromatographed on the capillary columns used for this study; therefore, the values listed are for the two compounds combined.

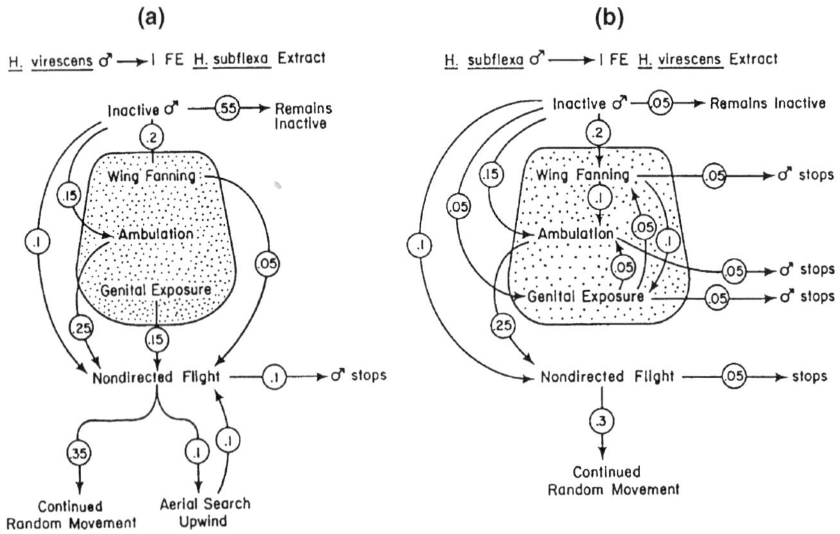

Figure 46.1. Responses of *Heliothis virescens* (**a**) and *H. subflexa* males (**b**) to one female equivalent (FE) of the extract of the sex pheromone glands of females of *H. subflexa* and *H. virescens* females, respectively. Probabilities of each behavioral transition were tabulated from data resulting from release of 20 individuals 1.5 m downwind from the pheromone source. When extracts of conspecific females were used, 72% of *H. virescens* and 68% of *H. subflexa* males landed on the source and attempted to copulate. Values in circles indicate the probability of a transition occurring.

(Teal 1981). In fact, the specificity of response to the pheromone blends is so great that in order to induce successful interspecific mating in the laboratory, we had to artificially stimulate males to mate by using lures impregnated with the conspecific pheromone blend.

This technique allowed us to conduct experiments on the mechanisms that regulate the sex pheromone communication systems of the two species as well as their hybrid and backcross progeny. These studies have included research on the mechanisms that control the structure of male hairpencil glands as well as chemical, biochemical, and physiological studies on production of both the male and female pheromones.

2. *Male Hairpencil Glands and Pheromone*

Hairpencil glands and other types of reversible epidermal glands are common in males of many species of noctuid moths. These structures are responsible for synthesis and release of pheromones that have been shown to have a variety of functions including short- and long-distance attraction of females and acceptance of males by females (Birch and Hefetz 1987). Males of *H. virescens* have two pair of hairpencil glands on the genital segments of the abdomen. The most prominent pair is situated at the base of the harpes and is composed of elongated hairs, while the second less prominent pair of tufts is situated laterally on the intersegmental membrane between abdominal segments 7 and 8 (Teal and Oostendorp 1993) (Fig. 46.2a). Both types consist of modified porous scales that are subtended by trichogenous gland cells. Males of *H. subflexa* have the less prominent tufts on the intersegmental membrane between abdominal segments 7 and 8 but lack the elongate scent scales at the base of the harpes (Fig. 46.2b). Males of both species are known to partially expose the genital segments and hairpencil glands when approaching females that are releasing pheromone and to fully expose the glands just prior to copulation (Teal et al. 1981; Cibrian-Tovar and Mitchell 1991). In *H. virescens*, effective hairpencil exposure during close-range courtship is of significant importance and less than 20% of the males that expose their hairpencils in an inappropriate manner successfully mate on the first or subsequent copulation attempts (Teal et al. 1981b).

Microscopic analysis of the genital segments of 150 F_1 hybrid males obtained from crosses between *H. subflexa* ♀ and *H. virescens* ♂ (S line) and 150 F_1 hybrid males from the reciprocal cross (*H. virescens* ♀ × *H. subflexa* ♂, V line) revealed that all males from both hybrid groups had hairpencil glands of similar structure to those present on the genital segments of males of *H. virescens* (Teal and Oostendorp 1993). This indicated that the *H. virescens* genome regulated the structure and presence of hairpencil glands in males. However, it did not indicate if the trait was sex-linked, because males of Lepidoptera are homogametic, having sex chromosomes denoted ZZ. Thus, dominant traits carried on the Z chromosome are expressed in males like dominant traits borne on the autosomes.

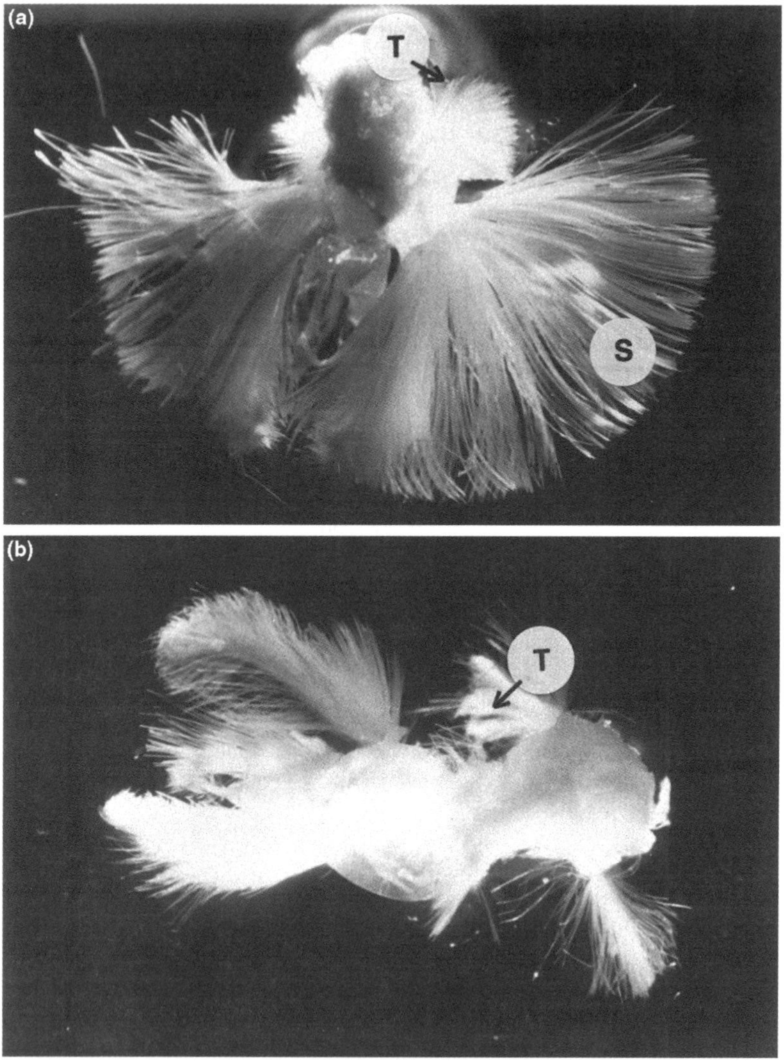

Figure 46.2. Exposed genital segments of males of *H. virescens* (**a**) and *H. subflexa* (**b**) showing the long scent scales (S) associated with the harpes of *H. virescens* and tufts of scent scales (T) on the intersegmental membrane between abdominal segments 7 and 8 of both species.

All males obtained from backcrossing females of either hybrid line to males of *H. virescens* had hairpencil glands identical to those of *H. virescens*. These results were predicted because males from these backcrosses obtained at least one Z chromosome and 75% of the autosomes from the dominant species, *H. virescens*. Males obtained from backcrosses between S-line F_1 females and males of *H. subflexa* (SD line, $n = 75$) also had hairpencil glands that were identical in structure to those of *H. virescens*. Alternatively, backcross males obtained from crosses between V-line F_1 females and *H. subflexa* (VA line, $n = 75$) had genital segments identical to those of *H. subflexa* males. These results did not support the action of autosomal genes in regulating the structure and presence of the *H. virescens*-type hairpencil glands because 50% of the populations of both SD- and VA-line backcross males had *H. virescens* autosomes. The results did, however, support the action of dominant genes linked to the Z chromosome of *H. virescens* because all SD line males obtained one Z chromosome from *H. virescens*, whereas both Z chromosomes of VA line males were from *H. subflexa*. The action of dominant genes linked to the Z chromosome of *H. virescens* was confirmed by crossing females from all backcross-1 generations to males of the parent species (Teal and Oostendorp 1993). In every case, males of the second backcross generations that obtained a single Z sex chromosome from *H. virescens* had hairpencil glands that were identical to those of *H. virescens*. However, males that obtained both Z chromosomes from *H. subflexa* had genital segments identical to those found in males of this species.

Close-range analyses of the reactions of females to courtship displays of conspecific males have shown that females accept courting males in response to full pulsed exposure of the hairpencil apparatus (Teal et al. 1981b). This suggests that the signal emanating from the hairpencils acts at close range. To determine if chemicals released from the hairpencils had an effect on female behavior, tests were conducted using hexane extracts of the hairpencils. During the peak period of reproductive activity, females were presented first with a solvent blank and then 10 min later, with either a sample of the extract of two male hairpencil glands or a solvent blank. Samples were delivered by rapidly depressing the plunger of a 1-ml gastight syringe attached to a Pasteur pipette containing the test material. The tip of the pipette was positioned 2–3 cm from the females. After delivery of the sample, the behaviors were monitored for a 2-min period. Significantly more females elevated the wings in an acceptance posture (Table 46.2; see also Teal et al. 1981b), exposed the terminal abdominal segments which house the pheromone gland, and bent the tip of the abdomen in a fashion associated with copulation.

Chemical analysis of the extracts and volatiles released by the exposed hairpencil glands of males of *H. virescens* resulted in the identification of a blend of components (Teal and Tumlinson 1989) composed principally (≥95%) of a 4:1 ratio of hexadecan-1-ol acetate (16:AC) and hexadecan-1-ol (16:OH). The mean amounts of these compounds were 286 ng and 70 ng, respectively. Bioassays

Table 46.2. Effect of exposure of virgin females of Heliothis virescens *to two gland equivalents of a hexane extract of the hairpencil glands of males*

Behavior	Hairpencil extract (n = 42)	Air Blank (n = 49)
Escape[a]	9	20[b]
Elevation of wings	21[b]	2
Exposure of pheromone gland	28[b]	1
Bend abdomen tip[c]	16[b]	0

[a]Escape attempts were indicated by wing fanning and ambulation away from the source of pheromone.

[b]Significant difference between the extract and air blank in a χ^2 test at p = 0.05.

[c]Abdomen tips were bent in the same manner as occurs during mating behavior.

showed that more than twice as many females elevated their wings in an acceptance posture and exposed the terminal abdominal segments when presented with two male equivalents of the synthetic blend of 16:AC and 16:OH than when presented with solvent blank controls (n = 80).

Capillary gas chromatographic analysis of the extracts obtained from the genital segments of males of *H. subflexa*, including the small tuft-like scent scales in the intersegmental membrane, revealed the presence of only minute amounts (mean = 1.4 ng, n = 50) of 16:OH. Only six extracts contained a compound having a retention index equivalent to that of 16:AC, but the presence of this compound could not be confirmed by mass spectroscopy.

Knowing that the presence and structure of the male hairpencil glands in hybrid and backcross males is regulated by dominant genes linked to the male Z sex chromosome of *H. virescens*, we were interested in determining if pheromone production was also regulated by sex-linked genes. Analysis of extracts obtained from 125 males of each of the F_1 hybrid lines revealed that only small amounts of 16:OH were present and that very few of the extracts contained the acetate analog. In fact, neither the amounts nor the ratios were different from those obtained from *H. subflexa* (Teal and Oostendorp 1994a). These results excluded the possibility that dominant genes on the Z chromosome of males of *H. virescens* regulated the production of hairpencil pheromone and suggested that the *H. subflexa* genome exerted a dominant influence.

Although extracts from backcross males obtained from crossing females of either F_1 hybrid line to *H. subflexa* males contained the same amounts and ratios of 16:OH to 16:AC as *H. subflexa* males, extracts from males obtained from backcrosses between either F_1 hybrid line and *H. virescens* males were variable. The majority of males from crosses between V-line F_1 females and *H. virescens* males (VB line, n = 75) and from S-line F_1 females crossed with *H. virescens* (SC line, n = 62) had ratios of 16:OH to 16:AC and also amounts of pheromone that were intermediate between those of the parent species. However, significant numbers of extracts from males of both the VB and SC lines contained amounts and ratios of pheromone that were identical to those present in males of either *H. subflexa* or *H. virescens* (Table 46.3). If the production of hairpencil pheromone was regulated by dominant sex-linked genes on the *H. subflexa* Z chromosome, then all of the VB-line backcross males should have been phenotypically the same as *H. subflexa* because they obtained one of their Z chromosomes from this species. Similarly, if the genes were sex-linked, then all SC-line males should have been phenotypically like *H. virescens* because they obtained both of the Z chromosomes from this species. Thus, the results excluded the action of sex-linked genes from regulation of production of either the ratio or amount of pheromone produced by hairpencil glands and implicated polygenic autosomal control of each trait with the *H. subflexa* genome having a dominant influence.

3. Regulation of Pheromone Production by Females

Although the pheromone produced and released by the hairpencil glands of males is of importance for successful mating, the most critical component of the chemical

Table 46.3. Effect of hybridization and backcrossing on ratio and total amount of pheromone in extracts of hairpencil glands on males[a]

Insect line	Male sex[b] chromosomes	Phenotypes if sex-linked, *H. subflexa* dominant	Observed percentage for total amount	Observed percentage for ratio of 16:AC and 16:OH
Parent, *H. virescens*	ZvZv	100% V	100% V	100% V
Parent, *H. subflexa*	ZsZs	100% S	100% S	100% S
S-line hybrid	ZvZs	100% S	100% S	100% S
V-line hybrid	ZvZs	100% S	100% S	100% S
BC1 SC line	ZvZv	100% V	27.4% V	19.4% V[c]
			62.9% I	54.8% I
			9.7% S	25.8% S
BC1 SD line	ZvZs	100% S	100% S	100% S
BC1 VA line	ZsZs	100% S	100% S	100% S
BC1 VB line	ZvZs	100% S	8.0% V[c]	17.3% V[c]
			82.7% I	48.0% I
			9.3% S	34.7% S

[a]Phenotype designations are: V, *H. virescens* type; S, *H. subflexa* type; I, intermediate between parental types; S-line hybrid, *H. subflexa* female × *H. virescens* male; V-line hybrid, *H. virescens* female × *H. subflexa* male; BC, backcross; SC, S-line female × *H. virescens* male; SD, S-line female × *H. subflexa* male; VA, V line female × *H. subflexa*; VB, V-line × *H. virescens*.

[b]Zv denotes Z chromosome from *H. virescens*, Zs denotes Z chromosome from *H. subflexa*.

[c]Here there exists a difference between the observed frequency and the frequency expected if the trait was linked to the Z sex chromosome.

communication system is the female-produced sex pheromone. Unlike some other species of closely related noctuid moths that have temporally discrete periods of pheromone release (calling) and reproduction, the calling and mating periods of *H. virescens* and *H. subflexa* overlap broadly (Cibrian-Tovar and Mitchell 1991; Heath et al. 1991). Consequently, females of these species utilize distinctly different blends or pheromone components to attract mates. Although both the oxygenated functional groups and ratios of components in the pheromone blends of the two species are different (Table 46.1), both species biosynthesize the pheromone components using common pathways (Teal et al., 1989b). Thus, the enzyme systems required to produce all of the pheromone components are possessed by females of both species. Therefore, we expected that females of interspecific hybrids would, at a minimum, produce the pheromone components that were common to the blends of both parent species.

In our initial experiments using female hybrids, we caged and observed females to determine the calling periodicity as described by Teal (1985). Females were observed during the first to fifth scotophases after adult emergence. As indicated in Fig. 46.3, females of the two hybrid lines were distinctly different. No females of the 20 S-line hybrids (Fig. 46.3a) were observed calling during the first scotophase after emergence. S-line females were observed calling during the subsequent four scotophases (Fig. 46.3a). Peak production of pheromone for this hybrid (sixth hour) intersected the periods of pheromone production of both parent species (see Teal et al. 1993). None of the 25 V-line individuals observed during the first to second scotophases were observed calling at any interval. During the third to fifth hour of the third scotophase, two V-line females were observed calling. One of the insects continued to call up to the seventh hour. During the fourth scotophase, one female was observed to be calling between hours 2 and 4 and another two were observed calling between hours five and eight. We observed two females calling between hours three and five during the fifth scotophase, but no females were observed calling at other intervals. The low incidence of calling exhibited by V-line hybrid females and lack of a distinct calling periodicity could not be ascribed to an abnormality stemming from hybridization between a single pair of individuals because equal numbers of females were selected from five separate groups of interspecific crosses.

To determine if V-line F_1 hybrids were producing pheromone, even though the majority did not appear to be assuming a classical calling posture, and to determine if calling by S-line females was accompanied by pheromone production, we obtained extracts from the pheromone glands of females of both hybrid lines at hourly intervals during the scotophase (Teal and Oostendorp 1994b). The extracts were then analyzed by capillary GC to determine the amount of pheromone present, as indicated by the quantity of (*Z*)-11-hexadecenal (Z11-16:AL), the principal component of the pheromone of both species. No distinct peak period of pheromone production was evident from analysis of extracts obtained from V-line females, and only 10–15% of the 20 extracts obtained at

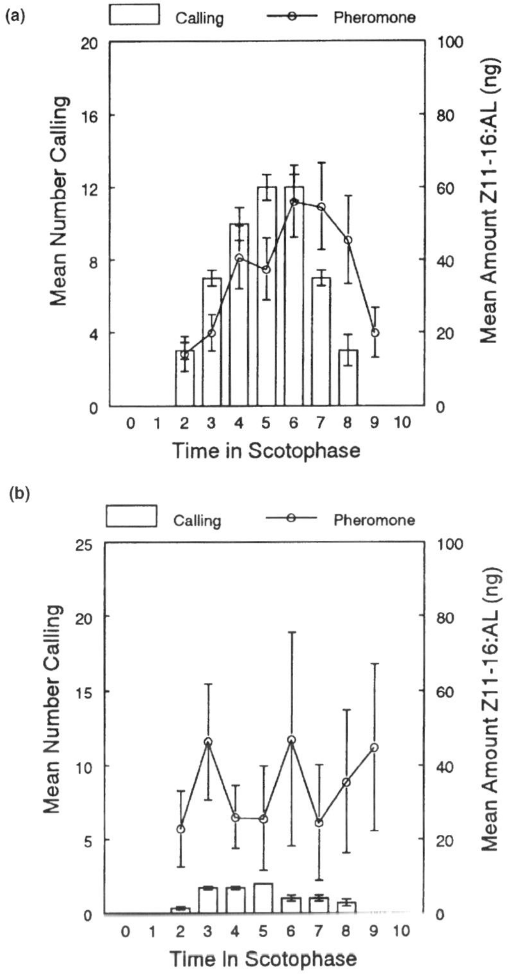

Figure 46.3. Comparison of numbers of hybrid S-line (3a, n = 20) and V-line (3b, n = 25) females exhibiting calling behavior with the amount of Z11-16:AL present in gland extracts (n = 20/hr, each species) obtained from females at hourly intervals during the scotophase.

each hourly interval contained more than 50 ng of Z11-16:AL. The majority of extracts contained fewer than 10 ng. However, as evidenced by the large standard errors (Fig. 46.3b), some extracts obtained at each interval contained very large amounts of pheromone (<150 ng). Thus, the majority of V-line hybrids neither exhibited calling behavior nor produced pheromone. Extracts obtained from S-line hybrid females indicated that these females produced pheromone in a diel periodic fashion. The greatest amounts of pheromone were obtained during the sixth hour of the scotophase and corresponded to the peak of calling (Fig. 46.3a).

The fact that only a small proportion of the V-line F_1 females produced pheromone posed a significant problem for our studies on the genetic regulation of blend production by females. Females of the parent species can be stimulated to produce pheromone during the photophase, when little or no pheromone is normally present in the pheromone gland, by injection of pheromone biosynthesis activating neuropeptide (PBAN) (Teal et al. 1993). Therefore, we injected female V-line hybrids with either brain extracts, which contain PBAN, or synthetic PBAN. Hexane extracts obtained from glands 1 hr after injection of any of the pheromonotropic substances all contained large amounts of pheromone (>50 ng, Z11-16:AL). Extracts obtained from females injected with only saline contained no more pheromone than was present in extracts obtained from glands from untreated females (≤3.5 ng, Z11-16:AL) (Teal and Oostendorp 1994b). These results indicated that all V-line hybrid females had the capacity to produce pheromone and that PBAN was present in the brains of females. Thus, hybridization adversely affected both the release of PBAN, which results in pheromone production, and calling by the majority of V-line females. However, the fact that all females from both V-line and S-line hybrid lines could be stimulated to produce pheromone by injection of PBAN (Teal and Oostendorp 1994b) provided an important tool for our studies on regulation of blend production.

To use PBAN as an effective tool to study the regulation of pheromone blend production by the hybrids, we first needed to determine the optimal dose and incubation period using females of the parent species. Detailed studies using the parent species indicated that injection of a 5-pmol dose and incubation for 1 hr prior to extraction of the pheromone gland resulted in production of the most uniform amount of Z11-16:AL in both species (Teal et al. 1993) (Fig. 46.4). Additionally, studies in which we compared the ratios of all pheromone components in extracts from both species obtained at 15-min intervals after injection of PBAN indicated that the blends of components did not vary between 15 and 120 min after injection. Similarly, extracts obtained from both F_1 hybrid lines at 15-min intervals after injection of PBAN indicated that no changes in blend ratios were evident between 15 and 90 min after injection. Therefore, we chose to inject hybrid and backcross females with a 5-pmol dose of synthetic PBAN during the fifth hour of the third photophase after adult emergence and to extract and analyze extracts obtained from females 1 hr after injection.

Although females of Lepidoptera are the heterogametic sex (sex chromosomes denoted "ZW"), only the Z chromosome is known to be functional. Therefore, sex-linked traits result from the action of genes borne on the Z chromosome. If the ratios were controlled by chromosomes on the Z sex chromosome, then the F_1 females should have produced the same pheromone blend that was produced by the paternal species. Analysis of 125 extracts obtained from both hybrid lines showed that the ratios of components were the same and that the populations

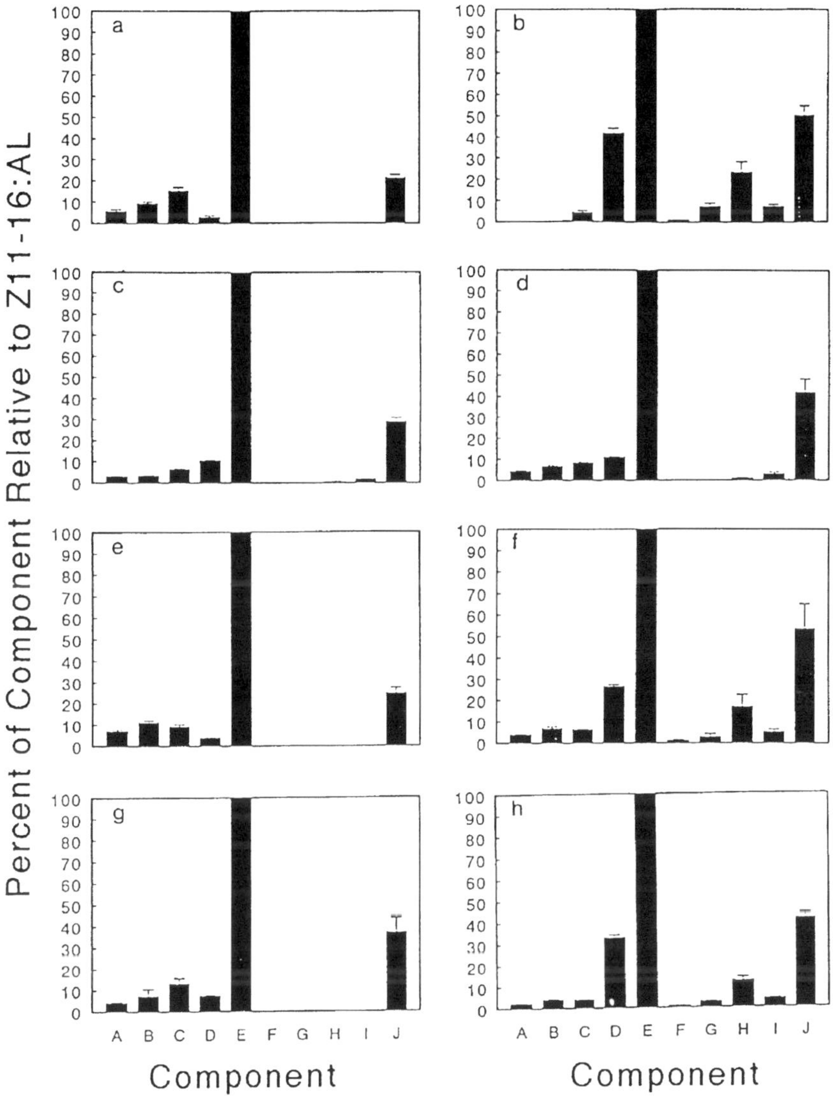

Figure 46.4. Comparison of ratios (±SE) of pheromone components in extracts obtained from pheromone glands of females 1 hr after intact females were injected with PBAN during the mid-photophase of the third day after adult emergence. The compounds (A through J) are the same as those listed in Table 46.1 and are compared to Z11-16:AL. (**a**) *H. virescens*, (**b**) *H. subflexa*, (**c**) V-line F_1 hybrid, (**d**) S-line F_1 hybrid, (**e**) V-line ♀ × *H. virescens* ♂, (**f**) S-line ♀ × *H. virescens* ♂, (**g**) V-line ♀ × *H. subflexa* ♂, (**h**) S-line ♀ × *H. subflexa* ♂.

showed little variation (Fig. 46.4). Therefore, genes regulating the ratio of pheromone components are not linked to the Z chromosome of either species. Although the ratios were more similar to those of *H. virescens* than to *H. subflexa*, ratios of (*Z*)-9-hexadecenal and (*Z*)-11-hexadecen-1-ol to Z11-16:AL were different from *H. virescens*. Additionally, extracts of hybrid females contained (*Z*)-11-hexadecen-1-ol acetate and (*Z*)-9-hexadecen-1-ol, while extracts from glands of *H. virescens* do not contain these compounds. This indicated that regulation of the pheromone blend was under polygenic control.

The mean ratios of components present in extracts obtained from females from backcrosses between F_1 females of either hybrid line with males of *H. virescens* ($n = 75$, each backcross line) were identical to those obtained from *H. virescens* females (Fig. 46.4). Additionally, the only components identified were those found in extracts of *H. virescens*. Thus, backcrossing the hybrid females one time diluted the genes obtained from *H. subflexa* to the point that no influence on blend regulation was observed. Extracts obtained from backcrosses between F_1 females and *H. subflexa* males ($n = 75$, each backcross line) contained all of the components common to the pheromone blend of *H. subflexa*, and the mean ratios of the 16-carbon compounds were quite similar to those of this species (Fig. 46.4). This suggested that genomic material inherited from *H. virescens* had a minimal effect on the ratios of the 16-carbon compounds in these backcross females. However, the 14-carbon aldehydes were present also in the extracts of these backcross females, and the ratios of these aldehydes were the same as those found in extracts from females of *H. virescens*. This indicated that autosomal genes inherited from *H. virescens* exert a dominant effect on production of the 14-carbon compounds and are probably not closely linked to genes regulating the production of the 16-carbon pheromone components.

4. Conclusion

Our studies on interspecific hybrids and backcrosses obtained from crosses between *H. virescens* and *H. subflexa* have provided information on the genetic mechanisms that control the pheromone communication systems of these insects. Although the structure of hairpencil glands of males is regulated by dominant genes on the male sex chromosome of *H. virescens*, the production of pheromone by males is under polygenic autosomal regulation with the *H. subflexa* genome exerting a dominant influence. The production of the female sex pheromone blend is also regulated by autosomal genes. However, the only genes which exert a dominant effect are those of the *H. virescens* genome that regulate the ratio of 14-carbon aldehydes. No dominant influence from either of the parent species was evident for the 16-carbon pheromone components.

The results of these studies are of significance for ongoing research aimed at control of *H. virescens* by release of laboratory-reared backcross insects into the wild population. This control strategy is based on the fact that backcross females are fertile whereas males are infertile. Thus, mating between wild *H. virescens* males and backcross females released into the field would result in production of fertile backcross females but infertile males. Continued mating between field-bred backcross females and wild *H. virescens* males for several generations would be expected to reduce the *H. virescens* population because fewer *H. virescens* males would be present with each successive generation (Laster 1972). However, the backcross females released into the field must compete effectively with wild *H. virescens* females for *H. virescens* mates in order for the technique to be effective (Laster 1972; Proshold and LaChance 1974). Our results indicate that progeny of hybrid females backcrossed to *H. virescens* produce the same sex pheromone as *H. virescens*. Therefore, they should compete effectively with females of the wild population for mates. This should result in a decline of the natural population of *H. virescens* due to introduction of male sterility.

47

Can Population Genetic Simulations Help to Interpret Pheromone Evolution?

Roger K. Butlin and Adam J. Trickett

1. Introduction

The evolutionary problem presented by insect pheromone systems is one that is common to sexual signaling systems in general: How is extensive divergence among species, even closely related species, reconciled with apparent conservatism within species? This pattern is a necessary feature of a signal system that is effective in prezygotic isolation in a speciose lineage. Many possible mechanisms have been suggested to explain its evolution (see Butlin and Ritchie 1994, for example) but most center on the special nature of signal evolution.

Sexual signaling systems are subject to two classes of selection pressure: those arising from the environment, including individuals of other species, and those arising within the system itself due to interactions between signal and receiver (Endler 1992). Signal–receiver interactions can generate evolutionary change that is largely independent of the environment and may be very rapid, as in "Fisherian" sexual selection (Lande 1981; Kirkpatrick and Ryan 1991).

Two asymmetries dominate thinking about the evolution of sexual signaling systems. The first is the asymmetry between signal and receiver: A signal must fall *within* the response window of the receiver in order to be effective, and a receiver must *include* the whole range of appropriate signals in its response envelope (De Jong and Sabelis 1991; Butlin 1993). The second asymmetry is the common difference between males and females in their parental investment per offspring which creates greater opportunity for sexual selection to operate on one sex, usually males, than on the other (Trivers 1972).

These relationships, and their genetic consequences, make sexual selection and signal evolution difficult questions to tackle. Moth pheromone evolution presents the added difficulty that the asymmetries are combined in the opposite way to most animal systems: high investing female moths signal and low investing males respond, unlike most well-studied systems. The mainstream of sexual selection research has, therefore, tended to ignore pheromones although there have been insightful discussions by pheromone biologists interested in evolution (e.g., Cardé 1986; Löfstedt 1990, 1993; Phelan 1992).

At the crux of pheromone evolution is the issue that we will consider in this chapter: the evolution of a simple two-component female-emitted pheromone and its associated preference function in the male. We are interested in the selection pressures that operate on mutant blend and preference alleles and we wish to find ways in which the system can evolve a sufficiently different blend/preference combination to contribute to prezygotic reproductive isolation and speciation.

Of course, isolation may be generated in other ways such as by divergence in timing of pheromone signaling (e.g., Greenfield and Karandinos 1979), by loss or gain of compounds (Löfstedt 1993), or by evolution of male pheromones (Phelan 1992). However, we feel that a shift in blend confronts the general problem of signal system evolution most clearly. It is also true that similar systems have been modeled extensively in the past and we hesitate to add a further crude representation of the real world. Much of what we conclude could probably be found in O'Donald's (1980) book by a diligent student. Our defense is that the simulation we present helped us to think clearly about the system. We hope it will help others to do so as well.

2. Observations on Which the Simulation is Based

Several sets of observations on real systems have influenced the structure of the simulation or prompted questions that the simulation can address. These are as follows:

1. **Narrow blend distributions** characterize pheromonal signals (Löfstedt 1993), especially mixtures of isomers (Linn and Roelofs 1994). This is in common with other sexual signals (Butlin 1994). They also share the observation made on other signal types that polygenic genetic variation is present within this narrow range. The best data for moth pheromones come from the selection experiment by Collins and Cardé (1989a) on *Pectinophora gossypiella* which produced a sustained response over 12 generations but nevertheless only a small shift in blend compared to the range of male response.
2. **Male preference matches pheromone blend** in that the maximum response is achieved with the mean blend, but males generally respond to a much wider range of blends than females produce (Löfstedt 1990). Capture–recapture experiments suggest that all males have wide response windows, rather than the overall wide window being due to

variation in preference among males Cardé et al. 1976; Haynes and Baker 1988; also see Haynes this volume). Evidence for genetic variation is stronger for the width of the response window than for the most preferred blend (Collins and Cardé 1989b,c; Löfstedt 1990), again in agreement with data on acoustic signals (Butlin 1993).

3. **Blend differences** between species are typically several times the within-population standard deviation so that there is no blend overlap and little overlap between the blend of one species and the preference envelope of the other. There is some evidence that major genes are responsible for the shifts in blend and preference. Certainly this is the case in the best-characterized example, the E and Z races of European corn borer *Ostrinia nubilalis* (Roelofs et al. 1987; Löfstedt et al. 1989a), and other similar cases appear to be accumulating (Löfstedt 1993; Haynes this volume; Foster et al. this volume). This is not parallel in acoustic systems (Butlin and Ritchie 1989). The European corn borer shows sex linkage for the male response locus, and this may be true of other species too (see Foster et al. this volume).
4. **Apparent coexistence** of races with divergent sexual signaling systems in the European corn borer (Bengtsson and Löfstedt 1990) provides a unique challenge to the theoretical expectation of positive frequency dependence (see below). Simulation of this system specifically may be very valuable.

3. Features of the Simulation

The simulation was written in Borland TurboPascal on an IBM-PC compatible microcomputer. It simulates a single population of moths with two loci of interest: Locus A controls the mean and variance of female pheromone blend, while locus B controls the mean and variance of male response. In other words, both loci have sex-limited effects. Each locus has two alleles: A/A′ and B/B′. The mean and variance of blend determine the probability with which individuals of a given genotype at the A locus produce any specified blend on the scale from 0% to 100% of component 1 in the two-component blend. The variance incorporates both within and among individual variation from any other source, genetic or environmental. Similarly the mean and variance of preference determine response probabilities. Thus the probability of mating can be determined for any genotypic combination by combining the distributions of female blend and male response.

Both blend and preference are considered in the simulation to be normally distributed on a logit scale (ln [proportion of component 1/proportion of component 2], amounts can be substituted for proportions with no effect on the result). We urge the use of this scale in the description of blends and preferences as well. These values cannot be normally distributed or additive on the percentage scale, but the logit scale gives a reasonable fit to observed distributions and should make genetic analysis simpler. Example distributions are given in Fig. 47.1.

The two loci may be unlinked; or they may be linked with any specified value of *r*, the recombination fraction. Either locus, but not both at once, may be considered sex-linked. The female is the heterogametic sex. Population size may be infinite (i.e. the simulation may be run deterministically) or may be specified.

Initial allele frequencies at each locus may be set, as may initial linkage disequilibrium. Alternatively, the program can simulate mixing of two multilocus genotypes, as in secondary contact between European corn borer races, for example. Linkage disequilibrium is a measure of the nonrandom association between alleles at different loci—in other words, the tendency for AB or A′B′ combinations to be more or less common in gametes than is expected from the allele frequencies. Here it is expressed as a proportion of its maximum value (Hedrick 1985). As a result, it varies from −1 (complete association between A and B′), to O (random association), to +1 (complete association between A and B). It is equivalent to genetic covariance in polygenic models (e.g. Lande 1981) and does not require that the loci involved are on the same chromosome.

There are two further parameters of importance. The first is the *cost* to females producing blends that have a low probability of attracting males (given the male genotypes available). This cost is equivalent to the Wallflower effect of De Jong and Sabelis (1991), who considered that, although females may all be mated eventually, unattractive females are likely to be mated late and this may expose them to extra predation risk or reduce the survival chances of their offspring. Here, we take a female genotype, sum the response probabilities for all male genotypes, and then discount the female's output of offspring by the product of our parameter cost and the total probability of response. Thus cost is the maximum

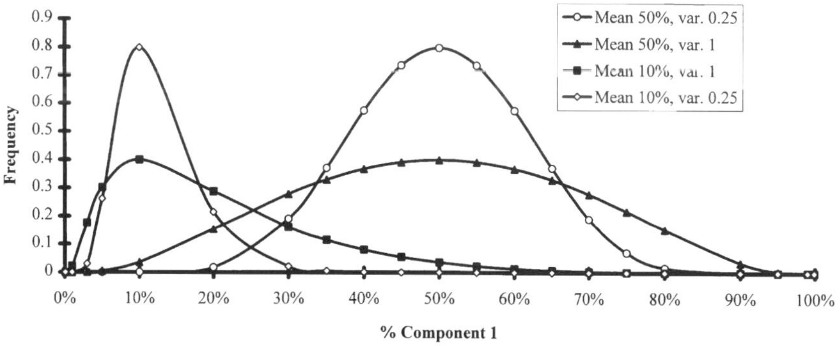

Figure 47.1. Examples of blend or preference distributions, plotted on the percentage scale but Gaussian on the logit scale. Note that for preference these distributions correspond to the maxpref = 0 option, and for the maxpref = 1 option they would be rescaled to peak at 1.0 (see text).

penalty a female can pay, as a proportion of her offspring, if her attractiveness is very low relative to that of other females in the population.

The final parameter relates to male response behavior. Two extreme assumptions can be made about widening male response windows (increasing the variance of response). At one extreme, a male with a wide response must sacrifice sensitivity to the optimum blend in such a way that his total responsiveness is unchanged. At the other extreme, it may be possible to respond to a wider range of blends without any cost in sensitivity at the peak. These two extremes can be selected in the simulation by setting "maxpref" to 0 or 1, respectively. The consequences will be considered below. The real situation must depend on the receptors present on the male antennae and the way in which their input is processed. It is probably nearer to the second extreme, but good empirical data do not seem to be available.

4. Preliminary Results

A series of runs of the simulation was used to establish the performance of the program and to demonstrate the primary effects before moving on to more complex runs. The results of these runs, in which genetic variation was present at only one locus at a time, are summarized briefly below, for blend followed by preference, and examples are given in Fig. 47.2.

1. **Cost.** In the absence of costs (cost = 0) there is no selection on alleles at the blend locus.
2. **Mean blend.** Where a cost is present, alleles specifying mean blends that do not correspond to the mean preference of males (fixed in these runs) are eliminated. The strength of selection against deviant blends depends on:
   a. the cost,
   b. the amount of deviation, with more divergent blends being eliminated more rapidly, and
   c. the variance in preference, with narrower preference windows imposing stronger selection on blend.

   Dominant alleles are eliminated more rapidly than alleles of additive effect, as expected. Where the blend is sex-linked, the system behaves like the introduction of a dominant mutation at an autosomal locus because the new allele is immediately and constantly exposed to selection in the hemizygous female.
3. **Variance of blend.** Alleles that elevate the variance in blend are eliminated while alleles that reduce the variance are favored by selection. The strength of selection is influenced by the same factors as for mean blend, with the effect of preference variance being particularly strong.

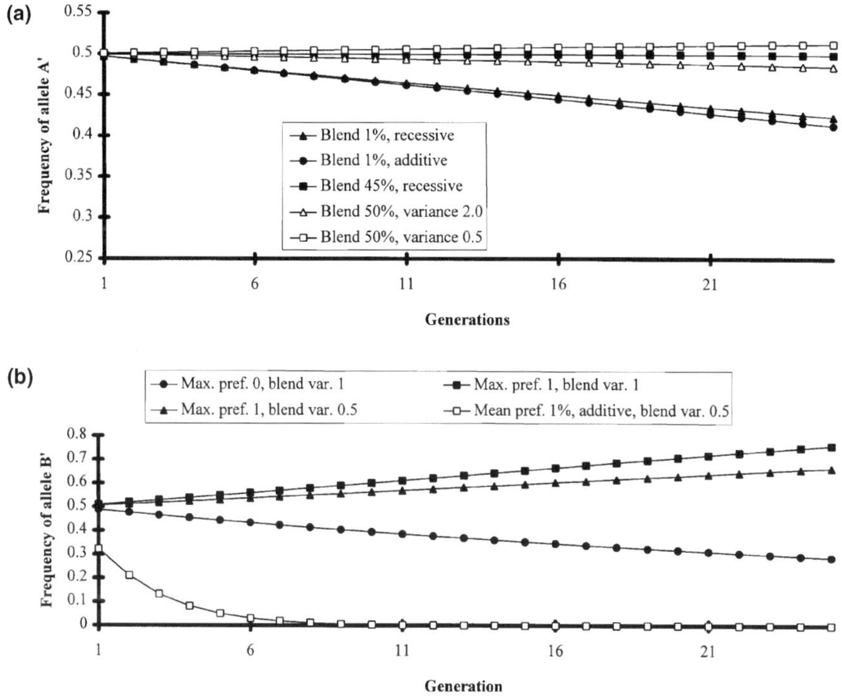

Figure 47.2. Examples of simulation runs in which only the blend locus (A/A′) (**a**) or the preference locus (B/B′) (**b**) was polymorphic. The caption specifies the characteristics that differ from the standard parameters, which were a mean of 50% and variance of 1.0 for both blend and preference, cost 0.5, and additive gene action. These runs were deterministic.

4. **Mean preference.** Alleles producing mean preferences that do not correspond to the mean blend are eliminated. Selection on males is much stronger than on females (even when cost is set high); that is an example of the effect of parental investment on the intensity of sexual selection since males can mate repeatedly or not at all while all females breed but may have reduced numbers of offspring. The rate of elimination depends on
   a. the amount of deviation from the mean blend and
   b. the preference variance and the blend variance with stronger selection where the variances are small.

   Dominant alleles are eliminated more rapidly than additive alleles, but sex-linkage of the preference locus has no effect since the trait is only expressed in the homogametic sex.
5. **Preference variance.** Here the behavior depends critically on the "max-pref" option. If there is no cost in reduced maximum response (maxpref =

1), selection always favors genotypes with greater response variance—that is, wider response windows. However, if there is a cost of reduced maximum response (maxpref = 0), selection always favors the narrower variance—in effect favoring maximum responsiveness to the mean female blend. Where selection favors wide preferences, the selection pressure declines as preference variance becomes wide relative to blend variance.

The simulation results are consistent with the observations on natural systems listed above. Selection maintains coordination between mean blend and mean preference. Narrow blend variance and wide preference variance are expected, if this does not have too much cost in terms of sensitivity which therefore appears to be the case. Once preference variance is wide, selection for narrowing of blend variance and for elimination of alleles with minor effects on blend mean becomes weak. Thus polygenic variation will be maintained by mutation-selection balance. Similarly, polygenic variation influencing the preference variance is expected, but the generally stronger selection on males makes this less likely for preference mean. This may explain the apparent widening of preference in Haynes' experiment (this volume) in response to an altered blend. However, in a large population a mutation producing a major shift in preference mean might reestablish coordination, rather than the slow change based on standing variation seen in Hayne's lines.

In reality, external selection pressures will also influence equilibrium variances of blend and preference. The risk of heterospecific mating may be the most important influence, although time and energy wasting by males who respond to a wide range of stimuli may also be important. It would be simple to incorporate costs like these into a simulation, but impossible to decide on realistic values for the parameters. In general, they will oppose the tendency for preference variances to become ever wider and may produce directional selection on means of both blend and preference.

5. Spread of a New Signal System

Summarizing the above observations, the principal selection pressures in the sexual signaling system act against novel blends and against novel preferences, apparently making shifts in signaling systems very unlikely. But this is not an accurate statement of the situation. More precisely, selection acts against blends for which preference is rare and against preferences whose corresponding blends are rare. The latter is the much stronger selection pressure. In addition, we know from Fisher's (1930) argument that when both signal and receiver vary, there will be assortative mating and that this assortative mating will generate nonrandom associations between alleles at signal and receiver loci—that is, linkage disequilibrium. When such associations exist, selection on one character produces a correlated response in the other. This is the basis of Fisher's runaway process.

Together, these arguments suggest a way in which new pheromone blend systems may spread. The process might operate as follows:

1. If cost is low, a mutant blend allele might spread into a small population by drift.
2. If the mutant blend allele reaches a sufficiently high frequency, then a corresponding preference mutant would be favored by selection. Because the new preference allele would initially be at a lower frequency than the corresponding blend, it would confer access to more females per male than the existing allele.
3. Assortative mating would generate linkage disequilibrium, and therefore the increase in frequency of the preference mutation would be accompanied by an increase in the blend frequency.
4. If this process were to go far enough, the new blend/preference combination may come to dominate the population, the old alleles being then at a minority disadvantage themselves and being removed.

This sequence contains a combination of the "asymmetric tracking" idea of Phelan (1992), an argument that Löfstedt also uses (1990, 1993), in that it relies on the relative weakness of selection on females to initiate the process. Step 3 is exactly the Fisher process: spread of preference as a correlated response to selection on the preferred trait, feeding back to increase selection on the trait. The difficulty with the verbal argument is that one cannot see how likely the sequence of events is. The simulation model can be used to address this problem.

The scenario appears most likely to work if the mutant blend allele is recessive, allowing it to drift more easily to high frequencies, and the mutant preference allele is dominant, immediately exposing it to selection. Since large shifts in blend are most interesting, we initially simulated a population with mean blend and preference of 50% into which mutant alleles determining a 1% blend were introduced. Blend variance was 0.25 and preference variance 1.0 for all genotypes. Recombination was 0.1 (neither locus sex-linked) and there was no initial disequilibrium. Cost was 0.01. The population size was 100 and maxpref was set to 1.

Fifty replicate runs of 100 generations were initiated with the starting frequency of the blend allele at 20%, and the preference allele is introduced at 1%. In 10 runs the original allele was replaced at both loci. Figure 47.3a shows the progress of a typical run of this type. The frequency of the new preference allele rises immediately due to its selective advantage in the presence of the novel blend. Strong disequilibrium is established rapidly and so the novel blend allele also rises in frequency. Since the blend allele stays well "ahead" of the preference allele, selection continues to favor its spread. The blend allele is fixed at generation 34 and then selection slowly eliminates the original preference allele (slowly because it is recessive relative to the new allele).

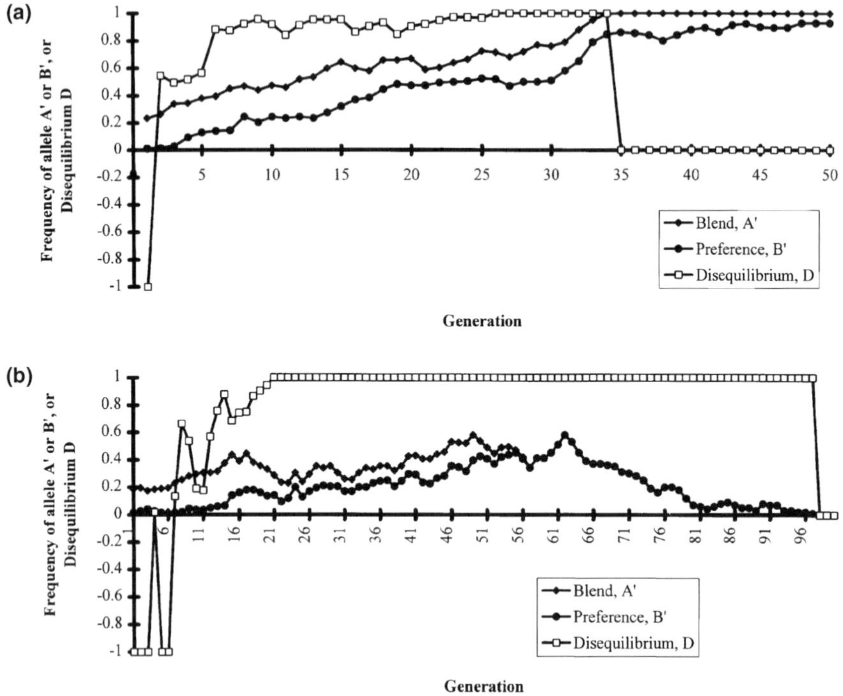

Figure 47.3. Examples of simulation runs in which new blend and preference alleles are introduced as described in the text. (**a**) A run in which the new alleles replaced the old alleles. (**b**) A run in which both new alleles were eliminated.

Deterministic runs (infinite population size and, therefore, no genetic drift) do not fix the mutant alleles from these starting conditions, nor do all runs at a population size of 100. The run shown in Fig. 47.3b started with conditions identical to those in Fig. 47.3a, but in this case disequilibrium built up more slowly; the correlated response of the blend allele was slower and thus the preference allele "caught up." Therefore, it lost its selective advantage and both mutant alleles were removed, despite effectively complete disequilibrium at this stage.

In other runs, one or both alleles remained polymorphic at generation 100. In those runs where the new preference allele was eliminated, the subsequent drift of the blend allele frequency helps to establish the likelihood of such a mutant reaching a frequency of 20%. With a population size of only 100 and the very weak selection implied by a cost of 0.01, drift took some populations to frequencies over 50% and values between 15% and 25% were common. In addition, some runs initiated with a blend frequency of only 10% did show the pattern of Fig. 47.3a and produce a switch of blend and preference.

The results for the original 50 runs are summarized in Table 47.1 together with results of simulations under modified assumptions. If the loci are unlinked, replacement of their pheromone system is less likely. This is because linkage disequilibrium builds up more slowly with freer recombination (cf. Trickett and Butlin 1994). Other things being equal, blend and preference loci that are linked would be expected to be involved in pheromone divergence more often than unlinked loci. The work of Grula and Taylor (1980) with *Colias* is the only supporting example of which we are aware.

If the new blend allele is additive in its effect, then the probability of replacement declines. This is because selection tends to eliminate the allele more rapidly and works against the correlated response to selection on the male preference. A similar effect was observed when female costs were increased to 0.1.

An additive preference allele was fixed only once in 50 simulation runs. Clearly the chances are that such an allele will be lost by drift because selection favors the heterozygote much less than it does in the dominant case.

Sex linkage of the blend locus produced a marked increased in the probability of replacement (compared with the free recombination autosomal case). This is because all of the new blend alleles are expressed, giving a much stronger selective advantage to the new preference allele. However, the result is somewhat artificial since sex-linkage makes drift of the new blend allele to a high level in the first place much less likely. A more thorough study will be needed to characterize the overall chances of replacement in the two systems.

Finally, the fates of alleles with much smaller effects on blend and preference (45% blend and preference means in the mutant homozygotes) were investigated. Most interestingly, this greatly reduces the chance of a replacement because the

Table 47.1. Outcomes of simulations of the introduction of both blend and preference mutations[a]

				Numbers of simulation runs, out of 50, showing the following outcomes:			
Blend[b]	Preference[b]	r^c	Cost	Both replaced	Both polymorphic	Both eliminated	Blend polymorphic
1%, rec.	1%, dom.	0.1	0.01	23	2	12	13
1%, rec.	1%, dom.	0.5	0.01	21	1	17	11
1%, add.	1%, dom.	0.1	0.01	5	7	23	15
1%, rec.	1%, add.	0.1	0.01	1	6	24	19
1%, SL	1%, dom.	0.5	0.01	24	0	21	5
1%, rec.	1%, dom.	0.1	0.1	5	0	33	12
45%, rec.	45%, dom.	0.1	0.1	0	1	16	33

[a]Parameters as described in the text.

[b]Mean blend/preference and mode of inheritance of the mutant allele: SL, sex-linked, rec., recessive autosomal; add., additive autosomal; dom., dominant autosomal.

[c]Recombination fraction.

much weaker assortative mating generates less linkage disequilibrium. In a series of simulation runs with different mutant blend and preference means, replacements were only observed where means were 5% or below. Thus sudden major switches in blend and preference appear more likely than accumulation of small changes, potentially explaining a tendency toward single locus control and counter to the arguments of Paterson (1985), who proposed that evolutionary changes in "specific-mate recognition systems" must occur in small steps.

It has often been suggested that pleiotropic effects of a single locus on both signal and receiver, "genetic coupling," would aid the evolution of new signal systems by maintaining coordination during the replacement of one system by another. In this simulation, and we believe in real pheromone systems, this is not the case. This is because a new mutation affecting both traits would be selected against by minority disadvantage on both counts. The mechanism described here, where the female trait becomes more common by drift and provides an initial selective advantage to the male trait, would not be possible.

A major limitation on this mechanism in natural populations is likely to be the supply of suitable mutations. The mechanism requires a mutation of large effect, in the right direction (perfect match is not necessary) in the male response machinery. This must happen in a small population in which a mutant blend allele is already segregating. Clearly this sequence of events is improbable. However, improbable events may make important contributions to evolution, partly because there are many replicate opportunities over space and time and partly because the events we are seeking to explain are, themselves, rare.

A further problem is the spread of the new variant, once it is established in a small local population. This may not be so difficult since the original deme can "seed" other demes with suitable alleles. However, this will happen in both directions and the dynamics of spread in a subdivided population certainly deserves attention in an extended version of the current simulation program.

6. European Corn Borer

The European corn borer, *Ostrinia nubilalis*, is unique in that two divergent sexual signaling systems appear to coexist in some natural populations (Bengtsson and Löfstedt 1990). It is also unique in that the genetics of the blend and preference differences are well characterized, as are the means and variances of the various possible genotypes. There is, therefore, the opportunity to tailor our simulation program specifically to the problem of the *O. nubilalis* races.

In *O. nubilalis*, the blend difference between the E and Z races is determined by a single autosomal locus while male behavioral response is determined by a sex-linked locus. The parameters used in our initial simulations are given in Table 47.2 and are derived from the distributions in Roelofs et al. (1987). Since the distributions were originally reported on the percentage scale, we adjusted

Table 47.2. Parameters for the European corn borer simulation[a]

Race	Mean blend	Blend variance	Mean preference	Preference variance
E	99%	0.04	99%	4.0 (9.0)
Hybrid	65%	0.04	30%	1.0
Z	3%	0.04	3%	0.25 (1.0)

[a]The preference locus is sex-linked. Parameters are those derived from Roelofs et al. (1987) with modified values in brackets where appropriate (see text). Blends and preferences are expressed as percentage of the E isomer.

variances on the logit scale by eye to achieve reasonable reproduction. This was possible for all except the F1 male response distribution which clearly has an unusual form.

Using these parameters in deterministic simulations introduction of a Z blend or preference into an E population, or vice versa, leads to rapid elimination whenever a cost is present. We then experimented with mixed populations, supposing secondary contact between the races, but this revealed a problem. The data in Bengtsson and Löfstedt (1990) imply symmetrical hybridization between pure E and Z individuals at a rate of about 4%. This cannot be achieved with the variances derived from Roelofs et al. (1987). Therefore, we increased the variances as shown in Table 47.2 and used these new values in all subsequent simulations.

Various runs simulating secondary contact (Fig. 47.4) at different initial frequencies of the two races indicate an unstable equilibrium point at about 75% Z. Runs started at 75% Z with a population size of 500 showed little change in

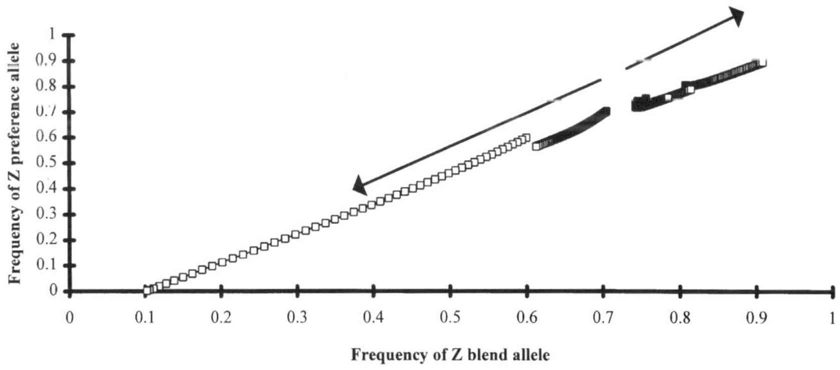

Figure 47.4. The result of simulating the mixing of European corn borer races at five different initial frequencies. Arrows indicate the direction of evolution. Note that some hybridization does occur and that selection favors the E preference allele moving the populations away from being simple mixtures of the two initial genotypes.

proportion initially but eventually moved to fixation for one or other race. This unstable equilibrium is biased away from the simple 50% expectation primarily because the high preference variance of E males (with maxpref = 1) gives them a selective advantage. At 75% Z this is approximately balanced by minority disadvantage. It is tempting to note that the apparently stable population discussed by Bengtsson and Löfstedt (1990) had a Z frequency very similar to this, but very little weight should be given to this coincidence.

Finally, we considered the spread of the E system into a Z population, or vice versa, by the mechanism described above. Runs were initiated with the allele frequency of the new blend at 10%, the new preference at 1%, and the population size at 100. The E system spread into a Z population in 3/10 runs, but in no run out of 10 did the Z system spread into an E population. Löfstedt (1990) suggested that spread of the blend allele before the preference allele was likely because of the lower intensity of selection on females, with which we agree. He further suggested that spread of the E system into the Z race was more likely because of the wider preference variance of E males. Our program also suggests that spread of the E system into the Z race is more likely, but that the major reason is because selection directly favors wide preference variance and thus aids the spread of the E preference allele. However, it is noteworthy that, when cost is low as in these runs, each blend variant can persist in the opposite race in small populations.

We hope that further development of the simulation program will help in the understanding of this fascinating example.

7. Conclusions

There is nothing entirely novel in the mechanism we propose for the transition between one pheromone blend and another. It is essentially a combination of the Fisher process of sexual selection with some of the realistic detail of pheromone communication systems. In the latter respect it simply extends the ideas of Phelan (1992) and Löfstedt (1990, 1993). We have also concentrated entirely on "internal" selection pressures. Clearly the signaling system must be subject to external pressures as well, especially the risk of heterospecific matings. In principle this risk could be incorporated into the simulation, but our current state of knowledge would make it very difficult to decide on suitable parameters to describe both the risks and the consequences.

In general, a simulation is always limited in its usefulness by the large number of parameters that is involved. It is very difficult to achieve generality by exploring the whole of parameter space, and yet accurate simulation of a specific system also tends to be difficult because the appropriate parameters have not been estimated. We feel that the greatest value of a simulation is to point to key parameters or assumptions that require further empirical study. Our discussion will, therefore, concentrate on this issue.

1. **Genetic analysis** is needed in more species. Is it true that major gene effects are common? If not, what features are shared by cases of polygenic determination that differ from examples of major gene effects? Can generalizations be made about sex linkage, dominance, or linkage between blend and preference loci? Within populations, the extent of genetic variation in male response behavior requires further work, in particular the difficult separation between variation in mean preference and variation in preference "windows."
2. **Male response** evolution was dominated in the simulation by the max-pref option. The real situation here is unknown. Although wide preference windows suggest that the sensitivity penalty is small, direct measurements are needed. Other potential costs to males with wide response windows should also be examined, especially the risk of wasting time in following heterospecific pheromone plumes, a cost that depends on the frequency with which homospecific plumes are encountered. The general idea of "forgiving receptors" (Vickers et al. 1991) is an important one, not just for the spread of mutations that alter blend but also for the spread of alleles that produce new compounds.
3. **Female costs** also represent a key feature of the model. De Jong and Sabelis (1991) identified the Wallflower effect and suggested that delayed mating was the major disadvantage to less attractive females. It is clear that males do respond to atypical blends (e.g. Löfstedt et al 1994a). Now we need to measure the delay experienced by females producing aberrant blends and measure the fitness consequences of such delays.
4. **Mutant blend alleles** should be segregating at low frequencies in natural populations if the mechanism proposed here is at all important. They should be common relative to the frequency of blend shifts because the supply of suitable male response mutations is expected to be the limiting factor. Haynes (this volume) has provided an excellent first example of the type of mutation expected, segregating at quite a high frequency in a laboratory line. It would be possible, though labor-intensive, to survey natural populations systematically for such mutations by collecting isofemale lines (Parsons 1980) and sib-mating within them to reveal recessive alleles. We think the return from such an experiment may well justify the effort.
5. **Divergent signaling systems** that interact in nature provide a very special opportunity for the study of evolutionary divergence and the characterization of the selection pressures at work. Study of hybrid zones has amply demonstrated this in recent years (Harrison 1993). The European corn borer thus deserves further attention, and other similar systems should be characterized (a further possible example is larch

budmoth, *Zeiraphera diniana*; Löfstedt 1990). In the European corn borer, two immediate questions to be answered are the spatial context of apparently stable mixed populations (are they maintained by gene flow?) and the fitness of hybrids. There is no doubt that marker loci close to the blend and response loci would greatly assist this venture. The tremendous background information available on pheromone production, reception, and subsequent behavior, illustrated by this collection of papers, makes moth pheromones superb systems to study from the evolutionary point of view. It is time they came into their own and had more input into the general field of sexual signaling system evolution.

48

Genetics and Phylogenetics in the Evolution of Sex Pheromones

P. Larry Phelan

1. Introduction

As with each area addressed by this symposium, understanding the evolution of sex pheromones has some unique challenges not shared by the others. Like the physiologist, the evolutionary biologist seeks rules that govern nature; but the special challenge is in understanding events that have transpired in the distant past, through processes that, although they continue in contemporary time, generally proceed on a time scale that is too protracted for us to observe. In the absence of a fossil record, this challenge is addressed by looking for patterns in extant species that suggest a historical sequence and processes that have governed it: patterns of trait divergence between related species, patterns of trait similarity among taxa only distantly related, and patterns of genetic control, which suggest modes of selection on the trait. By piecing together such fragments of an imperfect record, it is hoped that we can create a logical picture of how it all happened. The contributors to this symposium represent some of the field's best attempts at constructing such a picture for the evolution of sex pheromones.

A mate-signaling system is potentially the product of a number of evolutionary forces. A historical perusal of the literature will show that different factors have found favor among researchers for a period of time. During the 1940s until the 1970s, emphasis was placed on the importance of species fission, and field biologists looked for examples of reinforcement. During the 1970s and 1980s, there was a revival of interest in sexual selection, particularly with regard to mate choice and in the run-away process described by Fisher (1930). Presently, the greatest emphasis is on how signals evolve to match the predisposition of the receiver's sensory system [variously termed *sensory exploitation* (Ryan and Rand 1990), *sensory drive* (Endler 1992), and *sensory trap* (Christy 1995)].

Another view not restricted to a particular time period underscores the importance of stochastic factors leading to nonadaptive shifts in mate signaling. Although it is common for the focus of study to change as our understanding of an area evolves, unfortunately with regard to the evolution of mate signals, the historical shifts do not necessarily represent advances in resolving the relative role of these different processes. The accumulation of studies suggests that a number of adaptive forces and nonadaptive processes may be possible, and that may represent the ultimate answer; however, it is intellectually unsatisfying to simply state that all processes for which there is evidence are important. Recognizing at least some role for each and the potential for interactive effects among them, there is likely a hierarchical structure of relative importance. Unfortunately, there is little consensus at present for the nature of this hierarchy. What is needed is a unifying framework for understanding how these processes might interact, either synergistically or antagonistically, and within which we can predict the relative rank of various hypothesized evolutionary processes.

Despite the attention that mating behavior has attracted over the years, there are many key issues that remain unresolved (Table 48.1). These issues cannot be addressed in isolation but are highly interrelated. To provide a context for understanding the relative importance of different modes of evolution in mate signaling, I have previously proposed the asymmetric tracking hypothesis (Phelan 1992, 1997). The key premise of this framework is that differential parental investment plays a central role in the determination of mate-signal evolution. Arguments have been mounted for numerous processes and selective forces in the evolution of mate signaling. These arguments are often presented as mutually exclusive; for example, stochastic processes are more important than selection for reproductive isolation, or sensory exploitation is more important than sexual selection. In contrast, the asymmetric tracking hypothesis does not argue for one selective force at the expense of others. To the contrary, recognizing that the sexual-communication system will evolve as the interplay of a number of forces, the hypothesis suggests parental investment as a context for predicting the relative roles of other evolutionary processes. Although the hypothesis stresses differential strategies to maximize reproductive fitness as the primary selective agent for change in these signaling systems, expression of this process will be limited by other factors. For example, even in the absence of stabilizing pressures by male response, changes in female pheromone chemistry may be restricted by canalization in the biochemical pathways responsible for production. What the hypothesis does attempt to counter is the prevailing view of mate-signal evolution as a coevolutionary process, where changes are retarded by a mutually stabilizing process of "give and take" between signal and response. Recent emphasis on three aspects of pheromone biology has been particularly instructive in understanding the likelihood of these evolutionary scenarios: (1) measurements of geographic variability, (2) genetic architecture of control, and (3) heritability.

Table 48.1. Key issues for understanding the evolution of mate-signaling systems in animals

1. Determining adaptation
    - A. To what degree has sexual selection played a role in signal evolution?
        - 1. To the extent that sexual selection is involved, what is the relative importance of intrasexual competition versus mate preference?
            - a. If intersexual selection is involved, is the preference due to signaling of "good genes" or is the process governed by a Fisherian run-away due simply to linkage between a trait in one sex and preference for the trait in the other?
                - i. If a Fisherian run-away is involved, what, if anything, acted as the trigger of the run-away process?
    - B. To what degree has the environment either constrained or selected for a signal-receiver system?
    - C. How do properties of the signal-sensory system direct or constrain signal evolution? To what extent:
        - 1. does selection on the sensory system for other functions (e.g., food-finding, predation) affect evolution of mate signals—that is, sensory drive (Endler 1992)?
        - 2. do mate signals evolve to match the most sensitive characteristics of the sensory system, inherent characteristics that have not been selected for some other function—that is, sensory exploitation (Ryan and Rand 1990)?
        - 3. do signals evolve to mimic stimuli to which the receiver responds in other contexts, i.e., sensory trap (Christy 1995)?
    - D. What role do mate signals play in the speciation process?
        - 1. Is speciation largely effected by stochastic changes in the signal-response system (changes are incidental)?
        - 2. Can changes in the mate-signaling system result from selection for divergence in two or more interbreeding populations, thus effecting speciation (changes are an adaptive response via reinforcement)?
        - 3. Do changes in the mate-signaling system result from selection against interspecific mating attempts (changes are an adaptive response via reproductive character displacement)?

II. Phylogeny
    - A. What is the phylogenetic history of each particular mate-signaling system?

2. Genetic Control of Sex Pheromones

One of the greatest conundrums in the speciation process concerns changes in the mate-recognition system. The essential nature of coordination between signaler and responder in mate-finding would seem to be a strong drag on changes in the pheromone system. Moreover, the biological imperative of mating success is expected to heighten selection on such systems, limiting genetic variation and thus further restricting shifts in the communication system. Thus, when Roelofs and Comeau (1969) suggested that a single mutation controlling production of a pheromone could bring about "instant speciation," Lanier and Burkholder (1974) countered that this was highly unlikely since "in order for the mutant pheromone system to become established, it must convey more efficiency in mate-finding or be linked with genes for superior survival." Indeed, as a young student of pheromones, I was impressed with the apparent narrow tuning of component ratios in production and response, low variance that was attributed to "negative feedback inherent in closely co-adapted signaler and receiver (Miller and Roelofs 1980)." The assumption of "strong mutually stabilizing selection" has been a common theme shared by most discussions of pheromone evolution (Cardé and Baker 1984; Cardé 1986; Löfstedt 1993), and the apparent narrow variance in lepidopteran pheromone production has been cited as evidence of the inherent homeostatic properties of specific mate recognition systems (Lambert et al. 1987).

How then might shifts in pheromone blends be possible, which obviously have occurred given the species-specific nature of these signals? First, many of those who assume a system under strong reciprocal selection would argue that change occurs solely or at least largely by a slow stepwise coevolutionary process where the signal changes slightly followed by re-coordination by the responder (Paterson 1982). Alternatively, if signal and response are genetically coupled by some common mechanism or by pleiotropic effects in the genome, a change in one would be accompanied by a concomitant shift in the other. Such coupling could allow for even sudden drastic changes in the mate-signaling system, although coexistence between two such divergent populations would also require ecological separation (Cardé 1986). Evidence for coupling is seen in some acoustic systems (Hoy et al. 1977). In contrast to these scenarios, asymmetric tracking argues that in most mating systems, selection will be largely unidirectional between signaler and responder, instead of mutually stabilizing, with the low-investing sex feeling the brunt of the selection. Thus, for a typical moth species where males supply only sperm and in the absence of other selecting forces, this hypothesis predicts relatively high variability among females for pheromone production and only weak selection against mutations that produce new pheromone blends. But of course, other selection pressures do exist and wide variance in pheromone production will be countered by the presence of sympatric species sharing the pheromone channel, possibly tradeoffs between male sensitivity and breadth of chemical detection, and potentially even physicochemical considerations (Cardé and Baker

1984). Nevertheless, the important point is that under many circumstances, female-based pheromones should be free to change by nonadaptive processes.

There have been an increasing number of studies gauging the level of variation in pheromone blends in natural populations. These studies, which were reviewed previously (Löfstedt 1990, 1993; Phelan 1992, 1997), have shown that in contrast to early studies of low variance in sex-pheromone systems, not only can variability in female blends be broad, but many species possess pheromone polytypism, in either geographically separated or sympatric populations. In short, the number of such examples suggests that female pheromone blends may be under less intense selection than previously supposed.

With regard to the genetic architecture underlying pheromone systems, the genetic coupling hypothesis was given support by the earliest genetic analysis of lepidopteran pheromones. In *Colias* butterflies, Grula and Taylor (1979, 1980) showed that male pheromone production, male UV wing pattern, and female preference were largely controlled by genes on the X chromosome, in a "coadapted gene complex." However, a summary of the examples of genetic control from this symposium and other shows that this case is not representative of a widespread pattern of linkage (Table 48.2). In fact, although our elucidation of the genetic architecture of pheromone systems is in its infancy, evidence for control is given

Table 48.2. Genetic control of sex-pheromone systems in the Lepidoptera

Species	Female	Male	Reference
	Female pheromones		
Ctenopseustis	*Z5*/*Z8* ratio probably polygenic	Electrophysiological and behavioral responses: single gene and sex-linked	Foster et al. (this volume)
Heliothis	Polygenic autosomal for 14C and 16C compounds	?	Teal and Tumlinson (this volume)
Ostrinia nubilalis	*E*/*Z*11-14:Ac ratio single autosomal	Receptors: single autosomal; response: Z sex chromosome	Roelofs et al. (1987)
Trichoplusia ni	Polygenic involving both autosomes and sex chromosomes	?	Haynes (this volume)
	Male pheromones		
Colias	Response: X sex chromosome	UV wing pattern and pheromone: X sex chromosome, some autosomal control	Grula and Taylor (1979, 1980)
Heliothis	Response: ?	Hairpencil morphology: Z sex chromosome; pheromone: polygenic autosomal with dominance	Teal and Tumlinson (this volume)

for both autosomes and sex chromosomes; some aspects appear to be controlled by a single locus, while others are better characterized as polygenic. At this point, it is hard to discern any common features among these systems. Further conclusions await more detailed studies of these systems and others. It is possible that with more information, we will find that certain aspects of pheromone signal and response are inherited differently, reflecting the strength of selection and the probability for nonadaptive change. For example, genetic control by autosomal versus sex chromosomes may determine how likely a new mutation in the communication system is to persist in a population (Whitlock and Wade 1995; Butlin and Trickett this volume). Recessive mutants can be maintained for a considerable time, even when highly deleterious (Falconer 1980), and autosomal genes with sex-limited deleterious effects can also persist through their conveyance by the opposite sex where there is no selection against them. We simply do not have enough information to suggest how these processes might affect the evolution of sex pheromones. However, for a female pheromone system whose production is governed by a single locus and male response by a second major gene, the model of Butlin and Trickett (this volume) provides a first attempt to delineate the varied effects of dominance, linkage between blend production and preference, and sex linkage on the evolution of new blends.

Neither the strong reciprocal selection model nor the asymmetric-tracking hypothesis makes predictions for genetic architecture that will be applicable to each case; however, general trends can be stated. Relevant to the former hypothesis, strong stabilizing selection acts to reduce genetic variability in two ways: (1) by reducing heritable genetic variance at the loci controlling the traits under selection and (2) by reducing the probability of recombination (Falconer 1980). The latter can be brought about by consolidation of the controlling loci, giving rise to a tightly linked system or coadapted gene complex, such as is seen in genetic control of mimicry (Turner 1977). Given that most female lepidopteran pheromones consist of a blend of chemicals synthesized through the controlled action of a number of enzymes, most are likely under the control of a number of genes. If so, strong reciprocal selection generally should lead to coupling among these genes, as well as between signal and response. In contrast, asymmetric tracking predicts that for female sex pheromones, there should be weak linkage among components of the signaling/response system, while in male signaling systems, linkage is more likely. Particularly in those systems that have undergone runaway sexual selection, there is a greater probability of linkage (O'Donald 1962). Again recognizing the numerically limited data set, the combined results of genetic studies in pheromone systems are more consistent with the predictions of asymmetric tracking. Female-based pheromones appear to be generally polygenically controlled. Even in those cases where major genes have been disclosed, these genes control only parts of the pheromone blend. Furthermore, in all cases studied thus far, male response assorts independently from female blend production. Thus, there is little evidence of genetic linkage either through shared physiological mechanisms or due to selection to reduce recombination.

We do not have enough information about the genetic control of male pheromones to elucidate a pattern, but the strong linkage both among components of the signal and between signal and response in *Colias* is consistent with the predictions of asymmetric tracking. Analysis of the only other example, that of *Heliothis* male pheromones, is incomplete in this regard. The genetic control of female preference has yet to be determined, and therefore the level of linkage between male signal and female response is unknown; however, at least two aspects of the signal appear to be unlinked, pheromone production and hairpencil morphology.

The data presented by Haynes (this volume) on a *Trichoplusia ni* pheromone mutant represent the first direct test of the predictions of the asymmetric tracking hypothesis in female sex pheromones. Although these results were generated in the laboratory where preservation of a blend mutant might be more likely, they nonetheless demonstrate the sort of sexual asymmetry in selection envisioned by the hypothesis. In this case where a pheromone mutant arose, re-coordination of the communication system was achieved not by mutual selection but rather solely by change in male response without any change in the novel pheromone blend. Similarly, Butlin and Trickett (this volume) provide support for the hypothesis from theoretical treatment. To my knowledge, their model is the first attempt to put discussions of the evolution of sex pheromones on a more quantitative footing. In general, their simulations demonstrate the predicted asymmetric selection, the strength of whose expression depends on other factors. The model identifies two key parameters that act to stabilize pheromone blends: breadth of male response and cost to female in delayed matings. The latter is ultimately related back to parental investment, because the most significant male contribution to the mating is sperm. When the operational sex ratio (OSR) of the population is strongly skewed to males, sperm are not a valuable resource to females. This is usually the case for moths, whose males can typically mate many times; however, in an extreme case where a mutant blend approaches the limits of the male response window, this OSR bias would be diminished. It is a long-held assumption that breadth of male tuning is countered by a positive relationship between sensitivity and selectivity. Thus, males that are broadly tuned may be less competitive in finding females emitting the mean pheromone blend for the population. This possibility would certainly act to stabilize male response and (indirectly) female pheromone blends; however, there is not sufficient information as to the strength of this link and how important it might be as a selecting force. Those *T. ni* males of Haynes (this volume), whose response window had broadened to include the mutant pheromone blend, did not show a reduction in response to normal females, although it is possible that they would at longer distances from the female. More significantly, there are plentiful examples of cross-attraction in the field of males in species with polytypic pheromone systems (Baltensweiler and Priesner 1988; Klun and Huettel 1988) or among sibling species (Löfstedt et al. 1994b) to blend ratios very different from that of their own females. It is likely that a more significant force narrowing the response window of males is the avoidance of nonconspecifics with similar pheromone blends. When cross-matings between populations result in hybrids, even of reduced fitness, males that are cross-attracted are at a selective advantage (Hammerstein and Parker 1987); however, response to females with whom the male will produce no offspring represents (a) a loss of time and energy and (b) exposure to predation. Depending on the pheromone blend of the nonconspecific female, selection may be stabilizing, reducing variance in male preference, or it could be directional by selecting for a inhibitory response to novel compounds in the blend (Linn et al. 1988).

Summarizing for female pheromones, the asymmetric tracking framework allows a biologically meaningful role for both nonadaptive and adaptive evolution. In the absence of environmental or physiological constraints and competition in the pheromone channel with other species, mutant blends, within limits, are unlikely to face strong selection; recessive autosomal mutants are particularly likely to persist in a population. Under these circumstances, stochastic processes are likely to be important. Strong selection on male response will subsequently secure these changes in the pheromone signal in the population, leading to the broadening of male response, the evolution of a collection of pheromonally polytypic races, or, under some conditions, the displacement of the old pheromone blend with a new one (Butlin and Trickett this volume).

3. Male Pheromones and Sensory Drive

A mechanism for signal evolution that has gained attention recently has been variously termed *sensory drive* (Endler 1992), *sensory exploitation* (Ryan and Rand 1990), or *sensory trap* (Christy 1995). These concepts have the common feature that they describe the evolution of a signal due to the predisposition of the receiver to respond to it. I have suggested that the response of female moths to the odors of their host plants may predispose them to the evolution of male courtship pheromones (Phelan 1992). The evolution of male signals by such a mechanism is also consistent with the asymmetric-tracking hypothesis, which predicts that male signals will "track" female response. Support for this idea is suggested by male-pheromone chemistry, much of which is similar to that of plant volatiles. Thus, as Krasnoff (this volume) suggests, male pheromones may not function to signal either their species identity or "male quality." Rather these pheromones may simply mimic host odors to which females are responsive already. However, it is also true that predisposition of females to an odor in another context does not exclude the possibility of subsequent mating signal-response coevolution involving this odor (Phelan 1992). One aspect of moth mating systems that sensory exploitation fails to explain is female preference for the odor in the context of mating. Females of most moth species are sexually active during a species-specific diel period and usually engage in host finding at a different time. Presentation of a host odor does not elicit female ovipositional host finding outside that period. Likewise, females of most species must be mated before host odors will elicit orientation for oviposition. Moreover, aside from those species with sex-role reversal, where females are attracted long-range to males, the behaviors that the male pheromone elicits in courtship differ from those of oviposition. Thus, particularly in those cases where male pheromone is employed short range, after mate finding has occurred, there has been evolution of female mate-signal preference subsequent to, or concurrent with, the evolution of the male pheromone, and we return to the same question: Have male moth courtship pheromones evolved because they possess signal value of male identity or quality? More comparisons of male-pheromone blends among closely related species are needed to help address this question. Specificity of the pheromone signal should reflect the mode of selection. If these odors function to identify the male quality or species identity, differences in pheromone quality or quantity will have to reflect this difference among males. On the other hand, if male pheromones simply represent a tracking of the female sensory system, then the pheromone need be no more specific than required to trigger female response. If male odors have evolved as a sensory trap (Christy 1995) based on female host-finding, then the variability in male odor will be determined by the specificity of female response to plant hosts.

4. Phylogenetic Reconstruction and the Evolution of Pheromones

Phylogenetic reconstruction is one of the most significant recent tools to be advanced for studying evolutionary questions. A recurring theme coming from this symposium is the need to better integrate evolutionary issues of pheromone biology with phylogenetic methodology. This importance of drawing pheromone ecologists and systematists together has been emphasized previously (Phelan 1992, 1997), and it is clear from the contributions here that benefit from such collaborations can run both directions. For pheromone ecologists, a phylogenetic reconstruction of the taxa that they study can suggest an evolutionary path for traits of interest that otherwise can only be surmised. For their part, by understanding species patterns in variation in pheromone chemistry as well as the biosynthetic pathways responsible for that chemistry, systematists can gain a new and potentially independent set of characters for testing/refining their phylogenetic hypotheses (Dugdale this volume).

The value of phylogenetic methods is illustrated by recent advances in our understanding of the evolutionary sequence of the sex-pheromone system of the Lepidoptera (Löfstedt and Kozlov this volume; Phelan 1997). Thanks to the studies of pheromones in the Trichoptera and primitive Lepidoptera by Löfstedt and co-workers, as well as other researchers, a fascinating pattern of gradual evolution is beginning to emerge. Wagner and Rosovsky (1991) also effectively used phylogenetic reconstruction in the Hepialidae and other monotrysian taxa to provide evidence for the direction of evolution of pheromone communication in the primitive Lepidoptera, countering previous suggestions that male pheromones were a plesiomorphic trait. On the other hand, phylogenetic analysis of the sex-pheromone chemistry of the more advanced ditrysian Lepidoptera is less suggestive of a pattern of evolution. Foster et al. (this volume) correctly underscore the need for more comparative studies on enzymes and pathways to show how biochemical mechanisms are shared by related species. In addition to helping us better understand the genetic control of pheromone production, mapping enzyme systems on a robust phylogenetic tree of the Lepidoptera would suggest how these systems have evolved. Haynes (this volume) has clearly demonstrated in *T. ni* that even in a pheromone system under polygenic control, a single gene mutation can have profound effects on the resulting blend, changes that are attributed to the single effect on chain shortening. One of the problems in understanding the evolution of sex pheromones is that chemicals are more qualitative than quantitative characters. In acoustic mating signals, one can quantify a number of parameters that make up the call—for example, carrying frequency, chirp rate and pattern, and so on, which can be used to gauge species similarity. Such is not the case for sex pheromones. Are two aldehydes with different chain length more similar than an aldehyde and an alcohol with the same chain length? Are two compounds that differ only in the position of a double bond more similar than geometric isomers? Basing comparative studies on enzymes and pathways may allow a more quantitative analysis: Is a new blend the result of (a) the advent of a new enzyme system or (b) the loss of an old system, or has the order in which these enzymes enter the biosynthetic pathway changed? A phylogenetically dispersed database of pheromone biosynthesis, although time-consuming to accrue, would likely provide useful characters for cladists in their reconstruction efforts and would suggest specific cases of rapid evolution in pheromone communication for behavioral ecologists. These cases could provide the focus for further studies to test the relative importance of various agents of selection and/or stochastic processes.

Recently, much has been written concerning the value of phylogenetics to questions of historical ecology (Eggleton and Vane-Wright 1994), and its usefulness will probably increase as new methodologies are developed. Application of phylogenetic methods can provide evidence for (1) the historical sequence in the evolution of taxa and traits possessed by those taxa, (2) the number of independent origins and losses of a trait of interest, (3) the polarity in the transition sequence of a trait, and (4) the historical relationship between two traits or a trait and an ecological variable. In this symposium, we can distinguish the use of phylogenetics by Löfstedt and Kozlov and by Dugdale, which are primarily directed toward reconstructing the historical sequence of character evolution, from that of Krasnoff, whose application is meant more to test a hypothesis of adaptation. While the use of phylogenetics to reconstruct a historical record is largely unchallenged, significant debate is generated by some applications to test for adaptation [see Eggleton and Vane-Wright (1994) for a sampling of some of the varied views].

Krasnoff (this volume) is highly critical of Phelan and Baker's (1987) survey of male pheromones in the Lepidoptera, largely for its nonphylogenetic approach. Unlike the pheromone glands of female Lepidoptera, male pheromones are emitted by an extraordinarily diverse array of structures, examples of which can be found on almost any area of the body and which apparently have evolved independently many times throughout the order. Phelan and Baker (1987) provide the most species-inclusive test of a hypothesized function for male lepidopteran pheromones, that of male-species recognition by the female. This study measures the relative co-occurrence of androconia and sharing of host plants among congeners using >800 species of Lepidoptera. From my perspective, the study derives significance from its breadth, because trait associations were examined across five distantly related moth families. Its weakness stems from its dependence on an imperfect database to estimate the presence of species-specific pheromones and the potential for interspecific mating mistakes. Therefore, it is important that we consider the assumptions carefully to determine the reliability of the conclusions. First, since relatively few male pheromones have been characterized, we chose the presence of androconia (which have been well-described by taxonomists) as our character, with the only assumption being that species-specific pheromones are more likely to occur in species with androconia. This study did not assume that each occurrence of a particular androconium (abdominal hairpencils of the Phycitinae, for example) represented an independent evolution of that structure, as Krasnoff seems to suggest. Second, again working with the best available information, we chose the sharing of a host plant as an indicator of interspecific contact potentially leading to mating mistakes. Each of these measures has some error; however, the direction of error on the resulting analysis is predictable. As explained by Phelan and Baker (1987), if a relationship existed between male pheromones and interspecific contact, any shortcomings in the chosen parameters to measure this association would only lead to an underestimation of this relationship, just as adding noise to a communication channel can only weaken the signal. Likewise, in the absence of a relationship, noise added to noise is unlikely to produce a signal. Thus, ironically, the higher the error rate, the stronger would have to be the actual relationships to be detectable by our methods. Krasnoff's argument that this effect would not hold for inaccuracies due to nonfunctional androconia is in error, since all cells contribute to a χ^2 value.

Aside from this, Krasnoff raises three primary objections to this study. The first is his requirement that the male-pheromone system of each species involved be characterized. While more information is always better, presently there are no taxa that meet this criterion, as his own review of the literature aptly demonstrates. Thus, holding to this requirement would require waiting for such detailed studies to be conducted before we could address the functional evolution of male lepidopteran pheromones. Even so, without logical extrapolation, the conclusions of such a study would be limited to the groups in the study. Moreover, I reiterate the point that the degree to which androconia fail to properly index male pheromones reduces the probability of measuring the relationship. Secondly, Krasnoff argues that such analyses be limited to those pairs of species where we have direct knowledge of divergence between them. However, such studies could only be conducted on groups where the potential for cross-mating can be demonstrated under natural conditions. For this requirement, phylogenetic trees are of limited benefit since they provide only relative not absolute information. A phylogenetic reconstruction may suggest that species A and B are more likely to have separated than A and C, but A in fact may have diverged from species D, which is now extinct. On a related point that Krasnoff does not raise, I have modified my interpretation of the results of Phelan and Baker (1987). In light of the subsequent refinement made by Butlin (1987), limiting the definition of reproductive isolation to cases where postzygotic mechanisms are not in place, our data are still consistent with an adaptive function of reducing mating mistakes, but do not address whether these pressures were primarily between incipient or full species.

The third, and what Krasnoff considers the most damning, shortcoming of the survey is the use of nonindependent taxa in measuring the correspondence between androconia and interspecific contact. This problem of nonindependence for some comparative studies was discussed by Felsenstein (1985), who argued that correlations between traits or between a trait and environmental parameters may simple be due to common ancestry, not adaptation. Accordingly, for Phelan and Baker (1987), even if a significant association between androconia and host overlap is measured in a family of moths, it may be due to the incidental sharing of multiple traits or ecological requirements among related species. Although Felsenstein's logic is correct, common ancestry cannot explain the findings of Phelan and Baker (1987) for two reasons. First, although androconia are genetic-based characters and thus presence in one species is not independent of presence in related species, host–plant overlap is not a heritable character. In this study, host sharing is defined only in relation to congeneric species. While one might argue that host preference/utilization may be heritable and thus due to common ancestry, a species may be more likely to share a host with close relatives, the reverse is not true; that is, members of a clade cannot derive from their ancestor a trait called "lack of host sharing." If all members of a genus possess an androconial structure due to common ancestry and for the same reason all feed on apples, for example, one could erroneously conclude a relationship between the two. However, if this were the case in the taxa studied by Phelan and Baker (1987), host sharing should have been just as common in genera that did not have androconia, but our data showed that it was not. Thus, in the 2 × 2 contingency tables of androconia and host overlap, common ancestry could explain a large +/+ cell, but it cannot explain why the –/– cell would also be large.

The second reason Felsenstein's argument does not negate the conclusions of Phelan and Baker (1987) is the analysis was applied to five families of moths independently. By anyone's definition, the five families represent independent tests of the hypothesis since the androconia used to estimate the presence of unique male pheromones are not homologous. Because of the problems in using the χ^2 statistic in nonindependent sets, let us simplify the analysis. The hypothesis predicts that the frequency of androconia should be greater among host-sharing congeners than among non-host-sharing ones. For each of the five families, this prediction held [see Phelan and Baker (1987) or Phelan (1992) for actual numbers]. Taking a very conservative approach (n = 5, not considering the number of species in each family and not taking into account the magnitude of the differences), we can use a sign test to calculate the probability of getting this result by chance as $p = (0.5)^5$, or 0.03. The main criticism leveled against comparative studies such as that of Phelan and Baker (1987) is that basing the analysis on taxa with common ancestry can lead to inflated *df*'s. However, in this study, we can demonstrate that even with the most conservative analysis (reducing each family to $n = 1$), the argument that the association between moth androconia and host sharing could occur due to common descent can be rejected.

Krasnoff suggests an alternative approach for deriving the function of male lepidopteran pheromones, based on what I consider a rather extreme application of phylogenetic reconstruction. From this perspective, he then concludes that lepidopteran male scent structures, which most researchers see as possessing great diversity, actually "look more like variation on a surprisingly small number of themes." To make his argument, he picks one such structure, the pleated sacculus of the Notodontidae, and maps it on a phylogenetic tree constructed by Miller (1991). First, it is interesting that the conclusions that Krasnoff draws from this phylogenetic analysis appear so different from Miller's own characterization. Restricting the analysis to just this structure obscures the fact that a variety of nonhomologous androconia also exist in this family, occurring on the second, third, fourth, fifth, and eighth segments (Miller 1991). Likewise, Krasnoff's treatment of the pleated sacculus in a binary fashion—that is, presence or absence—overlooks the morphological diversity within this structure. Miller, on the other hand, classifies this trait using four character states, a system that because of the morphological diversity of the structure, he considers "was only marginally successful." In fact, he concludes that "Based on my results and those of Weller, [the pleated sacculus] exhibits a large amount of homoplasy." Plotting this trait as a 4-state character on the tree derived by Miller, there have been 20 evolutionary events among 50 genera, four leading to greater complexity and 16 to reduced complexity, giving a CI (consistency index) of 0.15, supporting a conclusion of high homoplasy (Sanderson and Donoghue 1989). With this information, it is hard to understand how Krasnoff finds support in the Notodontidae for his characterization of androconia.

Krasnoff finds significance in his phylogenetic interpretation that the pleated sacculus of the Notodontidae and the coremata of the Japanese Yponomeutidae each had a common origin in those groups; however, this is without dispute because no one would suggest that a structure as complex as the pleated sacculus of many Notodontidae or the coremata of the Yponomeutidae evolved independently numerous times within those families. Krasnoff seeks to trace the phylogenetic origin of each androconial system so as to focus our studies of adaptation in those taxa. This stricture does avoid the problem of confounding due to common descent; however, it comes at the great price of disposing of a lot of other information available to us. By way of analogy, there is no debate on the functional significance of the distal abdominal gland of female ditrysian moths, but we do not know in what group this gland originated. Should we set aside our notions of its adaptive function until we have had a chance to find that hypothesized monotrysian taxon in which its first arose? If not, why then should we restrict our studies of moth androconia and male pheromones to just their ancestral originators?

Moreover, how much confidence do we have that we can generally reconstruct the historical context in which androconia first arose? The primary assumption of this method is that traits and environmental conditions observed in contemporary time reflect those found at the time that the trait or character state appeared. In addition to the problem of missing data (extinct species, etc.) discussed above, there is also the potential problem of decay in "phylogenetic memory"; that is, "mapping [of traits of interest] onto a cladogram may bear little relation to the actual sequence of character state transitions between ancestral and derived taxa (Frumhoff and Reeve 1994)." The older the clade, the higher the probability that within-taxa state changes of the trait and species extinctions have occurred. Ironically, Krasnoff raises the issues of anagenetic change and current versus original function as a problem for studies of convergence, such as Phelan and Baker (1987). In actuality, the convergence approach is probably better suited to dealing with these issues than the form of analysis that Krasnoff espouses (Pagel 1994). Restricting one's analysis to single evolutionary events (e.g., point of origin for an androconial system) is more likely to lead to a conclusion based on an erroneous association than a convergence approach based on repeated instances of an association. In constructing a phylogenetic tree, cladists counter these historical inaccuracies by analyzing a large number of traits. One would have very little confidence in a tree constructed on a single trait, but these same uncertainties remain for a single trait mapped on a robust phylogenetic tree.

There are various sources of information to help us understand the evolutionary processes underlying the diversity of pheromone systems that we see, including phylogenetic reconstruction and patterns of repeated convergence; we need to recognize the strengths and weaknesses of each and use them complementarily. The most effective approach to understanding adaptation will be one that combines the robustness of the convergence approach, using large numbers of taxa, with the historical context provided by phylogenetic analysis and the precision of detailed natural-history and empirical studies of extant populations. A phylogenetic reconstruction can suggest which taxa should be included in the convergence study to avoid interdependence of the data set. In addition, phylogenetic evidence that a trait of interest evolved prior to a hypothesized selective factor can be used to argue against a given adaptive hypothesis. Finally, descriptive field studies and empirical work bring a third perspective and may reveal the mechanisms underlying the hypothesized association.

5. Conclusions and Future Directions

Although there is still much work to do, the recent studies presented at this symposium as well as others on the genetics and population variation of sex pheromones have indeed advanced our understanding of their evolution greatly. In order to understand the relative roles of different selective forces in the evolution of this communication system, I would hope that the asymmetric-tracking hypothesis could serve as a conceptual framework. The specific predictions of this hypothesis can be tested by future studies to determine its usefulness, particularly for suggesting the contextual dependence of different hypothesized processes. However, Butlin and Trickett (this volume) are correct in suggesting the need to put such conceptual models on a more quantitative foundation. As I have argued previously (Phelan 1992, 1997) that pheromone biologists have much to gain by incorporating phylogenetic information into their studies of pheromone evolution. Likewise, evolutionary biologists should make better use of the large database of information concerning the chemistry, genetics, and behavioral response of insect sex pheromones. This well-studied system provides an important opportunity for comparison with other mate-signaling systems.

Looking to the future, the papers presented here point out a number of areas that are deserving of attention.

1. Restating the above argument, putting comparative studies within a more robust phylogenetic context will greatly advance our understanding of the evolutionary processes underlying sex pheromones, as well as their evolutionary heritage. Fortunately, systematists are increasingly attending to some of the major insect-pest families, which have already been and will continue to be the focus of many pheromone biologists. We should continue to press for such collaborations, which will be mutually beneficial.
2. Mapping of pheromonal systems on phylogenetic tree of the Lepidoptera suggests certain answers and raises new questions. The female-based pheromone systems appears to have arisen early in the evolution of

the Lepidoptera, possibly from a gland shared by the Trichoptera and primitive Lepidoptera, a gland that may have originally or concurrently served a defensive role. Remaining to be resolved is how the well-known and evolutionarily conservative 8th/9th abdominal gland of the Ditrysia evolved from this primitive system. More cladistic analyses are needed to establish better the sister group of the Ditrysia. Moreover, almost nothing is known about the mating systems of the taxa that represent candidates as sister groups to the Ditrysia.

3. We have only a very few studies of the genetic control of pheromone-based signal-response systems, and none of these systems is completely described. Particularly wanting is our knowledge of the genetics of male courtship pheromones in the Lepidoptera. The recent investigations of female-produced moth pheromones measuring population variation in pheromone production and response, as well as their genetic architecture, have been extremely informative. They have understandably focused on ratios of paired geometric or positional isomers. The picture will undoubtedly become more complex as investigations broaden to identify genetic determination of additional components. Additionally, expanding these studies to include genetic control of the enzymes and biosynthetic pathways responsible for pheromone end-products will be immensely instructive, particularly when conducted in concert with cladistic analyses of the taxa involved. Given that a handful of enzymes are probably responsible for most of the chemical diversity in lepidopteran pheromones, charting the pathways involved should allow us to characterize quantitatively the diversity that we now only can describe in qualitative terms. Also I would hope to see a continuation of effort to use quantitative models in our analysis of the evolution of sex pheromones. Building on the work of Butlin and Trickett and models developed for other systems will help focus theoretical discussions.

4. Many of the questions require better descriptions of the pheromone chemistry and response behavior of populations undergoing divergence. The increasing emphasis on measuring degree and sources of intraspecific variation in pheromone production and response has changed the picture of this mate-signaling system as a spatially invariant one under strong mutually stabilizing selection. With more examples, we shall better see whether level of variability in pheromone blend and response is associated with certain physiological and/or environmental parameters. Particularly instructive will be additional studies of field populations of pheromone-communicating species that either are undergoing fission due to differential host utilization or some other adaptation or that interact with sibling species in hybrid zones. By doing so, we will better understand the importance of hybrid disadvantage or interspecific mating

mistakes in driving the evolution of new pheromone systems with greater species-specificity.

5. Finally, the contrasts between male- and female-based pheromone systems in the Lepidoptera provides many opportunities to test models of sexual-signal evolution. Some of the questions that we need to address in male systems are: (1) why did so many different androconia and the chemicals they emit evolve independently across the Lepidoptera, (2) what function did/do these structures and their odors serve, and (3) within the taxonomic distribution of homologous androconia, why were some maintained and others independently lost, often amongst such closely related taxa?

Acknowledgements

I wish to express my grateful appreciation to Dave Wagner and John Wenzel for their helpful discussions and review of the manuscript.

References

Abernathy R.L., R.J. Nachman, P.E.A. Teal, O. Yanashita and J.H. Tumlinson. 1995. Pheromonotropic activity of naturally occurring pyrokinin insect neuropeptides (FXP-RLamide) in *Helicoverpa zea*. *Peptides* **16**:215–219.

Ache, B.W. 1991. Phylogeny of smell and taste. In *Smell and Taste in Health and Disease*, ed. T.V. Getchell, pp. 3–18. Raven Press, New York.

Adams, R.G., K.D. Murray and L.M. Los. 1989. Effectiveness and selectivity of sex pheromone lures and traps for monitoring fall armyworm (Lepidoptera: Noctuidae) adults in Connecticut sweet corn. *Journal of Economic Entomology* **82**: 285–290.

Akers, R.P. and R.J. O'Connell. 1988. The contribution of olfactory receptor neurons to the perception of pheromone component ratios in male redbanded leafroller moths. *Journal of Comparative Physiology A* **163**:641–650

Allen, L.H. 1968. Turbulence and wind-speed spectra within a Japanese larch plantation. *Journal of Applied Meteorology* **7**: 73–78. Almaas, T.J. and H. Mustaparta. 1990. Pheromone reception in tobacco budworm moth, *Heliothis virescens*. *Journal of Chemical Ecology* **16**:1331–1347.

Almaas, T.J. and Mustaparta, H. 1990. Pheromone reception in tobacco budworm moth, *Heliothis virescens*. *Journal of Chemical Ecology* **16**:1331–1347.

Almaas, T.J. and H. Mustaparta, 1991. *Heliothis virescens*: response characteristics of receptor neurons in *sensilla trichodea* type 1 and type 2. *Journal of Chemical Ecology* **17**:953–972.

Almaas, T.J., T.A. Christensen and H. Mustaparta. 1991. Chemical communication in heliothine moths. I. Antennal receptor neurons encode several features of intra- and interspecific odorants in the male corn earworm moth *Helicoverpa zea*. *Journal of Comparative Physiology A* **169**:249–258.

Almaas, T.J., A.K. Amundgaard, B.G. Berg and H. Mustaparta. 1995. Functional subdivision of the macroglomerular complex in different heliothine moth species. In *Nervous Systems and Behaviour*, Proceedings International Congress of Neuroethology, p. 397. Thieme Medical Publishers, New York.

Altstein, M., M. Harel and E. Dunkelblum. 1989. Effect of a neuroendocrine factor on sex pheromone biosynthesis in the tomato looper, *Chrysodeixis chalcites* (Lepidoptera: Noctuidae). *Insect Biochemistry* **19**:645–649.

Altstein, M., Y. Gazit and E. Dunkelblum. 1990. Regulation of sex pheromone biosynthesis in *Chrysodeixis chalcites* and *Heliothis peltigera* by PBAN. In *Insect Neurochemistry and Neurophysiology*, eds. A.B. Borkovec and E. Masler, pp. 303–307. Humana Press, Clifton, New Jersey.

Altstein, M., Y. Gazit and E. Dunkelblum. 1993. Neuroendocrine control of sex pheromone biosynthesis in *Heliothis peltigera. Archives of Insect Biochemistry and Physiology* **22**:153–168.

Altstein, M., O. Benaziz and Y. Gazit. 1994. Pheromone biosynthesis activating neuropeptide (PBAN) and colour polymorphism—an immunochemical study in *Spodoptera littoralis. Journal of Insect Physiology* **40**:303–309.

Altstein, M., Y. Gazit, O. Ben Aziz, T. Gabay, R. Marcus, Z. Vogel and J. Barg. 1996. Induction of cuticular melanization in *Spodoptera littoralis* larvae by PBAN/MRCH: development of a quantitative bioassay and structure function analysis. *Archives of Insect Biochemistry and Physiology* **31**:355–370.

Alvarado-Rodriguez, B. and E. Rivera-Rubio. 1990. Tomato pinworm (Lepidoptera: Gelechiidae) and increasing pest on tomatoes in Sinaloa, Mexico. *Florida Entomologist* **73**:677–680.

Amman, G.D. 1994. Potential of verbenone for reducing lodgepole and ponderosa pine mortality caused by mountain pine beetle in high-value situations. In *Proceedings of the Symposium on Management of Western Bark Beetles with Pheromones—Recent Research and Development*, ed. P.J. Shea, pp. 33–37. United States Department of Agriculture, Forest Service, General Technical Report. PSW-GTR-150.

Amman, G.D., R.W. Thier, M.D. McGregor and R.F. Schmitz. 1989. Efficacy of verbenone in reducing lodgepole pine infestation by mountain pine beetles in Idaho. *Canadian Journal of Forest Research* **19**:60–64.

Amman, G.D., R.W. Thier, J.C. Weatherby, L.A. Rasmussen and A.S. Munson. 1991. Optimum dosage of verbenone to reduce infestation of mountain pine beetle in lodgepole pine stands of central Idaho. United States Department of Agriculture, Forest Service, Research Paper INT-446.

Anderson, P., B.S. Hansson and J. Löfqvist. 1995. Host odour specific receptor neurones on the antenna of female and male *Spodoptera littoralis. Physiological Entomology* **21**:189–198.

Ando, T., R. Arima, M. Uchiyama, H. Nagasawa, T. Inoue and A. Suzuki. 1988. Pheromone biosynthesis activating neuropeptide hormone in heads of the silkworm moth. *Agricultural and Biological Chemistry* **52**:881–883.

Ansteeg, O. and K. Dettner. 1991. Chemistry and possible biological significance of secretions from a gland discharging at the 5th abdominal sternit of adult caddisflies (Trichoptera). *Entomologia Generalis* **15**:303–312.

Anton, S. and B.S. Hansson. 1994. Central processing of sex pheromone, host odour, and oviposition deterrent information by interneurons in the antennal lobe of female *Spodoptera littoralis* (Lepidoptera: Noctuidae). *Journal of Comparative Neurology* **350**:199–214.

Anton, S. and B.S. Hansson. 1995. Sex pheromone and plant-associated odour processing in antennal lobe interneurons of male *Spodoptera littoralis* (Lepidoptera: Noctuidae). *Journal of Comparative Physiology A* **176**:773–789.

Arbas, E.A. and M.A. Willis. 1994. Pheromone-modulated flight behavior of the sphinx moth, *Manduca sexta*. In *Olfaction and Taste XI*, eds. K. Kurihara, N. Suzuki and H. Ogawa, pp. 835–837. Springer-Verlag, Tokyo.

Arbas, E.A., R. Kanzaki and J.G. Hildebrand. 1990. State-dependent activity changes in descending neurons of olfactory pathways in the moth *Manduca sexta*. *Society for Neuroscience Abstracts* **16**:758.

Arbas, E.A., M.A. Willis and R. Kanzaki. 1993. Organization of goal-oriented locomotion: pheromone-modulated flight behavior of moths. In *Biological Neural Networks in Invertebrate Neuroethology and Robotics*, eds. R.D. Beer, R.E. Ritzmann and T. McKenna, pp. 159–198, Academic Press, San Diego.

Arias, A., M. Bueno, J. Nieto, M. Valenzuela, A. Perez, B. Cuenda, F. Gallego, A. Alamada and M.A. Castillo. 1992. Essais de confusion sexuelle de *Lobesia botrana* Den. et Schiff., pendant 1989 et 1990 dans "Tierra de Barros" (Espagne). *International Organization for Biological Control—West Palearctic Regional Section Bulletin* **15**(2):18.

Arima, R., K. Takahara, T. Kadoshima, F. Numazaki, T. Ando, M. Uchiyama, H. Nagasawa, A. Kitamura and A. Suzuki. 1991. Hormonal regulation of pheromone biosynthesis in the silkworm moth, *Bombyx mori* (Lepidoptera: Bombycidae). *Applied Entomology and Zoology* **26**:137–147.

Armstrong, K.F. and D.M. Suckling. 1988. Investigations into the biochemical basis of azinphosmethyl resistance in the lightbrown apple moth *Epiphyas postvittana* (Lepidoptera: Tortricidae). *Pesticide Biochemistry and Physiology* **32**:62–73.

Arn, H. 1979. Developing insect control by disruption of sex pheromone communication: conclusions from programs on lepidopterous pests in Switzerland. In *Chemical Ecology: Odour Communication in Animals*, ed. F.L. Ritter, pp. 365–374. Elsevier, Amsterdam.

Arn, H. 1990. Pheromones: prophecies, economics, and the ground swell. In *Behavior-Modifying Chemicals for Insect Management. Applications of Pheromones and Other Attractants*, eds. R.L. Ridgway, R.M. Silverstein and M.N. Inscoe, pp. 717–722. Marcel Dekker, New York.

Arn, H. and S. Rauscher. 1987. Disruptant tests for *Eupoecilia ambiguella* and *Lobesia botrana*. *International Organization for Biological Control—West Palearctic Regional Section Bulletin* **10**(3):53–54.

Arn, H., S. Rauscher, P.M. Guerin and H.-R Buser. 1976. Sex pheromone of *Eupoecilia ambiguella: cis*-9-dococenyl acetate as a major component. *Zeitschrift für Naturforschung* **31c**:499–503.

Arn, H., S. Rauscher, A. Schmid, C. Jaccard and B. Bierl-Leonhardt. 1981. Field experiments to develop control of the grape moth, *Eupoecilia ambiguella*, by communication disruption. In *Management of Insect Pests with Semiochemicals: Concepts and Practice*, ed. E.R. Mitchell, pp. 327–338. Plenum Press, New York.

Arn, H., P. Esbjerg, R. Bues, M. Tóth, G. Szöcs, P. Guerin, and S. Rauscher. 1983. Field attraction of *Agrotis segetum* males in four European countries to mixtures containing three homologous acetates. *Journal of Chemical Ecology* **9**:267–276.

Arn, H., P.M. Guerin, H.R. Buser, S. Rauscher and E. Mani. 1985. Sex pheromone blend of the codling moth, *Cydia pomonella*: evidence for a behavioural role of dodecan-1-ol. *Experientia* **41**:1482–1484.

Arn, H., S. Rauscher, H.-R. Buser and P.M. Guerin. 1986. Sex pheromone of *Eupoecilia ambiguella* female: analysis and male response to ternary blend. *Journal of Chemical Ecology* **12**:1417–1429.

Arn, H., S. Rauscher, P.M. Guerin and H.-R. Buser. 1988. Sex pheromone blends of three tortricid pests in European vineyards. *Agriculture, Ecosystems and Environment.* **21**:111–117.

Arn, H., M. Tóth and E. Priesner. 1992. *List of Sex Pheromones of Lepidoptera and Related Attractants*, 2nd edition. International Organization for Biological Control, Montfavet, France.

Arn, H., M. Tóth and E. Priesner. 1995. *The Pherolist*. Internet Edition. http://www.nysaes.cornell.edu/pheronet

Astic, L., D. Saucier and A. Holley. 1987. Topographical relationships between receptor cells and glomerular foci in the rat olfactory bulb. *Brain Research* **424**:144–152.

Atkins, E.L., M.H. Frost, L.D. Anderson and A.S. Deal. 1957. The "omnivorous leaf roller" *Platynota stultana* Wlshm., on cotton in California: nomenclature, life history, and bionomics (Lepidoptera: Tortricidae). *Annals of the Entomological Society of America* **50**:251–259.

Atwood, M. 1990. *Cat's Eye*. Virago, London.

Audemard, H., E. Armand and C. Leblon. 1992. Essais de lutte par confusion sexuelle contre les lépidoptères nuisibles aux vergers de pommiers dans la basse vallée du Rhone. *International Organization for Biological Control—West Palearctic Regional Section Bulletin* **15**(5):94–97.

Augustynowicz, M., E. Malinski, Z. Warnke, J. Szafranek and J. Nawrot. 1987. Cuticular hydrocarbons of the German cockroach, *Blattella germanica* L. *Comparative Biochemistry Physiology* **86B**:519–523.

Bäädeı A., M. Schafer, C.H.F. Rowell. 1992. The perception of the visual flow field by flying locusts: a behavioural and neuronal analysis. *Journal of Experimental Biology* **165**:137–160.

Babson, A.L. 1963. Eradicating the gypsy moth. *Science* **142**:447–448.

Baker, K., M.A. Willis and E.A. Arbas. 1994. Spontaneous self-generated counterturns during tethered flight of moths, *Manduca sexta*. *Society for Neurosciences Abstracts* **20**:1025.

Baker, T.C. 1985. Chemical control of behavior. In *Behaviour*, Vol. 9, *Comprehensive Insect Physiology, Biochemistry and Pharmacology*, eds. G.A. Kerkut and L.I. Gilbert, pp. 621–672, Pergamon Press, Oxford.

Baker, T.C. 1986. Pheromone-modulated movements of flying moths. In *Mechanisms in Insect Olfaction*, eds. T.L. Payne, C.E.J. Kennedy and M.C. Birch, pp. 39–48, Clarendon Press, Oxford.

Baker, T.C. 1989a. Origin of courtship and sex pheromones of the oriental fruit moth and a discussion of the role of phytochemicals in the evolution of lepidopteran male scents. In *Phytochemical Ecology: Allelochemicals, Mycotoxins, and Insect Pheromones and Allomones*, ed. C.H. Chou and G.R. Waller, pp. 401–418. Monograph Series No. 9. Institute of Botany, Academia Sinica, Taipei.

Baker, T.C. 1989b. Pheromones and flight behavior. In *Insect Flight*, eds. G.T. Goldsworthy and C.H. Wheeler, pp. 231–255. CRC Press, Boca Raton, Florida.

Baker, T.C. 1989c. Sex pheromone communication in the Lepidoptera: new research progress. *Experientia* **45**:248–262.

Baker, T.C. 1990. Upwind flight and casting flight: complementary phasic and tonic systems used for location of sex pheromone sources by male moths. In *Proceedings of the X International Symposium on Olfaction and Taste*, ed. K.B. Døving, pp. 18–25. Graphic Communication Systems, Oslo.

Baker, T.C. and R.T. Cardé. 1979a. Courtship behavior of the oriental fruit moth (*Grapholitha molesta*): experimental analysis and consideration of the role of sexual selection in the evolution of courtship pheromones in the Lepidoptera. *Annals Entomological Society of America* **72**:173–188.

Baker, T.C. and R.T. Cardé. 1979b. Endogenous and exogenous factors affecting periodicities of female calling and male sex pheromone response in *Grapholitha molesta*. *Journal of Insect Physiology* **25**:943–950.

Baker, T.C. and K.F Haynes. 1987. Manoeuvres used by flying male oriental fruit moths to relocate a sex pheromone plume in an experimentally shifted wind-field. *Physiological Entomology* **12**:263–279.

Baker, T.C. and K.F. Haynes. 1989. Field and laboratory electroantennographic measurements of pheromone plume structure correlated with oriental fruit moth behaviour. *Physiological Entomology* **14**:1–12.

Baker, T.C. and L.P.S. Kuenen. 1982. Pheromone source location by flying moths: a supplementary nonanemotactic mechanism. *Science* **216**:424–427.

Baker, T.C. and N.J. Vickers. 1994. Behavioral reaction times of male moths to pheromone filaments and visual stimuli: determinants of flight track shape and direction. In *Olfaction and Taste XI*, ed. K. Kurihara, N. Suzuki and H. Ogawa, pp. 838–841. Springer-Verlag, Tokyo.

Baker, T.C. and R.G. Vogt. 1988. Measured behavioural latency in response to sex-pheromone loss in the large silkmoth *Antheraea polyphemus*. *Journal of Experimental Biology* **137**:29–38.

Baker. T.C., W. Meyer, and W.L. Roelofs. 1981. Sex pheromone dosage and blend specificity of response by oriental fruit moth males. *Entomologia experimentalis et applicata* **30**:269–276.

Baker, T.C., M.A. Willis and P.L. Phelan. 1984. Optomotor anemotaxis polarizes self-steered zigzagging in flying moths. *Physiological Entomology* **9**:365–376.

Baker, T.C., M.A. Willis, K.F. Haynes and P.L. Phelan. 1985. A pulsed cloud of sex pheromone elicits upwind flight in male moths. *Physiological Entomology* **10**:257–265.

Baker, T.C., B.S. Hansson, C. Löfstedt and J. Löfqvist. 1989. Adaptation of antennal neurons in moths is associated with cessation of pheromone-mediated upwind flight. *Proceedings of the National Academy of Sciences U.S.A.* **85**:9826–9830.

Baker, T.C., R.T. Staten and H.M. Flint. 1990. Use of pink bollworm pheromone in the southwestern United States. In *Behavior Modifying Chemicals for Insect Management*, eds. R.L. Ridgway, R.M. Silverstein and M.N. Inscoe, pp. 417–436. Marcel Dekker, New York.

Baker, T.C., W. Francke, J.G. Millar, C. Löfstedt, B. Hansson, J.-W. Du, P.L. Phelan, R.S. Vetter, R. Youngman and J.L. Todd. 1991. Identification and bioassay of sex pheromone components of carob moth, *Ectomyelois ceratoniae* (Zeller). *Journal of Chemical Ecology* **17**:1973–1987.

Bakke, A. 1981. Inhibition of the response in *Ips typographus* to the aggregation pheromone; field evaluation of verbenone and ipsenol. *Zeitschrift für angewandte Entomologie* **92**:172–177.

Bakke, A. 1987. Repression of *Ips typographus* infestations in stored logs by semiochemicals. *Scandinavian Journal of Forest Research* **2**:179–185.

Bakke, A. 1989. The recent *Ips typographus* outbreak in Norway—experiences from a control program. *Holarctic Ecology* **12**:515–519.

Baltensweiler, W. and E. Priesner. 1988. Studien zum Pheromon-Polymorphismus von *Zeiraphera diniana* Gn. (Lep., Tortricidae). 3. Anflugspezifität männlicher Falter zweier Wirtsrassen an synthetische Pheromonquellen. *Journal of Applied Entomology* **106**:217–231.

Bankes, E.R. 1890. *Opostega salaciella. Entomologist's Monthly Magazine* **26**:8.

Barlow, R. 1979. Circadian rhythms in the *Limulus* visual system. *Investigative Ophthalmology and Visual Science (Supplement)*, 108–109.

Barlow, R., S. Bolanowski, and M. Brachman. 1977. Efferent optic fibers mediate circadian rhythms in the *Limulus* eye. *Science* **197**:86–89.

Barnes, M.M., J.C. Millar, P.A. Kirsch and D.C. Hawks. 1992. Codling moth (Lepidoptera: Tortricidae) control by dissemination of synthetic female sex pheromone. *Journal of Economic Entomology* **85**:1274–1277.

Barry, M.W. and D.G. Nielsen. 1984. Behavior of adult peach tree borer. *Synanthedon exitiosa* (Lepidoptera: Sessiidae). *Annals Entomological Society of America* **77**:246–250.

Bartell, R.J. 1982. Mechanisms of communication disruption by pheromone in the control of Lepidoptera: a review. *Physiological Entomology* **7**:353–364.

Bartell, R.J. and L.A. Lawrence. 1973. Reduction in responsiveness of males of *Epiphyas postvittana* (Lepidoptera) to sex pheromone following brief pheromonal exposure. *Journal of Insect Physiology* **19**:845–855.

Barth, R.H., Jr. 1965. Insect mating behavior: endocrine control of a chemical communication system. *Science* **149**:882–883.

Barth, R.H., Jr. and L.J. Lester. 1973. Neuro-hormonal control of sexual behavior in insects. *Annual Review of Entomology* **18**:445–472.

Basolo, A.L. 1990. Female preference predates the evolution of the sword in swordtail fish. *Science* **250**:808–810.

Batllori, J.L., G.J. Brueren, V. Sarto i Monteys, J.J. de Vlieger and J.W. Klijnstra. 1992. Codling moth mating disruption with a new monolithic controlled release device. *International Organization for Biological Control—West Palearctic Regional Section Bulletin* **15**(5):34.

Bedard, W.D., P.E. Tilden, K.Q. Lindahl, Jr., D.L. Wood and P.A. Rauch. 1980. Effects of verbenone and *trans*-verbenol on the response of *Dendroctonus brevicomis* to natural and synthetic attractant in the field. *Journal of Chemical Ecology* **6**:997–1013.

Beehler, J.W., J.G. Millar and M.S. Mulla. 1994. Field evaluation of synthetic compounds mediating oviposition in *Culex* mosquitoes (Diptera: Culicidae). *Journal of Chemical Ecology* **20**:281–291.

Belanger, J.H. and E.A. Arbas. 1995. Behavioral strategies underlying pheromone-modulated flight in the moth Manduca sexta: lessons from simulation studies. *Society for Neuroscience Abstracts* **25**:187.5

Bellas, T.E., R.J. Bartell and A. Hill. 1983. Identification of the sex pheromone of *Epiphyas postvittana* (Lepidoptera: Tortricidae). *Journal of Chemical Ecology* **9**:503–512.

Bellés, X., J. Casas, A. Messeguer and M.D. Piulachs. 1987. *In vitro* biosynthesis of JH III by the corpora allata of adult females of *Blattella germanica* (L.). *Insect Biochemistry* **17**:1007–1010.

Bengtsson, B.O. and C. Löfstedt. 1990. No evidence for selection in a pheromonally polymorphic moth population. *American Naturalist* **136**:722–726.

Bengtsson, M., G. Karg, P.A. Kirsch, J. Löfqvist, A. Sauer and P. Witzgall. 1994. Mating disruption of pea moth *Cydia nigricana* F. (Lepidoptera: Tortricidae) by a repellent blend of sex pheromone and attraction inhibitors. *Journal of Chemical Ecology* **20**:871–887.

Benn, M., J. Degrave, C. Gnanasunderam and R. Hutchins. 1979. Host–plant pyrrolizidine alkaloids in *Nyctemera annulata* Boisduval: their persistence through the life cycle and transfer to a parasite. *Experientia* **35**:731–732.

Bentz, B., C.K. Lister, J.M. Schmid, S.A. Mata, L.A. Rasmussen and D. Haneman. 1989. Does verbenone reduce mountain pine beetle attacks in susceptible stands of ponderosa pine? United States Department of Agriculture, Forest Service, Research Note RM-495.

Berg, B.G. and H. Mustaparta. 1995. The significance of major pheromone components and interspecific signals as expressed by receptor neurons in the oriental tobacco budworm moth, *Helicoverpa assulta. Journal of Comparative Physiology A* **177**:683–694.

Berg., B.G., J.H. Tumlinson and H. Mustaparta. 1995. Chemical communication in heliothine moths. IV. Receptor neuron responses to pheromone compounds and formate analogues in the tobacco budworm moth *Heliothis virescens. Journal of Comparative Physiology A* **177**:527–534.

Berger, R.S. 1966. Isolation, identification, and synthesis of the sex attractant of the cabbage looper, *Trichoplusia ni. Annals of the Entomological Society of America* **59**:767–771.

Beroza, M., B.A. Bierl and H.R. Moffitt. 1974. Sex pheromones: (*E,E*)-8,10-dodecadien-1-ol in the codling moth. *Science* **183**:89–90.

Berryman, A.A. 1982. Population dynamics of bark beetles. In *Bark Beetles in North American Conifers*, eds. J.B. Mitton and K.B. Sturgeon, pp. 264–314. University of Texas Press, Austin, Texas.

Berson, D.R. 1992. Pheromones help save Torrey pines. United States Department of Agriculture. Forest Service, Forest Research West (September 1992), pp. 17–20.

Bertram, S.L. and T.D. Paine. 1994a. Response of *Dendroctonus brevicomis* LeConte (Coleoptera: Scolytidae) to different release rates and ratios of aggregation semiochemicals and the inhibitors verbenone and ipsdienol. *Journal of Chemical Ecology* **20**:1617–1629.

Bertram, S.L. and T.D. Paine. 1994b. Influence of aggregation inhibitors (verbenone and ipsdienol) on landing and attack behavior of *Dendroctonus brevicomis* (Coleoptera: Scolytidae). *Journal of Chemical Ecology* **20**:2931–2941.

Bestmann, H.J., A.B. Attygalle, T. Brosche, J. Erler, H. Platz, J. Schwarz, O. Vostrowsky and C.H. Wu. 1987a. Identification of three sex pheromone components of the female saturniid moth *Antheraea pernyi* (Lepidoptera: Saturniidae). *Zeitschrift für Naturforschung* **42c**:631–636.

Bestmann, H.J., M. Herrig and A.B. Attygalle. 1987b. Terminal acetylation in pheromone biosynthesis by *Mamestra brassicae* L. (Lepidoptera:Noctuidae). *Experientia* **43**:1033–1034.

Bestmann, H.J., M. Herrig, A.B. Attygalle and M. Hupe. 1989. Regulatory steps in sex pheromone biosynthesis in *Mamestra brassicae* L. (Lepidoptera: Noctuidae). *Experientia* **45**:778–781.

Birch, M.C. 1972. Male abdominal brush-organs in British noctuid moths and their value as a taxonomic character Part II. *The Entomologist* **105**:233–244.

Birch, M.C., ed. 1974. *Pheromones*. North-Holland, Amsterdam.

Birch, M.C. 1984. Aggregation in bark beetles. In *Chemical Ecology of Insects*, eds. W.J. Bell and R.T. Cardé, pp. 331–353. Chapman and Hall, London.

Birch, M.C. and A. Hefetz. 1987. Extrusible organs in male moths and their role in courtship behavior. *Bulletin of the Entomological Society of America* **33**:222–229.

Birch, M.C. and D.M. Light. 1977. Inhibition of the attractant pheromone response in *Ips pini* and *I. paraconfusus* (Coleoptera: Scolytidae): field evaluation of ipsenol and linalool. *Journal of Chemical Ecology* **3**:257–267.

Birch, M.C. and D.L. Wood. 1975. Mutual inhibition of the attractant pheromone response by two species of *Ips* (Coleoptera: Scolytidae). *Journal of Chemical Ecology* **1**:101–113.

Birch, M.C., D.M. Light and K. Mori. 1977. Selective inhibition of response of *Ips pini* to its pheromone by the (*S*)-(−)-enantiomer of ipsenol. *Nature* **270**:738–739.

Birch, M.C., D.M. Light, D.L. Wood, L.E. Browne, R.M. Silverstein, B.J. Bergot, G. Ohloff, J.R. West and J.C. Young. 1980. Pheromonal attraction and allomonal interruption of *Ips pini* in California by the two enantiomers of ipsdienol. *Journal of Chemical Ecology* **6**:703–717.

Birch, M.C., G.M. Poppy and T.C. Baker. 1990. Scents and eversible scent structures of male moths. *Annual Review of Entomology* **35**:25–58.

Birgersson, G., F. Schlyter, J. Löfqvist and G. Bergstrom. 1984. Quantitative variation of pheromone components in the spruce bark beetle *Ips typographus* from different attack phases. *Journal of Chemical Ecology* **10**:1029–1055.

Bjostad, L.B. and W.L. Roelofs. 1983. Sex pheromone biosynthesis in *Trichoplusia ni*: key steps involve delta-11 desaturation and chain-shortening. *Science* **220**:1387–1389.

Bjostad, L.B., L.K. Gaston, L.L. Noble, J.H. Moyer and H.H. Shorey. 1980. Dodecyl acetate, a second pheromone component of the cabbage looper, *Trichoplusia ni*. *Journal of Chemical Ecology* **6**:727–734.

Bjostad, L.B., C.E. Linn, J.-W. Du and W.L. Roelofs. 1984. Identification of new sex pheromone components in *Trichoplusia ni*, predicted from biosynthetic precursors. *Journal of Chemical Ecology* **10**:1309–1323.

Bjostad, L.B., W.A. Wolf and W.L. Roelofs. 1987. Pheromone biosynthesis in Lepidoptera: desaturation and chain shortening. In *Pheromone Biochemistry*, eds. G.D. Prestwich and G.J. Blomquist, pp. 77–120. Academic Press, Orlando, Florida.

Blight, M.M., N.J. Fielding, C.J. King, A.P. Ottridge, L.J. Wadhams and M.J. Wenham. 1983. Field response of the Dutch elm disease vectors, *Scolytus multistriatus* (Marsham) and *S. scolytus* (F.) (Coleoptera: Scolytidae) to 4-methyl-3-heptanol baits containing α-, β-, γ-, or δ-multistriatin. *Journal of Chemical Ecology* **9**:67–84.

Blomquist, G.J., J.A. Tillman-Wall, L. Guo, D.R. Quilici, P. Gu and C. Schal. 1993. Hydrocarbon and hydrocarbon-derived sex pheromones in insects: biochemistry and endocrine regulation. In *Insect Lipids: Chemistry, Biochemistry and Biology*, eds. D.W. Stanley-Samuelson and D.R. Nelson, pp. 317–351. University of Nebraska Press, Lincoln, Nebraska.

Blomquist, G.J., J.A. Tillman-Wall, J.R. Reed, P. Gu, D. Vanderwel, S. Choi and R.C. Reitz. 1995. Regulation of enzymatic activity involved in sex pheromone production in the housefly, *Musca domestica*. *Insect Biochemistry and Molecular Biology* **25**:751–757.

Bodnaryk, R.P. 1983. Regulation by Ca^{++} and calmodulin of brain adenylate cyclase from the moth, *Mamestra configurata* Wlk. *Insect Biochemistry* **13**:111–114.

Boeckh, J. 1962. Elektrophysiologische Untersuchungen an einzelnen Geruchsrezeptoren auf der Antenne des Totengräbers (*Necrophorus*, Coleoptera). *Zeitschrift für Vergleichende Physiologie* **46**:212–248.

Boeckh, J. and C.H. Boeckh. 1979. Theshold and odor specificity of pheromone-sensitive neurons in the deutocerebrum of *Antheraea pernyi* and *A. polyphemus* (Saturniidae). *Journal of Comparative Physiology A* **132**:235–242.

Boeckh, J. and L.P. Tolbert. 1993. Synaptic organization and development of the antennal lobe in insects. *Microscopy Research and Technique* **24**:260–280.

Boeckh, J., K.-E. Kaissling and D. Schneider. 1965. Insect olfactory receptors. *Cold Spring Harbor Symposium on Quantitative Biology* **30**:263–280.

Boeckh, J., P. Distler, K.D. Ernst, M. Hösl and D. Malun. 1990. Olfactory bulb and antennal lobe. In *Chemosensory Information Processing*, ed. D. Schild, pp. 201–227. Springer-Verlag, Berlin.

Boekhoff, I. and H. Breer. 1992. Termination of second messenger signalling in olfaction. *Proceedings of the National Academy of Sciences U.S.A.* **89**:471–474.

Boekhoff, I., K. Raming and H. Breer. 1990a. Pheromone-induced stimulation of inositol-trisphosphate formation in insect antennae is mediated by G-proteins. *Journal of Comparative Physiology* **160**:99–103.

Boekhoff, I., J. Strotmann, K. Raming, E. Tareilus and H. Breer. 1990b. Odorant-sensitive phospholipase C in insect antennae. *Cellular Signaling* **2**:49–56.

Boekhoff, I., S. Schleicher, J. Strotmann and H. Breer. 1992. Odor-induced phosphorylation of olfactory cilia proteins. *Proceedings of the National Academy of Sciences U.S.A.* **89**:11983–11987.

Boekhoff, I., E. Seifert, S. Göggerle, M. Lindemann, B.W. Krüger and H. Breer. 1993. Pheromone-induced second-messenger signaling in insect antennae. *Insect Biochemistry and Molecular Biology* **23**:757–762.

Boggs, C.L. 1990. A general model of the role of male-donated nutrients in female insects reproduction. *American Naturalist* **136**:598–617.

Boo, K.S., K.C. Park, D.R. Hall, A. Cork, B.G. Berg and Mustaparta, H. 1995 (*Z*)-9-Tetradecenal: a potent inhibitor of pheromone-mediated communication in the oriental tobacco budworm, *Helicoverpa assulta. Journal of Comparative Physiology* **177**:695–699.

Boppré, M. 1984. Chemically mediated interactions between butterflies. In *The Biology of Butterflies*, ed. R.I. Vane-Wright and P.R. Ackery, pp. 259–275. Academic Press, New York.

Borden, J.H. 1985. Aggregation pheromones. In *Behaviour*, Vol. 9, *Comprehensive Insect Physiology, Biochemistry and Pharmacology*, eds. G.A. Kerkut and L.I. Gilbert, pp. 257–285. Pergamon Press, Oxford.

Borden, J.H. 1990. Use of semiochemicals to manage coniferous tree pests in western Canada. In *Behavior-Modifying Chemicals for Insect Management*, eds. R.L. Ridgway, R.M. Silverstein and M.N. Inscoe, pp. 281–315. Marcel Dekker, New York.

Borden, J.H. 1992. Two tree baiting tactics for the management of bark beetles with semiochemicals. *Journal of Applied Entomology* **114**:201–207.

Borden, J.H. 1993. Strategies and tactics for the use of semiochemicals against forest insect pests in North America. In *Pest Management: Biologically Based Technologies*, eds. R.D. Lumsden and J.L Vaughn, pp. 265–279. American Chemical Society, Washington, D.C.

Borden, J.H. and B.S. Lindgren. 1988. The role of semiochemicals in IPM of the mountain pine beetle. In *Integrated Control of Scolytid Bark Beetles*, eds. T.L. Payne and H. Saarenmaa, pp. 247–255. Virginia Polytechnic Institute and State University, Blacksburg, Virginia.

Borden, J.H., J.R. Handley, J.A. McLean, R.M. Silverstein, L. Chong, K.N. Slessor, B.D. Johnston and H.R. Schuler. 1980. Enantiomer-based specificity in pheromone communication by two sympatric *Gnathotrichus* species (Coleoptera: Scolytidae). *Journal of Chemical Ecology* **6**:445–456.

Borden, J.H., L. Chong, K.N. Slessor, A.C. Oehlschlager, H.D. Pierce, Jr. and B.S. Lindgren. 1981. Allelochemic activity of aggregation pheromones of three sympatric species of ambrosia beetles (Coleoptera: Scolytidae). *Canadian Entomologist* **113**:557–563.

Borden, J.H., L.C. Ryker, L.J. Chong, H.D. Pierce, Jr., B.D. Johnston and A.C. Oehlschlager. 1987. Response of the mountain pine beetle, *Dendroctonus ponderosae* Hopkins (Coleoptera: Scolytidae), to five semiochemicals in British Columbia lodgepole pine forests. *Canadian Journal of Forest Research* **17**:118–128.

Borden, J.H., L.J. Chong and B.S. Lindgren. 1990. Redundancy in the semiochemical message required to induce attack on lodgepole pines by the mountain pine beetle, *Dendroctonus ponderosae* Hopkins (Coleoptera: Scolytidae). *Canadian Entomologist* **122**:769–777.

Borden, J.H., D.R. Devlin and D.R. Miller. 1992. Synomones of two sympatric species deter attack by the pine engraver, *Ips pini* (Coleoptera: Scolytidae). *Canadian Journal of Forest Research* **22**:381–387.

Boreham, B.W. and J.K. Harvey. 1984. Dipolar charged particles as markers for dispersion simulation in wind tunnels. *Journal of Physics E* **17**:994–998.

Boreham, B.W., C.D. Jones and R.F. Griffith. 1985. Field evaluation of a new Langmuir-type probe for atmospheric dispersion studies using ions as a tracer. *Journal of Physics E* **18**:218–223.

Borgo, M. 1992. Ultérieurs essais de lutte par confusion sexuelle contre les tordeuses de la grappe dans les vignobles de l'est de la Vénétie. *International Organization for Biological Control—West Palearctic Regional Section Bulletin* **15**(2):20.

Borroni, P.F. and R.J. O'Connell. 1992. Temporal analysis of adapataion in moth *(Trichoplusia ni)* pheromone receptor neurons. *Journal of Comparative Physiology A* **170**:691–700.

Boscheri, S., W. Rizzoli and N. Paoli. 1992. Experience with mating disruption for control of the codling moth and leafrollers at the Laimburg Experiment Station (South Tirol-Bolzano). *International Organization for Biological Control—West Palearctic Regional Section Bulletin* **15**(5):81–87.

Bossert, W.H. and E.O. Wilson. 1963. The analysis of olfactory communication among animals. *Journal of Theoretical Biology* **48**:443–469.

Bovey, R. 1979. *La Défense des Plantes Cultivées*, 7th edition, Payot, Lausanne, Switzerland.

Bradley, S.J., D.M. Suckling, K.G. McNaughton, C.H. Wearing and G. Karg. 1995. A temperature dependent model for predicting release rates of pheromones from polyethelene tubing dispensers. *Journal of Chemical Ecology* **21**:745–760.

Brady, J. 1974. The physiology of insect rhythms. *Advances in Insect Physiology* **10**:1–116.

Brady, J., G. Gibson and M.J. Packer. 1989. Odour movement, wind direction, and the problem of host finding by tsetse flies. *Physiological Entomology* **14**:369–380.

Brady, U.E. 1973. Isolation, identification and stimulatory activity of a second component of the sex pheromone system (complex) of the female almond moth, *Cadra cautella* (Walker). *Life Sciences* **13**:227–235.

Brady, U.E., J.H. Tumlinson III, R.G. Brownlee and R.M. Silverstein. 1971. Sex stimulant and attractant in the Indian meal moth and in the almond moth. *Science* **171**:802–804.

Brazzel, J.R. and D.F. Martin. 1957. Oviposition sites of the pink bollworm on the cotton plant. *Journal of Economic Entomology* **50**:122–124.

Brechbühler, C. and E. Meyer. 1992. Essais de lutte contre les tordeuses de la grappe (Eudémis, Cochylis, Eulia) par confusion sexuelle. *International Organization for Biological Control—West Palearctic Regional Section Bulletin* **15**(2):21.

Breeland, S.G. 1958. Biological studies on the armyworm, *Pseudaletia unipuncta* (Haworth) in Tennessee (Lepidoptera: Noctuidae). *Journal of Tennessee Academy of Science* **33**:263–347.

Breer, H. 1993. Second messenger signalling in olfaction. In *Molecular Basis of Smell and Taste Transduction*, eds. D. Chadwick, J. Marsh and J. Goode, pp. 97–114. Wiley, Chichester, United Kingdom.

Breer, H. 1994a. Odor recognition and second messenger signalling in olfactory receptor neurons. *Seminars in Cell Biology* **5**:25–32.

Breer, H. 1994b. Signal recognition and transduction in olfaction. In *Olfaction and Taste XI*, eds. K. Kurihara, N. Suzuki and H. Ogawa, pp. 132–134. Springer-Verlag, Tokyo.

Breer, H. and I. Boekhoff. 1992. Second messenger signalling in olfaction. *Current Opinion in Neurobiology* **2**:439–443.

Breer, H., K. Raming and I. Boekhoff. 1988. G-proteins in the antennae of insects. *Naturwissenschaften* **75**:627.

Breer, H., J. Krieger and K. Raming. 1990a. A novel class of binding proteins in the antennae of the silkmoth *Antheraea pernyi*. *Insect Biochemistry* **20**:735–740.

Breer, H., I. Boekhoff, J. Strotmann, K. Raming, E. Tareilus. 1990b. Molecular elements of olfactory signal transduction in insect antennae. In *Information Processing of Chemical Sensory Stimuli in Biological and Artificial Systems*, ed. D. Schild, pp 77–86. Springer-Verlag, Berlin.

Breer, H., I. Boekhoff and E. Tareilus. 1990c. Rapid kinetics of second messenger formation in olfactory transduction. *Nature* **345**:65–68.

Bretschneider, F. 1924. Über die Gehirne des Eichenspinners und des Seidenspinners (*Lasiocampa querqus* L. und *Bombyx mori* L.) *Jena Zeitschrift für Naturwissenschaften (Zoologie)* **60**:563–570.

Brewer, M.J. and J.T. Trumble. 1989. Field monitoring for insecticide resistance in the beet armyworm, *Spodoptera exigua* (Hübner). *Journal of Economic Entomology* **82**:1520–1526.

Brewer, M.J. and J.T. Trumble. 1991. Classifying resistance severity in field populations: sampling inspection plans for an insecticide resistance monitoring program. *Journal of Economic Entomology* **84**:379–389.

Brewer, M.J., D.J. Schuster, J.T. Trumble and B. Alvarado-Rodriguez. 1993. Tomato pinworm (Lepidoptera: Gelechiidae) resistance to fenvalerate from localities in Sinaloa, Mexico and California. *Tropical Agriculture* **70**:179–184.

Brooks, D.R. and D.A. McLennan. 1991. *Phylogeny, Ecology, and Behavior*. University of Chicago Press, Chicago.

Brooks, T.W., C.C. Doane and R.T. Staten. 1979. Experience with the first commercial pheromone communication disruption for suppression of an agricultural pests. In *Chemical Ecology: Odour Communication in Animals*, ed. F.M. Ritter, pp. 375–388. Elsevier/North-Holland, Amsterdam.

Brower, L.P. and M.A. Jones. 1965. Precourtship interaction of wing and abdominal sex glands in male *Danaus* butterflies. *Proceedings of the Royal Entomological Society of London (A)* **40**:147–151.

Brower, L.P., J.V.Z. Brower and F.P. Cranston. 1965. Courtship behavior of the Queen butterfly, *Danaus gilippus berenice. Zoologica* **50**:1–39.

Brown, C.S. and C. Nestler. 1985. Catecholamines and indolalkylamines. In *Pharmacology*, Vol. 11, *Comprehensive Insect Physiology Biochemistry and Pharmacology*, eds. G.A. Kerkut and L.I. Gilbert, pp. 435–497. Pergamon Press, Oxford.

Brusca, R.C. and G.J. Brusca. 1990. *Invertebrates*. Sinauer Associates, Sunderland, Massachusetts.

Buchner, E. 1984. Behavioral analysis of spatial vision in invertebrates. In *Photoreception and Vision in Invertebrates*, ed. M.A. Ali, pp 561–621. Plenum Press, New York.

Buck, L. and R. Axel. 1991. A novel multigene family may encode odorant receptors: a molecular basis for odor recognition. *Cell* **65**:175–187.

Bullock, T.H. 1983. Epilogue: neurobiological roots and neuroethological sprouts.In *Neuroethology and Behavioral Physiology*, eds. F. Huber and H. Markl, pp. 403–412. Springer-Verlag, Heidelberg, Germany.

Burghardt, G.M. 1970. Defining 'communication'. In *Communication by Chemical Signals*, eds. J.W. Johnson, D.G. Moulton and A. Turk, pp. 5–18. Appleton, New York.

Burkholder, W.E. 1990. Practical use of pheromones and other attractants for stored-product insects. In *Behavior-Modifying Chemicals for Insect Management*, eds. R.L. Ridgway, R.M. Silverstein and M.N. Inscoe, pp. 497–516. Marcel Dekker, New York.

Burns, E.L., A-S. Chiang, M. Gadot and C. Schal. 1991. Juvenile hormone regulation of the left colleterial gland in intact and ovariectomized *Blattella germanica* (L.) (Dictyoptera: Blattellidae). *Journal of Insect Physiology* **37**:401–405.

Buser, H.R., S. Rauscher and H. Arn. 1974. Sex pheromone of *Lobesia botrana*: (*E,Z*)-7,9-dodecadienyl acetate in the female grape vine moth. *Zeitschrift für Naturforschung* **29c**:781–783.

Buser, H.R., P.M. Guerin, M. Tóth, G. Szöcs, A. Schmid, W. Francke and H. Arn. 1985. (Z,Z)-6,9-nonadecadien-3-one and (Z,Z,Z)-3,6,9-nonadecatriene: Identification and synthesis of sex pheromone components of *Peribatodes rhomboidaria. Tetrahedron Letters* **26**:403–406.

Butenandt, A., R. Beckmann, D. Stamm and E. Hecker. 1959. Über den Sexual-Lockstoff des Seidenspinners *Bombyx mori*. Reindarstellung und Konstitutionsermittlung. *Zeitschrift für Naturforschung* **14b**:283–284.

Butlin, R. 1987. Speciation by reinforcement. *Trends in Ecology and Evolution* **2**:8–13.

Butlin, R. 1989. Reinforcement of premating isolation. In *Speciation and its Consequences*, eds. D. Otte and J. Endler, pp. 158–179. Sinauer Associates, Sunderland, Massachusetts.

Butlin, R.K. 1993. The variability of mating signals and preferences in the brown planthopper, *Nilaparvata lugens. Journal of Insect Behavior* **6**:125–140.

Butlin, R.K. 1994. Genetic variation in mating signals and responses. In *Speciation and the Recognition Concept: Theory and Application*, eds. D.M. Lambert and H.G. Spencer, pp. 327–366. Johns Hopkins University Press, Baltimore, Maryland.

Butlin, R.K. and M.G. Ritchie. 1989. Genetic coupling in mate recognition systems: what is the evidence? *Biological Journal of the Linnean Society* **37**:237–246.

Butlin, R.K. and M.G. Ritchie. 1994. Behaviour and speciation. In *Behaviour and Evolution*, eds. P.J.B. Slater and T.R. Halliday, pp. 43–79. Cambridge University Press, Cambridge.

Byers, J.A. 1982. Male-specific conversion of the host plant compound, myrcene, to the pheromone, (+)-ipsdienol, in the bark beetle, *Dendroctonus brevicomis. Journal of Chemical Ecology* **8**:363–371.

Byers, J.A. and D.L. Wood. 1980. Interspecific inhibition of the response of the bark beetles, *Dendroctonus brevicomis* and *Ips paraconfusus*, to their pheromones in the field. *Journal of Chemical Ecology* **6**:149–164.

Callahan, P.S. 1958. Behavior of the imago of the corn earworm, *Heliothis zea* (Boddie), with special reference to emergence and reproduction. *Annals of the Entomological Society of America* **51**:271–283.

Callahan, P.S. and J.B. Chapin. 1960. Morphology of the reproductive systems and mating in two representative members of the family Noctuidae, *Pseudaletia unipuncta* and *Peridroma margaritosa*, with comparison to *Heliothis zea. Annals of the Entomological Society of America* **53**:763–782.

Camacho, A.D., H.D. Pierce, Jr. and J.H. Borden. 1993. Geometrical and optical isomerism of pheromones in two sympatric *Dryocoetes* species (Coleoptera: Scolytidae) mediates species specificity and response level. *Journal of Chemical Ecology* **19**:2169–2182.

Camacho, A.D., H.D. Pierce, Jr. and J.H. Borden. 1994. Aggregation pheromones in *Dryocoetes affaber* (Mann.) (Coleoptera: Scolytidae): stereoisomerism and species specificity. *Journal of Chemical Ecology* **20**:111–124.

Camazine, S.M. and J.G. Hildebrand. 1979. Central projections of antennal sensory neurons in mature and developing *Manduca sexta. Society for Neuroscience Abstracts* **5**:492.

Camhi, J.M. 1984. Neuroethology. Sinauer Associates, Sunderland, Massachusetts.

Campbell, C.A.M., G.W. Dawson, D.C. Griffiths, J. Pettersson, J.A. Pickett, L.J. Wadhams and C.M. Woodcock. 1990. Sex attractant pheromone of damson-hop aphid *Phorodon humuli* (Homoptera, Aphididae). *Journal of Chemical Ecology* **16**:3455–3465.

Campbell, C.A.M., J. Pettersson, J.A. Pickett, L.J. Wadhams and C.M. Woodcock. 1993. Spring migration of damson-hop aphid, *Phorodon humuli* (Homoptera, Aphididae), and summer host plant-derived semiochemicals released on feeding. *Journal of Chemical Ecology* **19**:1569–1576.

Campbell, C.D., J.F. Walgenbach and G.G. Kennedy. 1992. Comparison of black light and pheromone traps for monitoring *Helicoverpa zea* (Boddie) (Lepidoptera: Noctuidae) in tomato. *Journal of Agricultural Entomology* **9**:17–24.

Campion, D.G. 1984. Survey of pheromone uses in pest control. In *Techniques in Pheromone Research*, eds. H.E. Hummel and T.A. Miller, pp. 405–450. Springer-Verlag, New York.

Campion, D.G., P. Hunter-Jones, L.J. McVeigh, D.R. Hall, R. Lester and B.F. Nesbitt. 1980. Modification of the attractiveness of the primary pheromone component of the Egyptian cotton leafworm, *Spodoptera littoralis* (Boisduval) (Lepidoptera: Noctuidae) by secondary pheromone components and related chemicals. *Bulletin of Entomological Research* **70**:417–434.

Cardé, R.T. 1984. Chemo-orientation in flying insects. In *Chemical Ecology of Insects*, eds. W.J. Bell and R.T. Cardé, pp. 111–124, Chapman and Hall, London.

Cardé, R.T. 1986. The role of pheromones in reproductive isolation and speciation of insects. In *Evolutionary Genetics of Invertebrate Behavior*, ed. M.D. Huettel, pp. 303–317. Plenum Press, New York.

Cardé, R.T. 1990. Principles of mating disruption. In *Behavior-Modifying Chemicals for Insect Management*, eds. R.L. Ridgway, R.M. Silverstein and M. Inscoe, pp. 47–71. Marcel Dekker, New York.

Cardé, R.T. and T.C. Baker. 1984. Sexual communication with pheromones. In *Chemical Ecology of Insects*, eds. W.J. Bell and R.T. Cardé, pp. 355–383. Chapman and Hall, London.

Cardé, R.T. and R.E. Charlton. 1984. Olfactory sexual communication in Lepidoptera: strategy, sensitvity and startegy. In *Insect Communication*, ed. T. Lewis, pp. 241–265. Academic Press, London.

Cardé, R.T. and T.E. Hagaman. 1979. Behavioral responses of the gypsy moth in a wind tunnel to airborne enantiomers of disparlure. *Environmental Entomology* **8**:475–484.

Cardé, R.T. and A.K. Minks. 1995. Control of moth pests by mating disruption: successes and constraints. *Annual Review of Entomology* **40**:559–585.

Cardé, R.T. and W.L. Roelofs. 1973. Temperature modification of male sex pheromone response and factors affecting female calling in *Holomelina immaculata* (Lepidoptera: Arctiidae). *Canadian Entomologist.* **105**:1505–1512.

Cardé, R.T. and R.P. Webster. 1981. Endogenous and exogenous factors controlling insect sex pheromone production and responsiveness, particularly among Lepidoptera. In *Regulation of Insect Development and Behaviour*, pp. 991–997. Wroclaw Technical University Press, Wroclaw, Poland.

Cardé, R.T., T.C. Baker and W.L. Roelofs. 1976. Sex attractant responses of male oriental fruit moths to a range of component ratios: pheromone polymorphism? *Experientia* **32**:1406–1407.

Cardé, R.T., A.M. Cardé A.S. Hill, and W.L. Roelofs. 1977. Sex pheromone specificity as a reproductive isolating mechanism among the sibling species *Archips agyrospilus* and *A. mortuanus* and other sympatric tortricine moths (Lepidoptera: Tortricidae). *Journal of Chemical Ecology* **1**:71–84.

Cardé, R.T., T.C. Baker and P. Castrovillo. 1978. Disruption of sexual communication in *Laspeyresia pomonella* (codling moth), *Grapholitha molesta* (oriental fruit moth) and *G. prunivora* (lesser appleworm) with hollow-fiber attractant sources. *Entomologia Experimentalis et Applicata* **22**:280–288.

Carlson, J. 1993. Molecular genetics of *Drosophila* olfaction. In *Molecular Basis of Smell and Taste Transduction*, eds. D. Chadwick, J. Marsh and J. Goode, pp. 150–166. Wiley, Chichester, United Kingdom.

Carpenter, J.E., S.D. Pair, and A.N. Sparks. 1984. Trapping of different noctuid moth species by one trap baited with two lures. *Journal of the Georgia Entomological Society* **19**:120–124.

Castrovillo, P.J. and R.T. Cardé. 1979. Environmental regulation of female calling and male pheromone response periodicities in the codling moth. *Journal of Insect Physiology* **25**:659–667.

Chapman, P.J. 1973. Bionomics of apple-feeding Tortricidae. *Annual Review of Entomology* **18**:73–96.

Chapman, P.J., S.E. Lienk and R.W. Dean. 1968. Bionomics of *Choristoneura rosaceana*. *Annals of the Entomological Society of America* **61**:285–290.

Charlton, R.E. and R.T. Cardé. 1982. Rate and diel periodicity of pheromone emission from female gypsy moths (*Lymantria dispar*) determined with a glass-adsorption collection system. *Journal of Insect Physiology* **28**:423–430.

Charlton, R.E. and R.T. Cardé. 1989. Factors mediating copulatory behavior and close-range mate recognition in the male gypsy moth, *Lymantria dispar* (L.). *Canadian Journal of Zoology* **68**:1995–2004.

Charlton, R.E. and R.T. Cardé. 1990a. Behavioral interactions in the courtship of *Lymantria dispar* (Lepidoptera: Lymantriidae). *Annals of the Entomological Society of America* **83**:89–96.

Charlton, R.E. and R.T. Cardé. 1990b. Orientation of male gypsy moth, *Lymantria dispar* (L.), to pheromone sources: the role of olfactory and visual cues. *Journal of Insect Behavior* **3**:443–469.

Charlton, R.E. and W.L. Roelofs. 1991. Biosynthesis of a volatile, methyl-branched hydrocarbon sex pheromone from leucine by arctiid moths (*Holomelina* spp.). *Archives of Insect Biochemistry and Physiology* **18**:81–97.

Charlton, R.E., J.A. Wayman, J.R. McLaughlin, J.-W. Du and W.L. Roelofs. 1991. Identification of sex pheromone of tomato pinworm, *Keiferia lycopersicella* (Wals.). *Journal of Chemical Ecology* **17**:175–183.

Charlton, R.E., H. Kanno, R.D. Collins, and R.T. Cardé. 1993a. Influence of pheromone concentration and ambient temperature on flight of the gypsy moth, *Lymantria dispar*, in a sustained-flight wind tunnel. *Physiological Entomology* **18**:349–362.

Charlton, R.E., F.X. Webster, A. Zhang, C. Schal, D. Liang, I. Sreng and W.L. Roelofs. 1993b. Sex pheromone of the brownbanded cockroach is an unusual dialkyl α-pyrone. *Proceedings of the National Academy of Sciences U.S.A.* **90**:10202–10205.

Charmillot, P.J. 1990. Mating disruption technique to control codling moth in western Switzerland. In *Behavior-Modifying Chemicals for Insect Management*, eds. R.L. Ridgway, R.M. Silverstein and M.N. Inscoe, pp. 165–182, Marcel Dekker, New York.

Charmillot, P.-J., D. Pasguler, N.J. Alipaz and U. Newman. 1995. Effet d'une ceinture de diffuseurs á attractif sexuel sur la distribution des captures de mâles, des accouplements de femelles attachées et de l'attaque de *Lobesia botrana* Den. & Schiff. en vignoble. *Journal of Applied Entomology* **199**:201–206.

Charmillot, P.J. and B. Bloesch. 1987. La technique de confusion sexuelle: Un moyen spécifique de lutte contre le carpocapse, *Cydia pomonella* L. *Revue Suisse de Viticulture, Arboriculture et Horticulture* **19**:129–138.

Charmillot, P.J. and D. Pasquier. 1992a. Deux ans de lutte par la technique de confusion appliquée simultanément contre *Eupoecilia ambiguella* Hb. et *Lobesia botrana* Den. et Schiff au moyen des ampoules BASF. *International Organization for Biological Control—West Palearctic Regional Section Bulletin* **15**(2):22

Charmillot, P.J. and D. Pasquier. 1992b. Lutte par confusion contre le carpocapse *Cydia pomonella* L. Expérimentation de longue durée en Suisse romande. *Revue Suisse de Viticulture, Arboriculture et Horticulture* **24**:213–220.

Chase, J., R.A. Jurenka, C. Schal, P.P. Halarnkar and G.J. Blomquist. 1990. Biosynthesis of methyl branched hydrocarbons of the German cockroach *Blattella germanica* (L.) (Orthoptera, Blattellidae). *Insect Biochemistry* **20**:149–156.

Chase, J., K. Touhara, G.D. Prestwich, C. Schal and G.J. Blomquist. 1992. Biosynthesis and endocrine control of the production of the German cockroach sex pheromone 3,11-dimethylnonacosan-2-one. *Proceedings of the National Academy of Sciences U.S.A.* **89**:6050–6054.

Chatwin, P.C. 1985. Interaction between turbulent and electrostatic effects in the atmospheric dispersion of electrically charged tracers. Report PTA/AZ/209. Chemical Defense Establishment, Porton Down, United Kingdom.

Cheung, W.Y. 1980. Calmodulin plays a pivotal role in cellular regulation. *Science* **207**:19–27.

Chiang, A-S. and C. Schal. 1994. Cyclic volumetric changes in corpus allatum cells in relation to juvenile hormone biosynthesis ovarian cycles in cockroaches. *Archives of Insect Biochemistry and Physiology* **27**:53–64.

Chiang, A-S., E.L. Burns and C. Schal. 1991a. Ovarian regulation of cyclic changes in size and activity of corpus allatum cells in *Blattella germanica. Journal of Insect Physiology* **37**:907–917.

Chiang, A-S., M. Gadot, E.L. Burns and C. Schal. 1991b. Sexual differentiation of nymphal corpora allata and the effects of ovariectomy on adult gland morphometrics in *Blattella germanica. Experientia* **47**:81–83.

Chon, T-S., D. Liang and C. Schal. 1990. Effects of mating and grouping on oocyte development and pheromone release activities in the female brown-banded cockroach, *Supella longipalpa* (F.) (Dictyoptera: Blattellidae). *Environmental Entomology* **19**:1716–1721.

Christensen, T.A. and J.G. Hildebrand. 1987a. Functions, organization and physiology of the olfactory pathways in the lepidopteran brain. In *Arthropod Brain: Its Evolution, Development, Structure and Functions*, ed. A.P. Gupta, pp. 457–484. John Wiley, New York.

Christensen, T.A. and J.G. Hildebrand. 1987b. Male-specific, sex pheromone-selective projection neurons in the antennal lobes of the moth *Manduca sexta. Journal of Comparative Physiology A* **160**:553–569.

Christensen, T.A. and J.G. Hildebrand. 1988. Frequency coding by central olfactory neurons in the sphinx moth *Manduca sexta. Chemical Senses* **13**:123–130.

Christensen, T.A. and J.G. Hildebrand. 1990. Representation of sex-pheromonal information in the insect brain. In *Proceedings X International Symposium on Olfaction and Taste*, ed. K.B. Døving, pp. 142–150. Graphic Communication Systems, Oslo.

Christensen, T.A. and J.G. Hildebrand. 1994. Neuroethology of sexual attraction and inhibition in heliothine moths. In *Neural Basis of Behavioural Adaptations*, eds. K. Schildberger and N. Elsner, pp. 37–46. Gustav Fischer Verlag, Stuttgart.

Christensen, T.A., J.G. Hildebrand, J.H. Tumlinson and R.E. Doolittle. 1989a. The sex pheromone blend of *Manduca sexta*: responses of central olfactory interneurons to antennal stimulation in male moths. *Archives of Insect Biochemistry and Physiology* **10**:281–291.

Christensen, T.A., H. Mustaparta and J.G. Hildebrand. 1989b. Discrimination of sex pheromone blends in the olfactory system of the moth. *Chemical Senses* **14**:463–477.

Christensen, T.A., H. Mustaparta and J.G. Hildebrand. 1991a. Chemical communication in heliothine moths. II. Central processing of intra- and interspecific olfactory messages in the male corn earworm moth *Helicoverpa zea*. *Journal of Comparative Physiology A* **169**:259–274.

Christensen, T.A., H. Itagaki, P.E.A. Teal, R.D. Jasensky, J.H. Tumlinson and J.G. Hildebrand. 1991b. Innervation and neural regulation of the sex pheromone gland in female *Heliothis* moths. *Proceedings of the National Academy of Sciences U.S.A.* **88**:4971–4975.

Christensen, T.A., H.K. Lehman, P.E.A. Teal, H. Itagaki, J.H. Tumlinson and J.G. Hildebrand. 1992. Diel changes in the presence and physiological actions of octopamine in the female sex-pheromone glands of heliothine moths. *Insect Biochemistry and Molecular Biology* **22**:841–849.

Christensen, T.A., B.R. Waldrop, I.D. Harrow and J.G. Hildebrand. 1993. Local interneurons and information processing in the olfactory glomeruli of the moth *Manduca sexta*. *Journal of Comparative Physiology A*. **173**:385–399.

Christensen, T.A., J.M. Lashbrook and J.G. Hildebrand. 1994. Neural activation of the sex-pheromone gland in the moth *Manduca sexta*: real-time measurement of pheromone release. *Physiological Entomology* **19**:265–270.

Christensen, T.A., Mustaparta, H. and Hildebrand, J.G. 1995. Chemical communication in heliothine moths. VI. Dendritic processing of pheromonal information in antennal lobe neurons in the male tobacco budworm moth *Heliothis virescens*. *Journal of Comparative Physiology A* **177**:545–557.

Christy, J.H. 1995. Mimicry, mate choice, and the sensory trap hypothesis. *American Naturalist* **139S**:125–153.

Cibrian-Tovar, J., and E.R. Mitchell. 1991. Courtship behavior of *Heliothis subflexa* (Gn.) (Lepidoptera: Noctuidae) and associated backcross insects obtained from hybridization with *H. virescens* (F.) *Environmental Entomology* **20**:419–426.

Cinelli, A.R. and J.S. Kauer. 1992. Voltage-sensitive dyes and functional activity in the olfactory pathway. *Annual Review of Neuroscience* **15**:321–351.

Clearwater, J.R., S.P. Foster, S.J. Muggleston, J.S. Dugdale and E. Priesner. 1991. Intraspecific variation and interspecific differences in the *Ctenopseustis obliquana* complex. *Journal of Chemical Ecology* **17**:413–429.

Cliff, D., P. Husbands, J.-A. Meyer and S.W. Wilson. 1994. *Animals to Animats 3*

Proceedings 3rd International Conference on Simulation of Adaptive Behavior. MIT Press, Cambridge, Massachusetts.

Coats, J.R. 1994. Risks from natural versus synthetic insecticides. *Annual Review of Entomology* **39**:489–515.

Cobb, F.W., Jr., E. Zavarin and J. Bergot. 1972. Effect of air pollution on the volatile oil from leaves of *Pinus ponderosa*. *Phytochemistry* **11**:1815–1818.

Coddington, J.A. 1988. Cladistic tests of adaptational hypotheses. *Cladistics* **4**:3–22.

Collett, T.S. 1978. Peering—a locust behaviour pattern for obtaining motion parallax information. *Journal of Experimental Biology* **76**:237–241.

Collett, T.S. 1992. Landmark learning and guidance in insects. *Philosophical Transactions of the Royal Society B* **337**:295–303.

Collett, T.S. and A.D. Blest. 1966. Binocular, directionally selective neurones, possibly involved in the optomotor response of insects. *Nature* **212**:1330–1333.

Collett, T.S. and M. Lehrer. 1993. Looking and learning: a spatial pattern in the orientation flight of the wasp *Vespula vulgaris*. *Proceedings of the Royal Society of London B* **252**:129–134.

Collins, R.D. and R.T. Cardé. 1985. Variation in and heritability of aspects of pheromone production in the pink bollworm moth, *Pectinophora gossypiella* (Lepidoptera: Gelechiidae). *Annals of the Entomological Society of America* **78**:229–234.

Collins, R.D. and R.T. Cardé. 1989a. Selection for altered pheromone-component ratios in the pink bollworm moth, *Pectinophora gossypiella* (Lepidoptera: Gelechiidae). *Journal of Insect Behavior* **2**:609–621.

Collins, R.D. and R.T. Cardé. 1989b. Heritable variation in pheromone response of the pink bollworm, *Pectinophora gossypiella* (Lepidoptera: Gelechiidae). *Journal of Chemical Ecology* **15**:2647–2659.

Collins, R.D. and R.T. Cardé. 1989c. Selection for increased pheromone response in the male pink bollworm moth, *Pectinophora gossypiella* (Lepidoptera: Gelechiidae). *Behavior Genetics* **20**:325–331.

Collins, R.D., S.L. Rosenblum, and R.T. Cardé. 1990. Selection for increased pheromone titre in the pink bollworm moth, *Pectinophora gossypiella* (Lepidoptera: Gelechiidae). *Physiological Entomology* **15**:141–147.

Collyer, E. and M. van Geldermalsen. 1975. Integrated control of apple pests in New Zealand. 1. Outline of experiment and general results. *New Zealand Journal of Zoology* **2**:101–134.

Common, I.F.B. 1975. Evolution and classification of the Lepidoptera. *Annual Review of Entomology* **20**:183–203.

Conner, W.E., T. Eisner, R.K. Vander Meer, A. Guerrero, D. Ghiringelli and J. Meinwald. 1980. Sex attractant of an arctiid moth (*Utetheisa ornatrix*): a pulsed chemical signal. *Behavioral Ecology and Sociobiology* **7**:55–63.

Conner, W.E., T. Eisner, R.K. Vander Meer, A. Guerrero and J. Meinwald. 1981. Precopulatory sexual interaction in an arctiid moth (*Utetheisa ornatrix*): role of a pheromone derived from dietary alkaloids. *Behavioral Ecology and Sociobiology* **9**:227–235.

Cork, A., K.S. Boo, E. Dunkelblum, D.R. Hall, K. Jee-Rajunga, M. Kehat, E. Kong Jie, K.C. Park, P. Teptidagarin and Liu Xun. 1991. Female sex pheromone of *Helicoverpa assulta* (Guenee) (Lepidoptera: Noctuidae): identification and field testing. *Journal of Chemical Ecology* **3**:403–418.

Corrent, G., D.J. McAdoo and A. Eskin. 1978. Serotonin shifts the phase of the circadian rhythm from the *Aplysia eye. Science* **202**:977–979.

Crow, T. and M.S. Bridge. 1985. Serotonin modulates photoresponses in *Hermissenda* type-B photoreceptors. *Neuroscience Letters* **60**:83–88.

Culvenor, C.C.J. and J.A. Edgar. 1972. Dihydropyrrolizine secretions associated with coremata of *Utetheisa* moths (family Arctiidae). *Experientia* **28**:627–628.

Cunningham, R.T., R.M. Kobayashi and D.H. Miyashita. 1990. The male lures of tephritid flies. In *Behavior-Modifying Chemicals for Insect Management*, eds. R.L. Ridgway, R.M. Silverstein and M.N. Inscoe, pp. 255–268. Marcel Dekker, New York.

Curtis, C.E., J.D. Clark, D.A. Carlson and J.A. Coffelt. 1987. A pheromone mimic: disruption of mating communication in the navel orangeworm, *Amyelois transitella*, with *Z,Z*-1,12,14-heptadecatriene. *Entomologia Experimentalis et Applicata* **44**:249–255.

Cusson, M. and J.N. McNeil. 1989a. Involvement of juvenile hormone in the regulation of pheromone release activities in a moth. *Science* **243**:210–212.

Cusson, M. and J.N. McNeil. 1989b. Ovarian development in female armyworm moths, *Pseudaletia unipuncta*: its relationship with pheromone release activities. *Canadian Journal of Zoology* **67**:1380–1385.

Cusson, M., J.N. McNeil and S.S. Tobe. 1990. *In vitro* biosynthesis of juvenile hormone by corpora allata of *Pseudaletia unipuncta* virgin females as a function of age, environmental conditions, calling behaviour and ovarian development. *Journal of Insect Physiology* **36**:139–146.

Cusson, M., K.J. Yagi, S.S. Tobe and J.N. McNeil. 1993. Identification of release products of corpora allata of male and female armyworm moths *Pseudaletia unipuncta. Journal of Insect Physiology* **39**:775–783.

Cusson, M., S.S. Tobe and J.N. McNeil. 1994a. Juvenile hormones: their role in the regulation of the pheromone communication system of the armyworm moth, *Pseudaletia unipuncta. Archives of Insect Biochemistry and Physiology* **25**:329–345.

Cusson, M., C.G. Yu, K. Carruthers, G.R. Wyatt, S.S. Tobe and J.N. McNeil. 1994b. Regulation of vitellogenin production in armyworm moths, *Pseudaletia unipuncta. Journal of Insect Physiology* **40**:129–136.

Cymborowski, B. 1970a. Investigations of the neurohormonal factors controlling circadian rhythm of locomotor activity in the house cricket I. The role of the brain and subesophageal ganglion. *Zoologia Poloniae* **20**:103–126.

Cymborowski, B. 1970b. Investigations of the neurohormonal factors controlling circadian rhythm of locomotor activity in the house cricket II. Daily histochemical changes in neurosecretory cells of the pars intercerebralis and subesophageal ganglion. *Zoologia Poloniae* **20**:127–153.

Davenport, A. and P. Evans. 1984a. Stress-induced changes in the octopamine levels of insect hemolymph. *Insect Biochemistry* **14**:135–143.

Davenport, A. and P. Evans. 1984b. Changes in hemolymph octopamine levels associated with food deprivation in the locust. *Physiological Entomology* **9**:269–274.

David, C.T. 1982. Competition between fixed and moving stripes in the control of orientation by flying *Drosophila. Physiological Entomology* **7**:151–156.

David, C.T. 1986. Mechanisms of directional flight in wind. In *Mechanisms in Insect Olfaction*, eds. T.L. Payne, M.C. Birch and C.E.J. Kennedy, pp. 49–57. Clarendon Press, Oxford.

David, C.T. and M.C. Birch. 1989. Pheromones and insect behaviour. In *Insect Pheromones in Plant Protection*, eds. A.R. Jutsum and R.F.S. Gordon, pp. 17–35. Wiley, New York.

David, C.T. and J.S. Kennedy. 1987. The steering of zigzagging flight by male gypsy moths. *Naturwissenschaften* **74**:194–196.

David, C.T., J.S. Kennedy, A.R. Ludlow, J.N. Perry, and C. Wall. 1982. A reappraisal of insect flight towards a distant point source of wind-borne odor. *Journal of Chemical Ecology* **8**:1207–1215.

David, C.T., J.S. Kennedy and A.R. Ludlow. 1983. Finding a sex pheromone source by gypsy moths, *Lymantria dispar*, released in the field. *Nature* **303**:804–806.

David, J.C. and H. Verron. 1982. Locomotor behavior in relation to octopamine levels in the ant *Lasius niger. Experientia* **38**:650–651.

Davis, D.R. 1975. Systematics and zoogeography of the family Neopseustidae with the proposal of a new superfamily (Lepidoptera: Neopseustoidea). *Smithsonian Contributions to Zoology* **210**:1–45.

Davis, D.R. 1978. A revision of the North American moths of the superfamily Eriocranioidea with the proposal of a new family Acanthopteroctetidae (Lepidoptera). *Smithsonian Contributions to Zoology* **251**:1–231.

Davis, D.R. 1989. Generic revision of the Opostegidae, with a synoptic catalog of the world's species (Lepidoptera: Nepticuloidea). *Smithsonian Contributions to Zoology* **478**:1–97.

Davis, D.R., O. Pellmyr and J.N. Thompson. 1992a. Biology and systematics of *Greya* Busck and *Tetragma*, new genus (Lepidoptera: Prodoxidae). *Smithsonian Contributions to Zoology* **524**:1–88.

Davis, M.B., V.N. Vakharia, J. Henry, T.G. Kempe and A.K. Raina. 1992b. Molecular cloning of the pheromone biosynthesis-activating neuropeptide in *Helicoverpa zea. Proceedings of the National Academy of Science U.S.A.* **89**:142–146.

Davis, N.T., H.K. Lehman, P.E.A. Teal and J.G. Hildebrand. 1993. Anatomy and peptide immunochemistry of the neurosecretory cells of the subesophageal ganglion of *Manduca sexta. Society of Neuroscience Abstract* **19**:931.

Dawson, G.W., D.C. Griffiths, A. Hassanali, J.A. Pickett, R.T. Plumb, B.J. Pye, L.E. Smart and C.M. Woodcock. 1986. Antifeedants: a new concept for control of barley yellow dwarf virus in winter cereals. *Proceedings Brighton Crop Protection Conference—Pests and Diseases 1986*, British Crop Protection Council, pp. 1001–1008.

Dawson, G.W., D.C. Griffiths, N.F. Janes, A. Mudd, J.A. Pickett, L.J. Wadhams and C.M. Woodcock. 1987. Identification of an aphid sex pheromone. *Nature* **325**:614–616.

Dawson, G.W., N.F. Janes, A. Mudd, J.A. Pickett, A.M.Z. Slawin, L.J. Wadhams and D.J. Williams. 1989. The aphid sex pheromone. *Pure and Applied Chemistry* **61**:555–558.

Dawson, G.W., D.C. Griffiths, L.A. Merritt, A. Mudd, J.A. Pickett, L.J. Wadhams and C.M. Woodcock. 1990. Aphid semiochemicals—a review, and recent advances on the sex pheromone. *Journal of Chemical Ecology* **16**:3019–3030.

de Groot, P. Biosystematics of *Conophthorus* Hopkins (Coleoptera: Scolytidae) in Eastern North America. 1991. Ph.D. thesis, Simon Fraser University, Burnaby, Canada.

De Jong, M.C.M. and M.W. Sabelis. 1991. Limits to runaway sexual selection: the wallflower paradox. *Journal of Evolutionary Biology* **4**:637–656.

De Kramer, J.J. and J. Hemberger. 1987. The neurobiology of pheromone reception. In *Pheromone Biochemistry*, eds. G.D. Prestwich and G.J. Blomquist, pp. 433–472. Academic Press, Orlando, Florida.

Delisle, J. 1992. Age-related changes in the calling behaviour and the attractiveness of obliquebanded leafroller virgin females, *Choristoneura rosaceana*, under different constant and fluctuating temperature conditions. *Entomologia Experimentalis et Applicata* **3**:55–62.

Delisle, J. and A. Bouchard. 1995. Male larval nutrition in *Choristoneura rosaceana* (Lepidoptera: Tortricidae): An important factor of reproductive success. *Oecologia* **104**:508–517.

Delisle, J. and M. Hardy. Male larval nutrition influences the reproductive success of both sexes in the spruce budworm, *Christoneura fumiferana* (Lepidoptera: Tortricidae). *Functional Ecology* (in press).

Delisle, J. and J.N. McNeil. 1986. The effect of photoperiod on the calling behaviour of virgin females of the true armyworm *Pseudaletia unipuncta* (Haw.) (Lepidoptera: Noctuidae). *Journal of Insect Physiology* **32**:199–206.

Delisle, J. and J.N. McNeil. 1987a. The combined effect of photoperiod and temperature on the calling behaviour of the true armyworm, *Pseudaletia unipuncta*. *Physiological Entomology* **12**:157–164.

Delisle, J. and J.N. McNeil. 1987b. Calling behaviour and pheromone titre of the true armyworm, *Pseudaletia unipuncta* (Haw.) (Lepidoptera: Noctuidae) under different temperature and photoperiodic conditions. *Journal of Insect Physiology* **33**:315–324.

Delisle, J., J.N. McNeil, E.W. Underhill and D. Barton. 1989. *Helianthus annuus* pollen, an oviposition stimulant for the sunflower moth, *Homoeosoma electellum*. *Entomologia Experimentalis et Applicata* **50**:53–60.

Den Otter, C.J. 1977. Single sensillum responses in the male moth *Adoxophyes orana* (F.v.R.) to female sex pheromone components and their geometrical isomers. *Journal of Comparative Physiology* **121**:205–222.

Denholm, I. and M.W. Rowland. 1992. Tactics for managing pesticide resistance in arthropods: theory and practice. *Annual Review of Entomology* **37**:91–112.

Dent, D. 1991. *Insect Pest Management*. CAB International, Wallingford, United Kingdom.

Dent, D.R. and C.S. Pawar. 1988. The influence of moonlight and weather on catches of *Helicoverpa armigera* (Hübner) (Lepidoptera: Noctuidae) in light and pheromone traps. *Bulletin of Entomological Research* **78**:365–377.

Derby, C.D., K.A. Hamilton and B.W. Ache. 1984. Processing of olfactory information at three neuronal levels in the spiny lobster. *Brain Research* **300**:311–319.

Derrick, M.E., J.W. v Duyn, C.E. Sorenson and G.G. Kennedy. 1992. Effect of pheromone trap placement on capture of male European corn borer (Lepidoptera: Pyralidae) in three North Carolina crops. *Environmental Entomology* **21**:240–246.

DeSalle, R. and A.R. Templeton. 1987. Comments on "The significance of asymmetrical sexual isolation". *Evolutionary Biology* **21**:21–27.

Descoins, C. 1990. Grape berry moth and grape vine moth in Europe. In *Behavior-Modifying Chemicals for Insect Management*, eds. R.L. Ridgway, R.M. Silverstein and M. Inscoe, pp. 213–222. Marcel Dekker, New York.

Devlin, D.R. and J.H. Borden. 1994. Efficacy of antiaggregants for the pine engraver, *Ips pini* (Say) (Coleoptera: Scolytidae). *Canadian Journal of Forest Research* **24**:2469–2476.

Devroetes, P.N. 1983. Cyclic necleotides and cell-cell communication in *Dictyostelium discoideum*. In *Advances in Cyclic Nucleotide Research*, eds. P. Greengard and G.A. Robinson, **15**:55–93. Raven Press, New York.

Dickens, J.C. 1989. Green leaf volatiles enhance aggregation pheromone of boll weevil, *Anthonomus grandis*. *Entomologia Experimentalis et Applicata* **52**:191–203.

Dickens, J.C. 1990. Specialized receptor neurons for pheromone and host plant odors in the boll weevil, *Anthonomus grandis* Boh. (Coleoptera: Curculionidae). *Chemical Senses* **15**:311–331.

Dickens, J.C. 1994. Sensitivity and responsiveness of neurons associated with male-specific trichoid sensilla in the beet armyworm, *Spodoptera exigua*, to volatile emissions of sympatric *Spodoptera* spp. and plant odors. In *Olfaction and Taste XI*, eds. K. Kurihara, N. Suzuki and H. Ogawa, pp. 847–848. Springer-Verlag, Tokyo.

Dickens, J.C. and F.C. Callahan. 1996. Antennal-specific protein in tarnished plant bug, *Lygus lineolaris*: Production and reactivity of antisera. *Entomologia Experimentalis et Applicata* **80**:19–22.

Dickens, J.C. and T.L. Payne. 1977. Bark beetle olfaction: pheromone receptor system in *Dendroctonus frontalis*. *Journal of Insect Physiology* **23**:481–489.

Dickens, J.C., E.B. Jang, D.M. Light and A.R. Alford. 1990. Enhancement of pheromone responses by green leaf volatiles. *Naturwissenschaften* **77**:29–31.

Dickens, J.C., R.F. Billings and T.L. Payne. 1992. Green leaf volatiles interrupt aggregation pheromone response in bark beetles infesting southern pines. *Experientia* **48**:523–524.

Dickens, J.C., J.W. Smith and D.M. Light. 1993. Green leaf volatiles enhance sex attractant pheromone of the tobacco budworm, *Heliothis virescens* (Lep.: Noctuidae). *Chemoecology* **4**:175–177.

Dickens, J.C., F.E. Callahan, W.P. Wergin and E.F. Erbe, 1995. Olfaction in a hemimetabolous insect: antennal-specific protein in adult *Lygus lineolaris* (Heteroptera: Miridae). *Journal of Insect Physiology* **41**:857–867.

Doane, C.C. 1968. Aspects of mating behavior of the gypsy moth. *Annals of the Entomological Society of America* **61**:768–773.

Dobzhansky, T. 1947. *Genetics and the Origin of Species*. Columbia University Press, New York.

Dobzhansky, T. 1951. *Genetics and the Origin of Species*. 3rd edition. Columbia University Press, New York.

Dohlman, H.G., M.G. Caron and R.J. Lefkowitz. 1987. A family of receptors coupled to guanine nucleotide regulatory proteins. *Biochemistry* **26**:2657–2664.

Donoghue, M.J. 1989. Phylogenies and the analysis of evolutionary sequences with examples from seed plants. *Evolution* **43**:1137–1156.

Downer, R.G.H. and R.J. Martin. 1987. Analysis of monoamines and their metabolites by high performance liquid chromatography with coulometric electrochemical detection. *Life Sciences* **41**:833–836.

Downer, R.G.H., B.A. Bailey and R.J. Martin. 1985. Estimation of biogenic amines in biological tissues. In *Neurobiology*, eds. R. Gilles and J. Balthazart, pp. 247–263. Springer-Verlag, Berlin.

Drapek, R.J., L.B. Coop, B.A. Croft and G.C. Fisher. 1990. *Heliothis zea* pheromone trapping: studies of trap and lure combinations and field placement in sweet corn. *Southwestern Entomologist* **15**:63–69.

Dreisig, H. 1980. The importance of illumination levels in the daily onset of flight activity in nocturnal moths. *Physiological Entomology* **5**:327–342.

Driesig, H. 1986. Timing of daily activities in adult Lepidoptera. *Entomologia Generalis.* **12**:24–43.

Drummond, B.A., III. 1984. Multiple mating and sperm competition. In *Sperm Competition and the Evolution of Animal Mating Systems*, ed. R.L. Smith, pp. 291–370. Academic Press, New York.

Du, G. and G.D. Prestwich. 1995. Protein structure encodes the ligand binding specificity in pheromone binding proteins. *Biochemistry* **34**:8726–8732.

Du, G., C.-S. Ng and G.D. Prestwich. 1994. Odorant binding by a pheromone binding protein: active site mapping by photoaffinity labeling. *Biochemistry* **33**:4812–4819.

Duffield, R.M. 1981. 2-Nonanol in the exocrine secretion of the nearctic caddisfly *Rhyacophila fuscula* (Walker) (Rhyacophilidae: Trichoptera). *Proceedings of the Entomological Society of Washington* **83**:61–63.

Duffield, R.M., M.S. Blum, J.B. Wallace, H.A. Lloyd and F.E. Regnier. 1977. Chemistry of the defensive secretion of the caddisfly *Pycnopsyche scabripennis* (Trichoptera: Limnephilidae), *Journal of Chemical Ecology* **3**:649–656.

Dugdale, J.S. 1974. Female genital configuration in the classification of Lepidoptera. *New Zealand Journal of Zoology* **1**:127–146.

Dugdale, J.S. 1990. Reassessment of *Ctenopseustis* Meyrick and *Planotortrix* Dugdale with descriptions of two new genera (Lepidoptera: Tortricidae). *New Zealand Journal of Zoology* **17**:437–465.

Dugdale, J.S. 1994. Hepialidae (Insecta: Lepidoptera). *Fauna of New Zealand* **30**:1–163.

Dulac, C. and R. Axel. 1995. A novel family of genes encoding putative pheromone receptors in mammals. *Cell* **83**: 195–206.

Dunkelblum, E. and M. Kehat. 1989. Female sex pheromone components of *Heliothis peltigera* (Lepidoptera: Noctuidae): chemical identification from gland extracts and male response. *Journal of Chemical Ecology* **15**:2233–2245.

Dunkelblum, E., H. Mamane, M. Altstein and Z. Goldschmidt. 1989. Acetylation of alcohols in the pheromone biosynthesis of the tomato looper, *Chrysodeixis chalcites* (Lepidoptera:Noctuidae). *Insect Biochemistry* **19**:523–526.

Dyer, E.D.A. and P.M. Hall. 1977. Effect of anti-aggregative pheromone 3,2-MCH and *trans*-verbenol on *Dendroctonus rufipennis* attacks on spruce stumps. *Journal of the Entomological Society of British Columbia* **74**:32–34.

Eberhard, W.G. 1985. *Sexual Selection and the Evolution of Animal Genitalia* Harvard University Press, Cambridge, Massachusetts.

Eckert, M., J. Rapus, A. Nürnberger, H. Penzlin. 1992. A new specific antibody reveals octopamine-like immunoreactivity in cockroach ventral nerve cord. *Journal of Comparative Neurology* **322**:1–15.

Edgar, J.A., C.C.J. Culvenor and T.E. Pliske. 1974. Coevolution of danaid butterflies with their host plants. *Nature* **259**:646–48.

Egelhaaf, M., and A. Borst. 1993. A look into the cockpit of the fly: visual orientation, algorithms, and identified neurons. *Journal of Neuroscience* **11**:4563–4574.

Eggleton, P. and R.I. Vane-Wright. 1994. *Phylogenetics and Ecology*. Academic Press, London.

Ehrman, L. and M. Wasserman. 1987. The significance of asymmetrical sexual isolation. *Evolutionary Biology* **21**:1–20.

Einhorn, J., J.F. Beauvais, M. Gallois, C. Descoins and R. Causse. 1984. Constituants secondaires de la phéromone sexuelle du Carpocapse des pommes, *Cydia pomonella* L. (Lepidoptera: Tortricidae). *Comptes Rendus de l'Académie de Sciences, Paris* **299**:773–777.

Eisenbraun, E.J., C.E. Browne, R.L. Irvin-Willis, D.J. McGurk, E.L. Eliel and D.L. Harris. 1980. Structure and stereochemistry of 4aβ, 7α, 7aβ-nepetalactone from *Nepta mussini* and its relationship to the 4aα, 7α, 7aα- and 4aα, 7α, 7aβ-neptetalactones from *N. cataria*. *Journal of Organic Chemistry* **45**:3811–3814.

Elizondo-Alapisco, D. and B. Avarado-Rodriguez. 1988. Uso de avispitas *Trichogramma pretiosum* en tomate industrial en Sinaloa. Taller Sobre el Manejo Integrado de Plagas en Tomate, Culiacan, Mexico, CAADES.

Elkinton, J.S. and R.T. Cardé. 1988. Effects of intertrap distance and wind direction on the interaction of gypsy moth (Lepidoptera: Lymantridae) pheromone-baited traps. *Environmental Entomology* **17**:764–769.

Elkinton, J.S., C. Schal, T. Ono and R.T. Cardé. 1987. Pheromone puff trajectory and upwind flight of the male gypsy moth in a forest. *Physiological Entomology* **12**:399–406.

Eltringham, H. 1915. Further observations on the structure of the scent organs in certain male danaine butterflies. *Transactions of the Entomological Society of London* **1915**:152–176.

Endler, J.A. 1992. Signals, signal conditions, and the direction of evolution. *American Naturalist* **139**:125–153.

Engelmann, F. 1960. Hormonal control of mating behavior in an insect. *Experientia* **16**:69–70.

Engelmann, F. 1970. *The Physiology of Insect Reproduction*. Pergamon Press, Oxford.

Erber, J., U. Homberg and W. Gronenberg. 1987. Functional roles of the mushroom bodies in insects. In *Arthropod brain: Its Evolution, Development, Structure and Functions*, ed. A.P. Gupta, pp. 485–511. Wiley, New York.

Erman, N.A. 1984. The mating behaviour of *Parthina linea* (Odontoceridae), a caddisfly of springs and seeps. In *Proceedings of the Fourth International Symposium on Trichoptera, Clemson, South Carolina*, ed. J.C. Morse, pp. 131–136, Series Entomologica 30. Junk, The Hague.

Eskin, A., G. Corrent, C.-Y. Lin, and D.J. McAdoo. 1982. Mechanism for shifting the phase of a circadian rhythm by serotonin: involvement of cAMP. *Proceedings of the National Academy of Science U.S.A.* **79**:660–664.

Ester, A., J. Gut, A.M. van Oosten and H.C.H. Pijnenburg. 1993. Controlling aphids in iceberg lettuce by alarm pheromone in combination with an insecticide. *Journal of Applied Entomology* **115**:432–440.

Evans, P. 1985. Octopamine. In *Pharmacology*, Vol. 11, *Comprehensive Insect Physiology, Biochemistry and Pharmacology*. eds. G. Kerkut and L. Gilbert, pp. 499–530, Pergamon Press, Oxford.

Ewert, J.P. 1980. *Neuroethology*. Springer-Verlag, Berlin.

Fabre, J.H. 1879. *Souveniers Entomologiques I–X*. Delagrave, Paris.

Fabriàs, G., R.A. Jurenka and W.L. Roelofs. 1992. Stimulation of sex pheromone production by proteinaceous extracts of the bursa copulatrix in the redbanded leafroller moth. *Archives of Insect Biochemistry and Physiology* **20**:75–86.

Fabriàs, G., M.-P. Marco and F. Camps. 1994. Effect of the pheromone biosynthesis activating neuropeptide on sex pheromone biosynthesis in *Spodoptera littoralis* isolated glands. *Archives of Insect Biochemistry and Physiology* **27**:77–87.

Faccioli, G., A. Antropoli and E. Pasqualini. 1993. Relationship between males caught with low pheromone doses and larval infestation of *Argyrotaenia pulchellana*. *Entomologia experimentalis et applicata* **68**:165–170.

Fackrell, J.E. 1978. A system for turbulent concentration measurement. *Journal of Physics E* **11**:1015–1022.

Fackrell, J.E. 1989. A flame ionization detector for measuring fluctuating concentration. *Journal of Physics E* **13**:888–893.

Fadool, D.A. and B.W. Ache. 1992. Plasma membrane inositol 1,4,5-triphosphate-activated channels mediate signal transduction in lobster olfactory neurons. *Neuron* **9**:907–918.

Fadool, D.A. and B.W. Ache. 1994. Mediation of opposing transduction cascades in lobster olfactory receptor neurons by cAMP and IP_3. In *Olfaction and Taste XI*, eds. K. Kurihara, N. Suzuki and H. Ogawa, pp. 178–182. Springer-Verlag, Tokyo.

Falconer, D.S. 1980. *Introduction to Quantitative Genetics*, 2nd edition. Longman Scientific and Technical, Essex, United Kingdom.

Fang, N., J.H. Tumlinson, P.E.A. Teal and H. Oberlander. 1992. Fatty acyl pheromone precursor in the sex pheromone gland of female tobacco hornworm moths, *Manduca sexta* (L.). *Insect Biochemistry and Molecular Biology* **22**:621–631.

Fang, N., P.E.A. Teal and J.H. Tumlinson. 1995a. PBAN regulation of pheromone biosynthesis in female tobacco hornworm moths, *Manduca sexta* (L.). *Archives of Insect Biochemistry and Physiology* **29**:35–44.

Fang, N., P.E.A. Teal, R.E. Doolittle and J.H. Tumlinson. 1995b. Biosynthesis of conjugated olefinic systems in the sex pheromone gland of female tobacco hornworm moths, *Manduca sexta* (L.). *Insect Biochemistry and Molecular Biology* **25**:39–48.

Farina, W.M., D. Kramer and D. Varju. 1995. The response of the hovering hawk moth *Macroglossum stellatarum* to translatory pattern motion. *Journal of Comparative Physiology A* **176**:551–562.

Farkas, S.R. and H.H. Shorey. 1972. Chemical trail-following by flying insects. A mechanism for orientation. *Science* **178**:67–68.

Fatzinger, C.W. 1973. Circadian rhythmicity of sex pheromone release by *Dioryctria abietella* (Lepidoptera: Pyralidae (Phycitinae)) and the effect of a diel light cycle on its precopulatory behavior. *Annals of the Entomological Society of America* **66**:1147–1153.

Feldhege, M., F. Louis and H. Schmutterer. 1995. Untersuchungen über Falterabundanzen des Bekreuzten Traubenwicklers *Lobesia botrana* Schiff. im Weinbau. *Anzeiger für Schädlingskunde, Pflanzenschutz, Umweltschutz* **68**:85–91.

Felsenstein, J. 1985. Phylogenies and the comparative method. *American Naturalist* **125**:1–15.

Fentress J.C. 1980. How can behavior be studied from a neuroethological perspective? In *Information Processing in the Nervous System*, eds. H.M. Pinsker and W.D. Willis, Jr., pp. 263–283. Raven Press, New York.

Fentress J.C. 1991. Analytical ethology and synthetic neuroscience. In *The Development and Integration of Behavior*, ed. P. Bateson, pp. 77–120. Cambridge University Press, Cambridge.

Ferguson, D.C. 1985. Contributions toward reclassification of the world genera of the tribe Arctiini, Part 1—Introduction and a revision of the *Neoarctia-Grammia* group (Lepidoptera: Arctiidae; Arctiinae). *Entomography* **3**:181–275.

Feyereisen, R. 1985. Regulation of juvenile hormone titer: synthesis. In *Endocrinology*, Vol. 7, *Comprehensive Insect Physiology, Biochemistry and Pharmacology*, eds. G.A. Kerkut and L.I. Gilbert, pp. 391–429. Pergamon Press, Oxford.

Fisher, R.A. 1930. *The Genetical Theory of Natural Selection*. Oxford University Press, Oxford.

Fisher, R.A. 1958. *The Genetical Theory of Natural Selection*. Dover, New York.

Fitzpatrick, S.M. and J.N. McNeil. 1988. Male scent in lepidopteran communication the role of male pheromone in mating behavior of *Pseudaletia unipuncta* Haw. (Lepidoptera: Noctuidae). *Memoirs of The Entomological Society of Canada* **146**:131–152.

Fitzpatrick, S.M. and J.N. McNeil. 1989. Lifetime mating potential and reproductive success in males of the true armyworm, *Pseudaletia unipuncta* (Haw.) (Lepidoptera: Noctuidae). *Functional Ecology* **3**:37–44.

Flint, H.M., B.S. Wright, H.A. Sallam, and B. Horn. 1975. Dispersal and mating in the field by male pink bollworm, *Pectinophora gossypiella*, labeled with ^{32}P. *Entomologia Experimentalis et Applicata* **18**:451–456.

Flint, H.M., J.R. Merkle and A. Yamamoto. 1985. Pink bollworm (Lepidoptera: Gelechiidae): field testing a new polyethylene tube dispenser for gossyplure. *Journal of Economic Entomology* **78**:1431–1436.

Flint, H.M., A.K. Yamamoto, N.J. Parks and K. Nyomura. 1993. Aerial concentrations of gossyplure, the sex pheromone of the pink bollworm (Lepidoptera Gelechiidae) within and above cotton fields treated with long lasting dispensers. *Environmental Ecology* **22**:43–48.

Fónagy, A., S. Matsumoto, K. Uchiumi, C. Orisaka and T. Mitsui. 1992a. Action of pheromone biosynthesis activating neuropeptide on pheromone glands of *Bombyx mori* and *Spodoptera litura. Journal of Pesticide Science* **17**:47–54.

Fónagy, A., S. Matsumoto, K. Uchiumi and T. Mitsui. 1992b. Role of calcium ion and cyclic nucleotides in pheromone production in *Bombyx mori. Journal of Pesticide Science* **17**:115–121.

Fónagy, A., S. Matsumoto, L. Schoofs, A. De Loof and T. Mitsui. 1992c. *In vivo* and *in vitro* pheromonotropic activity of two locustatachykinin peptides in *Bombyx mori. Bioscience Biotechnology and Biochemistry* **56**:1692–1693.

Fónagy, A., L. Schoofs, S. Matsumoto, A. De Loof and T. Mitsui. 1992d. Functional cross-reactivity of some *Locusta* myotropins and *Bombyx* pheromone biosynthesis activating neuropeptide. *Journal of Insect Physiology* **38**:651–657.

Food and Drug Administration. 1979. Results of FDA surveillance of Mexican produce for pesticide residues; Compliance Program 7305.008, April 23, 1979.

Forbush, E.H. and C.H. Fernald. 1896. *The Gypsy Moth*. Wright & Potter, Boston.

Foster, S.P. 1993. Neural inactivation of sex pheromone production in mated lightbrown apple moths, *Epiphyas postvittana* (Walker). *Journal of Insect Physiology* **39**:267–273.

Foster, S.P. and J.S. Dugdale. 1988. A comparison of morphological and sex pheromone differences in some New Zealand Tortricinae Moths. *Biochemical Systematics and Ecology* **16**:227–232.

Foster, S.P. and W.L. Roelofs. 1987. Sex pheromone differences in populations of the brownheaded leafroller, *Ctenopseustis obliquana. Journal of Chemical Ecology* **13**:623–629.

Foster, S.P. and W.L. Roelofs. 1988. Sex pheromone biosynthesis in the leafroller moth *Planotortrix excessana* by 10 desaturation. *Archives of Insect Biochemistry and Physiology* **8**:1–9.

Foster, S.P. and W.L. Roelofs. 1994. Regulation of pheromone production in virgin and mated females of two tortricid moths. *Archives of Insect Biochemistry and Physiology* **25**:271–285.

Foster, S.P., J.C. Bergh, S. Rose and M.O. Harris. 1991a. Aspects of pheromone biosynthesis in the hessian fly, *Mayetiola destructor* (SAY). *Journal of Insect Physiology* **37**:899–906.

Foster, S.P., J.S. Dugdale and C.S. White. 1991b. Sex pheromones and the status of greenheaded and brownheaded leafroller moths in New Zealand. *New Zealand Journal of Zoology* **18**:63–74.

Fowler, D.J. and C.J. Goodnight. 1966. Neurosecretory cells: daily rhythmicity in *Leiobunum longipes*. *Science* **152**:1978–1080.

Fowler, D.J., C.J. Goodnight, and M.M. LaBrie. 1972. Circadian rhythms of 5-hydroxtryptamine production in larvae, pupae, and adults of *Drosophila melanogaster*. *Annals of the Entomological Society of America* **65**:138–141.

Frérot, B., M. Renou, C. Descoins, M. Gallois. 1984. Les acétoxy-1 tétradécène-11Z et 11E (Z11 TDA et E11 TDA), constituants de la phéromone sexuelle de *Argyrotaenia pulchellana* (Haw.). (Lepid., Tortricianae, Archipini). *Agronomie* **4**:565–572.

Frumhoff, P.C. and H.K. Reeve. 1994. Using phylogenies to test hypotheses of adapatation: a critique of some current proposals. *Evolution* **48**:172–180.

Fukuda, S. 1951. The production of diapause eggs by transplanting the subesophageal ganglion in the silkworm. *Proceedings of the Japan Academy* **27**:672–677.

Furniss, M.M. and R.L. Livingston. 1979. Inhibition by ipsenol of pine engraver attraction in northern Idaho. *Environmental Entomology* **8**:369–372.

Furniss, M.M., L.N. Kline, R.F. Schmitz and J.A. Rudinsky. 1972. Tests of three pheromones to induce or disrupt aggregation of Douglas-fir beetles (Coleoptera: Scolytidae) on live trees. *Annals of the Entomological Society of America* **65**:1227–1232.

Furniss, M.M., G.E. Daterman, L.N. Kline, M.D. McGregor, G.C. Trostle, L.F. Pettinger and J.A. Rudinsky. 1974. Effectiveness of the Douglas-fir beetle antiaggregative pheromone methylcyclohexenone at three concentrations and spacings around felled host trees. *Canadian Entomologist* **106**:381–392.

Furniss, M.M., B.H. Baker and B.B. Hostetler. 1976. Aggregation of spruce beetles (Coleoptera) to seudenol and repression of attraction by methylcyclohexeanone in Alaska. *Canadian Entomologist* **108**:1297–1302.

Furniss, M.M., J.W. Young, M.D. McGregor, R.L. Livingston and D.R. Hamel. 1977. Effectiveness of controlled-release formulations of MCH for preventing Douglas fir-beetle (Coleoptera: Scolytidae) infestation in felled trees. *Canadian Entomologist* **109**:1063–1069.

Furniss, M.M., R.W. Clausen, G.P. Markin, M.D. McGregor and R.L. Livingston. 1981. Effectiveness of Douglas-fir beetle antiaggregative pheromone applied by helicopter. United States Department of Agriculture, Forest Service, General Technical Report INT-101.

Furniss, M.M., G.P. Markin, and Y.J. Hager. 1982. Aerial applications of Douglas-fir beetle antiaggregative pheromone: equipment and evaluation. United States Department of Agriculture, Forest Service, General Technical Report INT-137.

Futrelle, R.P. 1984. How molecules get to their detectors, the physics of diffusion of insect pheromones. *Trends in Neuroscience* **7**:116–120.

Gadenne, C. 1993. Effects of fenoxycarb, juvenile hormone mimetic, on female sexual behaviour of the black cutworm, *Agrotis ipsilon* (Lepidoptera: Noctuidae). *Journal of Insect Physiology* **39**:25–29.

Gadot, M., E.L. Burns and C. Schal. 1989a. Juvenile hormone biosynthesis and oocyte development in adult female *Blattella germanica*: effects of grouping and mating. *Archives of Insect Biochemistry and Physiology* **11**:189–200.

Gadot, M., A-S. Chiang and C. Schal. 1989b. Farnesoic acid-stimulated rates of juvenile hormone biosynthesis during the gonotrophic cycle of *Blattella germanica. Journal of Insect Physiology* **35**:537–542.

Gadot, M., A-S. Chiang, E.L. Burns and C. Schal. 1991. Cyclic juvenile hormone biosynthesis in the cockroach, *Blattella germanica*: effects of ovariectomy and corpus allatum denervation. *General and Comparative Endocrinology* **82**:163–171.

Gangavalli, R.R. and M.T. AliNiazee. 1985. Diapause induction in the oblique-banded leafroller *Choristoneura rosaceana* (Lepidoptera: Tortricidae): role of photoperiod and temperature. *Journal of Insect Physiology* **31**:831–835.

Garthwaite, J. 1991. Glutamate, nitric oxide and cell–cell signalling in the nervous system. *Trends in Neuroscience* **14**:60–67.

Gauld, I.D. and L.A. Mount. 1982. Homoplasy and the delineation of holophyletic genera in some inset groups. *Systematic Entomology* **7**:73–86.

Gayeski, T.E.J., R.J. Connet and C.R. Honig. 1985. Oxygen transport in rest-work transition illustrates new functions for myoglobin. *American Journal of Physiology* **248**:914–921.

Gazit, Y., E. Dunkelblum, M. Benichis and M. Altstein. 1990. Effect of synthetic PBAN and derived peptides on sex pheromone biosynthesis in *Heliothis peltigera* (Lepidoptera: Noctuidae). *Insect Biochemistry* **20**:853–858.

Gazit, Y., E. Dunkelblum, O. Ben-Aziz and M. Altstein. 1992. Immunochemical and biological analysis of pheromone biosynthesis activating neuropeptide in *Heliothis peltigera. Archives of Insect Biochemistry and Physiology* **19**:247–260.

Georghiou, G. 1990. Overview of insecticide resistance. In *Managing Resistance to Agrichemicals,* American Chemical Society Symposium Series 421. American Chemical Society, Washington, D.C.

Georghiou, G.P. and A. Lagunes-Tejeda. 1991. *The Occurrence of Resistance to Pesticides.* United Nations Food and Agriculture Organization, Rome.

Getchell, T.V., J.L. Margolis and M.L. Getchell. 1984. Perireceptor and receptor events in vertebrate olfaction. *Progress in Neurobiology* **23**:317–345.

Gewecke, M. 1975. The influence of the air-current sense organs on the flight behavior of *Locusta migratoria. Journal of Comparative Physiology* **103**:79–95.

Gibson, K.E., R.F. Schmitz, G.D. Amman and R.D. Oakes. 1991. Mountain pine beetle response to different verbenone dosages in pine stands of western Montana. United States Department of Agriculture, Forest Service, Research Paper INT-444.

Giebultowicz, J.M., A.K. Raina, E.C. Uebel and R.L. Ridgway. 1991a. Two-step regulation of sex-pheromone decline in mated gypsy moth females. *Archives of Insect Biochemistry and Physiology* **16**:95–105.

Giebultowicz, J.M., A.K. Raina and R.L. Ridgway. 1991b. Role of the spermatheca in the termination of sex pheromone production in mated gypsy moth females. In *Insect Chemical Ecology*, ed. I. Hrdy, pp. 101–104. SPB Academic Publishing, The Hague.

Giebultowicz, J.M., R.E. Webb, A.K. Raina and R.L. Ridgway. 1992. Effects of temperature and age on daily changes in pheromone titer in laboratory reared and wild gypsy moths, *Lymantria dispar* (L.). (Lepidoptera, Lymantriidae) *Environmental Entomology* **21**:822–826.

Gilbert, C., N.J. Strausfeld. 1992. Small-field neurons associated with oculomotor and optomotor control in muscoid flies: functional organization. *Journal of Comparative Neurology* **316**:72–86.

Gilbert, L.I., W.L. Combest, W.A. Smith, V.H. Meller and D.B. Rountree. 1988. Neuropeptides, second messengers and insect molting. *BioEssays* **8**:153–157.

Gittleman, J.L. and H.-K. Luh. 1992. On comparing comparative methods. *Annual Review of Ecology and Systematics* **23**:383–404.

Gosalbo, L., G. Fabriàs and F. Camps. 1994. Inhibitory effect of 10,11-methylenetetradec-10-enoic acid on a Z9-desaturase in the sex pheromone biosynthesis of *Spodoptera littoralis*. *Archives of Insect Biochemistry and Physiology* **26**:279–286.

Götz, B. 1941. Der Sexualduftstoff als Bekämpfungsmittel gegen die Traubenwickler im Freiland. *Wein und Rebe* **23**:75.

Grandi, R., U.M. Pagnoni, A. Pinetti and R. Trave. 1984. The possible role of photocitral-A in the biosynthesis of cyclopentane monoterpenes. *Journal of Chemical Research* (S), 194–195.

Grant, A.J., R.J. O'Connell and A.M. Hammond. 1988. A comparative study of pheromone perception in two species of noctuid moths. *Journal of Insect Behavior* **1**:75–96.

Grant, A.J., R.J. O'Connell and T. Eisner. 1989. Pheromone-mediated sexual selection in the moth *Utetheisa ornatrix* olfactory receptor neurons responsive to a male-produced pheromone *Journal of Insect Behavior* **2**:371–386.

Grant, G.G. 1978. Morphology of the presumed male pheromone glands on the forewings of tortricid and phycitid moths. *Annals of the Entomological Society of America* **71**:423–431.

Grant, G.G. 1987. Copulatory behavior of spruce budworm, *Choristoneura fumiferana* (Lepidoptera: Tortricidae): Experimental analysis of the role of sex pheromone and associated stimuli. *Annals Entomological Society of America* **80**:78–88.

Grant, G.G., E.B. Smithwick and U.E. Brady. 1975. Courtship behavior of phycitid moths. II. Behavioral and pheromonal isolation of *Plodia interpunctella and Cadra cautella* in the laboratory. *Canadian Journal of Zoology* **53**:827–832.

Gray, T.G., R.F. Shepard, D.L. Struble, J.B. Byers and T.F. Maher. 1991. Selection of pheromone trap and attractant dispenser load to monitor black army cutworm, *Actebia fennica*. *Journal of Chemical Ecology* **17**:309–316.

Greenbank, D.O., G.W. Schaefer and R.C. Rainey. 1980. Spruce budworm (Lepidoptera:

Tortricidae) moth flight and dispersal: new understanding from canopy observations, radar and aircraft. *Memoirs of the Entomological Society of Canada* **110**:1–48.

Greenfield, M.D. and M.G. Karandinos. 1979. Resource partitioning of the sex communication channel in clearwing moths (Lepidoptera: Sesiidae) of Wisconsin. *Ecological Monographs* **49**:403–426.

Greig, B.J.W. and J.N. Gibbs. 1983. Control of Dutch elm disease in Britain. In *Research on Dutch Elm Disease in Europe*, ed. D.A. Burdekin, pp. 10–16. Forestry Commission (U.K.) Bulletin No. 60.

Gries, G., H.D. Pierce, Jr., B.S. Lindgren and J.H. Borden. 1988. New techniques for capturing and analyzing semiochemicals for scolytid beetles (Coleoptera: Scolytidae). *Journal of Economic Entomology* **81**:1715–1720.

Gronenberg W. and N.J. Strausfeld. 1990. Descending neurons supplying the neck and flight motor of Diptera: physiological and anatomical characteristics. *Journal of Comparative Neurology* **302**:973–991.

Grula, J.W. and O.R. Taylor. 1979. The inheritance of pheromone production in the sulfur butterflies, *Colias eurytheme and C. philodice. Heredity* **42**:359–371.

Grula, J.W. and O.R. Taylor. 1980. The effect of X-chromosome inheritance on mate-selection behavior in the sulfur butterflies, *Colias eurytheme* and *C. philodice. Evolution* **34**:688–695.

Gu, X., D. Quilici, P. Juarez, G.J. Blomquist and C. Schal. 1995. Biosynthesis of hydrocarbons and contact sex pheromone and their transport by lipophorin in females of the German cockroach (*Blattella germanica*). *Journal of Insect Physiology* **41**:257–267.

Guerin, P.M., H.R. Buser, M. Tóth, E. Höbaus, A. Schmid and H. Arn. 1986. Sex pheromone of *Sparganothis pilleriana*: *E*- and *Z*-11-tetradecenyl acetates as essential components. *Entomologia Experimentalis et Applicata* **40**:137–140.

Guillon, G., B. Mouillac and A.L. Savage. 1992. Modulation of hormone-sensitive phospholipase C. *Cellular Signalling* **4**:11–23.

Guldemond, J.A., A.F.G. Dixon, J.A. Pickett, L.J. Wadhams and C.M. Woodcock. 1993. Specificity of sex pheromones, the role of host plant odour in the olfactory attraction of males, and mate recognition in the aphid *Cryptomyzus. Physiological Entomology* **18**:137–143.

Gut, L.J. and J.F. Brunner. 1992. Mating disruption as a control for codling moth and leafrollers. *Proceedings of the Oregon Horticultural Society* **83**:67–75.

Gut, L.J. and J.F. Brunner. 1994. Implementation of pheromone-based pest management programs in pear in Washington, U.S.A. *International Organization for Biological Control—West Palearctic Regional Section Bulletin* **17**(2):77–85.

Györgyi, T.K., A.J. Roby-Shemkovitz and M.R. Lerner. 1988. Characterization and cDNA cloning of the pheromone-binding protein from the tobacco hornworm *Manduca sexta*: a tissue-specific developmentally regulated protein. *Proceedings of the National Academy of Science U.S.A.* **85**:9851–9855.

Hadjitoffi, A. and M.J.G. Wilson. 1979. Fast response measurement of air pollution. *Atmospheric Environment* **13**:755–760.

Hagedorn, H.H. 1983. The role of ecdysteroids in adult insects. In *Endocrinology of*

Insects, eds. R.G.H. Downer and H. Laufer, pp. 271–304. Allen R. Liss Publishers, New York.

Haile, D.G., Snow J.W., and Goodenough J.L. 1973. Reduced captures of tobacco budworm and corn earworm males in the electric grid traps baited simultaneously with virgin females of both species. *Journal Economic Entomology* **66**:739–740

Hales, R.A. and M.D. Breed. 1983. Female calling and reproductive behavior in the brown-banded cockroach, *Supella longipalpa* (F.). *Annals of the Entomological Society of America* **76**:239–241.

Hallahan, D.L., G.W. Dawson, J.M. West and R.M. Wallsgrove. 1992. Cytochrome P-450 catalysed monoterpene hydroxylation in *Nepeta mussinii. Plant Physiology and Biochemistry* **30**:435–443.

Hallahan, D.L., S.-M.C. Lau, P.A. Harder, D.W.M. Smiley, G.W. Dawson, J.A. Pickett, R.E. Christofferson and D.P. O'Keefe. 1994. Cytochrome P-450-catalysed monoterpenoid oxidation in catmint (*Nepeta racemosa*) and avocado (*Persea americana*); evidence for related enzymes with different activities. *Biochimica et Biophysica Acta* **1201**:94–100.

Halliday, T. and S.J. Arnold. 1987. Multiple matings by females: a perspective from quantitative genetics. *Animal Behavior* **35**:939–941.

Hamilton, J.G.C., M.J. Dougherty and R.D. Ward. 1994. Sex pheromone activity in a single component of tergal gland extract of *Lutzomyia longipalpis* (Diptera: Psychodidae) from Jacobina, Northeastern Brazil. *Journal of Chemical Ecology* **20**:141–151.

Hammerstein, P. and Parker, G.A. 1987. Sexual selection: games between the sexes. In *Sexual Selection: Testing the Alternatives*, eds. J.W. Bradbury and M.B. Andersson, pp. 119–142. Wiley, Chichester, United Kingdom.

Hammond, P.M. 1981. The origin and development of reproductive barriers. In *The Evolving Biosphere*, ed. P.L. Florey, pp. 49–60. British Museum and Cambridge University Press.

Han, E.-N. and A.G. Gatehouse. 1991a. Effect of temperature and photoperiod on the calling behaviour of a migratory insect, the oriental armyworm *Mythimna separata. Physiological Entomology* **16**:419–427.

Han, E-N. and A.G. Gatehouse. 1991b. Genetics of precalling period in the oriental armyworm *Mythimna separata* (Walker) (Lepidoptera: Noctuidae), and implications for migration. *Evolution* **45**:1502–1510.

Hanna, S.R. 1984. The exponential PDF and concentration fluctuations in smoke plumes. *Boundary-Layer Meteorology* **29**:361–375.

Hansson, B.S., C. Löfstedt, and W.L. Roelofs. 1987. Inheritance of olfactory response to sex pheromone components in *Ostrinia nubilalis. Naturwissenschaften* **74**:497–499.

Hansson, B.S., C. Löfstedt and S.P. Foster. 1989. *Z*-linked inheritance of male olfactory response to sex pheromone components in two species of tortricid moths, *Ctenopseustis obliquana* and *Ctenopseustis* sp. *Entomologia Experimentalis et Applicata* **53**:137–145.

Hansson, B.S., M. Tóth, C. Löfstedt, G. Szöcs, M. Subchev, and J. Löfqvist. 1990. Pheromone variation among eastern European and a western Asian population of the turnip moth, *Agrotis segetum. Journal of Chemical Ecology* **16**:1611–1622.

Hansson, B.S., T.A. Christensen and J.G. Hildebrand. 1991. Functionally distinct subdivisions of the macroglomerular complex in the antennal lobe of the male sphinx moth *Manduca sexta. Journal of Comparative Neurology* **312**:264–278.

Hansson, B.S., H. Ljungberg, E. Hallberg and C. Löfstedt. 1992. Functional specialization of olfactory glomeruli in a moth. *Science* **256**:1313–1315.

Hansson, B.S., S. Anton and T.A. Christensen. 1994. Structure and function of antennal lobe neurons in the male turnip moth, *Agrotis segetum* (Lepidoptera: Noctuidae). *Journal of Comparative Physiology A* **175**:547–562.

Hansson, B.S., T.J. Almaas and S. Anthon. 1995. Chemical communication in heliothine moths. V. Antennal lobe projection patterns of pheromone-detecting olfactory receptor neurons in the male *Heliothis virescens* (Lepidoptera: Noctuidae) *Journal of Comparative Physiology A* **177**:535–543.

Hardie, J., S.F. Nottingham, W. Powell and L.J. Wadhams. 1991. Synthetic aphid sex pheromone lures parasitoids. *Entomologia Experimentalis et Applicata* **61**:97–99.

Hardie, J., S.F. Nottingham, G.W. Dawson, R. Harrington, J.A. Pickett and L.J. Wadhams. 1992. Attraction of field-flying aphid males to synthetic sex pheromone. *Chemoecology* **3**:113–117.

Hardie, J., R. Isaacs, F. Nazzi, W. Powell, L.J. Wadhams and C.M. Woodcock. 1993. Electroantennogram and olfactometer responses of aphid parasitoids to nepetalactone, a component of aphid sex pheromones. *Proceedings 5th International Symposium of the Global IOBC Working Group: Ecology of Aphidophaga*, September 1993, Antibes, France, p. 29.

Hardie, J., A.J. Hick, C. Höller, J. Mann, L.A. Merritt, S.F. Nottingham, W. Powell, L.J. Wadhams, J. Witthinrich and A.F. Wright. 1994a. The responses of *Praon* spp. parasitoids to aphid sex pheromone components in the field. *Entomologia Experimentalis et Applicata* **71**:95–99.

Hardie, J., R. Isaacs, J.A. Pickett, L.J. Wadhams and C.M. Woodcock. 1994b. Methyl salicylate and (–)-(1*R*,5*S*)-myrtenal are plant-derived repellents for the black bean aphid, *Aphis fabae* Scop. (Homoptera, Aphididae). *Journal of Chemical Ecology* **20**:2847–2855.

Hardie, J., R. Storer, S.F. Nottingham, L. Peace, R. Harrington, L.A. Merritt, L.J. Wadhams and D.K. Wood. 1994c. The interaction of sex pheromone and plant volatiles for field attraction of male bird-cherry-oat aphid, *Rhopalosiphum padi*. In *Proceedings Brighton Crop Protection Conference—Pests and Diseases 1994*, British Crop Protection Council, pp. 1223–1230.

Hardie, J., J.H. Visser and P.G.M. Piron. 1994d. Perception of volatiles associated with sex and food by different adult forms of the black bean aphid, *Aphis fabae. Physiological Entomology* **19**:278–284.

Harring, C.M. 1978. Aggregation pheromones of the European fir engraver beetles *Pityokteines curvidens, P. spinidens* and *P. vorontzovi* and the role of juvenile hormone in pheromone biosynthesis. *Zeitshrift für angewandte Entomologie* **85**:281–317.

Harris-Warrick, R.M., E. Marder, A.I. Selverston, and M. Moulins. 1992. *Dynamic Biological Networks: The Stomatogastric Nervous System.* MIT Press, Cambridge, Massachusetts.

Harrison, R.D., ed. 1993. *Hybrid Zones and the Evolutionary Process*. Oxford University Press, New York.

Hasegawa, K. 1951. Studies on the voltinism in the silkworm, *Bombyx mori* L., with special reference to the organs concerning determination of voltinism (a preliminary note). *Proceedings of the Japan Academy* **27**:667–672.

Hausen, K. and M. Egelhaaf. 1989. Neural mechanisms of visual course control in insects. In *Facets of Vision*, eds. D.G. Stavenga and R.C. Hardie, pp. 391–424, Springer-Verlag, Berlin.

Hayes, J.L. 1991. Dynamics of nocturnal activity of moths in the *Heliothis* complex (Lepidoptera: Noctuidae) in cotton. *Journal of Economic Entomology* **84**:855–865.

Hayes, J.L. and B.L. Strom. 1994. 4-Allylanisole as an inhibitor of bark beetle (Coleoptera: Scolytidae) aggregation. *Journal of Economic Entomology* **87**:1586–1594.

Hayes, J.L, B.L. Strom, L.M. Roton and L.L. Ingram, Jr. 1994. Repellent properties of the host compound 4-allyanisole to the southern pine beetle. *Journal of Chemical Ecology* **20**:1595–1615.

Haynes, K.F. and T.C. Baker. 1988. Potential for evolution of resistance to pheromones: worldwide and local variation in chemical communication system of pink bollworm moth, *Pectinophora gossypiella. Journal of Chemical Ecology* **14**:1547–1560.

Haynes, K.F. and T.C. Baker. 1989. An analysis of anemotactic flight in female moths stimulated by host odour and comparison with the males' response to sex pheromone. *Physiological Entomology* **14**:279–289.

Haynes, K.F. and M.C. Birch. 1984a. The periodicity of pheromone release and male responsiveness in the artichoke plume moth. *Physiological Entomology* **9**:287–295.

Haynes, K.F. and M.C. Birch. 1984b. Mate-locating and courtship beahviors of the artichoke plume moth, *Platyptilia carduidactyla* (Lepidoptera: Pterophoridae). *Environmental Entomology* **13**:399–408.

Haynes, K.F. and R.E. Hunt. 1990a. Interpopulational variation in the six-component pheromone blend of the cabbage looper moth, *Trichoplusia ni. Journal of Chemical Ecology* **16**:509–519.

Haynes, K.F. and R.E. Hunt. 1990b. A mutation in the pheromonal communication system of the cabbage looper moth, *Trichoplusia ni. Journal of Chemical Ecology* **16**:1249–1257.

Haynes, K.F., L.K. Gaston, M. Mistrot-Pope and T.C. Baker. 1984. Potential for evolution of resistance to pheromones: interindividual and interpopulation variation in chemical communication system of pink bollworm moth. *Journal of Chemical Ecology* **10**:1551–1565.

Haynes, K.F., M.P. Parrella, J.T. Trumble and T.A. Miller. 1986. Monitoring insecticide resistance with yellow sticky cards. *California Agriculture* **40**:1

Heath, R.R., J.R. McLaughlin, F. Proshold, and P.E.A. Teal. 1991. Periodicity of female sex pheromone titer and release in *Heliothis subflexa* and *H. virescens* (Lepidoptera: Noctuidae). *Annals of the Entomological Society of America* **84**:182–189.

Heath, R.R., P.J. Landolt, B.D. Dueben, R.E. Murphy, and R.E. Schneider. 1992. Identification of male cabbage looper sex pheromone attractive to females. *Journal of Chemical Ecology* **18**:441–453.

Hedden, R.L. and G.B. Pitman. 1978. Attack density regulation: a new approach to the use of pheromones in Douglas-fir beetle population management. *Journal of Economic Entomology* **71**:633–637.

Hedrick, P.W. 1985. *Genetics of Populations*. Jones and Bartlett Publishers, Boston.

Heisenberg, M. and R. Wolf. 1979. On the fine structure of yaw torque in visual flight orientation of *Drosophila melanogaster. Journal of Comparative Physiology A* **130**:113–130.

Heisenberg, M. and R. Wolf. 1984. Vision in *Drosophila*: *Genetics of Microbehavior*. Springer-Verlag, Berlin.

Heisenberg, M. and R. Wolf. 1988. Reafferent control of optomotor yaw torque in *Drosophila melanogaster. Journal of Comparative Physiology A* **163**:373–388.

Hendrikse, A. 1986. Intraspecific and interspecific sex-pheromone communication in the genus *Yponomeuta. Physiological Entomology* **11**:159–170.

Hendrikse, A. and E. Vos-Bunnemeyer. 1987. Role of host-plant stimuli in sexual behaviour of small ermine moths (*Yponomeuta*). *Ecological Entomology* **12**:363–371.

Hengstenberg, R., D.C. Sandemann and B. Hengstenberg. 1986. Compensatory head roll in the blowfly *Calliphora* during flight. *Proceedings of Royal Society of London B* **227**:455–482.

Hensler K and C.H.F. Rowell. 1990. Control of optomotor responses by descending deviation detector neurones in intact flying locusts. *Journal of Experimental Biology* **149**:191–205.

Henzell, R.F. and M.D. Lowe. 1970. Sex attractant of the grass grub beetle. *Science* **168**:1005–1006.

Herault, J-P.E. and J.P. Proux. 1987. Cyclic AMP: the second messenger of an antidiuretic hormone from the glandular lobes of the migratory locust corpora cardiaca. *Journal of Insect Physiology* **33**:487–491.

Higgings, C.F., S.C. Hyde, M.M. Mimmack, U. Gileadi, D.R. Gill and M.P. Gallagher. 1990. Binding dependent transport systems. *Journal of Bioenergetics and Biomembranes* **22**:571–592.

Higgins, L.G. 1975. *The Classification of European Butterflies*. Collins, London.

Hildebrand, J.G. 1985. Metamorphosis of the insect nervous system: influences of the periphery on the postembryonic development of the antennal sensory pathway in the brain of *Manduca sexta*. In *Model Neural Networks and Behavior*, ed. A.I. Selverston, pp. 129–148, Plenum Press, New York.

Hildebrand, J.G. 1995. Analysis of chemical signals by nervous systems. *Proceedings of the National Academy of Science U.S.A.* **92**:67–74.

Hildebrand, J.G. and T.A. Christensen. 1994. Olfactory mechanisms underlying processing of sex-pheromonal information in moths. In *Olfaction and Taste XI*, eds. K. Kurihara, N. Suzuki and H. Ogawa, pp. 827–830, Springer-Verlag, Heidelberg.

Hildebrand, J.G., T.A. Christensen, E.A. Arbas, J.H. Hayashi, U. Homberg, R. Kanzaki, and M. Stengl. 1992. Olfaction in *Manduca sexta*: cellular mechanisms of responses to sex pheromone. In *Proceedings of NEUROTOX 91—Molecular Basis of Drug and Pesticide Action*, ed. I.R. Duce, pp. 323–338. Elsevier, London.

Hille Ris Lambers, D. and A. Schepers. 1978. The effect of *trans*-β-farnesene, used as a repellent against landing aphid alatae in seed potato growing. *Potato Research* **21**:23–26.

Hillebrand, W., K.W. Eichhorn and D. Lorenz. 1990. *Rebschutz-Taschenbuch*. 9. Aufl. Fachverlag Dr. Fraund GmbH, Mainz.

Hinks, C. 1967. Relationship between serotonin and the circadian rhythm in some nocturnal moths. *Nature* **214**:386–387.

Hiruma, K., S. Matsumoto, A. Isogai and A. Suzuki. 1984. Control of ommochrome synthesis by both juvenile hormone and melanization hormone in the cabbage armyworm, *Mamestra brassicae*. *Journal of Comparative Physiology B* **154**:13–21.

Ho, H.Y., M.P. Tu, C.Y. Chang, C.M. Yin and R. Kou. 1995. Identification of *in vitro* products of corpora allata in female and male loreyi leafworms, *Leucania loreyi*. *Experientia* **51**:1–5.

Höbaus E. 1988. Zur Verbreitung der Traubenwicklerarten *Eupoecilia ambiguella* Hb. und *Lobesia botrana* Schiff. in Österreich. *Pflanzenschutzberichte* **49**:34–40.

Hobson, K.R. 1995. Host compounds as semiochemicals for bark beetles. In *Proceedings of the Symposium on Applications of Semiochemicals for Management of Bark Beetle Infestations*, eds. S.M. Salom and K.R. Hobson, pp. 48–51. United States Department of Agriculture, Forest Service, General Technical Report INT-GRT-318.

Hollander, A.L. and C.-M. Yin. 1982. Neurological influences on pheromone release by brain, corpora cardiaca, corpora allata and ovaries of the female gypsy moth, *Lymantria dispar* (L.). *Journal of Insect Physiology* **31**:159–163.

Höller, C., S.G. Micha, S. Schulz, W. Francke and J.A. Pickett. 1994. Enemy-induced dispersal in a parasitic wasp. *Experientia* **50**:182–185.

Holman, G.M., B.J. Cook and R.J. Nachman. 1986. Primary structure and synthesis of a blocked myotropic neuropeptide isolated from the cockroach, *Leucophaea maderae*. *Comparative Chemistry and Physiology* **85C**:219–224.

Homberg, U. 1987. Structure and functions of the central complex in insects. In *Arthropod Brain: Its Evolution, Development, Structure, and Functions*, ed. A.P. Gupta, pp. 347–363. Wiley, New York.

Homberg, U. and J.G. Hildebrand. 1989a. Serotonin-immunoreactive neurons in the median protocerebrum and subesophageal ganglion of the sphinx moth *Manduca sexta*. *Cell and Tissue Research* **258**:1–24.

Homberg, U. and J.G. Hildebrand. 1989b. Serotonin immunoreactivity in the optic lobes of the sphinx moth *Manduca sexta* and colocalization with FMRFamide and SCP_B Immunoreactivity. *Journal of Comparative Neurology* **288**:243–253.

Homberg, U., T.G. Kingan and J.G. Hildebrand. 1987. Immunocytochemistry of GABA in the brain and suboesophageal ganglion of *Manduca sexta*. *Cell and Tissue Research* **248**:1–24.

Homberg, U., R.A. Montague and J.G. Hildebrand. 1988. Anatomy of antenno-cerebral pathways in the brain of the sphinx moth *Manduca sexta. Cell and Tissue Research* **254**:255–281.

Homberg, U., T.A. Christensen, and J.G. Hildebrand. 1989. Structure and function of the deutocerebrum in insects. *Annual Review of Entomology* **34**:477–501.

Homberg, U., S. Wuden, H. Ircksen and K.R. Rao. 1991. Comparative anatomy of pigment-dispersing hormone-immunoreactive neuron in the brain of orthopteroid insects. *Cell and Tissue Research* **266**:343–357.

Hoos, D. 1989. Untersuchungen zur Populationsdynamik der Raubmilbe *Typhlodromus pyri* Scheuten 1857 (Acari, Phytoseiidae) im Agrarökosystem Weinberg. Dissertation Universität Kaiserslautern, Germany.

Horak, M. 1984. Assessment of taxonomically significant structures in Tortricinae (Lep., Tortricidae). *Mitteilungen der Schweizerischen Entomologischen Gesellschaft* **57**:3–64.

Horak, M. and R.L. Brown. 1991. Taxonomy and phylogeny. In *Tortricid Pests, Their Biology, Natural Enemies and Control*, eds. L.P.S. van der Geest and H.H. Evenhuis, pp. 23–48. Elsevier, Amsterdam.

Horak, M., C.P. Whittle, T.E. Bellas and E.R. Rumbo. 1988. Pheromone gland components of some Australian tortricids in relation to their taxonomy. *Journal of Chemical Ecology* **14**:1163–1175.

Hösl, M. 1990. Pheromone-sensitive neurons in the deutocerebrum of *Periplaneta americana*: receptive fields on the antenna. *Journal of Comparative Physiology A* **167**:321–327.

Howden, H.F. 1982. Larval and adult characteristics of *Frickius* Germain, its relationship to the Geotrupini, and a phylogeny of some major taxa in the Scarabbaeoidea (Insecta: Coleoptera). *Canadian Journal of Zoology* **60**:2713–2724.

Howell, J.F. 1992. Control of codling moth (*Cydia pomonella*) using pheromone to disrupt mating (Lepidoptera: Tortricidae). *Proceedings XIX International Congress of Entomology*, Beijing, p. 202.

Howell, J.F., A.L. Knight, T.R. Unruh, D.F. Brown, J.L. Krysan, C.R. Sell and P.A. Kirsch. 1992. Control of codling moth in apple and pear with sex pheromone mediated mating disruption. *Journal of Economic Entomology* **85**:918–925.

Howlader, M.A. and G.H. Gerber. 1986. Effects of age, egg development, and mating on calling behaviour of bertha armyworm, *Mamestra configurata* Walker (Lepidoptera: Noctuidae). *Canadian Entomologist* **118**:1221–1230.

Howse, P., I. Stevens and O.T. Jones. 1996. *Insect Pheromones and Their Use in Pest Management*. Chapman and Hall, London.

Hoy, C.W., A.J. Adams and F.R. Hall. 1990. Behavioral response of *Plutella xylostella* (Lepidoptera: Plutellidae) populations to permethrin deposits. *Journal of Economic Entomology* **83**:1216–1221.

Hoy, R.R., J. Hahn, and R.C. Paul. 1977. Hybrid cricket auditory behavior: evidence for genetic coupling in animal communciation. *Science* **195**:82–84.

Hoyle, G. 1970. Cellular mechanisms underlying behavior—neuroethology. *Advances in Insect Physiology* **7**:349–444.

Hoyle, G. 1984. The scope of neuroethology. *Brain and Behavioral Science* **7**:367–412.

Hoyt, C.P., G.O. Osborne and A.P. Mulcock. 1971. Production of insect sex attractant by symbiotic bacteria. *Nature* **230**:472.

Hughes, P.R. and J.A.A. Renwick. 1977. Neural and hormonal control of pheromone biosynthesis in the bark beetle, *Ips paraconfusus*. *Physiological Entomology* **2**:117–123.

Hunt, D.W.A. and J.H. Borden. 1988. Response of mountain pine beetle, *Dendroctonus ponderosae* Hopkins, and pine engraver, *Ips pini* (Say), to ipsdienol in southwestern British Columbia. *Journal of Chemical Ecology* **14**:277–293.

Hunt, R.E., B. Zhao and K.F. Haynes. 1990. Genetic aspects of interpopulational differences in the pheromone blend of the cabbage looper moth, *Trichoplusia ni*. *Journal of Chemical Ecology* **16**:2935–2946.

Ibbotson M.R., T. Maddess and R. DuBois. 1991. A system of insect neurons sensitive to horizontal and vertical image motion connects the medulla and midbrain. *Journal of Comparative Physiology A* **169**:355–367.

Ideses, R., J.T. Klug, A. Shani, S. Gothilf and E. Gurevitz. 1982. Sex pheromone of the European grapevine moth, *Lobesia botrana* Schiff. (Lepidoptera: Tortricidae): synthesis and effect of isomeric purity on biological activity. *Journal of Chemical Ecology* **8**:195–200.

Imai, K., T. Konno, Y. Nakazawa, T. Komiya, M. Isobe, K. Koga, T. Goto, T. Yaginuma, K. Sakakibara, K. Hasegawa and O. Yamashita. 1991. Isolation and structure of diapause hormone of the silkworm, *Bombyx mori*. *Proceedings of the Japanese Academy*. **67B**:98–101.

Ingle D. and D. Crews. 1985. Vertebrate neuroethology: Definitions and paradigms. *Annual Review of Neuroscience* **8**:457–494.

Ioriatti, C. and A. Dalla Serra. 1992. Five-year experience with mating disruption as a control method against codling moth and leafroller species in apple orchards of Trentino. *International Organization for Biological Control—West Palearctic Regional Section Bulletin* **15**(5):88–93.

Itagaki, H. and W.E. Connor. 1986. Physiological control of pheromone release behaviour in *Manduca sexta* (L.). *Journal of Insect Physiology* **32**:657–664.

Itagaki, H. and W.E. Connor. 1987. Neural control of rhythmic pheromone gland exposure in *Utetheisa ornatrix* (Lepidoptera: Arctiidae). *Journal of Insect Physiology* **33**:177–181.

Itagaki, H. and W.E. Connor. 1988. Calling behavior of Manduca sexta (L.) (Lepidoptera: Sphingidae) with notes on the morphology of the female sex pheromone gland. *Annals of the Entomological Society of America* **81**:798–807.

Iwao, S. 1962. Studies on the phase variation and related phenomena in some lepidopterous insects. *Memoirs of College of Agriculture, Kyoto University* **84**:1–80.

Jacquin, E., R.A. Jurenka, H. Ljungberg, P. Nagnan, C. Löfstedt, C. Descoins and W.L. Roelofs. 1994. Control of sex pheromone biosynthesis in the moth *Mamestra brassicae* by the pheromone biosynthesis activating neuropeptide. *Insect Biochemistry and Molecular Biology* **24**:203–211.

Jansson, R.K., L.J. Mason, R.R. Heath, K.A. Sorenson, A.M. Hammond and J.V. Robinson. 1992. Pheromone-trap monitoring system for sweetpotato weevil (Coleoptera: Apionidae) in the southern United States: effects of trap type and pheromone dose. *Journal of Economic Entomology* **85**:416–423.

Jastreboff, P.J., P.E. Pedersen, C.A. Greer, W.B. Stewart, J.S. Kauer, T.E. Benson and G.M. Shepherd. 1984. Specific olfactory receptor populations projecting to identified glomeruli in the rat olfactory bulb. *Proceedings of the National Academy of Science U.S.A.* **81**:5250–5254.

Jaynes, H.A. and C.F. Speers. 1949. Biological and ecological studies of the spruce budworm. *Journal of Economic Entomology* **42**:221–225.

Jenkins, J.W., C.C. Doane, D.J. Schuster, J.R. McLaughlin and M.J. Jiménez. 1990. Development and commercial application of sex pheromone for control of the tomato pinworm. In *Behavior Modifying Chemicals for Insect Management*. eds. R. Ridgway, R Silverstein and M. Inscoe, pp. 269–280. Marcel Deckker, New York.

Jiménez, M.J., N.C. Toscano, D.L. Flaherty, P. Ilic, F.G. Zalom and K. Kido. 1988. Controlling tomato pinworm by mating disruption. *California Agriculture* **42**:10–12.

Jones, C.D. 1978. Statistics of the concentration fluctuations in short range atmospheric dispersion. In *Mathematical Modelling of Turbulent Diffusion in the Environment*, ed. C.J. Harris, pp. 277–298. Academic Press, London.

Jones, C.D. 1983. On the structure on instantaneous plumes in the atmosphere. *Journal of Hazardous Materials* **7**:87–112.

Jones, C.D. and N.T. Gulliford. 1979. Developments in the use of ionized air as a wind tunnel tracer. *Journal of Physics E* **12**:321–325.

Jones, C.D. and R.F. Griffiths. 1984. Full scale experiments on dispersion round an isolated building using an ionized air tracer technique with very short averaging time. *Atmospheric Environment* **18**:903–916.

Jones, D.T. and R.R. Reed. 1987. Molecular cloning of five GTP-binding protein cDNA species from rat olfactory neuroepithelium. *Journal of Biological Chemistry* **262**:14241–14249.

Jones, G. 1995. Molecular mechanisms of action of juvenile hormone. *Annual Review of Entomology* **40**:147–169.

Jones, O.T. 1994. The current and future prospects for semiochemicals in the integrated management of insects pests. In *Proceedings of Brighton Crop Protection Conference—Pests and Diseases—1994*, Vol. 3, pp. 1213–1222.

Jönsson, P. and O. Anderbrandt. 1993. Weather factors influencing catch of *Neodiprion sertifer* (Hymenoptera: Diprionidae) in pheromone traps. *Environmental Entomology* **22**:445–452.

Juarez, P., J. Chase and G.J. Blomquist. 1992. A microsomal fatty acid synthetase from the integument of *Blattella germanica* synthesizes methyl-branched fatty acids, precursors to hydrocarbon and contact sex pheromone. *Archives of Biochemistry and Biophysics* **293**:333–341.

Judd, G.J.R. and H.L. McBrien. 1994. Pheromone-based mating disruption of pome fruit pests in British Columbia: a multiple-species approach. Proceedings of a Workshop "Research, development and commercialization of semiochemicals in insect pest management in Canada," Winnipeg, 19 October 1994, pp. 9–17. Pest Management Alternatives Office, Natural Resources, Ottawa, Canada.

Jurenka, R.A. and W.L. Roelofs. 1993. Biosynthesis and endocrine regulation of fatty acid derived sex pheromones in moths. In *Insect Lipids: Chemistry, Biochemistry and Biology*, eds. D.W. Stanley-Samuelson and D.R. Nelson, pp. 353–388. University of Nebraska Press, Lincoln.

Jurenka, R.A., C. Schal, E. Burns, J. Chase and G.J. Blomquist. 1989. Structural correlation between the cuticular hydrocarbons and the female contact sex pheromone of the German cockroach *Blattella germanica* (L.) *Journal of Chemical Ecology* **15**:939–949.

Jurenka, R.A., E. Jacquin and W.L. Roelofs. 1991a. Control of the pheromone biosynthesis pathway in *Helicoverpa zea* by the pheromone biosynthesis activating neuropeptide (PBAN). *Archives of Insect Biochemistry and Physiology* **17**:81–91.

Jurenka, R.A., E. Jacquin and W.L. Roelofs. 1991b. Stimulation of sex pheromone biosynthesis in the moth *Helicoverpa zea*: action of a brain hormone on pheromone glands involves Ca^{2+} and cAMP as second messengers. *Proceedings of the National Academy of Science U.S.A.* **88**:8621–8625.

Jurenka, R.A., G. Fabriàs and W.L. Roelofs. 1991c. Hormonal control of female sex pheromone biosynthesis in the redbanded leafroller moth, *Argyrotaenia velutinana*. Insect Biochemistry **21**:81–89.

Jurenka, R.A., G. Fabriàs, S. Ramaswamy and W.L. Roelofs. 1993. Control of pheromone biosynthesis in mated redbanded leafroller moths. *Archives of Insect Biochemistry and Physiology* **24**:129–137.

Jurenka, R.A., G. Fabriàs, L. DeVoe and W.L. Roelofs. 1994a. Action of PBAN and related peptides on pheromone biosynthesis in isolated pheromone glands of the redbanded leafroller moth, *Argyrotaenia velutinana. Comparative Biochemistry and Physiology* **108C**:153–160.

Jurenka, R.A. K.F. Haynes, R.O. Adlof, M. Bengtsson and W.L. Roelofs. 1994b. Sex pheromone component ratio in the cabbage looper moth altered by a mutation affecting the fatty acid chain-shortening reactions in the pheromone biosynthesis pathway. *Insect Biochemistry and Molecular Biology* **24**:373–381.

Jutsum, A.R. 1988. Commercial application of biological control. *Philosophical Transactions of the Royal Society London B* **318**:357–373.

Jutsum, A.R. and R.F.S. Gordon, eds. 1989a. *Insect Pheromones in Plant Protection*. Wiley, Chichester, United Kingdom.

Jutsum, A.R. and R.F.S. Gordon. 1989b. Epilogue. In *Insect Pheromones in Plant Protection*, eds. A.R. Jutsum and R.F.S. Gordon, pp. 353–355. Wiley, Chichester, United Kingdom.

Kafka, W.A. 1970. Molekuläre Wechselwirkungen bei der Erregung einzelner Riechzellen. *Zeitschrift für Vergleichende Physiologie* **70**:105–143.

Kaissling, K.-E. 1971. Insect olfaction. In *Handbook of Sensory Physiology*, Vol. IV, *Chemical Senses 1. Olfaction*, ed. L.M. Beidler, pp. 351–431. Springer-Verlag, Berlin.

Kaissling, K.-E. 1974. Sensory transduction in insect olfactory receptors. In *Biochemistry of Sensory Functions*, ed. L. Jaenicke, pp. 243–273. Springer-Verlag, Berlin.

Kaissling, K.-E. 1979. Recognition of pheromones by moths, especially in saturniids and *Bombyx mori*. In *Chemical Ecology: Odour Communication in Animals*, ed. F.J. Ritter, pp. 43–56. Elsevier/North-Holland, Amsterdam.

Kaissling, K.-E. 1986a. Chemo-electrical transduction in insect olfactory receptors. *Annual Review of Neuroscience* **9**:121–145.

Kaissling, K.-E. 1986b. Temporal characteristics of pheromone receptor cell responses in relation to orientation behaviour of moths. In *Mechanisms in Insect Olfaction*, eds. T.L. Payne, C.E.J. Kennedy and M.C. Birch, pp. 193–200, Clarendon Press, Oxford.

Kaissling, K.-E. 1987. R.H. *Wright-Lectures on Insect Olfaction*. Simon Fraser University. Burnaby, Canada.

Kaissling, K.-E. and E. Kramer. 1990. Sensory basis of pheromone-mediated orientation in moths. *Verhandlungen der Deutschen Zoologischen Gesellschaft* **83**:109–131.

Kaissling, K.-E. and E. Priesner. 1970. Die Riechschwelle des Seidenspinners. *Naturwissenschaften* **57**:23–28.

Kaissling, K.-E., L.Z. Meng and H.J. Bestmann. 1989a. Responses of bombykol receptor cells to (*Z,E*)-4,6-hexadecadiene and linalool. *Journal of Comparative Physiology A* **165**:147–154.

Kaissling, K.-E., J.G. Hildebrand and J.H. Tumlinson. 1989b. Pheromone receptor cells in the male moth *Manduca sexta. Archives of Insect Biochemistry Physiology* **10**:273–279.

Kaissling, K.-E., T.A. Keil and J.L.D. Williams. 1991. Pheromone stimulation in perfused sensory hairs of the moth *Antheraea polyphemus. Journal of Insect Physiology* **37**:71–78.

Kanaujia, L. and K.-E. Kaissling. 1985. Interactions of pheromone with moth antennae: adsorption, desoprtion and transport. *Journal of Insect Physiology* **31**:71–81.

Kandel, E.R., M. Klein, B. Hochner, S.A. Schuster, R.D. Siegelbaum, D.L. Hawkins, V. Glanzman, V. Castelluucci and T.W. Abrams. 1987. Synaptic modulation and learning: new insights into synaptic transmission from the study of behavior. In *Synaptic function*, eds. G.M. Edelman, W.S.E. Gall and W.M. Cowan, pp. 471–518. Wiley, New York.

Kaneshiro, K. 1980. Sexual isolation, speciation and the direction of evolution. *Evolution* **34**:437–444.

Kaneshiro, K.Y. 1983. Sexual selection and direction of evolution in the biosystematics of Hawaiian Drosophilidae. *Annual Review of Entomology* **28**:161–178.

Kaneshiro, K.Y. 1988. Speciation in the Hawaiian *Drosophila. Bioscience* **38**:258–263.

Kaneshiro, K.Y. and C.R.B. Boake. 1987. Sexual selection and speciation issues raised by Hawaiian *Drosophila. Trends in Ecology and Evolution* **2**:207–212.

Kaneshiro, K.Y. and L.V. Giddings. 1987. The significance of asymmetrical sexual isolation and the formation of new species. *Evolutionary Biology* **21**:29–44.

Kanost, M.R., J.K. Kawooya, J.H. Law, R.O. Ryan, M.C. Van Heusden and R. Ziegler. 1990. *Insect haemolymph proteins. Advances in Insect Physiology* **22**:299–396.

Kanzaki, R. and T. Mishima. 1996. Pheromone-triggered 'flipflopping' neural signals correlated with activities of neck motor neurons of a male moth, *Bombyx mori. Zoological Science* **13**:79–87.

Kanzaki, R. and T. Shibuya. 1986a. Identification of the deutocerebral neurons responding to the sexual pheromone in the male silkworm moth brain. *Zoological Science* **3**:409–418.

Kanzaki, R. and T. Shibuya. 1986b. Descending protocerebral neurons related to the mating dance of the male silkworm moth. *Brain Research* **377**:378–382.

Kanzaki, R. and T. Shibuya. 1992a. Long-lasting excitation of protocerebral bilateral neurons in the pheromone-processing pathways of the male moth *Bombyx mori. Brain Research* **587**:211–215.

Kanzaki, R. and T. Shibuya. 1992b. Personal computer-based processing technique for analyzing insect mating behavior in response to sex pheromone odor. *Sensors and Materials* **4**:1–9.

Kanzaki, R., E.A. Arbas, N.J. Strausfeld and J.G. Hildebrand. 1989. Physiology and morphology of projection neurons in the antennal lobe of the male moth *Manduca sexta. Journal of Comparative Physiology A* **165**:427–453.

Kanzaki, R., E.A. Arbas and J.G. Hildebrand. 1991a. Physiology and morphology of protocerebral olfactory neurons in the male moth *Manduca sexta. Journal of Comparative Physiology A* **168**:281–298.

Kanzaki, R., E.A. Arbas and J.G. Hildebrand. 1991b. Physiology and morphology of descending neurons in pheromone-processing olfactory pathways in the male moth *Manduca sexta. Journal of Comparative Physiology A* **169**:1–14.

Kanzaki, R., N. Sugi and T. Shibuya. 1992. Self-generated zigzag turning of *Bombyx mori* males during pheromone-mediated upwind walking. *Zoological Science* **9**:515–527.

Kanzaki, R., A. Ikeda and T. Shibuya. 1994. Morphological and physiological properties of pheromone-triggered flipflopping descending interneurons of the male silkworm moth, *Bombyx mori. Journal of Comparative Physiology A* **175**:1–14.

Karg, G., A.E. Sauer and U.T. Koch. 1990. Effects of the odour source distribution on the odour plume structure. In *Abstracts IX ECRO Symposium*, p. 41, Noordwijkerhout, The Netherlands.

Karg, G., D.M. Suckling and E.R. Rumbo. 1992. Measurement of sex pheromone in orchard air using electroantennogram apparatus. *Proceedings 45th New Zealand Plant Protection Conference* **1992**:304–306.

Karg, G., D.M. Suckling and S.J. Bradley. 1994. Absorption and release of pheromone of *Epiphyas postvittana* (Lepidoptera, Tortricidae) by apple leaves. *Journal of Chemical Ecology* **20**:1825–1841.

Kasang, G., L. Proff and M. Nichols. 1988. Enzymatic conversion and degradation of sex pheromones in antennae of the male silk worm *Antheraea polyphemus. Zeitschrift für Naturforschung*, **43c**:275–284.

Kass, L. and V. Barlow. 1984. Efferent neurotransmission of circadian rhythms in *Limulus* lateral eye. I. Octopamine-induced increases in retinal sensitivity. *Journal for Neuroscience* **4**:908–917.

Kass, L., J.L. Pelleitier, G. Renninger, and R. B. Barlow. 1988. Efferent neurotransmission of circadian rhythms in *Limulus* lateral eye. II. Intracellular recordings *in vitro. Journal of Comparative Physiology* **164**:95–105.

Kawano, T., H. Kataoka, H. Nagasawa, A. Isogai and A. Suzuki. 1992. cDNA cloning and sequence determination of the pheromone biosynthesis activating neuropeptide of the silkworm, *Bombyx mori. Biochemical and Biophysical Research Communications* **189**:221–226.

Kehat, M. and E. Dunkelblum. 1990. Behavioral responses of male *Heliothis armigera* (Lepidoptera: Noctuidae) moths in a flight tunnel to combinations of components identified from female sex pheromone glands. *Journal of Insect Behavior* **3**:75–83.

Kehat, M., S. Greenberg and Y. Tamaki. 1976. Field evaluation of the synthetic sex pheromone, as an attractant for males of the cotton leafworm, *Spodoptera littoralis* (Boisd.), in Israel. *Applied Entomology and Zoology* **11**:45–52.

Kehat, M., S. Gothilf, E. Dunkelblum and S. Greensberg. 1980. Field evaluation of female sex pheromone components of the cotton bollworm, *Heliothis armigera. Entomologia Experimentalis et Applicata* **27**:188–193.

Keil, T. 1982. Contacts of pore tubules and sensory dendrites in antennal chemosensilla of a silkmoth: demonstration of a possible pathway for olfactory molecules. *Tissue and Cell* **14**:451–462.

Keil, T. 1989. Fine strucute of the pheromone-sensitive sensilla on the antennae of the hawkmoth, *Manduca sexta. Tissue and Cell* **21**:139–151.

Kennedy, J.S. 1940. The visual responses of flying mosquitoes. *Proceedings of the Zoological Society London A* **109**:221–242.

Kennedy, J.S. 1951. The migration of the desert locust (*Schistocerca gregaria* Forsk.). *Philosophical Transactions of the Royal Society B* **253**:163–290.

Kennedy, J.S. 1977. Olfactory responses to distant plants and other odor sources. In *Chemical Control of Insect Behavior*, eds. H.H. Shorey and J.J. McKelvey, Jr., pp. 67–91. Wiley, New York.

Kennedy, J.S. 1978. The concept of olfactory 'arrestment' and 'attraction'. *Physiological Entomology* **3**:1–98.

Kennedy, J.S. 1982. Mechanism of moth sex attraction: a modified view based on wind-tunnel experiments with flying male *Adoxophyes. Les Colloques de L'INRA* **7**:189–192.

Kennedy, J.S. 1983. Zigzagging and casting as a programmed response to wind-borne odour: a review. *Physiological Entomology* **8**:109–120.

Kennedy, J.S. 1986. Some current issues in orientation to odour sources. In *Mechanisms in Insect Olfaction*, eds. T.L. Payne, M.C. Birch and C.E.J. Kennedy, pp. 11–25. Clarendon Press, Oxford.

Kennedy, J.S. 1992. *The New Anthropomorphism.* Cambridge University Press, Cambridge.

Kennedy, J.S. and D. Marsh. 1974. Pheromone-regulated anemotaxis in flying moths. *Science* **184**:999–1001.

Kennedy, J.S., A.R. Ludlow and C.J. Sanders. 1980. Guidance system used in moth sex attraction. *Nature* **288**:475–477.

Kennedy, J.S., A.R. Ludlow and C.J. Sanders. 1981. Guidance of flying male moths by wind-borne sex pheromone. *Physiological Entomology* **6**:395–412.

Kent K.S., Harrow I.D., P. Quartararo and J.G. Hildebrand. 1986. An accessory olfactory pathway in Lepidoptera: the labial pitorgan and its central projections in *Manduca sexta* and certain other sphinx moths and silk moths. *Cell and Tissue Research* **245**:237–245.

Khan, M.A. 1988. Brain-controlled synthesis of juvenile hormone in adult insects. *Entomologia Experimentalis et Applicata* **46**:3–17.

Kilias, G., G. Goulielmos and S. Alahiotis. 1989. Interspecific hybridization between *Drosophila* spp. group *melanogaster* sibling species: fitness components. *Hereditas* **110**:267–274.

Kingan, T.G., M. B. Blackburn and A.K. Raina. 1992. The distribution of pheromone-biosynthesis-activating neuropeptide (PBAN) immunoreactivity in the central nervous system of the corn earworm, *Helicoverpa zea. Cell and Tissue Research* **270**:229–240.

Kingan, T.G., P.A. Thomas-Laemont and A.K. Raina. 1993. Male accessory gland factors elicit change from 'virgin' to 'mated' behavior in the female corn earworm moth, *Helicoverpa zea. Journal of Experimental Biology* **183**:61–76.

Kingan, T.G., W.M. Bodnar, A.K. Raina and D.F. Hunt. 1994. The loss of female sex pheromone after mating in the corn earworm moth, *Heliocoverpa zea*: identification of a male pheromonostatic peptide. *Proceedings of the National Academy of Science U.S.A.* **92**:5082–5086.

Kinzer, G.W., A.F. Fentiman, Jr., R.L. Foltz and J.A. Rudinsky. 1971. Bark beetle attractants: 3-methyl-2-cyclohexen-1-one isolated from *Dendroctonus pseudotsugae. Journal of Economic Entomology* **64**:970–971.

Kirkpatrick, M. and M.J. Ryan. 1991. The paradox of the lek and the evolution of mating preferences. *Nature* **350**:33–38.

Kitamura, A., H. Nagasawa, H. Kataoka, T. Inoue, S. Matsumoto, T. Ando and A. Suzuki. 1989. Amino acid sequence of pheromone-biosynthesis-activating neuropeptide (PBAN) of the silkworm, *Bombyx mori. Biochemical and Biophysical Research Communications* **163**:520–526.

Kitamura, A., H. Nagasawa, H. Kataoka, T. Ando and A. Suzuki. 1990. Amino acid sequence of pheromone-biosynthesis-activating-neuropeptide II (PBAN-II) of the silkmoth, *Bombyx mori. Agricultural and Biological Chemistry Tokyo* **54**:2495–2497.

Klassen, W., R.L. Ridgway and M. Inscoe. 1982. Chemical attractants in integrated pest management programs. In *Insect Suppression with Controlled Release Pheromone Systems*, Vol. I, eds. A.F. Kydonieus and M. Beroza, pp. 13–130. CRC Press, Boca Raton, Florida.

Klein, M.G. 1981. Mass trapping for suppression of Japanese beetles. In *Management of Insect Pests with Semiochemicals*, ed. E.R. Mitchell, pp. 183–190. Plenum Press, New York.

Klein, U. 1987. Sensillum-lymph proteins from antennal olfactory hairs of the moth *Antheraea polyphemus* (Saturniidae) *Insect Biochemistry* **17**:1193–1204.

Klimetzek, D., J. Köhler, S. Krohn and W. Francke. 1989. Das Pheromon-System des Waldreben-Borkenkäfers, *Xylocleptes bispinus* Duft. (Coleoptera, Scolytidae). *Journal of Applied Entomology* **107**:304–309.

Kline, L.N., R.F. Schmitz, J.A. Rudinsky and M.M. Furniss. 1974. Repression of spruce beetle (Coleoptera) attraction by methylcyclohexenone in Idaho. *Canadian Entomologist* **106**:485–491.

Kloppenburg, P. and J. Erber. 1995. The modulatory effects of serotonin and octopamine in the visual system of the honey bee (*Apis mellifera* L.) II. Electrophysiological analysis of motion-sensitive neurons in the lobula. *Journal of Comparative Physiology* **176**:119–129.

Kluge, A. 1989. A concern for evidence and the phylogenetic hypothesis of relationships among *Epicrates* (Boidae, Serpentes). *Systematic Zoology* **38**:7–25.

Kluge, A.G. and A.J. Wolf. 1993. Cladistics: what's in a word? *Cladistics* **9**:183–199.

Klun, J.A. and Cooperators. 1975. Insect sex pheromones: intraspecific pheromonal variability of *Ostrinia nubilalis* in North America and Europe. *Environmental Entomology* **4**:891–894.

Klun, J.A. and M.D. Huettel. 1988. Genetic regulation of sex pheromone production and response: Interaction of sympatric pheromonal types of European corn borer, *Ostrinia nubilalis* (Lepidoptera: Pyralidae). *Journal of Chemical Ecology* **14**:2047–2061.

Klun, J.A., O.L. Chapman, K.C. Mattes, P.W. Wojtkowski, M. Beroza and P.E. Sonnet. 1973. Insect sex pheromones: minor amount of opposite geometrical isomer critical to attraction. *Science* **181**:661–663.

Klun, J.A., J.R. Plimmer, B.A. Bierl-Leonhardt, A.N. Sparks and O.L. Chapman. 1979. Trace chemicals: the essence of sexual communication systems in *Heliothis* species. *Science* **204**:1328–1330.

Klun, J.A., J.R. Plimmer, B.A. Bierl-Leonhardt, A.N. Sparks, M. Primiani, O.L. Chapman, G.H. Lee and G. Lepone. 1980a. Sex pheromone chemistry of female corn earworm moth, *Heliothis zea*. *Journal of Chemical Ecology* **6**:165–175.

Klun, J.A., Bierl-Leonhardt, B.A., Plimmer, J.R., Sparks, A.N., Primiani, M., Chapman, O.L., Lepone, G. and Lee, G.H. 1980b. Sex pheromone chemistry of the female tobacco budworm moth *Heliothis virescens*. *Journal of Chemical Ecology* **6**:177–183.

Koch, U.T. 1990. A portable EAG system for measurement of pheromones in the field. In *Abstracts American Chemical Society National Meeting*, Washington, D.C., August 26–30, 1990, AGFD 31.

Kochansky J., J. Tette, E.F. Taschenberg, R.T. Cardé, K.-E. Kaissling and W.L. Roelofs. 1975a. Sex pheromone of the moth *Antheraea polyphemus*. *Journal of Insect Physiology* **21**:1977–1983.

Kochansky, J., R.T. Cardé, J. Liebherr and W.L. Roelofs 1975b. Sex pheromone of the European corn borer, *Ostrinia nubilalis* (Lepidoptera: Pyralidae), in New York. *Journal of Chemical Ecology* **1**:225–231.

Kodadová, B. 1993. Effect of temperature on the electrophysiological response of moth olfactory sensilla. Dissertation, Max-Planck-Institut für Verhaltensphysiologie, Seewiesen, Germany.

Koehl, M.A.R. 1996a. Fluid flow through hair-bearing appendages: feeding, smelling and swimming at low and intermediate Reynolds numbers. *Journal of Experimental Biology* (in press).

Koehl, M.A.R. 1996b. Small-scale fluid dynamics of olfactory antennae. *Marine and Freshwater Behavior and Physiology* **27**:127–141.

Koehler, C.S., J.J. McKelvey, Jr., W.L. Roelofs, H.H. Shorey, R.M. Silverstein and D.L. Wood. 1977. Advancing towards operational behavior-modfying chemicals. In *Chemical Control of Insect Behavior*, eds. H.H. Shorey and J.J. McKelvey, Jr., pp. 395–400. Wiley, New York.

Kohnle, U. 1985. Untersuchungen über die Pheromonsysteme sekundärer Borkenkäfer (Col., Scolytidae). *Zeitshrift für angewandte Entomologie* **100**:197–218.

Kohnle, U., W. Francke, and A. Bakke. 1985. *Poligraphus poligraphus* (L.): response to enantiomers of beetle specific terpene alcohols and a bicyclic ketal. *Zeitshrift für angewandte Entomologie* **100**:5–8.

Kohnle, U., H. Schmutzenhofer, J. Bartels and W. Francke. 1988a. Oxygenated terpenes in the chemical communication system of the bark beetle, *Ips schmutzenhoferi* (Coleoptera: Scolytidae), a species recently described for the southeastern Himalaya *Journal of Applied Entomology* **106**:45–51.

Kohnle, U., J.P. Vité, C. Erbacher, J. Bartels and W. Francke. 1988b. Aggregation response of European engraver beetles of the genus *Ips* mediated by terpenoid pheromones. *Entomologica Experimentalis et Applicata* **49**:43–53.

Koontz, M.A. and D. Schneider. 1987. Sexual dimorphism in neuronal projections from the antennae of silk moths (*Bombyx mori*, *Antheraea polyphemus*) and the gypsy moth (*Lymantria dispar*). *Cell and Tissue Research* **249**:39–50.

Kostyk, B.C., J.H. Borden and G. Gries. 1993. Photoisomerism of antiaggregation pheromone verbenone: biological and practical implications with respect to the mountain pine beetle, *Dendroctonus ponderosae* Hopkins (Coleoptera: Scolytidae). *Journal of Chemical Ecology* **19**:1749–1759.

Koudele, K., J. Stout and D. Reichert. 1987. Factors which influence female crickets' (*Acheta domesticus*) phonotactic and sexual responsiveness to males. *Physiological Entomology* **12**:67–80.

Kozlov, M.V. 1982. New and little known species of Micropterigidae from Caucasus. *Vestnik Leningradskogo Universiteta Biologiya* **9**:116–118.

Kozlov, M.V. 1985. Precopulatory behaviour of primitive Lepidoptera. *Entomologicheskoje Obozrenie* **64**:493–504 (in Russian). English translation: *Entomological Review (Washington)* **64**:1–13.

Kozlov, M.V. 1988. Biology of *Eriocrania sparrmannella* Bosc. (Lepidoptera, Eriocraniidae). *Vestnik Zoologii* **4**:80–85.

Kozlov, M.V., J. Zhu, P. Philipp, W. Francke, E.L. Zvereva and C. Löfstedt. 1996. Pheromone specificity in *Eriocrania semipurpurella* (Stephens) and *E. sangii* (Wood) (Lepidoptera: Eriocraniidae) based on chirality of semiochemicals. *Journal of Chemical Ecology* **22**:431–457.

Kramer, E. 1975. Orientation of the male silkmoth to the sex attractant bombykol. In *Olfaction and Taste V*, eds. D.A. Denton and J.P. Coghlan, pp. 329–335. Academic Press, New York.

Kramer, E. 1986. Turbulent diffusion and pheromone-triggered anemotaxis. In *Mechanisms in Insect Olfaction*, eds. T.L. Payne, M.C. Birch and C.E.J. Kennedy, pp. 59–67. Clarendon Press, Oxford.

Kramer, E. 1992. Attractivity of pheromone surpassed by time-patterned application of two nonpheromone compounds. *Journal of Insect Behavior* **5**:83–97.

Krasnoff, S.B. and D.E. Dussourd. 1988. Dihydropyrrolizine attractants attractanct for arctiid moths that visit plants containing pyrrolizidine alkaloids. *Journal of Chemical Ecology* **15**:47–60.

Krasnoff, S.B. and W.L. Roelofs. 1989. Quantitative and qualitative effects of larval diet on male scent secretions of *Estigmene acrea*, *Phragmatobia fuliginosa*, and *Pyrrharctia isabella* (Lepidoptera: Arctiidae). *Journal of Chemical Ecology* **15**:1077:1093.

Krasnoff, S.B. and W.L. Roelofs. 1990. Evolutionary trends in the male pheromone systems of arctiid moths evidence from studies of courtship in *Phragmatobia fuliginosa* and *Pyrrharctia isabella* (Lepidoptera: Arctiidae). *Zoological Journal of The Linnean Society* **99**:319–338.

Krasnoff, S.B., L.B. Bjostad and W.L. Roelofs. 1987. Quantitative and qualitative variation in male pheromones of *Phragmatobia fuliginosa* and *Pyrrharctia isabella* (Lepidoptera: Arctiidae). *Journal of Chemical Ecology* **13**:807–822.

Krieger, J., K. Raming and H. Breer. 1991. Cloning of genomic and complementary DNA encoding insect pheromone binding proteins: evidence for micro diversity. *Biochimica et Biophysica Acta* **1088**:277–284.

Krieger, J., K. Raming, G.D. Prestwich, D. Frith, S. Stabel and H. Breer. 1992. Expression of a pheromone binding protein in insect cells using a baculovirus vector. *European Journal of Biochemistry* **203**:161–166.

Krieger, J., H. Gänßle, K. Raming and H. Breer. 1993. Odorant binding proteins of *Heliothis virescens*. *Insect Biochemistry and Molecular Biology* **23**:449–456.

Krieger, J., S. Schleicher, J. Strotmann, I. Wanner, I. Boekhoff, K. Raming, P. deGeus and H. Breer. 1994. Probing olfactory receptors with sequence-specific antibodies. *European Journal of Biochemistry* **219**:829–835.

Kristensen, N.P. 1981. Phylogeny of insect orders. *Annual Review of Entomology* **26**.135–157.

Kristensen, N.P. 1984a. Studies on the morphology and systematics of primitive Lepidoptera (Insecta). *Steenstrupia* **10**:141–191.

Kristensen, N.P. 1984b. The pregenital abdomen of the Zeugloptera (Lepidoptera). *Steenstrupia* **10**:113–136.

Kristensen, N.P. 1996. Lepidoptera of the world. Status and perspectives of the inventory of a major insect order. In *Biodiversity and Taxonomy of Terrestrial Arthropods: A Global Perspective*, ed. K.C. Kim, Universal Book Services, Leiden (in press).

Kristensen, N.P. and A.W. Skalski. 1996. Phylogeny and palaeontology. In *Handbook of Zoology/Handbuch der Zoologie. Lepidoptera. Moths and Butterflies 1*, ed. N.P. Kristensen. De Gruyter, Berlin (in press).

Kristensen, N.P. and E.S. Nielsen. 1979. A new subfamily of micropterigid moths from South America. A contribution to the morphology and phylogeny of the Micropterigidae, with a generic catalogue of the family (Lepidoptera, Zeugloptera). *Steenstrupia* **5**:69–147.

Kuenen, L.P.S. and T.C. Baker. 1981. Habituation versus sensory adaptation as the cause of reduced attraction following pulsed and constant sex pheromone preexposure by *Trichoplusia ni. Journal of Insect Physiology* **27**:721–726.

Kuenen, L.P.S. and T.C. Baker. 1982a. Optomotor regulation of ground velocity in moths during flight to sex pheromone at different heights. *Physiological Entomology* **7**:193–202.

Kuenen, L.P.S. and T.C. Baker. 1982b. The effects of pheromone concentration on the flight behaviour of the oriental fruit moth, *Grapholitha molesta. Physiological Entomology* **7**:423–434.

Kuenen, L.P.S. and T.C. Baker. 1983. A non-anemotactic mechanism used in pheromone source location by flying moths. *Physiological Entomology* **8**:277–289.

Kuenen, L.P.S. and R.T. Cardé. 1993. Effects of moth size on velocity and steering during upwind flight toward a sex pheromone source by *Lymantria dispar* (Lepidoptera: Lymantriidae). *Journal of Insect Behavior* **6**:177–193.

Kuenen, L.P.S. and R.T. Cardé. 1994. Strategies for recontacting a lost pheromone plume: casting and upwind flight in the male gypsy moth. *Physiological Entomology* **19**:15–29.

Kuenen, L.P.S., D.L. Wagner, W.E. Wallner and R.T. Cardé. 1994. Female sex pheromone in *Korscheltellus gracilis* (Grote) (Lepidoptera: Hepialidae). *Canadian Entomologist* **126**:31–41.

Kuniyoshi, H., A. Kitamura, H. Nagasawa, T. Chuman, K. Shimazaki, T. Ando and A. Suzuki. 1991. Structure—activity relationship of pheromone biosynthesis activating neuropeptide (PBAN) from the silkmoth, *Bombyx mori*. In *Peptide Chemistry* 1990, ed. Y. Shimonishi, pp. 251–254, Protein Research Foundation, Osaka, Japan.

Kuniyoshi, H., H. Nagasawa, T. Ando and A. Suzuki. 1992a. N-terminal modified analogs of C-terminal fragments of PBAN with pheromonotropic activity. *Insect Biochemistry and Molecular Biology* **22**:399–403.

Kuniyoshi, H., H. Nagasawa, T. Ando, A. Suzuki, R.J. Nachman and G.M. Holman. 1992b. Cross-activity between pheromone biosynthesis activating neuropeptide (PBAN) and myotropic pyrokinin insect peptides. *Bioscience Biotechechnology and Biochemistry* **56**:167–168.

Kuwahara, Y., C. Kitamura, S. Takahashi, H. Hara, S. Ishii and H. Fumaki. 1971. Sex pheromone of the almond moth and the Indian meal moth: *cis*-9, *trans*-12-tetradecadienyl acetate. *Science* **171**:801–802.

Kydonieus, A.F. and M. Beroza. 1978. Insect control with a multilayered luretrap dispenser. In *Proceedings of the Controlled Release Pesticides Symposium, Oregon State Review, August 22–24, 1977*, ed. H.B. Eber, pp. 78–89.

Kydonieus, A.F. and M. Beroza, eds. 1982. *Insect Suppression with Controlled Release Pheromone Systems*, Vols. I and II. CRC Press, Boca Raton, Florida.

Lacey, M.J. and C.J. Sanders. 1992. Chemical composition of sex pheromone of oriental fruit moth and rates of release by individual female moths. *Journal of Chemical Ecology* **18**:1421–1435.

Lal, L. 1989. Relationships between pheromone catches of adult moths, foliar larval populations and plant infestations by potato tuberworm in the field. *Tropical Pest Management* **35**:157–159.

Lambert, D.M., B. Michaux, and C.S. White. 1987. Are species self-defining? *Systematic Zoology* **36**:196–205.

Lancet, D. and N. Ben-Arie. 1993. Olfactory receptors. *Current Biology* **3**:668–674.

Lancet, D. N. Ben-Arie, S. Cohen, U. Gat, R. Grossisseroff, S. Hornsaban, M. Khen, H. Lehrach, M. Natochin, M. North, E. Seidemann, and N. Walker. 1993. Olfactory receptors–transduction, diversity, human psychophysics and genome analysis. In *Molecular Basis of Smell and Taste Transduction*, eds. D. Chadwick, J. Marsh and J. Goode, pp. 131–146. Wiley, Chichester, United Kingdom.

Land, M.F. 1969. Movements of the retinae of jumping spiders (Salticidae: Dendryphantinae) in response to visual stimuli *Journal of Experimental Biology* **51**:471–493.

Land, M.F. 1974. Head movements and fly vision. In *The Compound Eye and Vision of Insects*, eds. G.A. Horridge, pp. 469–489. Clarendon Press, Oxford.

Land, M.F. 1992. Visual tracking and pursuit: humans and arthropods compared. *Journal of Insect Physiology* **38**:939–951.

Land, M.F., N.J. Marshall, D. Brownless and T.W. Cronin. 1990. The eye-movements of the mantis shrimp *Odontodactylis scyllarus* (Crustacea: Stomatopoda). *Journal of Comparative Physiology A* **167**:155–166.

Lande, R. 1981. Models of speciation by sexual selection on polygenic traits. *Proceedings of the National Academy of Sciences U.S.A.* **78**:3721–3725.

Landolt, P.J. and R.R. Heath. 1989. Sexual role reversal in mate-finding strategies of the cabbage looper moth. *Science* **249**:1026–1028.

Lanier, G.N. 1990. Principles of attraction-annihilation. In *Behavior-Modifying Chemicals for Insect Management*, eds. R.L. Ridgway, R.M. Silverstein and M.N. Inscoe, pp. 25–45. Marcel Dekker, New York.

Lanier, G.N. and W.E. Burkholder. 1974. Pheromones in speciation of Coleoptera. In *Pheromones*, ed. M.C. Birch, pp. 161–189. North-Holland, Amsterdam.

Lanier, G.N., A. Classon, T. Stewart, J.J. Piston and R.M. Silverstein. 1980. *Ips pini*: the basis for interpopulational differences in pheromone biology. *Journal of Chemical Ecology* **6**:677–687.

Laster, M.L. 1972. Interspecific hybridization of *Heliothis virescens* and *H. subflexa*. *Environmental Entomology* **1**:682–687.

Latheef, M.A., J.A. Witz and J.D. Lopez. 1991. Relationships among pheromone trap catches of male corn earworm moths (Lepidoptera: Noctuidae), egg numbers, and phenology in corn. *Canadian Entomologist* **123**:271–281.

Latheef, M.A., J.D. Lopez and J.A. Witz. 1993. Capture of corn earworm (Lepidoptera: Noctuidae) in pheromone traps and hand nets: relationship to egg and adult densities in field corn, Texas Brazos River Valley. *Journal of Economic Entomolology* **86**:407–415.

Law, J.H., J.M. Ribeiro and M.A. Wells. 1992. Biochemical insights derived from insect diversity. *Annual Review of Biochemistry* **61**:87–111.

Lazard, D., K. Zupko, Y. Poria, P. Nef., J. Lazarovits, S. Horn., M. Khen and D. Lancet. 1991. Odorant signal termination by olfactory UDP glucuronosyl transferase. *Nature* **349**:790–793.

Leal, W.S. 1991. (*R*,*Z*)-5-(−)-(Oct-1-enyl)oxacyclopentan-2-one, the sex pheromone of the scarab beetle *Anomala cuprea. Naturwissenschaften* **78**:521–523.

Leal, W.S. 1993. (*Z*)- and (*E*)-Teteradec-7-en-2-one, a new type of sex pheromone from the oriental beetle. *Naturnwissenschaften* **80**:86–87.

Leal, W.S. and F. Mochizuki. 1993. Sex pheromone reception in the scarab beetle *Anomala cuprea*. Enantiomeric discrimination by sensilla placodea. *Naturwissenschaften* **80**:278–281.

Leal, W.S. and A. Shirata 1992a. Antimicrobial activity of some sex pheromones. In *Proceedings of the 9th Annual Meeting of the International Society of Chemical Ecology*, Kyoto, Japan, July 6–9, 1992, p. 71.

Leal, W.S. and A. Shirata 1992b. Antimicrobial activities of some scarab beetle pheromones and related compounds. Japanese patent 4-208371.

Leal, W.S., F. Mochizuki, S. Wakamura and T. Yasuda. 1992a. Electroantennographic detection of *Anomala cuprea* Hope (Coleoptera: Scarabaeidae) sex pheromone. *Applied Entomology and Zoology* **27**:289–291.

Leal, W.S., M. Hasegawa and M. Sawada. 1992b. Identification of *Anomala schonfeldti* sex pheromone by high-resolution GC-behavior bioassay. *Naturwissenschaften* **79**:518–519.

Leal, W.S., S. Matsuyama, Y. Kuwahara, S. Wakamura and M. Hasegawa. 1992c. An amino acid derivative as the sex pheromone of a scarab beetle. *Naturwissenschaften* **79**:184–185.

Leal, W.S., M. Sawada and M. Hasegawa. 1993a. The scarab beetle *Anomala cuprea* utilizes the sex pheromone of *Popillia japonica* as a minor component. *Journal of Chemical Ecology* **19**:1303–1313.

Leal, W.S., M. Sawada and M. Hasegawa. 1993b. The scarab beetle *Anomala daimiana* utilizes a blend of two other *Anomala* spp. sex pheromones. *Naturwissenschaften* **80**:181–183.

Leal, W.S., M. Sawada, S. Matsuyama, Y. Kuwahara and M. Hasegawa. 1993c. Unusual periodicity of sex pheromone production in the large black chafer *Holotrichia parallela. Journal of Chemical Ecology* **19**:1381–1319.

Leal, W.S., M. Hasegawa, M. Sawada and M. Ono. 1994a. Sex pheromone of the oriental beetle, *Exomala orientalis*: identification and field evaluation. *Journal of Chemical Ecology* **20**:1707–1718.

Leal, W.S., M. Hasegawa, M. Sawada, M. Ono and Y. Ueda. 1994b. The scarab beetle *Anomala octiescostata* (Coleoptera: Scarabaeidae) sex pheromone. *Journal of Chemical Ecology* **20**:1643–1655.

Leal, W.S., F. Kawamura and M. Ono. 1994c. The scarab beetle *Anomala albopilosa sakishimana* utilizes the same sex pheromone blend as a closely related and geographically isolated species, *Anomala cuprea. Journal of Chemical Ecology* **20**:1667–1676.

Leal, W.S., X. Shi, D. Liang, C. Schal and J. Meinwald. 1995. Application of chiral gas chromatography with electroantennographic detection to the determination of the stereochemistry of a cockroach sex pheromone. *Proceedings of the National Academy of Sciences U.S.A.* **92**:1033–1037.

Lehninger, A.L. 1979. *Biochemistry*. Worth Publishers, New York.

Lehrer, M. 1991. Bees which turn back and look. *Naturwissenschaften* **78**:274–276.

Lehrer, M. 1993. Why do bees turn back and look? *Journal of Comparative Physiology A* **172**:549–563.

Leonard, D.E. 1968. Diapause in the gypsy moth. *Journal of Economic Entomology* **61**:596–598.

Leonhardt, B.A. 1990. Controlled-release systems for the delivery of insect pheromones. In *Controlled Delivery of Crop Protection Agents*, ed. R.M. Wilkins, pp. 169–190. Taylor and Francis, London.

Leppla, N.C., E.W. Hamilton, R.H. Guy and F.L. Lee. 1979. Circadian rhythms of locomotion in six noctuid species. *Annals of the Entomological Society of America* **72**:209–215.

Lewis, T. 1984. Elements and frontiers of insect communication. In *Insect Communication*, ed. T. Lewis, pp 1–27. Academic Press, London.

Lewis, T. and E.D.M. Macaulay. 1976. Design and elevation of sex-attractant traps for pea moth, *Cydia nigricana* (Steph.) and the effect of plume shape on catches. *Ecological Entomology* **1**:175–187.

Liang, D. and C. Schal. 1993a. Volatile sex pheromone in the female German cockroach. *Experientia* **49**:324–328.

Liang, D. and C. Schal. 1993b. Female calling behavior of the German cockroach, *Blattella germanica* (Dictyoptera: Blattellidae). *Journal of Insect Behavior* **6**:603–614.

Liang, D. and C. Schal. 1993c. Ultrastructure and maturation of a sex pheromone gland in the female German cockroach, *Blattella germanica. Tissue and Cell* **25**:763–776.

Liang, D. and C. Schal. 1994. Neural and hormonal regulation of calling behavior in *Blattella germanica* females. *Journal of Insect Physiology* **40**:251–258.

Libbey, L.M., M.E. Morgan, T.B. Putnam and J.A. Rudinsky. 1976. Isomer of antiaggregative pheromone identified from male Douglas-fir beetle: 3-methylcyclohex-3-en-1-one. *Journal of Insect Physiology* **22**:871–873.

Lieke, E.E. 1993. Optical recording of neuronal activity in the insect central nervous system: odorant coding by the antennal lobes of honeybees. *European Journal of Neuroscience* **5**:49–55.

Light, D.M. 1986. Central integration of sensory signals: an exploration of processing of pheromonal and multimodal information in lepidopteran brains. In *Mechanisms in Insect Olfaction*, eds. T.L. Payne, M.C. Birch and C.E.J. Kennedy, pp. 287–301. Clarendon Press, Oxford.

Light, D.M. and M.C. Birch. 1979. Inhibition of the attractive pheromone response in *Ips paraconfusus* by (*R*)-(−)-ipsdienol. *Naturwissenschaften* **66**:159–160.

Light, D.M., R.A. Flath, R.G. Buttery, F.G. Zalom, R.E. Rice, J.C. Dickens and E.B. Jang. 1993. Host plant green leaf volatiles synergize the synthetic sex pheromones of the corn earworm and codling moth (Lepidoptera). *Chemoecology* **4**:145–152.

Lilley, R., J. Hardie and L.J. Wadhams. 1994. Field manipulation of *Praon* populations using semiochemicals. *Norwegian Journal of Agricultural Science Supplement* **16**:221–226.

Lin, S.Y.H. and J.T. Trumble. 1983. Bibliography of the tomato pinworm, *Keiferia lycopersicella* (Walsingham) (Lepidoptera: Gelechiidae). *Bibliography Entomogical Society of America* **1**:65–74.

Lin, S.Y.H. and J.T. Trumble. 1985. Influence of temperature and tomato maturity on development and survival of *Keifera lycopersicella* (Lepidoptera: Gelechiidae). *Environmental Entomology* **14**:855–858.

Lindgren, B.S. 1990. The research, development, and application continuum. In *Behavior-Modifying Chemicals for Insect Management. Applications of Pheromones and Other Attractants*, eds. R.L. Ridgway, R.M. Silverstein and M.N. Inscoe, pp. 149–164. Marcel Dekker, New York.

Lindgren, B.S. and J.H. Borden. 1993. Displacement and aggregation of mountain pine beetles, *Dendroctonus ponderosae* (Coleoptera: Scolytidae), in response to their antiaggregation and aggregation pheromones. *Canadian Journal of Forest Research* **23**:286–290.

Lindgren, B.S., J.H. Borden, G.H. Cushon, L.J. Chong and C.J. Higgins. 1989a. Reduction of mountain pine beetle (Coleoptera: Scolytidae) attacks by verbenone in lodgepole pine stands in British Columbia. *Canadian Journal of Forest Research* **19**:65–68.

Lindgren, B.S., M.D. McGregor, R.D. Oakes and H.E. Meyer. 1989b. Suppression of spruce beetle attacks by MCH released from bubble caps. *Western Journal of Applied Forestry* **4**:49–52.

Linn, C.E. and W.L. Roelofs. 1981. Modification of sex pheromone blend discrimination in male oriental fruit moths by pre-exposure to (*E*)-8-dodecenyl acetate. *Physiological Entomology* **6**:421–429.

Linn, C.E. and W.L. Roelofs. 1983. Effect of varying proportions of the alcohol component on sex pheromone blend discrimination in male oriental fruit moths. *Physiological Entomology* **8**:291–306.

Linn, C.E. and W.L. Roelofs. 1986a. Neuropharmacological effects on olfactory perception. In *Mechanisms in Insect Olfaction*, eds. T.L. Payne, M.C. Birch and C.E.J. Kennedy, pp. 209–215. Clarendon Press, Oxford.

Linn, C.E. and W.L. Roelofs. 1986b. Modulatory effects of octopamine and serotonin on thresholds for male sensitivity and periodicity of response to sex pheromone in the cabbage looper moth. *Archives of Insect Biochemistry and Physiology* **3**:161–171.

Linn, C.E. and W.L. Roelofs. 1987. Effects of octopaminergic agents on the sex pheromone mediated behavior of adult lepidoptera. In *Sites of Action for Neurotoxic Pesticides*, eds. R.M. Hollingworth and M.B. Green, ACS Symposium Series No. 356, pp. 162–173. American Chemical Society, Washington, D.C.

Linn, C.E. and W.L. Roelofs. 1989. Response specificity of male moths to multicomponent pheromones. *Chemical Senses* **14**:421–437.

Linn, C.E. and W.L. Roelofs. 1992. Role of photoperiod cues in regulating the modulatory action of octopamine on pheromone-response thresholds in the cabbage looper moth. *Archives of Insect Biochemistry and Physiology* **20**:285–302.

Linn, C.E. and W.L. Roelofs. 1994. Pheromone communication in the moths and its role in the speciation process. In *Speciation and the Recognition Concept: Theory and Application*, eds. D.M. Lambert and H.G. Spencer, pp. 263–300. Johns Hopkins University Press, Baltimore, Maryland.

Linn, C.E., L.B. Bjostad, J.W. Du and W.L. Roelofs. 1984. Redundancy in a chemical signal: behavioral responses of male *Trichoplusia ni* to a 6-component sex pheromone blend. *Journal of Chemical Ecology* **10**:1635–1658.

Linn, C.E., M.G. Campbell and W.L. Roelofs. 1986a. Male moth sensitivity to multicomponent pheromones: critical role of female-released blend in determining the functional role of components and active space of the pheromone. *Journal of Chemical Ecology* **12**:659–668.

Linn, C.E., M.G. Campbell and W.L. Roelofs. 1986b. Pheromone components and active spaces: what do moths smell and where do they smell it? *Science* **237**:650–652.

Linn, C.E., A. Hammond, J. Du and W.L. Roelofs. 1988. Specificity of male response to multicomponent pheromones in noctuid moths *Trichoplusia ni* and *Pseudoplusia includens. Journal of Chemical Ecology* **14**:47–57.

Linn, C.E., M.G. Campbell and W.L. Roelofs. 1992. Photoperiod cues and the modulatory action of octopamine and 5-hydroxytryptamine on locomotor and pheromone response in male gypsy moths, *Lymantria dispar. Archives of Insect Biochemistry and Physiology* **20**:265–284.

Linn, C.E., K.R. Poole and W.L. Roelofs. 1994a. Studies on biogenic amines and metabolites in nervous tissue and hemolymph of male cabbage looper moths: I. Quantitation of photoperiod changes. *Comparative Biochemistry and Physiology* **108**:73–85.

Linn, C.E., M.G. Campbell, K.R. Poole, and W.L. Roelofs. 1994b. Studies on biogenic amines and their metabolites in the nervous system and hemolymph of male cabbage looper moths: II. Photoperiod changes relative to random locomotor activity and pheromone-response thresholds. *Comparative Biochemistry and Physiology* **108**:87–98.

Linn, C.E., K.R. Poole, and W.L. Roelofs. 1994c. Studies on biogenic amines and their metabolites in the nervous system and hemolymph of male cabbage looper moths: III. Fate of injected octopamine, 5-hydroxytryptamine, and dopamine. *Comparative Biochemistry and Physiology* **108**:99–106.

Linn, C.E., K.R. Poole, W.-Q. Wu and W.L. Roelofs. 1994d. Circadian changes in levels of melatonin in the nervous system and hemolymph of the cabbage looper moth, *Trichoplusia ni. Journal of Comparative Physiology A* **176**:761–771.

Linn, C.E., M.G. Campbell, K.R. Poole, W.-Q. Wu and W.L. Roelofs. 1996. Effects of photoperiod on the circadian timing of pheromone response in male *Trichoplusia ni*: relationship to the modulatory action of octopamine. *Journal of Insect Physiology* **42**:881–891.

Lister C.K., J.M. Schmid, S.A. Mata, D. Haneman, C. O'Neil, J. Pasek and L. Sower. 1990. Verbenone bubble caps ineffective as a preventive strategy against mountain pine beetle attacks in ponderosa pine. United States Department of Agriculture, Forest Service, Research Note RM-501.

Liu, Y.-B. and K.F. Haynes. 1992. Filamentous nature of pheromone plumes protects integrity of signal from background chemical noise in cabbage looper moth, *Trichoplusia ni*. *Journal of Chemical Ecology* **18**:229–307.

Liu, Y.-B. and K.F. Haynes. 1993a. Impact of (*Z*)-7-dodecenol and turbulence on pheromone-mediated flight manoeuvres of male *Trichoplusia ni*. *Physiological Entomology* **18**:363–371.

Liu, Y.-B. and K.F. Haynes. 1993b. Pheromone-mediated responses of male cabbage looper moths, *Trichoplusia ni*, following various exposures to sex pheromone or (*Z*)-7-dodecenol. *Journal of Chemical Ecology* **19**:503–515.

Liu, Y.-B. and K.F. Haynes. 1994. Evolution of behavioral response to sex pheromone in mutant laboratory colonies of *Trichoplusia ni*. *Journal of Chemical Ecology* **20**:231–238.

Livingston, R.L. 1979. The pine engraver, *Ips pini* (Say), in Idaho. Life history, habits, and management recommendations. Idaho Department of Lands, Forest Insect and Disease Control Report 79-3.

Livingston, W.H., W.D. Bedard, A.C. Mangini and H.G. Kinzer. 1983. Verbenone interrupts attraction of the roundheaded pine beetle, *Dendroctonus adjunctus* (Coleoptera: Scolytidae), to sources of its natural attractant. *Journal of Economic Entomology* **76**:1041–1043.

Ljungberg, H., P. Anderson and B.S. Hansson. 1993. Physiology and morphology of pheromone-specific sensilla on the antennae of male and female *Spodoptera littoralis* (Lepidoptera: Noctuidae). *Journal of Insect Physiology* **39**:253–260.

Löfstedt, C. 1990. Population variation and genetic control of pheromone communication systems in moths. *Entomologia Experimentalis et Applicata* **54**:199–218.

Löfstedt, C. 1991. Evolution of moth pheromones. In *Insect Chemical Ecology*, ed. I. Hrdy, pp. 57–73. SPB Academic Publishing, The Hague.

Löfstedt, C. 1993. Moth pheromone genetics and evolution. *Philosophical Transactions of the Royal Society B* **340**:167–177.

Löfstedt, C. and W.L. Roelofs. 1985. Sex pheromone precursors in two primitive New Zealand Tortricid moth species. *Insect Biochemistry* **15**:729–734.

Löfstedt, C., J.N.C. Van Der Pers, J. Löfqvist, B.S. Lanne, M. Appelgren, G. Bergström and B. Thelin. 1982. Sex pheromone components of the turnip moth, *Agrotis segetum*: chemical identification, electrophysiological evaluation, and behavioral activity. *Journal of Chemical Ecology* **8**:1305–1322.

Löfstedt, C., W.M. Herrebout and J.-W. Du. 1986a. Evolution of the ermine moth pheromone tetradecyl acetate. *Nature* **323**:621–623.

Löfstedt, C., J. Löfqvist, B.S. Lanne, J.N.C. van der Pers and B.S. Hansson. 1986b. Pheromone dialects in European turnip moths *Agrotis segetum*. *Oikos* **46**:250–257.

Löfstedt, C., B.S. Hansson, W. Roelofs and B.O. Bengtsson. 1989a. No linkage between genes controlling female pheromone production and male pheromone response in the European corn borer, *Ostrinia nubilalis* Hübner (Lepidoptera: Pyralidae). *Genetics* **123**:553–556.

Löfstedt, C., N.J. Vickers, W.L. Roelofs and T.C. Baker. 1989b. Diet related courtship success in the Oriental fruit moth *Grapholita molesta* (Tortricidae). *Oikos* **55**:402–408.

Löfstedt, C., N. Vickers and T.C. Baker. 1990. Courtship pheromone titer and determination of the male mating success in the Oriental fruit moth *Grapholita molesta* (Lepidoptera: Tortricidae). *Entomologia Generalis* **15**:121–125.

Löfstedt, C., W.M. Herrebout and S.B.J. Menken. 1991. Sex pheromones and their potential role in the evolution of reproductive isolation in small ermine moths (Yponomeutidae). *Chemoecology* **2**:20–28.

Löfstedt, C., B.S. Hansson, M. Tóth, G. Szöcs, V. Buda, M. Bengtsson, N. Ryrholm, M. Svensson and E. Priesner. 1994a. Pheromone differences between sibling taxa *Diachrysia chrysitis* (Linaeus, 1758) and *D. tutti* (Kostrowicki, 1961) (Lepidoptera: Noctuidae). *Journal of Chemical Ecology* **20**:91–109.

Löfstedt, C., B.S. Hansson, E. Petersson, P. Valeur and A. Richards. 1994b. Pheromonal secretions from glands on the 5th abdominal sternite of hydrospychid and rhyacophilid caddisflies (Trichoptera). *Journal of Chemical Ecology* **20**:153–170.

Loher, W., T. Weber, H. Rembold and F. Huber. 1992. Persistence of phonotaxis in females of four species of crickets following allatectomy. *Journal of Comparative Physiology A* **171**:325–341.

Long, T. and L. Murdock. 1983. Stimulation of blowfly feeding behavior by octopaminergic drugs. *Proceedings of the National Academy of Science U.S.A.* **80**:4159–4163.

Longstaff, G.B. 1912. *Butterfly Hunting in Many Lands*. Longmans, Green, and Company, London.

Lopez, J.D. and J.A. Witz. 1988. Influence of *Heliothis vierscens* sex pheromone dispensers on captures of *H. zea* males in pheromone traps relative to distance and wind direction. *Journal of Chemical Ecology* **14**:265–276.

Lorenz, D. and F. Louis. 1993. Hinweise für den Rebschutz. *Der Deutsche Weinbau* **9**:21–23.

Louis, F. 1994. Ein Schädling hält alle auf Trab. *Der Deutsche Weinbau* **13**:18–20.

Louis, F. and K.J. Schirra. 1992. Mating disruption of *Lobesia botrana* and side effects of pheromones on beneficial arthropods in vineyards of the "Rheinpfalz" (Germany). *International Organization for Biological Control—West Palearctic Regional Section Bulletin* **15**(5):135.

Ludlow, A.R. 1984. Application of computer modelling to behavioral coordination. Ph.D. thesis, University of London.

Ma, P.W.K. 1992. Neuroendocrine control of sex-pheromone production in the European corn borer, *Ostrinia nubilalis* (Hübner) (Lepidoptera: Pyralidae). Ph.D. thesis, Cornell University, Ithaca, New York.

Ma, P.W.K. and W.L. Roelofs. 1995a. Calcium involvement in the stimulation of sex pheromone production by PBAN in the European corn borer, *Ostrinia nubilalis* (Lepidoptera: Pyralidae). *Insect Biochemistry and Molecular Biology* **25**:467–473.

Ma, P.W.K. and W.L. Roelofs. 1995b. Sites of synthesis and release of PBAN-like factor in female European corn borer, *Ostrinia nubilalis*. *Journal of Insect Physiology* **41**:339–350.

Ma, P.W.K. and W.L. Roelofs. 1995c. Anatomy of the neurosecretory cells in the cerebral and subesophageal ganglia of the female European corn borer moth, *Ostrinia nubilalis*

(Hübner)(Lepidoptera: Pyralidae). *International Journal of Insect Morphology and Embryology* **24**:343–359.

Ma, P.W.K., D.C. Knipple and W.L. Roelofs. 1994. Structural organization of the *Helicoverpa zea* gene encoding the precursor protein for pheromone biosynthesis-activating neuropeptide and other neuropeptides. *Proceedings of the National Academy of Science U.S.A.* **91**:6506–6510.

Ma, P.W.K., W.L. Roelofs and R.A. Jurenka. 1996. Characterization of PBAN and PBAN encoding gene neuropeptides in the central nervous system of the corn earworm moth, *Helicoverpa zea. Journal of Insect Physiology* **42**:257–266.

MacFerrin, K.D., L. Chen, M. Terranova, S.L. Schreiber and G.L. Verdine. 1993. Protein overproduction using the expression-cassette polymerase chain reaction. *Methods in Enzymology* **217**:79–102.

Maddison, W.P. 1990. A method for testing the correlated evolution of two binary characters: Are gains or losses concentrated on certain branches of a phylogenetic tree? *Evolution* **44**:539–557.

Maddison, W.P. and D.R. Maddison. 1992. *MacClade: Analysis of Phylogeny and Character Evolution.* Sinauer Associates, Sunderland, Massachusetts.

Mafra-Neto, A. 1993. Effects of the structure and composition of pheromone plumes on the response of the male almond moth, *Cadra cautella.* Ph.D. thesis, University of Massachusetts, Amherst.

Mafra-Neto, A. and R.T. Cardé. 1994. Fine-scale structure of pheromone plumes modulates upwind orientation of flying moths. *Nature* **369**:142.

Mafra-Neto, A. and R.T. Cardé. 1995a. Influence of plume structure and pheromone concentration on the upwind flight of *Cadra cautella* males. *Physiological Entomology* **20**:117–133.

Mafra-Neto, A. and R.T. Cardé. 1995b. Effect of the fine-scale structure of pheromone plumes: pulse frequency modulates activation and upwind flight of almond moth males. *Physiological Entomology* **20**:229–242.

Mafra-Neto, A. and R.T. Cardé. 1996. Dissection of the pheromone-modulated flight of moths using single-pulse response as a template. *Experientia* **52**:373–379.

Maida, R., A. Steinbrecht, G. Ziegelberger and P. Pelosi. 1993. The pheromone binding protein of *Bombyx mori*—purification, characterization and immunocytochemical localization. *Insect Biochemistry and Molecular Biology* **23**:243–253.

Mani, E. and F. Schwaller. 1992. Results of 12 years experience to control codling moth, *Cydia pomonella* L. by mating disruption. *International Organization for Biological Control—West Palearctic Regional Section Bulletin* **15**(5):76–80.

Mani, E., F. Schwaller and W. Riggenbach. 1984. Bekämpfung des Apfelwicklers (*Cydia pomonella* L.) mit der Verwirrungsmethode in einer Obstanlage im Bündner Rheintal: 1979–81. *Mitteilungen Schweizerischen Entomologischen Gesellschaft* **57**:341–348.

Marion-Poll, F. and T.R. Tobin. 1992. Temporal coding of pheromone pulses and trains in *Manduca sexta. Journal of Comparative Physiology A.* **171**:505–512.

Marsh, D., J.S. Kennedy and A.R. Ludlow. 1978. An analysis of anemotactic zigzagging flight in male moths stimulated by pheromone. *Physiological Entomology* **3**:221–240.

Marshall, L.D. and J.N. McNeil. 1989. Spermatophore mass as an estimate of male nutrient investment: a closer look in *Pseudaletia unipuncta* (Haworth) (Lepidoptera: Noctuidae). *Functional Ecology* **3**:605–612.

Martinez, T. and F. Camps. 1988. Stimulation of sex pheromone production by head extract in *Spodoptera littoralis* at different times of the photoperiod. *Archives of Insect Biochemistry and Physiology* **9**:211–220.

Martinez, T., G. Fabriás and F. Camps. 1990. Sex pheromone biosynthetic pathway in *Spodoptera littoralis* and its activation by a neurohormone. *Journal of Biological Chemistry* **265**:1381–1387.

Masler, E.P. and A.K. Raina. 1993. Pheromonotropic activity in the gypsy moth *Lymantria dispar*: evidence for a neuropeptide. *Journal of Comparative Physiology B* **163**:259–264.

Masler, E.P., A.K. Raina, R.M. Wagner and J.P. Kochansky. 1994. Isolation and identification of a pheromonotropic neuropeptide from the brain-subesophageal ganglion complex of *Lymantria dispar*: a new member of PBAN family. *Insect Biochemistry and Molecular Biology* **24**:829–836.

Masson, C. and H. Mustaparta. 1990. Chemical information processing in the olfactory system of insects. *Physiological Reviews* **70**:199–245.

Matsumoto S.G. and J.G. Hildebrand. 1981. Olfactory mechanisms in the moth *Manduca sexta*: response characteristics and morphology of central neurons in the antennal lobes. *Philosophical Transactions of the Royal Society, London B* **213**:249–277.

Matsumoto, S., A. Suzuki, H. Yamamoto, N. Ogura and H. Ikemoto. 1979. Cyclic nucleotides and hormonal control of cuticular melanization in the armyworm larva, *Leucania separata* (Lepidoptera: Noctuidae). *Applied Entomology and Zoology* **14**:159–163.

Matsumoto, S., A. Isogai, A. Suzuki, N. Ogura and H. Sonobe. 1981. Purification and properties of the melanization and reddish colouration hormone (MRCH) in the armyworm, *Leucania separata* (Lepidoptera). *Insect Biochemistry* **11**:725–733.

Matsumoto, S., A. Isogai and A. Suzuki. 1986. Isolation and amino terminal sequence of melanization and reddish colouration hormone (MRCH) from the silkworm, *Bombyx mori*. *Insect Biochemistry* **16**:775–779.

Matsumoto, S., A. Isogai and A. Suzuki. 1988. Purification and characterization of melanization and reddish coloration hormone (MRCH) in lepidopteran insects. In *Advances in Pigment Cell Research*, ed. J.T. Bagnara, pp. 437–451. Alan R. Liss, New York.

Matsumoto, S., A. Kitamura, H. Nagasawa, H. Kataoka, C. Orikasa, T. Mitsui and A. Suzuki. 1990. Functional diversity of neurohormone produced by the subesophageal ganglion: molecular identity of melanization and reddish colouration hormone and pheromone biosynthesis activating neuropeptide. *Journal of Insect Physiology* **36**:427–432.

Matsumoto, S., A. Fónagy, M. Kurihara, K. Uchiumi, T. Nagamine, M. Chijimatsu and T. Mitsui. 1992a. Isolation and primary structure of a novel pheromonotropic neuropeptide structurally related to leucopyrokinin from the armyworm larvae, *Pseudaletia separata*. *Biochemical and Biophysical Research Communications* **182**:534–539.

Matsumoto, S., O. Yamashita, A. Fónagy, M. Kurihara, K. Uchiumi, T. Nagamine and T. Mitsui. 1992b. Functional diversity of a pheromonotropic neuropeptide: induction of cuticular melanization and embryonic diapause in lepidopteran insects by *Pseudaletia* pheromonotropin. *Journal of Insect Physiology* **38**:847–851.

Matsumoto, S., A. Fónagy, L. Schoofs, A. De Loof, M. Kurihara, T. Nagamine and T. Mitsui. 1993. Induction of cuticular melanization in the armyworm larvae *Pseudaletia separata*, by insect myotropic neuropeptide possessing FXPRLamide at the C-terminus. *Journal of Pesticide Science* **18**:127–129.

Matsuyama, S., W.S. Leal, M. Hasegawa and M. Sawada. 1994. Is chirality essential for pheromonal activity of scarab beetles? The case of the large black chafer, *H. parallela*. Proceedings, 38th Annual Meeting of the Japan Society for Bioscience, Biotechnology and Agrochemistry. *Nogeikagakukaishi* **68**:58.

Mauchamp, B. and J.A. Pickett. 1987. Juvenile hormone-like activity of (**E**)-β-farnesene derivatives. *Agronomie* **7**:523–529.

Mayer, M.S. 1993. Responses of three antennal specialist neurons of male *Trichplusia ni* (Hübner) to sex pheromone components at and above naturally emitted levels. *Journal of Insect Physiology* **39**:401–412.

Mayer, M.S. 1995. The use of paired and single traps to assess perception and discrimination of sex pheromone mixtures in the field by *Trichoplusia ni* (Hübner). *Journal of Chemical Ecology* **21**:2057–2068.

Mayer, M.S. and J.R. McLaughlin. 1992. Discrimination of female sex pheromone by male *Trichoplusia ni* (Hübner). Chemical Senses **16**:699–710.

Mayer, M.S. and J.R. McLaughlin. 1993. Discrimination of mixtures of female sex pheromone components by male *Trichoplusia ni* (Hübner). *Chemical Senses* **18**:735–749.

Mayr, E. 1970. *Population, Species, and Evolution*. Harvard, Cambridge, Massachusetts.

McColl, H.P. 1969. The sexual scent organs of male Lepidoptera. M.S. thesis, University College, Swansea, Wales.

McDonough, L.M., H.G. Davis, P.S. Chapman and C.L. Smithhisler. 1993. Response of male codling moths (*Cydia molesta*) to components of conspecific female sex pheromone glands in flight tunnel tests. *Journal of Chemical Ecology* **19**:1737–1748.

McDonough, L.M., H.G. Davis, P.S. Chapman and C.L. Smithhisler. 1994. Codling moth (*Cydia pomonella*): disruptants of sex pheromonal communication. *Journal of Chemical Ecology* **20**:171–181.

McGregor, M.D., M.M. Furniss, R.D. Oaks, K.E. Gibson and H.E. Meyer. 1984. MCH pheromone for preventing Douglas-fir beetle infestation in windthrown trees. *Journal of Forestry* **82**:613–616.

McKenna, M.P., D.S. Hekmat-Scafe, P. Gaines and J.R. Carlson. 1994. Putative *Drosophila* pheromone-binding proteins expressed in a subregion of the olfactory system. *Journal of Biological Chemistry* **269**:16340–16347.

McLaughlin, J.R., E.R. Mitchell and J.H. Cross. 1981. Field and laboratory evaluation of mating disruptants of *Heliothis zea* and *Spodoptera frugiperda* in Florida. In *Management of Insect Pests with Semiochemicals: Concepts and Practice*, ed. E.R. Mitchell, pp. 243–251. Plenum Press, New York.

McNeil, J.N. 1986. Calling behavior: can it be used to identify species of moths? *Florida Entomologist* **69**:78–84.

McNeil, J.N. 1987. The true armyworm, *Pseudaletia unipuncta*: a victim of the pied piper or a seasonal migrant? *Insect Science and Its Application* **8**:591–597.

McNeil, J.N. 1991. Behavioral ecology of pheromone-mediated communication in moths and its importance in the use of pheromone traps. *Annual Review of Entomology* **36**:407–430.

McNeil, J.N. 1992. Evolutionary perspectives and insect pest control: an attractive blend for the deployment of semiochemicals in management programs. In *Insect Chemical Ecology: An Evolutionary Approach*, eds. B.D. Roitberg and M.B. Isman, pp. 334–351. Chapman and Hall, New York.

McNeil, J.N. and J. Delisle. 1989. Host plant pollen influences calling behavior and ovarian development of the sunflower moth, *Homoeosoma electellum. Oecologia* **80**:201–205.

McNeil, J.N., M. Cusson, J. Delisle, I. Orchard and S.S. Tobe. 1995. Physiological integration of migration in Lepidoptera. In *Migration: Physical Factors and Physiological Mechanisms*, eds. V.A. Drake and A.G. Gatehouse, pp. 279–302. Cambridge University Press, Cambridge.

Meng, L.Z., C.H. Wu, M. Wicklein, K.-E. Kaissling and H.J. Bestmann. 1989. Number and sensitivity of three types of pheromone receptor cells in *Antheraea pernyi* and *A. polyphemus. Journal of Comparative Physiology A* **165**:139–146.

Menken, S.B.J., W.M. Herrebout and J.T. Wiebes. 1992. Small ermine moths (Yponomeuta): their host relations and evolution. *Annual Review of Entomology* **37**:41–66.

Menzel, R. 1983. Neurobiology of learning and memory: the honeybee as a model system. *Naturwissenschaften* **70**:504–511.

Mercer, A. and R. Menzel. 1982. The effects of biogenic amines on conditioned and unconditioned responses to olfactory stimuli in the honeybee *Apis mellifera. Journal of Comparative Physiology A* **145**:363–368.

Mercer, A., P. Kloppenburg, and J.G. Hildebrand. 1993. Serotonin-induced changes in excitability of antennal-lobe neurons in *Manduca sexta. Society for Neuroscience Abstract* **19**:126.

Meyer, C.S., D.K. Lynn and D.J. Hamilton. 1968. *Analysis and Design of Integrated Circuits*. McGraw-Hill, New York.

Michel, W.C., D.A. Fadool and B.W. Ache. 1992. Cyclic nucleotides mediate an odor-evoked potassium conductance in lobster olfactory receptor cells. *Journal of Neuroscience* **12**:3979–3984.

Milde, J.J. 1989a. Neuronal basis for visually induced head movements in the sphingid moth *Manduca sexta*. In *Dynamics and Plasticity in Neuronal Systems*, eds. N. Elsner and W. Singer, Proceedings of the 17th Göttingen Neurobiology Conference, p. 81, Georg Thieme Verlag, Stuttgart, Germany.

Milde, J.J. 1989b. Neural basis of head movements evoked by optomotor stimuli in the moth *Manduca sexta. Society for Neuroscience Abstracts* **15**:1290.

Milde, J.J. 1993. Tangential medulla neurons in the moth *Manduca sexta*. Structure and responses to optomotor stimuli. *Journal of Comparative Physiology A* **173**:783–799.

Milde, J.J., H.S. Seyan and N.J. Strausfeld. 1987. The neck motor system of the fly

Calliphora erythrocephala II. Sensory organization. *Journal of Comparative Physiology A* **160**:225–238.

Miles, D.B. and A.E. Dunham. 1993. Historical perspectives in ecology and evolutionary biology: the use of phylogenetic comparative analyses. *Annual Review of Ecology and Systematics* **24**:587–619.

Miller, D.R. 1990. Reproductive and ecological isolation: community structure in the use of semiochemicals by pine bark beetles (Coleoptera: Scolytidae). Ph.D. thesis, Simon Fraser University, Burnaby, Canada.

Miller, D.R. and J.H. Borden. 1992. (*S*)-(+)-Ipsdienol: interspecific inhibition of *Ips latidens* (LeConte) by *Ips pini* (Say) (Coleoptera: Scolytidae) *Journal of Chemical Ecology* **18**:1577–1582.

Miller, D.R., G. Gries and J.H. Borden. 1990. *E*-Myrcenol: a new pheromone for the pine engraver, *Ips pini* (Say) (Coleoptera: Scolytidae). *Canadian Entomologist* **122**:401–406.

Miller, D.R., J.H. Borden, G.G.S. King and K.N. Slessor. 1991. Ipsenol: an aggregation pheromone for *Ips latidens* (LeConte) (Coleoptera: Scolytidae). *Journal of Chemical Ecology* **17**:1517–1527.

Miller, D.R., D.R. Devlin and J.H. Borden. 1994. The use of antiaggregation/inhibitory pheromones in controlling pine engravers in stands of lodgepole pine. In *Proceedings of the Symposium on Management of Western Bark Beetles with Pheromones—Recent Research and Development*, June 22–25, 1992, Kailua-Kona, ed. P.J. Shea, pp. 38–43. United States Department of Agriculture, Forest Service, General Technical Report PSW-GTR-150.

Miller, D.R., J.H. Borden and B.S. Lindgren. 1995. Verbenone: dose-dependent interruption of pheromone-based attraction of three sympatric species of pine bark beetles. *Environmental Entomology* **24**:692–696.

Miller, J.R. and R.S. Cowles. 1990. Stimulo-deterrent diversion: a concept and its possible application to onion maggot control. *Journal of Chemical Ecology* **16**:3197–3212.

Miller, J.R. and W.L. Roelofs. 1980. Individual variation in sex pheromone component ratios in two populations of the redbanded leafroller moth, *Argyrotaenia velutinana*. *Environmental Entomology* **9**:359–363.

Miller, J.S. 1991. Cladistics and classification of the Notodontidae (Lepidoptera: Noctuoidea) based on larval and adult morphology. *Bulletin of the American Museum of Natural History* **204**:1–230.

Miller, J.S. and J.W. Wenzel. 1994. Ecological characters and phylogeny. *Annual Review of Entomology* **40**:389–415.

Milli, R. 1990. Messung der Pheromonverteilung im Freiland zur Unterstützung der Paarungsstörmethode bei Lobesia botrana Schiff. Thesis submitted for Diploma, Universität Kaiserslautern, Germany.

Milli, R. and E. Dickler. 1992. Distribution of pheromones in apple orchards where codling moth is controlled by mating disruption. *International Organization for Biological Control—West Palearctic Regional Section Bulletin* **15**(5):140–145.

Milli, R. and J.J. de Kramer. 1989. An EAG detector for field pheromone measurements.

In *Dynamics and Plasticity in Neuronal Systems, Proceedings 17th Göttingen Neurobiology Conference*, p. 77, Thieme, Stuttgart, Germany.

Minet, J. 1984. Contribution à l'analyse phylogénétique des neolepidoptères (Lepidoptera, Glossata). *Nouvelle Revue d'Entomologie* **2**:139–149.

Minet, J. 1991. Tentative reconstruction of the ditrysian phylogeny. *Entomologica Scandinavica* **22**:69–95.

Minks, A.K. 1977. Trapping withe behavior-modifying chemicals: feasibility and limitations. In *Chemical Control of Insect Behavior*, eds. H.H. Shorey and J.J. McKelvey Jr., pp. 385–394. Wiley, New York.

Minks, A.K. 1990. Registration requirements and status for pheromones in Europe and other countries. In *Behavior-Modifying Chemicals for Insect Management*, eds. R.L. Ridgway, R.M. Silverstein and M.N. Inscoe, pp. 557–568. Marcel Dekker, New York.

Minks, A.K. and R.T. Cardé. 1988. Disruption of pheromone communication in moths: is the natural blend really most efficacious? *Entomologia Experimentalis et Applicata* **49**:25–36.

Mitchell, E.R., ed. 1981. *Management of Insect Pests with Semiochemicals*. Plenum Press, New York.

Mitchell, E.R., H.R. Agee and R.R. Heath. 1989. Influence of pheromone trap color and design on capture of male velvet bean caterpillar and fall armyworm moths (Lepidoptera: Noctuidae). *Journal of Chemical Ecology* **15**:1775–1784.

Mittelstaedt, H. 1962. Control systems of orientation in insects. *Annual Review of Entomology* **7**:177–198.

Moffitt, H.R. and P.H. Westigard. 1984. Suppression of the codling moth (Lepidoptera: Tortricidae) population on pears in southern Oregon. *Journal of Economic Entomology* **77**:1513–1519.

Möhl, B. 1988. Short-term learning during flight control in *Locusta migratoria. Journal of Comparative Physiology A* **156**:93–101.

Möhl, B. 1989. 'Biological noise' and plasticity of sensorimotor pathways in the locust flight system. *Journal of Comparative Physiology A* **166**:75–82.

Mole, N. and C.D. Jones. 1994. Concentration fluctuation data from dispersion experiments carried out in stable and unstable conditions. *Boundary-Layer Meteorology* **67**:41–74.

Montague, W. and J.R. Cook. 1971. The role of adenosine 3′:5′-cyclic monophosphate in the regulation of insuline release by isolated rat islets of Langerhans. *Biochemical Journal* **122**:115–120.

Moore, P.A., J. Atema and G.A. Gerhardt. 1991. Fluid dynamics and microscale chemical movement in the chemosensory appendages of the lobster, *Homarus americanus. Chemical Senses* **16**:663–674.

Moore, P.A., R.K. Zimmer-Faust, S.L. Bennet, M.J. Weissburg, J.M. Parrish and G.A. Gerhardt. 1992. Measurement of microscale patchiness in a turbulent aquatic odor plume using a semiconductor-based microprobe. *Biological Bulletin* **183**:138–142.

Moores, G.D., G.J. Devine and A.L Devonshire. 1994. Insecticide-insensitive acetylcholinesterase can enhance esterase-based resistance in *Myzus persicae* and *Myzus nicotianae. Pesticide Biochemistry and Physiology* **49**:114–120.

Mordue, A.J., A. Blackwell, B.S. Hansson, L.J. Wadhams and J.A. Pickett. 1992. Behavioural and electrophysiological evaluation of oviposition attractants for *Culex quinquefasciatus* Say (Diptera: Culicidae). *Experientia* **48**:1109–1111.

Morita, M., M. Hatakoshi and S. Tojo. 1988. Hormonal control of cuticular melanization in the common cutworm, *Spodoptera litura*. *Journal of Insect Physiology* **34**:751–758.

Moriuti, S. 1977. Fauna Japonica: Yponomeutidae S. Lat. (Insecta: Lepidoptera). Keigaku, Tokyo.

Muggleston, S.J. and S.P. Foster. 1989. Sustained-flight tunnel responses of male light-brown apple moth to synthetic sex pheromone. *Physiological Entomology* **14**:443–449.

Müller, F. 1877. On hair tufts, felted patches, and similar structures on the wings of male Lepidoptera. *Jenaische Zeitschrift für Naturwissenschaft* **11**:99–114.

Müller, F. 1878. Scent organs in butterflies. *Transactions of the Entomological Society (London)*, pp. 211–221.

Müller, M. and R. Wehner. 1988. Path integration in desert ants *Cataglyphis fortis*. *Proceedings of the National Academy of Sciences U.S.A.* **85**:5287–5290.

Murlis, J. 1986. The structure of odour plumes. In *Mechanisms in Insect Olfaction*, eds. T.L. Payne, M.C. Birch and C.E.J. Kennedy, pp. 27–38. Clarendon Press, Oxford.

Murlis, J. and B.W. Bettany. 1977. Night flight towards a sex pheromone source by male *Spodoptera littoralis* (Boisd.) (Lepidoptera, Noctuidae). *Nature* **268**:433–435.

Murlis, J. and C.D. Jones. 1981. Fine scale structure of odour plumes in relation to insect orientation to distant pheromone and other attractant sources. *Physiological Entomology* **6**:71–86.

Murlis, J., B.W. Bettany, J. Kelley and L. Martin. 1982. The analysis of flight paths of male Egyptian cotton leafworm moths, *Spodoptera littoralis*, to a sex pheromone source in the field. *Physiological Entomology* **7**:435–441.

Murlis, J., M.A. Willis and R.T. Cardé. 1989. Odour signals: pattern in time and space. In *Proceedings X International Symposium on Olfaction and Taste*, ed. K.B. Døving, pp. 6–17, Graphic Communication Systems, Oslo.

Murlis, J., J.S. Elkinton and R.T. Cardé. 1992. Odor plumes and how insects use them. *Annual Review of Entomology* **37**:505–532.

Mustaparta, H. 1989. Evolutionary aspects of pheromone perception. In *Proceedings X International Symposium on Olfaction and Taste*, ed. K.B. Døving, pp. 164–176, Graphic Communication Systems, Oslo.

Mustaparta, H., M.E. Angst and G.N. Lanier. 1977. Responses of single receptor cells in the pine engraver beetle, *Ips pini* (Say)(Coleoptera: Scolytidae) to its aggregation pheromone, ipsdienol, and its aggregation inhibitor, ipsenol, *Journal of Comparative Physiology* **121**:343–347.

Mustaparta, H., M.E. Angst and G.N. Lanier. 1979. Specialization of olfactory cells to insect and host-produced volatiles in the beetle, *Ips pini* (Say). *Journal of Chemical Ecology* **5**:109–123.

Mustaparta, H., M.E. Angst and G.N. Lanier. 1980. Receptor discrimination between enantiomers of the aggregation pheromone ipsdienol, in two species of *Ips*. *Journal of Chemical Ecology* **6**:689–70.

Muszynska-Pytel, M. and B. Cymborowski. 1978a. The role of serotonin in regulation of the circadian rhythms of locomotor activity in the cricket (*Acheta domesticus*) I. circadian variations in concentration in the brain and hemolymph. *Comparative Biochemistry and Physiology* **59C**:13–15.

Muszynska-Pytel, M. and B. Cymborowski. 1978b. The role of serotonin in regulation of the circadian rhythms of locomotor activity in the cricket (*Acheta domesticus*) II. Distribution of serotonin and variations in different brain structures. *Comparative Biochemistry Physiology* **59C**:17–20.

Mylne, K.R. 1988. Experimental measurements of concentration fluctuations. In *Air Pollution Modelling and Its Applications VII*, ed. H. van Dop, pp. 555–565. Plenum Press, New York.

Mylne, K.R. and P.J. Mason. 1991. Concentration fluctuation measurements in a dispersing plume at a range of up to 100 m. *Quarterly Journal of the Royal Meteorological Society* **117**:177–206.

Nachman, R.J., G.M. Holman and B.J. Cook. 1986. Active fragments and analogs of the insect neuropeptide leucopyrokinin: structure–function studies. *Biochemical and Biophysical Research Communications* **137**:936–942.

Nachman, R.J., G.M. Holman, L. Schoofs and O. Yamashita. 1993a. Silkworm diapause induction activity of myotropic pyrokinin (FXPRLamide) insect neuropeptides. *Peptides* **14**:1043–1048.

Nachman, R.J., H. Kuniyoshi, V. Roberts, M.G. Holman and A. Suzuki. 1993b. Promone biosynthesis in the silkworm. *Biochemical and Biophysical Research Communications* **193**:661–666.

Nagasawa, H., H. Kuniyoshi, R. Arima, T. Kawano, T. Ando and A. Suzuki. 1994. Structure and activity of *Bombyx* PBAN. *Archives of Insect Biochemistry and Physiology* **25**:261–270.

Nault, L.R. 1973. Alarm pheromones help aphids escape predators. *Ohio Report* **58**:16–17.

Nault, L.R. and M.E. Montgomery. 1977. Aphid pheromones. In *Aphids as Vectors*, eds. K.F. Harris and K. Maramorosch, pp. 527–545. Academic Press, London.

Neumann, U. 1992. La lutte contre les vers de la grappe (*Eupoecilia ambiguella* Hbn. et *Lobesia botrana* Schiff.) par la méthode de confusion: résultats et acquisitions. *Bulletin International Organization for Biological Control—West Palearctic Regional Section Bulletin* **15**(2):24.

Neumann, U., K. Kalter, P.-J. Charmillot, E. Mani, L. Blommers, M. Blanc, E Höbaus and G. Sterk. 1990. Experimentation and applications of sex pheromones with the mating disruption technique against the codling moth, *Cydia pomonella* (L.), and the summer fruit tortrix moth, *Adoxophyes orana* (F.v.R.). *Mededelingen van de Faculteit Landbouwwetenschappen Rijksuniversiteit Gent* **55**:379–386.

Neumann, U., A. Schmid, C. Ioriatti, M. Varner, R. Castillo, A. Lucas, J.L. Perez Marin and M.A. Castillo. 1993. La technique par confusion contre les "vers" de la grappe en Europe aujourd'hui. *Phytoma* **456**:15–17.

Newton, P.I. and C. Wilkinson. 1982. A taxonomic revision of the North American species of *Stigmella* (Lepidoptera: Nepticulidae). *Systematic Entomology* **7**:367–463.

Ngai, J., A. Chess, M.M. Dowling, N. Necles, E.R. Macgno and R. Axel. 1993. Coding of olfactory information: topography of odorant expression in the catfish olfactory epithelium. *Cell* **72**:667–680.

Nielsen, A. 1980. A comparative study of the genital segments and the genital chamber in female Trichoptera. *Biologiske Skrifter* **23**:1–200.

Nielsen, E.S. 1987. The recently discovered primitive (non-Ditrysian) family Palaephatidae (Lepidoptera) in Australia. *Invertebrate Taxonomy* **1**:201–209.

Nielsen, E.S. 1989. Phylogeny of the major lepidopteran groups. In *The Hierarchy of Life*, eds. B. Ferholm, K. Bremer and H. Jörnvall, pp. 281–294. Elsevier, New York.

Nielsen, E.S. and I.F.B. Common. 1991. Lepidoptera (moths and butterflies). In *The Insects of Australia*, Vol. 2, ed. I.D. Naumann, pp. 817–915. Melbourne University Press, Carlton, Victoria; and University College of London Press, London.

Nielsen, E.S. and D.R. Davis. 1985. The first southern hemisphere prodoxid and phylogeny of the Incurvarioidea (Lepidoptera). *Systematic Entomology* **10**:307–322.

Nishida, R. and H. Fukami. 1983. Female sex pheromone of the German cockroach, *Blattella germanica. Memoirs of the College of Agriculture, Kyoto University* **122**:1–24.

Nishizuka Y. 1992. Intracellular signalling by hydrolysis of phospholipids and activation of protein kinase C. *Science* **258**:607–614.

Noldus, L.P.J.J., R.P.J. Potting and H.E. Barendregt. 1991. Moth sex pheromone adsorption to leaf surface: bridge in time for chemical spies. *Physiological Entomology* **16**:329–344.

Nordlund, D.A., R.L. Jones and W.J. Lewis, eds. 1981. *Semiochemicals: Their Role in Pest Control.* Wiley, New York.

Oatman, E.R. 1970. Ecological studies of the tomato pinworm on tomato in southern California. *Journal of Economic Entomology* **63**:1531–1534.

Oatman, E.R., J.A. Wyman, and G.R. Platner. 1979. Seasonal occurrence and parasitization of the tomato pinworm on fresh market tomatoes in southern California. *Environmental Entomology* **8**:661–664.

Obara, Y. 1979. *Bombyx mori* mating dance: an essential in locating the female. *Applied Entomology and Zoology* **14**:130–132.

O'Carroll, D. 1993. Feature-detecting neurons in dragonflies. *Nature* **362**:541–543.

Ochieng, S.A., P. Anderson and B.S. Hansson. 1995. Antennal lobe projections of pheromone specific receptor neurons in male and female *Spodoptera littoralis. Tissue and Cell* **27**:221–232.

ODell, T.M. and V.C. Mastro. 1980. Crepuscular activity of gypsy moth adults. *Environmental Entomology* **9**:613–617.

O'Donald, P. 1962. The theory of sexual selection. *Heredity* **17**:541–552.

O'Donald, P. 1980. *Genetic Models of Sexual Selection.* Cambridge University Press, Cambridge.

Ogura, N. 1975a. Hormonal control of larval colouration in the armyworm, *Leucania separata. Journal of Insect Physiology* **21**:559–576.

Ogura, N. 1975b. Induction of cuticular melanization in larvae of armyworm, *Leucania separata* Walker (Lepidoptera: Noctuidae), by implantation of ganglia in the silkworm, *Bombyx mori* L. (Lepidoptera: Bombycidae). *Applied Entomology and Zoology* **10**:216–219.

Ogura, N., and T. Sato. 1973. Induction of embryonic diapause in the silkworm, *Bombyx mori* L. (Lepidoptera: Noctuidae). *Applied Entomology and Zoology* **8**:46–48.

Ohguchi, Y., S. Tatsuki, K. Usui, K. Arai, M. Kurihara, K. Uchiumi and J. Fukami. 1985. Hormone-like substance present in the cephalic organs of the female moth, *Chilo suppressalis* (Walker) (Lepidoptera: Pyralidae) and controlling sex pheromone production. *Japan Journal of Applied Entomology and Zoology* **29**:265–269.

Olberg, R.M. 1983a. Pheromone-triggered flip-flopping interneurons in the ventral nerve cord of the silkworm moth, *Bombyx mori. Journal of Comparative Physiology A* **152**:297–307.

Olberg, R.M. 1983b. Interneurons sensitive to female pheromone in the deutocerebrum of the male silkworm moth, *Bombyx mori. Physiological Entomology* **8**:419–428.

Olberg, R.M. and M.A. Willis. 1990. Pheromone-modulated optomotor response in male gypsy moths, *Lymantria dispar* L.: directionally selective visual interneurons in the ventral nerve cord. *Journal of Comparative Physiology A* **167**:707–714.

Orchard, I., J.-M. Ramirez and A.B. Lange. 1993. A multifunctional role for octopamine in locust flight. *Annual Review of Entomology* **38**:227–249.

Osborne, G.O. and C.P. Hoyt. 1969. A chemical attractant for males of the grass grub beetle, *C. zealandica* (White)(Col: Scarabaeidae). *Bulletin of Entomological Research* **59**:81–83.

Osborne, G.O. and C.P. Hoyt. 1970. Phenolic resins as chemical attractants for males of the grass grub beetle, *C. zealandica* (Col: Scarabaeidae). *Annals of the Entomological Society of America* **63**:1143–1147.

O'Shea, M. and R.K. Murphy. 1978. Octopamine modulates sensitivity of identified insect interneurons. *Society for Neuroscience Abstracts* **4**:203.

Otte, D. 1974. Effects and functions in the evolution of signaling systems. *Annual Review of Ecology and Systematics* **5**:385–417.

Outram, I. 1968. Polyandry in spruce budworm. *Bi-Monthly Research Notes (Ottawa, Canada)* **24**:6–7.

Owen, M.D., L. Pfaff and B.D. Sloley. 1987. The absence of diel change in the concentrations of dopamine, 5-hydroxytryptamine and their metabolites in the cerebral ganglia of the cockroach, *Periplaneta americana. Insect Biochemistry* **17**:723–729.

Ozawa, R.A., T. Ando, H. Nagasawa, H. Kataoka and A. Suzuki. 1993. Reduction of the acyl group: the critical step in bombykol biosynthesis that is regulated *in vitro* by the neuropeptide hormone in the pheromone gland of *Bombyx mori. Bioscience, Biotechnology and Biochemistry* **57**:2144–2147.

Page, T.L. 1987. Serotonin phase-shifts the circadian rhythm of locomotor activity in the cockroach. *Journal of Biological Rhythms* **2**:23–34.

Pagel, M.D. 1994. The adaptationist wager. In *Phylogenetics and Ecology*, eds. P. Eggleton and R.I. Vane-Wright, pp. 29–51. Academic Press, London.

Paine, T.D. and C.C. Hanlon. 1991. Response of *Dendroctonus brevicomis* and *Ips paraconfusus* (Coleoptera: Scolytidae) to combinations of synthetic pheromone attractants and inhibitors verbenone and ipsdienol. *Journal of Chemical Ecology* **17**:2163–2176.

Palaniswamy, P. and W.D. Seabrook. 1978. Behavioral response of the female eastern spruce budworm *Choristoneura fumiferana* (Lepidoptera: Tortricidae) to the sex pheromone of her own species. *Journal of Chemical Ecology* **4**:649–655.

Palaniswamy, P., E.W. Underhill, W.F. Steck and M.D. Chisholm. 1983. Responses of male redbacked cutworm, *Euxoa ochrogaster* (Lepidoptera: Noctuidae), to sex pheromone components in a flight tunnel. *Environmental Entomology* **12**:748–752.

Pandey, A. and M. Habibulla. 1980. Serotonin in the central nervous system of the cockroach. *Journal of Insect Physiology* **26**:1–6.

Pandey, A. and M. Habibulla. 1982. Circadian rhythms of serotonin and the electrical activity of the frontal ganglion of the cockroach. *Experientia* **38**:946–948.

Parker, G.A. 1970. Sperm competition and its evolutionary consequences in the insects. *Biological Reviews* **45**:525–567.

Parker, S.P. 1982. *Synopsis and Classification of Living Organisms*. McGraw-Hill, Philippines.

Parsons, P.A. 1980. Isofemale strains and evolutionary strategies in natural populations. *Evolutionary Biology* **13**:175–217.

Paterson, H.E.H. 1978. More evidence against speciation by reinforcement. *South African Journal of Science* **74**:369–371.

Paterson, H.E.H. 1982. Perspective on speciation by reinforcement. *South African Journal of Science* **78**:53–57.

Paterson, H.E.H. 1985. The recognition concept of species. In *Species and Speciation*, ed. E.S. Vrba, pp. 21–29. Transvaal Museum, Pretoria, South Africa.

Paterson, H.E.H. 1993. Variation and the specific-mate recognition system. In *Perspectives in Ethology*, ed. P.P.G. Bateson, pp. 209–227. Plenum Press, New York.

Payne, T.L. and R.F. Billings. 1989. Evaluation of (*S*)-verbenone applications for suppressing southern pine beetle (Coleoptera: Scolytidae) infestations. *Journal of Economic Entomology* **82**:1702–1708.

Payne, T.L. and J.V. Richerson. 1985. Pheromone-mediated competitive replacement between two bark beetle populations: influence on infestation suppression. *Zeitshrift für angewandte Entomologie* **99**:131–138.

Payne, T.L., J.E. Coster and P.C. Johnson. 1977. Effects of slow-release formulation of synthetic *endo*- and *exo*-brevicomin on southern pine beetle flight and landing behavior. *Journal of Chemical Ecology* **3**:133–141.

Payne, T.L., J.E. Coster, J.V. Richerson, L.J. Edson and E.R. Hart. 1978. Field response of the southern pine beetle to behavioral chemicals. *Environmental Entomology* **7**:578–582.

Payne, T.L., M.C. Birch and C.E.J. Kennedy, eds. 1986. *Mechanisms in Insect Olfaction*. Clarendon Press, London.

Payne, T.L., R.F. Billings, C.W. Berisford, S.M. Salom, D.M. Grossman, M.J. Dalusky and W.W. Upton. 1992. Disruption of *Dendroctonus frontalis* (Col., Scolytidae) infestations with an inhibitor pheromone. *Journal of Applied Entomology* **114**:341–347.

Pedersen, P.E., P.J. Jastreboff, W.B. Stewart and G.M. Shepherd. 1986. Mapping of an olfactory receptor population that projects to a specific region in the rat olfactory bulb. *Journal of Comparative Neurology* **250**:93–108.

Pellmyr O., L.B. Thien, G. Bergström and I. Groth. 1990. Pollination of New Caledonian Winteraceae: opportunistic shifts or parallel radiation with their pollinators? *Plant Systematics and Evolution* **173**:143–157.

Pelosi, P. 1994. Odorant-binding proteins. *Critical Review of Biochemistry and Molecular Biology* **29**:199–228.

Pelosi, P. and R. Maida. 1990. Odorant binding proteins in vertebrates and insects: similarities and possible common function. *Chemical Senses* **15**:205–215.

Penman, D.R., R.B. Chapman and M.H. Bowie. 1988. Selection for behavioral resistance in twospotted spider mite (Acari: Tetryanychidae) to flucythrinate. *Journal of Economic Entomology* **81**:40–44.

Percy-Cunningham, J.E. and J.A. MacDonald. 1987. Biology and ultrastructure of sex pheromone-producing glands. In *Pheromone Biochemistry*, eds. G.D. Prestwich and G.J. Blomquist, pp. 27–76. Academic Press, Orlando, Florida.

Perez Marin, J.L. 1992. Lutte contre *Lobesia botrana* de la vigne par la technique de confusion sexuelle en la Rioja (Espagne). *International Organization for Biological Control—West Palearctic Regional Section Bulletin* **15**(2):19.

Persoons, C.J., P.E.J. Verwiel, F.J. Ritter, E. Talman, P.E.J. Nooyen and W.J. Nooyen. 1976. Sex pheromone of the American cockroach, *Periplaneta americana*. Isolation and structure elucidation of periplanone-B. *Journal of Chemical Ecology* **5**:221–236.

Pettersson, J., J.A. Pickett, B.J. Pye, A. Quiroz, L.E. Smart, L.J. Wadhams and C.M. Woodcock. 1994. Winter host component reduces colonization by the bird-cherry–oat aphid, *Rhopalosiphum padi* (L.)(Homoptera, Aphididae), and other aphids in cereal fields. *Journal of Chemical Ecology* **20**:2265–2274.

Pevsner, J. and S.H. Snyder. 1990. Odorant-binding protein: odorant transport function in the vertebrate nasal epithelium. *Chemical Senses* **15**:217–222.

Pevsner, J., R.R. Reed, P.G. Feinstein and S.H. Snyder. 1988. Molecular cloning of odorant-binding protein: Member of a ligand carrier family. *Science* **241**:336–339.

Pevsner, J., V. Hou, A.M. Snowman and S.H. Snyder. 1990. Odorant-binding protein, characterization of ligand binding. *Journal of Biological Chemistry* **265**:6118–6125.

Pfeiffer, D.G., W. Kaakeh, J.C. Killian, M.W. Lachance and P. Kirsch. 1993. Mating disruption for control of damage by codling moth in Virginia apple orchards. *Entomologica Experimentalis et Applicata* **67**:57–64.

Pfeiffer, D.R., R.W. Taylor and H.A. Lardy. 1978. Ionophore A 23187: cation binding and transport properties. *Annals of New York Academy of Sciences* **307**:402–423.

Phelan, P.L. 1992. Evolution of sex pheromones and the role of asymmetric tracking. In *Insect Chemical Ecology*, eds. B.D. Roitberg and M.B. Isman, pp. 265–314. Chapman and Hall, New York.

Phelan, P.L. 1997. Evolution of mate signalling in moths: phylogenetic considerations and predictions from the asymmetric tracking hypothesis. In *Evolution of Mating Systems in Insects and Arachnids*, eds. J. Choe and B. Crespi. pp. 240–256. Cambridge University Press, Cambridge.

Phelan, P.L. and T.C. Baker. 1986. Male-size-related courtship success and intersexual selection in the tobacco moth *Ephestia elutella. Experientia* **42**:1291–1293.

Phelan, P.L. and T.C. Baker. 1987. Evolution of male pheromones in moths: reproductive isolation through sexual selection. *Science* **235**:205–207.

Phelan, P.L. and T.C. Baker. 1990a. Comparative study of courtship in twelve phycitine moths (Lepidoptera: Pyralidae). *Journal of Insect Behavior* **3**:303–326.

Phelan, P.L. and T.C. Baker. 1990b. Information transmission during intraspecific and interspecific courtship in *Ephestia elutella* and *Cadra figulilella. Journal of Insect Behavior* **3**:589–602.

Phillips, J.R. 1979. Migration of the bollworm, *Heliothis zea* (Boddie). In *Movement of Highly Mobile Insects: Concepts and Methodology in Research; Proceedings of a Conference*, eds. R.L. Rabb and G.G. Kennedy, pp. 409–411. University Graphics, North Carolina State University, Raleigh, North Carolina.

Philpott A. 1924. On an abdominal scent-organ (?) in *Sabatinca* and other primitive genera of Lepidoptera. *Transactions of the Entomological Society of London* pp. 457–461.

Picimbon, J.F., J.M. Becard, L.Sreng, J.L. Clement and C. Gadenne. 1995. Juvenile hormone stimulates release of pheromonotropic brain factor in the female black cutworm, *Agrotis ipsilon. Journal of Insect Physiology* **41**:377–382.

Pickett, J.A. 1992. Potential of novel chemical approaches for overcoming insecticide resistance. In *Resistance '91: Achievements and Developments in Combating Pesticide Resistance*, eds. I. Denholm, A.L. Devonshire and D.W. Hollomon, pp. 354–356. Elsevier, London

Pickett, J.A., G.R. Cayley, G.W. Dawson, D.C. Griffiths, S.H. Hockland, B. Marples, R.T. Plumb and C.M. Woodcock. 1986. Use of the alarm pheromone and derivatives against aphid mediated damage. *Abstracts 6th International Congress Pesticide Chemistry, Ottawa, 1986*, 2C-08.

Pickett, J.A., L.J. Wadhams, C.M. Woodcock and J. Hardie. 1992. The chemical ecology of aphids. *Annual Review of Entomology* **37**:67–90.

Pierce, H.D., Jr., P. de Groot, J.H. Borden, S. Ramaswamy and A.C. Oehlschlager. 1995. Identification and field activity of the pheromones of the red pine cone beetle, *Conophthorus resinosae* Hopkins, and its former sibling species, *C. banksianae* McPherson (Coleoptera: Scolytidae). *Journal of Chemical Ecology* **21**:169–185.

Pikielny, C.W., G. Hasan, F. Rouyer and M. Robash. 1994. Members of a family of *Drosophila* putative odorant-binding proteins are expressed in different subsets of olfactory hairs. *Neuron* **12**:35–49.

Pirngruber, O. 1944. Blütenökologische Untersuchungen über einige Falter der Gattung *Micropteryx. Verhandlungen der Zoologisch-Botaninischen Gesellschaft in Wien* **90/91**:129–220.

Pitcairn, M.J., F.G. Zalom and W.J. Bentley. 1990. Weather factors influencing capture of *Cydia pomonella* (Lepidoptera: Tortricidae) in pheromone traps during overwintering flight in California. *Environmental Entomology* **19**:1253–1258.

Pitman, G.B., M.W. Stock and R.C. McKnight. 1978. Pheromone application in mountain pine beetle/lodgepole pine management: theory and practice. In *Theory and Practice of Mountain Pine Beetle Management in Lodgepole Pine Forests*, eds. A.A. Berryman, G.D. Amman and R.W. Stark, pp. 165–173. University of Idaho, Moscow and United States Department of Agriculture, Forest Service, Washington.

Pivnick, K.A. and J.N. McNeil. 1985. Mate location and mating behavior of *Thymelicus lineola* (Lepidoptera: Hesperidae). *Annals of the Entomological Society of America* **78**:651–656.

Pivnick, K.A. and J.N. McNeil. 1987. Puddling in butterflies: sodium affects reproductive success in *Thymelicus lineola. Physiological Entomology* **12**:461–472.

Pollard, G.V. 1991. Constraints to IPM development and a strategy for management of tomato and cabbage pests in Trinidad, West Indies. *Journal of Tropical Pest Management* **37**:59–62.

Pop, L., H. Arn and H.R. Buser. 1993. Determination of release rates of pheromone dispensers by air sampling with C-18 bonded silica. *Journal of Chemical Ecology* **19**:2513–2519.

Pope, M.M., Gaston, L.K. and Baker, T.C. 1984. Composition, quantification, and periodicity of sex pheromone volatiles from individual *Heliothis zea* females. *Journal of Insect Physiology* **30**:943–945.

Powell, G., J. Hardie and J.A. Pickett. 1993. Effects of the antifeedant polygodial on plant penetration by aphids, assessed by video and electrical recording. *Entomologia Experimentalis et Applicata* **68**:193–200.

Preiss, R. 1980. Anemotaxis im Lauf und Flug beim Schwammspinner. Versuche zur Aufkärung des Wirkungsgefüges. Ph.D. thesis, Ludwig-Maximilians-Universität, Munich, Germany.

Preiss, R. 1992. Set point of retinal velocity of ground images in the control of swarming flight of desert locusts. *Journal of Comparative Physiology A* **171**:251–256.

Preiss, R. and E. Kramer. 1983. Stabilization of altitude and speed in tethered flying gypsy moth males: influence of (+) and (–)-disparlure. *Physiological Entomology* **8**:55–68.

Preiss, R. and L. Futschek. 1985. Flight stabilization by pheromone-enhanced optomotor responses. *Naturwissenschaften* **72**:435–436.

Preiss, R. and E. Kramer. 1986a. Pheromone-induced anemotaxis in simulated free flight. In *Mechanisms in Insect Olfaction*, eds. T.L. Payne, M.C. Birch and C.E.J. Kennedy, pp. 69–79. Clarendon Press, Oxford.

Preiss, R. and E. Kramer. 1986b. Mechanism of pheromone orientation in flying moths. *Naturwissenschaften* **73**:555–557.

Preiss, R. and M. Gewecke. 1991. Compensation of visually simulated wind drift in the swarming flight of the desert locust (*Schistocerca gregaria*) *Journal of Experimental Biology* **157**:461–481.

Prestwich, G.D. 1987. Chemical studies of pheromone reception and catabolism. In *Phero-*

mone Biochemistry, eds. G.D. Prestwich and G.J. Blomquist, pp. 473–527. Academic Press, Orlando, Florida.

Prestwich, G.D. 1991. Tritium-labeled hormones, pheromones, and photoaffinity labels. In *Isotopes in the Physical and Biomedical Sciences*, Vol. 1B, eds. J.R. Jones and E. Buncel, pp. 137–178, Elsevier, Oxford.

Prestwich, G.D. 1993a. Bacterial expression and photoaffinity labeling of a pheromone binding protein. *Protein Science* **2**:420–428.

Prestwich, G.D. 1993b. Chemical studies of pheromone receptors in insects. *Archives of Insect Biochemistry and Physiology* **22**:75–83.

Prestwich, G.D., F.A. Golec and N.H. Andersen. 1984. Synthesis of a highly tritiated photoaffinity labeled pheromone analog for the moth *Antheraea polyphemus*. *Journal of Labelled Compounds and Radiopharmaceuticals* **21**:593–601.

Prestwich, G.D., R.G. Vogt and L.M. Riddiford. 1986. Binding and hydrolysis of radiolabeled pheromone and several analogs by male-specific antennal proteins of the moth *Antheraea polyphemus*. *Journal of Chemical Ecology* **12**:323–333.

Prestwich, G.D., S.McG. Graham, B. Latli, M. Handley, L. Streinz and M.L. Tasayco. 1989. Enzymatic processing of pheromones and pheromone analogs. *Experientia* **45**:263–270.

Prestwich, G.D., G. Duand and S. LaForest. 1995. How is pheromone specificity encoded in proteins. *Chemical Senses* **20**:461–469.

Priesner, E. 1977. Evolutionary potential of specialized olfactory receptors. In: *Olfaction and Taste VI*, eds. J. Le Magnen and P. MacLeod. Information Retrieval, London.

Priesner, E. 1979a. Specificity studies on pheromone receptors of noctuid and tortricid Lepidoptera. In *Chemical Ecology: Odour Communication in Animals*, ed. F.J. Ritter, pp. 57–71. Elsevier/North-Holland, Amsterdam.

Priesner, E. 1979b. Progress in the analyses of pheromone receptor system. *Annales de Zoologies et Écologie Animale* **11**:533–546.

Priesner, E. 1986. Correlating sensory and behavioural responses in multichemical pheromone systems of Lepidoptera. In *Mechanisms in Insect Olfaction*, eds. T.L. Payne, M.C. Birch and C.E.J. Kennedy, pp. 225–233. Clarendon Press, Oxford.

Priesner, E. and P. Witzgall. 1984. Modification of pheromonal behaviour in wild *Coleophora laricella* male moths by (*Z*)-5-decenyl acetate, an attraction-inhibitor. *Journal of Applied Entomology* **98**:118–135.

Proctor, H.C. 1992. Sensory exploitation and the evolution of male mating behaviour a cladistic test using water mites (Acari: Parasitengona). *Animal Behaviour* **44**:745–752.

Prokopy, R.J. 1981. Oviposition deterring pheromone system of apple maggot flies. In *Management of Insect Pests with Semiochemicals*, ed. E.R. Mitchell, pp. 477–494. Plenum Press, New York.

Proshold, F.I., and L.E. LaChance. 1974. Analysis of sterility in hybrids from interspecific crosses between *Heliothis virescens* and *H. subflexa*. *Annals of the Entomological Society of America* **67**:445–449.

Proshold, F.I., J.L. Gonzalez, C. Asencio and R.R. Heath. 1986. A trap for monitoring the sweetpotato weevil (Coleoptera: Curculionidae) using pheromone or live females as bait. *Journal of Economic Entomology* **79**:641–647.

Rabb, R.L. 1966. Diapause in *Protoparce sexta* (Lepidoptera: Sphingidae). *Annals of the Entomological Society of America* **59**:160–165.

Rabb, R.L., F.E. Guthrie, H.E. Scott and C.F. Smith. 1959. Tobacco insects of North Carolina and their natural enemies. Bulletin 394, pp. 1–31. Agricultural Experiment Station, North Carolina State College, Raleigh.

Rafaeli, A. 1994. Pheromonotropic stimulation of moth pheromone gland cultures *in vitro. Archives of Insect Biochemistry and Physiology* **25**:287–299.

Rafaeli, A. and V. Soroker. 1989a. Influence of diel-rhythm and brain hormone on pheromone production in two lepidopteran species. *Journal of Chemical Ecology* **15**:447–455.

Rafaeli A. and V. Soroker. 1989b. Cyclic-AMP mediation of the hormonal stimulation of ^{14}C-acetate incorporation by *Heliothis armigera* pheromone glands *in vitro*. *Molecular and Cellular Endocrinology* **65**:43–48.

Rafaeli, A. and Z. Klein. 1994. Regulation of pheromone production by female pink bollworm moths, *Pectinophora gossypiella* (Sanders) (Lepidoptera: Gelechiidae). *Physiological Entomology* **19**:159–164.

Rafaeli, A. and V. Soroker. 1994. Second messenger interactions in response to PBAN stimulation of pheromone gland cultures. In *Insect Neurochemistry and Neurophysiology*, eds. A.B. Borkovec and M. Loeb, pp. 223–226. CRC Press, Boca Raton, Florida.

Rafaeli, A., V. Soroker, J.A. Klun and A.K. Raina. 1990a. Stimulation of *de novo* pheromone biosynthesis of (*Z*)-11 hexadecenal by *in vitro* pheromone glands of *Heliothis spp* In: *Insect Neurochemistry and Neurophysiology*, eds. A.B Borkovec and E. Masler, pp. 309–312. Humana Press, Clifton, New Jersey.

Rafaeli A., V. Soroker, B. Kamensky and A.K. Raina. 1990b. Action of pheromone biosynthesis activating neuropeptide on *in vitro* pheromone glands of *Heliothis armigera* females. *Journal of Insect Physiology* **36**:641–646.

Rafaeli, A., J. Hirsch, V. Soroker, B. Kamensky and A.K. Raina. 1991. Spatial and temporal distribution of pheromone biosynthesis-activating neuropeptide in *Helicoverpa (Heliothis) armigera* using RIA and *in vitro* bioassay. *Archives of Insect Biochemistry and Physiology* **18**.119–129.

Rafaeli A., V. Soroker, J. Hirsch, B. Kamensky and A.K. Raina. 1993. The influence of photoperiod and age on the competence of pheromone glands and the distribution of immunoreactive PBAN in *Helicoverpa* spp. *Archives of Insect Biochemistry and Physiology* **22**:169–180.

Raffa, K.F., T.W. Phillips and S.M. Salom. 1993. Strategies and mechanisms of host colonization by bark beetles. In *Beetle–Pathogen Interactions in Conifer Forests*, eds. T.D. Schowalter and G.M. Filip, pp. 103–128. Academic Press, New York.

Raina, A.K. 1988. Selected factors influencing neurohormonal regulation of sex pheromone production in *Heliothis* species. *Journal of Chemical Ecology* **14**:2063–2069.

Raina, A.K. 1989. Male-induced termination of sex pheromone production and receptivity in mated females of *Heliothis zea. Journal of Insect Physiology* **35**:821–826.

Raina, A.K. 1993. Neuroendocrine control of sex pheromone biosynthesis in Lepidoptera. *Annual Review of Entomology* **38**:329–349.

Raina, A.K. and G. Gade. 1988. Insect peptide nomenclature. *Insect Biochemistry* **18**:785–787.

Raina, A.K. and T.G. Kempe. 1990. A pentapeptide of the C-terminal sequence of PBAN with pheromonotropic activity. *Insect Biochemistry* **20**:849–851.

Raina, A.K. and T.G. Kempe. 1992. Structure activity studies of PBAN of *Helicoverpa zea* (Lepidoptera: Noctuidae). *Insect Biochemistry and Molecular Biology* **22**:221–225.

Raina, A.K. and J.A. Klun. 1984. Brain factor control of sex pheromone production in the female corn earworm moth. *Science* **225**:531–533.

Raina, A.K. and J.J. Menn. 1987. Endocrine regulation of pheromone production in Lepidoptera. In *Pheromone Biochemistry*, eds. G.D. Prestwich and G.J. Blomquist, pp. 159–174. Academic Press, Orlanda, Florida.

Raina, A.K., J.A. Klun and E.A. Stadelbacher. 1986. Diel periodicity and effect of age and mating on female sex pheromone titer in *Heliothis zea* (Lepidoptera: Noctuidae). *Annals of the Entomological Society of America* **79**:128–131.

Raina, A.K., H. Jaffe, J.A. Klun, R.L. Ridgway and D.K. Hayes. 1987. Characteristics of a neurohormone that controls sex pheromone production in *Heliothis zea*. *Journal of Insect Physiology* **33**:809–814.

Raina, A.K., H. Jaffe, T.G. Kempe, P. Keim, R.W. Blacher, H.M. Fales, C.T. Riley, J.A. Klun, R. Ridgway and D.K. Hayes. 1989a. Identification of a neuropeptide hormone that regulates pheromone production in female moths. *Science* **244**:796–798.

Raina, A.K., Stadelbacher, E.A. and Ridgway, R.L. 1989b. Comparison of sex pheromone composition and pheromone-mediated male behavior of laboratory reared and wild *Heliothis zea* (Lepidoptera: Noctuidae). *Journal of Chemical Ecology* **15**:1259–1265.

Raina, A.K., T.G. Kempe and H. Jaffe. 1991. PBAN: regulation of pheromone production in moths. In: *Insect Neuropeptides: Chemistry, Biology and Action*, eds. J.J. Menn, T.J. Kelly and E.P. Masler, pp. 100–109. American Chemical Society, Washington, D.C.

Raina, A.K., T.G. Kingan and A.K. Mattoo. 1992. Chemical signals from host plant and sexual behavior in a moth. *Science* **225**:592–594.

Raina, A.K., A. Rafaeli and T.G. Kingan. 1994a. Pheromonotropic activity of orally administered PBAN and its analogues in *Helicoverpa zea*. *Journal of Insect Physiology* **40**:393–397.

Raina, A.K., T.G. Kingan and J.M. Giebultowicz. 1994b. Mating-induced loss of sex pheromone and sexual receptivitiy in insects with emphasis on *Helicoverpa zea* and *Lymantria dispar*. *Archives of Insect Biochemistry and Physiology* **25**:317–327.

Raman, K.V. 1982. Field trials with sex pheromone of potato tuberworm. *Environmental Entomology* **11**:367–370.

Raman, K.V. 1988. Control of potato tuber moth *Pthorimaea operculella* with sex pheromones in Peru. *Agricultural, Ecosystems and Environonment* **21**:85–99.

Ramaswamy, S.B. and N.E. Cohen. 1992. Ecdysone: an inhibitor of receptivity in the moth, *Heliothis virescens*? *Naturwissenschaften* **79**:29–31.

Ramaswamy, S.B, G.N. Mbata, and N.E. Cohen. 1990. Necessity of juvenile hormone for choriogenesis in the moth, *Heliothis virescens* (Noctuidae). *Invertebrate Reproduction and Development* **17**:27–33.

Ramaswamy, S.B., G.N. Mbata, N.E. Cohen, A. Moore and N.M Cox. 1994. Pheromonotropic and pheromonostatic activity in moths. *Archives of Insect Biochemistry and Physiology* **25**:301–315.

Ramaswamy, S.B., R.A. Jurenka, C.E. Linn and W.L. Roelofs. 1995. Evidence for the presence of a pheromonotropic factor in hemolymph and regulation of sex pheromone production in *Helicoverpa zea*. *Journal of Insect Physiology* **41**:501–508.

Raming, K., J. Krieger and H. Breer. 1989. Molecular cloning of an insect pheromone-binding protein. *Federation of European Biochemical Societies Letters* **256**:215–218.

Raming, K., J. Krieger and H. Breer. 1990a. Primary structure of a pheromone binding protein from *Anthereae pernyi*: homologies with other ligand-carrying proteins. *Journal of Comparative Physiology B* **160**:503–509.

Raming, K., J. Krieger and H. Breer. 1990b. Molecular cloning, sequencing and expression of cDNA encoding a G_0-protein from insect. *Cellular Signalling* **2**:311–321.

Raming, K., J. Krieger, J. Strotmann, I. Boekhoff, S. Kubick, C. Baumstark and H. Breer. 1993a. Cloning and expression of odorant receptors. *Nature* **361**:353–356.

Raming, K., J. Freitag, J. Krieger, and H. Breer. 1993b. Arrestin-subtypes in insect antennae. *Cellular Signalling* **5**:69–80.

Rankin, L.J. and J.H. Borden. 1991. Competitive interactions between the mountain pine beetle and the pine engraver in lodgepole pine. *Canadian Journal of Forest Research* **21**:1029–1036.

Rasnitsyn, A.P. 1980. Superordo Papilionida. In *Historical Development of Insects*, eds. B.B. Rodendorf and A.P. Rasnitsyn, pp. 99–101. *Proceedings of the Palaeontological Institute* 178, Moscow.

Rau, P. and N.L. Rau. 1929. The sex attraction and rhythmic periodicity in the giant saturniid moths. *Transactions of the Academy of Science of Saint Louis* **26**:83–221.

Raupach, M.R. 1988. Canopy transport processes. In *Flow and Transport in the Natural Environment, Advances and Applications*, eds. W.L. Steffan and O.T. Denmead, pp. 95–127. Springer-Verlag, Berlin.

Rauscher, S., and H. Arn. 1979. Mating suppression in tethered females of *Eupoecilia ambiguella* by evaporation of (*Z*)-9-dodecenyl acetate in the field. *Entomologia Experimentalis et Applicata* **25**:16–20.

Razowski J. 1975. Abdominal scent organ in Homoneura (Lepidoptera). *Bulletin Entomologique de Pologne* **45**:3–7.

Reagan, J.D., W.H. Miller and S.J. Kramer. 1992. Allatotropin-induced formation of inositol phosphates in the corpora allata of the moth, *Manduca sexta*. *Archives of Insect Biochemistry and Physiology* **20**:145–155.

Reed, R.R. 1992. Mechanisms of sensitivity and specificity in olfaction. *Cold Spring Harbor Symposium on Quantitative Biology* **57**:501–504.

Reiter, R.J. 1987. The melatonin message: duration versus coincidence hypothesis. *Life Sciences* **40**:2119–2131.

Renwick, J.A.A. 1967. Identification of two oxygenated terpenes from the bark beetles *Dendrotonus frontalis* and *Dendroctonus brevicomis*. *Contributions of the Boyce Thompson Institute* **23**:355–360.

Renwick, J.A.A. and J.P. Vité. 1969. Bark beetle attractants: mechanism of colonization by *Dendroctonus frontalis*. *Nature* **224**:1222–1223.

Resh, V.H. and J.R. Wood. 1985. Site of sex pheromone production in three species of Trichoptera. *Aquatic Insects* **7**:65–71.

Ressler, K.J., S.L. Sullivan and L.B. Buck. 1993. A zonal organization of odorant receptor gene expression in the olfactory epithelium. *Cell* **73**:597–609.

Richerson, J.V. and T.L. Payne. 1979. Effects of bark beetle inhibitors on landing and attack behavior of the southern pine beetle and beetle associates. *Environmental Entomology* **8**:360–364.

Richerson, J.V., E.A. Cameron and G.L. Bernon. 1976. Sexual activity of the gypsy moth. *American Midland Naturalist* **95**:299–312.

Richerson, J.V., F.A. McCarty and T.L. Payne. 1980. Disruption of southern pine beetle infestations with frontalure. *Environmental Entomology* **9**:90–93.

Riddiford, L.M. 1974. The role of hormones in the reproductive behavior of female wild silk moths. In *Experimental Analysis of Insect Behaviour*, ed. L. Barton Browne, pp. 278–285. Springer-Verlag, New York.

Riddiford, L.M. and C.M. Williams. 1971. Role of corpora cardiaca in the behavior of saturniid moths I. Release of sex pheromone. *Biological Bulletin* **140**:1–7.

Ride, D.J. 1988. A model for the observed intermittency of a meandering plume. *Journal of Hazardous Materials* **19**:131–37.

Ridgway, R.L., R.M. Silverstein and M.N. Inscoe, eds. 1990. *Behavior-Modifying Chemicals for Insect Management*. Marcel Dekker, New York.

Ridley, M. 1990. The control and frequency of mating in insects. *Functional Ecology* **4**:75–84.

Riedl, H., A. Seaman and F. Henrie. 1985. Monitoring susceptibility to azinphosmethyl in field populations of the codling moth (Lepidoptera: Tortricidae) with pheromone traps. *Journal of Economic Entomology* **78**:692–699.

Riley, C.T., B.K. Barbeau, P.S. Keim, F.J. Kezdy, R.L. Heinrikson and J.H. Law. 1984. The covalent protein structure of insecticyanin, a blue biliprotein from the hemolymph of the tobacco hornworm, *Manduca sexta*. *Journal of Biological Chemistry* **259**:13159–13165.

Roach, S.H. 1975. *Heliothis zea* and *Heliothis virescens*: moth activity as measured by blacklight and pheromone traps. *Journal of Economic Entomology* **68**:17–21.

Roach, S.H. and P.L. Adkisson. 1970. Rôle of photoperiod and temperature in the induction of pupal diapause in the bollworm, *Heliothis zea*. *Journal of Insect Physiology* **16**:1591–1597.

Rodriguez, V. 1988. Spatial coding of olfactory information in the antennal lobe of *Drosophila melanogaster*. *Brain Research* **453**:299–307.

Roehrich, R., J.P. Carles, C. Trésor and M.A. de Vathaire. 1979. Essais de "confusion sexuelle" en vignoble contre les tordeuses de la grappe, l'eudímis *Lobesia botrana* Den et Schiff. et la cochylis *Eupoecilia ambiguella* Tr. *Annales de Zoologie et Écologie Animale* **11**:654–657.

Roelofs, W.L. 1984. Electroantennogram assays: rapid and convenient screening procedures for pheromones. In *Techniques in Pheromone Research*, eds. H.E. Hummel and T.A. Miller, pp. 131–159. Springer-Verlag, New York.

Roelofs, W.L. and L. Bjostad. 1984. Biosynthesis of lepidopteran pheromones. *Biorganic Chemistry* **12**:279–298.

Roelofs, W.L. and R.L. Brown. 1982. Pheromones and evolutionary relationships of Tortricidae. *Annual Review of Ecology and Systematics* **13**:395–422.

Roelofs, W.L. and R.T. Cardé. 1974. Sex pheromones in the reproductive isolation of lepidopterous species. In *Pheromones*, ed. M.C. Birch, pp. 96–114. North-Holland, Amsterdam.

Roelofs, W.L. and A. Comeau. 1969. Sex pheromone specificity: taxonomic and evolutionary aspects in Lepidoptera. *Science* **165**:398–400.

Roelofs, W.L. and Comeau, A. 1971. Sex pheromone perception: synergists and inhibitors for the red-banded leafroller attractant. *Journal of Insect Physiology* **17**:435–448.

Roelofs, W.L. and M.A. Novak. 1981. Small-plot disorientation tests for screening potential mating disruptants. In *Management of Insect Pests with Semiochemicals*, ed. E.R. Mitchell, pp. 229–242. Plenum Press, New York.

Roelofs, W.L. and W.A. Wolf. 1988. Pheromone biosynthesis in Lepidoptera. *Journal of Chemical Ecology* **14**:2019–2031.

Roelofs, W.L., A. Comeau, A. Hill and G. Milicevic. 1971. Sex attractant of the codling moth: characterization with electroantennogram technique. *Science* **174**:297–299.

Roelofs, W., J. Kochansky, R. Cardé, H. Arn and S. Rauscher. 1973. Sex attractant of the grapevine moth, *Lobesia botrana. Mitteilungen der Schweizerischen Entomologischen Gesellschaft* **46**:71–73.

Roelofs, W.L., A.S. Hill, R.T. Cardé and T.C. Baker. 1974. Two sex pheromone components of the tobacco budworm moth, *Heliothis virescens. Life Sciences* **14**:1555–1562.

Roelofs, W.L., J-W. Du, C. Linn, T.J. Glover and L.B. Bjostad. 1986. The potential for genetic manipulation of the redbanded leafroller moth sex pheromone blend. In *Evolutionary Genetics of Invertebrate Behavior*, ed. M.D. Huettel, pp. 263–281. Plenum, New York.

Roelofs, W.L., T. Glover, X-H. Tang, I. Sreng, P. Robbins, C. Eckenrode, C. Löfstedt, B.S. Hansson and B.O. Bengtsson. 1987. Sex pheromone production and perception in European corn borer moths is determined by both autosomal and sex-linked genes. *Proceedings of the National Academy of Science U.S.A.* **84**:7585–7589.

Ronacher, B. and R. Wehner. 1995. Desert ants *Cataglyphis fortis* use self-induced optic flow to measure distances travelled. *Journal of Comparative Physiology A* **177**:21–27.

Ronnett, G.V. and S.H. Snyder. 1992. Molecular messengers of olfaction. *Trends in Neuroscience* **15**:508–513.

Rospars, J.P. 1988. Structure and development of the insect antennodeutocerebral system. International *Journal of Insect Morphology and Embryology* **17**:243–294.

Rospars, J.P. and J.G. Hildebrand. 1992. Anatomical identification of glomeruli in the antennal lobes of the male sphinx moth *Manduca sexta. Cell and Tissue Research* **270**:205–227.

Ross, H.H. 1967. The evolution and past dispersal of the Trichoptera. *Annual Review of Entomology* **12**:169–206.

Rossiter, M., J. Gershenzon and T.J. Mabry. 1986. Behavioral and growth response of specialist herbivore, *Homoeosoma electellum*, to major terpenoid of its host, *Helianthus* spp. *Journal of Chemical Ecology* **12**:522–524.

Roth, L.M. 1962. Hypersexual activity induced in females of the cockroach *Nauphoeta cinerea. Science* **138**:1267–1269.

Roth, L.M. 1970. Evolution and taxonomic significance of reproduction in Blattaria. *Annual Review of Entomology* **15**:75–96.

Roth, L.M. and B. Stay. 1962. Oocyte development in *Blattella germanica and Blattella vaga* (Blattaria). *Annals of the Entomological Society of America* **55**:633–642.

Roth, L.M. and R.H. Barth, Jr. 1964. The control of sexual receptivity in female cockroaches. *Journal of Insect Physiology* **10**:965–975.

Rothschild, G.H.L. 1981. Mating disruption of lepidopterous pests: current status and future prospects. In *Management of Insect Pests with Semiochemicals: Concepts and Practice*, ed. E.R. Mitchell, pp. 207–228. Plenum Press, New York.

Rothschild, G.L. and A.K. Minks. 1974. Time of activity of male oriental fruit moth at pheromone sources in the field. *Environmental Entomology* **3**:1003–1007.

Rothschild, G.H.L., B.F. Nesbitt, P.S. Beevor, A. Cork, D.R. Hall, and R.A. Vickers. 1982. Studies of the female sex pheromone of the native budworm, *Heliothis punctiger*. *Entomologica Experimentalis et Applicata* **31**:395–401.

Rothschild, M., R.T. Aplin, P.A. Cockrum, J.A. Edgar, P. Fairweather and R. Lees. 1979. Pyrrolizidine alkaloids in arctiid moths (Lep.) with a discussion on host plant relationships and the role of these secondary plant substances in the Arctiidae. *Biological Journal of the Linnean Society* **12**:305–326.

Rowell, C.H.F. 1988. Mechanisms of flight steering in locusts. *Experientia* **44**:389–395.

Rowell, C.H.F. 1993. Intersegmental coordination of flight steering in locusts. *Seminar in the Neurosciences* **5**:59–66.

Royer, L. and J.N. McNeil. 1993a. Effect of relative humidity conditions on responsiveness of European corn borer (*Ostrinia nubilalis*) males to female sex pheromone in a wind tunnel. *Journal of Chemical Ecology* **19**:61–69.

Royer, L. and J.N. McNeil. 1993b. Male investment in the European corn borer, *Ostrinia nubilalis* (Lepidoptera: Pyralidae): impact on female longevity and reproductive performance. *Functional Ecology* **7**:209–215.

Rudinsky, J.A. 1968. Pheromone-mask by the female *Dendroctonus pseudotsugae* Hopk., an attraction regulator. *Pan-Pacific Entomologist* **44**:248–250.

Rudinsky, J.A. 1969. Masking of the aggregating pheromone in *Dendroctonus pseudotsugae* Hopk. *Science* **166**:884–885.

Rudinsky, J.A. 1973. Multiple functions of the Douglas fir beetle pheromone 3-methyl-2-cyclohexen-1-one. *Environmental Entomology* **2**:579–585.

Rudinsky, J.A. and L.C. Ryker. 1979. Field bioassay of male Douglas-fir beetle compound 3-methylcyclohex-3-en-1-one. *Experientia* **35**:1302.

Rudinsky, J.A., M.E. Morgan, L.M. Libbey and T.B. Putnam. 1974a. Additional compounds of the Douglas fir beetle (Col., Scolytidae) aggregative pheromone and their possible utility in pest control. *Zeitshrift für angewandte Entomologie* **76**:65–77.

Rudinsky, J.A., C. Sartwell, Jr., T.M. Graves and M.E. Morgan. 1974b. Granular formulation of methylcyclohexenone: an antiaggregative pheromone of the Douglas-fir beetle and spruce bark beetles. *Zeitshrift für angewandte Entomologie* **75**:254–263.

Rudinsky, J.A., L.N. Kline and J.D. Diekman. 1979. Response inhibition by four analogues of MCH, an antiaggregative pheromone of the Douglas-fir beetle. *Journal of Economic Entomology* **68**:527–528.

Rumbo, E.R. and K.-E. Kaissling. 1989. Temporal resolution of odour pulses by three types of pheromone receptor cells in *Antheraea polyphemus. Journal of Comparative Physiology A* **165**:281–291.

Rumbo, E.R., S.M. Deacon and L.P. Regan. 1993. Spatial discrimination between sources of pheromone and an inhibitor by the light-brown apple moth, *Epiphyas postvittana* (Walker) (Lepidoptera: Tortricidae). *Journal of Chemical Ecology* **19**:953–962.

Rumbo, E.R., D.M. Suckling and G. Karg. 1995. Measurement of airborne pheromone concentrations using electroantennograms: interactions between environment volatiles and pheromone. *Journal of Insect Physiology* **41**:465–471.

Ryan, M.J. and H.A. Keddy. 1992. Directional patterns of female mate choice and the role of sensory biases. *American Naturalist* **139S**:4–35.

Ryan, M.J. and A.S. Rand. 1990. The sensory basis of sexual selection for complex cells in the tungara frog, *Physalaemus pustulosus* (sexual selection for sensory exploitation). *Evolution* **44**:305–314.

Ryan, M.J., J.H. Fox, W. Wilczynski and A.S. Rand. 1990. Sexual selection for sensory exploitation in the frog *Physalaemus pustulosus. Nature* **343**:66–67.

Rybczynski, R., J. Reagan and M. Lerner. 1989. A pheromone-degrading aldehyde oxidase in the antennae of the moth *Manduca sexta. Journal of Neuroscience* **9**:1341–1353.

Ryker, L.C. and K.L. Yandell. 1983. Effect of verbenone on aggregation of *Dendroctonus ponderosae* Hopkins (Col., Scolytidae) to synthetic attractant. *Zeitshrift für angewandte Entomologie* **96**:452–459.

Ryker, L.C., L.M. Libbey and J.A. Rudinsky. 1979. Comparison of volatile compounds and stridulation emitted by the Douglas-fir beetle from Idaho and western Oregon populations. *Environmental Entomology* **8**:789–798.

Saglio, P., C. Descoins, M. Gallois, M. Lettéré, D. Jaouen and J. Mercier. 1977a. Etude de la phéromone sexuelle de la cochylis de la vigne *Eupoecilia (Clysia) ambiguella*, Hb. Lépidoptère *Tortricoidea Cochylidae. Annales de zoologie et écologie animale* **9**:553–562

Saglio, P., E. Priesner, C. Descoins and M. Gallois. 1977b. Un attractif sexuel de synthèse pour la pyrale de la vigne, *Sparganothis pilleriana* (Schiffermüller). *Comptes-rendus de l'Académie des Sciences (Paris) Série D* **284**:20007–2010.

Salom, S.M., R.F. Billings, W.W. Upton, M.J. Dalusky, D.M. Grossman, T.L. Payne, C.W. Berisford and T.N. Shaver. 1992. Effect of verbenone enantiomers and recemic *endo*-brevicomin on response of *Dendroctonus frontalis* (Coleoptera: Scolytidae) to attractant-baited traps. *Canadian Journal of Forest Research* **22**:925–931.

Sanders, C.J. 1971. Daily activity patterns and sex pheromone specificity as sexual isolating mechanisms in two species of *Choristoneura* (Lepidoptera: Tortricidae). *Canadian Entomologist* **103**:498–502.

Sanders, C.J. 1981. Disruption of spruce budworm mating—state of the art. In *Management of Insect Pests with Semiochemicals*, ed. E.R. Mitchell, pp. 339–349. Plenum Press, New York.

Sanders, C.J. 1984. Sex pheromone of the spruce budworm: evidence for a missing component. *Canadian Entomologist* **116**:93–100.

Sanders, C.J. 1985a. Disruption of spruce budworm, *Choristoneura fumiferana* (Lepidoptera: Tortricidae), mating in a wind tunnel: role of habituation. *Canadian Entomologist* **117**:391–393.

Sanders, C.J. 1985b. Flight speed of male spruce budworm moths in a wind tunnel at different wind speeds and at different distances from a pheromone source. *Physiological Entomology* **10**:83–88.

Sanders, C.J. 1986. The role of pheromone concentration in male moth flight behaviour. In *Mechanisms in Insect Olfaction*, eds. T.L. Payne, M.C. Birch and C.E.J. Kennedy, pp. 117–122. Clarendon Press, Oxford.

Sanders, C.J. 1989. The further understanding of pheromones: biological and chemical research for the future. In *Insect Pheromones in Plant Protection*, eds. A.R. Jutsum and R.F.S. Gordon, pp. 323–351. Wiley, Chichester, United Kingdom.

Sanders, C.J. 1990. Responses of male spruce budworm moths to sex pheromone released from filter paper and rubber septa. *Canadian Entomologist* **122**:263–269.

Sanders, C.J. and G.S. Lucuik. 1992. Mating behavior of spruce budworm moths, *Choristoneura fumiferana* (Lepidoptera: Tortricidae). *Canadian Entomologist* **124**:273–286.

Sanderson, M.J. and M.J. Donoghue. 1989. Patterns of variation in levels of homoplasy. *Evolution* **43**:1781–1795.

Sasaki, M. and L.M. Riddiford. 1984. Regulation of reproductive behaviour and egg maturation in the tobacco hawk moth, *Manduca sexta*. *Physiological Entomology* **9**:315–327.

Sasaki, M., L.M. Riddiford, J.W. Truman and J.K. Moore. 1983. Re-evaluation of the role of corpora cardiaca in calling and oviposition behaviour of giant silk moths. *Journal of Insect Physiology* **29**:695–705.

Sato, Y., M. Oguchi, N. Menjo, K. Imai, H. Saito, M. Ikeda, M. Isobe and O. Yamashita. 1993. Precursor polyprotein for multiple neuropeptides secreted from the suboesophageal ganglion of the silkworm *Bombyx mori*: characterization of the cDNA encoding the diapause hormone precursor and identification of additional peptides. *Proceedings of the National Academy of Science U.S.A.* **90**:3251–3255.

Satyanarayana K., J.H. Yu, G. Bhaskaran, K.H. Dahm, and R. Meola. 1992. Regulation of vitellogenin synthesis by juvenile hormone in the corn earworm, *Helicoverpa zea*. *Invertebrate Reproduction and Development* **21**:169–178.

Sauer, A.E. 1989. Ein System zur Messung von Pheromonkonzentrationen im Freiland mit Elektroantennogrammen. M.S. thesis, University of Kaiserslautern, Germany.

Sauer, A.E. 1991. Bestimmung von Pheromonkonzentrationen im Freiland mit Elektroantennogrammen zur Unterstützung der Paarungsstormethode bei Traubenwicklerarten *Lobesia botrana* und *Eupoecilia ambiguella*. Ph.D. thesis, University of Kaiserslautern, Germany.

Sauer, A.E., G. Karg, U.T. Koch and J.J. de Kramer. 1990. Measurements of pheromone concentrations in the field with EAGs. In *Abstracts IX ECRO Symposium*, p. 84, Noordwijkerhout, The Netherlands, Sept. 2–7, 1990.

Sauer, A.E., G. Karg, U.T. Koch, J.J. de Kramer and R. Milli. 1992. A portable EAG system for the measurement of pheromone concentrations in the field. *Chemical Senses* **17**:543–553.

Schal, C. and R.T. Cardé. 1986. Effects of temperature and light on calling in the tiger moth (Freeman) (Lepidoptera: Arctiidae). *Physiological Entomology* **11**:75–87.

Schal, C. and A-S. Chiang. 1995. Hormonal control of sexual receptivity in cockroaches. *Experientia* **51**:994–998.

Schal, C. and A.F. Smith. 1990. Neuroendocrine regulation of pheromone production in cockroaches. In *Cockroaches as Models for Neurobiology: Applications in Biomedical Research*, Vol. 2, eds. I. Huber, B.R. Rao and E.P. Masler, pp. 179–200. CRC Press, Boca Raton, Florida.

Schal, C., E.L. Burns and G.J. Blomquist. 1990a. Endocrine regulation of female contact sex pheromone production in the German cockroach, *Blattella germanica. Physiological Entomology* **15**:81–91.

Schal, C., E.L. Burns, R.A. Jurenka and G.J. Blomquist. 1990b. A new component of the sex pheromone of *Blattella germanica* (Dictyoptera: Blattellidae), and interaction with other pheromone components. *Journal of Chemical Ecology* **16**:1997–2008.

Schal, E.C., E.L. Burns, M. Gadot, J. Chase and G.J. Blomquist. 1991. Biochemistry and regulation of pheromone production in *Blattella germanica* (L.) (Dictyoptera, Blattellidae). *Insect Biochemistry* **21**:73–79.

Schal, C., D. Liang, L.K. Hazarika, R.E. Charlton and W.L. Roelofs. 1992. Site of pheromone production in female *Supella longipalpa* (F.) (Dictyoptera: Blattellidae): behavioral, electrophysiological and morphological evidence. *Annals of the Entomological Society of America* **85**:605–611.

Schal, C., A-S. Chiang, E.L. Burns, M. Gadot and R.A. Cooper. 1993. Role of the brain in juvenile hormone synthesis and oocyte development: effects of dietary protein in the cockroach *Blattella germanica* (L.). *Journal of Insect Physiology* **39**:303–313.

Schal, C., X. Gu, E.L. Burns, and G.J. Blomquist. 1994. Patterns of biosynthesis and accumulation of hydrocarbons and contact sex pheromone in the female German cockroach, *Blattella germanica. Archives of Insect Biochemistry and Physiology* **25**:375–391.

Scharrer, B. 1987. Insects as models in neuroendocrine research. *Annual Review of Entomology* **32**:1–16.

Schirra, K.J., K.W. Eichhorn, E. Tretzel and F. Louis. 1991. Integrierter Pflanzenschutz im Weinbau: Einfluss einer insektizidfreien Spritzfolge im Weinbau auf die Populationsentwicklung von Nützlingen und Gelegenheitsschädlingen. *Schriftenreihe des Bundesministers für Ernährung, Landwirtschaft und Forsten, Reihe A*, **396**:1–114.

Schleicher, S., I. Boekhoff, U. Konietzko and H. Breer. 1994. Pheromone-induced phosphorylation of antennal proteins from insects. *Journal of Comparative Physiology* **164**:76–80.

Schlyter, F., J.A. Byers and J. Löfqvist. 1987. Attraction to pheromone sources of different quantity, quality, and spacing: density-regulation mechanisms in bark beetle *Ips typographus. Journal of Chemical Ecology* **13**:1503–1523.

Schlyter, F., G. Birgersson and A. Leufven. 1989. Inhibition of attraction to aggregation pheromone by verbenone and ipsenol. Density regulation mechanisms in bark beetle *Ips typographus. Journal of Chemical Ecology* **15**:2263–2277.

Schmid, A. and G. Raboud. 1992. Essai de confusion contre *Lobesia botrana* dans le vignoble de Sierre (VS) 1989–1990. *International Organization for Biological Control—West Palearctic Regional Section Bulletin* **15**(2):23.

Schmidt, B.C. and B.W. Ache. 1979. Olfaction: response enhancement by flicking in a decapod crustacean. *Science* **205**:204–206.

Schmidt, J.O. and W.D. Seabrook. 1979. Mating of caged spruce budworm moths in pheromone environments. *Journal of Economic Entomology* **72**:509–511.

Schmidt, M. and B.W. Ache. 1992. Antennular projections to the midbrain of the spiny lobster. 2. Sensory innervation of the olfactory lobe. *Journal of Comparative Neurology* **318**:291–303.

Schmitz, R.F. and M.D. McGregor. 1990. Antiaggregative effect of verbenone on response of the mountain pine beetle to baited traps. United States Department of Agriculture, Forest Service, Research Paper INT-423.

Schmitz, V. 1992. Contribution à l'étude du mécanisme de la confusion sexuelle des mâles chez l'insecte. Application à la mise au point du procédé biotechnique de lutte contre l'eudémis de la vigne *Lobesia botrana* Den et Schiff. Thèse, Ecole Nationale Supérieure Agronomique, Rennes, France.

Schmutzenhofer, H. 1988. Mass outbreaks of *Ips* bark beetles in Bhutan and the revision of the genus *Ips* de Geer for the Himalayan region. In *Integrated Control of Scolytid Bark Beetles*, eds. T.L. Payne and H. Saarenmaa, pp. 345–355. Virginia Polytechnic Institute and State University, Blacksburg, Virginia.

Schneider, D. 1957. Elektrophysiologische Untersuchungen von Chemo- und Mechanorezeptoren der Antenne des Seidenspinners *Bombyx mori* L. *Zeitschrift für Vergleichende Physiologie* **40**:8–41.

Schneider, D. 1962. Electrophysiological investigation on the olfactory specificity of sexual attracting substances in different species of moths. *Journal of Insect Physiology* **8**:15–30.

Schneider, D. 1969. Insect olfaction: deciphering system for chemical messages. *Science* **163**:1031–1037.

Schneider, D. 1992. 100 years of pheromone research. *Naturwissenschaften* **79**:241–250.

Schneider, D. and E. Hecker. 1956. Zur Elektrophysiologie der Antenne des Seidenspinners *Bombyx mori* bei Reizung mit angereicherten Extrakten des Sexuallockstoffes. *Zeitschrift für Naturforschung* **11b**:121–124.

Schneider, D., V. Lacher, and K.-E. Kaissling. 1964. Die Reaktionsweisen und das Reaktionsspektrum von Riechzellen bei *Antherea pernyi* (Lepidoptera: Saturniidae). *Zeitschrift für Vergleichende Physiologie* **48**:632–662

Schneider, D., M. Boppré, H. Schneider, W.R. Thompson, C.J. Boriack, R.L. Petty and J. Meinwald. 1975. A pheromone precursor and its uptake in male *Danaus* butterflies. *Journal of Comparative Physiology A* **97**:245–256.

Schneider, D., M. Boppré, J. Zweig, B.B. Hjorsley, T.W. Bell, J. Meinwald, K. Hansen and E.W. Diehl. 1982. Scent organ development in *Creatonotos* moths: regulation by pyrrolizidine alkaloids. *Science* **215**:1264–1265.

Schneidermann, A.M., J.G. Hildebrand, M.M. Brennan, J.M. Tumlinson. 1986. Transsexually grafted antennae alter pheromone-directed behavior in a moth. *Nature* **323**:801–803.

Schoofs, L., J.V. Broeck and A. De Loof. 1993. The myotropic peptides of *Locusta migratoria*: Structures, distribution, functions and receptors. *Insect Biochemistry and Molecular Biology* **23**:859–881.

Schooley, D.A., F.C. Baker, L.W. Tsai, C.A. Miller, and G.C. Jamieson. 1984. Juvenile hormones O, I, and II exist only in Lepidoptera. In *Biosynthesis, Metabolism, and Mode of Action of Invertebrate Hormones*, eds. J. Hoffmann and M. Porchet, pp. 373–383. Springer-Verlag, Berlin.

Schoonhoven, L.M. 1990. Host-marking pheromones in Lepidoptera, with special reference to two *Pieris* spp. *Journal of Chemical Ecology* **16**:3043–3052.

Schreiber, S.L. 1991. Chemistry and biology of the immunophilins and their immunosuppressive ligands. *Science* **251**:283–288.

Schroeder, L.M. 1992. Olfactory recognition of nonhosts aspen and birch by conifer bark beetles *Tomicus piniperda and Hylurgops palliatus. Journal of Chemical Ecology* **18**:1583–1593.

Schulz, S., M. Boppré and R.I. Vane-Wright. 1993. Specific mixtures of secretions from male scent organs of African milkweed butterflies (Danainae). *Philosophical Transactions of the Royal Society London B* **342**:161–181.

Schwind, R. 1989. Size and distance perception in compound eyes. In: *Facets of Vision*, eds. D.G. Stavenga and R.C. Hardie, pp. 425–444, Springer-Verlag, Berlin.

Scoble, M. 1992. *The Lepidoptera: Form, Function and Diversity*. Oxford University Press, Oxford.

Seamon, K.B., W. Padgett and J.W. Daly. 1981. Forskolin: a unique diterpene activator of adenylate cyclase in membranes and insect cells. *Proceedings of the National Academy of Sciences U.S.A.* **78**:3363.

Sebeok, T.A. 1977. *How Animals Communicate*. Indiana University Press, Bloomington, Indiana.

Sevala, V.L. and K.G. Davey. 1989. Action of juvenile hormone on the follicle cells of *Rhodnius prolixus*: Evidence for a novel mechanism involving protein kinase C. *Experientia* **45**:355–356.

Shanbhag, S.R. and R.N. Singh. 1992. Functional implications of the projections of neurons from individual labellar sensillum of *Drosophila melanogaster* as revealed by neuronal-marker horseradish peroxidase. *Cell and Tissue Research* **267**:273–282.

Shapiro, J.P., J.H. Law and M.A. Wells. 1988. Lipid transport in insects. *Annual Review of Entomology* **33**:297–318.

Sharma, R.K., A.J. Mueller, H.T. Reynolds and N. Toscano. 1973. Techniques for trapping pink bollworm males. *California Agriculture* **27**:14–15.

Shaver, T.N., J.D. Lopez, Jr. and A.W. Hartstack, Jr. 1982. Effects of pheromone components and their degradation products on the response of *Heliothis* spp. to traps. *Journal of Chemical Ecology* **8**:755–762.

Shea, P.J. and M. Neustein. 1995. Protection of a rare stand of Torrey pine from *Ips paraconfusus*. In *Proceedings of the Symposium on Application of Semiochemicals for Management of Bark Beetle Infestations*, eds. S.M. Salom and K.R. Hobson. United States Department of Agriculture, Forest Service, General Technical Report INT-GTR-318.

Shea, P.J., M.D. McGregor and G.E. Daterman. 1992. Aerial application of verbenone reduces attack of lodgepole pine by mountain pine beetle. *Canadian Journal of Forest Research* **22**:436–441.

Shelton, A.M. and J.A. Wyman. 1979. Time of tuber infestation and relationships between pheromone catches of adult moths, foliar larval populations, and tuber damage by the potato tuberworm. *Journal of Economic Entomology* **72**:599–601.

Shirk, P.D., G. Baskaran and H. Roller. 1980. The transfer of juvenile hormone from male to female during mating in the *Cecropia* moth. *Experientia* **36**:682–683.

Shorey, H.H. 1973. Behavioral responses to insect pheromones. *Annual Review of Entomology* **18**:349–380.

Shorey, H.H. and L.K. Gaston. 1965. Sex pheromones of noctuid moths. V. Circadian rhythm of pheromone-responsiveness in males of *Autographa californica, Heliothis virescens, Spodoptera exigua*, and *Trichoplusia ni*. *Annals of the Entomological Society of America* **58**:597–600.

Shorey, H.H. and J.J. McKelvey, Jr., eds. 1977. *Chemical Control of Insect Behavior*. Wiley, New York.

Shorey, H.H., L.K. Gaston, and T.R. Fukuto. 1964. Sex pheromones of noctuid moths. I. A quantitative bioassay for the sex pheromone of *Trichoplusia ni* (Lepidoptera: Noctuidae). *Journal of Economic Entomology* **57**:252–254.

Shorey, H.H., R.S. Kaae, L.K. Gaston, and J.R. McLaughlin. 1972. Sex pheromones of Lepidoptera. XXX. Disruption of sex pheromone communication in *Trichoplusia ni* as a possible means of control. *Environmental Entomology* **1**:641–645.

Showers, W.B., R.B. Smelser, A.J. Keaster, F. Whitford, J.F. Robinson, J.D. Lopez and S.E. Taylor. 1989a. Recapture of marked black cutworm (Lepidoptera: Noctuidae) males after long-range transport. *Environmental Entomology* **18**:447–458.

Showers, W.B., F. Whitford, R.B. Smelser, A.J. Keaster, J.F. Robinson, J.D. Lopez and S.E. Taylor. 1989b. Direct evidence for meteorologically driven long-range dispersals of an economically important moth. *Ecology* **70**:987–992.

Silverstein, R.M. 1990. Practical use of pheromones and other behavior-modifying compounds: overview. In *Behavior-Modifying Chemicals for Insect Management. Applications of Pheromones and Other Attractants*, eds. R.L. Ridgway, R.M. Silverstein and M.N. Inscoe, pp. 1–8. Marcel Dekker, New York.

Silverstein, R.M., J.O. Rodin and D.L. Wood. 1966. Sex attractants in frass produced by male *Ips confusus* in ponderosa pine. *Science* **154**:509–510.

Singer, A.G. and F. Macrides. 1990. Aphrodisin: pheromone or transducer? *Chemical Senses* **15**:109–203.

Singh, P. 1974. A chemically-defined medium for rearing *Epiphyas postvittana* (Lepidoptera: Tortricidae). *New Zealand Journal of Zoology* **1**:241–243.

Smith, A.F. and C. Schal. 1990a. Corpora allata control of pheromone production and release in the female brown-banded cockroach, *Supella longipalpa* (F.) (Dictyopera: Blattellidae). *Journal of Insect Physiology* **36**:251–257.

Smith, A.F. and C. Schal. 1990b. The physiological basis for the termination of calling in the female brown-banded cockroach, *Supella longipalpa* (F.) *Journal of Insect Physiology* **36**:369–373.

Smith, A.F. and C. Schal. 1991. Circadian calling behavior in the adult female brown-banded cockroach, *Supella longipalpa* (F.) (Dictyoptera: Blattellidae). *Journal of Insect Behavior* **4**:1–14.

Smith, A.F., K. Yagi, S.S. Tobe and C. Schal. 1989. *In vitro* rates of juvenile hormone biosynthesis in adult virgin and mated female *Supella longipalpa* (F.) over two reproductive cycles. *Journal of Insect Physiology* **35**:781–785.

Smith, W.A. and W.L. Combest. 1985. Role of cyclic nucleotides in hormone action. In *Endocrinology II, Comprehensive Insect Physiology, Biochemistry and Pharmacology*, Vol. 8, eds. G.A. Kerkut and L.I. Gilbert, pp. 263–299. Pergamon Press, Oxford.

Smith, W.A. and L.I. Gilbert. 1986. Cellular regulation of ecdysone synthesis by the prothoracic glands of *Manduca sexta. Insect Biochemistry* **16**:143–147.

Smith, W.A., L.I. Gilbert and W.E. Bollenbacher. 1985. Calcium-cyclicAMP interactions in prothoracicotropic hormone stimulation of ecdysone synthesis. *Molecular and Cellular Endocrinology* **39**:71–78.

Sobel, E.C. 1990. The locust's use of motion parallax to measure distance. *Journal of Comparative Physiology A* **167**:579–588.

Solem, J.O. 1984. Adult behaviour of north European caddisflies. In *Proceedings of the 4th International Symposium on Trichoptera*, Series Entomologica 30, ed. J.C. Morse, pp. 375–382. Junk, The Hague.

Solem, J.O. 1985. Female sex pheromones in *Rhyacophila nubila* (Zetterstedt) (Trichoptera, Rhyacophilidae) and arrival pattern to sticky traps. *Fauna norvegica Series B* **32**:80–82.

Solem, J.O. 1986. Behaviour of adult caddisflies. *Fauna (Oslo)* **39**:58–64.

Solem, J.O. and E. Petersson. 1987. Demonstration of female sex pheromones and adult behaviour in *Molanna angustata* (Trichoptera: Molannidae). *Entomologia Generalis* **12**:115–118.

Soroker, V. and A. Rafaeli. 1989. *In vitro* hormonal stimulation of [^{14}C]acetate incorporation by *Heliothis armigera* pheromone glands. *Insect Biochemistry* **19**:1–5.

Soroker, V. and A. Rafaeli. 1995. Multi-signal transduction of the pheromonotropic response by pheromone gland incubations of *Helicoverpa armigera. Insect Biochemistry and Molecular Biology* **25**:1–9.

Sower, L.L, H.H. Shorey and L.K. Gaston. 1970. Sex pheromones of Noctuid moths. XXI. Light–dark cycle regulation and light inhibition of the sex pheromone release by females of *Trichoplusia ni. Annals of the Entomological Society of America* **63**:1090–1092.

Sreng, L. and A. Quennedey. 1976. Role of a temporary ciliary structure in the morphogenesis of insect glands. An electron microscope study of the tergal glands of male *Blattella germanica* L. (Dictyoptera, Blattellidae). *Journal of Ultrastructure Research* **56**:78–95.

Sreng, L., R. Moreau and A. Girardie. 1990. Locust neuropeptides stimulating sex pheromone production in female European corn borer moth, *Ostrina nubilalis. Journal of Insect Physiology* **36**:719–726.

Srinivasan, M.V. 1992. How bees exploit optic flow: behavioral experiments and neural models. *Philosophical Transactions of the Royal Society of London B* **337**:253–259.

Srinivasan, M.V., S.W. Zhang and B. Rolfe. 1993. Is pattern vision in insects mediated by 'cortical' processing? *Nature* **362**:539–540.

St. Clair, A., R.L. Goulding and J.A. Rudinsky. 1977. Controlled-release dispensing system for 3,2-MCH, antiaggregative pheromone of the Douglas-fir beetle. *Zeitschrift für Angewandte Entomologie* **83**:297–300.

Stange, G. 1992. High resolution measurements of atmospheric carbon dioxide concentration changes by the labial palp organ of the moth *Heliothis armigera* (Lepidoptera: Noctuidae). *Journal of Comparative Physiology A* **171**:317–324.

Starratt A.N., K.H. Dahm, N. Allen, J.G. Hildebrand, T.L. Payne and H. Roller. 1979. Bombykal, a sex pheromone of the sphinx moth, *Manduca sexta. Zeitschrift für Naturforschung* **34**:9–12.

Staten, R.T., H.M. Flint, R.C. Weddle, E. Quintero, R.E. Zarate, C.M. Finnell, M. Hernandes and A. Yamamoto. 1987. Pink bollworm (Lepidoptera: Gelechiidae): Large-scale field trials with a high-rate gossyplure formulation. *Journal of Economic Entomology* **80**:1267–1271.

Staten, R.T., R.W. Rosander, and D.F. Keaveney. 1993. Genetic control of cotton insects, the pink bollworm as a working program. In *Proceedings International Symposium on Management of Insect Pests: Nuclear and Related Molecular and Genetic Techniques.* IAEA and FAO, Vienna, pp. 269–283.

Stay, B. and L.M. Roth. 1958. The reproductive behavior of *Diploptera punctata* (Blattaria: Diplopteridae). *Proceedings X International Congress of Entomology, Montreal* **2**:547–552.

Stay, B. and S.S. Tobe. 1977. Control of juvenile hormone biosynthesis during the reproductive cycle of a viviparous cockroach. I. Activation and inhibition of corpora allata. *General and Comparative Endocrinology* **33**:531–540.

Steffan, A.W. 1987. Fern- und Nahorientierung geflügelter Gynoparae und Sexualis-Männchen bei Blattlöusen (Homoptera: Aphidinea: Aphididae). *Entomologia Generalis* **12**:235–258.

Steinbrecht, R.A. 1987. Functional morphology of pheromone-sensitive sensilla. In *Pheromone Biochemistry*, eds. G.D. Prestwich and G.J. Blomquist, pp. 353–383. Academic Press, Orlando, Florida.

Steinbrecht, R.A. 1992. Experimental morphology of insect olfaction—tracer studies, x-ray microanalysis, autoradiography, and immunocytochemistry with silkmoth antennae. *Microscopy Research and Technique* **22**:336–350.

Steinbrecht, R.A. and B. Müller. 1971. On the stimulus conducting structures in insect olfactory receptors. *Zeitschrift für Zellforschungen und Mikroskopische Anatomie* **117**:570–575.

Steinbrecht, R.A., M. Ozaki and G. Ziegelberger. 1992. Immunocytochemical localization of pheromone-binding protein in moth antennae. *Cell and Tissue Research* **270**:287–302.

Steinlen, S., S. Klumpp and J.E. Schultz. 1990. Guanylate cyclase in olfactory cilia from rat and pig. *Biochimica et Biophysica Acta* **1054**:60–72.

Stengl, M. 1993. Intracellular-messenger-mediated cation channels in cultured olfactory receptor neurons. *Journal of Experimental Biology* **178**:125–147.

Stengl, M. 1994. Inositol-trisphosphate-dependent calcium currents precede cation currents in insect olfactory receptor neurons *in vitro*. *Journal of Comparative Physiology A* **174**:187–194.

Stengl, M., H. Hatt and H. Breer. 1992. Peripheral processes in insect olfaction. *Annual Review of Physiology* **54**:665–681.

Stephen, F.M. and M.P. Lih. 1985. A *Dendroctonus frontalis* infestation growth model: organization, refinement, and utilization. In *Proceedings of Integrated Pest Management Research Symposium*, eds. S.J. Branham and R.C. Thatcher, pp. 186–194. United States Department of Agriculture, Forest Service, General Technical Report SO-56.

Stern, V. and V. Savacherean. 1978. Long range dispersal of pink bollworm into the San Joaquin Valley. *California Agriculture* **32**:4–5.

Stewart, R.C. and F.W. Dahlquist. 1987. Molecular components of bacterial chemotaxis. *Chemical Reviews* **87**:997–1025.

Still, W.C. 1979. (±)-Periplanone-B. Total synthesis and structure of the sex excitant pheromone of the American cockroach. *Journal of the American Chemical Society* **101**:2493–2495.

Stock, A.J., J.H. Borden, T.L. Pratt, H.D. Pierce, Jr. and B.D. Johnston. 1990. *endo*-Brevicomin: an antiaggregation pheromone for the western balsam bark beetle, *Dryocoetes confusus* Swaine (Coleoptera: Scolytidae). *Canadian Entomologist* **122**:935–940.

Stockel, J.P., V. Schmitz, P. Lecharpentier, R. Roehrich, M. Torres-Vila and U. Neumann. 1994. La confusion sexuelle chez l'eudémis *Lobesia botrana* (Lepidoptera Tortricidae). Bilan de 5 années d'expérimentation dans un vignoble bordelais. *Agronomie* **2**:71–82.

Stocker, R.F. 1994. The organization of the chemosensory system of *Drosophila melanogaster*: a review. *Cell and Tissue Research* **275**:3–26.

Stocker, R.F., R.N. Singh, M. Schorderet and O. Siddiqi. 1983. Projection patterns of different types of antennal sensilla in the antennal glomeruli of *Drosophila melanogaster*. *Cell and Tissue Research* **232**:237–248.

Stout, J., G. Atkins and D. Zacharias. 1992. Regulation of cricket phonotaxis through hormonal control of the threshold of an identified auditory neuron. *Journal of Comparative Physiology A* **169**:765–772.

Strausfeld, N.J. and W. Gronenberg. 1990. Descending neurons supplying the neck and flight motor of Diptera: Organization and neuroanatomical relationships with visual pathways. *Journal of Comparative Neurology* **302**:954–972.

Strausfeld, N.J. and T.A. Miller. 1980. *Neuroanatomical Techniques: Insect Nervous System*. Springer-Verlag, New York.

Strausfeld, N.J., H.S. Seyan and J.J. Milde. 1987. The neck motor system of the fly *Calliphora erythrocephala* I. Muscles and motor neurons. *Journal of Comparative Physiology A* **160**:205–224.

Strauss, R. and M. Heisenberg. 1993. A higher control center of locomotor behavior in the *Drosophila* brain. *Journal of Neuroscience* **13**:1852–1861.

Struble, D.L. 1983. Pheromone traps for monitoring moth (Lepidoptera) abundances: evaluation of cone-oriface and omni-directional designs. *Canadian Entomologist* **115**:59–65.

Struble, D.L. and H. Arn. 1984. Combined gas chromatography and electroantennogram recording of insect olfactory responses. In *Techniques in Pheromone Research*, eds. H.E. Hummel and T.A. Miller, pp. 161–178. Springer-Verlag, New York.

Stryer, L. 1986. Cyclic GMP cascade in vision. *Annual Review of Neuroscience* **9**:87–119.

Suckling, D.M. and J.R. Clearwater. 1990. Small scale trials of disruption of *Epiphyas postvittana* (Lepidoptera: Tortricidae) in New Zealand. *Environmental Entomology* **19**:1702–1709.

Suckling, D.M. and J.G.I. Khoo. 1993. Cline in frequency of azinphos-methyl resistance in lightbrown apple moth (Lepidoptera: Tortricidae). *Journal of Economic Entomology* **86**:1308–1316.

Suckling, D.M. and P.W. Shaw. 1990. Preliminary trials of mating disruption of lightbrown apple moth in Nelson. In *Proceedings of the 43rd New Zealand Weed and Pest Control Conference* pp. 311–316.

Suckling, D.M. and P.W. Shaw. 1991. Evaluation of mating disruption of the lightbrown apple moth in Nelson. In *Proceedings of the 44th New Zealand Weed and Pest Control Conference*, pp. 47–51.

Suckling, D.M. and P.W. Shaw. 1992. Conditions that favor mating disruption of *Epiphyas postvittana* (Lepidoptera: Tortricidae). *Environmental Entomology* **21**:949–956.

Suckling, D.M. and P.W. Shaw. 1995. Large scale trials of mating disruption of light brown apple moth. *New Zealand Journal of Crop and Horticultural Science* **23**:1–11.

Suckling, D.M., D.R. Penman, R.B. Chapman and C.H. Wearing. 1985. Pheromone use in insecticide resistance surveys of lightbrown apple moths (Lepidoptera: Tortricidae). *Journal of Economic Entomology* **78**:204–207.

Suckling, D.M., D.J. Rogers, P.W. Shaw, C.H. Wearing, W.P. Thomas, D.R. Penman and R.B. Chapman. 1987. Monitoring insecticide resistance in lightbrown apple moth. *Journal of Economic Entomology* **80**:733–738.

Suckling, D.M., D.J. Rogers and P.W. Shaw. 1990a. Disruption of lightbrown apple moth *Epiphyas postvittana* (Walker) (Lepidoptera: Tortricidae) trapping in Nelson, New Zealand. *Australian Journal of Zoology* **38**:363–373.

Suckling, D.M., P.W. Shaw, J.G.I. Khoo and V. Cruickshank. 1990b. Resistance management of lightbrown apple moth, *Epiphyas postvittana* (Lepidoptera: Tortricidae) by mating disruption. *New Zealand Journal of Crop and Horticultural Science* **18**:89–98.

Suckling, D.M., G. Karg, S.J. Bradley and C.R. Howard. 1994. Field electroantennogram and behavioral responses of *Epiphyas postvittana* (Lepidoptera: Tortricidae) under low pheromone and inhibitor concentrations. *Journal of Economic Entomology* **87**:1477–1487.

Suder, F., U. Dombrowski and G. Wendler. 1990. Control of flight orientation by wind and visual inputs in *Manduca sexta*. In: Brain–Perception–Cognition, eds. N. Elsner and G. Roth G, p. 79, Georg Thieme Verlag, Stuttgart, Germany.

Sun, X.J., L.P. Tolbert and J.G. Hildebrand. 1993. Ramification pattern and ultrastructural characteristics of the serotonin-immunoreactive neuron in the antennal lobe of the moth *Manduca sexta*: a laser scanning confocal and electron microscopic study. *Journal of Comparative Neurology* **338**:5–6.

Svärd, L. and J.N. McNeil. 1994. Female benefit, male risk: polyandry in the true armyworm *Pseudaletia unipuncta. Behavioral Ecology and Sociobiology* **35**:319–326.

Svärd, L. and C. Wiklund. 1986. Different ejaculate delivery strategies in first versus subsequent matings in the swallowtail butterfly *Papilio machaon* L. *Behavioral Ecology and Sociobiology* **18**:325–330.

Svärd, L. and C. Wiklund. 1989. Mass and production rate of ejaculates in relation to monoandry/polyandry in butterflies. *Behavioral Ecology and Sociobiology* **24**:395–402.

Tabashnik, B.E. 1994. Evolution of resistance to *Bacillus thuringiensis. Annual Review of Entomology* **39**:47–79.

Takeda, S. 1977. Induction of egg diapause in *Bombyx mori* by some cephalo-thoracic organs of the cockroach, *Periplaneta americana. Journal of Insect Physiology* **23**:813–816.

Tamaki, Y., H. Sugie and H. Noguchi. 1985. Methyl (*Z*)-5-tetradecenoate, sex-attractant pheromone of the soybean beetle, *Anomala rufocuprea* Motschulsky (Coleoptera: Scarabaeidae). *Applied Entomology and Zoology* **20**:359–361.

Tang, J.D., R.E. Charlton, R.T. Cardé and C.-M. Yin. 1987. Effect of allatectomy and ventral nerve cord transection on calling, pheromone emission and pheromone production in *Lymantria dispar. Journal of Insect Physiology* **33**:469–476.

Tang, J.D., R.E. Charlton, R.A. Jurenka, W.A. Wolf, P.L. Phelan, L., Sreng and W.L. Roelofs. 1989. Regulation of pheromone biosynthesis by a brain hormone in two moth species. *Proceedings of the National Academy of Science U.S.A.* **86**:1806–1810.

Tang, J.D., W.A. Wolf, W.L. Roelofs and D.C. Knipple. 1991. Development of functionally competent cabbage looper moth sex pheromone glands. *Insect Biochemistry* **21**:573–581.

Tauber, C.A. and M.J. Tauber. 1989. Sympatric speciation in insects: perception and perspective. In *Speciation and Its Consequences*, eds. D. Otte and J.A. Endler, pp. 307–344. Sinauer, Sunderland, Massachusetts.

Taylor, O.R. 1973. Reproductive isolation in *Colias eurytheme*, and *C. Philodice* (Lepidoptera: Pieridae): use of olfaction in mate selection. *Annals of the Entomological Society of America* **66**:621–626.

Teal, P.E.A. 1981. *The Role of Sex Pheromones in the Reproductive Isolation of* Heliothis *Species (Lepidoptera: Noctuidae)*. University of Florida Press, Gainesville.

Teal, P.E.A. 1985. The effect of hybridization on the periodicity of sex pheromone release of *Euxoa* species Lepidoptera: Noctuidae). *Physiological Entomology* **10**:463–466.

Teal, P.E.A. and A. Oostendorp. 1993. Effect of interspecific hybridization between *Heliothis virescens* and *H. subflexa* (Lepidoptera: Noctuidae) on the presence and structure of hairpencil glands of males. *Annals of the Entomological Society of America* **86**:322–326.

Teal, P.E.A. and A. Oostendorp. 1994a. Production of pheromone by hairpencil glands obtained from interspecific hybridization between *Heliothis virescens* and *H. subflexa*. *Journal of Chemical Ecology* **21**:59–67.

Teal, P.E.A. and A. Oostendorp. 1994b. Effect of interspecific hybridization between *Heliothis virescens* and *H. subflexa* (Lepidoptera: Noctuidae) on sex pheromone production by females. *Journal of Insect Physiology* **41**:519–525.

Teal, P.E.A. and J.H. Tumlinson. 1986. Terminal steps in pheromone biosynthesis by *Heliothis virescens* and *H. zea*. *Journal of Chemical Ecology* **12**:353–366.

Teal, P.E.A. and J.H. Tumlinson. 1987. The role of alcohols in pheromone biosynthesis by two noctuid moths that use acetate pheromone components. *Archives of Insect Biochemistry and Physiology* **4**:261–269.

Teal, P.E.A. and J.H. Tumlinson. 1989. Isolation, identification and biosynthesis of compounds produced by male hairpencil glands of *Heliothis virescens* (F.) (Lepidoptera: Noctuidae). *Journal of Chemical Ecology* **15**:413–427.

Teal, P.E.A. and J.H. Tumlinson. 1991. Sex pheromone production in moths: endogenous regulation of initiation and inhibition. In *Naturally Occurring Pest Bioregulators*, ed. P.A. Hedin, pp. 66–77. ACS Symposium Series 449, American Chemical Society, Washington, D.C.

Teal, P.E.A., J.R. Byers and B.J.R. Philogene. 1978. Differences in the female calling behavior of three interfertile sibling species of *Euxoa* (Lepidoptera: Noctuidae). *Annals of the Entomological Society of America* **71**:630–635.

Teal, P.E.A., R.R. Heath, J.H. Tumlinson, and J.R. McLaughlin. 1981a. Identification of sex pheromone of *Heliothis subflexa* (G.) (Lepidopera: Noctuidae) and field trapping studies using different blends of components. *Journal of Chemical Ecology* **7**:1011–1022.

Teal, P.E.A., J.R. McLaughlin and J.H. Tumlinson. 1981b. Analysis of the reproductive behavior of *Heliothis virescens* (F.) under laboratory conditions. *Annals of the Entomological Society of America* **74**:324–330.

Teal, P.E.A., J.H. Tumlinson, J.R. McLaughlin, R. Heath and R.A. Rush. 1984. (*Z*)-11-Hexadecen-1-o1: a behavioral modifying chemical present in the pheromone gland of female *Heliothis zea* (Lepidoptera: Noctuidae). *Canadian Entomologist* **116**:777–779.

Teal, P.E.A., J.H. Tumlinson and R.R. Heath. 1986. Chemical and behavioral analyses of volatile sex pheromone components released by calling *Heliothis virescens* (F.) females (Lepidoptera: Noctuidae). *Journal of Chemical Ecology* **12**:107–125

Teal, P.E.A., J.H. Tumlinson and H. Oberlander. 1989a. Neural regulation of sex pheromone biosynthesis in *Heliothis moths. Proceedings of the National Academy of Sciences U.S.A.* **86**:2488–2492.

Teal, P.E.A., J.H. Tumlinson, and A. Oostendorp. 1989b. Enzyme-catalyzed pheromone synthesis by *Heliothis* moths. In *Biocatalysis in Agricultural Biotechnology*, eds. J.R. Whitaker and P.E. Sonnet, pp. 332–343. American Chemical Society, Symposium Series 389. Washington, D.C.

Teal, P.E.A., A. Oostendorp, and J.H. Tumlinson. 1993. Induction of pheromone production in females of *Heliothis virescens* (F.) and *H. subflexa* (Gn.) (Lepidoptera: Noctuidae) during the photophase. *Canadian Entomologist* **125**:355–366.

Teicher, M.H., W.B. Stewart, J.S. Kauer and G.M. Shepherd. 1980. Suckling pheromone stimulation of a modified glomerular region in the developing rat olfactory bulb revealed by the 2-deoxyglucose method. *Brain Research* **194**:530–535.

Termer, A. 1992. Aufbau und Anwendung eines Systems zur Messung der Pheromonverteilung bei der Bekämpfung von *Lobesia botrana*, Schiff, durch Paarungsstörung. Diploma thesis, Universität Kaiserslautern, Germany.

Thomson, D.R., N.P.D. Angerilli, C. Vincent, and A.P. Gaunce. 1991. Evidence for regional differences in the response of obliquebanded leafroller (Lepidoptera: Tortricidae) to sex pheromone blends. *Environmental Entomology* **20**:935–938.

Thornhill, R. 1992. Female preference for the pheromone of males with low fluctuating asymmetry in the Japanese scorpionfly (*Panorpa japonica*, Mecoptera). *Behavioral Ecology* **3**:277–283.

Thornhill, R. and J. Alcock. 1983. *The Evolution of Insect Mating Systems*. Harvard University Press, Cambridge, Massachusetts.

Thorsteinson, A.J. 1960. Host selection in phytophagous insects. *Annual Review of Entomology* **5**:193–218.

Thyagaraja B.S.and A.K. Raina. 1994. Regulation of pheromone production in the gypsy moth, *Lymantria dispar*, and development of an *in vivo* bioassay. *Journal of Insect Physiology* **40**:969–974.

Tilden, P.E., W.D. Bedard, D.L. Wood and H.A. Stubbs. 1981. Interruption of response of *Dendroctonus brevicomis* to its attractive pheromone by components of the pheromone. *Journal of Chemical Ecology* **7**:183–196.

Timm, F. 1958. Zur Histochemie der Schwermetalle. Das sulfid-Silber-Verfahren. *Deutsche Zeitschrift für Gesamte-Gerichtliche Medizin* **46**:706–711.

Tips, A., L. Paemen, L. Schoofs, M. Ma and M. Blackburn. 1993. Co-localization of locustamyotropin- and pheromone biosynthesis activating neuropeptide-like immunoreactivity in the central nervous system of five insect species. *Comparative Biochemistry and Physiology A* **106**:195–207.

Tobe, S.S. and B. Stay. 1985. Structure and regulation of the corpus allatum. *Advances in Insect Physiology* **18**:305–432.

Todd, J.L. and T.C. Baker. 1993. Response of single antennal neurons of female cabbage loopers to behaviorally active attractants. *Naturwissenschaften* **80**:183–186.

Todd, J.L., K.F. Haynes, and T.C. Baker. 1992. Antennal neurones specific for redundant pheromone components in normal and mutant *Trichoplusia ni* males. *Physiological Entomology* **17**:183–192.

Todd, J.L., S. Anton, B.S. Hansson and T.C. Baker. 1995. Functional organization of the macroglomerular complex related to behaviorally-expressed olfactory redundancy in male cabbage looper moths. *Physiological Entomology* **20**:349–361.

Tolbert L.P. and Hildebrand J.G. 1981. Organization and synaptic ultrastructure of glomeruli in the antennal lobes of the moth *Manduca sexta*: a study using thin sections and freeze-structure. *Philosophical Transactions of the Royal Society, London B* **213**:279–301.

Tolbert, L.P. and L.A. Oland. 1990. Glial cells form boundaries for developing insect olfactory glomeruli. *Experimental Neurobiology* **109**:19–28.

Tomioka, K., M. Ikeda, T. Nagao and S. Tamotsu. 1993. Involvement of serotonin in the circadian rhythm of an insect visual system. *Naturwissenschaften* **80**:137–139.

Toscano, N.C., R.R. Youngman, E.R. Oatman, P.A. Phillips, M. Jiménez, and F. Munoz. 1987. Implementation of an integrated pest management program for fresh market tomatoes. *Applied Agricultural Research* **1**:315–324.

Tóth, M. and H.-R. Buser. 1992. Simple method for collecting volatile compounds from single insects and other point sources for gas chromatographic analysis. *Journal of Chromatography* **598**:303–308.

Tóth, M., C. Löfstedt, B.W. Blair, T. Cabello, A.I. Farag, B.S. Hansson, B.G. Kovalev, S. Maini, E.A. Nesterov, I. Pajor, A.P. Sazonov, I.V. Shamshev, M. Subchev, and G. Szöcs. 1992. Attraction of male turnip moths *Agrotis segetum* (Lepidoptera: Noctuidae) to sex pheromone components and their mixtures at 11 sites in Europe, Asia, and Africa. *Journal of Chemical Ecology* **18**:1337–1347.

Töth, M., G. Szöcs, E.J. van Nieukerken, P. Philipp, F. Schmidt and W. Francke. 1995. Novel type of sex pheromone structure identified from *Stigmella malella* (Stainton) (Lepidoptera: Nepticulidae). *Journal of Chemical Ecology* **21**:13–27.

Traynier, R.M.M. 1970. Sexual behavior of the mediterranean flour moth, *Anagasta kühniella*: some influences of age, photoperiod, and light intensity. *Canadian Entomologist* **102**:534–540.

Trickett, A.J. and R.K. Butlin. 1994. Recombination suppressors and the evolution of new species. *Heredity* **73**:339–345.

Trivers, R.L. 1972. Parental investment and sexual selection. In *Sexual Selection and the Descent of Man*, 1871–1971, ed. B. Campbell, pp. 136–179. Aldine, Chicago.

Troemel, E.R., J.H. Chou, N.D. Dwyer, H.A. Colbert and C.I. Bargmann. 1995. Divergent seven transmembrane receptors are candidate chemosensory receptors in *C. elegans*. *Cell* **83**:207–218.

Truman, J.W. and L.M. Riddiford. 1974. Hormonal mechanisms underlying insect behaviour. *Advances in Insect Physiology* **10**:297–352.

Trumble, J.T. 1985. Implications of changes in arthropod distribution following chemical application. *Researches on Population Ecology* **27**:277–285.

Trumble, J.T. 1989. Economic consequences of pesticide restriction. *Pesticide Use in a Technological Society*, Riverside, University of California, DNAR Series.

Trumble, J.T. 1990. Vegetable insect control with minimal use of insecticides. *Hort-Science* **25**:159–164.

Trumble, J.T. and T.C. Baker. 1984. Flight phenology and pheromone trapping of *Spodoptera exigua* (Hübner) (Lepidoptera: Noctuidae) in southern coastal California. *Environmental Entomology* **13**:1278–1282.

Trumble, J.T. and B. Alvarado-Rodriguez. 1993. Development and economic evaluation of an IPM program for fresh market tomato production in Mexico. *Agriculture, Ecosystems and Environment* **43**:267–284.

Tumlinson, J.H. 1988. Contemporary frontiers in insect semiochemical research. *Journal of Chemical Ecology* **14**:2109–2130.

Tumlinson, J.H., D.D. Hardee, R.C. Gueldner, A.C. Thompson, P.A. Hedin and J.P. Minyard. 1969. Sex pheromones produced by male boll weevils: isolation, identification, and synthesis. *Science* **166**:1010–1012.

Tumlinson, J.H., C.E. Yonce, R.E. Doolittle, R.R. Heath, C.R. Gentry and E.R. Mitchell. 1974. Sex pheromones and reproductive isolation of the lesser peachtree borer and the peachtree borer. *Science* **185**:614–616.

Tumlinson, J.H., P.E. Hendricks, E.R. Mitchell, R.E. Doolittle and M.M. Brennan. 1975. Isolation, identification and synthesis of the sex pheromone of the tobacco budworm. *Journal of Chemical Ecology* **1**:203–214.

Tumlinson, J.H., M.G. Klein, R.E. Doolittle, T.L. Ladd and A.T. Proveaux. 1977. Identification of the female Japanese beetle sex pheromone: inhibition of male response by an enantiomer. *Science* **197**:789–792.

Tumlinson, J.H., M.M. Brennan, R.E. Doolittle, E.R. Mitchell, A. Brabham, B.E. Mazomenos, A.H. Baumhover and D.M. Jackson. 1989. Identification of a pheromone blend attractive to *Manduca sexta* (L.) males in a wind tunnel. *Archives of Insect Biochemistry and Physiology* **10**:255–271.

Turgeon J.J. and J.N. McNeil. 1982. Calling behaviour of the armyworm, *Pseudaletia unipuncta. Entomologia Experimentalis et Applicata* **31**:402–408.

Turner, J.R.G. 1977. Butterfly mimicry: the genetical evolution of an adaptation. *Evolutionary Biology* **10**:163–206.

Vakharia, V.N., A.K. Raina, T.G. Kingan and T.G. Kempe. 1995. Synthetic pheromone biosynthesis activating neuropeptide gene expressed in a baculovirus expression system. *Insect Biochemistry and Molecular Biology* **25**:583–589.

Valencia, L. 1981. An egg trap to monitor ovipositional activity of the potato moth *Pthorimaea operculella* (Zeller) Lepidoptera: Gelechiidae. *Journal of the Australian Entomological Society* **20**:166.

Valles, S.M., J.L. Capinera and P.E.A. Teal. 1991. Evaluation of pheromone trap design, height, and efficiency for capture of male *Diaphania nitidalis* (Lepidoptera: Pyralidae) in a field cage. *Environmental Entomology* **20**:1274–1278.

Van den Berg, M.J. and G. Ziegelberger. 1991. On the function of the pheromone binding protein in the olfactory hairs of *Antheraea polyphemus. Journal of Insect Physiology* **37**:79–85.

Van der Horst, D.J. 1990. Lipid transport function of lipoproteins in flying insects. *Biochimica et Biophysica Acta* **1047**:195–211.

Van der Pers, J.N.C. and C.J. Den Otter. 1978. Single cell responses from olfactory receptors of small ermine moths to sex-attractants. *Journal of Insect Physiology* **24**:337–343.

Van der Pers, J.N.C. and C. Löfstedt. 1983. Continuous single sensillum recording as a detection method for moth pheromone components in the effluent of a gas chromatograph. *Physiological Entomology* **8**:203–211.

Van der Pers, J.N.C. and A.K. Minks. 1993. Pheromone monitoring in the field using single sensillum recording. *Entomologia Experimentalis et Applicata* **68**:237–245.

Van Deventer, P., A.K. Minks, L.H.M. Blommers, U. Neumann and K. Jilderda. 1992. Mating disruption utilizing lepidopterous sex pheromones: three years of testing in apple orchards in The Netherlands. *Brighton Crop Protection Conference—Pests and Diseases* **3**:1193–1198.

Van Deventer, P., A.K. Minks, J. Woets and K. Jilderda. 1994. Further tests on mating disruption of codling moth and leafrollers in apple orchards in The Netherlands. *Mededelingen van de Faculteit Landbouwwetenschappen Rijksuniversiteit Gent* **59**:671–680.

Van Lenteren, J.C. and J. Woets. 1988. Biological and integrated pest control in greenhouses. *Annual Review of Entomology* **33**:239–269.

Van Steenwyk, R.A. and E.R. Oatman. 1983. Mating disruption of the tomato pinworm (Lepidoptera: Gelechiidae) as measured by pheromone traps, foliage, and fruit infestation. *Journal of Economic Entomology* **76**:8–84.

Van Steenwyk, R.A., E.R. Oatman and J.A. Wyman. 1983. Density treatment level for tomato pinworm (Lepidoptera: Gelechiidae) based on pheromone trap catches. *Journal of Economic Entomology* **76**:440–445.

Vanderwel, D. 1994. Factors affecting pheromone production in beetles. *Archives of Insect Biochemistry and Physiology* **25**:347–362.

Vane-Wright, R.I. and M. Boppré. 1993. Visual and chemical signalling in butterflies: functional and phylogenetic perspectives. *Philosophical Transactions of the Royal Society London B* **340**:197–205.

Vane-Wright, R.I., S. Schulz and M. Boppré. 1992. The cladistics of *Amauris* butterflies: congruence, consensus, and total evidence. *Cladistics* **8**:125–138.

Vetter, R.S. and T.C. Baker. 1983. Behavioral responses of male *Heliothis virescens* in a sustained flight-tunnel to combinations of seven compounds identified from female sex pheromone glands. *Journal of Chemical Ecology* **9**:747–759.

Vetter, R.S. and T.C. Baker. 1984. Behavioral responses of male *Heliothis zea* moths in sustained flight-tunnel to combinations of four compounds identified from female sex pheromone gland. *Journal of Chemical Ecology* **10**:193–202.

Vickers, N.J. 1992. Pheromone-mediated olfactory and behavioral mechanisms in the tobacco budworm, *Heliothis virescens* (F.) Ph.D. thesis, University of California, Riverside.

Vickers, N.J. and T.C. Baker. 1992. Male *Heliothis virescens* maintain upwind flight in response to experimentally pulsed filaments of their sex pheromone (Lepidoptera: Noctuidae). *Journal of Insect Behavior* **5**:699–687.

Vickers, N.J. and T.C. Baker. 1994a. Visual feedback in the control of pheromone-mediated flight of *Heliothis virescens* males (Lepidoptera: Noctuidae). *Journal of Insect Behavior* **7**:605–632.

Vickers, N.J. and T.C. Baker. 1994b. Reiterative responses to single strands of odor promote sustained upwind flight and odor source location by moths. *Proceedings of the National Academy of Sciences of the United States of America.* **91**:5756–5760.

Vickers, N.J. and T.C. Baker. 1995. Neuroethology of attraction and inhibition of two species of heliothine moth in pheromone plume tainted with antagonistic compounds. In *Nervous Systems and Behaviour*, eds. M. Burrows, T. Matheson, P.L. Newland and H. Schuppe, p. 400. Thieme Medical Publishers, New York.

Vickers, N.J., T.A. Christensen, H. Mustaparta and T.C. Baker. 1991. Chemical communication in heliothine moths III. Flight behavior of male *Helicoverpa zea* and *Heliothis virescens* in response to varying ratios of intra- and interspecific sex pheromone components. *Journal of Comparative Physiology A* **169**:275–280.

Vickers, R.A. and G.H.L. Rothschild. 1991. Use of sex pheromones for control of codling moth. In *Tortricid Moths: Their Biology, Natural Enemies and Control*, eds. L.P.S. van der Geest and H.H. Evenhuis, pp. 339–354. Elsevier, Amsterdam.

Villet, R.H. 1978. Mechanisms of insect sex-pheromone sensory transduction: role of adenyl cyclase. *Comparative Biochemistry and Physiology* **61**:389–394.

Visser, J.H. 1986. Host odor perception in phytophagous insects. *Annual Review of Entomology* **31**:121–144.

Vité, J.P. and J.A.A. Renwick. 1971. Inhibition of *Dendroctonus frontalis* response to frontalin by isomers of brevicomin. *Naturwissenschaften* **58**:418.

Vité, J.P., A. Bakke and P.R. Hughes. 1974. Ein Populationslockstoff des Zwölfzähnigen Kiefernborkenkäfers *Ips sexdentatus. Naturwissenschaften* **61**:365–366.

Vité, J.P., P.R. Hughes and J.A.A. Renwick. 1976a. Southern pine beetle: effect of aerial pheromone saturation on orientation. *Naturwissenschaften* **63**:44.

Vité, J.P., D. Klimetzek, G. Loskant, R. Hedden and K. Mori. 1976b. Chirality of insect pheromones: response interruption by inactive antipodes. *Naturwissenschaften* **63**:582–583.

Vité, J.P., G. Ohloff and R.F. Billings. 1978. Pheromonal chirality and integrity of aggregation response in southern species of bark beetle *Ips* sp. *Nature* **272**:817–818.

Vité, J.P., R.F. Billings, C.W. Ware and K. Mori. 1985. Southern pine beetle: enhancement or inhibition of aggregation response mediated by enantiomers of *endo*-brevicomin. *Naturwissenschaften* **72**:99–100.

Vivien-Roels, B., and P. Pevet. 1993. Melatonin: presence and formation in invertebrates. *Experientia* **49**:642–647.

Vogel, S. 1981. *Life in Moving Fluids*. Princeton University Press, Princeton, New Jersey.

Vogel, S. 1983. How much air passes through a silkmoth's antenna? *Journal of Insect Physiology* **29**:597–602.

Vogt, H., A. Schropp, U. Neumann and K.W. Eichhorn. 1993. Field trials on mating disruption to control the European grape moth *Eupoecilia ambiguella* Hbn. *Journal of Applied Entomology* **115**:217–232.

Vogt, R.G. 1987. The molecular basis of pheromone reception: its influence on behavior. In *Pheromone Biochemistry*, eds. G.D. Prestwich and G.L. Blomquist, pp. 385–431. Academic Press, Orlando.

Vogt, R.G. and L.M. Riddiford. 1981. Pheromone binding and inactivation by moth antennae. *Nature* **293**:161–163.

Vogt, R.G. and L.M. Riddiford. 1986. Pheromone reception: a kinetic equilibrium. In *Mechanisms in Insect Olfaction*, eds. T.L. Payne, M.C. Birch and C.E.J. Kennedy, pp. 201–208. Clarendon Press, Oxford.

Vogt, R.G., L.M. Riddiford and G.D. Prestwich. 1985. Kinetic properties of a pheromone degrading enzyme: The sensillar esterase of *Antheraea polyphemus. Proceedings of the National Academy of Science U.S.A.* **82**:8827–8831.

Vogt, R.G., G.D. Prestwich and L.M. Riddiford. 1988. Sex pheromone receptor proteins: visualization using a radiolabeled photoaffinity analog. *Journal of Biological Chemistry* **263**:3952–3959.

Vogt, R.G., A.C. Kohne, J.T. Dubnau and G.D. Prestwich. 1989. Expression of pheromone binding proteins during antennal development in the gypsy moth *Lymantria dispar. Journal of Neuroscience* **9**:3332–3346.

Vogt, R.G., G.D. Prestwich and M.R. Lerner. 1991a. Odorant-binding-protein subfamilies associate with distinct classes of olfactory receptor neurons in insects. *Journal of Neurobiology* **22**:74–84.

Vogt, R.G., R. Rybcynski and M.R. Lerner. 1991b. Molecular cloning and sequencing of general odorant-binding proteins GOBP1 and GOBP2 from the tobacco hawk moth *Manduca sexta*: comparison with other insect OBPs and their signal peptides. *Journal of Neuroscience* **11**:2972–2984.

Von Keyserlingk, H.C. 1984. Close range orientation of flying Lepidoptera to pheromone sources in a laboratory wind tunnel and the field. *Mededelingen van de Faculteit Landbouwwetenschappen Rijksuniversiteit Gent* **49**:683–687.

Wadhams, L.J. 1984. The coupled gas chromatography–single cell recording technique. In *Techniques in Pheromone Research*, eds. H.E. Hummel and T.A. Miller, pp. 179–189. Springer-Verlag, New York.

Wadhams, L.J. 1990. The use of coupled gas chromatography electrophysiological techniques in the identification of insect pheromones. In *Chromatography and Isolation of Insect Hormones and Pheromones*, eds. A.R. McCaffery and I.D. Wilson, pp. 289–298. Plenum Press, New York.

Wagner, D.L. and J. Rosovsky. 1991. Mating systems in primitive Lepidoptera, with emphasis on the reproductive behaviour of *Korscheltellus gracilis* (Hepialidae). *Zoological Journal of the Linnean Society* **102**:277–303.

Waldner, W. 1991. Lockstoffe gegen Apfel- und Fruchtschalenwickler. Erfahrungen im praktischen Einsatz. *Obstbau-Weinbau* **12/91**:331–333.

Waldner, W. 1993a. Rückblick und Vorschau auf die Bekämpfung des Apfelwicklers. *Obstbau-Weinbau* **12/93**:355–357.

Waldner, W. 1993b. Verwirren—ein Schritt in die Zukunft. *Obstbau Weinbau* **30**:354.

Waldner, W. 1994a. Erfahrungen 1994 im Südtiroler Obstbau mit der Verwirrungsmethode gegen Apfel- und Fruchtschalenwickler. *Obstbau-Weinbau* **12/94**:276.

Waldner, W. 1994b. Verwirrung 1995—Unsere Empfehlungen. *Obstbau-Weinbau* **31**:298–299.

Wall, C. 1984. The exploitation of insect communication by Man: fact or fantasy? In *Insect Communication*, ed. T. Lewis, pp. 379–400. Academic Press, London.

Wall, C. 1989. Monitoring and spray timing. In *Insect Pheromones in Plant Protection*, eds. A.R. Jutsum and R.F.S. Gordon, pp. 39–66. Wiley, Chichester, United Kingdom.

Wall, C. and J.N. Perry. 1983. Further observations on the responses of male pea moth, *Cydia nigricana*, to vegetation previously exposed to sex-attractant. *Entomologia Experimentalis et Applicata* **33**:112–116.

Wang, Y. 1989. Turbulence structure, momentum and heat transport in the edge of broad leaf tree stands. Research in microclimate at the forest–urban interface, Progress Report 15. US Forest Service, Washington.

Wang, Y.-S., T.G. Kempe, A.K. Raina and P.H. Mazzocchi. 1994. Conformation of a biologically active C-terminal hexapeptide analog of the pheromone biosynthesis activating neuropeptide by NMR spectroscopy. *International Journal of Peptide and Protein Research* **43**:277–283.

Warren, R. and A.H. Wertheim. 1990. *Perception and Control of Self-Motion*. Lawrence Erlbaum Associates, Hillsdale, New Jersey.

Watterson, G.P., T.L. Payne and J.V. Richerson. 1981. The effects of verbenone and brevicomin on the within-tree populations of *Dendroctonus frontalis*. *Journal of the Georgia Entomological Society* **17**:118–126.

Wearing, C.H., W.P. Thomas, J.S. Dugdale and W. Danthanarayana. 1991. Tortricid pests of pome and stone fruits, Australian and New Zealand species. In *Tortricid Pests, Their Biology, Natural Enemies and Control*, eds. L.P.S. van der Geest and H.H. Evenhuis, pp. 453–472. Elsevier, Amsterdam.

Wearing, C.H., J.T.S. Walker and D.M. Suckling. 1993. Development of IPM for New Zealand apple production. Proceedings 2nd International Symposium on Integrated Fruit Production. *Acta Horticulturae* **347**:277–284.

Weatherston, I. and Minks, A.K. 1995. Regulation of semiochemicals—global aspects. *Integrated Pest Management Reviews* **1**:1–13.

Weaver, J.S., III. 1992a. Remarks on the evolution of the Trichoptera: a critique of Wiggins and Wichard's classification. *Cladistics* **8**:171–180.

Weaver, J.S., III. 1992b. Further remarks on the evolution of Trichoptera: a reply to Wiggins. *Cladistics* **8**:187–90.

Webb, R.E., B.A. Leonhardt, J.R. Plimmer, K.M. Tatman, V.K. Boyd, D.L. Cohen, C.P.

Schwalbe and L.W. Douglass. 1990. Effect of racemic disparlure released from grids of plastic ropes on mating success of gypsy moth (Lepidoptera: Lymantriidae) as influenced by dose and population density. *Journal of Economic Entomology* **83**:910–916.

Weber, D.C. and D.N. Ferro. 1991. Nontarget noctuids complicate integrated pest management monitoring of sweet corn with pheromone traps in Massachusetts. *Journal of Economic Entomology* **84**:1364–1369.

Webster, R.P. 1987. Modulation of the expression of calling by temperature in the omnivorous leafroller moth, *Platynota stultana* (Lepidoptera: Tortricidae) and other moths: a hypothesis. *Annals of the Entomological Society of America* **81**:138–151.

Webster, R.P. and R.T. Cardé. 1984. The effects of mating, exogenous juvenile hormone and a juvenile hormone analogue on pheromone titre, calling and oviposition in the omnivorous leafroller moth (*Platynota stultana*). *Journal of Insect Physiology* **30**:113–118.

Wegener, J. 1994. Elektrophysiologische Aspekte der olfaktorischen Signaltransduktion in Insekten. Ph.D. thesis, University Stuttgart—Hohenheim, Germany.

Weiss, G.B. 1974. Cellular pharmacology of lanthanium. *Annual Review of Pharmacology* **14**:343–354.

Wendler, G. and H. Scharstein. 1974. Physiology and systems analysis of gravity orientation in two insect species (*Carausius morosus, Calandra granaria*). *Fortschritte der Zoologie* **23**:33–44.

Wendler, G., M. Muller and U. Dombrowski. 1993. The activity of pleurodosal muscles during flight and at rest in the moth *Manduca sexta* (L.). *Journal of Comparative Physiology A* **173**:65–75.

Wenzel, J.W. 1992. Behavioral homology and phylogeny. *Annual Review of Ecology and Systematics* **23**:361–381.

Wenzel, J.W. and J.M. Carpenter. 1994. Comparing methods: adaptive traits and tests of adaptation. In *Phylogenetics and Ecology*, eds. P. Eggleton and R.I. Vane-Wright, pp. 79–101. Academic Press, London.

Werner, R.A. 1995. Toxicity and repellency of 4-allylanisole and monoterpenes from white spruce and tamarak to the spruce beetle and eastern larch beetle (Coleoptera: Scolytidae). *Environmental Entomology* **24**:372–379.

Werner, R.A., M.M. Furniss, L.C. Yarger and T. Ward. 1981. Effects on eastern larch beetle of its natural attractant and synthetic pheromones in Alaska. United States Department of Agriculture, Forest Service, Research Note PNW-371.

West-Eberhard, M.J. 1983. Sexual selection, social competition, and speciation. *Quarterly Review of Biology* **58**:155–234.

Wheeler, W.M. 1899. Anemotropisms and other tropisms in insects. *Archiv für Entwickelungsmechanik der Organismen* **8**:373–381.

White, C.S., D.M. Lambert and S.P. Foster. 1995. Chemical signals and the recognition concept. In *Speciation and the Recognition Concept: Theory and Application*, eds. D.M. Lambert and H. Spencer, pp. 301–326. Johns Hopkins University Press, Baltimore, Maryland.

Whitlock, M.C. and M.J. Wade. 1995. Speciation: founder events and their effects on X-linked and autosomal genes. *American Naturalist* **145**:676–685.

Wiesenborn, W.D., J.T. Trumble and V. Voth. 1988. Corn earworm outbreaks in strawberries. *California Agriculture* **42**:25–26.

Wiesenborn, W.D., J.T. Trumble and E.R. Oatman. 1990. Economic comparison of insecticide treatment programs for managing tomato pinworm (Lepidoptera: Gelechiidae) on fall tomatoes. *Journal of Economic Entomology* **83**:212–216.

Wiggins, G.B. and W. Wichard. 1989. Phylogeny of pupation in Trichoptera, with proposals the origin and higher classification of the order. *Journal of the North American Benthological Society* **8**:260–276.

Wiktelius, S. and J. Pettersson. 1985. Simulations of bird–cherry–oat aphid population dynamics: a tool for developing strategies for breeding aphid-resistant plants. *Agriculture, Ecosystems and Environment* **14**:159–170.

Wilkinson, C. and M.J. Scoble. 1979. The Nepticulidae (Lepidoptera) of Canada. *Memoirs of the Entomological Society of Canada* **107**:1–129.

Williams, C.M. 1946. Physiology of insect diapause. I. The role of the brain in the production and termination of pupal dormancy in the giant silkworm, *Platysamia cecropia. Biological Bulletin* **90**:234–243.

Williams, G.C. 1992. *Natural Selection: Domains, Levels and Challenges*. Oxford University Press, Oxford.

Williams, M. 1993. Control of LBAM, *Epiphyas postvittana* Walker, in apple orchards using the mating disruption technique: preliminary results. In *Pest Control in Sustainable Agriculture*, eds. S.A. Corey, D.J. Dall and W.M. Milne, pp. 284–286. CSIRO, Canberra, Australia.

Willis, J.H., G.P. Waldbauer and J.G. Sternburg. 1974. The initiation of development by the early and late emerging morphs *Hyalophora cecropia. Entomologia experimentalis et applicata* **17**:219–222.

Willis, M.A. and E.A. Arbas. 1991a. Odor-modulated upwind flight of the sphinx moth, *Manduca sexta* L. *Journal of Comparative Physiology A* **169**:427–440.

Willis, M.A. and E.A. Arbas. 1991b. Flight muscle activity underlying pheromone-modulated zigzagging flight in male moths, *Manduca sexta. Society for Neuroscience Abstract* **17**:492.15.

Willis, M.A. and E.A. Arbas. 1994. Motor patterns underlying pheromone-modulated flight in male moths, *Manduca sexta*. In *Olfaction and Taste XI*, eds. K. Kurihara, N. Suzuki and H. Ogawa, p. 850. Springer-Verlag, Tokyo.

Willis, M.A. and T.C. Baker. 1984. Effects of intermittent and continuous pheromone stimulation on the flight behaviour of the oriental fruit moth, *Grapholita molesta. Physiological Entomology* **9**:341–358.

Willis, M.A. and T.C. Baker. 1987. Comparison of manoeuvres used by walking versus flying *Grapholita molesta* males during the pheromone-mediated upwind movement. *Journal of Insect Physiology* **33**:875–883.

Willis, M.A. and T.C. Baker. 1988. Effects of varying sex pheromone component ratios on the zigzagging flight movement of the oriental fruit moth, *Grapholita molesta. Journal of Insect Behavior* **1**:357–371.

Willis, M.A. and T.C. Baker, 1994. Behaviour of flying oriental fruit moth males during approach to sex pheromone sources. *Physiological Entomology* **19**:61–69.

Willis, M.A. and M.C. Birch. 1982. Male lek formation and female calling in a population of the arctiid moth *Estigmene acrea. Science* **218**:168–170.

Willis, M.A. and R.T. Cardé. 1990. Pheromone-modulated optomotor response in male gypsy moths, *Lymantria dispar* L.: upwind flight in a pheromone plume in different wind velocities. *Journal of Comparative Physiology A* **167**:699–706.

Willis, M.A., J. Murlis and R.T. Cardé. 1991. Pheromone-mediated upwind flight of male gypsy moths, *Lymantria dispar*, in a forest. *Physiological Entomology* **16**:507–521.

Willis, M.A., C.T. David, J. Murlis and R.T. Cardé. 1994. Effects of pheromone plume structure and visual stimuli on the pheromone-modulated flight of male gypsy moths (*Lymantria dispar*) in a forest (Lepidoptera: Lymantriidea). *Journal of Insect Behavior* **7**:385–409.

Wilson, I.M. 1995. Antiaggregants for the mountain pine beetle. Master of Pest Management thesis, Simon Fraser University, Burnaby, Canada.

Winkelmann-Vogt, H. 1986. Untersuchungen zum Pheromoneinsatz bei der Bekämpfung des Einbindigen Traubenwicklers *Eupoecilia ambiguella* Hbn.) im Weinbau. Dissertation Universität Kaiserslautern, Germany.

Witz, J.A., J.J.D. Lopez and M.A. Latheef. 1992. Field density estimates of *Heliothis virescens* (Lepidoptera: Noctuidae) from catches in sex pheromone-baited traps. *Bulletin of Entomological Research* **82**:281–286.

Witzgall, P. 1990. Attraction of *Cacoecimorpha pronubana* male moths to synthetic sex pheromone blends in the wind tunnel. *Journal of Chemical Ecology* **16**:1507–1515.

Witzgall, P. and H. Arn. 1990. Direct measurement of the flight behavior of male moths to calling females and synthetic sex pheromones. *Zeitschrift für Naturforschung* **45c**:1067–1069.

Witzgall, P. and H. Arn. 1991. Recording flight tracks of *Lobesia botrana* in the wind tunnel. In *Insect Chemical Ecology*, ed. I. Hrdy, pp. 187–193. SPB Academic Publishers, The Hague.

Witzgall, P. and B. Frérot. 1989. Pheromone emission by individual females of carnation tortrix, *Cacoecimorpha pronubana. Journal of Chemical Ecology* **15**:707–717.

Witzgall, P. and E. Priesner. 1991. Wind tunnel study on an attraction inhibitor in male *Coleophora laricella* (Lepidoptera: Coleophoridae). *Journal of Chemical Ecology* **17**:1355–1362.

Witzgall, P., M. Bengtsson, C.R. Unelius and J. Löfqvist. 1993. Attraction of pea moth *Cydia nigricana* F. (Lepidoptera: Tortricidae) to female sex pheromone (*E,E*)-8,10-dodecadien-1-yl acetate, is inhibited by geometric isomers *E,Z, Z,E* and *Z,Z. Journal of Chemical Ecology* **19**:1917–1928.

Wiygul, G., J.C. Dickens and J.W. Smith. 1990. Effect of juvenile hormone III and beta-bisabolol on pheromone production in fat bodies from male boll weevils, *Anthonomus grandis* Boheman (Coleoptera: Curculionidae). *Comparative Biochemistry and Physiology* **95B**:489–491.

Wolf, R. and M. Heisenberg. 1990. Visual control of straight flight in *Drosophila melanogaster. Journal of Comparative Physiology A* **167**:269–283.

Wolf, R. and M. Heisenberg. 1991. Basic organization of operant behavior as revealed in *Drosophila* flight orientation. Journal of Comparative Physiology A **169**:699–705.

Wolf, R., A. Voss, S. Hein and M. Heisenberg. 1992. Can a fly ride a bicycle? *Philosophical Transactions of the Royal Society of London B* **337**:261–269.

Wong, S.S. 1993. *Chemistry of Protein Conjugation and Cross-Linking.* CRC Press, Boca Raton, Florida.

Wood D.E. and E.A. Arbas. 1994. Modulation of motor patterns in a tethered moth during multi-modal stimulation. *Society for Neuroscience Abstracts* **20**:1026.

Wood, J.R. and H.V. Resh. 1984. Demonstration of sex pheromones in caddisflies (Trichoptera). *Journal of Chemical Ecology* **10**:171–175.

Woodhead, A.P., M.A. Khan, B. Stay and S.S. Tobe. 1994. Two new allatostatins from the brains of *Diploptera punctata. Insect Biochemistry and Molecular Biology* **24**:257–263.

Woodring, J.P., O.W. Meirer and R. Rose. 1988. Effect of development, photoperiod, and stress on octopamine levels in the house cricket, *Acheta domesticus. Journal of Insect Physiology* **34**:759–765.

Wright, R.H. 1958. The olfactory guidance of flying insects. *Canadian Entomologist* **90**:81–89.

Wright, R.H. 1963. Chemical control of chosen insects. *New Scientist* **20**:598–600.

Wunderer, H., K. Hansen, T.W. Bell, D. Schneider and J. Meinwald. 1986. Sex pheromones of two Asian moths (*Creatonotos transiens, C. gangis*; Lepidoptera: Arctiidae): behavior morphology, chemistry, and electrophysiology. *Experimental Biology* **46**:11–27.

Yagi, S. 1980. Hormonal control of larval coloration associated with phase variation in the armyworm, *Spodoptera exempta. ICIPE Annual Report* **7**:41.

Yi, Q. and M. Cao. 1989. Mechanism of the control of sex pheromone production in female Asiatic corn borer (*Ostrinia furnacalis* Guenee). *Journal of the Xiamen University (Natural Science)* **28**:88–91.

Yin, C-M. 1994. Juvenile hormone III bisepoxide: new member of the insect juvenile hormone family. *Zoological Studies* **33**:237–245.

Young, H., R.A. Galbreath, M.H. Benn, V.A. Holt and D.L. Struble. 1985. Sex pheromone components in New Zealand brownheaded leafroller *Ctenopseustis obliquana* (Lepidoptera: Tortricidae). *Zeitschrift für Naturforschung* **40c**:262–265.

Yu, H.-S., and R.J. Reiter. 1992. *Melatonin: Biosynthesis, Physiological Effects, and Clinical Applications.* CRC Press, Boca Raton, Florida.

Zagulajev, A.K. and S. Yu. Sinev. 1988. Catapterigidae fam. n.—a new family of lower Lepidoptera (Dacnonypha). *Entomologicheskoe Obozrenie* **67**:593–601.

Zalokar, M. 1968. Effect of corpora allata on protein and RNA synthesis in colleterial glands of *Blattella germanica.* Journal of Insect Physiology **14**:1177–1134.

Zanen, P.O. 1993. A comparative study of the role of vision and olfaction during in-flight maneuvers by four species of insects to semiochemicals. Ph.D. thesis, University of Massachusetts, Amherst.

Zeil, J. 1993a. Orientation flights of solitary wasps (*Cerceris*; Sphecidae; Hymenoptera) I. Description of flight. *Journal of Comparative Physiology A* **172**:189–205.

Zeil, J. 1993b. Orientation flights of solitary wasps (*Cerceris*; Sphecidae; Hymenoptera) II. Similarities between orientation and return flights and the use of motion parallax. *Journal of Comparative Physiology A* **172**:207–222.

Zhu, J., M.V. Kozlov, P. Philipp., W. Francke and C. Löfstedt. 1995. Identification of a novel moth sex pheromone in *Eriocrania cicatricella* (Lepidoptera: Eriocraniidae) and its phylogenetic implications. *Journal of Chemical Ecology* **21**:29–43.

Zhu, J.-W. 1995. Diversity and conservatism in moth sex pheromone systems. Ph.D. thesis, Lund University, Sweden.

Ziegelberger, G. 1995. Redox shift of the pheromone-binding protein in the silkmoth *Antheraea polyphemus. European Journal of Biochemistry* **232**:706–711.

Ziegelberger, G., M.J. Van den Berg, K.-E. Kaissling, S. Klump and J.E. Schultz. 1990. Cyclic GMP levels and guanylate cyclase activity in pheromone-sensitive antennae of the silkmoths *Antheraea polyphemus and Bombyx mori. Journal of Neuroscience* **10**:1217–1225.

Zufall, F., M. Stengl, H. Hatt, C. Franke and J.G. Hildebrand. 1991. Ionic currents of cultured olfactory receptor neurons from antennae of male *Manduca sexta. Journal of Neuroscience* **11**:956–965.

Index